T0319268

BIOLOGY OF STRESS IN FISH

This is Volume 35 in the

FISH PHYSIOLOGY series

Edited by Anthony P. Farrell and Colin J. Brauner

Honorary Editors: William S. Hoar and David J. Randall

A complete list of books in this series appears at the end of the volume

BIOLOGY OF STRESS IN FISH

Fish Physiology

CARL B. SCHRECK

U.S. Geological Survey
Oregon Cooperative Fish and Wildlife Research Unit
Department of Fisheries and Wildlife
Oregon State University
Corvallis, Oregon
United States

LLUIS TORT

Department of Cell Biology, Physiology and Immunology
Universitat Autònoma de Barcelona
Bellaterra, Barcelona, Catalonia
Spain

ANTHONY P. FARRELL

Department of Zoology, and Faculty of Land and Food Systems
The University of British Columbia
Vancouver, British Columbia
Canada

COLIN J. BRAUNER

Department of Zoology
The University of British Columbia
Vancouver, British Columbia
Canada

ELSEVIER

AMSTERDAM • BOSTON • HEIDELBERG • LONDON
NEW YORK • OXFORD • PARIS • SAN DIEGO
SAN FRANCISCO • SINGAPORE • SYDNEY • TOKYO

Academic Press is an imprint of Elsevier

Academic Press is an imprint of Elsevier
125 London Wall, London EC2Y 5AS, United Kingdom
525 B Street, Suite 1800, San Diego, CA 92101-4495, United States
50 Hampshire Street, 5th Floor, Cambridge, MA 02139, United States
The Boulevard, Langford Lane, Kidlington, Oxford OX5 1GB, United Kingdom

British Library Cataloguing-in-Publication Data
A catalogue record for this book is available from the British Library

Library of Congress Cataloging-in-Publication Data
A catalog record for this book is available from the Library of Congress

ISBN: 978-0-12-802728-8
ISSN: 1546-5098

For Information on all Academic Press publications
visit our website at https://www.elsevier.com

Working together
to grow libraries in
developing countries

www.elsevier.com • www.bookaid.org

Publisher: Zoe Kruze
Acquisition Editor: Kristi Gomez and Kirsten Shankland
Editorial Project Manager: Pat Gonzalez and Hannah Colford
Production Project Manager: Lucía Pérez
Designer: Matthew Limbert

Typeset by MPS Limited, Chennai, India

CONTENTS

CONTRIBUTORS

N. ALURU *(113)*, *Woods Hole Oceanographic Institution, Woods Hole, MA, United States*

S.J. COOKE *(405)*, *Carleton University, Ottawa, ON, Canada*

M.R. DONALDSON *(405)*, *Carleton University, Ottawa, ON, Canada*

E. FAUGHT *(113)*, *University of Calgary, Calgary, AB, Canada*

G. FLIK *(75)*, *Radboud University, Nijmegen, The Netherlands*

M. GORISSEN *(75)*, *Radboud University, Nijmegen, The Netherlands*

E. HÖGLUND *(35)*, *Norwegian Institute of Water Research (NIVA), Oslo, Norway*

P.-P. HWANG *(207)*, *Academia Sinica, Taipei, Taiwan*

K.M.M. JONES *(333)*, *Cape Breton University, Sydney, NS, Canada*

M.L. KENT *(541)*, *Oregon State University, Corvallis, OR, United States*

C. LAWRENCE *(541)*, *Children's Hospital Boston, Aquatic Resources Program, Boston, MA, United States*

D.L.G. NOAKES *(333)*, *Oregon State University, Corvallis, OR, United States*

C.M. O'CONNOR *(405)*, *McMaster University, Hamilton, ON, Canada*

Ø. ØVERLI *(35)*, *Norwegian University of Life Sciences, Oslo, Norway*

N.W. Pankhurst *(295)*, *Australian Rivers Institute, Griffith University, Gold Coast, QLD, Australia*

J.V. Planas *(251)*, *Universitat de Barcelona, Barcelona, Spain*

K.J. Rodnick *(251)*, *Idaho State University, Pocatello, ID, United States*

B. Sadoul *(167)*, *University of Calgary, Calgary, AB, Canada*

C.B. Schreck *(1)*, *U.S. Geological Survey, Oregon State University, Oregon, United States*

L.U. Sneddon *(463)*, *University of Liverpool, Liverpool, United Kingdom*

N.M. Sopinka *(405)*, *Carleton University, Ottawa, ON, Canada*

S. Spagnoli *(541)*, *Oregon State University, Corvallis, OR, United States*

C.D. Suski *(405)*, *University of Illinois at Urbana-Champaign, Urbana, IL, United States*

Y. Takei *(207)*, *University of Tokyo, Chiba, Japan*

J.S. Thomson *(463)*, *University of Liverpool, Liverpool, United Kingdom*

L. Tort *(1, 365)*, *Universitat Autònoma de Barcelona, Barcelona, Spain*

M.M. Vijayan *(113, 167)*, *University of Calgary, Calgary, AB, Canada*

S. Winberg *(35)*, *Uppsala University, Uppsala, Sweden*

D.C.C. Wolfenden *(463)*, *Blue Planet Aquarium, Ellesmere Port, United Kingdom*

T. Yada *(365)*, *National Research Institute of Aquaculture, Nikko, Japan*

PREFACE

Stress is an inherent attribute of being alive. How fish respond to stressors of various types is critical to the individual's survival and the population's reproductive capacity. The response to stressors is also a critical variable that transcends all studies on the experimental biology of fish as well as many, if not most, management practices. Essentially, stress can affect all physiological processes and hence organismic performance. It is thus important for researchers across the entire suite of disciplines that could be called the study of fish or fishery biology to have an understanding of stress and what it means to the individual and the population. *Biology of Stress in Fish* is aimed at providing that understanding.

This book is intended to reach a diverse readership. It provides the most up-to-date information and thinking on the various topics covered, a greatly needed contribution to the field given that the last comprehensive book on stress in fish edited in 1981 by Alan D. Pickering is many decades old. *Biology of Stress in Fish* provides general understanding regarding stress biology, including most of the recent advances in the field. The book is intended to serve as a reference for professional fish physiologists as well as general fishery biologists who need to know about stress and its implications. Professional aquaculturists, ornamental fish hobbyists, and those interested in animal health and animal well-being should also find the book of interest. The book starts with a general discussion of stress—what is stress, the nature of the physiological stress response, and factors that affect the stress response. Moving forward, biotic and abiotic factors that cause variation in the stress response are considered. The book then discusses how the stress response is generated and controlled. Following chapters examine stress effects on physiological and organismic function and performance. The book concludes with applied chapters regarding assessment of stress, animal welfare, and stress as related to model species. To provide a true understanding of the state of knowledge regarding stress in fish, the book clearly denotes the boundaries of our understanding of the field. When venturing beyond what is understood, even strongly supported hypotheses become based more on subjective inference, untested models, and theories. Key unknowns are clearly identified and serve as topics for further research.

The *Biology of Stress in Fish* is the product of an initiative that started at least 6 years ago when, as fish stress specialists, we acknowledged the need for an updated book on this matter. The result is this book with 13 individual chapters authored by 30 experts in their respective fields. The final structure and content of the book is grounded in discussions with many individuals and on the outcome of two formal workshops, one at Oregon State University, Corvallis, and the other at Heriot-Watt University, Edinburgh. The editors are deeply indebted to the chapter authors who were a pleasure to work with and 25 outside reviewers of the chapters for their positive, constructive, and in-depth comments. Finally, we thank the production staff at Elsevier, in particular Pat Gonzalez and Kristi Gomez for their tireless efforts.

Carl B. Schreck
Lluis Tort

1

THE CONCEPT OF STRESS IN FISH

CARL B. SCHRECK

LLUIS TORT

1. Introduction
   1.1. What Is Stress?
   1.2. Dynamics of the Stress Response and Effects on Performance
   1.3. Contemporary View of the GAS: Eustress versus Distress
   1.4. Sensory Systems and Perception
   1.5. Adaptation versus Nonadaptation Aspects of the Stress Response
   1.6. Key Unknowns

The general physiological response of fish to threatening situations, as with all vertebrates, is referred to as *stress*. A stress response is initiated almost immediately following the perception of a stressor. Mildly stressful situations can have beneficial or positive effects (eustress), while higher severities induce adaptive responses but also can have maladaptive or negative consequences (distress). The stress response is initiated and controlled by two hormonal systems, those leading to the production of corticosteroids (mainly cortisol) and catecholamines (such as adrenaline and noradrenaline and their precursor dopamine). Together these regulate the secondary stress response factors that alter the distribution of necessary resources such as energy sources and oxygen to vital areas of the body, as well as compromise hydromineral imbalance and the immune system. If fish can resist death due to a stressor, they recover to a similar or somewhat similar homeostatic norm. Long-term consequences of repeated or prolonged exposures to stress are maladaptive by negatively affecting other necessary life functions (growth, development, disease resistance, behavior, and reproduction), in large part because of the energetic cost associated with mounting the stress response (allostatic load).

There is considerable variation in how fish respond to a stressor because of genetic differences among different taxa and also within stocks and

1

Biology of Stress in Fish: Volume 35
FISH PHYSIOLOGY

species. Variations within the stress response are introduced by the environmental history of the fish, present ambient environmental conditions, and the fish's present physiological condition. Currently, fish physiology has progressed to the point where we can easily recognize when fish are stressed, but we cannot always recognize when fish are unstressed because the lack of clinical signs of stress does not always correspond to fish being unstressed. In other words, we need to be aware of the possibility of false negatives regarding clinical signs of stress. In addition, we cannot use clinical data to precisely or accurately infer severity of a stressor.

1. INTRODUCTION

A fish's life is filled with overcoming, coping with, and recovering from threatening challenges. Threats to the well-being of vertebrates generally result in a physiological cascade of events that helps the organism react to and hopefully recover from or cope with stressors. This concept of how animals respond physiologically to such challenges was termed *stress* by Selye (1950). The response, he proposed, was similar irrespective of the nature of the stressor and referred to by him as the general adaptation syndrome (GAS). The GAS consists of a hormonal cascade that produces all the other responses to a stressor. He goes as far as to suggest that the complete absence of stress is death. The general contentions suggested by Selye (1976) are appropriate for fish but, as with other vertebrates, come with numerous caveats. Of course, the physiological stress response has consequences to essentially all physiological systems. Consequently, understanding the biology of stress in fish will provide a fundamental understanding important for the science of fish biology as well as fish research, management, and husbandry.

This book provides a detailed account of the state of the art of stress physiology in fishes. It builds upon and expands the seminal publication by Pickering (1981) on *Stress and Fish*. While providing a holistic review of stress, it does not focus on any particular stressor or environmental tolerance limits. Instead, we recognize that many environmental factors (eg, temperature, pH, turbidity, toxicants, pathogens, predators, handing by people) can lead to stress and dysfunction in the fish if they are encountered at levels approaching or beyond the normal tolerance capacity of fish.

The intent of this chapter is to provide an overview of the stress concept as it relates to fish. We discuss stress in fish in general terms; the other chapters will provide the necessary detail regarding what is known about their respective topics. Understanding the limits of our knowledge is also important from both a basic and applied perspective; the mission of this book is to also provide this insight.

1.1. What Is Stress?

The word "stress" is surprisingly difficult to define. It is a physiological response of the organism. It is not the environmental variable that causes the response; that would be referred to as the *stressor*.

1.1.1. STRESS DEFINITION

The word "stress" has its roots in the physiological definition proposed by Selye (1950, 1973): "stress is the nonspecific response of the body to any demand placed upon it." However, the definition we prefer and the one used throughout this book is "The physiological cascade of events that occurs when the organism is attempting to resist death or reestablish homeostatic norms in the face of insult" (Schreck, 2000).

Our preference is dictated in part by Selye's unfortunate selection of the word "stress" with reference to the GAS, which has led to some confusion in clearly defining the term. Classically, in terms of physics, stress refers to a force to which an object is exposed (the stressor in GAS). *Strain* would have been a better choice for GAS, being the distortion of an object due to a force (ie, the response to the stressor). To be complete, we list many numerous, differing definitions of the word "stress" that have appeared in the literature:

- *A state produced by an environmental or other factor that extends the adaptive responses beyond the normal range* (Brett, 1958).
- *The sum of all physiological responses that occur when animals attempt to establish or maintain homeostasis* (Wedemeyer and McLeay, 1981).
- *The alteration of one or more physiological variables to the point that long-term survival may be impaired* (Bayne, 1985).
- *Stress is a state caused by a stress factor, or stressor, that deviates from a normal resting or homeostatic state* (Barton and Iwama, 1991).
- *The cascade of biological events that occur when the organism faces a challenge out of the normal range and the attempt to reestablish homeostatic values* (Barton, 1997).
- *A state of threatened homeostasis that is re-established by a complex suite of adaptive responses* (Chrousos, 1998).
- *The reaction of the organism aimed at regaining homeostasis* (Chrousos, 2009).
- *Stress is a condition where an environmental demand exceeds the natural regulatory capacity of an organism* (Koolhaas et al., 2011).

A general theme does run through all definitions: stress is the physiological response to a stressor. Some definitions restrict the stress response to imply a neuroendocrine-induced cascade. Irrespective of definition, this cascade tends to be nonspecific, being qualitatively similar irrespective of the nature, type, and severity of stressor. That said, the quantitative magnitude of the stress response differs widely for a variety of reasons that are discussed and reviewed in detail by Winberg and colleagues (2016; Chapter 2 in this volume). A general discussion of the definition of stress can be found in Levine (1985), while McEwen and Lasley (2002) provide a very readable, more contemporary discussion of what constitutes stress.

Stressors can range anywhere from very brief (acute)—for example, being caught in a net or escaping a predator—to those that are prolonged and even more or less permanent (chronic)—for example, being over-crowded in a tank or at the bottom of a social hierarchy. The terms "acute" and "chronic" are context-dependent and so not easily defined. When considering whether something results in an acute or chronic stress, it should not be based on the duration of the stressor but rather on "… the duration of its consequences on the physiology of the animal" (Boonstra, 2013). Stressors also vary in severity.

1.1.2. THE PHYSIOLOGICAL STRESS RESPONSE

The GAS concept (Selye, 1950) embraced the notion that stress is a generalized response. We describe this physiological stress response for fishes in the following, and others expand on this in other chapters of this book. While in a very general sense the GAS concept is a good way to think about the stress response, we emphasize that the response is not in actuality all that general, something Schreck (1981) pointed out for fish many years ago. For example, fish can be killed by certain toxicants, such as cadmium, or by anesthesia without evoking a GAS-like response. Different stressors and different severities of stressors also result in considerably different response dynamics of elements of the GAS (Winberg et al., 2016; Chapter 2 in this volume) (Barton, 2002).

Events that extend the normal daily bounds of a fish's experiences can lead to our perception that a fish is in some way in danger, which then leads to a stress response. Detection of a real or perceived stressor is essential to initiate the stress response (Schreck, 1981). The classical stress response is not initiated if the fish does not perceive a real life-threatening stressor (Schreck, 1981). Therefore, the psychogenic aspect of the stress response is extremely important to the animal's well-being, as proposed by Ellis et al. (2012).

There are three main stages to the stress response: alarm, resistance, and either compensation or exhaustion (death) (Selye, 1950; Schreck, 2000).

The nature (magnitude and duration) of the stress response is dependent on the severity and duration of the stressor. In essentially all cases, the alarm phase consists of the upregulation of systems involved in flight, fight, and importantly, coping. During the resistance stage the fish either (1) fully overcomes the stressor, allowing for reestablishment of homeostatic norms, (2) sufficiently overcomes the stressor to allow it to nearly recover (compensate), or (3) starts down a trajectory leading to death. As we discuss more fully later, very low levels of stress (eustress) are actually adaptive, while higher levels of stress (distress, a term also difficult to define precisely; Holden, 2000) have maladaptive and adaptive elements. Here we are considering stress from the perspective of distress.

Basically, the primary reaction to the perception of a stressor involves the induction of a neuroendocrine cascade response involving the secretion and synthesis of the corticosteroid hormone (cortisol and related compounds) and the catecholamines (mainly adrenaline and noradrenaline, also called epinephrine and norepinephrine). The endocrinology of stress has been reviewed by Wendelaar Bonga (1977), Sumpter (1997), and recently by Pankhurst (2011). The main corticosteroid responding to stress is 11-deoxycortisol in lamprey (Close et al., 2010; Roberts et al., 2014), while in elasmobranchs it is 1α-hydroxycorticosterone (Idler and Truscott, 1966; Idler et al., 1967a) and cortisol and other related steroids in chondrosteans and teleosts (Idler and Sangalang, 1970; Webb et al., 2007). There are also large circulating concentrations of cortisone, a cortisol metabolite, consequent to stress in fish (Patino et al., 1987). It is thought that this conversion to cortisone serves to downregulate cortisol into a nonactive metabolite, but this is a subject little studied. It is evident that fish also can produce a host of other corticosteroids in response to stress; often these are at low concentrations, but we know almost nothing about their function or potency. A main role of these hormones is to make energy available for systems involved in fight, flight, or coping. Fig. 1.1 depicts the sequence of events that comprise the stress response. Gorissen and Flik (2016; Chapter 3 in this volume) provide a contemporary discussion of the endocrine stress axis and control of the stress response.

These secondary responses include the cardiovascular and respiratory responses (Rodnick and Planas, 2016; Chapter 7 in this volume), which increase distribution of oxygen as well as energy substrates that are liberated into the circulation also as a result of the stress response. Other accompanying secondary responses include a hydromineral dysfunction because adrenaline alters the gill blood flow patterns and gill permeability, both of which favor water flowing down its osmotic gradient, either in or out of the fish depending on environmental salinity. Thus, a logical role of cortisol in this regard would be the restoration of osmotic equilibrium

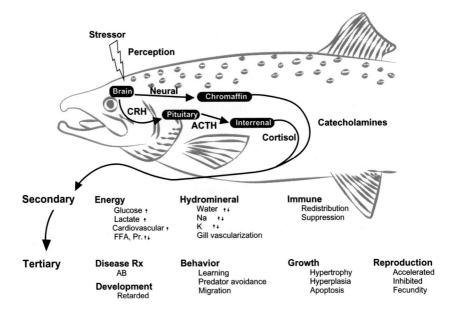

Figure 1.1. The primary (depicted inside fish), secondary, and tertiary responses of fish during distress. CRH, corticotropin releasing hormone; ACTH, adrenocorticotropic hormone; FFA, free fatty acids; P, proteins; AB, antibodies.

(Takei and Hwang, 2016; see Chapter 6 in this volume). Another response to cortisol, and perhaps other control factors, is immunosuppression. While the hormones ultimately have a positive effect on the organism by making energy available where it is needed during exposure to a stressor, many of the tertiary responses, those at the whole animal level, are maladaptive. For example, health and disease resistance (Yada and Tort, 2016; Chapter 10 in this volume), reproduction (Pankhurst, 2016; Chapter 8 in this volume), growth (Sadoul and Vijayan, 2016; Chapter 5 in this volume), learning, and other behaviors such as predator avoidance (Noakes and Jones, 2016; Chapter 9 in this volume) are all impaired.

1.1.3. STRESS AND HOMEOSTASIS

The stress response is clearly a necessary mechanism of fish to overcome severe challenges and, if possible, restore homeostasis. All the same, it is equally important to understand that under normal, nonstressful conditions a low circulating level of stress hormones is essential for maintenance of routine life functions such as growth (Schreck, 1992), the immune system (Schreck, 1996; Yada and Nakanishi, 2002; Dhabhar, 2008; Verburg-van Kemenade et al., 2009; Tort, 2011; Dhabhar and McEwen, 2001), development (Schreck, 1981), and learning (Sorensen et al., 2013; Martins et al., 2013). That is, the effects of levels of cortisol and catecholamines at low severities of stress are positive on routine life functions, while the effects become negative at higher and more prolonged elevations in concentration (Schreck, 1992). In other words, the individual elements involved in the stress response are part of homeostasis.

"Homeostasis" (Cannon, 1926, 1932) is another difficult word to define and concept to describe. The word is derived from the Greek roots *homeo* meaning "like" or "similar" used as a prefix and s*tasis*, meaning "stability" or "standing still," and is used to imply the relative constancy of the internal environment of an organism "despite continual changes in the organisms' external environment." This condition of stasis is maintained by feedback systems (Tortora and Grabowski, 2000; Guyton and Hall, 2011). Thus, homeostasis is a dynamic process and so it is very difficult to recognize a stable zone, which adds difficulty to the definition. Selye (1973) recognized this difficulty in terms of describing stress within a concept related to homeostasis. He suggested the term *heterostasis*, meaning that a new steady state is established by exogenous factors stimulating adaptive (nongenetic, or compensatory) mechanisms. This happens via development and maintenance of defensive tissue reactions that heretofore were dormant. The fluctuations experienced by the internal environment of a fish are small in contrast to the fluctuations resultant from stress. So, we can conclude that fish with elements of their internal environments beyond the ranges experienced when in homeostasis would suggest that the fish are under stress, which fits our definition. However, just as the variations experienced by a fish's internal environment during stress are quite large, there can be considerable variation as well during normal times when the fish is at homeostasis if we consider fish only during this state, excluding stress and other challenges that push the fish beyond this scope.

Stress results when a fish's system experiences conditions that result in homeostatic overload, the situation where routine physiological mechanisms cannot maintain a fish's internal environment within normal bounds. A conceptual integration of concepts involving stress, allostasis (see Section 1.1.4), and homeostasis led Romero et al. (2009) to propose a "reactive scope model" as a way of considering the physiological bounds between lack of stress and stress.

Recovery from stress returns the internal environment to within homeostatic ranges. If the stressor is acute and not overly injurious, full recovery can occur. However, in the face of a prolonged, chronic stressor or a very severe acute stressor, the fish may never reestablish the exact same characteristics of homeostasis, a condition referred to as compensation (Schreck, 2000). A good example of a temporary state of compensation in fish is acclimation to a new external environment; for example, the compensatory responses to thermal acclimation. However, there are two important distinctions: compensatory responses when acclimating are typically beneficial and reversible. Thus, while many clinical signs of stress will return to normal after an acute stressor, other physiological systems may be affected for a long time. For example, cortisol levels in a fish may return to normal within a few hours to a day following a brief (seconds to minutes) handling stressor, while the immune system can take at least a up to a week to do so (Maule et al., 1989). Consequently, the recovery process is very dynamic and not described by unimodal trajectories. There can be overshoots followed by corrections as systems return to unstressed conditions. In the example cited earlier, stress was followed by an immediate suppression of the acquired immune system followed the next day by an apparent enhancement over a prestress condition with a return to prestressed levels within a week (Maule et al., 1989). In addition, fish may never recover from severe acute or chronic stressors despite recovery of primary and secondary stress factors. Indeed, there can be a complete lack of concordance between clinical indicators (components of the primary and secondary stress responses) of stress that suggest recovery and the delayed mortality that starts to happen a week or two later (eg, see Davis et al., 2001 and Davis and Schreck, 2005).

Some of the earliest work in this regard found that forced activity across all species tested resulted in significant elevations in circulating lactic acid concentrations and that elevated concentrations correlated with the ability of the fish to perform under differing environmental conditions such as temperature (Black, 1955; Black et al., 1960). Further, fish can experience delayed mortality following severe muscular activity that first becomes apparent about 30 min after the cessation of the activity, perhaps due to disturbance to acid–base balance (Black, 1958). While the classical GAS indicators of the stress response may not suggest that the fish are out of homeostatic ranges, the best predictor of impending death was actually not a measurement of something classically thought of as being part of the GAS but related to reflex actions of the fish (Davis and Ottmar, 2006; Davis, 2007, 2010). Chronic stressors may also lack concordance between concentrations of stress response factors such as cortisol and whether or not fish are stressed. It is the mechanisms associated with homeostasis and heterostasis that appear to cause lack of concordance between primary and tertiary stress response factors. For example, Patino et al. (1986) found that cortisol levels can be similar for individuals that are unstressed or under chronic moderate stress due to crowding, while the capacity of the latter to resist other stressors such as low oxygen levels, elevated salinity, or disease challenge is severely lowered. This example of chronically stressed fish not having elevated plasma cortisol (termed a false negative) can be explained by the fact that they had elevated their cortisol clearance rates to manage the circulating concentrations down to a tolerable level. Spagnoli and colleagues (2016; Chapter 13 in this volume) describe indicators of stress and interpretation of data in detail.

The nonspecific stress response initiated by the central nervous system releases hormones that operate via highly specific receptors on or in the cells. Therefore, the specificity of target organs and cell types and whether the effect is an up- or downregulation are determined at the cellular level. For example, the nature of the receptor can determine the rates of reactions by a cell; membrane receptors mediate more rapid effects than those affecting genes located in the nucleus. Once set in motion, the stress response is regulated by various processes as the fish attempts to restore a homeostatic equilibrium:

1. Principally the negative feedback mechanisms that concern virtually all endocrine factors involved.
2. Hormonal up- or downregulation of their own receptors, thereby altering the magnitude of cellular responses.
3. Hormonal (or their end products) effects on metabolism and clearance of themselves. For example, induction of hepatic enzymes specific for each hormone can inactivate the hormone into a form that can be excreted, preventing prolonged exposure of tissues to extreme hormone concentrations that could eventually lead to death if allowed to persist.
4. Hormone-binding plasma carrier proteins that are somewhat specific and reduce hormone availability for receptor binding at target tissues. Indeed, these carrier proteins may have a large binding capacity relative to the freely dissolved (ie, unbound) hormone in the blood, which is the case with cortisol, but the affinity for the steroid is relatively low when compared the comparable protein in mammals (Caldwell et al., 1991). The high capacity suggests that a high proportion of steroids such as cortisol that are bound to such proteins are not available to bind to their specific receptors; the consequence of this would be a reduction in effects of the hormone. However, the low affinity suggests that receptors could perhaps strip the hormones off of the proteins, hence reducing the regulatory efficiency of the proteins. While we know that they are present in elasmobranchs (Idler et al., 1967b; Idler and Freeman, 1968) and teleosts (Freeman and Idler, 1966; Idler and Freeman, 1968), there is a paucity of information on carrier proteins of hormones concerned with stress in fish. We thus do not know how important such a protein is in fish as a regulatory mechanism for downregulating stress hormone action. Such hormone-specific binding proteins may also be protective of eggs being developed in females that are experiencing stress. Binding proteins could help keep elevated circulating levels of hormones such as the

hydrophobic cortisol in the maternal circulation, greatly reducing how much could cross into lipophilic eggs (Schreck et al., 2001).

It should also be noted that because of all the regulatory mechanisms involved, that the response patterns of onset and duration for each stress response factor can be quite different from each other. Gorissen and Flik (2016; Chapter 3 in this volume) provide a more thorough review of this area.

1.1.4. STRESS AND ENERGETICS

The physiological response to stressors ultimately results in an effect on the entire body; all physiological processes are impacted to some extent largely because there is a need to provide energy needed to overcome stressors. Fight or flight (Cannon, 1926, 1932) requires additional energy for the acute nervous and muscular emergency responses, while chronic exposure to stressors that render energy unavailable for other life processes must be rectified. These costs can be viewed as allostasis, the price that a fish pays in recovering from or coping with a stressor (Schreck, 2010). Sterling and Eyer (1988) proposed the paradigm allostasis as "the ability to achieve stability through change." Others have elaborated on the process in endothermic animals (eg, McEwen, 1998; McEwen and Wingfield, 2003; Ashley and Wingfield, 2012), while Schreck (2010) discussed this concept for fish. Also relevant to energetics are Sadoul and Vijayan (2016; Chapter 5 in this volume) on metabolism, and Rodnick and Planas (2016; Chapter 7 in this volume) on the physiological impacts and on swimming and cardiovasacular responses.

The energetic cost of resistance to, recovery from, and coping (attempting to restore homeostasis) with a stressor can be energetically costly and quantified in terms of calories or another metabolic currency (eg, mg/L oxygen consumed/hour/kg of fish). There are two types of allostatic load, type 1 and type 2 (McEwen and Wingfield, 2003). Type 1 is more acute and is associate more with emergency-type of responses, such as fleeing from a predator, while type 2 refers more to the coping-type of responses, such as being in a constantly overcrowded rearing environment, and is more typical of chronic stressors (Schreck, 2010). Moreover, accommodation to a stressor diverts energy from other life processes. Mathematically, the energetic cost of an allostatic load ($E_{\text{allostasis}}$) can be viewed as:

$$E_{\text{Somatic Growth}} + E_{\text{Gametic Growth Maturation}} + E_{\text{Activity}}$$
$$= (E_{\text{Stored}} + E_{\text{Food}}) - (E_{\text{STD}} + E_{\text{SDA}} + E_{\text{Waste}} + E_{\text{Allostasis}})$$

with $E_{\text{Somatic Growth}}$ and $E_{\text{Gametic Growth/Maturation}}$ representing the energy used for the two types of growth, the body and reproduction (gonads and secondary and tertiary sex characteristics).

$E_{Activity}$ includes energy needed for all general activities such as moving (eg, feeding, swimming, and those associated with reproduction).

($E_{Stored} + E_{Food}$) represents mobilizable forms of energy available to the fish. It can be used for positive functions such as both types of growth, activity, and standard metabolism (STD). STD in the context used here is instantaneous standard metabolism, which can be thought of as the instantaneous standard metabolic rate (SMR). We use the label STD to avoid confusion that could arise by the use of SMR, which has a time component.

E_{STD} represents the instantaneous energetic costs associated with standard metabolism (the minimum amount of expenditures needed just to keep the fish alive such as for respiration, circulation, and osmoregulation).

E_{SDA} represents the energy unavailable due to the energy cost of feeding and food processing. Metabolic inefficiencies as part of this can be thought of as entropy and is unavailable to the fish, being lost to the environment in the form of heat.

E_{Waste} represents energy in excreted wastes.

$E_{Allostasis}$ represents the gross energy needed to resist and recover from a stressor. That includes the energy that goes directly into processes of resistance and recovery as well as the energy lost due to specific dynamic action.

There is a surprising scarcity of literature regarding the energetic cost of stress in general. Most of the papers relate to stress due to infection and energy demands of the immune system (for reviews see Sheldon and Verhulst, 1996; Lochmiller and Deerenberg, 2000; Sandland and Minchella, 2003; Martin et al., 2003, 2008; Demas et al., 2012). Only a few papers discuss energetics of stress in fish, but we know that it is metabolically costly (Barton and Schreck, 1987). For example, stressors such as brief handling can cost between 12% and 30% of the entire scope for metabolic activity (Davis and Schreck, 1997). There are also a few papers that consider environmental stressors from an energetic perspective (eg, Glencross and Bermudes, 2011; Ogoshia et al., 2012).

The capacity of fish to perform any of life's tasks such as growth, development, reproduction, disease resistance, and general activities are ultimately determined by the animal's genetics. However, a fish's realized performance capacity, the phenotype that it expresses, is delimited by its ambient environment (how near or far it is from optimum conditions for each of these activities) and the environments the fish has had earlier in its life, such as nutritional history (Schreck, 1981; Schreck and Li, 1991) or behavioral experiences (Schjolden et al., 2005). Eustress can increase and distress can decrease this performance scope (Fig. 1.2). This conceptualization of energetic cost associated with stress allows us to suggest that the $E_{allostasis}$ needed during acutely stressful situations may not by sufficiently great enough to interfere to any large extent with normal life processes such as growth, reproduction, migration, and the like. Indeed, during acute stress a fish relies almost solely on stored forms of energy, in part because feeding behavior is disrupted (eg, reduced feeding, digestion, assimilation) and blood flow to certain organs is diverted for use by those needed in fight or flight. Also, glycolysis generates adenosine triphosphate more rapidly than oxidative phosphorylation. Death resulting from acute stress would likely be due to cardiovascular failure and happens during the time the fish is exposed to the stressor or soon thereafter (delayed mortality). If death happens

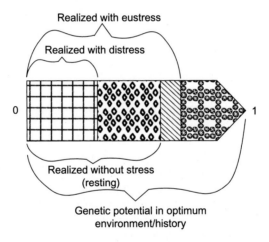

Figure 1.2. Conceptual depiction of the performance capacity of a fish as represented by a vector that ranges from death, or 0 capacity, to the maximum capacity that a fish could achieve based on its genetics and if it lived in an optimum environment in every way, valued at 1. This vector could describe any performance trait. The concept illustrates that the maximum performance can be viewed as a magnitude ranging from no performance (0) to the maximum potentially achievable performance (1), determined by the organism's genetics in a perfect environment. But given that the environment is never perfect, in the absence of stress (realized without stress) the actual performance realized is less than the maximum potential performance. The realized performance can be viewed as the scope to actually perform work. A small amount of stress (eustress) would increase the magnitude of the actual performance capacity while more severe stress (distress) would diminish it. Therefore, given that the environment and magnitude of a stressor can change through time, so does the realized performance capacity. The fish's realized capacity in the absence of any stressor is delimited by the fish's environment and history (and epigenetics). Stress further increases or decreases the realized performance capacity.

within a week or more following a stressor, then it is likely due to secondary disease caused by replicative pathogens (those that can complete their entire life cycle in a host) with which the fish are infected because of the immunosuppressive effects of cortisol. However, there are situations where the cause of mortality starting about a week after an acute stressor cannot be determined (see Section 1.1.3).

In terms of total energetic costs we propose that chronic stress would impose a significantly greater allostatic load than could be experienced during acute stress, even those that are quite severe but not injurious; that is because of the prolonged energy demand under chronic conditions. Under severe chronic stress, from which the fish could not either fully compensate or recover, energetic costs would be sufficiently large enough that stored energy and food acquisition would be insufficient for growth, reproductive development, and ontogenetic processes. They would be negatively impacted or completely turned off. At the highest levels of allostatic load there would be insufficient energy for general activities, and death due to other challenges becomes more probable; for example, they cannot mount a response to a random encounter with some threat such as a predator or an infection because of immune failure. In this event, even replicative and nonreplicative parasites (those requiring another host to complete their life cycles) could cause death. These paradigms are supported by the syntheses proposed by Schreck (2010).

1.2. Dynamics of the Stress Response and Effects on Performance

The response of the primary, secondary, and tertiary stress response factors to stressors are dependent on the severity and the duration of the stressor. They are also contingent on the genetic heritage of the fish, the environment in which the fish are located, their prior experiences, and the ontogenetic stage they are in when exposed to a stressor. These variables also shape the impact of the stressor on each individual (Schreck, 1981; Schreck and Li, 1991). Fig. 1.3 depicts conceptual responses of the primary, secondary, and tertiary stress response factors to stressors of different types, severities, and exposures. Because of the inherent variability associated with the dynamics of the stress responses and because of differences in experimental animals and research designs and procedures, studies on stress in fish can differ in major ways from each other relative to the dynamics of the stress responses. This variability between studies is a function of differing initial states of the fish (genetics, developmental stage, history, general health, etc.), nature of the stressors involved, duration of the stressful events, and sampling periodicities. Thus, interpretation of results using a single sampling time are particularly problematic. This confounds

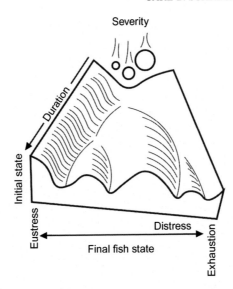

Figure 1.3. A conceptualization of how the final state (well-being in terms of stress response) of a fish would differ when exposed to stressors of differing severity for various periods of time and under differing potential starting physiological conditions of the fish using a landscape perspective. Envision balls (representing the physiologic state of the fish) of differing mass (depicted as size) falling onto a downward sloping landscape of uneven terrain; where the balls would end up at the bottom would be different. How far down the hill a particular ball lands would represent the duration of a stressor, with a shorter duration being closer to the end (downhill). The shape of the landscape is affected by the initial state of the fish: the better the initial condition, the better the final state would be at the bottom of the slope. If the ball were to take a trajectory ending near the far left (near the left side of the double-headed arrow), the fish would experience eustress. A mild stressor of short duration would create a landscape where the state of the fish would be somewhat similar to what it was prior to the stressor (somewhere toward the left but not as far as for eustress along the double-headed arrow). Increased stressor severity, durations of stress, and decreased quality of initial fish condition would result in a trajectory down landscapes leading to various degrees of poor final state of the fish and at extreme stress exhaustion (toward the right end of the double-headed arrow). The double-pointed arrow represents a continuum ranging from maximum eustress on the far left to extremely poor health and condition on the far right.... This conceptualization is depicted somewhat similar to Waddington's (1977) epigenetic landscape.

comparisons from which generalizations can be precisely drawn regarding descriptions of the temporal dynamics of the responses involved with stress. While we can describe the shapes of responses, we cannot overlay these with accurate times or magnitudes.

There is a strong genetic component to the dynamics of the stress response (Fevolden et al., 2002; Øverli et al., 2005), with large differences between taxa. Thus, the heritability must be quite high. Although not quite as large, individuals within a population differ considerably in their stress responses (Ellis et al., 2012). There can also be sexual dimorphisms in how fish respond to stressors ranging from gene activation (eg, see Momoda et al., 2007) to the magnitude of responses by GAS factors. Therefore, resting values for stress response factors, magnitudes of the response, and temporal dynamics of the responses can all vary considerably with genotype. These differences are reflected in how well different taxa and individuals within taxa can tolerate different stressors; some have small scopes of tolerance, others have quite broad ranges of tolerance. In addition, there is lack of concordance of how different taxa and different individuals of a population respond to different stressors. We speculate that variability exists between different ontogenetic stages of fish within a taxon, but there is an extreme paucity of data in this area. Winberg et al. (2016; Chapter 2 in this volume) and Faught and colleagues (2016; Chapter 4 in this volume) review variation the genetic aspects of the stress response in detail.

The ambient environment in which fish experience a stressor influences the physiological responses, the tolerance levels, and their rates. Fish live in a variable environment and therefore potentially stressful (Schulte, 2014), and environmental variables can affect the stress response of fish differently because they operate via different mechanisms. Some examples include (1) temperature can affect the dynamics of the stress response elements by affecting the rate of reactions because of the Q_{10} phenomenon; (2) oxygen concentrations can affect metabolic capacity, thereby influencing the magnitude and duration of the allostatic load; (3) the social environment can cause fish hierarchies that respond differently to a common stressor; (4) state of digestion (full vs empty stomach) can affect stress tolerance; and (5) pathogen infection can alter the responses and tolerances compared with healthy individuals. In this volume, Chapters 2, 5, 7, 10 and 12, all consider effects of the environment, including fish welfare, in some fashion.

A fish's prior experience, especially the environments in which they have been residing, can greatly affect how they respond to and during stress. An easy way to visualize this is to consider differences between starved and well-fed fish. Also, prior experiences with stress can potentially either debilitate stress responsiveness and/or tolerance, or could physically and psychologically "harden" the fish to better handle another stressor. Chapters 2, 7 and 9 discuss the importance of fish condition further.

An emerging area is how the experiences of its parents and even perhaps grandparents can influence the response to a stressor though an epigenetic effect. There is considerable literature on mammals concerning epigenetics as related to stress (Fish et al., 2004), including environmental stressors (Feil and Fraga, 2012). Some of the epigenetic effects of stress appear to be mediated by the corticosteroids (Lee et al., 2010). While there is a paucity of literature regarding epigenetics in fishes, stress must also have profound epigenetic consequences. For example, Mommer and Bell (2014) and McGhee and Bell (2014) found maternal effects of stress in fish on their progeny, as in mammals, and the latter demonstrated that the effect appears to operate via a DNA and histone methylation processes.

Different developmental stages of a fish's life cycle have differing responses to and tolerances of stressors. It appears that when fish are going through some ontogenetic transition in their lives they are more sensitive to stressors (Barton et al., 1985). Thus, the magnitude of the stress response is greater and their tolerance to stressors are lower than at times when they are not undergoing such a transition. Transitional stages when the fish are less resistant to stressors include the embryo before the eye pigments are formed, hatching, the time of onset of per os feeding, metamorphosis (or smolting), and near the end of reproduction (Fig. 1.4) (Feist and Schreck, 2002). The ability to respond to stressors appears quite early in a fish's development. The embryo is capable of producing cortisol starting around the time when the eye pigmentation becomes apparent. Elevated crowding of embryos at this stage impaired cortisol secretion and feedback (Ghaedi et al., 2013, 2014).

Stress can negatively affect the reproductive fitness of adult females (Schreck et al., 2001), by either lowering fecundity (number of eggs ovulated) or lowering the quality of the eggs that are ovulated. We conceptualize that the reproductive responses to stress as a series of tradeoffs. Females under stress must partition available energy between recovery from, or tolerance of a stressor and energy investment into maturing eggs to an acceptable level of quality before ovulation. The ultimate tradeoff is between probabilities of female survival versus net lifetime reproductive success (number and quality of offspring). However, a stressed female may not invest in maturing eggs and spawning, delay spawning and potentially lose any investment already made in egg production, mature as many eggs as possible that are nutritionally deficient and result in progeny with lower fitness but in higher numbers, or allow some eggs to become atretic and maximize reproductive investment into a smaller number of eggs that have higher nutritional content and fitness. Given that energy availability is a common currency for these various tradeoffs, nutritional stress is obviously of fundamental importance in setting a female off along one of these trajectories. Examples of all of these tactics for coping with a stressor during egg maturation can be found among fishes. Maternal transfer of biologically active substances associated with the stress response such as cortisol can occur in eggs prior to ovulation. We know very little regarding how stress affects the male's stress response or its consequences. Chapters 2, 8 and 12 review various roles of ontogeny and the stress response in more detail.

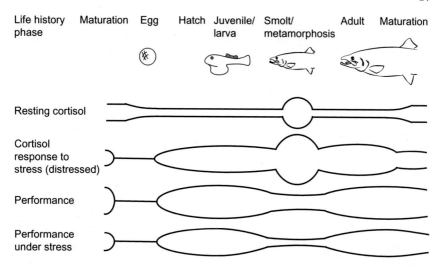

Figure 1.4. Conceptual model depicting circulating resting and distressed fish cortisol levels and performance capacity in the rested (performance) and stressed (performance under stress) fish over the life history of a fish. The space between the lines represents the magnitude of either the plasma cortisol concentration or the performance capacity (ability to perform work). This conceptualization can be likened to the dimensions of the n-dimensional niche hypervolume concept of Hutchinson (1958).

1.2.1. THE RESPONSE OF THE PHYSIOLOGICAL SYSTEMS

Perception of a stressor initiates an essentially immediate chemically mediated cascade (the primary stress response) that results in qualitatively more or less similar secondary and tertiary stress responses. Neuronal signaling, within a fraction of a second, triggers chromaffin tissue to produce and secrete catecholamines, which control the secondary and tertiary responses. Other neuropeptides such as dopamine and gamma-aminobutyric acid (GABA) are also affected. The rapidity of catecholamine release into the bloodstream (within seconds) makes it difficult to study in fishes, in part also because of the difficulty of obtaining a blood sample from free-swimming fish without further stress (blood vessel cannulation techniques are required). For a review of the catecholamne stress response in fish, see Perry and Bernier (1999). The other hormonal release is slightly slower and more prolonged (within minutes and lasting hours) and is initiated by the production of corticotropic hormone releasing hormone (corticotropin releasing hormone, or CRH; also referred to as corticotropic releasing factor, or CRF) in hypothalamic cells of the brain. The CRH travels to the adrenocorticotropic hormone (ACTH) producing cells in the anterior pituitary, which are specialized to produce and secrete ACTH into the circulation. ACTH targets interrenal cells in the head kidney, causing synthesis and secretion of cortisol in the circulation. Both CRH and ACTH can directly affect secondary stress responses, with ACTH operating centrally directly on the brain (eg, see Clements et al., 2001) and both CRH and ACTH acting peripherally as on the immune system (eg, see Schreck and Maule, 2001). In fact, there is considerable cross-talk between chemical mediators of the endocrine and immune systems related to stress (for a review of psychoneuroimmunology, see Ader, 2007).

There is considerable variation between species in how long it takes before significant elevations in circulating cortisol levels can be seen following onset of a stressor, with most fish studied requiring several minutes. Winberg et al. (2016; Chapter 2 in this volume) review the variation seen in the dynamics of various stress response factors and species, and Gorissen and Flik (2016; Chapter 3 in this volume) cover the endocrinology involved.

The main role of the stress response is to make energy available for processes involved in fight, flight, or coping. Stored forms of energy are metabolized and liberated from organs such as the liver into the circulation in forms more readily utilized by tissues. For example, both the catecholamines and cortisol cause hepatic glycogenolysis, making glucose readily available for use by skeletal muscles. In addition, cortisol can downregulate glyconeogenic systems, thereby keeping the glucose available to the tissues. Free fatty acid and protein concentrations in blood may also be affected, as explained by Sadoul and Vijayan (2016; Chapter 5 in this volume). The catecholamines trigger cardiorespiratory responses (recruit gill lamellae; increase cardiac output, blood pressure, and gill ventilation) that help provide more oxygen to tissues; Rodnick and Planas (2016; Chapter 7 in this volume) review this aspect of the stress response. However, an increase in functional gill surface area enhances water as well as oxygen flow, termed the osmorespiratory compromise. Hence, stressed fish in freshwater take on water down its osmotic gradient, while those in seawater lose water; there can thus be weight gain in the former due to stress and weight loss in fish in the latter (Stevens, 1972). Blood ion fluxes are also evident during osmo-respiratory compromise, and any resulting change in ionic concentrations must be restored. Concentrations of differing electrolytes may not follow similar response patterns during and after exposure to a stressor; this is likely explainable by differing cellular processes relative to each electrolyte (Stewart et al., 2016). One of the main roles of cortisol is the activation of pumps involved in transporting ions into and out of the fish at the gill, gut, and kidney. Restoration of osmotic and ionic equilibrium must be one of the main roles of cortisol following the hydromineral disequilibrium consequent to stress. Takei and Hwang (2016; Chapter 6 in this volume) review the effects of stress on hydromineral balance in fish. The time-course for measurable hydromineral disturbances is extremely variable.

Stress affects the immune system. The initial phase of an immune response is the activation of certain nonspecific responses, increasing processes such as apoptosis. However, even very acute stressors can suppress the acquired arm of the immune system, which is highly specific. Cortisol suppresses the capacity of lymphocytes to synthesize antibodies and the processes leading to leukocyte production and mobilization. An inevitable consequence of this suppression is that exposure to a stressor makes fish more susceptible to infection and disease until the immune status of the fish oscillates back to prestress immune competence.

Stress also can affect biological processes by shifting the normal timing of such processes. Schreck et al. (2001) and Schreck (2010) discussed this phenomenon relative to reproduction. It appears highly likely that other "biological clocks" can be interfered with when fish are under stress (Sánchez et al., 2009).

There are also other physiological responses to stressors that are somewhat general but that do not fit neatly into the GAS paradigm. For example, heat shock proteins (HSPs) are part of the cellular stress response. These proteins, also referred to as stress proteins, can be classified into several families of proteins based on their molecular weights. Each family can have somewhat different functions. Basically, these proteins are protective. One of their main functions is to serve as chaperone proteins, helping preserve the tertiary structure of other proteins by preventing their deformation by stressors. These proteins are produced by cells in response to a variety of stressors including heat, hypoxia, oxidative stressors, and pathogens, and they are found in most teleosts (Iwama et al., 1998; Basua et al., 2002; Kayhan and Duman, 2010; Roberts et al., 2010; Currie, 2011; LeBlanc et al., 2012; Templeman et al., 2014; Stitt et al., 2014). They appear to have similar functions in elasmobranchs (Renshawa et al., 2012).

While we know that many processes at the level of the whole organism such as growth, disease resistance, reproduction, and development are impaired to some degree by stress, the nature of the pattern of this impairment or the duration of the impairment are not well studied. Of course, under severe stress, there would be no recovery and death would result.

1.2.2. ACUTE, CHRONIC, AND MULTIPLE STRESSORS

The magnitude and duration of the individual components of the physiological stress response are highly dependent on both duration and severity of a stressor such that the magnitude of the response to an acute, severe stressor could be as great as that due to a prolonged, mild stressor. Even so, the dynamics and magnitude of the response to the alarm phase tend to be more stereotypical irrespective of the magnitude or duration of the stressor, unlike the resistance and compensation/recovery phases (Schreck, 2000). The physiological responses to acute stressors and the subsequent effect on the capacity of fish to perform necessary life functions (eg, grow, develop, reproduce, and resist pathogens) can vary with exposure to acute stressors (Fig. 1.5) and to chronic stressors (Fig. 1.6).

Regarding chronic stressors, Boonstra (2013) argues that chronic stressors in nature do not lead to pathology in wild animals, basing his views on the functionality of the hypothalamic–pituitary–adrenal (HPA) axis (hypothalamic–pituitary–interrenal (HPI) axis in fishes). We view this slightly differently and suggest that other coping mechanisms may maintain life and even a functional HPI axis in the face of a common stressor (eg, Patino et al., 1986). That is not to say that under chronic stress the stress response is not adaptive, it is. However, there are bioenergetic costs associated with physiological resisting and coping mechanisms that can lead to poorer capacity to perform other of life's functions (eg, resist pathogens) (see Section 1.1.4). How we define stress also can affect how we interpret effects of chronic stressors. The question really is, does chronic stress lower an animal's fitness? It does not need to be lowered to zero (as implied by Boonstra, 2013), in fact not even very much, to have the potential for major evolutionary consequences. It has been suggested that acute stressors are much more common than chronic stressors in nature, at least for terrestrial animals (Boonstra, 2013). This may not be the case for fishes, particularly those in freshwater environments that can experience prolonged elevations in temperature, reduced flows, or contaminant exposure. Even marine environments can experience prolonged cooling or warming, ocean acidification, and hypoxia that at times can push some species of fish beyond their tolerance limits. The role of social status related to chronic stress and its effects is important (see Sapolsky (2005) for an interesting analysis of this in primates). In any case, there is an extreme paucity of information on the physiology of fish in nature, and much of what we think is inferred from studies conducted in laboratory or hatchery environments.

Fish can often be exposed to the same acute stressor in some sequential order, such as in mark-recapture experiments. We know that such situations can lead to cumulative effects. However, how stress response factors respond temporally and in magnitude to multiple stressors is very poorly understood. We do know that acute stressors experienced a few hours apart lead to nearly additive effects on stress response elements, for at least three

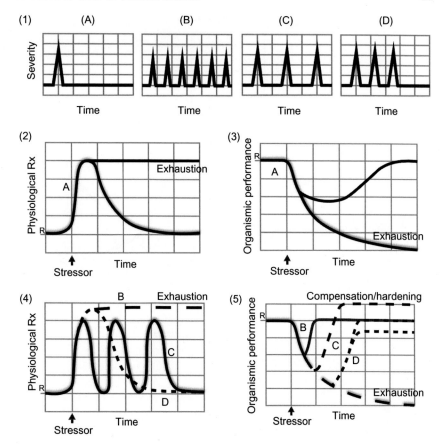

Figure 1.5. Four temporal patterns of a stressor (1): (A) a single acute stressor, (B) numerous sequential stressors with relatively short time intervals in between each, (C) sequential stressors with relatively longer time intervals in between each, and (D) sequential acute stressors with relatively short time interval between each. Temporal dynamics of primary or secondary stress factors (2) and performanceor physiological response (Rx) (ie, ability to perform work) (3) when exposed to a stress pattern A. Stressor effects can be mild, resulting in recovery to the resting state, or severe, resulting in exhaustion (ie, death). Temporal dynamics of primary or secondary stress factors (4) and performance (5) when exposed to stress patterns B, C, and D. R represents the resting state. Single or sequential stress patterns (A, B, C, D) are initiated starting at the Stressor arrow. Stressor effects range from those having no long-term performance consequences to those that are mild, resulting in compensation/hardening (ie, eustress), or those that are severe, resulting in exhaustion (ie, death).

exposures, but we do not know how many more could continue this trend. We also do not know the minimum amount of time between stressors before two or more stressors are perceived as one event rather than multiple events. Similarly, we do not know how much time needs to elapse between

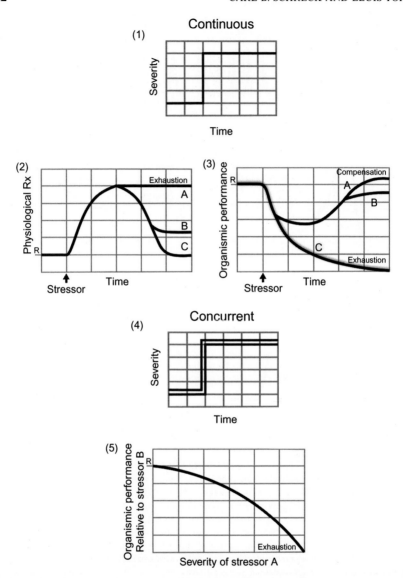

Figure 1.6. Temporal pattern of a chronic, continuous stressor of various severity (1). Temporal dynamics of primary or secondary stress factors over time (2) and performance or physilogical response (Rx) (ability to perform work) (3) when exposed to a continuous stressor initiated at the time of Stressor arrow. A mild stressor leading to compensation and eustress (A), severe stressor leading to distress (B), and an extremely severe stressor leading to exhaustion and death (C) are represented. R represents the resting state. Stressors may also occur concurrently and continuously (4). The effects of increasing severity of stressor A on the ability of the fish to perform or tolerate stressor B to which it is exposed concurrently (5).

sequential stressors before the presence of a former stressor is no longer apparent (Schreck, 2000) (Fig. 1.5).

When two different stressors are encountered simultaneously (eg, crowding in low oxygen environments), the presence of one stressor can lower the tolerance to another stressor (Schreck, 2000) (Fig. 1.6). However, a gradual downregulation of the stress response in a fish that is coping and slowly compensating for a chronic stressor should not be confused with an apparent exhaustion, attributable perhaps to lack of capacity to physiologically respond to more stressors. The diminished capacity to physiologically respond to other stressful encounters (Huntingford et al., 2006; Pankhurst, 2011; Zuberi et al., 2014) during chronic stress is covered by Sneddon et al. (Sneddon et al., 2016; Chapter 12 in this volume). During compensation for stress, fish purposely downregulate the stress response, while a diminished stress response can also occur because of an overload of stressors and a response cannot be mounted.

Care is also needed to distinguish when fish condition themselves to stressors. After prior experiences to other mild, sequential (perhaps daily) stressors for a period of time, the magnitude and duration of the stress response can become attenuated. This is likely due to the physical conditioning that occurred during the earlier exposures to the earlier stressors. A mental component can play a role in the enhanced performance of fish that have had prior experiences with mild stressors. Habituation to repeated stressors is a good example. For example, positive conditioning where food (a reward) was given to fish following brief and mild daily stressful experiences improved performance better to subsequent severe challenges compared with no reward. The positively conditioned fish had lower physiological responses to subsequent stressful events and were also better at tolerating other challenges such as lower dissolved oxygen, increased salinity, and disease (Schreck et al., 1995). Schreck (2010) referred to this as stress hardening. Winberg et al. (2016; Chapter 2 in this volume) discuss variation in how fish respond to and handle stressful situations.

Conceptually, we view the response dynamics of both physiological stress factors and animal performances to stressful situations as a multidimensional family of curves that cannot simply be described or displayed. The nature of the response curve due to a stressor is contingent upon multiple factors including those endogenous to the fish such as the fish's genetics, developmental stage, and prior experiences by the fish (environments and stressors experienced in the past). They are also contingent upon external factors such as severity and duration of the stressor, developmental stage of the fish, and the present environment of the fish. All of these factors operate in concert to affect the way fish respond physiologically to stressors. They also affect the performance capacity of the fish; that is, they affect how well the fish can perform all other life functions (Schreck and Li, 1991). Fig. 1.3 illustrates conceptually the pattern and duration of responses to acute and chronic stressors in a much simplified manner.

1.3. Contemporary View of the GAS: Eustress versus Distress

We need to revisit the concepts of eustress and distress to properly acknowledge that the responses of fish to various stressors or durations are not necessarily unimodal. Bimodal responses are common, such as a low severity of a stressor has one effect, while a higher level of severity has the opposite effect. This concept, hormesis, was discussed more fully by Schreck (2010). It can be considered an adaptive response that induces compensatory biological processes following an initial disruption of homeostasis. How the immune system of fish responds to various severities of stress is a classic example of this phenomenon. Low levels of stress result in just slightly elevated levels of cortisol that have a positive effect on the immune capacity by upregulating leukocyte redeployment, innate immunity, effector cell function, and cell-mediated immunity. The net effect of this is enhanced resistance to infections and cancers. This positive level of stress is referred to as eustress (Dhabhar, 2008; Dhabhar and McEwen, 2001; Schreck, 2010).

As the severity of a stressor increases and hormones such as cortisol become somewhat more elevated over eustress states, the immune system teeters in a state where the positive effects on the immune system start to disappear and those that are all negative appear. This condition, resilience, is followed as the severity of the stressor increases yet more by a suppression of all the immune factors mentioned earlier. The net effect of this is suppressed resistance to infections and cancers. This state of stress where effects are negative is referred to as distress.

1.4. Sensory Systems and Perception

As mentioned earlier, a classic stress response is only initiated consequent to the perception of a threat. Most of the literature concerning the perception of stressors in fish follows the paradigm that stressors are sensed through activation of the classic senses (vision, hearing, touch, smell, and likely taste in fishes). These activate the neuroendocrine response in the hypothalamus of fish. Schreck (1981) suggested that from an anthropogenic perspective that those events in a fish's life that can cause fright, discomfort, or pain result in a stress response. However, other sensory systems also appear to play a role in stress perception and a subsequent stress response. For example, exposure to environmentally relevant levels of acidity (Barton et al., 1985), hypoxia (O'Connor et al., 2011), hypercapnia (Schreck, 1972; Sandblom et al., 2013), osmostress (Takei and Hwang, 2016; Chapter 6 in this volume), and pathogens/parasites (Suzumoto et al., 1977) can result in elevated cortisol. These may affect the stress response system via neural routes not part of the classical senses. Stressors affecting mucosal surfaces, such as gills, stomach–intestine, or skin, can generate responses at these sites that first develop as a local response that subsequently leads to a neuroendocrine or systemic response. This is the case with stressors such as unbalanced or deficient diets that affect the gut and may become chronic stressors (Montero et al., 2001). Similarly, contaminants or pollutants at subacute levels can affect gills locally. Clearly pathogens that act at as the portals of entry such that gill, skin, or gut can generate local responses that involve cytokine expression and production and consequent activation of the HPI axis. It is difficult to precisely determine when the HPI axis will be activated after the entry of a pathogen or a parasite as it contingent upon the virulence of the pathogen and infection intensity. Once the HPI axis is activated, their messenger elements perform their actions in the target tissues through their specific receptors. Gorissen and Flik (2016; Chapter 3 in this volume) and Yada and Tort (2016; Chapter 10 in this volume) discuss the roles of the stress hormones and their receptors within this context.

1.5. Adaptation versus Nonadaptation Aspects of the Stress Response

Physiological adaptation is the sum of mechanisms that allow the organism to overcome (adjust to or recover from) the challenge posed by the stressor. These mechanisms could involve compensatory responses and/or habituation. This involves all physiological response mechanisms, including energetics, immune responses, and also learning and coping strategy.

Nonadaptation is the result of not meeting the resources to overcome the stressor, and/or the uncontrollability of the reaction. In fact, fishes, like other vertebrates, have a number of abilities and mechanisms to cope with stressor stimuli and overcome the consequences of such stressors. As mentioned earlier, the stress reaction depends on the type of stressor, its intensity, and its eventual repetition. However, two factors, uncontrollability and unpredictability, will always make very difficult the coping possibilities, since in such conditions the allostatic load will increase (Korte et al., 2007; Schreck, 2010).

Fish in aquaculture are subjected to a number of constrictions, in particular the movement restriction and the regular environment. This means that the learning features and hardening will be different from wild fish.

For an adequate fish stress management under captivity conditions, it is very relevant to know the main behavioral responses, the capacity of coping, and therefore a certain predictability of the fish responses. In fact, most of the work done in the area of fish welfare over recent years has investigated the behavioral traits that determine the capacity of coping with stressors and the limits of welfare. Model fish should provide a more complete framework (Spagnoli et al., 2016; Chapter 13 in this volume).

1.6. Key Unknowns

Contemporary research approaches have provided major new insights into the stress physiology of fish and have considerably broadened the paradigms that describe how fish respond to stressors (Table 1.1). However, there are numerous unknowns that must be resolved for us to have a more comprehensive understanding of the stress in fish. The following describe some of these key unknowns and questions.

Our ability to predict when a fish is stressed is relatively good. For example, stress hormone levels that are above resting concentrations allow reasonable inference that fish are stressed. Similarly, elevated concentrations of HSPs also allow fairly robust speculation that fish are stressed to near the point of cellular damage, which needs to be managed. However, there is little basis upon which to predict when fish have recovered from exposure to either acute or chronic stressors. In other words, circulating concentrations of stress hormones or other stress factors do not necessarily mean that the fish have recovered fully; they just indicate that either the fish have recovered or they have been able to lower these concentrations.

Information is needed to allow conclusions regarding the magnitude of stress response factors over resting levels. For example, slight elevations in magnitude of response may be adaptive if they lead to a eustress condition while slightly higher magnitudes of response may be maladaptive in either the short or long term. We do not know where this tipping point is. Further, we do not know if higher magnitudes of a stress response indicate that a fish is more stressed than at lower levels of response once magnitudes are above those associated with eustress; similarly, we do not know if the magnitude of a response is the same or differs over a range of severities of the same or different stressors. Another dilemma is that we do not know if stressors perceived through different sensory systems lead to similar or dissimilar stress responses. Information is needed to inform the question: Is the response to a stressor operating under a graded or a threshold control system? For example, do increasingly elevated concentrations of stress hormones cause continuously increasing levels of responses, or are there minimum concentrations needed for a response with higher concentrations having no additional effect, until potentially some other set point is reached?

Table 1.1
Classical and contemporary approaches and thinking regarding stress characteristics in fish

Subject	Classical	Contemporary
Behavioral response	Species-specific	Species-specific Coping strategy
Biological hierarchy	Organismic	Organismic Cellular Molecular
Endocrine stress response	Hierarchical (eg, hypothalamus, pituitary, interrenal)	Nonhierarchical (multiple directions)
Learning and adaptation	Stereotyped	Learning and adaptation
Physiological regulation	Nervous system Endocrine system	Nervous system Endocrine system Immune system
Regulation	Homeostasis	Homeostasis Allostasis
Response dynamics	Unimodal	Unimodal Bimodal
Response variables	Physiological Metabolic	Physiological Metabolic Energetics/resource allocation
Sensing	Central (nervous system)	Multiple (central and peripheral)
Signaling	Hormones	Hormones Cytokines
Stress effects	Organismic Physiological	Organismic Physiological Pain Psychological
Stressor effects	Distress	Distress Eustress
Stressor traits	Nature of stressor Intensity of stressor Duration of stressor	Nature of stressor Intensity of stressor Duration of stressor Control of stressor Predictability of stressor
Variation	Interspecific	Interspecific Individual

Information on how fish in the wild respond to stress is extremely rare. It is important to validate that what is learned about stress from fish in captivity can be extended to their wild counterparts. It would also be exceptionally useful to determine how to study the initial phases of stress in wild fish that avoid potential error introduced due to stressors associated with the fish and sample collection process.

A large number of studies in fish have provided data that should allow for the establishment of normal, nonstress ranges in values for stress factors and for their functions. However, this is actually quite problematic because the numerous factors affecting resting and response levels and their interactions coupled with intra- and interspecific differences basically defy the proposals of generalities. Because of this, interpretation of data concerning the GAS is not simple. Further confounding interpretation of data is the likelihood that some GAS factors operate more via threshold rather than continuous control, and there is no hard data on this phenomenon that would allow for establishment of globally applicable thresholds. The bottom line is that it would be exceptionally useful if a validated "cookbook" could be developed that allowed for interpretation of stress measurements.

REFERENCES

Ader, R. (2007). *Psychoneuroimmunology*, vols. 1 and 2. Burlington: Academic Press.

Ashley, N. T. and Wingfield, J. C. (2012). Sickness behavior in vertebrates: allostasis, life-history modulation, and hormonal regulation. In *Ecoimmunology* (eds. G. Demas and R. Nelson), pp. 45–91. New York, NY: Oxford University Press.

Barton, B. A. (1997). Stress in finfish: past, present, and future – a historical perspective. In *Fish Stress and Health in Aquaculture* (eds. G. K. Iwama, A. D. Pickering, J. P. Sumpter and C. B. Schreck), pp. 1–33. Cambridge: Cambridge University Press.

Barton, B. A. (2002). Stress in fishes: a diversity of responses with particular reference changes in circulating corticosteroids. *Integr. Comp. Biol.* **42**, 517–525.

Barton, B. A. and Iwama, G. (1991). Physiological changes in fish from stress in aquaculture with emphasis on the response and effects of corticosteroids. *Ann. Rev. Fish Dis.* **1**, 3–26.

Barton, B. A. and Schreck, C. B. (1987). Metabolic cost of acute physical stress in juvenile steelhead. *Trans. Am. Fish. Soc.* **116**, 257–263.

Barton, B. A., Schreck, C. B., Ewing, R. D., Hemmingsen, A. R. and Patino, R. (1985). Changes in plasma cortisol during stress and smoltification in coho salmon, Oncorhynchus kisutch. *Gen. Comp. Endocrinol.* 59, 468–471.

Basua, N., Todghama, A. E., Ackermana, P. A., Bibeaub, M. R., Nakanoa, K., Schulteb, P. M., et al. (2002). Heat shock protein genes and their functional significance in fish. *Gene* **295**, 173–183.

Bayne, B. L. (1985). Responses to environmental stress: tolerance, resistance and adaptation. In *Marine Biology of Polar Regions and Effects of Stress on Marine Organisms* (eds. J. S. Gray and M. E. Christiansen), pp. 331–349. New Jersey: Wiley.

Black, E. C. (1955). Blood levels of haemoglobin and lactic acid in some freshwater fishes following exercise. *J. Fish. Res. Board Can.* **12**, 917–929.

Black, E. C. (1958). Hyperactivity as a lethal factor in fish. *J. Fish. Res. Board Can.* **15**, 573–586.

Black, E. C., Robertson, A. C., Hanslip, A. R. and Chiu, W.-G. (1960). Alterations in glycogen, glucose and lactate in rainbow and Kamloops trout, *Salmo gairdneri* following muscular activity. *J. Fish. Res. Board Can.* **17**, 487–500.

Boonstra, R. (2013). Reality as the leading cause of stress: rethinking the impact of chronic stress in nature. *Funct. Ecol.* **27**, 11–23.

Brett, J. R. (1958). Implications and assessment of environmental stress. In *The Investigation of Fish-Power Problems* (ed. P. A. Larkin), pp. 69–83. Vancouver: H.R. MacMillan Lectures in Fisheries, University of British Columbia.

Caldwell, C. A., Kattesh, H. G. and Strange, R. J. (1991). Distribution of cortisol among its free and protein-bound fractions in rainbow trout (*Oncorhynchus mykiss*): evidence of control by sexual maturation. *Comp. Biochem. Physiol.* **99A**, 593–595.

Cannon, W. B. (1926). Physiological regulation of normal states: some tentative postulates concerning biological homeostatics. In *A Charles Richet: ses amis, ses collègues, ses élèves* (ed. A. Pettit), p. 91. Paris: Les Éditions Médicales.

Cannon, W. B. (1932). *The Wisdom of the Body*. New York, NY: W.W. Norton & Company, Inc.

Chrousos, G. P. (1998). Stressors, stress, and neuroendocrine integration of the adaptive response. *Ann. N.Y. Acad. Sci.* **851**, 311–335.

Chrousos, G. P. (2009). Stress and disorders of the stress system. *Nat. Rev. Endocrinol.* **5**, 374.

Clements, S., Schreck, C. B., Larsen, D. A. and Dickhoff, W. W. (2001). Central administration of corticotrophin-releasing hormone stimulates locomotor activity in juvenile chinook salmon (*Oncorhynchus tshawytscha*). *Gen. Comp. Endocrinol.* **125**, 319–327.

Close, D. A., Yun, S.-S., McCormick, S. D., Wildbill, A. J. and Li, W. (2010). 11-Deoxycortisol is a corticosteroid hormone in the lamprey. *Proc. Natl. Acad. Sci. U. S. A.* **107**, 13942–13947.

Currie, S. (2011). Temperature. Heat shock proteins and temperature. In *Encyclopedia of Fish Physiology*, vol. 1 (eds. A. P. Farrell, E. D. Stevens, J. J. Cech, Jr. and J. G. Richards), pp. 1732–1737. San Diego: Academic Press.

Davis, L. E. and Schreck, C. B. (1997). The energetic response to handling stress in juvenile coho salmon. *Trans. Am. Fish. Soc.* **126**, 248–258.

Davis, M. W. (2007). Simulated fishing experiments for predicting delayed mortality rates using reflex impairment in restrained fish. *ICES J. Mar. Sci.* **64**, 1535–1542.

Davis, M. W. (2010). Fish stress and mortality can be predicted using reflex impairment. *Fish Fisher* **11**, 1–11.

Davis, M. W. and Ottmar, M. L. (2006). Wounding and reflex impairment may be predictors for mortality in discarded or escaped fish. *Fish. Res.* **82**, 1–6.

Davis, M. W. and Schreck, C. B. (2005). Responses by Pacific halibut to air exposure: lack of correspondence among plasma constituents and mortality. *Trans. Am. Fish. Soc.* **134**, 991–998.

Davis, M. W., Olla, B. L. and Schreck, C. B. (2001). Stress induced by hooking, net towing, elevated sea water temperature and air in sablefish: lack of concordance between mortality and physiological measures of stress. *J. Fish Biol.* **58**, 1–15.

Demas, G., Greive, T. J., Chester, E. and French, S. S. (2012). The energetics of immunity. In *Ecoimmunology* (eds. G. Demas and R. Nelson), pp. 259–296. New York, NY: Oxford University Press.

Dhabhar, F. S. (2008). Enhancing versus suppressive effects of stress on immune function: implications for immunoprotection versus immunopathology. *Allergy Asthma Clin. Immunol.* **4**, 2–11.

Dhabhar, F. S. and McEwen, B. S. (2001). Bidirectional effects of stress and glucocorticoid hormones on immune function: possible explanations for paradoxical observations. In *Psychoneuroimmunology*, vol. 1 (eds. R. Ader, D. L. Felten and N. Cohen), p. 301. San Diego: Academic Press.

Ellis, T., Yildiz, H. Y., López-Olmeda, J., Spedicato, M. T., Tort, L., Øverli, Ø., et al. (2012). Cortisol and finfish welfare. *Fish Physiol. Biochem.* **38**, 163–188.

Faught, E., Aluru, N. and Vijayan, M. M. (2016). The Molecular Stress Response. In *Fish Physiology - Biology of Stress in Fish*, Vol. 35 (eds. C. B. Schreck, L. Tort, A. P. Farrell and C. J. Brauner), San Diego, CA: Academic Press.

Feil, R. and Fraga, M. F. (2012). Epigenetics and the environment: emerging patterns and implications. *Nat. Rev. Genet.* **13**, 97–109.

Feist, G. and Schreck, C. B. (2002). Ontogeny of the stress response in chinook salmon, *Oncorhynchus tshawytscha*. *Fish Physiol. Biochem.* **2**, 31–40.

Fevolden, S. E., Røed, K. H. and Gjerde, B. (2002). Genetic components of post-stress cortisol and lysozyme activity in Atlantic salmon; correlations to disease resistance. *Fish Shellfish Immunol.* **4**, 507–519.

Fish, E. W., Shahrokh, D., Bagot, R., Caldji, C., Bredy, T., Szyf, M., et al. (2004). Epigenetic programming of stress responses through variations in maternal care. *Ann. N. Y. Acad. Sci.* **1036**, 167–180.

Freeman, H. C. and Idler, D. R. (1966). Transcortin binding in Atlantic salmon (*Salmo salar*) plasma. *Gen. Comp. Endocrinol.* **7**, 37–43.

Ghaedi, G., Yavari, V., Falahatkar, B., Nikbakht, G., Sheibani, M. T. and Salati, A. P. (2013). Whole egg and alevin changes of cortisol and interrenal tissue differences in rainbow trout *Oncorhynchus mykiss* exposed to different stocking densities during early development. *Zoolog. Sci.* **30**, 1102–1109.

Ghaedi, G., Falahatkar, B., Yavari, V., Sheibani, M. T. and Broujeni, G. N. (2014). The onset of stress response in rainbow trout *Oncorhynchus mykiss* embryos subjected to density and handling. *Fish Physiol. Biochem.* **41**, 485 493.

Glencross, B. D. and Bermudes, M. (2011). Adapting bioenergetic factorial modelling to understand the implications of heat stress on barramundi (*Lates calcarifer*) growth, feed utilisation and optimal protein and energy requirements – potential strategies for dealing with climate change? *Aquacult. Nutr.* **18**, 411–422.

Gorissen, M. and Flik, G. (2016). Endocrinology of the Stress Response in Fish. In *Fish Physiology - Biology of Stress in Fish*, Vol. 35 (eds. C. B. Schreck, L. Tort, A. P. Farrell and C. J. Brauner), San Diego, CA: Academic Press.

Guyton, A. C. and Hall, J. E. (2011). *Guyton and Hall Textbook of Medical Physiology*. Philadelphia: Saunders Elsevier.

Holden, C. (2000). Laboratory animals. Researchers pained by effort to define distress precisely. *Science* 290, 1474–1475.

Huntingford, F. A., Adams, C., Braithwaite, V. A., Kadri, S., Pottinger, T. G., Sandoe, P., et al. (2006). Current issues in fish welfare. *J. Fish Biol.* **68**, 332–372.

Idler, D. R. and Freeman, H. C. (1968). Binding of testosterone, 1α-hydroxycorticosterone and cortisol by plasma proteins of fish. *Gen. Comp. Endocrinol.* **11**, 366–372.

Idler, D. R. and Sangalang, G. B. (1970). Steroids of a chondrostean: in-vitro steroidogenesis in yellow bodies isolated from kidneys and along the posterior cardinal veins of the American Atlantic sturgeon. Acipenser oxyrhynchus Mitchell. *J. Endocrinol.* **48**, 627–637.

Idler, D. R. and Truscott, B. (1966). 1α-hydroxycorticosterone from cartilaginous fish: a new adrenal steroid in blood. *J. Fish. Res. Bd. Can.* **23**, 615–619.

Idler, D. R., Freeman, H. C. and Truscott, B. (1967a). A preliminary communication on the biological activity of 1α-hydroxycorticosterone isolated from cartilaginous fish. *J. Fish. Res. Bd. Can.* **24**, 205–206.

Idler, D. R., Freeman, H. C. and Truscott, B. (1967b). Biological activity and protein-binding of 1α-hydroxycorticosterone: an interrenal steroid in elasmobranch fish. *Gen. Comp. Endocrinol.* **9**, 207–213.

Iwama, G. K., Thomas, P. T., Forsyth, R. and Vijayan, M. M. (1998). Heat shock proteins (chaperones) in fish and shellfish and their potential role in relation to fish health: a review. *Rev. Fish Biol. Fish* **8**, 35–56.

Kayhan, F. E. and Duman, B. S. (2010). Heat shock protein genes in fish. *Turkish J. Fish. Aquat. Sci.* **10**, 287–293.

Koolhaas, J. M., Bartolomucci, A., Buwalda, B., de Boer, S. F., Flügge, G., Korte, S. M., et al. (2011). Stress revisited: a critical evaluation of the stress concept. *Neurosci. Biobehav. Rev.* **35**, 1291–1301.

Korte, S. M., Olivier, B. and Koolhaas, J. M. (2007). A new animal welfare concept based on allostasis. *Physiol. Behav.* **92**, 422–428.

LeBlanc, S., Hoglund, E., Gilmour, K. M. and Currie, S. (2012). Hormonal modulation of the heat shock response: insights from fish with divergent cortisol stress responses. *Am. J. Physiol. Regul. Integr. Comp. Physiol.* **302**, 184–192.

Lee, R. S., Tamashiro, K. L. K., Yang, X., Purcell, R. H., Harvey, A., Willour, V. L., et al. (2010). Chronic corticosterone exposure increases expression and decreases deoxyribonucleic acid methylation of Fkbp5 in mice. *Endocrinology* **151**, 4332–4343.

Levine, S. (1985). A definition of stress? In *Animal Stress* (ed. G. P. Moberg), pp. 51–69. Bethesda: American Physiological Society.

Lochmiller, R. L. and Deerenberg, C. (2000). Trade-offs in evolutionary immunology: just what is the cost of immunity? *Oikos* **88**, 87–98.

Martin, L. B., Scheuerlein, A. and Wikelski, M. (2003). Immune activity elevates energy expenditure if house sparrows: a link between direct and indirect costs? *Proc. R. Soc. B* **270**, 153–158.

Martin, L. B., Weil, Z. M. and Nelson, R. J. (2008). Seasonal changes in vertebrate immune activity: mediation by physiological trade-offs. *Proc. R. Soc. B* **363**, 321–339.

Martins, C. I. M., Galhardo, L., Noble, C., Damsgard, B., Spedicato, M. T., Zupa, W., et al. (2013). Behavioural indicators of welfare in farmed fish. *Fish Physiol. Biochem.* **38**, 17–41.

Maule, A. G., Tripp, R. A., Kaattari, S. L. and Schreck, C. B. (1989). Stress alters immune function and disease resistance in chinook salmon (*Oncorhynchus tshawytscha*). *J. Endocrinol.* **120**, 135–142.

McEwen, B. S. (1998). Protective and damaging effects of stress mediators. *N. Engl. J. Med.* **338**, 171–179.

McEwen, B. S. and Lasley, E. N. (2002). *The End of Stress As We Know It.* Washington, D.C.: Joseph Henry Press.

McEwen, B. S. and Wingfield, J. C. (2003). The concept of allostasis in biology and biomedicine. *Horm. Behav.* **43**, 2–15.

McGhee, K. E. and Bell, A. M. (2014). Paternal care in a fish: epigenetics and fitness enhancing effects on offspring anxiety. *Proc. R. Soc. B* **281**, 20141146.

Mommer, B. C. and Bell, A. M. (2014). Maternal experience with predation risk influences genome-wide embryonic gene expression in threespined sticklebacks (*Gasterosteus aculeatus*). *PLoS One* **9** (6), e98564. ⟨http://dx.doi.org/10.1371/journal.pone.0098564⟩

Momoda, T. S., Schwindt, A. R., Feist, G. W., Gerwick, L., Bayne, C. J. and Schreck, C. B. (2007). Gene expression in the liver of rainbow trout, Oncorhynchus mykiss, during the stress response. *Comp. Biochem. Physiol. Part D Genomics Proteomics* **2**, 303–315.

Montero, D., Tort, L., Robaina, L., Vergara, J. M. and Izquierdo, M. S. (2001). Low vitamin E in diet reduces stress resistance of gilthead sea bream (*Sparus aurata*) juveniles. *Fish Shellfish Immunol.* **11**, 473–490.

Noakes, D. L. G. and Jones, K. M. M. (2016). Cognition, Learning, and Behavior. In *Fish Physiology - Biology of Stress in Fish*, Vol. 35 (eds. C. B. Schreck, L. Tort, A. P. Farrell and C. J. Brauner), San Diego, CA: Academic Press.

O'Connor, E. A., Pottinger, T. G. and Sneddon, L. U. (2011). The effects of acute and chronic hypoxia on cortisol, glucose and lactate concentrations in different populations of three-spined stickleback. *Fish Physiol. Biochem.* **37**, 461–469.

Ogoshia, M., Katoa, K., Takahashia, H., Ikeuchid, T., Abea, T. and Sakamotoa, T. (2012). Growth, energetics and the cortisol-hepatic glucocorticoid receptor axis of medaka (*Oryzias latipes*) in various salinities. *Gen. Comp. Endocrinol.* **178**, 175–179.

Øverli, O., Winberg, S. and Pottinger, T. G. (2005). Behavioral and neuroendocrine correlates of selection for stress responsiveness in rainbow trout—a review. *Integr. Comp. Biol.* **45**, 463–474.

Pankhurst, N. W. (2011). The endocrinology of stress in fish: an environmental perspective. *Gen. Comp. Endocrinol.* **170**, 265–275.

Pankhurst, N. W. (2016). Reproduction and Development. In *Fish Physiology - Biology of Stress in Fish*, Vol. 35 (eds. C. B. Schreck, L. Tort, A. P. Farrell and C. J. Brauner), San Diego, CA: Academic Press.

Patino, R., Schreck, C. B., Banks, J. L. and Zaugg, W. S. (1986). Effects of rearing conditions on the developmental physiology of smolting of coho salmon. *Trans. Am. Fish. Soc.* **115**, 828–837.

Patino, R., Redding, J. M. and Schreck, C. B. (1987). Interrenal secretion of corticosteroids and plasma cortisol and cortisone concentrations after acute stress and during seawater acclimation in juvenile coho salmon (*Oncorhynchus kisutch*). *Gen. Comp. Endocrinol.* **68**, 431–439.

Perry, S. F. and Bernier, N. J. (1999). The acute humoral and adrenergic stress response in fish: facts and fiction. *Aquaculture* 177, 285–295.

Pickering, A. D. (ed.). (1981). *Stress and Fish.* London: Academic Press.

Renshawa, G. M. C., Kuteka, A. K., Grantb, G. D. and Anoopkumar-Dukieb, S. (2012). Forecasting elasmobranch survival following exposure to severe stressors. *Comp. Biochem. Physiol. Part A Mol. Integr. Physiol.* **162**, 101–112.

Roberts, B. W., Didier, W., Rai, S., Johnson, N. S., Libants, S., Yun, S.-S., et al. (2014). Regulation of a putative corticosteroid, 17,21-dihydroxypregn-4-ene,3,20-one, in sea lamprey, *Petromyzon marinus. Gen. Comp. Endocrinol.* **196**, 17–25.

Roberts, R. J., Agius, C., Saliba, C., Bossier, P. and Sung, Y. Y. (2010). Heat shock proteins (chaperones) in fish and shellfish and their potential role in relation to fish health: a review. *J. Fish Dis.* **33**, 789–801.

Rodnick, K. J. and Planas, J. V. (2016). The Stress and Stress Mitigation Effects of Exercise: Cardiovascular, Metabolic, and Skeletal Muscle Adjustments. In *Fish Physiology - Biology of Stress in Fish*, Vol. 35 (eds. C. B. Schreck, L. Tort, A. P. Farrell and C. J. Brauner), San Diego, CA: Academic Press.

Romero, L. M., Dickens, M. J. and Cyr, N. E. (2009). The reactive scope model—a new model integrating homeostasis, allostasis, and stress. *Horm. Behav.* **55**, 375–389.

Sadoul, B. and Vijayan, M. M. (2016). Stress and Growth. In *Fish Physiology - Biology of Stress in Fish*, Vol. 35 (eds. C. B. Schreck, L. Tort, A. P. Farrell and C. J. Brauner), San Diego, CA: Academic Press.

Sánchez, J. A., López-Olmeda, J. F., Blanco-Vives, B. and Sánchez-Vázquez, F. J. (2009). Effects of feeding schedule on locomotor activity rhythms and stress response in sea bream. *Physiol. Behav.* **98**, 125–129.

Sandblom, E., Seth, H. and Sundh, H. (2013). Stress responses in Arctic char (*Salvelinus alpinus* L.) during hyperoxic carbon dioxide immobilization relevant to aquaculture. *Aquaculture* **414**, 254–259.

Sandland, G. J. and Minchella, D. J. (2003). Cost of immune defense: an enigma wrapped in an environmental cloak? *Trends Parasitol.* **19**, 571–574.

Sapolsky, R. M. (2005). The influence of social hierarchy on primate health. *Science* **308**, 648–652.

Schjolden, J., Backström, T., Pulman, K. G. T., Pottinger, T. G. and Winberg, S. (2005). Divergence in behavioural responses to stress in two strains of rainbow trout (*Oncorhynchus mykiss*) with contrasting stress responsiveness. *Horm. Behav.* **48**, 537–544.

Schreck, C. B. (1972). Steroid assays and their usefulness in fisheries research. *Proc. Annu. Conf. Southeast Assoc. Fish Wildl. Agencies* **26**, 649–652.

Schreck, C. B. (1981). Stress and compensation in teleostean fishes: response to social and physical factors. In *Stress and Fish* (ed. A. D. Pickering), pp. 295–321. London: Academic Press.

Schreck, C. B. (1992). Glucocorticoids: metabolism, growth, and development. In *The Endocrinology of Growth, Development and Metabolism in Vertebrates* (eds. M. P. Schreibman, C. G. Scanes and P. K. T. Pang), pp. 367–392. New York, NY: Academic Press.

Schreck, C. B. (1996). Immunomodulation: endogenous factors. In *Fish Physiology - The Fish Immune System: Organism, Pathogen, and Environment*, Vol. 15 (eds. G. Iwama and T. Nakanishi), pp. 311–337. New York, NY: Academic Press.

Schreck, C. B. (2000). Accumulation and long-term effects of stress in fish. In *The Biology of Animal Stress: Basic Principles and Implications for Animal Welfare* (eds. G. P. Moberg and J. A. Mench), pp. 147–158. Wallingford: CAB International.

Schreck, C. B. (2010). Stress and reproduction: the roles of allostasis and hormesis. *Gen. Comp. Endocrinol.* **165**, 549–556.

Schreck, C. B. and Li, H. W. (1991). Performance capacity of fish: stress and water quality. In *Aquaculture and Water Quality* (eds. D. E. Brune and J. R. Tomasso), pp. 21–29. Baton Rouge: World Aquaculture Society, Advances in World Aquaculture 3.

Schreck, C. B. and Maule, A. G. (2001). Are the endocrine and immune systems really the same thing? In *International Symposium on Comparative Endorinology* (eds. H. J. T. Goos, R. K. Rostogi, H. Vaudry and R. Pierantoni), pp. 351–357. Naples: Monduzzi Editore.

Schreck, C. B., Jonsson, L., Feist, G. and Reno, P. (1995). Conditioning improves performance of juvenile chinook salmon, *Oncorhynchus tshawytscha*, to transportation stress. *Aquaculture* **135**, 99–110.

Schreck, C. B., Contreras-Sanchez, W. and Fitzpatrick, M. S. (2001). Effects of stress on fish reproduction, gamete quality, and progeny. *Aquaculture* **197**, 3–24.

Schulte, P. M. (2014). What is environmental stress? Insights from fish living in a variable environment. *J. Exp. Biol.* **217**, 23–34.

Selye, H. (1950). Stress and the general adaptation syndrome. *Br. Med. J.* **1**, 1383–1392.

Selye, H. (1973). Homeostasis and heterostasis. *Perspect. Biol. Med.* 441–445.

Selye, H. (1976). *Stress in Health and Disease*. Boston: Butterworth.

Sheldon, B. C. and Verhulst, S. (1996). Ecological immunology: costly parasite defenses and trade-offs in evolutionary ecology. *Trends Ecol. Evol.* **11**, 317–321.

Sneddon, L. U., Wolfenden, D. C. C. and Thomson, J. S. (2016). Stress Management and Welfare. In *Fish Physiology - Biology of Stress in Fish*, Vol. 35 (eds. C. B. Schreck, L. Tort, A. P. Farrell and C. J. Brauner), San Diego, CA: Academic Press.

Sorensen, C., Johansen, I. B. and Øverli, Ø. (2013). Neural plasticity and stress coping in teleost fishes. *Gen. Comp. Endocrinol.* **181**, 25–34.

Spagnoli, S., Lawrence, C. and Kent, M. L. (2016). Stress in Fish as Model Organisms. In *Fish Physiology - Biology of Stress in Fish*, Vol. 35 (eds. C. B. Schreck, L. Tort, A. P. Farrell and C. J. Brauner), San Diego, CA: Academic Press.

Sterling, P. and Eyer, J. (1988). Allostasis: a new paradigm to explain arousal pathology. In *Handbook of Life Stress, Cognition and Health* (eds. S. Fisher and J. Reason), pp. 629–649. New York, NY: John Wiley & Sons.

Stevens, E. D. (1972). Change in body weight caused by handling and exercise. *J. Fish. Res. Board Can.* **29**, 202–203.

Stewart, H. A., Noakes, D. L. G., Cogliati, K. M., Peterson, J. T., Iverson, M. H. and Schreck, C. B. (2016). Salinity effects on plasma ion levels, cortisol, and osmolality in Chinook Salmon following lethal sampling. *Comp. Biochem. Physiol. A Mol. Integr. Physiol.* **192**, 38–43.

Stitt, B. C., Burness, G., Burgomaster, K. A., Currie, S., McDermid, J. L. and Wilson, C. C. (2014). Intraspecific ariation in thermal tolerance and acclimation capacity in brook trout (*Salvelinus fontinalis*): physiological implications for climate change. *Physiol. Biochem. Zool.* **87**, 15–29.

Sumpter, J. P. (1997). The endocrinology of stress. In *Fish Stress and Health in Aquaculture* (eds. G. K. Iwama, A. D. Pickering, J. P. Sumpter and C. B. Schreck), pp. 95–118. Cambridge: Cambridge University Press.

Suzumoto, B. K., Schreck, C. B. and McIntyre, J. D. (1977). Relative resistances of three transferrin genotypes of coho salmon (*Oncorhynchus kisutch*) and their hematological responses to bacterial kidney disease. *J. Fish. Res. Board Can.* **34**, 1–8.

Takei, Y. and Hwang, P.-P. (2016). Homeostatic Responses to Osmotic Stress. In *Fish Physiology - Biology of Stress in Fish*, Vol. 35 (eds. C. B. Schreck, L. Tort, A. P. Farrell and C. J. Brauner), San Diego, CA: Academic Press.

Templeman, N. M., LeBlanc, S., Perry, S. F. and Currie, S. (2014). Linking physiological and cellular responses to thermal stress: β-adrenergic blockade reduces the heat shock response in fish. *J. Comp. Physiol. B* **184**, 719–728.

Tort, L. (2011). Stress and immune modulation in fish. *Dev. Comp. Immunol.* **35**, 1366–1375.

Tortora, G. J. and Grabowski, R. R. (2000). *Principles of Anatomy and Physiology*. New York, NY: John Wiley & Sons.

Verburg-van Kemenade, B. M. L., Stolte, E. H., Metz, J. R. and Chadzinska, M. (2009). Neuroendocrine-immune interactions in teleost fish. In *Fish Physiology - Fish Neuroendocrinology*, Vol. 28 (eds. N. J. Bernier, G. van der Kraak, A. P. Farrell and C. J. Brauner), pp. 313–364. London: Academic Press.

Waddington, C. H. (1977). *Tools for Thought: How to Understand and Apply the Latest Scientific Techniques of Problem Solving*. New York, NY: Basic Books.

Webb, M. A. H., Allert, J. A., Kappenman, K. M., Marcos, J., Feist, G. W., Schreck, C. B., et al. (2007). Identification of plasma glucocorticoids in pallid sturgeon in response to stress. *Gen. Comp. Endocrinol.* **154**, 98–104.

Wedemeyer, G. A. and McLeay, D. J. (1981). Methods for determining the tolerance of fishes to environmental stressors. In *Stress and Fish* (ed. A. D. Pickering), pp. 295–321. London: Academic Press.

Wendelaar Bonga, S. E. (1977). The stress response in fish. *Physiol. Rev.* **77**, 591–625.

Winberg, S., Höglund, E. and Øverli, Ø. (2016). Variation in the Neuroendocrine Stress Response. In *Fish Physiology - Biology of Stress in Fish*, Vol. 35 (eds. C. B. Schreck, L. Tort, A. P. Farrell and C. J. Brauner), San Diego, CA: Academic Press.

Yada, T. and Nakanishi, T. (2002). Interaction between endocrine and immune systems in fish. *Int. Rev. Cytol.* **220**, 35–92.

Yada, T. and Tort, L. (2016). Stress and Disease Resistance: Immune System and Immunoendocrine Interactions. In *Fish Physiology - Biology of Stress in Fish*, Vol. 35 (eds. C. B. Schreck, L. Tort, A. P. Farrell and C. J. Brauner), San Diego, CA: Academic Press.

Zuberi, A., Brown, C. and Ali, S. (2014). Effect of confinement on water-borne and whole body cortisol in wild and captive-reared rainbowfish (*Melanotaenia duboulayi*). *Int. J. Agric. Biol.* **16**, 183–188.

VARIATION IN THE NEUROENDOCRINE STRESS RESPONSE

SVANTE WINBERG

ERIK HÖGLUND

ØYVIND ØVERLI

Between and within species, individuals vary in how and when they respond to stress and environmental perturbations. Neuroendocrine and physiological mechanisms mediating this flexibility are to a large degree conserved throughout the vertebrate subphylum, but are reviewed here with particular reference to adaptive variation in teleost fish. The influence of both genetic and environmental factors, such as the social environment and nutrient availability, can be exploited to reveal underlying proximate mechanisms. Associations between stress responsiveness, behavior, and life history traits likewise illuminate how natural selection apprehends and maintains individual variation.

Biology of Stress in Fish: Volume 35
FISH PHYSIOLOGY

1. INTRODUCTION

Behavioral, autonomic, and neuroendocrine stress responses all show considerable interspecific as well as intraspecific variation (Barton and Iwama, 1991; Barton, 2002; Cockrem, 2013). Species differences appear to be in part related to differences in habitat and lifestyle (Wendelaar Bonga, 1997; Mommsen, 1999; Barton, 2002; Pankhurst, 2011). However, this chapter will focus mainly on differences in stress responses between individual fish of the same species.

Such differences could relate to quality (eg, individuals showing different behavioral and neuroendocrine profiles) as well as to magnitude (eg, divergent elevation of plasma levels of cortisol and catecholamines in individuals subjected to the same stressor). Moreover, an animal's stress response is often related to its health, nutritional, and general physiological status. Furthermore, factors like previous experience, age, and sex are also important in shaping individual stress response profiles. Neuronal and neuroendocrine systems are highly plastic, and the experience of stress may affect the subsequent responses to stressors, processes known as habituation and sensitization. Usually, these effects are most pronounced at early life stages and at certain life stages, like during smoltification in anadromous salmonids (Schreck and Tort, 2016; Chapter 1 in this volume).

Activation of the sympathetic–chromaffin (SC) axis and the hypothalamic–pituitary–interrenal (HPI) axis are often used as indices of stress (Wendelaar Bonga, 1997; Mommsen, 1999; Barton, 2002). However, it is important to realize that these systems are also activated in response to a variety of stimuli, such as sexual behavior and social victory, which are not normally thought of in the context of stress. Glucocorticoids, as well as sympathetic tone and circulating catecholamines, are important for energy mobilization and metabolic control. Thus, energetically demanding behavior, even if not induced by aversive stimuli, will result in an activation of these systems. It is becoming increasingly clear that it is not the physical nature of an aversive stimulus that results in stress, but rather the degree to which this stimulus could be predicted and controlled. In mammals the cognitive functions of appraisal are important in creating affective states resulting in stress responses (Paul et al., 2005; Mendl et al., 2010). Recent studies suggest that this is true also in teleost fish and other nonmammalian vertebrates (Oliveira, 2013).

Moreover, stress response profiles are correlated with behavioral traits, forming what has been described as coping styles (Koolhaas et al., 1999). Fish displaying contrasting stress coping styles respond differently to stressors and also differ in behavioral patterns including levels of aggression, neophobia, and general locomotor activity and the tendency to develop behavioral routines

(Øverli et al., 2007). Stress coping styles in part are controlled by heritable factors, but individual coping styles are also affected by previous experiences. In addition, coping styles may also be related to life history strategies, such as time to initiate endogenous feeding, age of sexual maturation, and age of migration (Mommsen, 1999; Wendelaar Bonga, 1997).

In this chapter we review research results on the ontogenetic development of stress responses and the effects of early stress on subsequent stress response profiles. We focus especially on the effects of early life events on subsequent stress responsiveness in fish, and through what mechanisms such effects may act. We also review studies on the mechanism of sensitization and habituation in teleosts as well as studies on divergent stress coping strategies in fish. In addition, present knowledge on the effects of nutritional factors, social status, and sexual maturation is reviewed.

2. ONTOGENY OF THE TELEOST STRESS RESPONSE

In teleosts the developmental maturity at hatching varies between species. Marine, pelagic fishes, such as those of the families Gadidae, Pleuronectidae, and Sparidae, lay small eggs from which larvae hatch with immature organs. In these species the HPI axis does not appear until several days after hatching (Tanaka et al., 1995). Salmonidae, on the contrary, lay large eggs from which larvae hatch a more developed stage with a functional HPI axis (Barry et al., 1995a).

The ontogenetic development of the HPI axis has been most extensively studied in the zebrafish (*Danio rerio*). Using whole-mount in situ hybridization, Chandrasekar et al. (2007) detected preoptic corticotropin releasing hormone (CRH) positive cells in this species at 24 hours postfertilization (hpf). Using reverse transcription polymerase chain reaction, the same authors obtained limited evidence for an upregulation of CRH expression between days 1 and 5 postfertilization. However, it is not clear if this reflects an increased sensitivity to stressor exposure during this time.

Adrenocorticotrophic hormone (ACTH) is detected in the developing zebrafish pituitary at 24 hpf (Herzog et al., 2003; Liu et al., 2003). ACTH is derived from the precursor pro-opiomelanocortin (POMC), which also gives rise to other biologically active peptides such as α-melanocyte stimulating hormone (α-MSH) (Wendelaar Bonga, 1997). At 48 to 64 hpf an organization of the pars distalis ACTH and neurointermediate lobe (α-MSH) is observed (Liu et al., 2003). However, even though the HPI axis appears already functional at hatching (48 hpf), it does not respond to stressors by elevating cortisol production and release. In fact, the genes coding for enzymes involved in cortisol biosynthesis as well as the interrenal melanocortin 2 receptor (MC2R) genes are all expressed at hatching, at a time when endogenous cortisol starts to rise, but the HPI axis does not appear to react to stressors, as indicated by an elevated cortisol release, until 2 days after hatching (Alsop and Vijayan, 2009). Cortisol concentration in the embryo decreases during the initial development but starts increasing again prior to hatching (Alsop and Vijayan, 2008). The time when basal cortisol levels begin to increase most likely corresponds to the start of endogenous cortisol biosynthesis. The cortisol observed during early development probably is of maternal origin.

There appear to be large interspecific differences in the time at which the HPI axis becomes stress responsive. In some species, for example, common carp (*Cyprinus carpio*), the HPI axis is responsive to stress immediately at hatching (Stouthart et al., 1998; Flik et al., 2002), whereas in others like yellow perch (*Perca flavescens*), rainbow trout (*Oncorhynchus mykiss*), Chinook salmon (*Oncorhynchus tshawytscha*) (Feist and Schreck, 2002), and Mossambique tilapia (*Oreochromis mossambicus*), stress does not result in increased cortisol secretion until several days or weeks after hatching (Barry et al., 1995a; Jentoft et al., 2002; Pepels and Balm, 2004). In rainbow trout, basal cortisol levels increased from 4 weeks postfertilization, but a stress-induced elevation of cortisol concentrations was not detected until 2 weeks posthatch (Barry et al., 1995a). Still, trout interrenal cells (or adrenal homolog) were responsive to ACTH stimulation already before hatching (Barry et al., 1995b). Thus, possibly hypothalamic factors are limiting, or sensory input necessary for stress perception is not reaching the hypothalamus (Alsop and Vijayan, 2009).

At least in zebrafish it appears as if cortisol signaling during embryogenesis is mainly mediated through mineralocorticoid receptors (MRs) (Alsop and Vijayan, 2009). Throughout zebrafish development glucocorticoid receptor (GR) transcripts decrease whereas MR mRNA increases (Alsop and Vijayan, 2008). Cortisol binds with high affinity to MR receptors, and in species where other MR ligands, such as aldosterone in mammals, are present MR is usually coexpressed with 11β-hydroxysteroid dehydrogenase 1(1β-HSD2), an enzyme that converts cortisol to corticosterone, eliminating MR binding. However, in the zebrafish embryo 11β-HSD2 shows very low expression (Alsop and Vijayan, 2008). Interestingly, even though in most species fish do not respond to stress with elevated cortisol secretion until some time after hatching, early stress may affect stress responses displayed later in life (Auperin and Geslin, 2008). For example, in rainbow trout subjected to a brief stress exposure (1 min in 0°C water, 1 min out of water) during early developmental stages as 5-month-old fingerlings showed reduced stress-induced plasma cortisol.

Effects of stressors experienced early in ontogeny on neuroendocrine stress responses in adulthood have been extensively studied in mammals. For example, prenatal stress affects brain monoaminergic neurotransmission (Hayashi et al., 1998; Ishiwata et al., 2005) and hypothalamic–pituitary–adrenal (HPA) axis activity (for references, see review by Glover et al., 2010), and poor parental care induces consistent behavioral and neuroendocrine changes. Moreover, social isolation during the period between postweaning and adolescence has been demonstrated to have a persistent effect on the neuroendocrine stress response, and also to affect brain architecture and neurotransmission in rats (Fone and Porkess, 2008; Lukkes et al., 2009). Some of these changes are associated with alterations in the activity of the brain 5-hydroxytryptamine (5-HT) system.

In fish, the level of parental care varies from just depositing the eggs after fertilization to nest building and feeding the offspring (Smith and Wootton, 1995). However, the knowledge on the effects of stressors early in ontogeny on the neuroendocrine stress response so far, to our knowledge, is limited to species that just deposit their offspring without showing parental care. In the study by Auperin and Geslin (2008) rainbow trout embryos (eyed egg stage) and newly hatched larvae subjected to cold stress (0°C) in combination with air exposure displayed a reduction in their cortisol response to a stressor when tested as juveniles 5 months after hatching. This dampening of HPI axis reactivity did not become apparent after ACTH treatment, suggesting that early stress exerts its effects on mechanisms upstream of the pituitary.

Maternal stress may affect behavior and stress responses of the progeny, and there are indications that maternal cortisol is part of the mechanism transferring information about the environment over generations. For example, increased levels of cortisol in plasma and eggs of female three-spined sticklebacks (*Gasterosteus aculeatus*) have been associated with a higher predator pressure. Moreover, this predator-induced rise in maternal cortisol concentrations was associated with a more pronounced predator avoidance in the offspring (Giesing et al., 2011). Similarly, in Atlantic salmon (*Salmo salar*), maternal cortisol treatment induced a more pronounced behavioral inhibition to novelty in the offspring (Eriksen et al., 2006; Espmark et al., 2008). Furthermore, cortisol treatments of eggs affect learning (Sloman, 2010) and cortisol responses to stress (Auperin and Geslin, 2008) in juvenile rainbow trout. This suggests that parental stress could induce epigenetic effects, which in turn modify stress coping in the offspring. Information about the mechanisms involved in these effects of parental cortisol endowment is just emerging. Factors involved in the epigenetic silencing of gene transcription, including multiple DNA-methyltransferases and two histone proteins, are expressed differentially in embryos originating from stressed three-spined stickleback mothers (Mommer and Bell, 2014).

In addition to the hormonal endowment described earlier, the parental diet may affect the stress response in the offspring. For example, dietary fatty acid composition affects HPI axis reactivity (discussed in Section 6.2) and muscle FA composition. Moreover, muscle FA composition has been shown to be reflected in the oocytes (Garrido et al., 2007), suggesting effects of parental diet on offspring early in the ontogeny of the HPI axis.

3. NEURONAL SUBSTRATE FOR STRESS AND VARIATION IN STRESS RESPONSES

In fish as in other vertebrates, autonomic, neuroendocrine, and behavioral stress responses are closely coordinated, but even though recent results have shed some light on the neural mechanisms involved, our knowledge of the mechanisms responsible for this coordination is still rudimentary. The neuroendocrine control of the SC and HPI axes has been extensively reviewed (Gorissen and Flik, 2016; Chapter 3 in this volume) (Wendelaar Bonga, 1997; Mommsen, 1999), and in this section we will focus on how these mechanisms relate to individual variation in stress responses.

Prolonged elevation of plasma cortisol levels has been shown to suppress HPI axis reactivity (Øverli et al., 1999a,b; Jeffrey et al., 2014). Thus, chronically stressed fish respond to an acute stressor with lower plasma cortisol levels than unstressed control fish (Barton et al., 2005). Moreover, the time course of the stress-induced cortisol response appeared to be slower in chronically stressed fish. This effect of chronic stress on HPI axis function is obvious in socially subordinate fish. For instance, Øverli et al. (1999a,b) reported that socially subordinate Arctic charr (*Salvelinus alpinus*) show elevated basal plasma levels of cortisol but respond to an acute netting stress with a smaller and more sluggish cortisol response than dominant charr. Jeffrey et al. (2014) obtained similar results in rainbow trout. However, it is still not clear at what level of the HPI axis and through what mechanisms these differences originate.

The HPI axis is controlled by a series of peptides acting at different levels of the axis. Secretion of cortisol from teleost interrenal cells is under stimulatory control of ACTH, which in turn is secreted by corticotropic cells of the pituitary pars distalis. Pituitary ACTH secretion is stimulated by corticotropin releasing factor (CRF) originating from the hypothalamic preoptic area (POA). The major cortisol secretagogue, ACTH, is produced from POMC, a preprohormone also giving rise to αMSH and β-endorphin, two peptides that may act together with ACTH in the control of cortisol secretion. In addition, several other hormones and neuroendocrine factors, such as urotensin I (UI), melanin-concentrating hormone (MCH), arginine vasotocin (AVT), isotocin (IST), and neuropeptide Y (NPY), could be important in the control of the HPI axis. The peptides UI, AVT, IT, and NPY have been reported to stimulate pituitary ACTH release whereas MCH appears to have the opposite effect (Wendelaar Bonga, 1997). The peptide UI, which is produced by hypothalamic neurons as well as by the teleost caudal neurosecretory organ, the urophysis, has a strong stimulatory effect on pituitary ACTH release (Gorissen and Flik, 2016; Chapter 3 in this volume).

At the level of the interrenal tissue the sensitivity to ACTH may vary. For instance, in brook trout (*Salmo fontinalis*) the interrenal tissue of fish held at high stocking density is less sensitive to ACTH than the interrenals of fish held at a lower density, suggesting an effect of chronic stress on ACTH-induced cortisol secretion (Vijayan and Leatherland, 1990). Similarly, confinement stress resulted in decreased ACTH sensitivity of the interrenal tissue in rainbow trout (Balm and Pottinger, 1995). Sloman et al. (2002) demonstrated that when the interrenal cells were stimulated in situ with ACTH, cells from socially subordinate rainbow trout displayed a lower rate of cortisol secretion than those of dominants. Moreover, interrenals of rainbow trout selectively bred for low poststress plasma cortisol (low responsive (LR)) showed lower sensitivity to ACTH than those of trout bred for high poststress plasma cortisol (high responsive (HR)) (Pottinger and Carrick, 2001a,b). A possible hypothesis is that reduced interrenal ACTH sensitivity is due to a downregulation of melanocortin II (MCR II, the ACTH receptor) expression and/or desensitization of these receptors. In addition, elevated plasma levels of cortisol may increase interrenal GR expression and by that potentiate the ultrashort negative feedback by cortisol on the interrenal tissue.

The synthesis and secretion of ACTH at the pituitary may also be affected by chronic stress. Winberg and Lepage (1998) reported that social subordination results in an upregulation of pituitary POMC mRNA expression in rainbow trout. However, the increase in POMC mRNA appeared most obvious in the neurointermediate lobe where POMC is processed to α-MSH.

Gonadal steroids may also have effects on interrenal ACTH sensitivity, providing a mechanism for gender differences and modulation of stress responsiveness by sexual maturation. In rainbow trout, Young et al. (1996) reported that 11-ketotestosterone (11-KT) suppresses interrenal ACTH secretion.

At the level of the hypothalamus, social rank affects CRF expression, an effect through which socially induced effects on stress responses could also be mediated. However, the time course of socially induced alteration in CRF expression appears complex. In rainbow trout, Doyon et al. (2003) reported that after 72 h subordinate fish showed elevated CRF expression in the preoptic area. Bernier et al. (2008) found that the increase of CRF expression in the POA of subordinate rainbow trout was temporary since it was elevated after 8 h but not after 24 h of social interaction. Further, Jeffrey et al. (2012) reported that following 5 days of social interaction, there was no difference in telencephalic or POA, CRF, or CRF binding protein expression between dominant and subordinate rainbow trout.

4. DIVERGENT STRESS COPING STYLES, ANIMAL PERSONALITIES, AND BEHAVIORAL SYNDROMES

Even when all other factors such as age and previous experience are kept constant, individuals within any given group of fish will vary in their behavioral patterns and physiological responses to stressful stimuli. This type of variation has received increasing attention in both animal studies and human biomedicine (Koolhaas et al., 1999, 2010; Korte et al., 2005; Øverli et al., 2007; Carere et al., 2010; Coppens et al., 2010; Cockrem, 2013). Emphasizing the large degree of conserved mechanisms and patterns throughout the vertebrate subphylum, while reviewing studies in fish in particular, is the point of this section. As an introduction, we briefly dwell on terminology. Individual differences in how animals respond to environmental perturbations are described using many terms, such as behavioral syndromes (Sih et al., 2004a,b), temperaments (Francis, 1990; Clarke and Boinski, 1995; Réale et al., 2007), animal personalities (Gosling, 2001), or coping styles (Koolhaas et al., 1999, 2010). All these terms ultimately seek to describe consistent links between traits (behavioral, physiological, or both) and fitness. Such links are widespread in the animal kingdom, and have been described in invertebrates (Riechert and Hedrick, 1993; Sinn et al., 2006; Wilson and Krause, 2012; Kralj-Fišer and Schuett, 2014), mammals (Réale et al., 2000, 2009; Boon et al., 2007), birds (Dingemanse et al., 2004; Both et al., 2005; David et al., 2011), reptiles (Stapley and Keogh, 2005), and fish (eg, Smith and Blumstein, 2010; Wilson et al., 2010).

Andrew Sih and coworkers coined the term behavioral syndromes to describe suites of correlated behaviors reflecting consistency between individuals across multiple situations (Sih et al., 2004a,b). Behavioral syndromes are stable if the same association between behaviors is seen at different life stages (Bell and Stamps, 2004). Importantly, all behaviors are preceded by neurological and neuroendocrine processes, so consistent variations in behavior are to some extent derived from variation in neuroendocrine activity. Koolhaas et al. (1999) introduced the term "stress coping style" to describe correlated suites of behavioral and physiological responses to challenges. Those authors defined stress coping style as "a coherent set of behavioral and physiological stress responses, which is consistent over time and which is characteristic to a certain group of individuals." Typically, contrasting coping styles are referred to as proactive and reactive, each being identified by a distinct set of behavioral and physiological characteristics. The behavioral characteristics of a proactive stress coping style largely correspond to the fight–flight response originally described by Cannon (1929); that is, high levels of activity and aggression, greater risk-taking and boldness, and acts that are indicative of active prevention or manipulation of the stressful stimulus. Proactive individuals are also typically more prone to routine-based behavior with low flexibility, relying more on previous experience than information from the actual current surroundings (Benus et al., 1990; Bolhuis et al., 2004; Ruiz-Gomez et al., 2011). Reactive coping is characterized by avoidance, immobility, low levels of aggression, shyness, and higher behavioral flexibility, with animals being more perceptive to changes in the surroundings and reacting accordingly. These differences in behavioral characteristics are associated with consistent physiological differences; proactive individuals typically show higher sympathetic reactivity and catecholamine release while reactive individuals show higher poststress levels of plasma cortisol or corticosterone (Koolhaas et al., 1999; Øverli et al., 2005).

4.1. Conserved Physiology of Contrasting Stress Coping Styles

Individual variation indicative of divergent coping styles appear to be well conserved between fish and mammals (Øverli et al., 2005, 2007; Silva et al., 2010; Martins et al., 2011a,b; Castanheira et al., 2013; Sørensen et al., 2013; Tudorache et al., 2013). Divergent corticosteroid secretion is one of the major characteristics of proactive and reactive coping styles, with proactive individuals displaying lower poststress plasma concentrations compared with reactive individuals. The physiological stress response in teleost fishes resembles that of mammals and other terrestrial vertebrates (Wendelaar Bonga, 1997), suggesting that it is evolutionarily conserved.

In this context, it is interesting to observe that differences in behavior of rainbow trout selected for divergent poststress cortisol responsiveness (HR and LR trout), such as decreased appetite and increased locomotor activity, are consistent with reported effects of cortisol in other nonmammalian vertebrates (Cash and Holberton, 1999; Gregory and Wood, 1999; Øverli et al., 2002a). Selective breeding demonstrated the heritability of the

magnitude of the cortisol stress response and allowed for the creation of HR
and LR lines of trout (Pottinger and Carrick, 1999). The patterns of
behavioral, physiological, and neurobiological responses found in salmonid
model systems are generally consistent with the proactive and reactive
coping styles found in mammals (Table 2.1), but sensitivity to specific
environmental factors and experience, such as nutritional status, has been
demonstrated (Ruiz-Gomez et al., 2008). Also underlining the context-
dependent nature of physiological–behavioral correlations, several studies
show no relationship between HPI axis reactivity and behavioral profile in
wild salmonids (Brelin et al., 2008; Thörnqvist et al., 2015). Nonetheless, the
distinct physiological and behavioral profiles of the HR and LR trout lines
reported in numerous studies suggest that they reflect selection for reactive
and proactive stress coping styles (Schjolden et al., 2005, 2006a), comparable
to those identified in other animal groups. Pottinger and Carrick (2001a,b)
initially examined the behavioral differences between HR and LR rainbow
trout lines in relation to the outcome of dyadic fights for social dominance
between juveniles. Juvenile rainbow trout, like other predominantly stream-
resident salmonids, are highly territorial animals. When introduced into a
territory they will engage in agonistic activity until a dominance hierarchy is
established (Jönsson et al., 1998; Winberg and Lepage, 1998; Øverli et al.,
1999a,b). Pottinger and Carrick (2001a,b) reported that LR fish became
dominant in the majority of HR–LR pairings, suggesting a link between
competitive ability and stress responsiveness. In comparison to fish from the
HR strain, LR fish are typically bolder, in the sense of showing faster
resumption of feeding after transfer to a novel environment (Øverli et al.,
2002b; but see Ruiz-Gomez et al., 2008). One further distinct behavioral
feature was reported by Øverli et al. (2002b): while both lines of fish
increased their activity in response to a territorial intruder, the activity of
HR trout was significantly higher compared with LR fish.

The HR–LR strain selection for consistent divergence in cortisol response
to stress also led to divergence in other physiological traits. Behavior and
brain monoamine activity was initially investigated in adult females in HR
and LR rainbow trout in relation to selection for stress responsiveness (Øverli
et al., 2001). The brain monoamine neurotransmitters dopamine (DA),
norepinephrine (NE), and serotonin (5-HT) are deeply involved in the control
and integration of behavioral and physiological stress responses in both
teleost fish and mammals (Blanchard et al., 1993; Stanford and Salmon, 1993;
Winberg and Nilsson, 1993; Winberg et al., 1997, 2001; Øverli et al., 1998,
1999a,b; Höglund et al., 2001, 2002a,b, 2005; Clements et al., 2003; Larson et
al., 2003; Lepage et al., 2005; Carpenter et al., 2007; Lillesaar, 2011; Medeiros
and McDonald, 2013; Chabbi and Ganesh, 2015). Notably, HR trout showed
an increase in the concentrations of both 5-HT (brain stem), DA (brain stem),

Table 2.1

Physiological stress responses and behavioral patterns in proactive and reactive phenotypes in salmonid model species

Trait	Proactive	Reactive	Model species	References
Cortisol basal	×	×	HR–LR rainbow trout	Pottinger and Carrick (1999)
Cortisol stress response	Low	High	HR–LR rainbow trout	Pottinger and Carrick (1999)
	Low	High	Rainbow trout aquaculture strain	Øverli et al. (2004)
	Low	High	Atlantic salmon aquaculture strain	Kittilsen et al. (2009)
	Low	High	Arctic charr aquaculture strain	Backström et al. (2014)
Sympathetic activity	High	Low	HR–LR rainbow trout	Schjolden et al. (2006a,b)
	High	Low	Rainbow trout aquaculture strain	van Raaij et al. (1996)
	High	Low	Brown trout wild type	Brelin et al. (2005)
Social dominance	Dominant	Subordinate	HR–LR rainbow trout	Pottinger and Carrick (2001a,b)
	Dominant	Subordinate	Rainbow trout aquaculture strain	Øverli et al. (2004)
	Subordinate	Dominant	HR–LR rainbow trout	Ruiz-Gomez et al. (2008)
Locomotor activity basal	×	×	HR–LR rainbow trout	Øverli et al. (2002a,b)
Locomotor activity acute stress	Low	High	HR–LR rainbow trout	Øverli et al. (2002a,b)
	Low	High	Atlantic salmon aquaculture strain	Kittilsen et al. (2009)
Feeding in novel environment	Quick	Slow	HR–LR rainbow trout	Øverli et al. (2002a,b)
	Quick	Slow	Rainbow trout aquaculture strain	Øverli et al. (2004)
	Quick	Slow	Atlantic salmon aquaculture strain	Kittilsen et al. (2009)
	Slow	Quick	HR–LR rainbow trout	Ruiz-Gomez et al. (2008)
Struggling, net restraint	High	Low	Arctic charr aquaculture strain	Magnhagen et al. (2015)
Routine formation	High	Low	HR–LR rainbow trout	Ruiz-Gomez et al. (2011)
Retention of conditioned response	High	Low	HR–LR rainbow trout	Moreira et al. (2004)

Traits that show large consistency between studies and with other model systems are shown (eg, newly developed restraint test applied to European sea bass; see text). Note, however, context-dependent effects that are evident in one study (Ruiz-Gomez et al., 2008), where compensatory growth after stress and starvation was associated with proactive behavior in stress-sensitive HR trout. Life history traits, energy metabolism, and several other behavioral patterns, such as boldness in novel object test and attack latency, for example, show inconsistent and largely unresolved associations with coping style.

and NE (optic tectum, telencephalon) in response to stress, whereas LR fish did not (Øverli et al., 2001). Concentrations of monoamine metabolites were elevated in the brain stem and optic tectum of HR fish following confinement stress, but not in LR fish. The simultaneous increase in the concentration of monoamines and their metabolites suggests that both synthesis and metabolism of these transmitters were elevated after stress in HR trout. A different pattern was observed in the hypothalamus, however, where LR trout displayed higher levels of the serotonin metabolite, 5-hydroxyindoleacetic acid (5-HIAA), and the norepinephrine metabolite, 3-methoxy-4-hydroxy-phenylglycol. There were also differences in the telencephalon between the two lines, in that a higher baseline ratio of 5-HIAA/5-HT was seen in LR fish, but no differences were observed between the stressed LR and HR trout. These differences in neurotransmitter metabolism likely reflect a true difference in functional transmitter release in the two strains, as there was no significant difference in brain monoamine oxidase (MAO) activity between the LR and HR fish (Schjolden et al., 2006b). These results clearly suggest that selection for stress responsiveness in rainbow trout is also associated with changes in the function of brain monoaminergic systems.

In the HR–LR rainbow trout model, it seems likely that neuropeptides such as CRF are also involved in determining the behavioral profile (Backström et al., 2011). Intracerebroventricular injections of CRF in juvenile Chinook salmon induced hyperactivity, an effect that was shown to depend on concurrent 5-HT activation (Clements et al., 2003). CRF administration in newts (*Taricha granulosa*) has been shown to increase DA concentrations in the dorsal medial hypothalamus (Lowry et al., 2001). An increase in stress-induced DA concentrations and turnover in brain stem and optic tectum of HR rainbow trout is one of the main neurochemical differences between the HR and LR strains, selected for their poststress cortisol levels (Øverli et al., 2001). The increase in DA synthesis and release may be an effect of acute elevations of glucocorticoid concentrations (Barrot et al., 2001); therefore it is not known whether the differences in HR–LR dopaminergic systems are a cause or a consequence of hormone dynamics.

Genetically determined variability in the brain serotonin system is associated with variability in personality and temperament, as well as a propensity to develop mood disorders. Polymorphisms in MAO and the serotonin transporter (5-HTT) gene or promoter regions (Lesch et al., 1996; Caspi et al., 2003; Fernandez et al., 2003) are among the most prominent examples in humans and other mammals, and variants in these genes have also been reported in fishes (Elipot et al., 2014). The expression of 5-HTT is regulated by glucocorticoid hormones, but the response to these hormones is attenuated in the 5-HTT "short" type, suggesting that the functional link between 5-HTT polymorphisms and personality traits, to some degree, may depend on interaction with glucocorticoid hormones. To our knowledge, this potential mechanism for altered brain monoaminergic signaling systems has not been investigated in fish models of contrasting stress coping styles.

In addition to membrane receptors mediating rapid effects on behavior (Moore and Orchinik, 1994), two receptor types mediating the effects of corticosteroid hormones in vertebrates are MR and GR receptors. In mammals, there is an extensive literature on the role of corticosteroid receptors in the brain. Glucocorticoids bind to MRs predominantly localized in limbic brain structures (eg, hippocampus) with a 10-fold higher affinity than to GRs, which are widely distributed in the brain (De Kloet, 2004). The high expression of MRs and GRs in limbic brain structures reflects their involvement in memory, learning, and general alertness. First, a proactive mode is mediated through MRs and involves maintenance of tonic HPA axis reactivity and neuronal excitability. In contrast to MR activation, extensive GR activation mediates a decreased level of neuronal excitation. In this way corticosterone exerts biphasic effects on excitability of neurons.

In teleost fishes, cortisol receptor types differ in glucocorticoid affinity and cortisol actions are mediated through GR1, GR2, and MR (Colombe et al., 2000; Bury et al., 2003). In rainbow trout MRs have higher affinity for cortisol than do GRs (Bury and Sturm, 2007). Furthermore, the two different GR paralogs also differ in glucocorticoid sensitivity (Stolte et al., 2006). Since teleost fishes largely lack the conventional MR ligand, aldosterone, cortisol is likely the most important MR ligand in fish (Wendelaar Bonga, 1997; Stolte et al., 2008), including rainbow trout (Colombe et al., 2000). The rainbow trout MR is activated at lower cortisol levels than the GRs, indicating that the trout MR has a similar role as the mammalian MR in the brain (Sturm et al., 2005). Hypothesizing that different expression of MRs and GRs could be involved in coping style, Johansen et al. (2011) quantified the mRNA expression of GR1, GR2, and MR in different brain regions of stressed and nonstressed LR and HR rainbow trout. The expression of MR was higher in LR trout than in HR fish in all brain parts investigated. Altered GR–MR balance is thus likely to be associated with the phenotypic expression of coping style, including reduced anxiety and enhanced memory retention in LR fish. However, transcriptional patterns of the receptors are regulated by cortisol exposure in a time- and organ-specific manner (Teles et al., 2013); hence, it is not straightforward to determine which traits are causes and which are consequences of altered hormone levels.

4.2. Stress, Neuroplasticity, and Coping Style

It is becoming increasingly clear that individual variation in stress responsiveness both affects and is affected by brain function. Predictability and control and psychological processes associated with how animals assess a stressful situation and the outcome they expect might be factors equally important as the physical challenge in determining the magnitude of the stress response. Expectancy of the outcome of stimuli is a powerful cognitive modulator of the stress response, and if we select for a divergence in cognitive ability, differences in stress coping style appear (Giorgi et al., 2003; Aguilar et al., 2004; Steimer and Driscoll, 2005). The opposite is also true; that is, selection for stress responsiveness yields animals with diverging cognitive abilities. Using the HR–LR trout lines, Moreira et al. (2004) showed that LR rainbow trout retained a conditioned response longer than did HR fish. This observation raised the question whether low poststress cortisol production is associated with generally better performance in learning and memory tasks. Later experiments, however, suggest that this is likely not the case. Instead, the selected lines seem to encompass a general difference in behavioral flexibility in response to environmental changes, as opposed to routine formation (Ruiz-Gomez et al., 2011). The latter study showed that two different manipulations, relocating food from a previously learned location and introducing a novel object, yielded contrasting responses in HR and LR fish. No difference was seen in the rate of learning the original food location; however, proactive LR fish were markedly slower than reactive HR fish in altering their food-seeking behavior when food was relocated. In contrast, LR fish largely ignored a novel object that disrupted feeding in HR fish (Ruiz-Gomez et al., 2011). Thus, it would seem that the major difference between the lines is not the capacity to learn the features of completely novel environments, but rather to which degree more subtle changes in the environment affect the animal and provoke a response, in this case deviation from already learned routines. This conclusion is largely in line with studies performed in pigs, rodents, and birds, suggesting no differences in the learning capabilities, but differences in routine strength between proactive and reactive individuals (Benus et al., 1991; Bolhuis et al., 2004).

It is not known to what extent the link between cognition and stress coping style reflects acute effects of circulating glucocorticoids, organizational and structural effects of chronic exposure to different hormone levels throughout ontogeny, or inherent differences in brain function. In other words, is an altered cognitive function a cause or a consequence of a variable stress response? The complex interaction between stress coping, personality, and plasticity and diversity in behavior and brain function was the topic of a special issue of the journal, *Brain, Behavior, and Evolution*, in 2007. Several examples of both heritable and apparently consistent trait associations (Schjolden and Winberg, 2007; Veenema and Neumann, 2007) and rapidly modifiable responses (Burmeister, 2007) are given there. Furthermore, sensitivity to social challenge varies during an animal's life history (Wommack and Deville, 2007), and events early in hierarchy formation may be more important for neurobiological impact than what happens during daily social interactions (Sørensen et al., 2007). Finally, preexisting differences in physiology may determine social rank (Korzan and Summers, 2007); hence, it is not necessarily straightforward to disentangle causes and consequences of particular events in an animal's life.

Sørensen et al. (2011) found that reduced proliferative activity seen in brains of socially stressed rainbow trout is likely mediated by cortisol and that there is a similar suppressive effect of cortisol on brain cell proliferation in the teleost forebrain as in the mammalian hippocampus. These results warranted investigating whether rates of brain cell proliferation and neurogenesis and other aspects of neuroplasticity differ between HR and LR trout lines, as divergence in the rate or total amount of structural changes in key brain areas may well explain apparent differences in cognitive ability, behavior, and stress responsiveness, irrespective of the mechanisms that lead to this divergent neuronal plasticity.

4.3. Genetic Basis for Individuality

Selection studies in several species show that physiological and behavioral responses to stress are often correlated in a coherent and heritable pattern (van Oers et al., 2005; Øverli et al., 2005). The heritability of a variety of traits, including personality traits, varies as a function of conditions (Charmantier and Garant, 2005; Dingemanse et al., 2009). In other words, experiences change genetically determined behavioral patterns, but genes also determine what situations an individual is likely to experience. As a further complication, genes affect how sensitive individuals are to environmental changes (Ruiz-Gomez et al., 2011).

A fundamental genetic basis for behavior is conclusively demonstrated not only in selection and cloning studies, but also in the fact that behavioral variation occurs both within and between fish populations. One example is a population of Trinidad guppies (*Poecilia reticulata*) that are sympatric with predatory charachids and cichlids. These guppies exhibit broadly different behavioral patterns from guppy populations allopatric to these predators (Seghers, 1974), being more shy and exhibiting higher schooling tendencies. This might be explained by the difference in predation pressure in the two habitats, a hypothesis supported by a study on stickleback populations showing heritable, adaptive antipredator variation in behavior (Huntingford et al., 1994; Bell et al., 2010).

Genetic, epigenetic, and environmental factors interact to shape individual neuroendocrine and behavioral profiles, conferring variable vulnerability to stress, disease, and environmental changes. How alternative behavioral syndromes and stress coping styles evolve and are maintained by natural selection remains debatable. In this context it should be mentioned that individual variation in stress responsiveness is reflected in the visual appearance of two species of teleost fish: rainbow trout and Atlantic salmon (Kittilsen et al., 2009). Salmon and trout skin vary from nearly immaculate to densely spotted, with black spots formed by eumelanin-producing chromatophores. In rainbow trout, selection for divergent HPI axis responsiveness in the HR–LR regime has led to a change in dermal pigmentation patterns, with low cortisol-responsive fish being consistently more spotted. This trait correlation is not restricted to the selection regime, since in an aquaculture population of Atlantic salmon individuals with more spots showed a reduced physiological and behavioral response to stress (Kittilsen et al., 2009) (see Fig. 2.1). These data demonstrate a heritable behavioral–physiological and

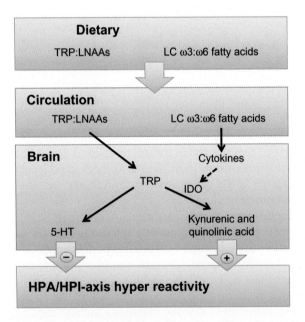

Figure 2.1. Hypothetical pathways for the interaction between dietary fatty acids, amino acids, and HPA/HPI axis activity. The ratio between tryptophan (TRP) and other large neutral amino acids (LNAAs) in the diet affects plasma concentrations of these amino acids, which in turn affect brain availability of TRP. The dietary ratio between long-chain omega-3 ($\omega 3$) and omega-6 ($\omega 6$) fatty acids (LC $\omega 3$:$\omega 6$) affects plasma concentrations of these fatty acids, influencing cytokine activity. This may have an impact on the activity of the enzyme indoleamine dioxygenase (IDO), which converts TRP to kynurenic and quinolinic acid. Serotonin (5-HT) has been suggested to dampen the HPA/HPI axis hyperreactivity caused by kynurenic and quinolinic acid. The dashed line indicates a pathway not shown in fish.

morphological trait correlation that may be specific to alternative coping styles, and the link between poststress cortisol production and eumelanin-based pigmentation was indeed later reported to be present in other species (Almasi et al., 2010). This observation may illuminate the evolution of contrasting coping styles and behavioral syndromes, as occurrence of phenotypes in different environments and their response to selective pressures can be precisely and easily recorded.

4.4. Stress Coping and Life History

In salmonids, stress coping style has been suggested to be associated with life history traits, such as time alevin emergence from the nest, growth, and smoltification (Metcalfe and Thorpe, 1992; Metcalfe et al., 1995; Einum and Fleming, 2000). In this family of fishes, the female typically buries eggs in gravel nests on the stream bottom. After hatching, the larvae remain within the gravel nests and feed on their yolk until they emerge and start defending territories for exogenous feeding. The timing of this event may vary by several weeks among individuals within the same spawning nest (Mason and Chapman, 1965; Brännäs, 1988) and is related to behavioral trait composition. Early emerging fish display proactive traits, such as being more aggressive, bold, and socially dominant (Metcalfe and Thorpe, 1992; Metcalfe et al., 1995). In contrast, individuals that emerge later are generally shy, less aggressive, and socially subordinate (Metcalfe and Thorpe, 1992; Metcalfe et al., 1995), traits characteristic of a reactive stress coping style (Koolhaas et al., 1999). This relationship between timing of emergence and stress coping styles was supported by Åberg-Andersson et al. (2013), who showed that LR rainbow trout generally emerged earlier than HR fish from artificial spawning nests. However, even if recent studies demonstrate differences in boldness (Vaz-Serrano et al., 2011; Thörnqvist et al., 2015) and forebrain gene expression (Thörnqvist et al., 2015) between late and early emerging fish, time of emergence does not seem to be related to HPI axis reactivity (Vaz-Serrano et al., 2011; Thörnqvist et al., 2015).

5. AGONISTIC INTERACTIONS: STRESS AND AGGRESSION

Agonistic interactions play a major role in establishing social hierarchies and access to resources in gregarious animals (Bernstein and Gordon, 1974; Huntingford and Turner, 1987; Francis, 1988). In these gregarious animals, dominant/subordinate relationships usually take the form of a hierarchy, where the outcome of the aggressive encounters is the main factor in determining individual rank order (Huntingford and Turner, 1987). These aggressive interactions consist of behaviors such as displays, charging, nipping, and chasing. In fish, agonistic behavior in a dyadic contest will typically lead to the establishment of a dominant and a subordinate individual, in which the latter will suppress further aggression and retreat from the dominant individual (Øverli et al., 1999a,b; Larson et al., 2006). Behavioral effects of social defeat include appetite inhibition (Meerlo et al., 1997; Øverli et al., 1998; Kramer et al., 1999; Montero et al., 2009), reduced aggression (Blanchard et al., 1995; Höglund et al., 2001), and decreased reproductive behavior (D'Amato, 1988; Perret, 1992).

Initially the stress response (especially the HPI axis) for both individuals becomes activated. As a result, circulating cortisol levels increase in both dominant and subordinate individuals during initial stages of social interaction, but usually return to baseline in the dominant within hours of establishment of the hierarchy (Ejike and Schreck, 1980; Øverli et al., 1999a,b). An increase in plasma cortisol in response to stress is an adaptive response that prepares the organism for changes in the environment and/or physiological changes, as cortisol helps to maintain homeostasis. In subordinate individuals, cortisol levels are typically elevated for days and possibly weeks after hierarchy establishment (Winberg and Lepage, 1998; Øverli et al., 1999a,b; Sloman et al., 2001). Chronically elevated cortisol levels are associated with pathologies, and are thus harmful to the animal. There is a negative correlation between plasma cortisol levels and ACTH, where increased levels of cortisol lead to a decrease in ACTH concentration. This applies for both subordinates and cortisol-fed trout (Jeffrey et al., 2012), and is in agreement with another study where fish fed cortisol had lower ACTH than did control-fed fish (Balm and Pottinger, 1995). Cortisol has been shown to negatively influence performance in social competition (Øverli et al., 2002b). In rainbow trout bred for high or low cortisol response to stress, the individuals responding with high plasma cortisol are more prone to becoming socially subordinate (Pottinger and Carrick, 2001a,b).

In fish, several behavioral and physiological characteristics differ between dominant and subordinate individuals, in addition to plasma cortisol levels. Dominant individuals typically perform more aggressive acts (Larson et al., 2006; Pavlidis et al., 2011; Dahlbom et al., 2012). Subordinate individuals, however, experience freezing behavior (Pavlidis et al., 2011) and retreating behavior (Larson et al., 2006), higher 5-HIAA/5-HT levels (Winberg and Lepage, 1998; Sørensen et al., 2011; Dahlbom et al., 2012), higher POMC A and POMC B levels (precursor for ACTH) (Winberg and Lepage, 1998), reduced cell proliferation (Sørensen et al., 2011), and reduced feed intake (Sørensen et al., 2011).

Social interactions are one of the best-studied factors that can modify both behavior and physiology. However it can be difficult to discern whether the physiological and behavioral differences between dominant and subordinate individuals are causes or consequences of social rank. It is a complex subject where several factors contribute: for example, sensitivity to social challenge varies during an animal's life history (Wommack and Deville, 2007) and events early in hierarchy formation are as important for neurobiological impact as what happens during social interaction (Sørensen et al., 2007). Furthermore, preexisting differences in physiology may determine social rank (Korzan and Summers, 2007), hence it is not necessarily straightforward to disentangle causes and consequences of particular events in an animal's life. Both in the wild and in captivity, a range of factors such as age, body size, sex, kinship, secondary sexual characteristics, prior dominance experience, or residence in a particular territory have been shown to affect social rank (Abbott et al., 1985; Beacham, 1988; Beaugrand and Cotnoir, 1996; Sprague, 1998; Cote, 2000; Renison et al., 2002). Some studies suggest that in predictable and stable environments, a difference in behavior and physiology predisposes an animal for a certain social rank (Morgan et al., 2000; McCarthy, 2001; Plusquellec et al., 2001; Pottinger and Carrick, 2001a,b). In most cases the stress response in the dominant (winner) individual is downregulated. In a stable hierarchy it is stress relieving to be the dominant individual. However, when the environment shifts from stable to unstable, the dominant individual now responds with a higher stress level than the subordinate.

In a dominance battle a number of factors can influence social hierarchy position. Higher innate aggressiveness (Holtby et al., 1993; Adams and Huntingford, 1996; Adams et al., 1998; Cutts et al., 1999b), higher feeding motivation (Johnsson et al., 1996), and a larger size (Holtby et al., 1993; Rhodes and Quinn, 1998; Cutts et al., 1999a) are associated with becoming socially dominant. In other studies, no correlation between size and tendency to become dominant was observed (Huntingford et al., 1990; Adams and Huntingford, 1996; Yamamoto et al., 1998). In the case of the cichlid, *Cichlasoma dimerus*, size determines social position in males, but not in females (Alonso et al., 2012). Previous social experience with being dominant or subordinate (winner/loser effect) are strong indicators of the outcome of social interactions (Abbott and Dill, 1985; Dugatkin, 1997; Rhodes and Quinn, 1998; Hsu and Wolf, 1999; Johnsson et al., 1999; Oliveira, 2009; Oliveira et al., 2011). It is likely that this is caused by neurochemical or endocrine changes brought on by prior social interaction (Winberg and Nilsson, 1993; Øverli et al., 1999a,b; Höglund et al., 2001; Winberg et al., 2001). Several studies report a change in the likelihood of attaining dominance after neurochemical or endocrine modulations. Treatment with growth hormones has been shown to increase aggressive behavior and feeding motivation, which may in turn lead to social dominance (Johnsson and Björnsson, 1994; Jönsson et al., 1998, 2003), and elevation of brain dopamine activity through treatment with the DA precursor L-DOPA ensured social dominance in Arctic charr (Winberg and Nilsson, 1992). A genetically low stress response predisposes an individual for dominance (Ruiz-Gomez et al., 2008); however, this is not always the case. In a study in which rainbow trout from two different stress reacting lines (HR and LR) were moved and starved, the HR individuals became dominant over the LR ones that normally became dominant (Ruiz-Gomez et al., 2008).

Subordinate individuals display a general inhibition of behavior both in fishes (Abbott and Dill, 1985; Franck and Ribowski, 1993; Huntingford et al., 1993; Winberg and Nilsson, 1993; Winberg et al., 1993; Nakano, 1994, 1995a,b; Gómez-Laplaza and Morgan, 2003; Desjardins et al., 2012) and in other vertebrates (Raab et al., 1986; Blanchard et al., 1993; Albonetti and Farabollini, 1994; Meerlo et al., 1997; Engh et al., 2005; Hsu et al., 2006; Korzan and Summers, 2007). This is characterized by suppressed aggressive and/or reproductive behavior, reduced feeding, and low spontaneous locomotor activity and exploration. The behavioral inhibition in subordinates can be viewed as a passive coping strategy to avoid costly interaction with dominants (Leshner, 1980; Benus et al., 1991).

Aggression can be both stressful and stress relieving. Stress can induce aggression (briefly) or inhibit it (prolonged). A good stress management capability can predispose an individual for aggression and dominance, and in this way, confound studies on the effect of aggression on stress (since winner and loser individuals have a different stress response in the first place). However, a good stress management capability can also predispose an individual for a loss, in the case where hunger and relocation is involved.

6. NUTRITIONAL FACTORS AFFECTING STRESS RESPONSES

The amount as well as the composition of the food will have effects on stress responses and could be additional factors generating intraspecific variation in the stress response profile. Especially, the amino acid (AA) and FA composition of the food seem to be important since AAs and FAs have effects on neurotransmitter synthesis, membrane composition, and neuronal excitability.

6.1. Amino Acids

There are a number of studies showing behavioral and neuroendocrine effects of dietary manipulation of the AAs tryptophan (TRP) and tyrosine (TYR) (for reviews see Fernstrom, 1983; Fernstrom and Fernstrom, 2007). The AAs TRP and TYR are the precursors of 5-HT and catecholamines (dopamine, norepinephrine and epinephrine), respectively, and the effects of TRP and TYR on stress responses and behavior are believed to be mediated by these monoaminergic neurotransmitters. Both TRP and TYR cross the blood–brain barrier through the same carrier in competition with other large neutral amino acids (LNAAs) such as phenylalanine, leucine, isoleucine, valine, and methionine. Thus, in addition to dietary TYR and TRP content, the amount of these other LNAAs in the food will also have effects on brain uptake of TYR and TRP (Fernstrom, 1983, 1990), and thus on neuroendocrine processes, including integration of the behavioral and endocrine responses to stress (reviewed by Markus, 2008).

The rate-limiting step in 5-HT biosynthesis is catalyzed by tryptophanhydroxylase (TPH), an enzyme that occurs in two different isoforms, TPH1 and TPH2 (Lillesaar, 2011). The Km value of TPH1 is lower than that of TPH2. In fact, the Km for TRP of TPH2 is in the range of the free concentration of brain TRP, making the rate of 5-HT synthesis drastically affected by TRP availability (McKinney et al., 2001, 2005). In the teleost brain 5-HT cell bodies are located in the hindbrain raphe nuclei as well as in several diencephalic areas. Interestingly, TPH1 and TPH2 show divergent expression, with TPH2 being expressed in the raphe whereas TPH1 dominates in diencephalic and peripheral 5-HT cells (eg, along the gustatory tract) (Lillesaar, 2011). Thus, the rate of raphe 5-HT synthesis is strictly restricted by TRP availability whereas 5-HT synthesis in other 5-HT neuronal populations is not. However, raphe 5-HT neurons show an extremely divergent projection pattern, sending projections to forebrain areas as well as down the medulla and spinal cord. Plasma TRP concentrations and thus brain TRP availability are affected by factors like dietary TRP content, stress, and immune responses, and it has been hypothesized that changes in plasma TRP levels act as a peripheral signal modifying central 5-HT release (Russo et al., 2009).

In contrast to other LNAAs, TRP is transported in the plasma bound to albumin, at least in mammals (Fernstrom, 1983; Fernstrom and Fernstrom, 2007). As a consequence, dietary carbohydrates will also have effects on brain TRP availability since insulin secretion results in an uptake of AAs to muscle. However, since TRP is bound to albumin, plasma TRP concentrations are not affected by insulin. Instead, insulin secretion results in an increase in TRP uptake by the brain since plasma levels of other LNAAs that compete with TRP for the same carrier are reduced. However, rainbow trout albumin appears to lack the binding site for TRP (Fuller and Roush, 1973), and it is not clear how TRP is transported in the plasma of other teleosts.

A number of studies in fish have demonstrated the effects of dietary TRP supplementation on brain 5-HT metabolism (Winberg et al., 2001; Lepage et al., 2002; Höglund et al., 2005, 2007; Basic et al., 2013b; Martins et al., 2013). Moreover, these studies show that TRP supplementation results in behavioral and physiological effects, suggesting changes in 5-HT signaling. For instance, dietary supplementation of TRP has been shown to suppress aggression and to have anxiolytic effects, attenuating stress-induced anorexia (Winberg et al., 2001; Hseu et al., 2003; Höglund et al., 2005, 2007; Wolkers et al., 2012, 2014). Furthermore, TRP-enriched feed has been shown to reduce poststress plasma cortisol levels in a dose-dependent way in juvenile rainbow trout (Lepage et al., 2002), Atlantic cod (*Gadus morhua*) (Basic et al., 2012), and Atlantic salmon (Basic et al., 2013b). Moreover, Lepage et al. (2003) showed that this suppressive effect on HPI reactivity was present after 7 days of treatment with TRP-enriched feed, but not after treatment periods of 3 and 28 days. This suggests a rather narrow time span of the TRP treatment period for obtaining the stress-reducing effects of TRP. It is not clear if treatment with feed supplemented with TRP has any persistent effects on the behavior and stress responses of the fish since in most cases the effects of TRP supplementation has been tested while still feeding the fish this feed (Lepage et al., 2002, 2003; Höglund et al., 2005, 2007). However, recent studies demonstrate that the effect of dietary TRP may persist for up to 10 days after ending the dietary treatment period (Basic et al., 2013a,b).

The mechanism involved in mediating the stress-reducing effect of TRP is currently not fully understood. Interestingly, in rainbow trout treatment with the selective serotonin reuptake inhibitor (SSRI) citalopram reduced aggressive behavior and poststress plasma cortisol in the same manner as TRP-enriched feed (Lepage et al., 2005). In mammalian models of depression the antidepressive effects of SSRI, such as suppression of HPA axis hyperactivity, has been attributed to increased extracellular 5-HT levels and altered 5-HT receptor expression. However, 5-HT may also act through neuroplasticity mechanisms by stimulating neurogenesis via stimulatory effects on brain-derived neurotrophic factor (BDNF) and neural growth factor (Dong-Ryulu et al., 1999; van Donkelaar et al., 2009). Effects of 5-HT on neuroplasticity appear to be most expressed in the hippocampus, a brain area that has an inhibitory action on HPA axis activity (for references, see review by Mahar et al., 2014).

Aside from 5-HT synthesis, TRP may also be a rate-limiting factor in other metabolic pathways, which potentially can interact with the neuroendocrine stress response. In mammals, the kynurenic pathway convert TRP into other bioactive substances, such as kynurenic acid and quinolinic acid, by the enzymes indoleamine dioxygenase (IDO) and tryptophan dioxygenase (TDO) (Maes et al., 2009; Le Floc'h et al., 2011). The enzymes TDO and IDO are activated by glucocorticoids and proinflammatory cytokines, respectively. The kynurenic pathway shunts TRP away from 5-HT production (reviewed by Russo et al., 2009). Furthermore, kynurenic and quinolinic acid are neuroactive substances acting on excitatory AA receptors (AMPA, NMDA, and kainite glutamate receptors) (Zadori et al., 2009) and may have effects on monoaminergic signaling (Okuno et al., 2011). Thus, stress and inflammatory processes are likely to interact with the effect of dietary TRP supplementation, shunting available TRP toward the kyneric pathway. However, while extensively studied in mammals, there is less information on the kynurenic pathway in fish and how it interacts with the HPI axis.

The rate-limiting enzyme in catecholamine biosynthesis, tyrosine hydroxylase, appears fully saturated with its substrate, Tyr, in vivo. This suggests that dietary TYR should not affect production and release of catecholamines, even though TYR, like TRP, is an essential AA. However, there are some studies reporting behavioral and endocrine effects of TYR administration (Fernstrom and Fernstrom, 2007). Research on physiological effects of dietary TYR in fish is relatively limited (Li et al., 2009). In a recent study Costas et al. (2012) reported that a diet containing high levels of AAs, including TYR and TRP, increased brain DA and 5-HT concentrations. This finding led the authors to suggest that dietary TYR enrichment can stimulate DA production in teleost fish.

6.2. Fatty Acids

Fish, like other vertebrates, require the omega-3 FAs, docosahexaenoic acid (DHA) and eicosapentaenoic acid (EPA), and the omega-6 FA, arachidonic acid (ARA), for their normal growth and development (for references, see Bell et al., 1986). Generally, the physiological functions of these three long-chain polyunsaturated FAs seem to be similar within the vertebrate lineage, where they have a generalized role in maintaining the structural and functional integrity of cell membranes and a more specific role as eicosanoid precursors. The latter are a group of paracrine hormones, including prostaglandins, thromboxanes, leukotrienes, and a variety of hydroxy and hydroperoxy FAs (for references, see Wainwright, 2002). They are involved in a number of physiological functions, and almost all tissues in the body are able produce them. Generally, they are synthesized both at the cellular and whole-body levels in response to stressful situations.

As in mammals, the major eicosanoid precursor in fish is ARA. Furthermore, the biological activity of eicosanoids with EPA as a precursor is generally lower compared with those synthesized from ARA. Consequently, eicosanoid activity is affected by the available tissue content of FAs. Concurrent to this, there is considerable evidence that eicosanoid activity and production can be modified by changes in the relative proportions of the dietary omega-6 and omega-3 FAs (for references, see review by Tocher, 2003). Yet, it is important to keep in mind that many vertebrate species, including most of the freshwater fishes, are capable of producing DHA and EPA from linolenic acid (LNA) as well as ARA from linoleic acid (LA). In such species, LNA and LA are dietary essential FAs and the dietary ratios of these FAs are major determinants of the final tissue ratios of DHA:EPA:ARA. In marine fishes, however, the conversion rate of LNA and LA is very low, and DHA, EPA, and ARA are considered as being dietary essential FAs (Tocher, 2003).

The dietary FA composition may affect the neuroendocrine stress response at different levels of the stress axis. This has been demonstrated in rat studies where prostaglandins, eicosanoids that are converted from ARA by the enzyme cyclooxygenases (COX), stimulate HPA axis activity at the pituitary and adrenal levels (Mulla and Buckingham, 1999). Indications of similar mechanisms in fish have been demonstrated in European sea bass (*Dicentrarchus labrax*). This species showed higher poststress plasma cortisol levels when being fed a diet where some of the omega-3 rich fish oil was substituted with omega-3 deficient vegetable oil (Montero et al., 2003). Moreover, sea bass larvae fed a diet supplemented with ARA showed higher cortisol levels when exposed to a regimen of daily repeated stress (Koven et al., 2003), and perfusion studies suggested that both EPA and ARA are able to stimulate cortisol production at the interrenal level of the HPI axis in this species. Moreover, the fact that COX inhibitors reduced this stimulating effect suggests that COX-derived eicosanoids mediate the cortisol-stimulating effects of EPA and ARA (Ganga et al., 2006).

Dietary DHA seems to be especially important during early development. This has been demonstrated in a number of studies where it promotes survival, stress resistance (Lund and Steenfeldt, 2011; Lund et al., 2014), proper brain development, neuronal migration, and neurophysiological functioning in fish larvae (Benítez-Santana et al., 2012). However, the mechanism underlying the link between dietary DHA and HPI axis activity is currently not known. Low levels of omega-3 FAs have been associated with depression (Hibbeln, 1998, 2002), including being accompanied by chronically elevated plasma cortisol levels. Interestingly, a diet rich in fish oil containing high levels of DHA and EPA has been shown to affect serotonergic transmission (Vancassel et al., 2008) and to stimulate expression of genes involved in trophic processes (Wu et al., 2004) in the same manner as antidepressants. The mechanisms for the antidepressive action of omega-3 FAs are currently not fully understood. It is possible that a diet with a high omega-3:omega-6 ratio results in a decrease in the eicosanoids that activate IDO. This, in turn, shunts the kyrenic pathway from the TRP catabolites kynurenic acid and quinolinic acid toward 5-HT production (Fig. 2.1). Since quinolinic acid has been shown to have neurodegenerative effects (Maes et al., 2009) and 5-HT stimulates neurogenesis, it can be hypothesized that this AA and FA interplay results in structural processes in the brain, which in turn have suppressive effects on HPA axis hyperactivity. Whether such a relationship between AA and FA and activity of the HPI axis exists in fish is not known. But in the event that it does, TRP-enriched feed may offer a way to enhance the low stress tolerance observed in fish fed with a high inclusion rate of vegetable oil containing low levels of omega-3 FAs in the feed (Fig. 2.1).

The nutritional composition of the diet may be especially important during early ontogeny since it could induce long-term effects on the neuroendocrine stress response. This has been shown in rats where the FA composition of maternal milk affects the HPA axis function in the adult (D'Asti et al., 2010). A recent study demonstrated long-term effects of dietary FAs also in fish (Lund et al., 2014). In the latter study, pikeperch (*Sander lucioperca*) was fed feeds with different FA compositions from day 7 to day 27 after hatching. When tested 85 days after termination of the dietary treatment, pikeperch fed a feed containing low levels of DHA showed reduced locomotor activity in response to novel treatment. However, even though this suggests that the behavioral stress response was affected by dietary FA composition, no effects on the HPI axis have been reported.

7. DIRECTIONS FOR FUTURE RESEARCH

Variation in neuroendocrine stress responses in fish is related to genetic factors, current environment, and previous experiences. Divergent stress responses are usually correlated to specific behavioral profiles forming different stress coping styles. Our knowledge on how heritable factors interact with the environment for controlling the development of these stress coping styles, and how coping style relates to life history variation, is still very limited. Agonistic interactions are known to have drastic behavioral effects (ie, classic winner–loser effects). As discussed earlier, social stress as well as other challenges has been shown to modify at least the behavioral traits associated with specific stress coping styles. These effects are likely to be mediated by epigenetic effects, mechanisms that have so far not been explored in any detail. It is also clear that stress coping is related to life history traits like time of sexual maturation, migration, and so on. However, again our knowledge on both ecological implications and mechanisms involved is limited. Dietary composition is also known to affect stress response profiles and behavior in fish. Feed supplementation could be an interesting strategy for producing stress-resistant nonaggressive fish for aquaculture. Another strategy to produce stress-resistant fish for aquaculture is selective breeding. Fish displaying a proactive stress coping style appear to be best suited for rearing in confined aquaculture settings. However, as it comes to aquaculture production proactive fish have one clear disadvantage—they are aggressive, and it is not known whether aggression could be separated from other traits of the proactive phenotype by targeted selection. A research field of both fundamental and applied interest is thus to what degree and under which conditions correlated traits can diverge under artificial or natural selection.

REFERENCES

Abbott, J. C. and Dill, L. M. (1985). Patterns of aggressive attack in juvenile steelhead trout (*Salmo gairdneri*). *Can. J. Fish. Aquat. Sci.* **42**, 1702–1706.

Abbott, J. C., Dunbrack, R. L. and Orr, C. D. (1985). The interaction of size and experience in dominance relationships of juvenile steelhead trout (*Salmo gairdneri*). *Behaviour* **92**, 241–253.

Åberg-Andersson, M., Wahid Khanb, U., Øverli, Ø., Gjøen, H. M. and Höglund, E. (2013). Coupling between stress coping style and time of emergence from spawning nests in salmonid fishes: evidence from selected rainbow trout strains (*Oncorhynchus mykiss*). *Physiol. Behav.* **116**, 30–34.

Adams, C. E. and Huntingford, F. A. (1996). What is a successful fish? Determinants of competitive success in Arctic char (*Salvelinus alpinus*) in different social contexts. *Can. J. Fish. Aquat. Sci.* **53**, 2446–2450.

Adams, C. E., Huntingford, F. A., Turnbull, J. F. and Beattie, C. (1998). Alternative competitive strategies and the cost of food acquisition in juvenile Atlantic salmon (*Salmo salar*). *Aquaculture* **167**, 17–26.

Aguilar, R., Gil, L., Fernández-Teruel, A. and Tobeña, A. (2004). Genetically-based behavioral traits influence the effects of shuttle box avoidance overtraining and extinction upon intertrial responding: a study with the Roman rat strains. *Behav. Processes* **66**, 63–72.

Albonetti, M. E. and Farabollini, F. (1994). Social stress by repeated defeat: effects on social behaviour and emotionality. *Behav. Brain. Res.* **62**, 187–193.

Almasi, B., Jenni, L., Jenni-Eiermann, S. and Roulin, A. (2010). Regulation of stress response is heritable and functionally linked to melanin-based coloration. *J. Evol. Biol.* **23**, 987–996.

Alonso, F., Honji, R. M., Moreira, R. G. and Pandolfi, M. (2012). Dominance hierarchies and social status ascent opportunity: anticipatory behavioral and physiological adjustments in a Neotropical cichlid fish. *Physiol. Behav.* **106**, 612–618.

Alsop, D. and Vijayan, M. M. (2008). Development of the corticosteroid stress axis and receptor expression in zebrafish. *Am. J. Physiol. Regul. Integr. Comp. Physiol.* **294**, R2021–R2021.

Alsop, D. and Vijayan, M. M. (2009). The zebrafish stress axis: molecular fallout from the teleost-specific genome duplication event. *Gen. Comp. Endocrinol.* **161**, 62–66.

Auperin, B. and Geslin, M. (2008). Plasma cortisol response to stress in juvenile rainbow trout is influenced by their life history during early development and by egg cortisol content. *Gen. Comp. Endocrinol.* **158**, 234–239.

Backström, T., Schjolden, J., Øverli, Ø., Thörnqvist, P.-O. and Winberg, S. (2011). Stress effects on AVT and CRF systems in two strains of rainbow trout (*Oncorhynchus mykiss*) divergent in stress responsiveness. *Horm. Behav.* **59**, 180–186.

Backström, T., Brännäs, E., Nilsson, J. and Magnhagen, C. (2014). Behaviour, physiology and carotenoid pigmentation in Arctic charr (*Salvelinus alpinus*). *J. Fish. Biol.* **84**, 1–9.

Balm, P. H. M. and Pottinger, T. G. (1995). Corticotrope and melanotrope POMC-derived peptides in relation to interrenal function during stress in raibow trout (*Oncorhynchus mykiss*). *Gen. Comp. Endocrinol.* **98**, 279–288.

Barrot, M., Abrous, D. N., Marinelli, M., Rougé-Pont, F., Le Moal, M. and Piazza, P. V. (2001). Influence of glucocorticoids on dopaminergic transmission in the rat dorsolateral striatum. *Eur. J. Neurosci.* **13**, 812–818.

Barry, T. P., Malison, J. A., Held, J. A. and Parrish, J. J. (1995a). Ontogeny of the cortisol stress response in larval rainbow trout. *Gen. Comp. Endocrinol.* **97**, 57–65.

Barry, T. P., Ochiai, M. and Malison, J. A. (1995b). In-vitro effects of acth on interrenal corticosteroidogenesis during early larval development in rainbow-trout. *Gen. Comp. Endocrinol.* **99**, 382–387.

Barton, B. A. (2002). Stress in fishes: a diversity of responses with particular reference to changes in circulating corticosteroids. *Integr. Comp. Biol.* **42**, 517–525.

Barton, B. A. and Iwama, G. K. (1991). Physiological changes in fish from stress in aquaculture with emphasis on the response and effects of corticosteroids. *Annu. Rev. Fish Dis.* **1**, 3–26.

Barton, B. A., Ribas, L., Acerete, L. and Tort, L. (2005). Effects of chronic confinement on physiological responses of juvenile gilthead sea bream, *Sparus aurata* L., to acute handling. *Aquacult. Res.* **36**, 172–179.

Basic, D., Winberg, S., Schjolden, J., Krogdahl, A. and Höglund, E. (2012). Context-dependent responses to novelty in rainbow trout (*Oncorhynchus mykiss*), selected for high and low post-stress cortisol responsiveness. *Physiol. Behav.* **105**, 1175–1181.

Basic, D., Krogdahl, A., Schjolden, J., Winberg, S., Vindas, M. A., Hillestad, M., et al. (2013a). Short- and long-term effects of dietary ι-tryptophansupplementation on the neuroendocrine stress response in seawater-reared Atlantic salmon (*Salmo salar*). *Aquaculture* **388**, 8–13.

Basic, D., Schjolden, J., Krogdahl, A., von Krogh, K., Hillestad, M., Winberg, S., et al. (2013b). Changes in regional brain monoaminergic activity and temporary down-regulation in stress response from dietary supplementation with ʟ-tryptophan in Atlantic cod (*Gadus morhua*). *Br. J. Nutr.* **109**, 2166–2174.

Beacham, J. L. (1988). The relative importance of body size and aggressive experience as determinants of dominance in pumpkinseed sunfish, *Lepomis gibbosus. Anim. Behav.* **36**, 621–623.

Beaugrand, J. P. and Cotnoir, P. A. (1996). The role of individual differences in the formation of triadic dominance orders of male green swordtail fish (*Xiphophorus helleri*). *Behav. Processes* **38**, 287–296.

Bell, A. M. and Stamps, J. A. (2004). Development of behavioural differences between individuals and populations of sticklebacks, *Gasterosteus aculeatus. Anim. Behav.* **68**, 1339–1348.

Bell, A. M., Henderson, L. and Huntingford, F. A. (2010). Behavioral and respiratory responses to stressors in multiple populations of three-spined sticklebacks that differ in predation pressure. *J. Comp. Physiol. B.* **180**, 211–220.

Bell, M. V., Henderson, R. J. and Sargent, J. R. (1986). The role of polyunsaturated fatty-acids in fish. *Comp. Biochem. Physiol. B* **83**, 711–719.

Benítez-Santana, T., Juárez-Carrillo, E., Beatriz Betancor, M., Torrecillas, S., José Caballero, M. and Soledad Izquierdo, M. (2012). Increased mauthner cell activity and escaping behaviour in seabream fed long-chain PUFA. *Br. J. Nutr.* **107**, 295–301.

Benus, R., Den Daas, S., Koolhaas, J. and van Oortmerssen, G. (1990). Routine formation and flexibility in social and non-social behaviour of aggressive and non-aggressive male mice. *Behaviour* **112**, 176–193.

Benus, R. F., Bohus, B., Koolhaas, J. M. and van Oortmerssen, G. A. (1991). Heritable variation for aggression as a reflection of individual coping strategies. *Experientia* **47**, 1008–1019.

Bernier, N. J., Alderman, S. L. and Bristow, E. N. (2008). Heads or tails? Stressor-specific expression of corticotropin-eleasing factor and urotensin I in the preoptic area and caudal neurosecretory system of rainbow trout. *J. Endocrinol.* **196**, 637–648.

Bernstein, I. S. and Gordon, T. P. (1974). The function of aggression in primate societies: uncontrolled aggression may threaten human survival, but aggression may be vital to the establishment and regulation of primate societies and sociality. *Am. Sci.* **62**, 304–311.

Blanchard, D. C., Sakai, R. R., McEwen, B., Weiss, S. M. and Blanchard, R. J. (1993). Subordination stress: behavioral, brain, and neuroendocrine correlates. *Behav. Brain Res.* **58**, 113–121.

Blanchard, D. C., Spencer, R. L., Weiss, S. M., Blanchard, R. J., McEwen, B. and Sakai, R. R. (1995). Visible burrow system as a model of chronic social stress: behavioral and neuroendocrine correlates. *Psychoneuroendocrinology* **20**, 117–134.

Bolhuis, J. E., Schouten, W. G., De Leeuw, J. A., Schrama, J. W. and Wiegant, V. M. (2004). Individual coping characteristics, rearing conditions and behavioural flexibility in pigs. *Behav. Brain Res.* **152**, 351–360.

Boon, A. K., Réale, D. and Boutin, S. (2007). The interaction between personality, offspring fitness and food abundance in North American red squirrels. *Ecol. Lett.* **10**, 1094–1104.

Both, C., Dingemanse, N. J., Drent, P. J. and Tinbergen, J. M. (2005). Pairs of extreme avian personalities have highest reproductive success. *J. Anim. Ecol.* **74**, 667–674.

Brännäs, E. (1988). Emergence of Baltic salmon, *Salmo salar* L., in relation to temperature—a laboratory study. *J. Fish. Biol.* **33**, 589–600.

Brelin, D., Petersson, E. and Winberg, S. (2005). Divergent stress coping styles in juvenile brown trout (*Salmo trutta*). *Ann. N. Y. Acad. Sci.* **1040**, 239–245.

Brelin, D., Petersson, E., Dannewitz, J., Dahl, J. and Winberg, S. (2008). Frequency distribution of coping strategies in four populations of brown trout (*Salmo trutta*). *Horm. Behav.* **53**, 546–556.

Burmeister, S. S. (2007). Genomic responses to behavioral interactions in an African cichlid fish: mechanisms and evolutionary implications. *Brain Behav. Evol.* **70**, 247–256.

Bury, N. R., Sturm, A., Le Rouzic, P., Lethimonier, C., Ducouret, B., Guiguen, Y., et al. (2003). Evidence for two distinct functional glucocorticoid receptors in teleost fish. J. Mol. Endocrinol. **31**, 141–156.

Bury, N. R. and Sturm, A. (2007). Evolution of the corticosteroid receptor signalling pathway in fish. *Gen. Comp. Endocrinol.* **153**, 47–56.

Cannon, W. B. (1929). Bodily changes in pain, fear, hunger, and rage. New York: Appleton.

Carere, C., Caramaschi, D. and Fawcett, T. W. (2010). Covariation between personalities and individual differences in coping with stress: converging evidence and hypotheses. *Curr. Zool.* **56**, 728–740.

Carpenter, R. E., Watt, M. J., Forster, G. L., Øverli, Ø., Bockholt, C., Renner, K. J., et al. (2007). Corticotropin releasing factor induces anxiogenic locomotion in trout and alters serotonergic and dopaminergic activity. *Horm. Behav.* **52**, 600–611.

Cash, W. B. and Holberton, R. L. (1999). Effects of exogenous corticosterone on locomotor activity in the red-eared slider turtle, *Trachemys scripta* elegans. *J. Exp. Zool.* **284**, 637–644.

Caspi, A., Sugden, K., Moffitt, T. E., Taylor, A., Craig, I. W., Harrington, H., et al. (2003). Influence of life stress on depression: moderation by a polymorphism in the 5-HTT gene. *Science* **301**, 386–389.

Castanheira, M. F., Herrera, M., Costas, B., Conceição, L. E. and Martins, C. I. (2013). Linking cortisol responsiveness and aggressive behaviour in gilthead seabream *Sparus aurata*: indication of divergent coping styles. *Appl. Anim. Behav. Sci.* **143**, 75–81.

Chabbi, A. and Ganesh, C. (2015). Evidence for the involvement of dopamine in stress-induced suppression of reproduction in the cichlid Fish *Oreochromis mossambicus. J. Neuroendocrinol.* **27**, 343–356.

Chandrasekar, G., Lauter, G. and Hauptmann, G. (2007). Distribution of corticotropin-releasing hormone in the developing zebrafish brain. *J. Comp. Neurol.* **505**, 337–351.

Charmantier, A. and Garant, D. (2005). Environmental quality and evolutionary potential: lessons from wild populations. *Proc. R. Soc. Lond. B Biol. Sci.* **272**, 1415–1425.

Clarke, A. S. and Boinski, S. (1995). Temperament in nonhuman primates. *Am. J. Primatol.* **37**, 103–125.

Clements, S., Moore, F. L. and Schreck, C. B. (2003). Evidence that acute serotonergic activation potentiates the locomotor-stimulating effects of corticotropin-releasing hormone in juvenile Chinook salmon (*Oncorhynchus tshawytscha*). *Horm. Behav.* **43**, 214–221.

Cockrem, J. F. (2013). Individual variation in glucocorticoid stress responses in animals. *Gen. Comp. Endocrinol.* **181**, 45–58.

Colombe, L., Fostier, A., Bury, N., Pakdel, F. and Guiguen, Y. (2000). A mineralocorticoid-like receptor in the rainbow trout, *Oncorhynchus mykiss*: cloning and characterization of its steroid binding domain. *Steroids* **65**, 319–328.

Coppens, C. M., De Boer, S. F. and Koolhaas, J. M. (2010). Coping styles and behavioural flexibility: towards underlying mechanisms. *Phil. Trans. R. Soc. Lond. B Biol. Sci.* **365**, 4021–4028.

Costas, B., Aragao, C., Soengas, J. L., Miguez, J. M., Rema, P., Dias, J., et al. (2012). Effects of dietary amino acids and repeated handling on stress response andbrain monoaminergic neurotransmitters in Senegalese sole (*Solea senegalensis*) juveniles. *Comp. Biochem. Physiol. A Mol. Integr. Physiol.* **161**, 18–26.

Cote, S. D. (2000). Dominance hierarchies in female mountain goats: stability, aggressiveness and determinants of rank. *Behaviour* **137**, 1541–1566.

Cutts, C. J., Metcalfe, N. B. and Taylor, A. C. (1999a). Competitive asymmetries in territorial juvenile Atlantic salmon, *Salmo salar. Oikos* **86**, 479–486.

Cutts, C. J., Brembs, B., Metcalfe, N. B. and Taylor, A. C. (1999b). Prior residence, territory quality and life-history strategies in juvenile Atlantic salmon (*Salmo salar* L.). *J. Fish. Biol.* **55**, 784–794.

Dahlbom, S. J., Backström, T., Lundstedt-Enkel, K. and Winberg, S. (2012). Aggression and monoamines: effects of sex and social rank in zebrafish (*Danio rerio*). *Behav. Brain. Res.* **228**, 333–338.

D'Amato, F. R. (1988). Effects of male social status on reproductive success and on behavior in mice (*Mus musculus*). *J. Comp. Psychol.* **102**, 146–151.

D'Asti, E., Long, H., Tremblay-Mercier, J., Grajzer, M., Cunnane, S. C., Di Marzo, V., et al. (2010). Maternal dietary fat determines metabolic profile and the magnitude of endocannabinoid inhibition of the stress response in neonatal rat offspring. *Endocrinology* **151**, 1685–1694.

David, M., Auclair, Y. and Cézilly, F. (2011). Personality predicts social dominance in female zebra finches, *Taeniopygia guttata*, in a feeding context. *Anim. Behav.* **81**, 219–224.

De Kloet, E. R. (2004). Hormones and the stressed brain. *Ann. N. Y. Acad. Sci.* **1018**, 1–15.

Desjardins, J. K., Hofmann, H. A. and Fernald, R. D. (2012). Social context influences aggressive and courtship behavior in a cichlid fish. *PLoS One* **7**, e32781.

Dingemanse, N. J., Both, C., Drent, P. J. and Tinbergen, J. M. (2004). Fitness consequences of avian personalities in a fluctuating environment. *Proc. R. Soc. Lond. Ser. B Biol. Sci.* **271**, 847–852.

Dingemanse, N. J., Van Der Plas, F., Wright, J., Réale, D., Schrama, M., Roff, D. A., et al. (2009). Individual experience and evolutionary history of predation affect expression of heritable variation in fish personality and morphology. *Proc. R. Soc. Lond. B Biol. Sci.* **282**. http://dx.doi.org/10.1098/rspb.2008.1555.

Dong-Ryulu, L., Semba, R., Kondo, H., Goto, S. and Nakano, K. (1999). Decrease in the levels of NGF and BDNF in brains of mice fed a tryptophan-deficient diet. *Biosci. Biotechnol. Biochem.* **63**, 337–340.

Doyon, C., Gilmour, K. M., Trudeau, V. L. and Moon, T. W. (2003). Corticotropin-releasing factor and neuropeptide Y mRNA levels are elevated in the preoptic area of socially subordinate rainbow trout. *Gen. Comp. Endocrinol.* **133**, 260–271.

Dugatkin, L. A. (1997). Winner and loser effects and the structure of dominance hierarchies. *Behav. Ecol.* **8**, 583–587.

Einum, S. and Fleming, I. A. (2000). Selection against late emergence and small offspring in Atlantic salmon (*Salmo salar*). *Evolution* **54**, 628–639.

Ejike, C. and Schreck, C. B. (1980). Stress and social hierarchy rank in coho salmon. *Trans. Am. Fish. Soc.* **109**, 423–426.

Elipot, Y., Hinaux, H., Callebert, J., Launay, J.-M., Blin, M. and Rétaux, S. (2014). A mutation in the enzyme monoamine oxidase explains part of the Astyanax cavefish behavioural syndrome. *Nat. Commun.* **5**, 3647.

Engh, A. L., Siebert, E. R., Greenberg, D. A. and Holekamp, K. E. (2005). Patterns of alliance formation and post conflict aggression indicate spotted hyaenas recognize third-party relationships. *Anim. Behav.* **69**, 209–217.

Eriksen, M. S., Bakken, M., Espmark, A., Braastad, B. O. and Salte, R. (2006). Prespawning stress in farmed Atlantic salmon *Salmo salar*: maternal cortisol exposure and hyperthermia during embryonic development affect offspring survival, growth and incidence of malformations. *J. Fish. Biol.* **69**, 114–129.

Espmark, A. M., Eriksen, M. S., Salte, R., Braastad, B. O. and Bakken, M. (2008). A note on pre-spawning maternal cortisol exposure in farmed Atlantic salmon and its impact on the behaviour of offspring in response to a novel environment. *Appl. Anim. Behav. Sci.* **110**, 404–409.

Feist, G. and Schreck, C. B. (2002). Ontogeny of the stress response in Chinook salmon, *Oncorhynchus tshawytscha*. *Fish Physiol. Biochem.* **25**, 31–40.

Fernandez, F., Sarre, S., Launay, J. M., Aguerre, S., Guyonnet-Dupérat, V., Moisan, M. P., et al. (2003). Rat strain differences in peripheral and central serotonin transporter protein expression and function. *Eur. J. Neurosci.* **17**, 494–506.

Fernstrom, J. D. (1983). Role of precursor availability in control of monoamine biosynthesis in brain. *Physiol. Rev.* **63**, 484–546.

Fernstrom, J. D. (1990). Aromatic-amino-acid and monoamine syntesis in the central-nervous-system—influence of the diet. *J. Nutr. Biochem.* **1**, 508–517.

Fernstrom, J. D. and Fernstrom, M. H. (2007). Tyrosine, phenylalanine, and catecholamine synthesis and function in the brain. *J. Nutr.* **137**, 1539–1547.

Flik, G., Stouthart, X. J. H. X., Spanings, F. A. T., Lock, R. A. C., Fenwick, J. C. and Wendelaar Bonga, S. E. (2002). Stress response to waterborne Cu during early life stages of carp, *Cyprinus carpio*. *Aquat. Toxicol.* **56**, 167–176.

Fone, K. C. F. and Porkess, M. V. (2008). Behavioural and neurochemical effects of post-weaning social isolation in rodents—relevance to developmental neuropsychiatric disorders. *Neurosci. Biobehav. Rev.* **32**, 1087–1102.

Francis, R. C. (1988). On the relationship between aggression and social dominance. *Ethology* **78**, 223–237.

Francis, R. C. (1990). Temperament in a fish: a longitudinal study of the development of individual differences in aggression and social rank in the Midas cichlid. *Ethology* **86**, 311–325.

Franck, D. and Ribowski, A. (1993). Dominance hierarchies of male green swordtails (*Xiphophorus helleri*) in nature. *J. Fish. Biol.* **43**, 497–499.

Fuller, R. W. and Roush, B. W. (1973). Binding of tryptophan to plasma-proteins in several species. *Comp. Biochem. Physiol. B.* **46**, 273–276.

Ganga, R., Tort, L., Acerete, L., Montero, D. and Izquierdo, M. S. (2006). Modulation of ACTH-induced cortisol release by polyunsaturated fatty acids in interrenal cells from gilthead seabream, *Sparus aurata. J. Endocrinol.* **190**, 39–45.

Garrido, S., Rosa, R., Ben-Hamadou, R., Cunha, M. E., Chicharo, M. A. and van der Lingen, C. D. (2007). Effect of maternal fat reserves on the fatty acid composition of sardine (*Sardina pilchardus*) oocytes. *Comp. Biochem. Physiol. B* **148**, 398–409.

Giesing, E. R., Suski, C. D., Warner, R. E. and Bell, A. M. (2011). Female sticklebacks transfer information via eggs: effects of maternal experience with predators on offspring. *Proc. R. Soc. B* **278**, 1753–1759.

Giorgi, O., Lecca, D., Piras, G., Driscoll, P. and Corda, M. (2003). Dissociation between mesocortical dopamine release and fear-related behaviours in two psychogenetically selected lines of rats that differ in coping strategies to aversive conditions. *Eur. J. Neurosci.* **17**, 2716–2726.

Glover, V., O'Connor, T. G. and O'Donnell, K. (2010). Prenatal stress and the programming of the HPA axis. *Neurosci. Biobehav. Rev.* **35**, 17–22.

Gómez-Laplaza, L. M. and Morgan, E. (2003). The influence of social rank in the angelfish, *Pterophyllum scalare*, on locomotor and feeding activities in a novel environment. *Lab. Anim.* **37**, 108–120.

Gorissen, M. and Flik, G. (2016). Endocrinology of the Stress Response in Fish. In *Fish Physiology - Biology of Stress in Fish*, Vol. 35 (eds. C. B. Schreck, L. Tort, A. P. Farrell and C. J. Brauner), San Diego, CA: Academic Press.

Gosling, S. D. (2001). From mice to men: what can we learn about personality from animal research? *Psychol. Bull.* **127**, 45–86.

Gregory, T. R. and Wood, C. M. (1999). The effects of chronic plasma cortisol elevation on the feeding behaviour, growth, competitive ability, and swimming performance of juvenile rainbow trout. *Physiol. Biochem. Zool.* **72**, 286–295.

Hayashi, A., Nagaoka, M., Yamada, K., Ichitani, Y., Miake, Y. and Okado, N. (1998). Maternal stress induces synaptic loss and developmental disabilities of offspring. *Int. J. Dev. Neurosci.* **16**, 209–216.

Herzog, W., Zeng, X., Lele, Z., Sonntag, C., Ting, J.-W., Chang, C.-Y., et al. (2003). Adenohypophysis formation in the zebrafish and its dependence on sonic hedgehog. *Dev. Biol.* **254**, 36–49.

Hibbeln, J. R. (1998). Fish consumption and major depression. *Lancet* **351**, 1213.

Hibbeln, J. R. (2002). Seafood consumption, the DHA content of mothers' milk and prevalence rates of postpartum depression: a cross-national, ecological analysis. *J. Affect. Disord.* **69**, 15–29.

Höglund, E., Kolm, N. and Winberg, S. (2001). Stress-induced changes in brain serotonergic activity, plasma cortisol and aggressive behavior in Arctic charr (*Salvelinus alpinus*) is counteracted by L-DOPA. *Physiol. Behav.* **74**, 381–389.

Höglund, E., Balm, P. H. and Winberg, S. (2002a). Behavioural and neuroendocrine effects of environmental background colour and social interaction in Arctic charr (*Salvelinus alpinus*). *J. Exp. Biol.* **205**, 2535–2543.

Höglund, E., Balm, P. H. and Winberg, S. (2002b). Stimulatory and inhibitory effects of 5-HT 1A receptors on adrenocorticotropic hormone and cortisol secretion in a teleost fish, the Arctic charr (*Salvelinus alpinus*). *Neurosci. Lett.* **324**, 193–196.

Höglund, E., Bakke, M. J., Øverli, Ø., Winberg, S. and Nilsson, G. E. (2005). Suppression of aggressive behaviour in juvenile Atlantic cod (*Gadus morhua*) by ɪ-tryptophan supplementation. *Aquaculture* **249**, 525–531.

Höglund, E., Sørensen, C., Bakke, M. J., Nilsson, G. E. and Øverli, Ø. (2007). Attenuation of stress-induced anorexia in brown trout (*Salmo trutta*) by pre-treatment with dietary ɪ-tryptophan. *Br. J. Nutr.* **97**, 786–789.

Holtby, L. B., Swain, D. P. and Allan, G. M. (1993). Mirror-elicited agonistic behavior and body morphology as predictors of dominance status in juvenile coho salmon (*Oncorhynchus kisutch*). *Can. J. Fish. Aquat. Sci.* **50**, 676–684.

Hseu, J. R., Lu, F. I., Su, H. M., Wang, L. S., Tsai, C. L. and Hwang, P. P. (2003). Effect of exogenous tryptophan on cannibalism, survival and growth in juvenile grouper, *Epinephelus coioides*. *Aquaculture* **218**, 251–263.

Hsu, Y. and Wolf, L. L. (1999). The winner and loser effect: integrating multiple experiences. *Anim. Behav.* **57**, 903–910.

Hsu, Y., Earley, R. L. and Wolf, L. L. (2006). Modulation of aggressive behaviour by fighting experience: mechanisms and contest outcomes. *Biol. Rev. Camb. Philos. Soc.* **81**, 33–74.

Huntingford, F. A. and Turner, A. K. (1987). *Animal Conflict.* London: Chapman & Hall/CRC.

Huntingford, F. A., Metcalfe, N. B., Thorpe, J. E., Graham, W. D. and Adams, C. E. (1990). Social dominance and body size in Atlantic salmon parr, *Salmo salar* L. *J. Fish. Biol.* **36**, 877–881.

Huntingford, F. A., Metcalfe, N. B. and Thorpe, J. E. (1993). Social status and feeding in Atlantic salmon *Salmo salar* parr: The effect of visual exposure to a dominant. *Ethology* **94**, 201–206.

Huntingford, F. A., Wright, P. and Tierney, J. (1994). *Adaptive Variation in Antipredator Behaviour in Threespine Stickleback. The Evolutionary Biology of the Threespine Stickleback.* Oxford: *Oxford University Press.*

Ishiwata, H., Shiga, T. and Okado, N. (2005). Selective serotonin reuptake inhibitor treatment of early postnatal mice reverses their prenatal stress-induced brain dysfunction. *Neuroscience* **133**, 893–901.

Jeffrey, J. D., Esbaugh, A. J., Vijayan, M. M. and Gilmour, K. M. (2012). Modulation of hypothalamic–pituitary–interrenal axis function by social status in rainbow trout. *Gen. Comp. Endocrinol.* **176**, 201–210.

Jeffrey, J. D., Gollock, M. J. and Gilmour, K. M. (2014). Social stress modulates the cortisol response to an acute stressor in rainbow trout (*Oncorhynchus mykiss*). *Gen. Comp. Endocrinol.* **196**, 8–16.

Jentoft, S., Held, J. A., Malison, J. A. and Barry, T. P. (2002). Ontogeny of the cortisol stress response in yellow perch (*Perca flavescens*). *Fish Physiol. Biochem.* **26**, 371–378.

Johansen, I. B., Sandvik, G. K., Nilsson, G. E., Bakken, M., Øverli, Ø., et al. (2011). Cortisol receptor expression differs in the brains of rainbow trout selected for divergent cortisol responses. *Comp. Biochem. Physiol. Part D Genomics Proteomics* **6**, 126–132.

Johnsson, J. I. and Björnsson, B. T. (1994). Growth hormone increases growth rate, appetite and dominance in juvenile rainbow trout, *Oncorhynchus mykiss. Anim. Behav.* **48**, 177–186.

Johnsson, J. I., Jonsson, E. and Björnsson, B. T. (1996). Dominance, nutritional state, and growth hormone levels in rainbow trout (*Oncorhynchus mykiss*). *Horm. Behav.* **30**, 13–21.

Johnsson, J. I., Nöbbelin, F. and Bohlin, T. (1999). Territorial competition among wild brown trout fry: effects of ownership and body size. *J. Fish. Biol.* **54**, 469–472.

Jönsson, E., Johnsson, J. I. and Björnsson, B. T. (1998). Growth hormone increases aggressive behavior in juvenile rainbow trout. *Horm. Behav.* **33**, 9–15.

Jönsson, E., Johansson, V., Björnsson, B. T. and Winberg, S. (2003). Central nervous system actions of growth hormone on brain monoamine levels and behavior of juvenile rainbow trout. *Horm. Behav.* **43**, 367–374.

Kittilsen, S., Schjolden, J., Beitnes-Johansen, I., Shaw, J., Pottinger, T. G., Sørensen, C., et al. (2009). Melanin-based skin spots reflect stress responsiveness in salmonid fish. *Horm. Behav.* **56**, 292–298.

Koolhaas, J. M., Korte, S. M., De Boer, S. F., Van Der Vegt, B. J., Van Reenen, C. G., Hopster, H., et al. (1999). Coping styles in animals: current status in behavior and stress-physiology. *Neurosci. Biobehav. Rev.* **7**, 925–935.

Koolhaas, J., De Boer, S., Coppens, C. and Buwalda, B. (2010). Neuroendocrinology of coping styles: towards understanding the biology of individual variation. *Front Neuroendocrinol.* **31**, 307–321.

Korte, S. M., Koolhaas, J. M., Wingfield, J. C. and Mcewen, B. S. (2005). The Darwinian concept of stress: benefits of allostasis and costs of allostatic load and the trade-offs in health and disease. *Neurosci. Biobehav. Rev.* **29**, 3–38.

Korzan, W. J. and Summers, C. H. (2007). Behavioral diversity and neurochemical plasticity: selection of stress coping strategies that define social status. *Brain Behav. Evol.* **70**, 257–266.

Koven, W., van Anholt, R., Lutzky, S., Ben Atia, I., Nixon, O., Ron, B., et al. (2003). The effect of dietary arachidonic acid on growth, survival, and cortisol levels in different-age gilthead seabream larvae (*Sparus auratus*) exposed to handling or daily salinity change. *Aquaculture* **228**, 307–320.

Kralj-Fišer, S. and Schuett, W. (2014). Studying personality variation in invertebrates: why bother? *Anim. Behav.* **91**, 41–52.

Kramer, M., Hiemke, C. and Fuchs, E. (1999). Chronic psychosocial stress and antidepressant treatment in tree shrews: time-dependent behavioral and endocrine effects. *Neurosci. Biobehav. Rev.* **23**, 937–947.

Larson, E., Norris, D. and Summers, C. (2003). Monoaminergic changes associated with socially induced sex reversal in the saddleback wrasse. *Neuroscience* **119**, 251–263.

Larson, E. T., O'Malley, D. M. and Melloni, R. H. (2006). Aggression and vasotocin are associated with dominant–subordinate relationships in zebrafish. *Behav. Brain Res.* **167**, 94–102.

Le Floc'h, N., Otten, W. and Merlot, E. (2011). Tryptophan metabolism, from nutrition to potential therapeutic applications. *Amino. Acids* **41**, 1195–1205.

Lepage, O., Tottmar, O. and Winberg, S. (2002). Elevated dietary intake of ʟ-tryptophan counteracts the stress-induced elevation of plasma cortisol in rainbow trout (*Oncorhynchus mykiss*). *J. Exp. Biol.* **205**, 3679–3687.

Lepage, O., Vilchez, I. M., Pottinger, T. G. and Winberg, S. (2003). Time-course of the effect of dietary ʟ-tryptophan on plasma cortisol levels in rainbow trout *Oncorhynchus mykiss*. *J. Exp. Biol.* **206**, 3589–3599.

Lepage, O., Larson, E. T., Mayer, I. and Winberg, S. (2005). Serotonin, but not melatonin, plays a role in shaping dominant-subordinate relationships and aggression in rainbow trout. *Horm. Behav.* **48**, 233–242.

Lesch, K.-P., Bengel, D., Heils, A., Sabol, S. Z., Greenberg, B. D., Petri, S., et al. (1996). Association of anxiety-related traits with a polymorphism in the serotonin transporter gene regulatory region. *Science* **274**, 1527–1531.

Leshner, A. I. (1980). The interaction of experience and neuroendocrine factors in determining behavioural adaptions to aggression. *Prog. Brain Res.* **53**, 427–438.

68 SVANTE WINBERG *ET AL.*

Li, P., Mai, K. S., Trushenski, J. and Wu, G. Y. (2009). New developments in fish amino acid nutrition: towards functional and environmentally oriented aquafeeds. *Amino. Acids* **37**, 43–53.

Lillesaar, C. (2011). The serotonergic system in fish. *J. Chem. Neuroanat.* **41**, 294–308.

Liu, N.-A., Huang, H., Yang, Z., Herzog, W., Hammerschmidt, M., Lin, S., et al. (2003). Pituitary corticotroph ontogeny and regulation in transgenic zebrafish. *Mol. Endocrinol.* **17**, 959–966.

Lowry, C. A., Burke, K. A., Renner, K. J., Moore, F. L. and Orchinik, M. (2001). Rapid changes in monoamine levels following administration of corticotropin-releasing factor or corticosterone are localized in the dorsomedial hypothalamus. *Horm. Behav.* **39**, 195–205.

Lukkes, J., Vuong, S., Scholl, J., Oliver, H. and Forster, G. (2009). Corticotropin-releasing factor receptor antagonism within the dorsal raphe nucleus reduces social anxiety-like behavior after early-life social isolation. *J. Neurosci.* **29**, 9955–9960.

Lund, I. and Steenfeldt, S. J. (2011). The effects of dietary long–chain essential fatty acids on growth and stress tolerance in pikeperch larvae (*Sander lucioperca* L.). *Aquacult. Nutr.* **17**, 191–199.

Lund, I., Höglund, E., Ebbesson, L. O. E. and Skov, P. V. (2014). Dietary LC-PUFA deficiency early in ontogeny induces behavioural changes in pike perch (*Sander lucioperca*) larvae and fry. *Aquaculture* **432**, 453–461.

Maes, M., Yirmiya, R., Noraberg, J., Brene, S., Hibbeln, J., Perini, G., et al. (2009). The inflammatory & neurodegenerative (I&ND) hypothesis of depression: leads for future research and new drug developments in depression. *Metab. Brain Dis.* **24**, 27–53.

Magnhagen, C., Backström, T., Øverli, Ø., Winberg, S., Nilsson, J., Vindas, M., et al. (2015). Behavioural responses in a net restraint test predict interrenal reactivity in Arctic charr (*Salvelinus alpinus*). *J. Fish. Biol.* **87**, 88–99.

Mahar, I., Bambico, F. R., Mechawar, N. and Nobrega, J. N. (2014). Stress, serotonin, and hippocampal neurogenesis in relation to depression and antidepressant effects. *Neurosci. Biobehav. Rev.* **38**, 173–192.

Markus, C. R. (2008). Dietary amino acids and brain serotonin function; implications for stress-related affective changes. *Neuromol. Med.* **10**, 247–258.

Martins, C. I., Castanheira, M. F., Engrola, S., Costas, B. and Conceição, L. E. (2011a). Individual differences in metabolism predict coping styles in fish. *Appl. Anim. Behav. Sci.* **130**, 135–143.

Martins, C. I., Silva, P. I., Conceição, L. E., Costas, B., Höglund, E., Øverli, Ø., et al. (2011b). Linking fearfulness and coping styles in fish. *PLoS One* **6**, e28084.

Martins, C. I. M., Silva, P. I. M., Costas, B., Larsen, B. K., Santos, G. A., Conceicao, L. E. C., et al. (2013). The effect of tryptophan supplemented diets on brain serotonergic activity and plasma cortisol under undisturbed and stressed conditions in grouped-house Nile tilapia *Oreochromis niloticus*. *Aquaculture* **400**, 129–134.

Mason, J. C. and Chapman, D. W. (1965). Significance of early emergence, environmental rearing capacity and behavioral ecology of juvenile coho salmon in stream channels. *J. Fish. Res. Fish. Res. Board Can.* **22**, 173–190.

McCarthy, I. D. (2001). Competitive ability is related to metabolic asymmetry in juvenile rainbow trout. *J. Fish. Biol.* **59**, 1002–1014.

McKinney, J., Teigen, K., Froystein, N. A., Salaun, C., Knappskog, P. M., Haavik, J., et al. (2001). Conformation of the substrate and pterin cofactor bound to human tryptophan hydroxylase. Important role of Phe313 in substrate specificity. *Biochemistry* **40**, 15591–15601.

McKinney, J., Knappskog, P. M. and Haavik, J. (2005). Different properties of the central and peripheral forms of human tryptophan hydroxylase. *J. Neurochem.* **92**, 311–320.

Medeiros, L. R. and McDonald, M. D. (2013). Cortisol-mediated downregulation of the serotonin 1A receptor subtype in the Gulf toadfish, *Opsanus beta*. *Comp. Biochem. Physiol. A Mol. Integr. Physiol.* **164**, 612–621.

Meerlo, P., Overkamp, G. J. F. and Koolhaas, J. M. (1997). Behavioural and physiological consequences of a single social defeat in Roman high- and low-avoidance rats. *Psychoneuroendocrinology* **22**, 155–168.

Mendl, M., Burman, O. H. and Paul, E. S. (2010). An integrative and functional framework for the study of animal emotion and mood. *Proc. R. Soc. B* **277**, 2895–2904.

Metcalfe, N. B. and Thorpe, J. E. (1992). Early predictors of life-history events—link between 1st feeding date, dominance and seaward migration in Atlantic salmon, *Salmo salar* L. *J. Fish. Biol.* **41**, 93–99.

Metcalfe, N. B., Taylor, A. C. and Thorpe, J. E. (1995). Metabolic-rate, social-status and life-history strategies in Atlantic salmon. *Anim. Behav.* **49**, 431–436.

Mommer, B. C. and Bell, A. M. (2014). Maternal experience with predation risk influences genome-wide embryonic gene expression in threespined sticklebacks (*Gasterosteus aculeatus*). *PLoS One* **9**, e98564. http://dx.doi.org/10.1371/journal.pone.0098564.

Mommsen, R. P. (1999). Cortisol in teleosts: dynamics, mechanisms of action, and metabolic regulation. *Rev. Fish Biol. Fish.* **9**, 211–268.

Montero, D., Kalinowski, T., Obach, A., Robaina, L., Tort, L., Caballero, M. J., et al. (2003). Vegetable lipid sources for gilthead seabream (*Sparus aurata*): effects on fish health. *Aquaculture* **225**, 353–370.

Montero, D., Lalumera, G., Izquierdo, M. S., Caballero, M. J., Saroglia, M. and Tort, L. (2009). Establishment of dominance relationships in gilthead sea bream (*Sparus aurata*) juveniles during feeding: effects on feeding behaviour, feed utilization, and fish health. *J. Fish. Biol.* **74**, 1–16.

Moore, F. L. and Orchinik, M. (1994). Membrane receptors for corticosterone: a mechanism for rapid behavioral responses in an amphibian. *Horm. Behav.* **28**, 512–519.

Moreira, P., Pulman, K. and Pottinger, T. (2004). Extinction of a conditioned response in rainbow trout selected for high or low responsiveness to stress. *Horm. Behav.* **46**, 450–457.

Morgan, D., Grant, K. A., Prioleau, O. A., Nader, S. H., Kaplan, J. R. and Nader, M. A. (2000). Predictors of social status in cynomolgus monkeys (*Macaca fascicularis*) after group formation. *Am. J. Primatol.* **52**, 115–131.

Mulla, A. and Buckingham, J. C. (1999). Regulation of the hypothalamo-pituitary-adrenal axis by cytokines. *Best Pract. Res. Clin. Endocrinol. Metab.* **13**, 503–521.

Nakano, S. (1994). Variation in agonistic encounters in a dominance hierarchy of freely interacting red-spotted masu salmon (*Oncorhynchus masouishikawai*). *Ecol. Freshw. Fish* **3**, 153–158.

Nakano, S. (1995a). Individual differences in resource use, growth and emigration under the influence of a dominance hierarchy in fluvial red-spotted masu salmon in a natural habitat. *J. Anim. Ecol.* **64**, 75–84.

Nakano, S. (1995b). Competitive interactions for foraging microhabitats in a size-structured interspecific dominance hierarchy of two sympatric stream salmonids in a natural habitat. *Can. J. Zool.* **73**, 1845–1854.

Okuno, A., Fukuwatari, T. and Shibata, K. (2011). High tryptophan diet reduces extracellular dopamine release via kynurenic acid production in rat striatum. *J. Neurochem.* **118**, 796–805.

Oliveira, R. F. (2009). Social behavior in context: hormonal modulation of behavioral plasticity and social competence. *Integr. Comp. Biol.* **49**, 423–440.

Oliveira, R. F. (2013). Mind the fish: zebrafish as a model in cognitive social neuroscience. *Front. Neural Circuits.* **7**, 131. ⟨ http://dx.doi.org/10.3389/fncir.2013.00131 ⟩.

Oliveira, R. F., Silva, J. F. and Simões, J. M. (2011). Fighting zebrafish: characterization of aggressive behavior and winner-loser effects. *Zebrafish* **8**, 73–81.

Øverli, Ø., Winberg, S., Damsård, B. and Jobling, M. (1998). Food intake and spontaneous swimming activity in Arctic char (*Salvelinus alpinus*): role of brain serotonergic activity and social interactions. *Can. J. Zool.* **76**, 1366–1370.

Øverli, Ø., Olsen, R. E., Løvik, F. and Ringø, E. (1999a). Dominance hierarchies in Arctic charr, *Salvelinus alpinus* L.: differential cortisol profiles of dominant and subordinate individuals after handling stress. *Aquacult. Res.* **30**, 259–264.

Øverli, Ø., Harris, C. A. and Winberg, S. (1999b). Short-term effects of fights for social dominance and the establishment of dominant–subordinate relationships on brain monoamines and cortisol in rainbow trout. *Brain Behav. Evol.* **54**, 263–275.

Øverli, Ø., Pottinger, T. G., Carrick, T. R., Øverli, E. and Winberg, S. (2001). Brain monoaminergic activity in rainbow trout selected for high and low stress responsiveness. *Brain Behav. Evol.* **57**, 214–224.

Øverli, Ø., Kotzian, S. and Winberg, S. (2002a). Effects of cortisol on aggression and locomotor activity in rainbow trout. *Horm. Behav.* **42**, 53–61.

Øverli, Ø., Pottinger, T. G., Carrick, T. R., Øverli, E. and Winberg, S. (2002b). Differences in behaviour between rainbow trout selected for high-and low-stress responsiveness. *J. Exp. Biol.* **205**, 391–395.

Øverli, Ø., Korzan, W. J., Höglund, E., Winberg, S., Bollig, H., Watt, M., et al. (2004). Stress coping style predicts aggression and social dominance in rainbow trout. *Horm. Behav.* **45**, 235–241.

Øverli, Ø., Winberg, S. and Pottinger, T. G. (2005). Behavioral and neuroendocrine correlates of selection for stress responsiveness in rainbow trout—a review. *Integr. Comp. Biol.* **45**, 463–474.

Øverli, Ø., Sørensen, C., Pulman, K. G., Pottinger, T. G., Korzan, W., Summers, C. H., et al. (2007). Evolutionary background for stress-coping styles: relationships between physiological, behavioral, and cognitive traits in non-mammalian vertebrates. *Neurosci. Biobehav. Rev.* **31**, 396–412.

Pankhurst, N. W. (2011). The endocrinology of stress in fish: an environmental perspective. *Gen. Comp. Endocrinol.* **170**, 265–275.

Paul, E. S., Harding, E. J. and Mendl, M. (2005). Measuring emotional processes in animals: the utility of a cognitive approach. *Neurosci. Biobehav. Rev.* **29**, 469–491.

Pavlidis, M., Sundvik, M., Chen, C. Y. and Panula, P. (2011). Adaptive changes in zebrafish brain in dominant–subordinate behavioral context. *Behav. Brain Res.* **225**, 529–537.

Pepels, P. P. L. M. and Balm, P. H. M. (2004). Ontogeny of corticotropin-releasing factor and of hypothalamic–pituitary–interrenal axis responsiveness to stress in tilapia (*Oreochromis mossambicus*; Teleostei). *Gen. Comp. Endocrinol.* **139**, 251–265.

Perret, M. (1992). Environmental and social determinants of sexual function in the male lesser mouse lemur (*Microcebus murinus*). *Folia Primatol.* **59**, 1–25.

Plusquellec, P., Bouissou, M. F. and Le Pape, G. (2001). Early predictors of dominance ability in heifers (*Bos taurus* L.) of the Herens breed. *Behaviour* **138**, 1009–1031.

Pottinger, T. and Carrick, T. (1999). Modification of the plasma cortisol response to stress in rainbow trout by selective breeding. *Gen. Comp. Endocrinol.* **116**, 122–132.

Pottinger, T. G. and Carrick, T. R. (2001a). ACTH does not mediate divergent stress responsiveness in rainbow trout. *Comp. Biochem. Physiol. A: Mol. Integr. Physiol.* **129**, 399–404.

Pottinger, T. G. and Carrick, T. R. (2001b). Stress responsiveness affects dominant—subordinate relationships in rainbow trout. *Horm. Behav.* **40**, 419–427.

Raab, A., Dantzer, R., Michaud, B., Mormede, P., Taghzouti, K., Simon, H., et al. (1986). Behavioral, physiological and immunological consequences of social status and aggression in chronically coexisting resident intruder dyads of male rats. *Physiol. Behav.* **36**, 223–228.

Réale, D., Gallant, B. Y., Leblanc, M. and Festa-Bianchet, M. (2000). Consistency of temperament in bighorn ewes and correlates with behaviour and life history. *Anim. Behav.* **60**, 589–597.

Réale, D., Reader, S. M., Sol, D., Mcdougall, P. T. and Dingemanse, N. J. (2007). Integrating animal temperament within ecology and evolution. *Biol. Rev.* **82**, 291–318.

Réale, D., Martin, J., Coltman, D., Poissant, J. and Festa-Bianchet, M. (2009). Male personality, life-history strategies and reproductive success in a promiscuous mammal. *J. Evol. Biol.* **22**, 1599–1607.

Renison, D., Boersma, D. and Martella, M. B. (2002). Winning and losing: causes for variability in outcome of fights in male Magellanic penguins (*Spheniscus magellanicus*). *Behav. Ecol.* **13**, 462–466.

Rhodes, J. S. and Quinn, T. P. (1998). Factors affecting the outcome of territorial contests between hatchery and naturally reared coho salmon parr in the laboratory. *J. Fish. Biol.* **53**, 1220–1230.

Riechert, S. E. and Hedrick, A. V. (1993). A test for correlations among fitness-linked behavioural traits in the spider Agelenopsis aperta (*Araneae, Agelenidae*). *Anim. Behav.* **46**, 669–675.

Ruiz-Gomez, M. D., Kittilsen, S., Hoglund, E., Huntingford, F. A., Sørensen, C., Pottinger, T. G., et al. (2008). Behavioral plasticity in rainbow trout (*Oncorhynchus mykiss*) with divergent coping styles: when doves become hawks. *Horm. Behav.* **54**, 534–538.

Ruiz-Gomez, M. D., Huntingford, F. A., Øverli, Ø., Thörnqvist, P.-O. and Höglund, E. (2011). Response to environmental change in rainbow trout selected for divergent stress coping styles. *Physiol. Behav.* **102**, 317–322.

Russo, S., Kema, I. P., Bosker, F., Haavik, J. and Korf, J. (2009). Tryptophan as an evolutionarily conserved signal to brain serotonin: molecular evidence and psychiatric implications. *World J. Biol. Psychiatry* **10**, 258–268.

Schjolden, J. and Winberg, S. (2007). Genetically determined variation in stress responsiveness in rainbow trout: behavior and neurobiology. *Brain Behav. Evol.* **70**, 227–238.

Schjolden, J., Backström, T., Pulman, K. G., Pottinger, T. G. and Winberg, S. (2005). Divergence in behavioural responses to stress in two strains of rainbow trout (*Oncorhynchus mykiss*) with contrasting stress responsiveness. *Horm. Behav.* **48**, 537–544.

Schjolden, J., Pulman, K., Pottinger, T., Metcalfe, N. and Winberg, S. (2006a). Divergence in locomotor activity between two strains of rainbow trout (*Oncorhynchus mykiss*) with contrasting stress responsiveness. *J. Fish. Biol.* **68**, 920–924.

Schjolden, J., Pulman, K. G., Pottinger, T. G., Tottmar, O. and Winberg, S. (2006b). Serotonergic characteristics of rainbow trout divergent in stress responsiveness. *Physiol. Behav.* **87**, 938–947.

Schreck, C. B. and Tort, L. (2016). The Concept of Stress in Fish. In *Fish Physiology - Biology of Stress in Fish*, Vol. 35 (eds. C. B. Schreck, L. Tort, A. P. Farrell and C. J. Brauner), San Diego, CA: Academic Press.

Seghers, B. H. (1974). Schooling behavior in the guppy (*Poecilia reticulata*): an evolutionary response to predation. *Evolution* **28**, 486–489.

Sih, A., Bell, A. and Johnson, J. C. (2004a). Behavioral syndromes: an ecological and evolutionary overview. *Trends Ecol. Evol.* **19**, 372–378.

Sih, A., Bell, A. M., Johnson, J. C. and Ziemba, R. E. (2004b). Behavioral syndromes: an integrative overview. *Q. Rev. Biol.* **79**, 241–277.

Silva, P. I. M., Martins, C. I., Engrola, S., Marino, G., Øverli, Ø. and Conceição, L. E. (2010). Individual differences in cortisol levels and behaviour of Senegalese sole (*Solea senegalensis*) juveniles: evidence for coping styles. *Appl. Anim. Behav. Sci.* **124**, 75–81.

Sinn, D., Apiolaza, L. and Moltschaniwskyj, N. (2006). Heritability and fitness-related consequences of squid personality traits. *J. Evol. Biol.* **19**, 1437–1447.

Sloman, K. A. (2010). Exposure of ova to cortisol pre-fertilisation affects subsequent behaviour and physiology of brown trout. *Horm. Behav.* **58**, 433–439.

Sloman, K. A., Metcalfe, N. B., Taylor, A. C. and Gilmour, K. M. (2001). Plasma cortisol concentrations before and after social stress in rainbow trout and brown trout. *Physiol. Biochem. Zool.* **74**, 383–389.

Sloman, K. A., Montpetit, C. J. and Gilmour, K. M. (2002). Modulation of catecholamine release and cortisol secretion by social interactions in the rainbow trout, *Oncorhynchus mykiss*. *Gen. Comp. Endocrinol.* **127**, 136–146.

Smith, C. and Wootton, R. J. (1995). The costs of parental care in teleost fishes. *Rev. Fish Biol. Fish.* **5**, 7–22.

Smith, B. R. and Blumstein, D. T. (2010). Behavioral types as predictors of survival in Trinidadian guppies (*Poecilia reticulata*). *Behav. Ecol.* **21**, 919–926. http://dx.doi.org/10.1093/beheco/arq084.

Sørensen, C., Øverli, Ø., Summers, C. H. and Nilsson, G. E. (2007). Social regulation of neurogenesis in teleost. *Brain Behav. Evol.* **70**, 239–246.

Sørensen, C., Bohlin, L. C., Øverli, Ø. and Nilsson, G. E. (2011). Cortisol reduces cell proliferation in the telencephalon of rainbow trout (*Oncorhynchus mykiss*). *Physiol. Behav.* **102**, 518–523.

Sørensen, C., Johansen, I. B. and Øverli, Ø. (2013). Neural plasticity and stress coping in teleost fishes. *Gen. Comp. Endocrinol.* **181**, 25–34.

Sprague, D. S. (1998). Age, dominance rank, natal status, and tenure among male macaques. *Am. J. Phys. Anthropol.* **105**, 511–521.

Stanford, S. and Salmon, P. E. (1993). *Stress: From Synapse to Syndrome*. London: Academic Press.

Stapley, J. and Keogh, J. S. (2005). Behavioral syndromes influence mating systems: floater pairs of a lizard have heavier offspring. *Behav. Ecol.* **16**, 514–520.

Steimer, T. and Driscoll, P. (2005). Inter-individual vs line/strain differences in psychogenetically selected Roman High-(RHA) and Low-(RLA) Avoidance rats: neuroendocrine and behavioural aspects. *Neurosci. Biobehav. Rev.* **29**, 99–112.

Stolte, E. H., van Kemenade, B. M. L. V., Savelkoul, H. F. J. and Flik, G. (2006). Evolution of glucocorticoid receptors with different glucocorticoid sensitivity. *J. Endocrinol.* **190**, 17–28.

Stolte, E. H., Nabuurs, S. B., Bury, N. R., Sturm, A., Flik, G., Savelkoul, H. F. J., et al. (2008). Stress and innate immunity in carp: corticosteroid receptors and pro-inflammatory cytokines. *Mol. Immunol.* **46**, 70–79.

Stouthart, A. J. H. X., Lucassen, E. C. H. E. T., van Strien, F. J. C., Balm, P. H. M., Lock, R. A. C. and Wendelaar Bonga, S. E. (1998). Stress responsiveness of the pituitary-interrenal axis during early life stages of common carp (*Cyprinus carpio*). *J. Endocinol.* **157**, 127–137.

Sturm, A., Bury, N., Dengreville, L., Fagart, J., Flouriot, G., Rafestin-Oblin, M., et al. (2005). 11-deoxycorticosterone is a potent agonist of the rainbow trout (*Oncorhynchus mykiss*) mineralocorticoid receptor. *Endocrinology* **146**, 47–55.

Tanaka, M., Tanangonan, J. B., Tagawa, M., de Jesus, E. G., Nishida, H., Isaka, M., et al. (1995). Development of the pituitary, thyroid and interrenal glands and applications of endocrinology to the improved rearing of marine fish larvae. *Aquaculture* **135**, 111–126.

Teles, M., Tridico, R., Callol, A., Fierro-Castro, C. and Tort, L. (2013). Differential expression of the corticosteroid receptors GR1, GR2 and MR in rainbow trout organs with slow release cortisol implants. *Comp. Biochem. Physiol. A Mol. Integr. Physiol.* **164**, 506–511.

Thörnqvist, P.-O., Höglund, E. and Winberg, S. (2015). Natural selection constrains personality and brain gene expression differences in Atlantic salmon (*Salmo salar*). *J. Exp. Biol.* **218**, 1077–1083.

Tocher, D. R. (2003). Metabolism and functions of lipids and fatty acids in teleost fish. *Rev. Fish. Sci.* **11**, 107–184.

Tudorache, C., Schaaf, M. J. and Slabbekoorn, H. (2013). Covariation between behaviour and physiology indicators of coping style in zebrafish (*Danio rerio*). *J. Endocrinol.* **219**, 251–258.

Vancassel, S., Leman, S., Hanonick, L., Denis, S., Roger, J., Nollet, M., et al. (2008). n-3 Polyunsaturated fatty acid supplementation reverses stress-induced modifications on brain monoamine levels in mice. *J. Lipid. Res.* **49**, 340–348.

van Donkelaar, E. L., van den Hove, D. L. A., Blokland, A., Steinbusch, H. W. M. and Prickaerts, J. (2009). Stress-mediated decreases in brain-derived neurotrophic factor as potential confounding factor for acute tryptophan depletion-induced neurochemical effects. *Eur. Neuropsychopharmacol.* **19**, 812–821.

van Oers, K., De Jong, G., Van Noordwijk, A. J., Kempenaers, B. and Drent, P. J. (2005). Contribution of genetics to the study of animal personalities: a review of case studies. *Behaviour* **142**, 1185–1206.

van Raaij, M. T., Pit, D. S., Balm, P. H., Steffens, A. B. and Van Den Thillart, G. E. (1996). Behavioral strategy and the physiological stress response in rainbow trout exposed to severe hypoxia. *Horm. Behav.* **30**, 85–92.

Vaz-Serrano, J., Ruiz-Gomez, M. L., Gjøen, H. M., Skov, P. V., Huntingford, F. A., Øverli, Ø., et al. (2011). Consistent boldness behaviour in early emerging fry of domesticated Atlantic salmon (*Salmo salar*): decoupling of behavioural and physiological traits of the proactive stress coping style. *Physiol. Behav.* **103**, 359–364.

Veenema, A. H. and Neumann, I. D. (2007). Neurobiological mechanisms of aggression and stress coping: a comparative study in mouse and rat selection lines. *Brain Behav. Evol.* **70**, 274–285.

Vijayan, M. M. and Leatherland, J. F. (1990). High stocking density affects cortisol secretion and tissue distribution in brook charr, *Salvelinus fontinalis*. *J. Endocrinol.* **124**, 311–318.

Wainwright, P. E. (2002). Dietary essential fatty acids and brain function: a developmental perspective on mechanisms. *Proc. Nutr. Soc.* **61**, 61–69.

Wendelaar Bonga, S. E. (1997). The stress response in fish. *Physiol. Rev.* **77**, 591–624.

Wilson, A. D. and Krause, J. (2012). Metamorphosis and animal personality: a neglected opportunity. *Trends Ecol. Evol.* **27**, 529–531.

Wilson, A. D., Godin, J. G. J. and Ward, A. J. (2010). Boldness and reproductive fitness correlates in the eastern mosquitofish, *Gambusia holbrooki*. *Ethology* **116**, 96–104.

Winberg, S. and Lepage, O. (1998). Elevation of brain 5-HT activity, POMC expression, and plasma cortisol in socially subordinate rainbow trout. *Am. J. Physiol.* **274**, 645–654.

Winberg, S. and Nilsson, G. E. (1992). Induction of social dominance by L-dopa treatment in Arctic charr. *Neuroreport* **3**, 243–246.

Winberg, S. and Nilsson, G. E. (1993). Roles of brain monoamine neurotransmitters in agonistic behavior and stress reactions, with particular reference to fish. *Comp. Biochem. Physiol. C Pharmacol. Toxicol. Endocrinol.* **106**, 597–614.

Winberg, S., Nilsson, G. E., Spruijt, B. M. and Höglund, U. (1993). Spontaneous locomotor activity in Arctic charr measured by a computerized imaging technique—role of brain serotonergic activity. *J. Exp. Biol.* **179**, 213–232.

Winberg, S., Nilsson, A., Hylland, P., Söderstöm, V. and Nilsson, G. E. (1997). Serotonin as a regulator of hypothalamic-pituitary-interrenal activity in teleost fish. *Neurosci. Lett.* **230**, 113–116.

Winberg, S., Øverli, Ø. and Lepage, O. (2001). Suppression of aggression in rainbow trout (*Oncorhynchus mykiss*) by dietary L-tryptophan. *J. Exp. Biol.* **204**, 3867–3876.

Wolkers, C. P. B., Serra, M., Hoshiba, M. A. and Urbinati, E. C. (2012). Dietary ι-tryptophan alters aggression in juvenile matrinxa *Brycon amazonicus*. *Fish Physiol. Biochem.* **38**, 819–827.

Wolkers, C. P. B., Serra, M., Szawka, R. E. and Urbinati, E. C. (2014). The time course of aggressive behaviour in juvenile matrinxa *Brycon amazonicus* fed with dietary ι-tryptophan supplementation. *J. Fish. Biol.* **84**, 45–57.

Wommack, J. C. and Deville, Y. (2007). Stress, aggression, and puberty: neuroendocrine correlates of the development of agonistic behavior in golden hamsters. *Brain Behav. Evol.* **70**, 267–273.

Wu, A., Ying, Z. and Gomez-Pinilla, F. (2004). Dietary omega-3 fatty acids normalize BDNF levels, reduce oxidative damage, and counteract learning disability after traumatic brain injury in rats. *J. Neurotrauma* **21**, 1457–1467.

Yamamoto, T., Ueda, H. and Higashi, S. (1998). Correlation among dominance status, metabolic rate and otolith size in masu salmon. *J. Fish. Biol.* **52**, 281–290.

Young, G., Thorarensen, H. and Davie, P. S. (1996). 11-ketotestosterone suppresses interrenal activity in rainbow trout (*Oncorhynchus mykiss*). *Gen. Comp. Endocrinol.* **103**, 301–307.

Zadori, D., Klivenyi, P., Vamos, E., Fulop, F., Toldi, J. and Vecsei, L. (2009). Kynurenines in chronic neurodegenerative disorders: future therapeutic strategies. *J. Neural. Transm.* **116**, 1403–1409.

3

THE ENDOCRINOLOGY OF THE STRESS
RESPONSE IN FISH
An Adaptation-Physiological View

MARNIX GORISSEN

GERT FLIK

1. Introduction
   1.1. The Fish Forebrain
   1.2. Stress
2. Stress and the Brain: The (Neuro-)Endocrine Hypothalamus
   2.1. Fundamental Axes Interact
   2.2. The CRF System
   2.3. Ontogeny of the CRF System
   2.4. Control Over the Pituitary Gland
   2.5. CRF and Behavior
3. Stress and the Pituitary Gland
   3.1. Adrenocorticotropic Hormone (ACTH)
   3.2. Alpha-MSH
4. Stress and the Head Kidney
   4.1. Catecholamine-Producing Cells
   4.2. Steroid-Producing Cells
   4.3. Communication Within the Head Kidney
   4.4. Stress and Energy
5. Synthesis and Perspective

For any organism dealing with environmental challenges, proper handling of stressful conditions is key to survival. Extant fishes represent the earliest vertebrates on earth and must have been masters in doing so, given their vast and sometimes fast radiation. Ancestral genome expansions (two or three whole genome duplication rounds) and stable water conditions contributed to their great ability to evolve and the eventual rise of tetrapods. An elaborate endocrine machinery provides the chemical mediation of a hypothalamically integrated signal to properly spend energy and allow for fight or flight when confronted with stressful conditions. We discuss developments in fish

Biology of Stress in Fish: Volume 35
FISH PHYSIOLOGY

forebrain and (nonexhaustively) hypothalamic lay-out from the newest insights, obtained mostly from zebrafish studies. Corticotropin releasing factor, adrenocorticotropic hormone (ACTH), α-melanocyte-stimulating hormone, adrenaline, and cortisol, the key chemical mediators in the hypothalamic–pituitary–interrenal (HPI) axis, are passed in review and in the context of allostatic regulation of stress responses. We dedicate this chapter to Sjoerd E. Wendelaar Bonga, friend and teacher, who introduced us to the concept of stress and taught us to deal with it.

1. INTRODUCTION

It is extremely important to recognize the great diversity among fishes and the consequence of that variation on the physiological stress response. While much of what we know about the physiology of the stress response in mammals is based on model organisms, such as mouse and rat, which were (in-) bred to eliminate between-animal variation, the study of stress in fishes is based on animals representing extreme taxonomic variation, as well as extreme variation between individuals within a species and populations. We may forget that the many things we know about human stress physiology concerns the physiology of a single species, *Homo sapiens*.

The extant bony fishes[1] are representatives of the first vertebrates that inhabited the world's seas and oceans. The great success of extant fishes, estimated at about 35,000 species, is often ascribed to at least two whole genome duplication events (WGDs) in early vertebrate evolution (Dehal and Boore, 2005). WGDs yielded an enriched genetic library coding for new proteins, for elaborate and increasingly specific adaptations to the wealth of niches available in the aquatic world. The niches found required evolution of more or less variegated behaviors (from isolated life forms to shoaling and sophisticated parental care), depending on their biological complexity. Indeed, we find fishes in almost every niche on earth, aquatic and beyond.

---

[1]Extant bony fishes, *Osteichthyes*, are gill-breathing vertebrates with a bony skeleton, bony fin rays (lepidotrichia), and paired lungs or swim bladder(s) derived thereof, and comprise lobe-finned fish, *Sarcopterygii* (vertebrates with fins with a humerus-equivalent in their pectoral fins and a femur-equivalent in their pelvic fins), and ray-finned fish, *Actinopterygii* (vertebrates with a single dorsal fin, rhombic scales, an endoskeleton and muscles that do not extend far into the fins, and elongated flexible fin rays that form a fin). The sarcopterygians comprise the coelacanths (two species) and three species of *Ceratodontiformes* (lung fishes). An astounding approximately 35,000 actinopterygian species have been described to date, roughly 70% of all vertebrate species known (see the Berkeley website: http://evolution.berkeley.edu/evolibrary/article/fishtree_01).

Remarkably, after the earliest fishes arose around 500 million years ago (Mya) during the Ordovicium, at least five mass extinctions took place, worldwide disasters that apparently were survived by fishes (Long, 1995). Of note here is that all conodonts and also several lines of fishes became extinct—ostracoderms, acanthodians (spiny sharks), and placoderms (eg, *Dunkleosteus*). We may speculate that the relative constancy of a seawater milieu provided a buffering medium and protected against rapid and life-threatening environmental changes. Through its buffering role, the seawater environment, when it changed slowly, provided the time for the adaptations seen in the lobe- and ray-finned bony fishes. The strong ability of fishes to evolve—that is, their capacity to generate heritable phenotypic variation and ample physiological diversification—is well illustrated by the phenomenal radiation (since about 110 Mya) of the *Perciformes* (10,033 extant perch species). In more recent times (since 1 Mya) over 500 species of Lake Malawi cichlids have radiated from a single ancestor (Albertson et al., 1999).

An adaptation considered of crucial importance for the success of early and present-day vertebrates is the capacity to deal with a dynamically changing environment that may present any form of stress, be it eustress or distress. Stimuli from a continuously changing environment need to be dealt with and learned from to cope with recurring situations. In an allostasis frame (Schreck and Tort, 2016; Chapter 1 in this volume), we consider distress the condition that arises when a stressor negatively impacts an organism. As a consequence, it then cannot mount the proper, effective endocrine response normally mediated by adrenaline and cortisol, the rapid mobilization of sugars (adrenaline) and restoration of prestress conditions by adjusting energy allocation to new demands (cortisol). It follows that a successful response to a stressor, a surge in adrenaline and cortisol to timely and adequately counter the stressor and reset physiology, accordingly will be beneficial and has survival value. We want to stress that proper coping with stress requires memory and learning, the capacity to discriminate risky and safe conditions, appraisal of aversive and nonaversive stimuli, and conscious perception of the world around. Do fish have those qualities?

In 1997, our colleague Wendelaar Bonga authored a canonical paper, "The Stress Response in Fish" (Wendelaar Bonga, 1997). In the current chapter, we focus on developments in fish stress endocrine research since that seminal paper, in which the recent advancement provided by zebrafish studies takes a prominent position. Wendelaar Bonga's main conclusions were that the messengers in the catecholaminergic and steroidal stress pathways and their functions (stimulation of oxygen uptake and transfer, mobilization of energy and reallocation of energy away from growth and reproduction, immunosuppression) were largely similar for all vertebrates, from fish to humans. Conspicuous differences relate to stress-related disturbances of the systemic hydromineral balance that result from adrenaline disturbing epithelial (skin and gills) permeability to water and ions and actions of cortisol (acting in its targets as glucocorticoid and mineralocorticoid via specific receptor characteristics) in interplay with, among others, prolactin and growth hormone (GH). Important to recall, too, is that seawater fish (and euryhaline fish in seawater) have to drink to maintain hydromineral balance. But also fish in fresh water will drink when stressed (Takei, 2002; Huising et al., 2003; Nobata et al., 2013), a rather unfavorable behavior from an osmoregulation point of view, considering a significant osmotic water influx in a freshwater fish (ie, they have no need to drink). It follows that the water fish inhabit, its ionic composition, including its potential pollutants, affects not only the extensive gill surface, but often also the elaborate and delicate intestinal tract. Notably, pollutants in the water may act both as toxicant and as stressor evoking endocrine responses. Water temperature, pH, ionic composition, and calcium levels may mitigate or exacerbate hydromineral disturbances and all have to be considered to understand stressor intensity and quality. The aquatic milieu and lifestyle clearly contribute to the complexity of the stress responsiveness of an organism.

Since 1997, ongoing inventories have increased the estimate of the number of fish species from around 20,000 to 35,000, which will shed greater insight into additional adaptations to stress especially in light of the increased number of annotated genomes and modern molecular biological tools. It is also important to keep in mind that "going on land/leaving the aquatic niche" may have resulted in differences in the stress responses in fishes as compared to the terrestrial vertebrates.

1.1. The Fish Forebrain

To appreciate the stress response of a vertebrate it is important to recognize and appreciate the role of complex behaviors in that response. As indicated earlier, it follows that a proper stress response, when memory, learning, appraisal, and prediction are crucial in coping with a dynamic environment, requires brain structures that must have facilitated such behaviors. From an evolutionary point of view, considering the success of fish, a vertebrate Bauplan for a brain fitting these faculties must have been developed in them to great complexity, and therefore it seems justified to search for the entities that allow such complex behaviors. The layman in fish physiology may be and will be surprised by what modern biology has discovered in the simple brains of fish; key roles were played by studies on the zebrafish, which is now the best studied vertebrate model in developmental biology and physiology (Kalueff et al., 2014; Stewart et al., 2014).

As early as the late 19th century (Gage, 1893; Studnička, 1896) it was appreciated that the forebrain of fishes develops differently from that in higher vertebrates, by eversion (outward bending) of the most rostral part of the neural tube, resulting in a pair of solid lobes; lateral ventricles surrounded by nervous tissue are completely lacking. This holds for all actinopterygian species, the superorders of teleosts (the bony fishes), cladistians (bichirs and reedfishes), chondrosteans (sturgeons and paddle-fishes), ginglymodes (gars), and halecomorphs (bowfins). Eversion may be considered the principal and dominant morphogenetic event in the forebrain of actinopterygian fishes (Nieuwenhuys et al., 1998; Nieuwenhuys, 2009). Until the 1960s, the prevailing supposition was that fish (in fact all prereptilian vertebrates) forebrain development and organization is dominated by olfactory-input wiring (Herrick, 1933). However, we now know from a continuing flow of studies that the development of the teleostean telencephalon is very much more than that and highly varied. Input from visual, auditory, electrosensory, olfactory, and somatosensory systems is relayed to subdomains of the forebrain via preglomerular and thalamic nuclei (Ito and Yamamoto, 2009). A consistently emerging picture suggests convergent evolution of forebrain homologues in all vertebrates, not only to integrate essential physicochemical input (from the *milieu extérieur*) with signals on internal conditions (eg, energy status, oxygen requirements; *milieu intérieur*), but also to mount increasingly sophisticated behaviors in response to multivariate cues from the biome. Such cues may be beneficial and hold the promise of reward, or be disadvantageous and imply punishment inherent to risk taking and come with anxiety, fear, and lasting, chronic stress. Indeed, evidence is accruing that the fish (actinopterygian and sarcopterygian) forebrain contains regions homologous to the mammalian hippocampus, amygdala, piriform cortex, and isocortex (Portavella et al., 2002, 2004; Northcutt, 2006; Mueller et al., 2011; Mueller, 2012). We infer, again from an evolutionary point of view, that dealing with positive and negative environmental reinforcement cues, dealing with eustress and distress, has been quintessential to survival in the competitive vertebrate world. Also the endocrine systems related to stress handling must already have been developed to a great sophistication in the earliest of vertebrates. We stress the importance of an eco-physiological view on stress physiology. Just consider the ecology of any organism, include the transition from water to land or the ever-increasing complexity in vertebrate social communities, from fish to humans.

1.2. Stress

In any paper on stress the definition of stress will recur, as the term is used in many disciplines of life sciences. The original term "stress" comes from physics and describes the resistance of a steel bar to bending force, or in other words the resilience of a steel bar to a force imposed on it. The discipline of psychology took over the term to describe the pressure on a biological system and included human emotional components in the pressure on the organism. This of course introduced a world of complexity and confusion alike. Schreck and Tort (2016; Chapter 1 in this volume) define stress as "the physiological cascade of events that occurs when the organism is attempting to resist death or reestablish homeostatic norms in the face of insult." In this chapter on chemical mediation we narrow the definition of stress: the condition induced by a factor (stressor) that evokes an endocrine response that could be beneficial as well as disadvantageous. When the stress response results in timely return to prestress conditions and restoration of homeostatic regulations we discern what is called eustress (indeed, not all stress is bad). When the stressor becomes chronic, cortisol does not restore prestress conditions. If the endocrine response does not result in a timely recovery to prestress conditions a situation of distress arises, homeostatic regulations remain disturbed, and pathologies will follow. The concept of eustress and distress is crucial to the allostasis concept on which we further elaborate at the end of this chapter.

Importantly, we consider the adrenergic section of the stress response as the stress response proper, as adrenaline immediately frees energy for fight or flight (Selye, 1950, 1973). Following the adrenaline boost, the corticosteroid component (cortisol) of the response in fact concerns adaptation: cortisol secures proper and more lasting energy distribution following fight or flight and reestablishment of prestress conditions including resetting of set points for homeostatic controls (Korte et al., 2007; Koolhaas et al., 2011). Indeed, cortisol actions can be best understood (cortisol as a permissive factor for prolactin and GH actions) through its antagonism with adrenaline in the restoring disturbance of permeability control and the apoptosis inherent to adrenaline surges. Eventually, proper energy balance is what stress regulation is all about. Let us elaborate briefly on this with two examples: the energetic costs of osmoregulation and of feeding.

An aquatic niche imposes on fish, which have nonkeratinized skin, often significant energy-consuming stress. Most fishes maintain a plasma osmolality of around 300 mOsmol/kg (milliosmoles/kg) and this means that in fresh water (osmolality around 20 mOsmol/kg) and in seawater (osmolality around 1000 mOsmol/kg) osmotic gradients of 280 and 700 mOsmol/kg exist, respectively. Living in fresh water or seawater requires the maintenance of osmotic balance, an energy-costly process as it depends on active ion transports (mediated by adenosine triphosphate (ATP)-consuming enzymes and secondary active exchange mechanisms); obviously, brackish waters (osmolality around 300 mOsmol/kg) minimize the osmotic challenge and fish adapted to such waters may be predicted to have a larger budget for coping with stress. Indeed, the cost of osmoregulation was shown to be lowest in isosmotic media (Febry and Lutz, 1987). This notion is further corroborated by very efficient finfish production in (highly fertile and well-protected) estuarine mangrove locations in Asia, where isosmotic conditions prevail (Primavera, 2005). Locations with brackish water empirically and historically are preferred sites/locations for aquaculture. Also, euryhaline fish often spend part of their life in estuarine niches, in particular during the vulnerable stages of larval-to-adult transition and rapid growth. Those critically important stages may benefit from a minimal energy expenditure on osmoregulation and prioritize investment in energy management for growth and coping with stress. In particular, the consequences of osmoregulation for the energy budget of the fish are considered in our view on stress physiology. For detailed discussions on stress and osmoregulation we refer to Takei and Hwang (2016; Chapter 6 in this volume).

The energetic cost of feeding is termed specific dynamic action (SDA), which is the sum of all costs of ingestion, digestion, absorption, and assimilation of a meal. Characteristic of SDA is the rapid postprandial increase in metabolic rate that upon peaking returns more slowly to prefeeding levels. The increase in metabolic rate due to digestion may be as high as 136% for fishes (Fitzgibbon et al., 2007). Indeed, obtaining energy comes with a price, not in the least because digestion proper is so energy-costly. Moreover, the delicate and precise regulation of food intake involves a plethora of endocrine responses that show cross-talk and bidirectional communication with components of the endocrine stress axis. For this chapter, we will concentrate on the endocrine stress axis components with the aforementioned and fish-specific aspects in mind.

2. STRESS AND THE BRAIN: THE (NEURO-)ENDOCRINE HYPOTHALAMUS

Early on fishes spread over the two major zones available in the oceans and seas, the pelagic and the benthic zones. In the benthic zone, bottom-dwelling sarcopterygians thrived (and laid the basis for all later tetrapods); in the pelagic zone actinopterygian species radiated quickly and to great success. We find in fishes roughly two different hypothalamic–pituitary anatomies, one with a portal system (*eminentia mediana*; eg, coelacanths and lungfishes[1]), and one without (all modern bony fishes, teleosts). The modern bony fishes lost the more original shark-type hypophyseal portal vessel system as seen in some early actinopterygian species and the sarcopterygians (and indeed in all higher vertebrates, ie, the tetrapods that descended from the sarcopterygians).

The consequences of such different anatomies is that the hypothalamus in fish *with* a hypothalamic portal system for feedforward control over pituitary output acts as a true endocrine organ, whereas in those *without* a portal system the signals are of a neurotransmitter nature. This will have consequences for the responsiveness of pituitary targets, as receptors in a synaptic cleft generally have lower affinities than classical receptors for endocrine signals with high affinity. Humoral signals become (highly) diluted in the circulation and require adaptation of the receptor accordingly. The vessel bed volume determines the extent of dilution of the hormone and thus the circulating concentration after release; consequently, receptor affinities will be adjusted and may be predicted to be high in an endocrine setting. In a neurotransmitter setting the hormone is delivered next to its target in a synaptic cleft and the presumed higher hormone concentration in the cleft is a determining factor for receptor quality. Moreover, in the latter setting the axon-target cell construction anatomically determines the control; in an endocrine setting the receptor profile of the target cell determines control.

In actinopterygian teleostean fishes (see earlier) the preoptic neurons send axons containing corticotropin releasing factor (CRF) directly to the pituitary ACTH cells (and melanophore stimulating hormone (MSH) cells; see a later section). In contemporary teleostean fishes, therefore, two separate neuroendocrine pathways descend from the hypothalamus, the CRF pathway to the *pars distalis* and a pathway carrying arginine vasopressin (AVP), isotocin (IST), and CRF to the *pars nervosa* and MSH cells. The latter pathway subserves direct neuronal control over the MSH cells (comparable to that seen for the ACTH cells) and a neuroendocrine control over peripheral targets with the *pars intermedia* acting as a neurohemal organ. Besides release of hypothalamic CRF via the *pars nervosa*, fish are unique in having one peculiar, nonbrain site of CRF (and urotensin I (UI)) expression, *viz.* the caudal neurosecretory system (CNSS) (Lu et al., 2004; Craig et al., 2005; Bernier et al., 2008). This neuroendocrine organ is located at the caudal end of the spinal cord, and CRF and UI released from the CNSS are believed to contribute substantially to control over the pituitary gland and the regulation of the stress response in teleostean fish (Bernier et al., 2008). In addition, the chromaffin cells in the head kidney produce and release CRF and CRF-binding protein (CRFBP) (Huising et al., 2007; see Section 4.3). There is an urgent need to show, by in situ hybridization, the cellular localization of CRF receptors in head kidney tissue, in order to establish whether (ultra-)short loop feedback mechanisms may be present. A series of peripheral organs express CRF_1 receptor, including gills, (head) kidney, spleen, and heart (Huising et al., 2007) and these tissues may therefore be the target for humoral CRF and related peptides.

In addition to CRF, CRF-related peptides may play a large role in peripheral control. It is noteworthy to mention semelparous Pacific masou salmon (*Oncorhynchus masou*) that come to full sexual maturation during their upstream migration, spawn once, and (then) rapidly senesce and die. This process is characterized by a significant hypercortisolinemia that is considered causal to the demise of the fish. Cortisol levels in the spawning salmon may show a five- to sevenfold increase compared to nonstress levels (typically < 10 ng mL^{-1}) seen in juvenile and nonspawning fish (Westring et al., 2008). Westring and colleagues (2008) suggest that UI rather than CRF performs a more direct role in this hypercortisolinemia as UI expression levels consistently rise and correlate well with enhanced cortisol secretion, while CRF expression peaks twice and correlates poorly. In other words, in the maturing masou salmon, control over cortisol production may be differential, *viz.* constitutive via UI as well as (stress-)regulated via CRF. This situation of differential control over cortisol production is reminiscent of the independent control over the head kidney as seen in chronically stressed Mozambique tilapia kept in acidified water (Lamers et al., 1994).

The external milieu is monitored continuously by numerous sensors. Upon registration of an external threat to the internal milieu (eg, the presence of a predator), a change in water quality, or physical stressors related to aquaculture practices, this perception of potential threats leads to an activation of hypothalamic nuclei of which the *nucleus preopticus* (*npo*) and *nucleus lateralis tuberis* (*nlt*) are the most important when considering the stress axis. It is in these nuclei that the key hypophysiotropic signals are produced (Cerdá-Reverter and Canosa, 2009).

More specifically, in studies on larval zebrafish (Fernandes et al., 2013) it was shown that paralogous so-called orthopedia transcription factors (Otps; products of the *otpa* and *otpb* genes in zebrafish), among others, specify the diencephalic, hypothalamic subsets of neurons that produce CRF and arginine vasotocin (AVT) and ventral diencephalic dopamine (DA)-producing cells. This process is not restricted to zebrafish but is conserved in all vertebrates, including humans. The Otp-dependence of CRF/AVT and DA neurons in all vertebrates corroborates the notion of a shared Bauplan, at the fine level of hypothalamic functional anatomy. Indeed, in the zebrafish hypothalamus we find parvocellular and magnocellular clusters. The former project to the *pars distalis* and there release CRF, somatostatin, and thyrotropin-releasing hormone (TRH) and thus play a key role in the endocrine stress axis. The latter produce IST (now often called oxytocin in zebrafish, as the gene is the oxytocin orthologue) and AVT (in zebrafish now often called vasopressin or AVP, as the gene is the *vasopressin* orthologue) for release in neurohemal areas mainly situated in the *pars intermedia*. It should be kept in mind that cells often coexpress hypophysiotropic signals (notably CRF *plus* AVT and cholecystokinin *plus* enkephalin-B; Herget and Ryu, 2015), and this may impede our understanding of the individual signals. Production of a series of hypophysiotropic signals by closely intermingled subsets of cells in nuclei is even more common, and our understanding of multifactorial cocktails of signals awaits characterization of the individual cells by analysis of their homeobox gene product-dependence, in situ hybridization of multiple genes, their protein products, and the receptor profiles at the basis of differential signal release. Such exciting studies in zebrafish are now rapidly appearing in the literature; their contribution to comparative vertebrate endocrinology is invaluable (Machluf et al., 2011; Fernandes et al., 2013; Herget et al., 2014; Herget and Ryu, 2015).

2.1. Fundamental Axes Interact

The hypothalamus integrates control over the fundamental processes of stress handling, growth, and reproduction. Four tripartite (hypothalamus, hypophysis, peripheral endocrine glands) neuroendocrine axes participate in that control and are found in all vertebrates. We want to emphasize that the stress axis shows crossover with the other nonstress axes. This notion may help to explain complex phenomena such as growth spurts, (seasonal or diurnal) metabolic suppression, and sexual maturation (masculinization and feminization). Life means energy consumption, the strongly energy-consuming processes of coping with stress and growth therefore take top hierarchical positions in the priority list of controls that guarantee survival.

For growth control the hypothalamic key players are growth hormone-releasing hormone, ghrelin, and somatostatin, and these determine pituitary GH and eventually hepatic insulin-like growth factor output. For stress handling the hypothalamic key players are CRF, AVT, melanin-concentrating hormone (MCH), and DA, which determine pituitary ACTH, and interrenal (adrenal) cortisol output; the actions of cortisol are essentially the distribution of energy to organs in need of energy, as dictated by environmental conditions that challenge the organism and require redistribution of energy to restore prestress conditions. A second axis that parallels the stress axis is the thyroidal axis, in which hypothalamic TRH, CRF, and DA steer pituitary thyroid-stimulating hormone that controls thyroid output of tetraiodothyronine (T4) and triiodothyronine (T3). Thyroid hormones are essential regulators of basal metabolism and metamorphosis related processes and often play permissive roles in energy-consuming activities. In the reproductive axes, gonadotropin-releasing hormone, kisspeptin, and gonadotropin-inhibiting hormone from the hypothalamus target pituitary luteinizing hormone and follicle-stimulating hormone producing cells that eventually control testicular testosterone and ovarian estradiol output.

Another important notion to keep in mind when discussing stress/ energy availability is that fish, like most animal life forms, are ectotherms that consume up to threefold less energy in early life growth than endotherms. Although the energy cost of offspring production (estimated to 250 kJ per day and per kg egg mass, hatchlings, or litter) may be similar for ectotherms and endotherms; fish may spend 35% of the metabolizable energy on this versus 2–6% in endotherms (Wieser, 1985). It follows that regulation of energy-consuming processes including dealing with stress may differ in fishes greatly from what we know from studies on mammalian species and that it is important to consider life stage and reproductive status when addressing stress coping. Another consideration in the context of ectothermy is that fish may show both behavioral and emotional fever (Boltaña et al., 2013; Rey et al., 2015). Briefly, zebrafish given a thermal preference choice and exposed to viral infection or stress spend more time at higher temperatures (+2 to 4°C). This response is a real fever in the sense that it activates the same immuno-physiological and metabolic pathways as seen in mammals. The response is immediate and lasts up to 8 h. We must consider the consequences of carrying out stress-related experiments in less time than this 8-h window, thus preventing the animal from fully utilizing its adaptive qualities, as it may function only suboptimally.

With regard to the stress axis of fishes we will now focus on CRF and AVT as the dominant positive feedforward signals, and briefly consider DA and MCH as negative feedforwards. Indeed, the situation is far more complex as the rostral *pars distalis*, which harbors the ACTH and prolactin cells, receives in addition to CRF, AVT, DA, and MCH a plethora of other ACTH and MSH release-modulating factors including IST, growth hormone-releasing hormone, pituitary adenylate cyclase activating polypeptide (PACAP), somatostatin, and galanin. An extensive overview of these hypothalamic, hypophysiotropic factors was provided by Cerdá-Reverter and Canosa (2009), and Bernier et al. (2009).

2.2. The CRF System

In fish, as in mammals, CRF (in this chapter we adhere to the conventional nomenclature as defined by the international union of pharmacology; Hauger et al., 2003) is the principal hypothalamic factor that controls the stress axis (Vale et al., 1981; Wendelaar Bonga, 1997; Flik et al., 2006). Release of CRF from the preoptic area results in release of ACTH from the pituitary gland *pars distalis* (see Section 3.1) and the release of cortisol from the head kidney (see Section 4.2). CRF is a 41 amino acid peptide derived from a 160 amino acid precursor (Vale et al., 1981) and is remarkably conserved across vertebrates (Huising et al., 2004; Huising, 2006). Next to CRF, the CRF family consists of UI (Pearson et al., 1980), urocortin-1 (Ucn1) in mammals (Vaughan et al., 1995) and the urocortins-2 and -3 (Pearson et al., 1980; Coulouarn et al., 1998; Lovejoy and Balment, 1999; Lewis et al., 2001). We refer to Boorse and Denver (Boorse et al., 2005) for an excellent phylogenetic overview of the CRF family in vertebrates. In most vertebrate species there are two CRF receptors, *viz.* the CRF_1 and the CRF_2 receptors (a third CRF receptor has been described only in catfish; Arai et al., 2001), which vary considerably in their affinities for the different members of the CRF family. Whereas CRF_1 receptor has equal affinity for CRF and UI/Ucn1 (Arai et al., 2001; Coste et al., 2002; Manuel et al., 2014c), CRF_2 receptor has higher affinity for UI and urocortins than for CRF (Wei et al., 1998; Hsu and Hsueh, 2001; Manuel et al., 2014c).

An interesting and key modulator of CRF bioavailability is CRFBP. CRFBP is a 322 amino acid protein, structurally nonrelated to the CRF receptors (Potter et al., 1991) and present also in invertebrates (Huising and Flik, 2005). Invertebrates do not make CRF proper, and this suggests that CRFBP may have different roles in chaperoning proteins and related functions in invertebrates and vertebrates. Indeed, in insects CRFBP may chaperone diuretic hormone (DH), a hormone that controls water balance. The terrestrial environment and its potential drought may have put a serious stress on the water balance of land-invading insects, and the DH-CRFBP tandem may have played a role in the control of this stressor. We consider the role in the vertebrate endocrine stress axis as a convergent neo-functionalization of CRFBP with CRF as partner. Interestingly, DH and CRF and related proteins share an ancestor predating the Cambrium (*viz.* between 700 and 993 Mya; Huising and Flik, 2005), which is consistent with a very early origin of CRFBP. CRFBPs have distinct binding sites for both CRF and UI/Ucn1 (Huising et al., 2008) and bind to these ligands with a higher affinity than CRF receptors do (Sutton et al., 1995; Manuel et al., 2014c). Such an affinity profile means that the equilibrium between CRFBP and its ligands is shifted toward the binding protein-ligand complex and that CRFBP plays an important role in determining the free, bioactive portion of CRF (and its relatives). Little to nothing is known about the conditions/microenvironments at receptor sites and interactions among the three protein groups. In physiological studies on stress that evaluate gene expression of the proteins, it often appears that the expression of CRFBP best explains the anticipated endocrine stress axis activity (Geven et al., 2006): suppression of *crfbp* expression in stressful conditions, upregulation under basal nonstress conditions.

2.3. Ontogeny of the CRF System

There is a seemingly enigmatic condition with the demonstration of CRF mRNA in fish eggs. Chandrasekar and colleagues (Chandrasekar et al., 2007) showed this signal via in situ hybridization from 24 h postfertilization (hpf) onward in zebrafish; with a more sensitive semiquantitative RT-PCR technique mRNAs for *crf*, its receptors *crf1&2 receptor*, and *crfbp* were demonstrated even in unfertilized eggs (Alderman and Bernier, 2009). The question arises then what such an apparently maternally deposited signal means for the embryo. Thus, although we do not know the distribution of the signal *in ovo* (ie, yolk or blastodisc), we do know that after fertilization and first cleavages all cells and yolk are in open contact through cytoplasmatic bridges and small molecules can migrate between yolk and cells at the animal pole (Kimmel and Law, 1985; Kimmel et al., 1995). Indeed, for several maternally deposited RNAs it was shown that these may unite after fertilization in the blastodisc at the animal pole (eg, Theusch et al., 2006; Kosaka et al., 2007). Apparently, the maternal CRF signal is already functional in early development, not necessarily in embryonic stress physiology: although CRF_1 receptor is also present in unfertilized eggs, this signal disappears between 6 and 12 hpf (Alderman and Bernier, 2009). This may explain the mechanism underlying the window of unresponsiveness to stress (measured as cortisol production) around hatching in several species of fish, including rainbow trout (*Oncorhynchus mykiss*) (Barry et al., 1995a, b), chinook salmon (*Oncorhynchus tshawytscha*) (Feist and Schreck, 2002), zebrafish (Alsop and Vijayan, 2009), and European sea bass (*Dicentrarchus labrax*) (Tsalafouta et al., 2014). However, this may not hold for all fish, as common carp (*Cyprinus carpio*) (Stouthart et al., 1998) and Mozambique tilapia (*Oreochromis mossambicus*) (Pepels et al., 2004) do show stress responsiveness before hatching. It should be noted that stress responsiveness (ie, cortisol production) is not necessarily CRF-dependent (Fuzzen et al., 2011; Tsalafouta et al., 2014). The vast numbers of fish species and survival strategies possibly underlie the dichotomy in stress responsiveness of fish early life stages.

2.4. Control Over the Pituitary Gland

CRF and CRFBP are found in the vicinity of pituitary ACTH cells and in massive amounts in the *pars intermedia* when these proteins are detected by specific immunohistochemistry (Huising et al., 2004; Fig. 3.1). In the stress

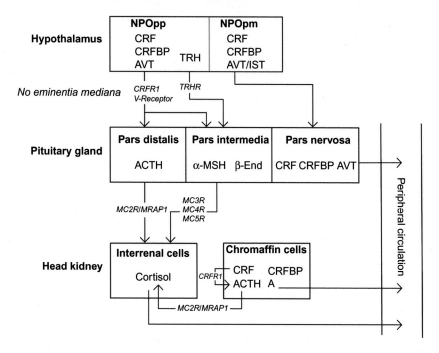

Figure 3.1. Diagram of the actinopterygian/teleostean hypothalamic–pituitary–interrenal (HPI) axis and its chemical mediators underlying the endocrine stress response. In the hypothalamus neuroendocrine cells of the ventral, parvocellular section of the *nucleus preopticus* (NPOpp) send corticotrophin releasing factor (CRF)-containing axons to pituitary adrenocorticotropic hormone (ACTH)-cells (no *eminentia mediana*) and, by doing so, convey the major feedforward to the pituitary gland to initiate the endocrine stress response; an as yet poorly defined (situated at the borders of the NPO) group of cells projects, in parallel and similarly, to send CRF-binding protein (CRFBP) to the ACTH cells. The equilibrium between CRF and CRFBP determines CRF bioactivity, that is, activation of the pituitary CRF-receptor (CRFR$_1$). The NPOpp cells coexpress arginine vasotocin (AVT) and possibly isotocin (IST). The *pars intermedia* melanophore stimulating hormone (α-MSH) producing cells are under hypothalamic control by CRF/CRFBP and thyrotropin-releasing hormone (TRH, via its receptor TR). ACTH signals in its targets via the MC2R-MRAP1-complex (2 MC2Rs and 4 MRAP1s), and this receptor is highly specific for ACTH (no or little promiscuity with other α-MSH-sequences). Alpha-MSH released upon stress may stimulate constitutive release of interrenal cortisol and act as (stress-related) anorexic signal (assuming that the circular α-MSH molecule may easily pass the blood–brain barrier). Neuroendocrine cells in the dorsal magnocellular section of the NPO (NPOpm) project to the neurohemal *pars intermedia* to signal to the general circulation. This diagram is illustrated by immunohistochemistry in Fig. 3.2.

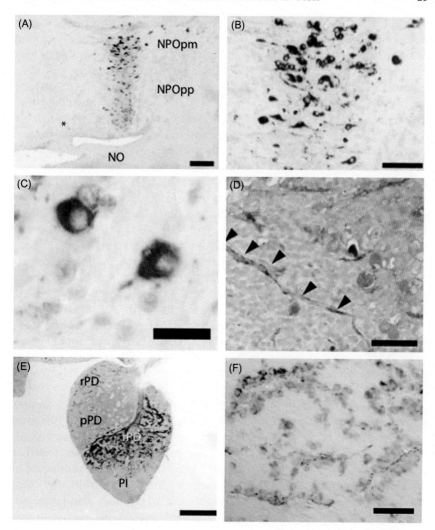

Figure 3.2. Immunohistochemistry of the hypothalamo-pituitary complex of common carp, *Cyprinus carpio*. (A) Sagittal section of the nucleus preopticus stained with a rabbit antiserum to sheep corticotrophin releasing factor (CRF) (Incstar), showing the dorsal magnocellular (NPOpm; enlarged in (B) and ventral parvocellular (NPOpp) sections of the nucleus; the NPO is found above the ventricular space dorsal of the nervus opticus (NO). The asterisk in (A) indicates the area where CRF-binding protein (CRFBP) expressing cells are found (C). In (D) a bundle of CRFBP-containing axons (arrow heads) projecting to the *pars distalis* (PD; see (E)); CRFBP was demonstrated with a rabbit antihuman CRFBP (#5144, 1:1000; generous gift from late Dr. Wylie Vale). Picture (E) shows a sagittal section of a carp pituitary gland stained with anti-CRF (as in A). Note profound staining of the *pars nervosa* (PN) projecting to the *pars intermedia* (PI) as well as staining of fibers in the rostral *pars distalis* (rPD) but not in the

axis CRF and CRFBP concur (not necessarily in higher, telencephalic regions and optic lobe; Alderman and Bernier, 2007). At present we lack any procedure or technique to observe and quantify CRF and its binding protein in vivo and we can only speculate on the physiological meaning of these molecules in stress physiology, although molecular analyses (gene up- and downregulation in stress paradigms) give strong evidence for involvement (Bernier et al., 1999; Huising et al., 2004; Bernier and Craig, 2005; Madaro et al., 2015, 2016) of these signals in stress responses; nonetheless, actual data on protein actions are virtually absent in the pertinent literature. Differences in half-life time of mRNA and protein molecules further complicate our understanding (Maier et al., 2009; Schwanhäusser et al., 2011). However, perifusion studies with (recombinant) proteins/peptides on pituitary tissue of a few species do support this role, but in all cases the studies are oversimplifications of real life with its highly complex multifactorial control over stress steroid production. One study illustrating this complexity is that of carp pituitary ACTH and MSH release. In line with the DA master control over ACTH cells in vivo (Metz et al., 2005; Section 3 of this chapter), pituitary explants in perifusion (ie, when removed from hypothalamic dopaminergic control) show an increased release of ACTH. In such a setup we could not show CRF-stimulated ACTH release unless the tissue operated under a DA tonus (Metz et al., 2005). Clearly, the dopaminergic cyclic-AMP second messenger pathway and that of CRF (also a cyclic-AMP pathway) interdependently determine ACTH exocytosis; this is likely to occur in vivo.

The ACTH-releasing function of CRF was established for several species, including goldfish (Fryer et al., 1984), trout (Baker et al., 1996), sea bream (Rotllant et al., 2000), Mozambique tilapia (van Enckevort et al., 2000), and carp (Metz et al., 2005). However, when this was tested in common carp, Metz and colleagues (2005) initially failed to demonstrate an ACTH-releasing effect of CRF using in vitro perifusion. In their experiments, basal ACTH secretion by the isolated pituitary gland increased slowly and steadily over time. They concluded that ACTH release in vivo must be under tonic inhibitory control (which is lost in vitro when

proximal *pars distalis* (pPD). In (F) we see CRF axons in blue projecting to *pars distalis* corticotropes (brown, adrenocorticotropic hormone (ACTH)); the ACTH antibody was raised against carp cys-ACTH[10–23]; Metz et al., 2004). Primary antibodies were detected by goat antirabbit IgG-biotin (1:200, Bio-rad) amplified with the Vectastain ABC amplification kit. Pictures (A–E) are taken from Huising, M. O., Metz, J. R., van Schooten, C., Taverne-Thiele, A. J., Hermsen, T., Verburg-van Kemanade, B. M. L., et al. (2004). Structural characterisation of a cyprinid (*Cyprinus carpio* L.) CRH, CRH-BP and CRH-R1, and the role of these proteins in the acute stress response. *J. Mol. Endocrinol.* **32**, 627–648, with permission.

hypothalamic connections are severed), providing a mechanism to keep circulating ACTH levels low in the nonstressed animal. Dopaminergic neurons in the preoptic area project to the pituitary gland (Kah et al., 1986) and there inhibit ACTH secretion via a DA-receptor (D2 type) (Selbie et al., 1989; Civelli et al., 1993; Stefaneanu et al., 2001). Accordingly, CRF-stimulated ACTH-release could be demonstrated from isolated pituitary glands when DA was coadministered at concentrations that are known to activate D2 receptors (Lamers et al., 1997). So at least in carp, activation of both CRF_1 and D2 receptor is necessary to elicit an intracellular pathway leading to ACTH secretion (Metz et al., 2005).

In Mozambique tilapia, a persistent mild stress (or in fact eustress) imposed by 5 days' exposure to acidified (pH 4.0) water produced a dramatically different and surprising control over pituitary output. A second, stimulatory DA-receptor (D1 type) becomes expressed and TRH appears to take over as the main and mild regulator of cortisol production, with di-acetylated α-MSH from the *pars intermedia* as pituitary mediator (Lamers et al., 1997). Such a scenario would leave the CRF/ACTH/cortisol pathway free for a more rapid/acute stress response in an already challenging multistressor environment. This dual organization of the stress axis seems dictated by the aquatic environment and provides the fish with a wonderful window of physiological adaptation strategies.

Baker and colleagues (Baker et al., 1996) showed that in trout CRF and AVT act both independently and synergistically as stimulators of ACTH in vitro. This was then confirmed by Pierson and coworkers (Pierson et al., 1996); the study also included IST as yet another stimulator. CRF was shown to be the most potent (ED_{50}: 8×10^{-14} M), followed by AVT (ED_{50}: 2×10^{-10} M) and IST (ED_{50}: 10^{-7} M). Coadministration of AVT (to CRF) did not affect pituitary output of MSH; however, AVT did potentiate cortisol release when coadministered with ACTH, in line with its main humoral action after release from the (neurohemal) *pars nervosa* associated with the *pars intermedia* MSH and somatolactin cells (Jerez-Cepa et al., 2016).

CRF- and UI-expressing neurons may be recruited to action in a similar fashion in response to different stressors (Bernier et al., 2008). Indeed, hypothalamic UI-expressing neurons are thought to innervate the pituitary gland in fish (Lederis et al., 1994), but whether UI released from these cells directly contributes to control of ACTH release is not yet clear (Bernier and Craig, 2005). In the latter study, Bernier and Craig found that hypoxia stimulates CRF and UI gene expression in the preoptic hypothalamus of rainbow trout, but no causal link between UI and ACTH was established. Intracerebroventricular injection of UI in trout is over five times more potent as an anorexigenic signal than CRF (Bernier, 2006; Ortega et al., 2013). It is well known that stressed fish lose appetite and stop feeding. Such a dual wiring of the stress and feeding controlling networks in vertebrate brains emphasizes not only the importance of bidirectional communication, but also the time required for the mutual independence of energy-consuming (stress) and providing (feeding) physiologies. The exact neural circuitry controlling stress-related suppression of feeding (Volkoff et al., 2005; Bernier, 2006; Gorissen et al., 2006) is at present only poorly understood and awaits detailed functional characterization of CRF receptors in hypothalamus, higher brain centers, and brainstem.

2.5. CRF and Behavior

Not surprisingly, in addition to its well-characterized role in the regulation of the stress axis, CRF (seemingly independently) impacts behavior in fish. CRF attenuates competitive ability, and both CRF and UI induce anxiety (Backstrom et al., 2011) and hyperventilation in rainbow trout (Le Mevel et al., 2009). On the other hand, social subordination increased CRF mRNA levels in the preoptic hypothalamus of trout (Doyon et al., 2005), indicating that behavior itself also impacts hypothalamic CRF activity. For an extensive overview on CRF and behavior see Noakes and Jones (2016; Chapter 9 in this volume).

3. STRESS AND THE PITUITARY GLAND

The pituitary gland, often referred to as the master endocrine gland, is the interface between the brain and peripheral endocrine regulatory systems. Information from higher brain centers (including telencephalic pallial centers) reaches the pituitary gland from where chemical signals are sent via the blood to peripheral endocrine tissues; by doing so these signals carry an integrated message for control over growth, stress handling, energy metabolism, reproduction, and immunity. In the early development of (zebra-) fish, *gh* (first and most abundantly) and *pro-opiomelanocortin* (*pomc*) genes (second in time and abundance) are expressed (Pogoda and Hammerschmidt, 2009) in line with the physiologies they control: growth initially and a functional stress axis secondarily to cope with a dynamic and dangerous environment.

The specificity of the signal is determined eventually by the receptor profile of the peripheral targets. Multiple hormones and interactions between receptor pathways need consideration; understanding those interactions requires precise definition of the receptor profiles in targets. In model species with well-annotated genomes such as zebrafish and Japanese ricefish (*Oryzias latipes*, Temminck and Schlegel), we should now reverse our approach and, knowing the hormones, focus on the receptor profile to understand what is going on in the targets and cross-talk and bidirectional communication in second messenger pathways.

Stress coping is an energy-demanding process in which two players are key: adrenaline and cortisol. Whereas adrenaline, *inter alia*, mobilizes energy as glucose to prepare for fight or flight in a matter of seconds to minutes, cortisol is in fact the adaptation hormone that directs the energy flows within the body to restore prestress conditions through a redistribution of energy in the body over a time span from hours to days (Korte et al., 2007; Koolhaas et al., 2011). From this notion it follows that cortisol production needs regulation over such time frames and requires steering both for basic maintenance, a constitutive release at various levels depending on the allostatic load, and for more acute needs, so-called regulated peak surges. The pituitary gland harbors two cell populations that express pro-opiomelanocortin (POMC) and that play key roles in relation to stress and energy household; POMC is processed either to ACTH and β-lipotropic hormone (LPH) in the corticotropes, or to α-MSH, corticotropin-like intermediate peptide (CLIP), and β-endorphin(s) in the melanotropes. ACTH and α-MSH are just two signals from the highly pleiotropic and multifactorial melanocortin system, including multiple ligands and multiple well-defined receptors (Metz et al., 2006; Takahashi et al., 2006). Interestingly, it is the acetylation of α-MSH and β-endorphins that determines their bioactivity; acetylated α-MSH (Lamers et al., 1992) tends to be more potent than the nonacetylated peptide, while acetylation quiets β-endorphins (Wilkinson, 2006). The posttranslational acetylation machinery thus plays a key role in the *pars intermedia* output. The roles of ACTH and acetylated α-MSH in stress responses of fish are discussed later. In zebrafish, during development two POMC cell populations that find their origin in the hypophysial placode settle in the pituitary gland: first the ACTH cells and second the α-MSH cells (Herzog et al., 2003). We may wonder what information is laid down in the POMC molecule that is at the basis of this distribution. The coexpression of prohormone convertases (aka subtilisins, serine-proteases produced in *Bacillus subtilis*) eventually determines to which degree and what POMC is processed (Zhou et al., 1993; Tanaka, 2003). Information on LPH and CLIP in relation to stress physiology is exiguous and the signals will not be addressed in this chapter. More is known about β-endorphins and, recently, the opioid receptor repertoire in brain and peripheral organs (eg, Chadzinska et al., 2009). In mammals β-endorphins are elevated during acute and chronic stress and have an important role in the termination of the stress response (Drolet et al., 2001). Such a role in fish stress physiology is not yet established. The presynaptic inhibitory modulation exerted by the various, more or less

94 MARNIX GORISSEN AND GERT FLIK

truncated, β-endorphins, as seen in common carp (van den Burg et al., 2001), seems to be a vertebrate property.

3.1. Adrenocorticotropic Hormone (ACTH)

Upon arrival in the anterior pituitary, the CRF-CRFBP complex must dissociate so that free CRF can activate its cognate CRF_1 receptor expressed in the corticotropic cells (Wynn et al., 1983; Chen et al., 1993; Huising et al., 2004) to stimulate synthesis (Bruhn et al., 1984) and release (Rivier et al., 1982) of POMC-derived peptides, such as ACTH, β-endorphin, and others (Benjannet et al., 1991; Castro and Morrison, 1997). The target of circulating ACTH is the head kidney interrenal cell that synthesizes and secretes the glucocorticoid cortisol (Flik et al., 2006). This is the endocrine cascade that we call the endocrine stress axis.

What is ACTH doing, beyond the firmly established control of cortisol production and release (Dores et al., 2014)? To answer this we ask the question, where do we find the ACTH receptor, the so-called melanocortin receptor 2 (MC2R)? In fish, the MC2R is predominantly, in fact almost exclusively, found in the head kidney (Metz et al., 2005, 2006), and within that organ presumably in the interrenal cells, but even that is not known for sure. Head kidney tissue harbors a multitude of different cell types, including cell lines from the immune system that interact with the stress axis (Verburg-van Kemenade and Schreck, 2007). In fugu (*Takifugu rubripes*) MC2R was found in telencephalic and hypothalamic tissue (Klovins, 2003), along with MC4R and MC5R, as in other fish (Metz et al., 2006).

The paucity of studies on ACTH function in fish derives from the lack of suitable antibody/antisera to assess plasma levels of the hormone. Moreover, analysis of *pomc*-transcript abundance conveys only limited information as the diversification of the signal occurs posttranslationally. Tools to discriminate ACTH from α-MSH (and POMCs) are needed both at the mRNA and protein levels. Fortunately, immunohistochemistry and in situ hybridization have shown that ACTH and α-MSH cells are anatomically and topologically separated, which means that proper separation of the cell groups is feasible, yet it is seldom applied. In Section 4.4, we elaborate on the effects of leptin on the stress axis and show that this signal of energy status may attenuate stress axis activity through direct effects on ACTH- and cortisol production and secretion (Gorissen et al., 2012).

3.2. Alpha-MSH

During chronic stress, α-MSH-levels in plasma rise (Lamers et al., 1994; Arends et al., 2000; Metz et al., 2005). Studies show that α-MSH is corticotropic in chronically stressed Mozambique tilapia (Lamers et al., 1997) and in barfin flounder (*Verasper moseri*) when kept in seawater (Kobayashi et al., 2011); seawater with its high osmolality (1000 mOsmol/kg) imposes an unescapable osmotic stress load to the animal. In common carp, however, any stress-related effects of α-MSH could not be demonstrated, neither for α-MSH alone nor with any combination of coreleased POMC-derived peptides, in particular the truncated β-endorphins (Metz et al., 2005; van den Burg et al., 2005). This suggests not only species-specificity, but also life-stage dependency (eg, fresh water vs seawater life stages) of specific actions of α-MSH on cortisol release. Several explanations can be advanced for an absence of α-MSH corticotropic activity. Alpha-MSH is a circular peptide, while ACTH (which includes the α-MSH sequence at its first 13 positions) is a linear peptide and thus the receptor-binding moieties may well be different. Both have the binding motif HFRW, of which the W is assumed to be the most important amino acid for ligand receptor interaction (Liang et al., 2015); it is postulated that ACTH has an additional unique RKRRP binding motif (Dores et al., 2014). Also, melanocortin receptors (MCRs) seem to have radiated into great diversity and specificity alike, so why allow promiscuity with ligands? Most importantly, ACTH needs the coexpressed, coregulatory protein melanocortin 2 receptor accessory protein 1 (MRAP1) to be able to bind to its cognate MC2R (Agulleiro et al., 2010; Liang et al., 2015; Dores et al., 2016) (Faught et al., 2016; Chapter 4 in this volume). This seems to exclude α-MSH actions on the ACTH receptor pathway. However, potentially other α-MSH-sensitive MCRs (MC1R, MC3R, MC4R, and MC5R) found in head kidney tissue of rainbow trout (Haitina et al., 2004), carp (Metz et al., 2005), and barfin flounder (Kobayashi et al., 2011) may also be present in Mozambique tilapia to explain corticotropic actions of α-MSH.

4. STRESS AND THE HEAD KIDNEY

Head kidney tissue arose in the fish lineage (Hofmann et al., 2010) and harbors chromaffin cells, interrenal cells, erythropoietic and leucopoietic cell lines, the secondary lymphoid moiety of the immune system. As in all vertebrates, the spleen watches over the blood circulation with its potential pathogens and is the site where red blood cells mature and, after becoming apoptotic, will be removed. Thymocytes from head kidney origin mature in the thymus and the antibody-producing B cells are produced and mature in the head kidney. The former describes the situation in adult fish; during early (zebrafish) development, two waves of migration of immune-cell progenitors occur before the adult situation is reached (Chen and Zon, 2009).

An abundance of recent studies addresses hypothalamic and pituitary endocrinology. Recent studies on the interrenals proper of fish are scarce. Interrenal tissue is often mentioned as the final location of the stress axis cascade and as the production site of cortisol, and some work has been done on early development of steroid-producing cells and pituitary-interrenal interaction (Hsu et al., 2004; Liu, 2007; To et al., 2007). More often studies describe the anatomical variability of the interrenal tissue found in different fish species, but no significant new endocrine insights have been found. For proper understanding of activation of the stress axis, the consequences of communication within the HPI axis components (in particular communication between adrenaline- and cortisol-producing cells) must be addressed. What do we know about effects of chromaffin cell on cortisol-producing cells and vice versa?

4.1. Catecholamine-Producing Cells

The teleostean homologue of the mammalian adrenal medulla are the chromaffin cells found intermingled with the interrenal tissue in the head kidney. The chromaffin cells originate from neural crest cells, and are arranged around postcardinal veins, together with cortisol-producing cells (this varies among teleostean species). In cyprinids such as common carp and zebrafish, the interrenal tissue forms rather compact and separate organs; in most others the endocrine tissue is loosely arranged around the cardinal veins, in between nephric ducts of the pronephros, dorsally in the body cavity just behind the gill chamber. Chromaffin cells derive their name from the property to bind chromate or bi-chromate in classical histochemical stains.

4.2. Steroid-Producing Cells

The teleostean homologue of the adrenal cortex resides along the postcardinal veins in the head kidney, a unique organ found in (bony) fishes that originates from the pronephros. The main product of the corticosteroid-producing cells is cortisol, as in humans. However, fish lack the aldosterone synthase to produce a specific mineralocorticoid. Cortisol functions either as a glucocorticoid or as a mineralocorticoid signal; it is the receptor makeup of target cells that determines which section of the genome is read by the cortisol receptor complex (F-GR or F-MR). Also the receptor/transcription factor determines, through its interaction with the cortisol-responsive elements of genes, whether activation or suppression of the gene occurs. Primary target tissues for actions of cortisol include the gills, liver, intestinal epithelia, and the brain.

In head kidney of fish we therefore find, in close proximity, cells producing catecholamines and cortisol, as well as the full repertoire of immune-cell cytokines and related signals that may locally and peripherally interact and synergize, as well as show cross-talk. We refer to Yada and Tort (2016; Chapter 10 in this volume) for detailed discussions on the interaction with the immune system.

4.3. Communication Within the Head Kidney

There is one canonical review on catecholamine storage and release in fish (Reid et al., 1998) addressing the biosynthesis and exocytotic release of the afferent limb of the adrenergic response. They conclude that cortisol is affecting adrenaline storage and release only on the longer term (3–7 days following injection of coconut deposits with cortisol; Reid et al., 1998). Cortisol may sensitize the adrenaline release, but does not stimulate catecholamine secretion in situ. From studies on trout (Reid et al., 1996) it was concluded that cortisol does not affect phenyl ethanolamine- N-methyl transferase, the enzyme that methylates noradrenaline to adrenaline (in contrast with what is found in mammals (Betito et al., 1992)). Whether this holds true for all species of fish requires a further search for cortisol-responsive elements in the genes coding for the enzymes in the Blaschko pathway (Blaschko, 1939).

In carp, CRF and CRFBP are present in a subset of the chromaffin cells, in line with their neural crest origin (Huising et al., 2007), yet the significance of CRF and CRFBP in these cells remains enigmatic. Following stimulation in vitro with 8-bromo-AMP, CRF was released from head kidney tissue. The carp system is reminiscent of the intraadrenal CRF system in mammals involved in modulation of the *fasciculata* glucocorticoid release and adrenal blood flow control. The absence of CRF_1-receptors on cortical cells calls for identification of a medullary component to explain this indirect CRF effect. Indeed, ACTH is produced and released in conjunction with CRF (Suda et al., 1984; Ehrhart-Bornstein et al., 2000). We suggest that CRF, also in fish, could act then as an autocrine feedforward that results in production and secretion of ACTH; ACTH then targets the MC2R on interrenal cells to produce and secrete cortisol. ACTH directly and dose-dependently stimulates adrenaline release in trout (Reid et al., 1996), a phenomenon likely common to all fish and in accordance with the close anatomical organization of the two cell types. Possibly this system allows for local fine-tuning of the outcome of the centrally initiated stress response, a trait apparently already present in the common ancestor of fish and mammals (~ 450 Mya) (Huising et al., 2007).

4.4. Stress and Energy

Overall the stress response culminates in a transient hyperglycemia to provide the organism/fish with fuel for fight or flight. Direct neural actions from the hypothalamus to the chromaffin cell result in release of glycogenolytic adrenaline on a second-to-minute basis. Then cortisol takes over to sustain hyperglycemia over minutes to hours and allows for adaptive redistribution and reallocation of energy. Indeed, adrenaline is the true stress hormone, cortisol the adaptive hormone (Korte et al., 2007; Koolhaas et al., 2010, 2011).

In a series of studies on rainbow trout, the Soengas laboratory (Vigo, Spain) showed that glucose modulates ACTH-induced cortisol release (Conde-Sieira et al., 2013). Cortisol release in vitro was potentiated by high glucose levels and decreased by cytochalasin-B (a blocker of glucose transport). This provided the first evidence in fish for a link between glycemic state and the release of a nonpancreatic hormone. It follows that the hyperglycemia is relevant for cortisol synthesis and release under acute stress conditions. In this setting adrenaline not only frees glucose but also performs as a hyperglycemic signal, actions that are prerequisite for ACTH-mediated cortisol release. A glucose sensor (specifically glucokinase/hexokinase IV) was demonstrated in in vitro studies on head kidney slices and confirmed by immunohistochemistry in the head kidney interrenal cells. These data corroborate the notion that interrenal glucose sensitivity must be an intrinsic property of cortisol-producing cells. The stress axis and energy metabolism are intimately entangled, further illustrated by the effects of one of the major cytokines in control of food/energy intake, leptin. Leptin was shown to inhibit stress axis activity via direct inhibition of both pituitary constitutive and CRF-regulated ACTH release and interrenal cortisol release (Gorissen et al., 2012). One mode of elevating leptin levels in fish (cyprinids, salmonids) is via exposure to stressful hypoxia (Bernier et al., 2012; MacDonald et al., 2014). Depending on the degree of hypoxia, energy expenditure decreases more or less dramatically. Under such conditions launching a large stress response could be counterproductive, as stress and energy metabolism are intimately linked. It was proposed that leptin conveys to central brain regions information on energy status and downplays the stress response. In this way leptin contributes to the coordination of the delicate balance of eustress and distress (Gorissen and Flik, 2014).

5. SYNTHESIS AND PERSPECTIVE

Allostasis (McEwen and Wingfield, 2003) is defined as "the active process of maintaining or re-establishing homeostasis" (Romero et al.,

2009). It follows that we can refer to allostasis as the ability of an organism to produce the chemical mediators for that process, *viz.* stress-related hormones (adrenaline and cortisol), cytokines, and parasympathetic activity (Romero et al., 2009), in order to adapt to a new situation or an environmental challenge. There is a large variation in individual perception (ie, appraisal) and allostasis allows customization of the individual physiological response to a changing environment (McEwen and Wingfield, 2010). Deviations from normal environmental states generate a situation called an allostatic state, in which the regulatory capacity of the animal changes (Koob and Le Moal, 2001; Korte et al., 2007; McEwen and Wingfield, 2010). This condition is characterized by increases in the primary chemical mediators of allostasis, as defined earlier, and can only be maintained for a relatively short period. Although the stress response serves as an adaptive response at first, prolonged exposure to stressors leads to a situation named allostatic overload, in which pathologies and diseases occur, as a result of wear-and-tear of the body (McEwen and Wingfield, 2003, 2010; Korte et al., 2007; Schreck, 2010). The ability to predict the outcome of a current situation, based on prior experience, is fundamental to the allostasis principle (McEwen and Wingfield, 2003) and allows the organism to reset behavioral and physiological set points to (better) cope with the challenge or changed environmental conditions. It is clear that the concept of allostasis takes a key role in our understanding of individual variation in stress responsiveness, coping styles, animal characteristics, and the underlying chemical mediation.

The power of the allostasis model is well illustrated by a series of studies by Manuel and coworkers (Manuel et al., 2014a,b; Gorissen et al., 2015; Manuel et al., 2016), who used an inhibitory fear-avoidance paradigm (task variables: shock number and shock intensity) with zebrafish of different age, different strains (AB and Tupfel Long-Fin), prior enrichment, and prior unpredictable chronic stress (subject variables). The capacity of an animal to cope with the stress imposed is determined by the interaction between subject characteristics and environmental characteristics. When environmental conditions are favorable (eg, enriched environment during development) the coping capacity of the fish increases, while unfavorable environmental conditions (chronic stress, barren conditions, etc.) will decrease coping capacity (Manuel, 2015; Fig. 3.3). Such a multidimensional, allostatic view on stress coping deepens our insight into fish resilience in particular, and vertebrates in general. Indeed in fish, representatives of the earliest vertebrates, there is much more involved than just the classical top-down regulation seen in the textbook stress axis that we know from studies on mammals and humans. A number of factors are involved, including locally produced components of the HPI axis (CRF and ACTH) in the

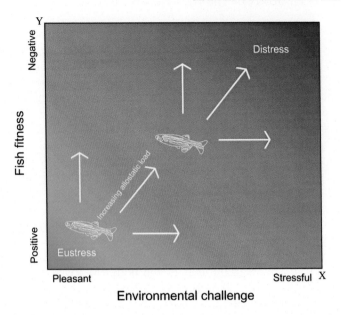

Figure 3.3. Diagram showing the allostatic landscape represented by the fitness of the fish (Y-axis) and the environmental challenge (X-axis). Green reflects conditions of eustress, red indicates distress. Transition from eustress to distress is multifactorial and two-dimensional. Obviously, a stressful environment has a larger impact on fish with poor fitness than on those with high fitness. Fish with high fitness can handle more environmental challenges. In real life both environment and fish characteristics form continua and together determine the allostatic load. A fish with increased allostatic load (eg, by stressful procedures such as frequent sorting and transportation) more rapidly enters the distress corner. Stress means energy expenditure and reallocation and thus perception of their environment is key to launching the proper stress response: adrenaline and cortisol mediate the new requirements and allow for resetting of physiological set points.

interrenal tissue and the demonstration of interrenal glucose-sensors, the machinery for energy-sensing and use of glucose. This illustrates the complexity, pleiotropy, and even redundancy in stress physiology. The complex regulation must be an early vertebrate invention and at the basis of the success of fish and all tetrapods.

ACKNOWLEDGMENTS

MG received funding from the European Community's Seventh Framework Program (FP7/2010–2014) under grant agreement number 265957 (COPEWELL). Tina Reilink and Gonny Kremers of the Library of Science

(Radboud University) are heartily thanked for carrying out a dedicated and systematic literature search. Present and former members and students of the department of Animal Physiology, as well as our (inter-)national collaborators on stress physiology and fish welfare are acknowledged for numerous discussions. Mark Huising kindly provided Fig. 3.2F. Jim and Julie Fenwick are thanked wholeheartedly for proofreading and correcting the second-to-last version of this manuscript.

REFERENCES

Agulleiro, M. J., Roy, S., Sanchez, E., Puchol, S., Gallo-Payet, N. and Cerda-Reverter, J. M. (2010). Role of melanocortin receptor accessory proteins in the function of zebrafish melanocortin receptor type 2. *Mol. Cell. Endocrinol.* **320**, 145–152.

Albertson, R. C., Markert, J. A., Danley, P. D. and Kocher, T. D. (1999). Phylogeny of a rapidly evolving clade: the cichlid fishes of Lake Malawi, East Africa. *Proc. Natl. Acad. Sci. U.S.A.* **96**, 5107–5110.

Alderman, S. L. and Bernier, N. J. (2007). Localization of corticotropin-releasing factor, urotensin I, and CRF-binding protein gene expression in the brain of the zebrafish, *Danio rerio. J. Comp. Neurol.* **502**, 783–793. ⟨http://dx.doi.org/10.1002/cne.21332⟩.

Alderman, S. L. and Bernier, N. J. (2009). Ontogeny of the corticotropin-releasing factor system in zebrafish. *Gen. Comp. Endocrinol.* **164**, 61–69. ⟨http://dx.doi.org/10.1016/j.ygcen.2009.04.007⟩.

Alsop, D. and Vijayan, M. M. (2009). Molecular programming of the corticosteroid stress axis during zebrafish development. *Comp. Biochem. Physiol. Mol. Integr. Physiol.* **153**, 49–54. ⟨http://dx.doi.org/10.1016/j.cbpa.2008.12.008⟩.

Arai, M., Assil, I. Q. and Abou-Samra, A. B. (2001). Characterization of three corticotropin-releasing factor receptors in catfish: a novel third receptor is predominantly expressed in pituitary and urophysis. *Endocrinology* **142**, 446–454. ⟨http://dx.doi.org/10.1210/endo.142.1.7879⟩.

Arends, R. J., Rotllant, J., Metz, J. R., Mancera, J. M., Wendelaar Bonga, S. E. and Flik, G. (2000). alpha-MSH acetylation in the pituitary gland of the sea bream (*Sparus aurata* L.) in response to different backgrounds, confinement and air exposure. *J. Endocrinol.* **166**, 427–435.

Backstrom, T., Pettersson, A., Johansson, V. and Winberg, S. (2011). CRF and urotensin I effects on aggression and anxiety-like behavior in rainbow trout. *J. Exp. Biol.* **214**, 907–914. ⟨http://dx.doi.org/10.1242/jeb.045070⟩.

Baker, B. I., Bird, D. J. and Buckingham, J. C. (1996). In the trout, CRH and AVT synergize to stimulate ACTH release. *Regul. Pept.* **67**, 207–210.

Barry, T. P., Malison, J. A., Held, J. A. and Parrish, J. J. (1995a). Ontogeny of the cortisol stress response in larval rainbow trout. *Gen. Comp. Endocrinol.* **97**, 57–65. ⟨http://dx.doi.org/10.1006/gcen.1995.1006⟩.

Barry, T. P., Ochiai, M. and Malison, J. A. (1995b). In vitro effects of ACTH on interrenal corticosteroidogenesis during early larval development in rainbow trout. *Gen. Comp. Endocrinol.* **99**, 382–387. ⟨http://dx.doi.org/10.1006/gcen.1995.1122⟩.

Benjannet, S., Rondeau, N., Day, R., Chrétien, M. and Seidah, N. G. (1991). PC1 and PC2 are proprotein convertases capable of cleaving proopiomelanocortin at distinct pairs of basic residues. *Proc. Natl. Acad. Sci. U. S. A.* **88**, 3564–3568.

Bernier, N. J. (2006). The corticotropin-releasing factor system as a mediator of the appetite-suppressing effects of stress in fish. *Gen. Comp. Endocrinol.* **146**, 45–55. ⟨http://dx.doi.org/10.1016/j.ygcen.2005.11.016⟩.

Bernier, N. J. and Craig, P. M. (2005). CRF-related peptides contribute to stress response and regulation of appetite in hypoxic rainbow trout. *Am. J. Physiol. Regul. Integr. Comp. Physiol.* **289**, R982–R990. ⟨http://dx.doi.org/10.1152/ajpregu.00668.2004⟩.

Bernier, N. J., Lin, X. W. and Peter, R. E. (1999). Differential expression of corticotropin-releasing factor (CRF) and urotensin I precursor genes, and evidence of CRF gene expression regulated by cortisol in goldfish brain. *Gen. Comp. Endocrinol.* **116**, 461–477. ⟨http://dx.doi.org/10.1006/gcen.1999.7386⟩.

Bernier, N. J., Alderman, S. L. and Bristow, E. N. (2008). Heads or tails? Stressor-specific expression of corticotropin-releasing factor and urotensin I in the preoptic area and caudal neurosecretory system of rainbow trout. *J. Endocrinol.* **196**, 637–648. ⟨http://dx.doi.org/10.1677/JOE-07-0568⟩.

Bernier, N. J., Flik, G. and Klaren, P. H. M. (2009). Regulation and Contribution of The Corticotropic, Melanotropic and Thyrotropic Axes to The Stress Response in Fishes. In *Fish Physiology - Fish Neuroendocrinology*, Vol. 28 (eds. N. J. Bernier, G. Van der Kraak, A. P. Farrell and C. J. Brauner), pp. 235–311. London: Academic Press.

Bernier, N. J., Gorissen, M. and Flik, G. (2012). Differential effects of chronic hypoxia and feed restriction on the expression of leptin and its receptor, food intake regulation and the endocrine stress response in common carp. *J. Exp. Biol.* **215**, 2273–2282. ⟨http://dx.doi.org/10.1242/jeb.066183⟩.

Betito, K., Diorio, J., Meaney, M. J. and Boksa, P. (1992). Adrenal phenylethanolamine *N*-methyltransferase induction in relation to glucocorticoid receptor dynamics: evidence that acute exposure to high cortisol levels is sufficient to induce the enzyme. *J. Neurochem.* **58**, 1853–1862.

Blaschko, H. (1939). The specific action of L-Dopa decarboxylase. *J. Physiol. (Lond.)* **96**, 50P–51P.

Boltaña, S., Rey, S., Roher, N., Vargas, R., Huerta, M., Huntingford, F. A., et al. (2013). Behavioural fever is a synergic signal amplifying the innate immune response. *Proc. Biol. Sci.* **280**, 20131381. ⟨http://dx.doi.org/10.1098/rspb.2013.1381⟩.

Boorse, G. C., Crespi, E. J., Dautzenberg, F. M. and Denver, R. J. (2005). Urocortins of the South African clawed frog, *Xenopus laevis*: conservation of structure and function in tetrapod evolution. *Endocrinology* **146**, 4851–4860. ⟨http://dx.doi.org/10.1210/en.2005-0497⟩.

Bruhn, T. O., Plotsky, P. M. and Vale, W. W. (1984). Effect of paraventricular lesions on corticotropin-releasing factor (CRF)-like immunoreactivity in the stalk-median eminence: studies on the adrenocorticotropin response to ether stress and exogenous CRF. *Endocrinology* **114**, 57–62. ⟨http://dx.doi.org/10.1210/endo-114-1-57⟩.

Castro, M. G. and Morrison, E. (1997). Post-translational processing of proopiomelanocortin in the pituitary and in the brain. *Crit. Rev. Neurobiol.* **11**, 35–57.

Cerdá-Reverter, J. M. and Canosa, L. F. (2009). Neuroendocrine Systems of the Fish Brain. In *Fish Physiology - Fish Neuroendocrinology*, Vol. 28 (eds. N. J. Bernier, G. Van der Kraak, A. P. Farrell and C. J. Brauner), pp. 3–74. London: Academic Press.

Chadzinska, M., Hermsen, T., Savelkoul, H. F. J. and Kemenade, B. M. L. V.-V. (2009). Cloning of opioid receptors in common carp (*Cyprinus carpio* L.) and their involvement in regulation of stress and immune response. *Brain Behav. Immun.* **23**, 257–266. ⟨http://dx.doi.org/10.1016/j.bbi.2008.10.003⟩.

Chandrasekar, G., Lauter, G. and Hauptmann, G. (2007). Distribution of corticotropin-releasing hormone in the developing zebrafish brain. *J. Comp. Neurol.* **505**, 337–351. ⟨http://dx.doi.org/10.1002/cne.21496⟩.

Chen, A. T. and Zon, L. I. (2009). Zebrafish blood stem cells. *J. Cell. Biochem.* **108**, 35–42. ⟨http://dx.doi.org/10.1002/jcb.22251⟩.

Chen, R., Lewis, K. A., Perrin, M. H. and Vale, W. W. (1993). Expression cloning of a human corticotropin-releasing-factor receptor. *Proc. Natl. Acad. Sci. U.S.A.* **90**, 8967–8971.

Civelli, O., Bunzow, J. R. and Grandy, D. K. (1993). Molecular diversity of the dopamine receptors. *Annu. Rev. Pharmacol. Toxicol.* **33**, 281–307. ⟨http://dx.doi.org/10.1146/annurev.pa.33.040193.001433⟩.

Conde-Sieira, M., Alvarez, R., López-Patiño, M. A., Míguez, J. M., Flik, G. and Soengas, J. L. (2013). ACTH-stimulated cortisol release from head kidney of rainbow trout is modulated by glucose concentration. *J. Exp. Biol.* **216**, 554–567. ⟨http://dx.doi.org/10.1242/jeb.076505⟩.

Coste, S. C., Quintos, R. F. and Stenzel-Poore, M. P. (2002). Corticotropin-releasing hormone-related peptides and receptors: emergent regulators of cardiovascular adaptations to stress. *Trends Cardiovasc. Med.* **12**, 176–182.

Coulouarn, Y., Lihrmann, I., Jegou, S., Anouar, Y., Tostivint, H., Beauvillain, J. C., et al. (1998). Cloning of the cDNA encoding the urotensin II precursor in frog and human reveals intense expression of the urotensin II gene in motoneurons of the spinal cord. *Proc. Natl. Acad. Sci. U.S.A.* **95**, 15803–15808.

Craig, P. M., Al-Timimi, H. and Bernier, N. J. (2005). Differential increase in forebrain and caudal neurosecretory system corticotropin-releasing factor and urotensin I gene expression associated with seawater transfer in rainbow trout. *Endocrinology* **146**, 3851–3860. ⟨http://dx.doi.org/10.1210/en.2005-0004⟩.

Dehal, P. and Boore, J. L. (2005). Two rounds of whole genome duplication in the ancestral vertebrate. *PLoS Biol.* **3**, 1700–1708. ⟨http://dx.doi.org/10.1371/journal.pbio.0030314⟩.

Dores, R. M., Londraville, R. L., Prokop, J., Davis, P., Dewey, N. and Lesinski, N. (2014). Molecular evolution of GPCRS: melanocortin/melanocortin receptors. *J. Mol. Endocrinol.* **52**, T29–T42. ⟨http://dx.doi.org/10.1530/JME-14-0050⟩.

Doyon, C., Trudeau, V. L. and Moon, T. W. (2005). Stress elevates corticotropin-releasing factor (CRF) and CRF-binding protein mRNA levels in rainbow trout (*Oncorhynchus mykiss*). *J. Endocrinol.* **186**, 123–130. ⟨http://dx.doi.org/10.1677/joe.1.06142⟩.

Dores, R. M., Liang, L., Hollmann, R. E., Sandhu, N. and Vijayan, M. M. (2016). Identifying the activation motif in the N-terminal of rainbow trout and zebrafish melanocortin-2 receptor accessory protein 1 (MRAP1) orthologs. *General and Comparative Endocrinology* In press⟨http://dx.doi.org/10.1016/j.ygcen.2015.12.031⟩.

Drolet, G., Dumont, E. C., Gosselin, I., Kinkead, R., Laforest, S. and Trottier, J. F. (2001). Role of endogenous opioid system in the regulation of the stress response. *Prog. Neuropsychopharmacol. Biol. Psychiatry* **25**, 729–741.

Ehrhart-Bornstein, M., Haidan, A., Alesci, S. and Bornstein, S. R. (2000). Neurotransmitters and neuropeptides in the differential regulation of steroidogenesis in adrenocortical-chromaffin co-cultures. *Endocr. Res.* **26**, 833–842.

Faught, E., Aluru, N. and Vijayan, M. M. (2016). The Molecular Stress Response. In *Fish Physiology - Biology of Stress in Fish*, Vol. 35 (eds. C. B. Schreck, L. Tort, A. P. Farrell and C. J. Brauner), San Diego, CA: Academic Press.

Febry, R. and Lutz, P. (1987). Energy partitioning in fish: the activity-related cost of osmoregulation in a euryhaline cichlid. *J. Exp. Biol.* **128**, 63–85.

Feist, G. and Schreck, C. B. (2002). Ontogeny of the stress response in chinook salmon, *Oncorhynchus tshawytscha*. *Fish Physiol. Biochem.* **25**, 31–40.

Fernandes, A. M., Beddows, E., Filippi, A. and Driever, W. (2013). Orthopedia transcription factor otpa and otpb paralogous genes function during dopaminergic and neuroendocrine cell specification in larval zebrafish. *PLoS One* **8**, e75002. ⟨http://dx.doi.org/10.1371/journal.pone.0075002⟩.

Fitzgibbon, Q. P., Seymour, R. S. and Ellis, D. (2007). The energetic consequence of specific dynamic action in southern bluefin tuna *Thunnus maccoyii*. *J. Exp. Biol.* **210** (Pt 2), 290–298.

Flik, G., Klaren, P. H. M., Van den Burg, E. H., Metz, J. R. and Huising, M. O. (2006). CRF and stress in fish. *Gen. Comp. Endocrinol.* **146**, 36–44. ⟨http://dx.doi.org/10.1016/j.ygcen.2005.11.005⟩.

Fryer, J., Lederis, K. and Rivier, J. (1984). Cortisol inhibits the ACTH-releasing activity of urotensin-I, CRF and sauvagine observed with superfused goldfish pituitary-cells. *Peptides* **5**, 925–930.

Fuzzen, M. L. M., Alderman, S. L., Bristow, E. N. and Bernier, N. J. (2011). Ontogeny of the corticotropin-releasing factor system in rainbow trout and differential effects of hypoxia on the endocrine and cellular stress responses during development. *Gen. Comp. Endocrinol.* **170**, 604–612. ⟨http://dx.doi.org/10.1016/j.ygcen.2010.11.022⟩.

Gage, S. P. (1893). *The Brain of Diemyctilus Viridescens From Larval to Adult Life and Comparison With the Brain of Amia and Petromyzon.* Ithaca: The Wilder Quarter Century Book.

Geven, E. J. W., Verkaar, F., Flik, G. and Klaren, P. H. M. (2006). Experimental hyperthyroidism and central mediators of stress axis and thyroid axis activity in common carp (*Cyprinus carpio* L.). *J. Mol. Endocrinol.* **37**, 443–452.

Gorissen, M. and Flik, G. (2014). Leptin in teleostean fish, towards the origins of leptin physiology. *J. Chem. Neuroanat.* **61–62**, 200–206. ⟨http://dx.doi.org/10.1016/j.jchemneu.2014.06.005⟩.

Gorissen, M., Marnix, H. A. G., Flik, G. and Huising, M. O. (2006). Peptides and proteins regulating food intake: a comparative view. *Anim. Biol.* **56**, 447–473.

Gorissen, M., Bernier, N. J., Manuel, R., de Gelder, S., Metz, J. R., Huising, M. O., et al. (2012). Recombinant human leptin attenuates stress axis activity in common carp (*Cyprinus carpio* L.). *Gen. Comp. Endocrinol.* **178**, 75–81. ⟨http://dx.doi.org/10.1016/j.ygcen.2012.04.004⟩.

Gorissen, M., Manuel, R., Pelgrim, T. N. M., Mes, W., de Wolf, M. J. S., Zethof, J., et al. (2015). Differences in inhibitory avoidance, cortisol and brain gene expression in TL and AB zebrafish. *Genes Brain Behav.* **14**, 428–438. ⟨http://dx.doi.org/10.1111/gbb.12220⟩.

Haitina, T., Klovins, J., Andersson, J., Frederiksson, R., Lagerström, M. C., Larhammar, D., et al. (2004). Cloning, tissue distribution, pharmacology and three-dimensional modelling of melanocortin receptors 4 and 5 in rainbow trout suggest close evolutionary relationship of these subtypes. *Biochemical Journal* **380**, 475–486.

Hauger, R. L., Grigoriadis, D. E., Dallman, M. F., Plotsky, P. M., Vale, W. W. and Dautzenberg, F. M. (2003). International Union of Pharmacology. XXXVI. Current status of the nomenclature for receptors for corticotropin-releasing factor and their ligands. *Pharmacol. Rev.* **55**, 21–26. ⟨http://dx.doi.org/10.1124/pr.55.1.3⟩.

Herget, U. and Ryu, S. (2015). Coexpression analysis of nine neuropeptides in the neurosecretory preoptic area of larval zebrafish. *Front. Neuroanat.* **9**, 2. ⟨http://dx.doi.org/10.3389/fnana.2015.00002⟩.

Herget, U., Wolf, A., Wullimann, M. F. and Ryu, S. (2014). Molecular neuroanatomy and chemoarchitecture of the neurosecretory preoptic-hypothalamic area in zebrafish larvae. *J. Comp. Neurol.* **522**, 1542–1564. ⟨http://dx.doi.org/10.1002/cne.23480⟩.

Herrick, C. J. (1933). The functions of the olfactory parts of the cerebral cortex. *Proc. Natl. Acad. Sci. U. S. A.* **19**, 7–14.

Herzog, W., Zeng, X., Lele, Z., Sonntag, C., Ting, J.-W., Chang, C.-Y., et al. (2003). Adenohypophysis formation in the zebrafish and its dependence on sonic hedgehog. *Dev. Biol.* **254**, 36–49.

Hofmann, J., Greter, M., Pasquier Du, L. and Becher, B. (2010). B-cells need a proper house, whereas T-cells are happy in a cave: the dependence of lymphocytes on secondary lymphoid tissues during evolution. *Trends Immunol.* **31**, 144–153. ⟨http://dx.doi.org/10.1016/j.it.2010.01.003⟩.

Hsu, S. Y. and Hsueh, A. J. (2001). Human stresscopin and stresscopin-related peptide are selective ligands for the type 2 corticotropin-releasing hormone receptor. *Nat. Med.* **7**, 605–611. ⟨http://dx.doi.org/10.1038/87936⟩.

Hsu, H. J., Lin, G. and Chung, B. C. (2004). Parallel early development of zebrafish interrenal glands and pronephros: differential control by wt1 and ff1b. *Endocr. Res.* **30**, 803.

Huising, M. O. (2006). Communication in the Endocrine and Immune Systems, pp. 1–298. Nijmegen.

Huising, M. O. and Flik, G. (2005). The remarkable conservation of corticotropin-releasing hormone (CRH)-binding protein in the honeybee (*Apis mellifera*) dates the CRH system to a common ancestor of insects and vertebrates. *Endocrinology* **146**, 2165–2170. ⟨http://dx.doi.org/10.1210/en.2004-1514⟩.

Huising, M. O., Guichelaar, T., Hoek, C., Verburg-van Kemenade, B. M. L., Flik, G., Savelkoul, H., et al. (2003). Increased efficacy of immersion vaccination in fish with hyperosmotic pretreatment. *Vaccine* **21**, 4178–4193. ⟨http://dx.doi.org/10.1016/S0264-410X(03)00497-3⟩.

Huising, M. O., Metz, J. R., van Schooten, C., Taverne-Thiele, A. J., Hermsen, T., Verburg-van Kemanade, B. M. L., et al. (2004). Structural characterisation of a cyprinid (*Cyprinus carpio* L.) CRH, CRH-BP and CRH-R1, and the role of these proteins in the acute stress response. *J. Mol. Endocrinol.* **32**, 627–648.

Huising, M. O., van der Aa, L. M., Metz, J. R., de Fátima Mazon, A., Verburg-van Kemenade, B. M. L. and Flik, G. (2007). Corticotropin-releasing factor (CRF) and CRF-binding protein expression in and release from the head kidney of common carp: evolutionary conservation of the adrenal CRF system. *J. Endocrinol.* **193**, 349–357. ⟨http://dx.doi.org/10.1677/JOE-07-0070⟩.

Huising, M. O., Vaughan, J. M., Shah, S. H., Grillot, K. L., Donaldson, C. J., Rivier, J., et al. (2008). Residues of corticotropin releasing factor-binding protein (CRF-BP) that selectively abrogate binding to CRF but not to urocortin 1. *J. Biol. Chem.* **283**, 8902–8912. ⟨http://dx.doi.org/10.1074/jbc.M709904200⟩.

Ito, H. and Yamamoto, N. (2009). Non-laminar cerebral cortex in teleost fishes? *Biol. Lett.* **5**, 117–121. ⟨http://dx.doi.org/10.1098/rsbl.2008.0397⟩.

Jerez-Cepa, I., Mancera, J. M., Flik, G. and Gorissen, M. (2016). Vasotinergic and isotonergic co-regulation in stress response of common carp (*Cyprinus carpio* L.). *Adv. Comparative Endocrinol.* **VII** (in press).

Kah, O., Dubourg, P., Onteniente, B., Geffard, M. and Calas, A. (1986). The dopaminergic innervation of the goldfish pituitary. An immunocytochemical study at the electron-microscope level using antibodies against dopamine. *Cell Tissue. Res.* **244**, 577–582.

Kalueff, A. V., Echevarria, D. J. and Stewart, A. M. (2014). Gaining translational momentum: more zebrafish models for neuroscience research. *Progress Neuro-Psychopharmacol. Biol. Psychiatry* **55**, 1–6. ⟨http://dx.doi.org/10.1016/j.pnpbp.2014.01.022⟩.

Kimmel, C. B. and Law, R. D. (1985). Cell lineage of zebrafish blastomeres. I. Cleavage pattern and cytoplasmic bridges between cells. *Dev. Biol.* **108**, 78–85.

Kimmel, C. B., Ballard, W. W., Ullmann, B. and Schilling, T. F. (1995). Stages of embryonic-development of the zebrafish. *Dev. Dynam.* **203**, 253–310. ⟨http://dx.doi.org/10.1002/aja.1002030302⟩.

Klovins, J. (2003). The melanocortin system in fugu: determination of POMC/AGRP/MCR gene repertoire and synteny, as well as pharmacology and anatomical distribution of the MCRs. *Mol. Biol. Evol.* **21**, 563–579. ⟨http://dx.doi.org/10.1093/molbev/msh050⟩.

Kobayashi, Y., Chiba, H., Yamanome, T., Schiöth, H. B. and Takahashi, A. (2011). Melanocortin receptor subtypes in interrenal cells and corticotropic activity of alpha-melanocyte-stimulating hormones in barfin flounder, *Verasper moseri. Gen. Comp. Endocrinol.* **170**, 558–568. ⟨http://dx.doi.org/10.1016/j.ygcen.2010.11.019⟩.

Koob, G. F. and Le Moal, M. (2001). Drug addiction, dysregulation of reward, and allostasis. *Neuropsychopharmacology* **24**, 97–129. ⟨http://dx.doi.org/10.1016/S0893-133X(00)00195-0⟩.

Koolhaas, J. M., de Boer, S. F., Coppens, C. M. and Buwalda, B. (2010). Neuroendocrinology of coping styles: towards understanding the biology of individual variation. *Front Neuroendocrinol.* **31**, 307–321. ⟨http://dx.doi.org/10.1016/j.yfrne.2010.04.001⟩.

Koolhaas, J. M., Bartolomucci, A., Buwalda, B., de Boer, S. F., Flügge, G., Korte, S. M., et al. (2011). Stress revisited: a critical evaluation of the stress concept. *Neurosci. Biobehav. Rev.* **35**, 1291–1301. ⟨http://dx.doi.org/10.1016/j.neubiorev.2011.02.003⟩.

Korte, S. M., Olivier, B. and Koolhaas, J. M. (2007). A new animal welfare concept based on allostasis. *Physiol. Behav.* **92**, 422–428. ⟨http://dx.doi.org/10.1016/j.physbeh.2006.10.018⟩.

Kosaka, K., Kawakami, K., Sakamoto, H. and Inoue, K. (2007). Spatiotemporal localization of germ plasm RNAs during zebrafish oogenesis. *Mech. Dev.* **124**, 279–289. ⟨http://dx.doi.org/10.1016/j.mod.2007.01.003⟩.

Lamers, A. E., Flik, G., Atsma, W. and Wendelaar Bonga, S. E. (1992). A role for di-acetyl alpha-melanocyte-stimulating hormone in the control of cortisol release in the teleost *Oreochromis mossambicus. J. Endocrinol.* **135**, 285–292.

Lamers, A. E., Flik, G. and Wendelaar Bonga, S. E. (1994). A specific role for Trh in release of diacetyl alpha-MSH in tilapia stressed by acid water. *Am. J. Physiol. Regul. Integr. Comp. Physiol.* **267**, R1302–R1308.

Lamers, A. E., Brugge Ter, P. J., Flik, G. and Wendelaar Bonga, S. E. (1997). Acid stress induces a D1-like dopamine receptor in pituitary MSH cells of *Oreochromis mossambicus. Am. J. Physiol.* **273**, R387–R392.

Lederis, K., Fryer, J. N., Okawara, Y., Schonrock, C. and Richter, D. (1994). Cortico-tropinreleasing factors acting on the fish pituitary: experimental and molecular analysis. In *Fish Physiology - Molecular Endocrinology of Fish*, Vol. 13 (eds. N. M. Sherwood and C. L. Hew), pp. 67–100. San Diego, CA: Academic Press.

Le Mevel, J. C., Lancien, F., Mimassi, N. and Conlon, J. M. (2009). Central hyperventilatory action of the stress-related neurohormonal peptides, corticotropin-releasing factor and urotensin-I in the trout *Oncorhynchus mykiss. Gen. Comp. Endocrinol.* **164**, 51–60. ⟨http://dx.doi.org/10.1016/j.ygcen.2009.03.019⟩.

Lewis, K., Li, C., Perrin, M. H., Blount, A., Kunitake, K., Donaldson, C., et al. (2001). Identification of urocortin III, an additional member of the corticotropin-releasing factor (CRF) family with high affinity for the CRF2 receptor. *Proc. Natl. Acad. Sci. U. S. A.* **98**, 7570–7575. ⟨http://dx.doi.org/10.1073/pnas.121165198⟩.

Liang, L., Schmid, K., Sandhu, N., Angleson, J. K., Vijayan, M. M. and Dores, R. M. (2015). Structure/function studies on the activation of the rainbow trout melanocortin 2 receptor. *Gen. Comp. Endocrinol.* **210**, 145–151.

Liu, Y.-W. (2007). Interrenal organogenesis in the zebrafish model. *Organogenesis* **3**, 44–48.

Long, J. A. (1995). *The Rise of Fishes*. London: Johns Hopkins Press Ltd.

Lovejoy, D. A. and Balment, R. J. (1999). Evolution and physiology of the corticotropin-releasing factor (CRF) family of neuropeptides in vertebrates. *Gen. Comp. Endocrinol.* **115**, 1–22. ⟨http://dx.doi.org/10.1006/gcen.1999.7298⟩.

Lu, W., Dow, L., Gumusgoz, S., Brierley, M. J., Warne, J. M., McCrohan, C. R., et al. (2004). Coexpression of corticotropin-releasing hormone and urotensin I precursor genes in the caudal neurosecretory system of the euryhaline flounder (*Platichthys flesus*): a possible shared role in peripheral regulation. *Endocrinology* **145**, 5786–5797. ⟨http://dx.doi.org/10.1210/en.2004-0144⟩.

MacDonald, L. E., Alderman, S. L., Kramer, S., Woo, P. T. K. and Bernier, N. J. (2014). Hypoxemia-induced leptin secretion: a mechanism for the control of food intake in diseased fish. *J. Endocrinol.* **221**, 441–455. ⟨http://dx.doi.org/10.1530/JOE-13-0615⟩.

Machluf, Y., Gutnick, A. and Levkowitz, G. (2011). Development of the zebrafish hypothalamus. *Ann. N. Y. Acad. Sci.* **1220**, 93–105. ⟨http://dx.doi.org/10.1111/j.1749-6632.2010.05945.x⟩.

Madaro, A., Olsen, R. E., Kristiansen, T. S., Ebbesson, L. O. E., Nilsen, T. O., Flik, G., et al. (2015). Stress in Atlantic salmon: response to unpredictable chronic stress. *J. Exp. Biol.* **218**, 2538–2550. ⟨http://dx.doi.org/10.1242/jeb.120535⟩.

Madaro, A., Olsen, R. E., Kristiansen, T. S., Ebbesson, L. O. E., Flik, G. and Gorissen, M. (2016). A comparative study of the response to repeated chasing stress in Atlantic salmon (*Salmo salar* L.) parr and post-smolts. *Comp. Biochem. Physiol. A Mol. Integr. Physiol.* **192**, 7–16. ⟨http://dx.doi.org/10.1016/j.cbpa.2015.11.005⟩.

Maier, T., Güell, M. and Serrano, L. (2009). Correlation of mRNA and protein in complex biological samples. *FEBS Lett.* **583**, 3966–3973. ⟨http://dx.doi.org/10.1016/j.febslet.2009.10.036⟩.

Manuel, R. (2015). *Biology of welfare in fish (Ph.D. thesis)*. Nijmegen: Radboud University.

Manuel, R., Gorissen, M., Roca, C. P., Zethof, J., van de Vis, H., Flik, G., et al. (2014a). Inhibitory avoidance learning in zebrafish (*Danio rerio*): effects of shock intensity and unraveling differences in task performance. *Zebrafish* **11**, 341–352. ⟨http://dx.doi.org/10.1089/zeb.2013.0970⟩.

Manuel, R., Gorissen, M., Zethof, J., Ebbesson, L. O. E., van de Vis, H., Flik, G., et al. (2014b). Unpredictable chronic stress decreases inhibitory avoidance learning in Tuebingen long-fin zebrafish: stronger effects in the resting phase than in the active phase. *J. Exp. Biol.* **217**, 3919–3928. ⟨http://dx.doi.org/10.1242/jeb.109736⟩.

Manuel, R., Metz, J. R., Flik, G., Vale, W. W. and Huising, M. O. (2014c). Corticotropin-releasing factor-binding protein (CRF-BP) inhibits CRF- and urotensin-I-mediated activation of CRF receptor-1 and -2 in common carp. *Gen. Comp. Endocrinol.* **202**, 69–75. ⟨http://dx.doi.org/10.1016/j.ygcen.2014.04.010⟩.

Manuel, R., Gorissen, M. and van den Bos, R. (2016). Relevance of test- and subject-related factors on inhibitory avoidance (performance) of zebrafish for psychopharmacology studies. *Current Psychopharmacology* **5** (2), epub.

McEwen, B. S. and Wingfield, J. C. (2003). The concept of allostasis in biology and biomedicine. *Horm. Behav.* **43**, 2–15.

McEwen, B. S. and Wingfield, J. C. (2010). What is in a name? Integrating homeostasis, allostasis and stress. *Horm. Behav.* **57**, 105–111. ⟨http://dx.doi.org/10.1016/j.yhbeh.2009.09.011⟩.

Metz, J. R., Huising, M. O., Meek, J., Taverne-Thiele, A. J., Wendelaar Bonga, S. E. and Flik, G. (2004). Localization, expression and control of adrenocorticotropic hormone in the nucleus preopticus and pituitary gland of common carp (Cyprinus carpio L.). *The Journal of Endocrinology* **182**, 23–31.

Metz, J. R., Geven, E. J. W., van den Burg, E. H. and Flik, G. (2005). ACTH, alpha-MSH, and control of cortisol release: cloning, sequencing, and functional expression of the melanocortin-2 and melanocortin-5 receptor in *Cyprinus carpio*. *Am. J. Physiol. Regul. Integr. Comp. Physiol.* **289**, R814–R826. ⟨http://dx.doi.org/10.1152/ajpregu.00826.2004⟩.

Metz, J. R., Peters, J. J. M. and Flik, G. (2006). Molecular biology and physiology of the melanocortin system in fish: a review. *Gen. Comp. Endocrinol.* **148**, 150–162. ⟨http://dx.doi.org/10.1016/j.ygcen.2006.03.001⟩.

Mueller, T. (2012). What is the thalamus in zebrafish? *Front. Neurosci.* **6**, 1–14. ⟨http://dx.doi.org/10.3389/fnins.2012.00064⟩.

Mueller, T., Dong, Z., Berberoglu, M. A. and Guo, S. (2011). The dorsal pallium in zebrafish, *Danio rerio* (Cyprinidae, Teleostei). *Brain Res.* **1381**, 95–105. ⟨http://dx.doi.org/10.1016/j.brainres.2010.12.089⟩.

Nieuwenhuys, R. (2009). The forebrain of actinopterygians revisited. *Brain Behav. Evol.* **73**, 229–252. ⟨http://dx.doi.org/10.1159/000225622⟩.

Nieuwenhuys, R., ten Donkelaar, H. J. and Nicholson, C. (1998). *The Central Nervous System of Vertebrates*. Berlin Heidelberg: Springer-Verlag.

Noakes, D. L. G. and Jones, K. M. M. (2016). Cognition, Learning, and Behavior. In *Fish Physiology - Biology of Stress in Fish*, Vol. 35 (eds. C. B. Schreck, L. Tort, A. P. Farrell and C. J. Brauner), San Diego, CA: Academic Press.

Nobata, S., Ando, M. and Takei, Y. (2013). Hormonal control of drinking behavior in teleost fishes; insights from studies using eels. *Gen. Comp. Endocrinol.* **192**, 214–221. ⟨http://dx.doi.org/10.1016/j.ygcen.2013.05.009⟩.

Northcutt, R. G. (2006). Connections of the lateral and medial divisions of the goldfish telencephalic pallium. *J. Comp. Neurol.* **494**, 903–943. ⟨http://dx.doi.org/10.1002/cne.20853⟩.

Ortega, V. A., Lovejoy, D. A. and Bernier, N. J. (2013). Appetite-suppressing effects and interactions of centrally administered corticotropin-releasing factor, urotensin I and serotonin in rainbow trout (*Oncorhynchus mykiss*). *Front. Neurosci.* **7**, 196. ⟨http://dx.doi.org/10.3389/fnins.2013.00196⟩.

Pearson, D., Shively, J. E., Clark, B. R., Geschwind, I. I., Barkley, M., Nishioka, R. S., et al. (1980). Urotensin II: a somatostatin-like peptide in the caudal neurosecretory system of fishes. *Proc. Natl. Acad. Sci. U. S. A.* **77**, 5021–5024.

Pepels, P., van Helvoort, H., Wendelaar Bonga, S. E. and Balm, P. H. M. (2004). Corticotropin-releasing hormone in the teleost stress response: rapid appearance of the peptide in plasma of tilapia (*Oreochromis mossambicus*). *J. Endocrinol.* **180**, 425–438. ⟨http://dx.doi.org/10.1677/joe.0.1800425⟩.

Pierson, P. M., Guibbolini, M. E. and Lahlou, B. (1996). A V1-type receptor for mediating the neurohypophysial hormone-induced ACTH release in trout pituitary. *J. Endocrinol.* **149**, 109–115.

Pogoda, H.-M. and Hammerschmidt, M. (2009). How to make a teleost adenohypophysis: molecular pathways of pituitary development in zebrafish. *Mol. Cell. Endocrinol.* **312**, 2–13. ⟨http://dx.doi.org/10.1016/j.mce.2009.03.012⟩.

Portavella, M., Vargas, J. P., Torres, B. and Salas, C. (2002). The effects of telencephalic pallial lesions on spatial, temporal, and emotional learning in goldfish. *Brain Res. Bull.* **57**, 397–399.

Portavella, M., Torres, B. and Salas, C. (2004). Avoidance response in goldfish: emotional and temporal involvement of medial and lateral telencephalic pallium. *J. Neurosci.* **24**, 2335–2342. ⟨http://dx.doi.org/10.1523/JNEUROSCI.4930-03.2004⟩.

Potter, E., Behan, D. P., Fischer, W. H., Linton, E. A., Lowry, P. J. and Vale, W. W. (1991). Cloning and characterization of the cDNAs for human and rat corticotropin releasing factor-binding proteins. *Nature* **349**, 423–426. ⟨http://dx.doi.org/10.1038/349423a0⟩.

Primavera J. H. (2005). Mangroves and Aquaculture in Southeast Asia.

Reid, S. G., Vijayan, M. M. and Perry, S. F. (1996). Modulation of catecholamine storage and release by the pituitary interrenal axis in the rainbow trout, *Oncorhynchus mykiss*. *J. Comp. Physiol. B* **165**, 665–676.

Reid, S. G., Bernier, N. J. and Perry, S. F. (1998). The adrenergic stress response in fish: control of catecholamine storage and release. *Comp. Biochem. Physiol. C Pharmacol. Toxicol. Endocrinol.* **120**, 1–27.

Rey, S., Huntingford, F. A., Boltaña, S., Vargas, R., Knowles, T. G. and MacKenzie, S. (2015). Fish can show emotional fever: stress-induced hyperthermia in zebrafish. *Proc. R. Soc. B Biol. Sci.* **282**, 20152266–20152267. ⟨http://dx.doi.org/10.1098/rspb.2015.2266⟩.

Rivier, C., Brownstein, M., Spiess, J., Rivier, J. and Vale, W. (1982). In vivo corticotropin-releasing factor-induced secretion of adrenocorticotropin, beta-endorphin, and corticosterone. *Endocrinology* **110**, 272–278. ⟨http://dx.doi.org/10.1210/endo-110-1-272⟩.

Romero, L. M., Dickens, M. J. and Cyr, N. E. (2009). The reactive scope model–a new model integrating homeostasis, allostasis, and stress. *Horm. Behav.* **55**, 375–389. ⟨http://dx.doi.org/10.1016/j.yhbeh.2008.12.009⟩.

Rotllant, J., Balm, P. H., Ruane, N. M., Pérez-Sánchez, J., Wendelaar Bonga, S. E. and Tort, L. (2000). Pituitary proopiomelanocortin-derived peptides and hypothalamus-pituitary-interrenal axis activity in gilthead sea bream (*Sparus aurata*) during prolonged crowding stress: differential regulation of adrenocorticotropin hormone and alpha-melanocyte-stimulating hormone release by corticotropin-releasing hormone and thyrotropin-releasing hormone. *Gen. Comp. Endocrinol.* **119**, 152–163. ⟨http://dx.doi.org/10.1006/gcen.2000.7508⟩.

Schreck, C. B. (2010). Stress and fish reproduction: the roles of allostasis and hormesis. *Gen. Comp. Endocrinol.* **165**, 549–556. ⟨http://dx.doi.org/10.1016/j.ygcen.2009.07.004⟩.

Schreck, C. B. and Tort, L. (2016). The Concept of Stress in Fish. In *Fish Physiology - Biology of Stress in Fish*, Vol. 35 (eds. C. B. Schreck, L. Tort, A. P. Farrell and C. J. Brauner), San Diego, CA: Academic Press.

Schwanhäusser, B., Busse, D., Li, N., Dittmar, G., Schuchhardt, J., Wolf, J., et al. (2011). Global quantification of mammalian gene expression control. *Nature* **473**, 337–342. ⟨http://dx.doi.org/10.1038/nature10098⟩.

Selbie, L. A., Hayes, G. and Shine, J. (1989). The major dopamine D2 receptor: molecular analysis of the human D2A subtype. *DNA (Mary Ann Liebert, Inc.)* **8**, 683–689.

Selye, H. (1950). Stress and the general adaptation syndrome. *Br. Med. J.* **1**, 1383–1392.

Selye, H. (1973). The evolution of the stress concept. *Am. Sci.* **61**, 692–699.

Stefaneanu, L., Kovacs, K., Horvath, E., Buchfelder, M., Fahlbusch, R. and Lancranjan, L. (2001). Dopamine D2 receptor gene expression in human adenohypophysial adenomas. *Endocrine* **14**, 329–336.

Stewart, A. M., Braubach, O., Spitsbergen, J., Gerlai, R. and Kalueff, A. V. (2014). Zebrafish models for translational neuroscience research: from tank to bedside. *Trends Neurosci.* **37**, 264–278. ⟨http://dx.doi.org/10.1016/j.tins.2014.02.011⟩.

Stouthart, A. J. H. X., Lucassen, E., van Strien, F., Balm, P., Lock, R. A. C. and Wendelaar Bonga, S. E. (1998). Stress responsiveness of the pituitary-interrenal axis during early life stages of common carp (*Cyprinus carpio*). *J. Endocrinol.* **157**, 127–137.

Studnička, F. K. (1896). Beiträge zur Anatomie und Entwickelungsgeschichte des Vorderhirns der Cranioten. *Sitzungsber K-Bohm Gesell- Sch Wissensch Mathem Naturw Kl* **15**, 1–32.

Suda, T., Tomori, N., Tozawa, F., Demura, H. and Shizume, K. (1984). Effects of corticotropin-releasing factor and other materials on adrenocorticotropin secretion from pituitary glands of patients with cushing's disease in vitro. *J. Clin. Endocrinol. Metabol.* **59**, 840–845. ⟨http://dx.doi.org/10.1210/jcem-59-5-840⟩.

Sutton, S. W., Behan, D. P., Lahrichi, S. L., Kaiser, R., Corrigan, A., Lowry, P., et al. (1995). Ligand requirements of the human corticotropin-releasing factor-binding protein. *Endocrinology* **136**, 1097–1102. ⟨http://dx.doi.org/10.1210/endo.136.3.7867564⟩.

Takahashi, A., Amano, M., Amiya, N., Yamanome, T., Yamamori, K. and Kawauchi, H. (2006). Expression of three proopiomelanocortin subtype genes and mass spectrometric identification of POMC-derived peptides in pars distalis and pars intermedia of barfin flounder pituitary. *Gen. Comp. Endocrinol.* **145**, 280–286. ⟨http://dx.doi.org/10.1016/j. ygcen.2005.09.005⟩.

Takei, Y. (2002). Hormonal control of drinking in eels: an evolutionary approach. *Symp. Soc. Exp. Biol.* 61–82.

Takei, Y. and Hwang, P.-P. (2016). Homeostatic Responses to Osmotic Stress. In *Fish Physiology - Biology of Stress in Fish*, Vol. 35 (eds. C. B. Schreck, L. Tort, A. P. Farrell and C. J. Brauner), San Diego, CA: Academic Press.

Tanaka, S. (2003). Comparative aspects of intracellular proteolytic processing of peptide hormone precursors: studies of proopiomelanocortin processing. *Zool. Sci.* **20**, 1183–1198.

Theusch, E. V., Brown, K. J. and Pelegri, F. (2006). Separate pathways of RNA recruitment lead to the compartmentalization of the zebrafish germ plasm. *Dev. Biol.* **292**, 129–141. ⟨http://dx.doi.org/10.1016/j.ydbio.2005.12.045⟩.

To, T. T., Hahner, S., Nica, G., Rohr, K. B., Hammerschmidt, M., Winkler, C., et al. (2007). Pituitary-interrenal interaction in zebrafish interrenal organ development. *Mol. Endocrinol. (Baltimore, Md.)* **21**, 472–485. ⟨http://dx.doi.org/10.1210/me.2006-0216⟩.

Tsalafouta, A., Papandroulakis, N., Gorissen, M., Katharios, P., Flik, G. and Pavlidis, M. (2014). Ontogenesis of the HPI axis and molecular regulation of the cortisol stress response during early development in *Dicentrarchus labrax*. *Sci. Rep.* **4**.⟨http://dx.doi.org/10.1038/ srep05525⟩.

Vale, W., Spiess, J., Rivier, C. and Rivier, J. (1981). Characterization of a 41-residue ovine hypothalamic peptide that stimulates secretion of corticotropin and beta-endorphin. *Science (New York, N.Y.)* **213**, 1394–1397.

van den Burg, E. H., Metz, J. R., Arends, R. J., Devreese, B., Vandenberghe, I., Van Beeumen, J., et al. (2001). Identification of beta-endorphins in the pituitary gland and blood plasma of the common carp (*Cyprinus carpio*). *J. Endocrinol.* **169**, 271–280.

van den Burg, E. H., Metz, J. R., Spanings, F. A. T., Wendelaar Bonga, S. E. and Flik, G. (2005). Plasma alpha-MSH and acetylated beta-endorphin levels following stress vary according to CRH sensitivity of the pituitary melanotropes in common carp, *Cyprinus carpio*. *Gen. Comp. Endocrinol.* **140**, 210–221. ⟨http://dx.doi.org/10.1016/j.ygcen.2004.11. 010⟩.

van Enckevort, F. H., Pepels, P. P., Leunissen, J. A., Martens, G. J., Wendelaar Bonga, S. E. and Balm, P. H. (2000). *Oreochromis mossambicus* (tilapia) corticotropin-releasing hormone: cDNA sequence and bioactivity. *J. Neuroendocrinol.* **12**, 177–186. ⟨http://dx. doi.org/10.1046/j.1365-2826.2000.00434.x⟩.

Vaughan, J., Donaldson, C., Bittencourt, J., Perrin, M. H., Lewis, K., Sutton, S., et al. (1995). Urocortin, a mammalian neuropeptide related to fish urotensin I and to corticotropin-releasing factor. *Nature* **378**, 287–292. ⟨http://dx.doi.org/10.1038/378287a0⟩.

Verburg-van Kemenade, B. M. L. and Schreck, C. B. (2007). Immune and endocrine interactions. *Gen. Comp. Endocrinol.* **152**, 352. ⟨http://dx.doi.org/10.1016/j. ygcen.2007.05.025⟩.

Volkoff, H., Canosa, L. F., Unniappan, S., Cerda-Reverter, J. M., Bernier, N. J., Kelly, S. P., et al. (2005). Neuropeptides and the control of food intake in fish. *Gen. Comp. Endocrinol.* **142**, 3–19. ⟨ http://dx.doi.org/10.1016/j.ygcen.2004.11.001 ⟩.

Wei, E. T., Thomas, H. A., Christian, H. C., Buckingham, J. C. and Kishimoto, T. (1998). D-amino acid-substituted analogs of corticotropin-releasing hormone (CRH) and urocortin with selective agonist activity at CRH1 and CRH2beta receptors. *Peptides* **19**, 1183–1190.

Wendelaar Bonga, S. E. (1997). The stress response in fish. *Physiol. Rev.* **77**, 591–625.

Westring, C. G., Ando, H., Kitahashi, T., Bhandari, R. K., Ueda, H., Urano, A., et al. (2008). Seasonal changes in CRF-I and urotensin I transcript levels in masu salmon: correlation with cortisol secretion during spawning. *Gen. Comp. Endocrinol.* **155**, 126–140. ⟨ http://dx. doi.org/10.1016/j.ygcen.2007.03.013 ⟩.

Wieser, W. (1985). A new look at energy conversion in ectothermic and endothermic animals. *Oecologia* **66** (4), 506–510.

Wilkinson, C. W. (2006). Roles of acetylation and other post-translational modifications in melanocortin function and interactions with endorphins. *Peptides* **27**, 453–471. ⟨ http://dx. doi.org/10.1016/j.peptides.2005.05.029 ⟩.

Wynn, P. C., Aguilera, G., Morell, J. and Catt, K. J. (1983). Properties and regulation of high-affinity pituitary receptors for corticotropin-releasing factor. *Biochem. Biophys. Res. Commun.* **110**, 602–608.

Yada, T. and Tort, L. (2016). Stress and Disease Resistance: Immune System and Immunoendocrine Interactions. In *Fish Physiology - Biology of Stress in Fish*, Vol. 35 (eds. C. B. Schreck, L. Tort, A. P. Farrell and C. J. Brauner), San Diego, CA: Academic Press.

Zhou, A., Bloomquist, B. T. and Mains, R. E. (1993). The prohormone convertases PC1 and PC2 mediate distinct endoproteolytic cleavages in a strict temporal order during proopiomelanocortin biosynthetic processing. *J. Biol. Chem.* **268**, 1763–1769.

4

THE MOLECULAR STRESS RESPONSE

ERIN FAUGHT
NEEL ALURU
MATHILAKATH M. VIJAYAN

1. Introduction
2. Molecular Regulation of the Hypothalamic–Pituitary–Interrenal (HPI) Axis
    2.1. Hypothalamus
    2.2. Pituitary
    2.3. Head Kidney (Interrenal Tissue)
3. Genomic Cortisol Signaling
    3.1. Glucocorticoid Receptor
    3.2. Mineralocorticoid Receptor
4. Genomic Effects of Cortisol
    4.1. Development of the Stress Axis
    4.2. Molecular Adjustments During Stress
    4.3. Cellular Adjustments
5. Significance of Molecular Responses
6. Approaches to Study Molecular Responses to Stress
    6.1. Mechanistic Studies Using Targeted Mutagenesis
    6.2. Epigenetic Regulation of Stress Response
7. Concluding Remarks and the Unknowns

In this chapter we summarize the key patterns observed in the transcript abundances of genes involved in the hypothalamic–pituitary–interrenal (HPI) axis regulation, as well as regulation of stress-responsive genes that are modulated by corticosteroid signaling in fish. Plasma levels of cortisol, the primary corticosteroid in teleosts, rise within minutes after a stressor encounter, and this steroid action is mediated primarily by genomic signaling involving the family of nuclear receptors including the glucocorticoid receptor (GR) and mineralocorticoid receptor (MR). Specific molecular responses in target tissues associated with activation of these corticosteroid receptors have become apparent by increased use of receptor antagonists and gene knockdown tools. Although molecular adjustments to stress are

113

Biology of Stress in Fish: Volume 35
FISH PHYSIOLOGY

dependent on species, developmental stage, type, and duration of stressor, overall cortisol action limits energy-demanding processes and enhances energy mobilization and reallocation. Recent work has also underscored the necessity of maternal cortisol and GR signaling as essential for fish development, but the underlying mechanisms are far from clear. While transcript abundance is an excellent indicator of pathway modulation by cortisol, this data by itself lacks physiological relevance unless accompanied by downstream protein and/or metabolite changes. As the field of molecular biology continues to advance, approaches including next-generation sequencing and gene editing tools will allow production of transgenic animals, even in nonmodel fish species, to reveal molecular mechanisms essential for stress adaptation. This fundamental knowledge may have relevance to improving the welfare of the organism in aquaculture and for protecting ecosystem health and biodiversity.

1. INTRODUCTION

The hormonal stress response has been extensively studied in teleosts and is a key mediator of the physiological adjustments essential to reestablish homeostasis (Wendelaar Bonga, 1997; Mommsen et al., 1999; Barton, 2002; Vijayan et al., 2010). The primary endocrine response is the rapid elevation of plasma epinephrine, the predominant catecholamine, and cortisol, the principal glucocorticoid, levels in response to stressor exposure (Reid et al., 1998; Mommsen et al., 1999). From a functional standpoint of physiological adjustments, stressor-induced plasma cortisol and epinephrine responses are temporally distinct; epinephrine increases within seconds and is also cleared rapidly from circulation, whereas cortisol is elevated within minutes to hours in response to a stressor challenge (Fig. 4.1A; Vijayan et al., 2010). The epinephrine action involves the rapid mobilization of glucose to fuel the stressor-induced energy demand (Fabbri and Moon, 2015). Target tissue response to cortisol stimulation, on the other hand, is slower and usually involves the synthesis of effector proteins that facilitate energy substrate mobilization and reallocation, including replenishment of depleted glycogen stores in fish (Mommsen et al., 1999; Vijayan et al., 2010).

Epinephrine signals through β-adrenoreceptors, a family of g-protein coupled receptors (GPCRs), and modulates cellular responses by phosphorylation/dephosphorylation of target proteins (Fabbri and Moon, 2015; Fig. 4.1B). Although epinephrine affects transcript changes, including genes involved in metabolism (Ings et al., 2012) and immune response (Castillo et al., 2009; Chadzinska et al., 2012), the predominant role of this hormone

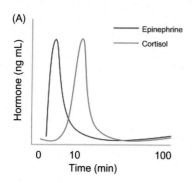

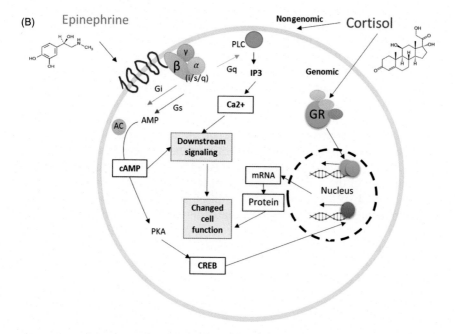

Figure 4.1. Stress hormone regulation of molecular responses to stress. (A) The stressor-mediated plasma epinephrine response is faster (seconds to minutes) than cortisol response (minutes to hours), and the former brings about rapid cellular responses compared to the steroid response, which mostly involves genomic signaling. (B) Epinephrine signals through g-protein coupled adrenergic receptors (α,β), which are expressed in a tissue-specific manner. The g-protein coupled receptors (GPCRs) cause secondary signaling cascades either by activation (Gs) or by inhibition (Gi) of adenylyl cyclase, which produces cAMP. Subsequent phosphorylation of PKA will cause activation of the transcription factor cAMP response element binding protein (CREB), leading to transcriptional regulation by binding CREB response elements (CREs). The genomic cortisol signaling is the most widely studied stress signaling pathway; however nongenomic signaling also occurs during stress, but the mechanisms are far from clear. In genomic cortisol signaling, the steroid binds to its cytosolic located nuclear receptor, which is present as a heterocomplex with other proteins, and two such ligand-bound receptors dimerize, translocate to the nucleus, and induce targeted gene transcription by binding to the glucocorticoid response elements (GREs).

is independent of transcriptional regulation in fish (Fabbri et al., 1998). In mammalian models, epinephrine activates the cAMP response binding protein (CREB), a key transcription factor involved in the metabolic regulation during stress (Thiel et al., 2005). However, the role of CREB in epinephrine-mediated stress signaling in piscine models is not clear (Fabbri and Moon, 2015).

Cortisol action is predominantly, but not exclusively, genomic and involves gene transcription and protein synthesis (Fig. 4.1B; Mommsen et al., 1999; Vijayan et al., 2005). The genomic action of cortisol is mediated by its activation of GRs and/or MRs in fish (see Section 3). Targeted gene expression as a molecular marker of stress and/or cortisol stimulation in teleosts has been widely used since 2000, and this was mostly driven by the advent of quantitative tools for genome analysis, including quantitative real-time polymerase chain reaction (PCR) and microarrays (Aluru and Vijayan, 2009). The majority of studies examined the steady-state transcript levels, while a mechanistic underpinning for the observed responses is still largely lacking in fish (Aluru and Vijayan, 2009). Despite this, the stressor-mediated molecular response underscore metabolic changes, including energy substrate repartitioning, as essential for stress adaptation (Aluru and Vijayan, 2009). However, this may occur at the expense of other energy-demanding pathways, including growth, health, and reproduction (Vijayan et al., 2010; Philip and Vijayan, 2015). As elevated plasma cortisol is a key characteristic of the primary stress response in teleosts, and since most of its action studied so far involves transcriptional regulation, this chapter will focus on the molecular responses mediated by this stress steroid in fish. Also, we will highlight the latest developments in genetic manipulations, genome and epigenome analysis, including reverse genetics techniques and next-generation sequencing, as a means to understand the mode of action of hormones and the associated stress-coping mechanisms.

2. MOLECULAR REGULATION OF THE HYPOTHALAMIC–PITUITARY–INTERRENAL (HPI) AXIS

The HPI axis in teleosts is synonymous to the hypothalamic–pituitary–adrenal (HPA) axis in mammals (Wendelaar Bonga, 1997). Although fish lack a discrete adrenal gland, the steroidogenic cells are distributed around the postcardinal vein in the head kidney region (interrenal tissue; Wendelaar Bonga, 1997). The stressor-induced HPI axis activation is highly conserved and culminates in the release of cortisol, the principle circulating corticosteroid in teleosts (Wendelaar Bonga, 1997; Mommsen et al., 1999;

Barton, 2002). Briefly, stressor input stimulates the release of corticotropin-releasing factor (CRF) from the hypothalamus, and this peptide hormone stimulates the release of adrenocorticotropic hormone (ACTH) from the anterior pituitary corticotrophs (Wendelaar Bonga, 1997). ACTH is produced by posttranslational modification of the precursor protein pro-opiomelanocortin (POMC), which is cleaved to produce multiple stress-related peptides, including ACTH, alpha-melanocyte-stimulating hormone (α-MSH), and β-endorphins (Takahashi and Kawauchi, 2006; Dores, 2013). These peptide levels in circulation have been used as markers of acute stress response in fish (Wendelaar Bonga, 1997). Although several hormones stimulate cortisol release from the head kidney in fish, ACTH is the primary secretagogue for corticosteroidogenesis (Wendelaar Bonga, 1997; Mommsen et al., 1999). ACTH binds to melanocortin 2 receptor (MC2R) on the steroidogenic cells in the interrenal tissue to activate the signaling cascade leading to cortisol production (Aluru and Vijayan, 2008). The molecular responses, including transcript changes of key genes involved in HPI axis regulation, as well as target tissue response to cortisol stimulation, have been extensively studied in teleosts (Fig. 4.2), and is outlined in the following sections.

2.1. Hypothalamus

The stressor-mediated upregulation of CRF in the hypothalamus constitutes the primary step in HPI axis activation (Flik et al., 2006). As in mammals, this neuropeptide is widely expressed not only in the central nervous system of teleosts (Alderman and Bernier, 2007), but also in peripheral tissues, including the head kidney (Huising et al., 2007). However, the principal site of hypophysiotropic CRF is the preoptic area (POA), which is analogous to the hypophysiotropic CRF neurons of the paraventricular nucleus in mammals (Pepels et al., 2002; Bernier et al., 2008). Teleosts also have an additional source of CRF-related peptides, the caudal neurosecretory system (CNSS), located in the terminal segments of the spinal cord (Bernier et al., 2008). While a physiological role for the caudal neurosecretory peptides during stress remains to be determined, urotensin I (UI) produced by the CNSS plays a role in HPI axis regulation (Craig et al., 2005; Bernier et al., 2008). UI stimulates cortisol production from the interrenals either directly (Arnold-Reed and Balment, 1994; Kelsall and Balment, 1998; Huising et al., 2007) or indirectly by increasing ACTH secretion from the pituitary (Fryer et al., 1983). These studies suggest a role for CRF-related peptides in the organismal stress response, while the majority of studies have focused on *CRF* transcript changes in relation to HPI axis function in fish (Flik et al., 2006).

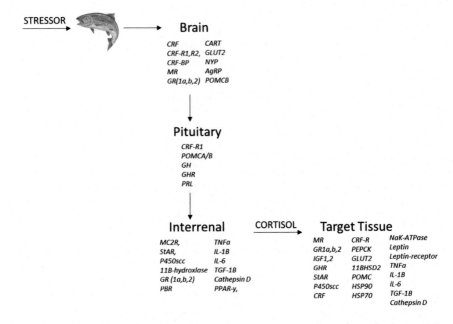

Figure 4.2. Stress-related genes regulated in response to HPI axis activation. *CRF*, corticotropin-releasing factor; *CRF-R1,R2*, CRF-receptors; CRF-BP, CRF-binding protein; *MR*, mineralocorticoid receptor; *GR(1a,b,2)*, glucocorticoid receptor; *CART*, cocaine and amphetamine-regulated transcript; *GLUT2*, glucose transporter 2; *NYP*, neuropeptide Y; *AgRP*, agouti-related protein; *POMCA/B*, pro-opiomelanocortin; *GH*, growth hormone; *GHR*, growth hormone receptor; *PRL*, prolactin; *MC2R*, melanocortin 2 receptor; *StAR*, steroidogenic acute regulatory protein; *P450scc*, P450 side chain cleavage; *PBR*, peripheral-type benzodiazepine receptor; *TNFa*, tumor necrosis factor a; *IL*, interleukin; *TGF*, transforming growth factor; *PPAR*, peroxisome proliferator-activated receptors.

CRF transcript levels are modulated by stress, but the changes have been anything but consistent and may reflect the variability in the stressor model utilized. For instance, while studies showed stressor-mediated elevation in *CRF* mRNA levels in the POA (Ando et al., 1999; Huising et al., 2004; Doyon et al., 2005, 2006; Craig et al., 2005; Fuzzen et al., 2010; Aruna et al., 2012; Wunderink et al., 2012), others failed to observe any changes in *CRF* transcript levels (Bernier et al., 2012; Ghisleni et al., 2012). Similarly, elevated plasma cortisol levels either increased (Madison et al., 2015), caused no change (Doyon et al., 2005), or suppressed *CRF* transcript levels (Bernier and Peter, 2001; Bernier et al., 2004; Yeh, 2015). However, a study showed that GR signaling is essential for maintaining *CRF* transcript abundance and stress-mediated cortisol production in rainbow trout (*Oncorhynchus mykiss*) (Doyon et al., 2006; Alderman et al., 2012). For instance, blocking GR signaling with mifepristone reduced *CRF* transcript abundance in the POA of trout (Alderman et al., 2012) and zebrafish (Yeh, 2015), leading to the proposal that other control mechanisms may be in place, including a role for MR signaling in the negative feedback regulation to restore cortisol homeostasis (Alderman et al., 2012; Alderman and Vijayan, 2012). Studies also implicate a role for CRF-binding proteins in regulating the bioavailability of CRF during stress (Flik et al., 2006; Bernier, 2006; Alderman and Bernier, 2009; Bernier et al., 2012; Wunderink et al., 2012), but this remains to be verified empirically.

Nutritional status also influences *CRF* transcript levels in the fish hypothalamus. For instance, elevated CRF and UI reduces feed intake and this involves CRF receptor activation (Bernier and Peter, 2001; Ortega et al., 2013). Also, CRF-binding protein decreased in response to food deprivation, suggesting an increase in CRF bioavailability to suppress feeding (Bernier et al., 2012; Wunderink et al., 2012). Although plasma cortisol levels did not correlate to gene changes in these studies, it is reasonable to hypothesize that elevated CRF reflects increased HPI axis activation. As elevated plasma cortisol level modulates genes encoding feeding-related peptides, including POMC, neuropeptide Y, and cocaine- and amphetamine-regulated transcript abundance in the hypothalamus (Gesto et al., 2014) and leptin in the liver (Madison et al., 2015), the results point to an interplay between appetite-related peptides and CRF in the stress axis functioning in fish. Recently we showed that treatment with venlafaxine, a potent selective serotonin norepinephrine reuptake inhibitor, elevated serotonin, norepinephrine, and dopamine content in the midbrain of trout (Melnyk-Lamont et al., 2014). This treatment led to a higher *CRF* transcript abundance in the POA along with reduced feed intake (Melnyk-Lamont et al., 2014). A similar increase in POA *CRF* transcript level was also seen with fluoxetine, a selective serotonin reuptake inhibitor, in goldfish (*Carassius auratus*) (Mennigen et al., 2010). Together, these results suggest a role for monoamines along with other neuropeptides in modulating *CRF* transcript levels in response to stress in fish (Bernier et al., 2012; Ortega et al., 2013).

2.2. Pituitary

The anterior pituitary is the primary source of ACTH production in response to stressor-mediated CRF stimulation (Flik et al., 2006; Dores, 2013). CRF binds to CRF-receptors (CRF-Rs), a GPCR, in the adenohypophysis and stimulates the release of ACTH (Baker et al., 1996). There are only two isoforms of CRF-Rs in fish—CRF-R1 and CRF-R2—but their distinct functional roles in the stress response is not clear (Dores, 2013; Manuel et al., 2014). CRF signaling leads to the transactivation of *POMC* gene, encoding the precursor protein that is posttranslationally modified to produce the melanocortin family of hormones, including ACTH and α, β, γ, and δ-MSH (Dores and Baron, 2011). Both elevated plasma cortisol and ACTH levels modulate *POMC* transcript levels in the pituitary (Nakano et al., 2013; Laiz-Carrión et al., 2009; Aruna et al., 2012; Madison et al., 2015), supporting a negative feedback control of the hormone (Mommsen et al., 1999). Interestingly, in response to chronic stressors, including hypoxia and fasting, pituitary *POMC* transcript remained elevated for days poststressor exposure (Bernier et al., 2012). Long-term fasting also consistently upregulated *POMC* mRNA abundance in the trout hypothalamus, suggesting a role for this protein/peptide as satiety signals to reduce energy expenditure (Jorgensen et al., 2015). Consequently, an important role for this gene transcript changes during the stress response may involve appetite control and regulation of energy homeostasis.

Zebrafish (*Danio rerio*) represent an excellent model to address mechanistic questions pertaining to hypothalamus and pituitary functions. This is because of the ease of manipulating the embryos for forward and reverse genetics studies. To this end, we recently showed that knockdown of GR (via morpholino oligonucleotides) in zebrafish affected several molecular pathways, including those involved in stress axis function, suggesting a key role for cortisol signaling in regulating *CRF* and *POMC* transcript levels (Nesan and Vijayan, 2013a,b). Also, manipulation of zygotic cortisol content, mimicking maternal transfer of cortisol, modulated *CRF* and *POMCA* transcript levels in zebrafish embryos (Nesan and Vijayan, 2016). Specifically, lower zygotic cortisol content led to a higher *CRF* and *POMCA* transcript levels in zebrafish embryos, while both these transcripts were lower with elevated cortisol content. Changes in these key stress-responsive genes during early development corresponded with altered cortisol response to a stressor in zebrafish larvae, suggesting a role for these transcripts in the programming of stress axis function posthatch in zebrafish (Nesan and Vijayan, 2016).

The importance of the pituitary in stress regulation was also evident from the rx3 zebrafish mutant (Table 4.1), which carries a mutation in the retinal homeobox gene 3 (encoded for by the chokh gene; Dickmeis et al., 2007). They have severe impairment of eye and retinal development and a lack of the corticotrope/melanotrope cells in the pituitary (Dickmeis et al., 2007). This mutant exhibits lower cortisol levels, which is in agreement with a lack of corticotropes and the associated absence of *POMC* expression. The use of this mutant has underscored the importance of the HPI axis activation and cortisol in the regulation of peripheral circadian rhythms (Dickmeis et al., 2007; Dickmeis, 2009). Specifically, cortisol is playing a key role in the regulation of peripheral clocks to generate cell cycle rhythms (Dickmeis et al., 2007). Also, the circadian clock in the adrenals may be regulating the

Table 4.1

Pharmacological and genomic techniques employed to analyze the molecular response to stress

Agent/method	Method/antagonist	Target	Sources
Pharmacological:			
RU486 (Mifepristone)	Cortisol	GR	Medeiros et al. (2014), Philip et al. (2012), Alderman et al. (2012), Li et al. (2012), Tipsmark et al. (2009), Aluru and Vijayan (2007)
Spironolactone	Cortisol, 11-deoxycorticoterone	MR	Kiilerich et al. (2007)
18β-Glycyrrhetinic acid	Catalyzes (Cortisol—cortisone)	11βHSD2	Alderman et al. (2012)
Metyrapone	Catalyzes (11-deoxycortosol—cortisol)	11β-hydroxylase	Rodela et al. (2011)
Genomic:			
Morpholino	GR morpholino	GR	Pikulkaew et al. (2011), Nesan and Vijayan (2012), Chatzopoulou et al. (2015)
TALENS	SR4G transgenic line	GRE reporter line	Krug et al. (2014)
Zinc finger nucleases			No publications to date
Large-scale mutagenesis (ethylnitrosourea)	GR^{s357} (previously utouto)	GR disruption (allele 357)	Muto et al. (2005), Ziv et al. (2013), Muto et al. (2013), Griffiths et al. (2012)
Large-scale mutagenesis (ethylnitrosourea)	Chokh/rx3	Reduction of corticotropes in the anterior pituitary (allele s399 of the *chk* locus)	Loosli et al. (2003), Dickmeis et al. (2007), Dickmeis et al. (2011)
CRISPR/Cas9			No publications to date

sensitivity of steroidogenic cells to ACTH stimulation, leading to the rhythmicity in cortisol release (Dickmeis, 2009). While the rhythmicity in cortisol release has been shown in zebrafish larvae (Yeh, 2015), the rx3 mutants appear to be an excellent model to study the role of POMC in stress axis functioning during early development. Surprisingly, studies on HPI axis activation and cortisol regulation is still lacking with this mutant and may be in part because of the short lifespan (3–4 weeks), precluding long-term studies. Overall, zebrafish is an excellent model for teasing out the mechanisms of action of CRF and POMC in the regulation of HPI axis development and function.

2.3. Head Kidney (Interrenal Tissue)

The head kidney is the primary source for corticosteroids in lower vertebrates, and is analogous to the mammalian adrenal cortex (Wendelaar Bonga, 1997; Mommsen et al., 1999). ACTH binds to MC2R, a GPCR, on the steroidogenic cells dispersed throughout the head kidney (Aluru and Vijayan, 2008; Dores et al., 2014). While other melanocortin families of proteins bind multiple ligands, MC2R is unique in only having a single ligand (Aluru and Vijayan, 2008; Dores et al., 2014). Recent studies have shown that MC2R function in fish is associated with the availability of two melanocortin receptor accessory proteins (MRAPs), MRAP1 and MRAP2 (Cerdá-Reverter et al., 2013; Dores and Garcia, 2015). The receptor activation in response to ACTH stimulation requires the presence of MRAP1, but not MRAP2 (Liang et al., 2013; Agulleiro et al., 2013; Dores et al., 2016). MRAP-independent signaling of MC2R has only been reported for the elephant shark (*Callorhynchus milii*) MC2R gene expressed in CHO cells (Reinick et al., 2012). Overall, these results underscore the specificity of ACTH in MC2R activation, and a critical role for the coevolution of MC2R and MRAPs in regulating stress axis functioning (Cerdá-Reverter et al., 2013; Dores and Garcia, 2015). All these studies on fish MC2R activation were carried out in vitro using reporter assays with mammalian cells (Liang et al., 2015; Dores and Garcia, 2015; Dores et al., 2016), while little is known about the regulation of MRAPs in response to stress in fish in vivo.

Recently, we showed that head kidney *MRAP1*, but not *MRAP2*, transcript levels are transiently elevated in response to an acute stress in rainbow trout (N. Sandhu, L. Liang, B. Dores and M.M. Vijayan, in preparation). The *MRAP1* transcript level closely followed the transient increase in plasma cortisol levels following stress in trout, leading to the proposal that MRAP1 regulation may be a critical part of the acute stress response for MC2R activation and cortisol production (Sandhu et al., in preparation). Studies have shown *MC2R* transcript levels to be impacted by stressors, including contaminants in fish (Aluru and Vijayan, 2008; Hontela and Vijayan, 2009; Sandhu et al., 2014; Jeffrey et al., 2014). This correlated with attenuation of stressor-mediated cortisol production in vivo and ACTH-stimulated cortisol production in vitro (Hontela and Vijayan, 2009; Aluru and Vijayan, 2008; Sandhu and Vijayan, 2011). However, the mechanisms leading to disruption of MC2R signaling and impaired cortisol response to contaminants are unclear. Given the emerging picture that

MC2R signaling is tightly regulated by their accessory proteins, it is likely that disruption of MC2R activation may involve multiple targets, including MRAPs, but this remains to be determined. The role of MRAP2 in MC2R signaling is still unclear. However, a recent study showed that MC4R may act as an ACTH receptor by coexpression of MRAP2a in zebrafish (Agulleiro et al., 2013), suggesting multiple control mechanisms that may modulate corticosteroidogenesis in fish.

Although cortisol production in response to ACTH signaling has been examined both in vivo and in vitro extensively, the signaling pathways involved in the modulation of interrenal steroidogenesis are far from clear (Hontela and Vijayan, 2009). Both cAMP analogs and calcium modulate corticosteroid synthesis (Sandhu and Vijayan, 2011; Alsop et al., 2009; Lacroix and Hontela, 2006; Hontela and Vijayan, 2009), suggesting a cross-talk between multiple signaling pathways in the modulation of cortisol biosynthesis. There is no known antagonist for MC2R, making it difficult to elucidate key mechanisms involved in downstream signaling postligand binding. There have been some reports in mammals that a truncated ACTH molecule will act to partially antagonize the function of the receptor (Bouw et al., 2014); however, this has yet to be tested in fish. Most studies have employed the use of either a cAMP analog (8-bromo-cyclic AMP) or the activator of adenylyl cyclase, forskolin (Alsop et al., 2009) to stimulate ACTH-induced cortisol production in the interrenal tissues (Hontela and Vijayan, 2009). In fish, ACTH activation of MC2R is correlated with transcript abundance of the steroidogenic acute regulatory protein (StAR), a rate-limiting step responsible for transporting cholesterol across the outer to the inner mitochondrial membrane; P450 side chain cleavage (P450scc), the rate-limiting enzyme in steroidogenesis catalyzing the conversion of cholesterol to pregnenolone; and 11β-hydroxylase (cyp11A1), the terminal step catalyzing the conversion of 11-deoxycortisol to cortisol (Kusakabe et al., 2009; Geislin and Auperin, 2004; Hagen et al., 2006; Aluru and Vijayan, 2006, 2008; Castillo et al., 2008; Fuzzen et al., 2010). All these steroidogenic genes are generally well conserved (Arukwe, 2008; Dores, 2013), and StAR shows 100% homology for the predicted protein kinase A phosphorylation sites between rainbow trout and humans (Arukwe, 2008).

The steroidogenic gene expression patterns are not always consistent, as studies have failed to see an increase in transcript abundance of steroidogenic genes in response to acute stressors in vivo or ACTH stimulation in vitro (Geislin and Auperin, 2004; Hagen et al., 2006). While the reason for the inconsistency in these transcript responses to stress is not known, there may be several factors including the steroid levels that may be modulating the transcript abundance (Castillo et al., 2008). Most corticosteroids in teleosts reach their peak concentration in plasma around 1–4 h after an acute stressor, and return to basal concentrations within 24 h (Vijayan et al., 2010). There exist an ultrashort-loop control of cortisol production at the level of the interrenal tissue (Bradford et al., 1992), and this may involve regulation of steroidogenic genes. In support of this, we observed a direct involvement of GR signaling in the regulation of head kidney *StAR* transcript levels in rainbow trout (Alderman et al., 2012). However, it is unclear how GR signaling regulates *StAR* transcript abundance, as a putative glucocorticoid response element (GRE) in the promotor region of this gene has not been reported yet. One possibility is that cortisol may be acting through rapid nongenomic pathways, including regulation of CREB, to affect *StAR* transcription (Manna et al., 2003; Clark and Cochrum, 2007). In support of this notion, cortisol exposure rapidly increases CREB phosphorylation in trout hepatocytes (L. Dindia and M.M. Vijayan, in preparation), but whether this is involved in the regulation of *StAR* transcription in response to cortisol stimulation remains to be determined.

A consistent suppression of steroidogenic genes, including StAR and P450scc, was observed in response to adrenotoxicants suppression of corticosteroidogenesis in fish (Hontela and Vijayan, 2009). These genes are considered key targets for chemical stressors-mediated suppression of cortisol production in fish (Aluru et al., 2005; Aluru and Vijayan, 2006; Gravel and Vijayan, 2006; Sandhu and Vijayan, 2011; Best et al., 2014; Sandhu et al., 2014). The mode of action of contaminants in affecting these key steps in steroidogenesis is not known. Recently we showed that cadmium disruption of steroidogenesis occurs at the level of MC2R signaling, as a cAMP analog and forskolin were able to rescue *StAR* and *P450scc* transcript levels in response to this metal exposure (Sandhu and Vijayan, 2011). While contaminant effect on upstream targets may limit downstream protein activity, we cannot discount a direct effect on target proteins, especially given that chemical stressors directly impact *StAR* transcription (Clark and Cochrum, 2007).

Longer-term exposure studies with contaminants revealed a consistent activation of the cortisol biosynthetic machinery, as evidenced by the higher *StAR* and *p450scc* transcript levels in rainbow trout (Ings et al., 2011a; Best et al., 2014; Sandhu et al., 2014). However, this did not correlate with elevated cortisol levels, suggesting a mismatch in steroid production and the steroidogenic capacity. A major limiting factor currently is the lack of tools to measure protein expression for these target genes in fishes, given that transcript changes may not necessarily reflect the functional capacity of the tissue. This is because gene expression may be dependent on factors that influence mRNA stability and turnover (Roy and Jacobson, 2013), and this is one area where studies are lacking in fish. Another possible scenario for the mismatch between steroidogenic capacity and steroid production may involve modulation of 11β-hydroxysteriod dehydrogenase type 2 (*11βHSD2*), a key enzyme that catalyzes the conversion of cortisol into the biologically inactive cortisone, during long-term exposure to contaminants. Indeed, in mammalian choriocarcinoma cells treated with cadmium, a lower cortisol concentration was observed in the culture media and this correlated with higher *11βHSD2* mRNA abundance (Ronco et al., 2010). Overall, stressors activate the HPI axis, leading to molecular responses supportive of enhanced corticosteroidogenic capacity in the head kidney tissues of fish. Contaminants disrupt cortisol production and this appears to be a targeted response at multiple sites along the corticosteroidogenic pathway. While there is value in using transcript responses as molecular markers of steroidogenesis, the type and intensity of stressors and the duration of exposure may influence their expression patterns. The mechanisms leading to consistent upregulation of steroidogenic transcripts over long periods, as well as the consequence to corticosteroid biosynthetic capacity, remains to be determined.

3. GENOMIC CORTISOL SIGNALING

Several reviews have been published on corticosteroid receptors and cortisol signaling in fish (Vijayan et al., 2005; Prunet et al., 2006; Stolte et al., 2006; Bury and Sturm, 2007; Alsop and Vijayan, 2009; Aluru and Vijayan, 2009; Schaaf et al., 2009; Nesan and Vijayan, 2013a). In general, cellular action of cortisol involves a combination of genomic and nongenomic signaling (Fig. 4.1B). While GR and MR are involved in genomic cortisol signaling, the mechanism of action and the cellular consequences in teleosts are far from clear (Borski, 2000; Dindia et al., 2012, 2013). Recent studies point to multiple modes of action for rapid cortisol-mediated cellular effects, including membrane biophysical alterations to rapidly activate stress signaling pathways (Dindia et al., 2012, 2013) and membrane receptor-mediated activation of stress signaling pathways (Borski, 2000). However, a membrane corticosteroid receptor is yet to be discovered (Dindia et al., 2013). Overall, the majority of cortisol effects reported are mediated by genomic signaling involving GR and MR activation.

3.1. Glucocorticoid Receptor

GR is the primary receptor for glucocorticoid action in teleosts, and was first cloned and sequenced in rainbow trout and the receptor is expressed in every tissue (Ducouret et al., 1995). Subsequently, multiple isoforms of GR have been cloned and sequenced in a number of species, including a cichlid fish (*Haplochromis burtoni*; Greenwood et al., 2003), sea bass (*Dicentrarchus labrax*; Terova et al., 2005), fathead minnow (*Pimephales promelas*; Filby and Tyler, 2007), gilthead seabream (*Sparus aurata*; Acerete et al., 2007), common carp (*Cyprinus carpio*; Stolte et al., 2008), and midshipman fish (*Porichthys notatus*; Arterbery et al., 2010). The multiple isoforms are thought to have risen from the whole genome duplication ~350 million years ago (Alsop and Vijayan, 2009). Teleosts carry two genes, *GR1* and *GR2*, encoding GR (Bury et al., 2003; Vijayan et al., 2005; Prunet et al., 2006), with the exception of zebrafish, which has only one GR in the genome (Alsop and Vijayan, 2008, 2009; Schaaf et al., 2008). A splice variant of GR2, α and β, have been characterized in *Haplochromis burtoni* (Greenwood et al., 2003), as well as a GRβ splice variant in zebrafish (Schaaf et al., 2008; Alsop and Vijayan, 2009). The affinity between GR1 and GR2 differ, with GR2 being more sensitive to lower concentrations of both cortisol and dexamethasone (Bury et al., 2003). Although the GR antagonist mifepristone inhibited transactivation activity of both GR1 and GR2, GR1 was more sensitive to this antagonist (Bury et al., 2003). In spite of these differences in receptor affinities and ligand specificity, which were mostly determined by in vitro reporter assays (Bury et al., 2003), a functionally distinct role for the two GRs during stress has not been clearly established in fish (Prunet et al., 2006). A recent study using a GR knockdown approach demonstrated for the first time the transcriptome and metabolome changes specific to GRα and GRβ activation in zebrafish embryos (Chatzopoulou et al., 2015). This study underscores a key role for GRβ in the negative regulation of GRα signaling in zebrafish (Chatzopoulou et al., 2015).

In mammals, GR exists in the cytoplasm as a heterocomplex bound to a dimer of Hsp90, one molecule each of Hsp70, immunophilin, and p23 (Pratt et al., 2006). In steroid free cells the receptor will shuttle continuously between the cytoplasm and the nucleus (Madan and DeFranco, 1993); however, binding of a steroid to its receptor will cause a rapid translocation and binding to GREs. Classical unliganded GR resides in the cytoplasm, while MR is seen both in the cytoplasm and the nucleus. Cortisol diffuses through the plasma membrane and binds to the receptor-chaperone complex in the cytoplasm. The chaperones will then dissociate from the receptor, and the ligand-receptor complex translocates to the nucleus and binds to the GRE on target genes to elicit a molecular response (Pratt et al., 2006; Fig. 4.1). While little information is available on the GR heterocomplex in fish, the GRE, a specific 15 bp, imperfect, palindromic DNA motif (AGAACA nnnn TGTTCT), is highly conserved (Esbaugh and Walsh, 2009). The GRE acts as an allosteric activator by providing a scaffold to bind GR in the correct position (Schoneveld et al., 2004). GR will also exert repression of gene transcription by modulating the transactivating properties of other transcription factors, including NF-kB and/or AP-1 (Schoneveld et al., 2004). While few studies have characterized target gene regulation by glucocorticoid (Esbaugh and Walsh, 2009), bioinformatic analysis identified several genes with upstream GREs in fish (Aluru and Vijayan, 2007; Philip and Vijayan, 2015; Alderman et al., 2012; Table 4.2). This, along with the availability of new tools for genome analysis (Section 6), provides possible avenues in the near future for mechanistic characterization of target gene regulation by cortisol during stress in fish.

The transcript abundance of GR is differentially regulated by stress, and most studies have shown an upregulation of *GR* transcript levels in response to stress and/or cortisol treatment (Vijayan et al., 2005, 2010; Prunet et al., 2006; Bury and Sturm, 2007; Stolte et al., 2008; Alderman et al., 2012).

Table 4.2

Fish genes with putative glucocorticoid response elements (GREs) in the promoter region

Gene	Species	Source
11βHSD2	*Danio rerio*	Alderman et al. (2012)
Lactate dehydrogenase B	*Fundulus heteroclitus*	Schulte et al. (2000)
a-Amylase	*Lates calcarifer*	Ma et al. (2004)
Growth hormone 2[a]	*Oncorhynchus mykiss*	Yang et al. (1997)
Prolactin	*Oncorhynchus mykiss*	Argenton et al. (1996)
Suppressors of cytokine signaling 1, 2	*Oncorhynchus mykiss*	Philip and Vijayan (2015)
Metallothionein 1	*Oncorhynchus mykiss*	Olsson et al. (1995)
PEPCK[b]	*Oncorhynchus mykiss*	Aluru and Vijayan (2007)
Glutamine synthase1,2,4[b]		
Arginase[b]		
Glutamine decarboxylase 65[b]		
Cysteine-rich protein[b]		
Hsp90[b]		
Thyroid hormone receptor a[b]	*Oncorhynchus mykiss*	Aluru and Vijayan (2007)
Metallothionein[b]		
P4502M1[b]		
Fibroblast growth factor[b]		
Lipoprotein receptor[b]		
Aryl hydrocarbon receptor-α+β[b]		
Vitelline envelope protein-B[b]		
Glucocorticoid receptor	*Haplochromis burtoni*	Greenwood et al. (2003)
Glutamine synthetase	*Opsanus beta*	Esbaugh and Walsh (2009)
Cytochrome P450 1A (CYP1A)	*Platichthys flesus*	Williams et al. (2000)

[a]Also contain cAMP response elements (CRE) in the promoter region.
[b]Regulated genes considered GR-dependent as a combination of cortisol and RU486 abolished cortisol-mediated transcription (bioinformatics analysis of the promoter region was not carried out).

While in general the two GR paralogs in fishes show similar expression patterns, differential regulation of the paralogs in a tissue-specific manner in response to stressors has also been reported (Stolte et al., 2008). However, the functional relevance associated with the differential GR isoform expression remains to be determined. The *GR* transcript upregulation seen with stress may be a direct result of elevated cortisol levels and the associated activation of the GR transcriptional machinery (Sathiyaa and Vijayan, 2003). Studies have shown that while *GR* transcript levels are elevated, GR protein expression is downregulated in response to cortisol stimulation (Vijayan et al., 2003; Sathiyaa and Vijayan, 2003). The GR protein downregulation is due to cortisol-mediated increase in proteosomal degradation (Sathiyaa and Vijayan, 2003). Interestingly, inhibition of proteosomal breakdown abolishes cortisol-mediated downregulation of GR protein and upregulation of *GR* transcript levels. The mismatch between *GR* transcript levels and GR protein expression in response to cortisol treatment was seen both in vivo in trout liver and in vitro in trout hepatocytes, suggesting autoregulation of GR as a key cellular adaptation to stress in fish (Vijayan et al., 2003; Sathiyaa and Vijayan, 2003). Overall, GR expression in target tissues may be tightly regulated by a negative feedback mechanism, given the importance of this receptor in wide-ranging functions, including growth and metabolism, immune, stress, and osmotic and ionoregulatory performances (Vijayan et al., 2010). However, at the mechanistic level, GR regulation in target cells in response to stress and glucocorticoid stimulation, and the associated cellular responses remains to be determined in piscine models.

3.2. Mineralocorticoid Receptor

Teleosts lack aldosterone, the primary ligand for MR signaling in mammals (Funder, 1997). Although MR binds cortisol with high affinity, the target tissue for MR action in mammals will have $11\beta HSD2$, which degrades cortisol to cortisone, thereby allowing aldosterone-MR signaling (Funder, 1997). As in mammals, $11\beta HSD2$ is thought to play an important role in regulating cortisol bioavailability in fishes (Kusakabe et al., 2003). This gene contains GRE in its promoter region, and is upregulated in response to excess cortisol stimulation (Alderman and Vijayan, 2012; Li et al., 2012). This enzyme is also important in regulating cortisol levels during early development in zebrafish (Alsop and Vijayan, 2008), as well as controlling maternal cortisol incorporation into the yolk during oogenesis in trout (Li et al., 2012).

MR is widely expressed in fish tissues, and studies suggest the possibility that 11-deoxcorticosterone may act as an MR agonist (Milla et al., 2008; Stolte et al., 2008; Pippal et al., 2011). However, the physiological relevance of this hormone action is still unclear in fish (Stolte et al., 2008). While MR has been cloned and sequenced in a number of species, including rainbow trout (Colombe et al., 2000), zebrafish (Alsop and Vijayan, 2008), carp (Stolte et al., 2008), and *Haplochromis* (Greenwood et al., 2003), the action of MR and the physiological effects, especially on ion regulation, are far less clear (McCormick et al., 2008; Cruz et al., 2013). Most species seem to have a single MR so far in spite of genome duplication, suggesting loss of the extra receptor from the teleost genome (Alsop and Vijayan, 2008). While the genomic actions of cortisol via GR and the associated target genes regulated have been studied, less is known about MR-mediated cortisol signaling in fish (Takahashi and Sakamoto, 2013). The overriding evidence seems to suggest a dual role for cortisol as both a glucocorticoid and mineralocorticoid, and some of these effects may be mediated by MR signaling especially in tissues, and under conditions, where maintaining osmotic balance is paramount (Takahashi and Sakamoto, 2013; Kiilerich et al., 2007; McCormick et al., 2008; Kelly and Chasiotis, 2011). A recent study showed that at least in vitro GR and MR interact to modulate transcriptional control, suggesting the possibility that both these receptors may also be involved in modulating target tissue cortisol responsiveness during stress in fish (Kiilerich et al., 2015). This seems plausible in vivo given that both these receptors are expressed in most tissues in fish (Aruna et al., 2012).

Recent studies also propose a role for cortisol signaling via MR activation on the negative feedback regulation of cortisol during stress in fish (Alderman et al., 2012; Alderman and Vijayan, 2012), as well as a role for this receptor in early development (Alsop and Vijayan, 2008; Nesan and Vijayan, 2013a) and behavior (Manuel et al., 2014, 2015). While GR transcripts are maternally deposited in zebrafish embryos and their levels do not increase until hatch (Alsop and Vijayan, 2008), *MR* transcripts are deposited in very low amounts and gradually increase during embryogenesis. This mismatch in *GR* and *MR* transcript levels during embryogenesis led to the proposal that MRs may have an important role in early development (Alsop and Vijayan, 2008; Nesan and Vijayan, 2013a), but this remains to be tested. Overall, the role of MR signaling, the ligands that mediate the response, and the target genes that are affected are still unclear, but so far the evidence seems to suggest cortisol as a key mediator of MR effects in fish.

4. GENOMIC EFFECTS OF CORTISOL

The genomic cortisol signaling has wide-ranging effects in fish (Aluru and Vijayan, 2009); in this book it is discussed in the context of osmoregulation and acid base balance (Takei and Hwang, 2016; Chapter 6 in this volume), reproduction (Pankhurst, 2016; Chapter 8 in this volume), and health (Yada and Tort, 2016; Chapter 10 in this volume). To limit overlap with other chapters, this chapter will focus on the molecular responses that are cortisol-mediated and essential in the context of early development and acute stress performance.

4.1. Development of the Stress Axis

Recent studies clearly point to a key role for cortisol in the molecular regulation essential for early development in teleosts (Nesan and Vijayan, 2013a). Much of the work has been in zebrafish due to their rapid development, transparent embryos, and the utility of this species for forward and reverse genetics studies (Nesan and Vijayan, 2013a). The chronology of HPI gene expression, including *CRF* (Chandrasekar et al., 2007), *POMC* (Herzog et al., 2003; Liu et al., 2003), *StAR*, *MC2R* (To et al., 2007), *11β-hydroxylase*, *MR*, *GR*, and *11βHSD2* (Alsop and Vijayan, 2008), during early development has been described in zebrafish (Alsop and Vijayan, 2009; Nesan and Vijayan, 2013a). Despite all the molecular machinery for HPI axis function in place by hatch (48 hpf) in zebrafish (Alsop and Vijayan, 2009), the earliest stress response occurs only at around 72 hpf, pointing to a delay in HPI axis activation (Alderman and Bernier, 2009; Alsop and Vijayan, 2009; Nesan and Vijayan, 2013a). This supports the concept of a stress hyporesponsive period during embryogenesis as essential for proper developmental programming in animal models (see review by Nesan and Vijayan, 2013a).

Both *GR* and cortisol are maternally transferred and de novo synthesis of this protein and steroid commences during gastrulation (>8 hpf) and posthatch, respectively (Nesan and Vijayan, 2013a). Indeed, the maternal contribution of GR and cortisol are essential for developmental programming (Nesan et al., 2012; Nesan and Vijayan, 2012, 2013a,b; Pikulkaew et al., 2011). The importance of GR signaling was established by transiently knocking down GR using morpholino oligonucleotides and studying the associated changes in genotype and phenotype (Pikulkaew et al., 2011; Nesan and Vijayan, 2012, 2013b). GR signaling, presumably by maternal cortisol stimulation, affects early genome changes and is also playing a role in the mid-blastula transition (MBT), a key period for transcriptional regulation of developmental genes (Pikulkaew et al., 2011; Nesan et al., 2012). Also, GR signaling is involved in the regulation of a number of key developmental genes post-MBT stage in zebrafish as indicated by changes in genome profiles using microarray (Nesan and Vijayan, 2013b). The transcriptome analysis during embryogenesis (36 hpf) revealed several pathways regulated by GR signaling, including nervous system develop-ment, cellular movement, cell-to-cell signaling, cardiovascular development, organ morphology, and skeletal/muscular system development (Nesan and Viayan, 2013a,b). While mechanistic studies to demonstrate direct regula-tion of these pathways by cortisol signaling is still wanting, there are indications that bone morphogenetic proteins (BMPs) may be a key morphogen regulated by maternal cortisol during early zebrafish develop-ment (Nesan and Vijayan, 2012). BMPs are involved in all aspects of early development, including organogenesis, anteroposterior endoderm pattern-ing (Tiso et al., 2002; Poulain et al., 2006), mesoderm patterning (Neave et al., 1997), and inner ear and lateral line formation (Mowbray et al., 2001), supporting a key role for maternal cortisol in developmental programming (Nesan and Vijayan, 2013a,b). Beyond the transient knockdown of morpholinos, stable transgenic zebrafish knockout lines have also recently been established (Table 4.1). Specifically, GRs357 mutants have substitution of arginine to cysteine in the second zinc finger motif of the DNA binding domain of GR, which abolishes the ability of the mutant GR to bind to the GREs of target genes (Ziv et al., 2013). This mutant has chronically elevated CRF and cortisol, essentially mimicking a chronic stress scenario, and exhibit behavioral phenotype akin to depression, including decreased exploratory behavior and impaired habituation to repeated exposure to an anxiogenic environment (Ziv et al., 2013). This suggests that cortisol is playing an important role in brain development and behavioral phenotypes, while the molecular mechanisms linking genotype to phenotype remains to be determined.

While maternal cortisol is essential for developmental programming, abnormal levels of zygotic cortisol content, mimicking a maternal stress and excess cortisol transfer scenario, is detrimental to the offspring (Nesan and Vijayan, 2012, 2016; Wilson et al., 2013, 2015). Maternal stress and increased cortisol levels led to excess transfer of cortisol to the embryos in fish (Geising et al., 2011; Stratholt et al., 1997; Mccormick, 1999; Eriksen et al., 2006; Kleppe et al., 2013; Faught et al., 2016). Recent studies suggest that maternal cortisol is extremely important for proper heart development (Nesan and Vijayan, 2012; Wilson et al, 2013, 2015). For instance, cortisol microinjected into one-cell embryos to elevate basal cortisol levels mimicking maternal stress cause heart deformities, which was correlated to suppression of key cardiac genes, including nkx2.5 cardiac myosin light chain 1, cardiac troponin type T2A, and calcium transporting ATPase (Nesan and Vijayan, 2012). These results suggest a role for cortisol in cardiac development; however, it is unclear how these genes are inhibited by excess cortisol during early development (Nesan and Vijayan, 2012). We are also seeing that excess zygotic cortisol content, mimicking maternal stress, increases forebrain neurogenesis in zebrafish embryos, and this corresponded with increased spatial abundance of neurogenic markers, neuroD and orthopedia b in zebrafish embryos (C. Best, D. Kurassch and M.M. Vijayan, in preparation). Also, these fish were anxiolytic leading to the proposal that maternal cortisol-mediated neurogenesis may be playing a key role in the development of behavioral phenotypes. These results underscore the critical role for tight regulation of maternal cortisol in proper developmental programming, because stress and excess cortisol deposition affect brain function, leading to abnormal behavioral and cardiac phenotypes.

Recent studies also highlight a role for 20β-hydroxysteriod dehydrogenase type 2 (20βHSD2) in the breakdown of cortisone to 20β-hydroxycortisone in zebrafish (Tokarz et al., 2012, 2013). This enzyme transcript is upregulated in response to stress/cortisol treatment, and knockdown of *20βHSD2* led to distinct stress phenotype, suggesting a role for this enzyme in the cortisol stress response (Tokarz et al., 2013). Given the importance of tight regulation of maternal cortisol deposition for proper development, it is not surprising that cortisol degrading enzymes, including 11βHSD2 and 20βHSD2, may be playing a key role in dampening the effect of excess cortisol in zebrafish embryos (Tokarz et al., 2012, 2013). Also, we showed that excess cortisol, mimicking maternal stress, upregulated *11βHSD2* transcript abundance in the ovarian follicles in vitro and may be a mechanism to tightly regulate embryo cortisol content in zebrafish (Faught et al., 2016). Overall, these results suggest that a tight regulation of zygotic cortisol content is essential for proper developmental programming in zebrafish, including muscle growth and development (Nesan and Vijayan, 2012, 2013a,b). The use of GR knockdown mutants in zebrafish highlights the role for cortisol in regulating molecular mechanisms essential for development (Nesan and Vijayan, 2013b; Ziv et al., 2013; Chatzopoulou et al., 2015).

A recent study detailed the transcriptome and metabolome response to GR activation in zebrafish embryos (Chatzopoulou et al., 2015). Using GRα and GRβ knockdown approaches, this study for the first time showed specific molecular and metabolic responses associated with the activation of the two GR splice variants. The authors examined the differential regulation by α and β GR signaling by using splice blocking morpholinos and subsequent treatment with the synthetic glucocorticoid dexamethasone. From the transcriptomics and metabolomics data two distinct gene clusters were activated under either basal or stressed conditions (Chatzopoulou et al., 2015). The increase in glucose in the dexamethasone treated group was linked to several key gluconeogenic genes, including pck1, pck2, pfkrb41, and g6pca (Chatzopoulou et al., 2015). The pck genes encode for phosphoenolpyruvate carboxykinase (PEPCK), and are expressed in multiple cell types prior to hepatogenesis (Jurczyk et al., 2011). As the pck genes have GREs in the promoter regions (Table 4.2), they are regulated by changes in cortisol levels during development. Consequently, the increase in cortisol content posthatch in zebrafish suggests a key role for the GR receptors in energy repartitioning during the transition from endogenous (yolk) nutrient utilization to exogenous feeding. As cortisol also increases genes associated with protein catabolism (Wiseman et al., 2007), including increased transcript abundance of genes involved in serine proteases and the proteasome pathway (ube2a/b, myliqb, foxred2; Chatzopoulou et al., 2015), the interplay between the two GR receptors in regulating energy metabolism may become significant. Interestingly, the activation of HPI axis in zebrafish corresponds with the reduced endogenous nutrient resources (yolk), leading us to hypothesize that elevation in cortisol levels posthatch may be a stimulus for energy substrate allocations and the commencement of exogenous feeding.

4.2. Molecular Adjustments During Stress

The liver is a prime target for hormonal action that allows animals to metabolically cope with stress. Glucose is the key energy fuel essential to reestablish homeostasis in the face of stressor insult (Mommsen et al., 1999; Enes et al., 2009), and this is produced in the liver either rapidly by glycogenolysis or by delayed activation involving gluconeogenesis (Vijayan et al., 2010). The classical flight-or-fight response involves an epinephrine-mediated increase in glycogenolysis, leading to rapid glucose output into the circulation (Mommsen et al., 1999; Fabbri and Moon, 2015). The delayed glucose response to stress is usually mediated by glucocorticoid activation of GR leading to upregulation of metabolic genes involved in gluconeogenesis, which not only increases stress-mediated circulating glucose levels, but also repletes the depleted glycogen stores poststressor exposure (Mommsen et al., 1999; Leung and Woo, 2010; Vijayan et al., 2010). A number of metabolic pathways are activated in response to stress and cortisol stimulation in fish, including enhanced gluconeogenesis, glycolysis, and fatty acid and amino acid metabolism (Mommsen et al., 1999). At the level of the liver, the main metabolic effect of cortisol revolves around glucose production and the associated substrate reallocation to achieve production of this metabolite. Several studies showed that cortisol increases glucose production, and this can be abolished with the GR antagonist mifepristone, confirming a direct effect of GR signaling (Aluru and Vijayan, 2009; Philip et al., 2012; Philip and Vijayan, 2015; Fig. 4.3). Although studies point to a role for cortisol in glycogen mobilization to increase glucose production, this is thought to be more a rapid nongenomic response rather than a genomic effect (Mommsen et al., 1999; Vijayan et al., 2010). The glycogen depletion

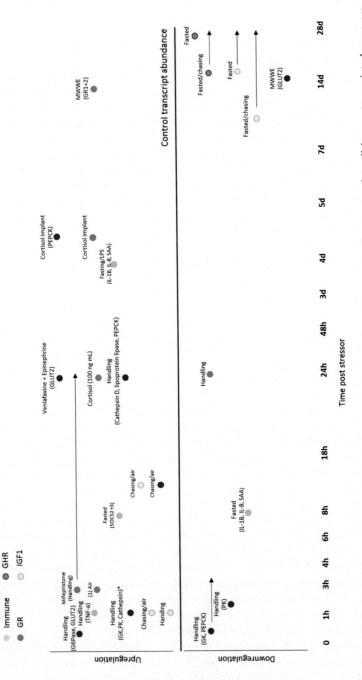

Figure 4.3. Temporal gene transcript changes in the liver in response to stressor exposure. Stressors used to elicit gene transcript changes are described above as color-specific dot; for specific species, see text. StAR, steroidogenic acute regulatory protein; P450scc, P450 side chain cleavage; MC2R, Melanocortin 2 receptor; GR, glucocorticoid receptor; GLUT2, glucose transporter 2; GH, growth hormone; GHR, growth hormone receptor; PRL, prolactin; MC2R, melanocortin 2 receptor; IGF-1, insulin-like growth factor 1; IGF-R, insulin-like growth factor receptor; PEPCK, phosphenolpyruvate carboxykinase; G6Pase, glucose-6-phosphatase; GK, glucokinase; GS, glutamine synthase. *See Wiseman et al., 2007 for a list of metabolic genes. Note this is not an exhaustive list of all gene changes due to stress in the liver; see text for further details. Arrows denote duration of time that gene remained up- or downregulated over the course of the experiment.

poststress is mostly epinephrine mediated (Mommsen et al., 1999; Reid et al., 1998; Fabbri et al., 1998); however, we cannot discount a role for corticosteroid in priming the hepatocyte for epinephrine action during stress (Reid et al., 1992; Mommsen et al., 1999). Other hormones, notably glucagon, may also be playing a role in facilitating glycogen metabolism during stress and enhancing glucose output. However, little is known about the interplay between glucoregulatory hormonal dynamics or the signaling mechanisms that modulate glycogen depletion during acute stress, and its repletion during recovery from stress (Mommsen et al., 1999; Vijayan et al., 2010).

A commonly seen molecular response to cortisol is the upregulation of transcript abundance and enzyme activity for the key rate-limiting step in gluconeogenesis, activation of PEPCK, in fish liver (Sathiyaa and Vijayan, 2003; Vijayan et al., 2003; Aluru and Vijayan, 2007; Fig. 4.3). Studies have shown that amino acids may be preferentially used for gluconeogenesis in fish liver, while glycerol and lactate have also been shown to be utilized for glucose production in response to cortisol stimulation (reviewed in Mommsen et al., 1999). In addition to PEPCK, hexokinase (HK and glucokinase (GK) in the liver) and pyruvate kinase (PK) are often measured, although the latter two enzymes are rate-limiting steps in glycolysis, and unlike PEPCK are an indication of glucose utilization. HK catalyzes the phosphorylation of glucose to glucose-6-phosphate, and it is the first rate-limiting step in glycolysis (Mommsen et al., 1999), whereas PK catalyzes the transfer of phosphoenolpyruvate to pyruvate. The regulation of these transcripts during stress is far from clear, but changes in nutritional status, usually associated with stressor exposure, will modulate mRNA abundance and activities of HK and GK in fish (Soengas et al., 2006; Pérez-Jiménez et al., 2007; Fig. 4.3).

Other notable tissues that are affected by cortisol stimulation during stress include white muscle, as this tissue provides the amino acids and lactate for enhancing gluconeogenesis in the liver (Milligan, 1997, 2003; Frolow and Milligan, 2004). Studies have confirmed that GR signaling is involved in this response; however, mechanistic studies that characterize the regulation of transcripts and proteins involved in muscle catabolism and energy substrate regulation during stress is still lacking in piscine models. To achieve a more comprehensive understanding of transcriptional regulation during stress, studies have recently employed genome sequencing, including microarrays and next-generation sequencing in metabolically active tissues, including the liver and muscle (Table 4.3). Wiseman et al. (2007), using targeted trout microarray, showed increases in transcript abundance of genes involved in protein catabolism, including cathepsin D, arginase, and glutamine synthetase. Cathepsin D, an aspartic protease that has been

Table 4.3

Effect of stress/cortisol on gene changes assessed by large-scale transcriptome analysis (cDNA microarray or RNA-sequencing)

Stressor	Species	Tissue	Method	Genes affected	Source
Handling/crowding	Oncorhynchus mykiss	Liver	RNAseq [Illumina HiSeq 2000]	Global gene changes	Liu et al. (2014)
LPS (fasting; salinity)	Lates calcarifer	Intestine	RNAseq [Roche 454]	Global gene changes (unspecific)	Xia et al. (2013)
Aluminum	Salmo salar	Muscle	RNAseq [Illumina Genome Analyzer IIx]	Alterations in miRNA	Kure et al. (2013)
Handling	Coilia nasus	Liver	RNAseq [Illumina HiSeq (2000)]	Global gene changes	Du et al. (2014)
Cortisol implant	Sparus aurata	Liver	Microarray/qPCR	Global gene changes [GR, alcohol dehydrogenase, MHC II antigen—associated invariant chain, bloodthirsty, MHC class II antigen beta chain, T cell receptor, CIDE3, pitrilysin metalloproteinase 1, cdc48, phosphoglucomutase, zinc-binding1]	Teles et al. (2013)
Cortisol	Oncorhynchus mykiss	Hepatocytes	Microarray/qPCR	Global gene changes	Aluru and Vijayan (2007)
Cortisol	Danio rerio	Embryos	Microarray/qPCR	Global gene changes	Nesan and Vijayan (2013b)
Cortisol implant	Gadus morhua	Embryos	Microarray/qPCR	Global gene changes [stxbp6, fbxw2, capn12, thbs4, sytl2, coco1c, sel1l3, ipo7]	Kleppe et al. (2013)
Handling/crowding	Oncorhynchus mykiss	Liver	Microarray/qPCR	Global gene changes measured at 1 h post stressor and 24 h poststressor	Wiseman et al. (2007)
Handling/air exposure	Oncorhynchus mykiss	Liver	Microarray/qPCR	Global gene changes	Momoda et al. (2007)
	Oncorhynchus mykiss	Liver	Microarray/qPCR	Haptoglobin, complement factor H, CIHBP, B-fibrinogen TC8422, anti-trypsin, ceruloplasmin, serum albumin, TC8200, EST10729, orosomucoid, 14-3-3	Cairns et al. (2008)

Stressor	Species	Tissue	Method	Genes/gene changes	Reference
Handling	*Oncorhynchus mykiss*	Brain and kidney	Microarray/qPCR	Global gene changes	Krasnov et al. (2005)
Fasting/fed	*Sparus aurata*	Heart and skeletal muscle	Microarray/qPCR	Makroin1, serpin h1 precursor, delta-9-desaturase1, skeletal muscle sodium channel, neural cell adhesion molecule, membrane-type matrix metalloproteinase, annexina2a, pyruvate dehydrogenase isoenzyme2, adiponectin receptor protein 1	Calduch-Giner et al. (2014)
LPS	*Oncorhynchus mykiss*	Monocyte/macrophage primary culture	Microarray/qPCR	Hsp70, Jun-B, iNFkBm, MMP13	MacKenzie et al. (2006a)
LPS	*Oncorhynchus mykiss*	Ovary	Microarray/qPCR	Global gene changes	MacKenzie et al. (2006b)
Lepeophtheirus salmonis/cortisol	*Salmo salar*	Skin	Microarray/qPCR	Global gene changes	Krasnov et al. (2012)
Heat stress	*Oncorhynchus mykiss*	RBCs	Microarray/qPCR	Global gene changes	Lewis et al. (2010)
Heat/salinity	*Oncorhynchus mykiss*	Gill, brain, liver, spleen, kidney, and muscle	Microarray/PCR	Global gene changes [casp3, casp8,p53,nupr, Hsp70b,nakatpa1a, nakatpa1b, nakatpa1c, nakatpa3]	Sánchez et al. (2011)
Heat	*Gadus morhua*	Spleen	Microarray/qPCR	Global gene changes [DHX58, STAT1, IRF7, ISG15, RSAD2, IkBa]	Hori et al. (2012)
Propranolol	*Pimpephales promelas*	Brain	Microarray/qPCR	Global gene changes	Lorenzi et al. (2012)
Flyoxetine, nelafaxin, carbamazepine	*Pimpephales promelas*	Whole body	Microarray/qPCR	Genes associated with human diseases (Alzheimer's, bipolar, schizophrenia, ADHD, Parkinson's)	Thomas and Klaper (2012)
MWWE	*Oncorhynchus mykiss*	Liver	Microarray/qPCR	Global gene changes	Ings et al. (2011b)
Atrazine, nonylpenol	*Oncorhynchus mykiss*	Liver	Microarray/qPCR	Global gene changes	Shelley et al. (2012)

LPS, lipopolysaccharide; MWWE, municipal wastewater effluent.

implicated in nonspecific protein degradation, is upregulated and may explain discrepancies between gene upregulation and protein attenuation, as protein stability may be nonspecifically affected during times of stress (Mommsen et al., 1999; Fig. 4.3). Genes that commonly change poststressor exposure can be broadly categorized to include enzymes and transporters involved in tissue metabolism, including glucose, protein and amino acid metabolism, and immune response, including major histocompatibility complex-2, interleukins, acute phase proteins, and tumor necrosis factor-α (Table 4.3). The transcript responses in general are variable and there are some gene transcripts that are expressed within a short-term period, while the majority are modulated over a 24 h period and several are also seen to be regulated over a longer time period in response to stress or cortisol stimulation (Fig. 4.3). Overall, a general sense of the role of these transcript changes in stress adaptation or maladaptation is unclear given that most studies have examined the steady-state level of transcript changes. Also, the mechanisms involved or if the gene changes are transcriptionally and translationally regulated during stress and/or by cortisol stimulation are far from clear.

The target tissue molecular responses, including enhanced gene transcription and translation, may contribute to the increased energy demand associated with stress. Indeed stress increases the metabolic rate in fish (Barton and Iwama, 1991), while the contribution of gene and protein expression to the metabolic demand in fish is not known. The energy demand associated with protein synthesis contributes predominantly to the total increase in cellular metabolic rate (Iwama et al., 2006; Vijayan et al., 2010), but the contribution of transcriptional machinery to the overall energy budget is unknown. This is important as gene transcription does not always correlate with the corresponding protein synthesis in fish (Sathiyaa and Vijayan, 2003; Table 4.4). The general thinking is that upregulation of targeted gene and protein expression is critical to allow animals to cope with stressor(s), which would lead to increased metabolic demand during stress. However, there are stressors that reduce metabolism as an adaptive response, including hypoxia/anoxia and cold shock (Storey, 1997). Under such circumstances, even though the overall protein synthetic capacity is curtailed, there are targeted increases in gene transcription and protein synthesis that are essential for coping with the stressors. This suggests a control mechanism at the level of target cells that modulates energy-demanding pathways during stress. The factor(s) dictating such energy substrate repartitioning during stress adaptation has not been thoroughly investigated (Vijayan et al., 2010).

With energy demands during stress being increased, the amount of energy that is allotted to other pathways, including growth and immune

Table 4.4

Cortisol-responsive gene (G) and the corresponding protein (P) changes

Stressor	Species	Tissue	Gene (G)	Protein (P)	G-P correlate?	Source
Hypoxia/chasing/confinement/cortisol implants	Oncorhynchus mykiss	Pineal organ	Aanat2	AANAT2 enzyme activity	Y	Lopez-Patino et al. (2014)
Handling stress	Oncorhynchus mykiss	Brain	GR	GR	N	Alderman et al. (2012)
		Liver	GR	GR	N	
Cortisol implants	Oncorhynchus mykiss	Liver	PEPCK	PEPCK enzyme	N	Vijayan et al. (2003)
			GR	GR	N	
			Hsp90	Hsp90	Y	
Handling stress	Oncorhynchus mykiss	Liver	GK	PEPCK activity	N	Lopez-Patino et al. (2014)
			PEPCK			
			G6Pase			
			GLUT2			
			PK	PK	Y	
MWWE	Oncorhynchus mykiss	Liver	GR1, GR2	GR	Y/N	Ings et al. (2011b)
			Hsp90	Hsp90 [D 100%]	N	
Cortisol/urea	Opsanus beta	Gill	GS Ubiquitous-GS [U unfused groups (not F)]	GS activity [U crowded]	N	McDonald et al. (2009)
Cortisol	Oncorhynchus mykiss	Hepatocytes	GR [U F; D RU groups]	GR [D F; D RU groups]	N	Aluru and Vijayan (2007)
Cortisol-induced fasting	Oreochromis mossambicus	Stomach	Ghrelin [D F]	Plasma Ghrelin [D F]	Y	Janzen et al. (2012)
Cortisol injection	Danio rerio	Embryos	GR [NC]	GR [D48 h]	N	Nesan and Vijayan (2012)
GR knockdown	Danio rerio	Embryos	GR [NC]	GR [D Mo treated]	N	Nesan and Vijayan (2012)
Fasting (no stress hormones measured)	Sparus aurata	Liver	GK [U]	GK activity [U]	Y	Caseras et al. (2000)

The G-P represents correlation (Y=positive and N=negative) between gene and protein expression or enzyme activity in that study.

response, may be downregulated at the transcriptional level (Tort, 2011; Reindl and Sheridan, 2012). In response to stress, growth hormone receptors and IGF-1 transcript levels are downregulated; this is observed both in vivo (Small et al., 2006; Nakano et al., 2013) and in vitro in fish hepatocytes (Leung et al., 2008; Philip and Vijayan, 2015). While studies have examined the effects of chronic stress and cortisol treatment on growth (Bernier and Peter, 2001; Bernier et al., 2004), the mechanisms of action are far from clear. Myostatin is known to negatively regulate muscle growth in vertebrates. In mammals it is known that myostatin has GRE in its promotors, suggesting a direct role of cortisol in regulating growth. However, this response of cortisol is not conserved in teleosts (Biga et al., 2004; Galt et al., 2014), and cortisol treatment downregulated myostatin expression in the tilapia (Rodgers et al., 2003). Cortisol treatment is also known to attenuate the GH-induced IGF-1 mRNA abundance in the liver, suggesting a link between stress and growth suppression (Leung et al., 2008; Philip and Vijayan, 2015). However, IGF-2 transcript level is actually stimulated by cortisol in coho salmon (*Oncorhynchus kisutch*; Pierce et al., 2010) and tilapia (*Oreochromis mossambicus*; Pierce et al., 2011). A recent study also showed that chronic cortisol treatment increased liver igfbp1, suggesting a potential role in growth suppression by modifying IGFs action (Madison et al., 2015). Both IGF-1 and IGF-2 are the primary mediators of GH-promoting effects, and elicit downstream effects on other tissues, including gonads and heart. For instance, Igf2b is upregulated in the heart during zebrafish regeneration (Huang et al., 2013), and a positive correlation between cardiosomatic index and cortisol levels were reported in adult European brown trout (*Salmo trutta*), suggesting a role for cortisol in myocardial remodeling (Johansen et al., 2011). The interaction between the somatotropic axis and the cortisol-mediated effects on cardiac development requires further work, especially given the fact that excess cortisol during development is having a very pronounced effect on cardiac development and function in zebrafish (Nesan and Vijayan, 2012; Wilson et al., 2013).

Recent work in rainbow trout hepatocytes has started to explain the mechanistic underpinnings between the growth and immune axis suppression during stress. Suppressors of cytokine signaling (SOCS) that contain GRE in their promoter region negatively regulate growth hormone and cytokine signaling by suppression of JAK-STAT pathways (Philip and Vijayan, 2015; Fig. 4.3). Lipopolysaccharide (LPS) stimulation upregulated transcript abundance of cytokine genes, including *IL-6* and *IL-8* in trout hepatocytes; however, cortisol treatment suppressed only LPS-induced *IL-6* and not *IL-8* transcript abundance (Philip et al., 2012; Philip and Vijayan, 2015). As LPS-mediated *IL-6* but not *IL-8* expression involves the JAK-STAT pathway, the results suggest a role for cortisol-induced upregulation

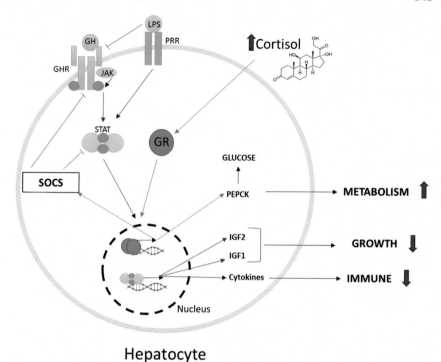

Hepatocyte

Figure 4.4. Intracellular cross-talk between stress, growth, and immune pathways in teleost hepatocytes. Cortisol diffuses through the plasma membrane and binds to the glucocorticoid receptor (GR). Once binding occurs it translocates to the nucleus and induces gene transcription by binding to the glucocorticoid response elements (GREs) of target genes, including suppressors of cytokine signaling (SOCS) and phosphoenolpyruvate carboxykinase (PEPCK). Growth hormone (GH) signals through the growth hormone receptor (GHR) and activates the JAK-STAT pathway. Upon ligand to GHR, JAK (janus kinase) is phosphorylated and the signal transducers and activators of transcription (STAT) bind to the phospho-tyrosine of JAK. The JAK phosphorylates STAT and they dimerize and translocate to the nucleus where it will initiate target gene transcription, including insulin-like growth factor 1 (IGF-1) and 2 (IGF-2). LPS binds to pattern recognition receptors, which also activate JAK-STAT signaling, leading to the synthesis of cytokines. Green arrow denotes cortisol signaling; red arrow denotes GH signaling; blue arrow denotes LPS signaling.

of SOCS in this cytokine suppression (Philip and Vijayan, 2015). Indeed, cortisol treatment increased *SOCS-1* and *SOCS-2* transcript abundance; this was abolished in the presence of mifepristone in trout hepatocytes (Philip et al., 2012; Fig. 4.4). Cortisol treatment also suppressed acute growth hormone-stimulated IGF-1 transcript abundance, which correlated with reduced JAK-STAT activation in trout liver slices (Philip and Vijayan, 2015). Consequently, the changes in growth and immune modulation with cortisol seem to correlate with the upregulation of SOCS and suppression of the JAK-STAT pathway (Philip and Vijayan, 2015; Fig. 4.4). Together these results suggest a potential control mechanism at the level of the liver to regulate energy-demanding pathways during stress in fish. We propose SOCS regulation by cortisol as a key mediator of cellular energy reallocation during stress in fish (Philip and Vijayan, 2015). Overall, the target tissue molecular effects of cortisol are only starting to emerge and the evidence suggests a key role for this steroid in the stressor-mediated effects on growth and immune function.

4.3. Cellular Adjustments

There have been a number of reviews on the cellular stress response and specifically the heat shock response in fish (Iwama et al., 1998; Ackerman et al., 2000a,b; Basu et al., 2002; Iwama, 2004; Deane and Woo, 2011). At the cellular level, studies have examined the regulation of several classes of heat shock proteins (HSPs), generally categorized as the heat shock response (Richter et al., 2010). The Hsps are ubiquitously expressed in almost every cell type and they provide cytoprotection to offset the proteotoxicity in response to stressors (Iwama et al., 1998; Deane and Woo, 2011), making them excellent indicators of cellular stress response. Hsps can be broadly classified into seven categories based on function, including the traditional molecular chaperones, components of the proteolytic system, RNA- and DNA-modifying enzymes, metabolic enzymes, transcription factors, and kinases, as well as transport, detoxifying, and membrane-modulating proteins (Richter et al., 2010). The heat shock response is conserved among species; however, the composition and rate of upregulation of each class of proteins vary (Richter et al., 2010). The class of proteins whose primary role is to act as molecular chaperones are divided into five families based on molecular mass, including Hsp 27, 60, 70, 90, and 110 (Kregel, 2002; Richter et al., 2010). These stress proteins comprise 5–10% of total protein in an unstressed cell, but can be induced to encompass up to 15% of total protein during times of stress (Calderwood et al., 2007; Pockley et al., 2008). The regulation of the heat shock response and synthesis of Hsps occurs through the activation of heat shock factors (HSFs), which will bind to the promoter region of heat shock genes during stress, initiating transcription. There are four possible HSFs, while HSF1 is the principal transcription factor in mammals and lower vertebrates, including fish (Deane and Woo, 2011). In the absence of stress, HSF1 is found within the cytoplasm in a latent monomeric state bound to Hsp70 (Kregel, 2002). Upon stimulation by stressors affecting the protein machinery, the Hsp70 or ubiquitin is removed from HSF1, leading to homotrimerization and subsequent transport to the nucleus where it is hyperphosphorylated by several kinases (Richter et al., 2010). The trimer binds to the heat shock element on the promoter region of the Hsp70 gene, inducing transcription. Regulation of HSF1 binding to the heat shock element is mediated by heat shock binding protein 1, which can be subjected to negative feedback by Hsp70, thereby repressing synthesis (Shi et al., 1998).

The intracellular chaperoning roles of Hsps in fish, specifically Hsp70, have been well characterized (Iwama et al., 1998; Deane and Woo, 2011). However, the majority of studies examined the temporal expression pattern of these proteins under different stressor scenarios, and in different species of fish and cell systems (Vijayan et al., 2005; Deane and Woo, 2011). Surprisingly few studies have actually looked at the mechanism by which these proteins provide cellular protection to stressors. While proteotoxicity is a major stimulus for the expression of Hsps (Iwama et al., 1999; Basu et al., 2002), the chaperoning role for Hsps in regaining protein homeostasis in fish are far from clear. Studies have established linkages between the organismal stress response and the corresponding modulation of the cellular stress response (Vijayan et al., 2005; Iwama et al., 2006). Cortisol attenuates the heat shock response, in gill and liver of rainbow trout (Basu et al., 2001; Boone and Vijayan, 2002b) and in common carp exposed to copper (De Boeck et al., 2003); however, exogenous cortisol exposure did not affect hepatic levels of Hsp70 expression or *Hsp70* mRNA levels in silver sea bream (*Sparus sarba*) in vitro or in cutthroat trout (*Oncorhynchus clarkii*) in vivo (Ackerman et al., 2000a,b). While cortisol also decreased *Hsp90* mRNA abundance in primary cultures of trout hepatocytes (Sathiyaa et al., 2001), Hsp90 protein expression was elevated in vivo in the trout liver (Vijayan et al., 2003). In addition to cortisol both GH and prolactin also reduced *Hsp70* transcript abundance and protein levels in the silver sea bream liver (Deane et al., 1999).

These variations highlight the need for establishing the mode of action of how the organismal stress response affects the cellular stress response (Deane and Woo, 2011). For instance, it is unknown how cortisol modulates the heat shock response in cells, because a putative GRE has not been reported on the promoter of *Hsp70* or *Hsp90* genes. In rainbow trout, the glucocorticoid-mediated attenuation of Hsp70 expression involves the proteasome (Boone and Vijayan, 2002a). Also, a recent work suggests that glucocorticoid-induced modulation of microRNAs (miRNAs) reduces Hsp70 protein abundance in a mammalian cell system (Kukreti et al., 2013). The Hsps may also influence GR signaling, as both Hsp90 and Hsp70 are important components of the GR heterocomplex, essential for maintaining the receptor in a ligand-binding conformation (Collingwood et al., 1999; Morishima et al., 2000). These studies highlight the possibility of a multifactorial control of cellular stress response by stressor-induced cortisol stimulation. Most of our knowledge on the mode of action of cortisol-GR binding and signaling is based on mammalian models (Vijayan et al., 2005), while there is a dearth of mechanistic research linking organismal stress response to cellular stress adaptation in lower vertebrates.

5. SIGNIFICANCE OF MOLECULAR RESPONSES

While this review has focused primarily on the molecular responses to stress, it is equally important to determine whether transcript abundance translates to protein expression during stress. The protein expression is the best indication of functional relevance of the transcript response. Studies have shown a positive match between transcript level and protein expression and/or enzyme activity in response to stress or cortisol stimulation (Table 4.4), including Hsp70 and Hsp90 (Vijayan et al., 2003), PK (Lopez-Patino et al., 2014), ghrelin (Janzen et al., 2012), and GK (Caseras et al., 2000). While studies have correlated gene transcript abundance and their associated protein expression/enzyme activity, there are instances when there is a mismatch between transcript abundance and the corresponding protein expression (Sathiyaa and Vijayan, 2003). For instance, GR transcript level and protein expression show an inverse relationship in trout liver in response to cortisol treatment (Vijayan et al., 2003). This mismatch was also seen with stressor exposure. For instance, an increase in *GR* transcript abundance was seen in both the liver and brain of rainbow trout exposed to a handling stress (Alderman et al., 2012) or cortisol treatment (hepatocytes; Aluru and Vijayan, 2007); however, protein levels were significantly lower in these studies. Enzymes show a similar lack of correlation. While there was a downregulation of the transcript abundance of GK and PEPCK as quickly as 15 min posthandling stress in rainbow trout (Lopez-Patino et al., 2014), there was an upregulation of PEPCK activity at 240–480 min poststressor (Lopez-Patino et al., 2014). The lack of correlation between protein expression and mRNA level may be due to changes in mRNA turnover, as was proposed as a mechanism for GR autoregulation in trout hepatocytes (Sathiyaa and Vijayan, 2003). It appears that GR activation by cortisol increases the receptor degradation through the proteasomal pathway, which in turn triggers the GR transcriptional machinery, leading to a mismatch in GR transcript levels and protein expression (Sathiyaa and Vijayan, 2003). This sort of feedback control mechanism at the level of the target cells may be adaptive to prolong target tissue cortisol responsiveness.

Transcript levels of genes are commonly used as markers of exposure to stressors. This is increasingly used in toxicology where specific gene transcript levels signals the exposure to specific contaminants, including polychlorinated biphenyls (PCBs) (cytochrome P450 1A) and metals (metallothionein), but is not indicative of effects unless accompanied by other protein and biochemical markers (Sarkar et al., 2006). Although measuring mRNA transcript abundance provides an accurate indication of pathway activation, the results must be interpreted with caution when extrapolating the information to assign specific effects. We know that there are several regulators of gene transcription, including coactivators and proteins involved in histone modification and methylation factors (Sexton and Cavalli, 2015; Venkatesh and Workman, 2015), but few studies have looked at the regulation of these proteins in piscine models. Also, factors regulating mRNA stability and turnover (Roy and Jacobson, 2013) may play a major role in influencing gene expression and protein synthesis in response stress but remain poorly understood in the context of stress and adaptation. However, several interesting avenues are starting to emerge with respect to transcript regulation, including epigenome modification and generational inheritance through nonmendalian methods (Zhang et al., 2013). Indeed, the addition of hormones, including cortisol, has illuminated alterations in methylation patterns of CpG islands in mammals (Zhang et al., 2013). The effect of stress or cortisol treatment on epigenetic changes has not yet been studied in lower vertebrates. To date, the vast majority of work regarding molecular responses in fish has been the quantification of gene transcript abundance using quantitative PCR or by large-scale genome analysis techniques, including microarray or next-generation sequencing. The use of knockouts and knockdowns to determine the functional relevance of stress-related proteins has been limited to use in zebrafish and medaka (Pikulkaew et al., 2011; Nesan and Vijayan, 2012; Wilson et al., 2013; Cruz et al., 2013; Benato et al., 2014; Chatzopoulou et al., 2015; Ishikawa et al., 2013) and mostly from the standpoint of early development. The availability of new technologies, including next-generation sequencing and clustered regularly interspaced short palindromic repeat (CRISPR)-Cas9 gene-editing tool (Hwang et al., 2013; Auer et al., 2014; Ablain et al., 2015), that can be used for targeted knockout in a tissue-specific manner with nonmodel organisms will allow for mechanistic characterization of stress axis functioning, and its relevance to animal fitness.

6. APPROACHES TO STUDY MOLECULAR RESPONSES TO STRESS

Until recently most of our understanding on molecular responses to stress in nonmodel fish species came from quantifying the expression of individual genes or proteins using quantitative PCR or western blotting,

respectively. This changed considerably with the advent of oligonucleo-
tide and cDNA microarrays, which enabled quantification of global
changes in gene expression (Aluru and Vijayan, 2009; Rise et al., 2004;
Villeneuve et al., 2008; Whitehead et al., 2011). However, these methods
are not readily and widely available for nonmodel species because of the
lack of genomic information. These limitations have been addressed with
the advent of new sequencing technologies. Major transformation came in
the past decade with the availability of next-generation or deep
sequencing technologies for sequencing the genomes and transcriptomes
of nonmodel species. These technologies offer considerably cheaper and
cost-effective ways for quantifying gene expression patterns in an
unbiased manner. Using these approaches, transcriptomes of several fish
species have been sequenced (Kolmakov et al., 2008; Oleksiak et al., 2011;
Salem et al., 2015; Salem et al., 2010; Wang et al., 2014). Transcriptome
studies, if properly designed, can test a hypothesis and also help in
discovering pathways essential for stress adaptation, as well as leading to
maladaptation in fish.

6.1. Mechanistic Studies Using Targeted Mutagenesis

In order to investigate the mechanistic basis of stress response,
knockdown and knockout approaches are widely used in mammalian
systems. In fish, morpholino oligonucleotide-based knockdown has been
extensively used. Until recently knockout or reverse genetic approaches
(inactivating a gene and evaluating the phenotypic consequence) were not
widely available for research involving fish species. That has changed with
the advent of programable nucleases, which enabled targeted mutagenesis in
a cost-effective and efficient manner (Kim and Kim, 2014). These targeted
genome-editing tools include zinc finger nucleases, transcription activator-
like effector nucleases, and RNA-guided engineered nucleases (widely
known as CRISPR-Cas9). These technologies have enabled generation of
mutants in a wide range of fish species other than zebrafish, a biomedical
model species (Aluru et al., 2015a; Ansai et al., 2012, 2013, 2014; Dong
et al., 2011; Edvardsen et al., 2014; Li et al., 2013, 2014; Wang and Hong,
2014; Yano et al., 2014; Table 4.1 for stress-specific knockouts). So far
researchers have taken advantage of these methods in species with short
generation time, including zebrafish, tilapia, and medaka, due to the
minimum amount of time necessary to raise potential founders and
heterozygous mutants. This has changed with the development of the
highly efficient CRISPR-Cas9 approach as it allows generation of biallelic
knockouts in the injected embryo itself (Jao et al., 2013). This method allows
studying gene function in any species of interest. In addition to generation of

gene knockouts, the CRISPR-Cas9 approach also allows gene insertion, gene correction, and chromosomal rearrangements in any region of interest (Kim and Kim, 2014). Overall, targeted mutagenesis can be used for understanding gene functions, particularly the ones that are duplicated in the fish-specific lineage. The emergence of targeted genome editing, together with high throughput sequencing technology, including RNAseq, has opened the door for conducting mechanistic studies in any fish species of interest (Table 4.1).

6.2. Epigenetic Regulation of Stress Response

There is a growing amount of evidence from biomedical research suggesting that epigenetic mechanisms play a very important role in gene regulation. In some cases these mechanisms are shown to be responsible for transgenerational inheritance of phenotypic traits. Environmental factors that are well studied in mammals and other species include nutrition, behavior, stress, and chemical, and they influence the epigenome during critical periods of development (Turecki and Meaney, 2014; Yan et al., 2014; Yano et al., 2014; Zannas and West, 2015; Zheng et al., 2014). In comparison very little is known about the epigenome modification in response to stress in lower vertebrates. One of the widely tested hypotheses is that environmental influences on early development alter the epigenome and lead to complex phenotypes and disease susceptibility throughout the course of life. Understanding the environmental effects on epigenetic mechanisms is essential for predicting the changes in gene expression as well as phenotypes. In this section, we outline the widely studied epigenetic mechanisms and illustrate the potential environmental factors that influence them. Epigenetics is the study of heritable changes in gene function that occur independently of alterations to primary DNA sequence (Laird and Jaenisch, 1996). The best-studied epigenetic modifications are DNA methylation, histone modifications, and noncoding RNAs. It is well demonstrated that early embryonic developmental events are tightly regulated by epigenetic mechanisms (Reik et al., 2003). In mammals, epigenetic modifications are key regulators of important developmental events, including X-inactivation, genomic imprinting, and terminal differentiation (Callinan, 2006; Reik et al., 2003).

6.2.1. DNA METHYLATION

DNA methylation is a covalent modification of DNA by the addition of a methyl group. DNA-methyltransferases catalyze the reaction by transferring a methyl group to the 5 position of cytosine on CpG dinucleotides. DNA methylation is considered to be a repressive mark by directly interfering with transcription factor binding to DNA recognition sites or by recruiting corepressor complexes that harbor histone deacetylases or histone methyl-transferases (Cedar and Bergman, 2012; Fuks, 2005). In vertebrates, heritable methylation mainly occurs at the CpG dinucleotide, and methylation of DNA can change the functional state of regulatory regions. Given the importance of DNA methylation in development, any exogenous factors that alter these mechanisms could influence the developmental trajectory and phenotypes. Most of our current understanding on DNA methylation comes from mammalian studies and very little information is available from lower vertebrates. So far very few studies have investigated the impact of stressors on DNA methylation in fish (Aluru et al., 2015b; Liu et al., 2014; Pierron et al., 2014a,b). Chemical stressors (particularly toxicants including PAHs, PCBs, and xenoestrogens) are shown to alter DNA methylation patterns of certain target genes (Aluru et al., 2015b; Corrales et al., 2014; Fang et al., 2013a,b). Similar studies need to be conducted in order to determine the role of other stressors in modulating methylation patterns. For example, the effects of environmental factors, including temperature, pH, and dissolved oxygen on DNA methylation patterns need to be determined. It is well known these environmental factors have an influence on the growth and reproductive outcomes in vertebrates. However, the mechanistic basis of these changes is not well understood. Recent studies point to the involvement of genetic and epigenetic mechanisms in determining these complex phenotypes. Furthermore, DNA methylation is an important epigenetic mechanism that is of relevance to aquaculture based on the evidence that there is a developmental basis for later onset phenotypes (Li et al., 2010). For instance, hatchery and larval rearing practices, larval nutrition, and chemicals in the environment can influence adult growth and disease. Mammalian studies have determined that various environmental factors can influence behavior and health of the adults not only in the exposed generation, but also in subsequent generations (Skinner, 2014). These studies have shown evidence that these long-term and/or transgenerational effects have an epigenetic basis, particularly DNA methylation. These kinds of studies in fish should shed light on the impact of early life-rearing practices on growth, disease resistance, and reproductive capacity. With regard to the HPA axis in mammals, it is very well demonstrated that promoter methylation of GR in the brain is sensitive and responsive to stressors (Weaver et al., 2004; McGowan et al., 2009). It remains to be determined if a similar mechanism exists in teleosts.

6.2.2. NONCODING RNAs

Until recently, noncoding RNAs (ncRNAs) were known to play generic functions in cells, including mRNA translation (rRNAs and tRNAs), splicing (small nuclear RNAs; snRNAs), and modification of ribosomal RNAs (small nucleolar RNAs; snoRNAs) (Mattick, 2006). This notion has changed in the past decade with the discovery of a variety of ncRNAs that are shown to play critical roles in development, physiology, and disease (Berezikov, 2011). Noncoding RNA is a nonspecific term used to describe any RNA species that does not code for functional protein and is <200–300 nucleotides long. Based on the size and function, they are further classified into miRNAs, short-interfering RNAs (siRNAs), piwi-interacting RNAs (piRNAs), double-stranded RNAs (dsRNAs), and long noncoding RNAs (lncRNAs). Since ncRNAs regulate several epigenetic phenomena that establish long-term effects of gene expression, they are considered to be important players in epigenetic regulation (Berezikov, 2011). There is very little information available on the role of noncoding RNAs in regulating the stress response in fish. With the advent of high throughput sequencing approaches, there is enormous interest in cataloging various noncoding RNAs. Future studies should focus on sequencing the noncoding RNA in species of aquaculture importance and environmental relevance under different environmental conditions. Once the important groups of noncoding RNA are identified, reverse genetic tools can be used to knockout specific ncRNAs to characterize their function. With respect to stress response, identifying the temporal and tissue-specific changes in the expression of ncRNAs, and their regulatory mechanisms, is of paramount importance.

7. CONCLUDING REMARKS AND THE UNKNOWNS

While many of the broad physiological responses to stress have been characterized, the emerging picture from the abundance of gene transcript data is that cortisol is playing a key role in regulating molecular mechanisms not only during stress adaptation, but also in other aspects of animal performance, including growth and development. Indeed when fish are perturbed by a stressor, the effect of cortisol is ubiquitous, affecting various aspects of animal performance; however, the mode of action and the specific signaling pathways involved in bringing about various physiological changes are less clear. Most studies relate the physiological responses to stress in terms of plasma cortisol levels. This may not be particularly useful in all instances unless we know the dynamics of the hormone and the associated effect on target tissue corticosteroid receptor regulation. Understanding target tissue corticosteroid receptor regulation and activation, as well as the regulation of cortisol degrading enzymes, is of paramount importance when interpreting physiological responses associated with elevated cortisol levels. For instance, chronically elevated plasma cortisol levels may not necessarily reflect a negative effect on metabolism if the target tissue receptors are downregulated or if there is a corresponding increase in cortisol degrading enzymes. This is supported by the observation that elevated cortisol levels increase *11βHSD2* and *20βHSD2* transcript levels (Li et al., 2012; Alderman and Vijayan, 2012; Tokarz et al., 2013), as well as reduce GR protein expression in fish (Vijayan et al., 2003, 2005).

Although there is an abundance of transcript data available, there is still much that is unclear. The specific mechanism by which stress hormones affect target tissues during stress remains largely unknown. This is, in part, due to the fact that fish research employing pharmacological or molecular techniques (knockouts/morpholinos) to understand the underlying mechanisms are few and far between. The majority of current research often demonstrates correlative relationships between changes in stress hormone levels and associated gene changes. Knockdown and rescue studies with mutant strains, morpholinos, or more extensive use of known antagonists for receptors and other signaling molecules can help to avoid speculative conclusions and more concretely define the role of stress hormones in modulating stress signaling pathways and target tissue responses. For instance, a recent study (Chatzopoulou et al., 2015) using gene knockdown, along with transcriptome and metabolome approaches, identified GR splice variant-specific responses in zebrafish embryos, providing insights into the functional relevance of the activation of these receptors during stress. Although such studies require model organisms, the stress-responsive pathways and candidate genes identified from those studies can be tested in nonmodel native and aquaculture-relevant species. The use of mutants to answer mechanistic questions is currently limited to the teleost developmental models (zebrafish and medaka). However, recent advances in genome editing, including CRISPR/Cas9, will allow creation of targeted knockouts and knockins, and will pave the way for more precise characterization of mechanisms of action during stress in nonmodel animals, including aquaculture-relevant and/or ecologically-sensitive species.

While this chapter highlights a key role for cortisol in the molecular adjustments to stress, there are many unknowns, including:

- How does cortisol elicit a rapid, nongenomic response? Specifically, is there a putative membrane receptor for GR signaling?
- What is the role of GR and MR signaling in the stress adaptation?
- Does maternal stress affect subsequent generations and what are the mechanisms?
- What is the role of environmental factors (pH, temperature, oxygen) in altering epigenetic machinery?
- What is the role of epigenetic mechanisms in the long-term effects of developmental exposure to stress?

- Does DNA methylation regulate expression of genes associated with the HPI axis?
- What are the candidate microRNAs and long noncoding RNAs that are involved in the regulation of genes associated with the HPI axis?
- Does acute or chronic stress alter DNA methylation patterns and/or noncoding RNAs in the steroidogenic tissues/cells and the brain?

REFERENCES

Ablain, J., Durand, E. M., Yang, S., Zhou, Y. and Zon, L. I. (2015). A CRISPR/Cas9 vector system for tissue-specific gene disruption in zebrafish. *Dev. Cell* **32**, 756–764.

Acerete, L., Balasch, J. C., Castellana, B., Redruello, B., Roher, N., Canario, A. V., et al. (2007). Cloning of the glucocorticoid receptor (GR) in gilthead seabream (*Sparus aurata*). Differential expression of GR and immune genes in gilthead seabream after an immune challenge. *Comp. Biochem. Physiol. B Biochem. Mol. Biol.* **148**, 32–43.

Ackerman, P. A., Forsyth, R. B., Mazur, C. F. and Iwama, G. K. (2000a). Stress hormones and the cellular stress response in salmonids. *Fish. Physiol. Biochem.* **23**, 327–336.

Ackerman, P. A., Forsyth, R. B., Mazur, C. F. and Iwama, G. K. (2000b). Stress hormones and the cellular stress response in salmonids. *Stress Int. J. Biol. Stress* 327–336.

Agulleiro, M. J., Sánchez, E., Leal, E., Cortés, R., Fernández-Durán, B., Guillot, R., et al. (2013). Molecular characterization and functional regulation of melanocortin 2 receptor (MC2R) in the sea bass. A putative role in the adaptation to stress. *PLoS One* **8**, e65450.

Alderman, S. L. and Bernier, N. J. (2007). Localization of corticotropin-releasing factor, urotensin I and CRF-binding protein gene expresion in the brain of the zebrafish, *Danio rerio*. *J. Comp. Neurol.* **502**, 783–793.

Alderman, S. L. and Bernier, N. J. (2009). Ontogeny of the corticotropin-releasing factor system in zebrafish. *Gen. Comp. Endocrinol.* **164**, 61–69.

Alderman, S. L. and Vijayan, M. M. (2012). 11β-Hydroxysteroid dehydrogenase type 2 in zebrafish brain: a functional role in hypothalamus-pituitary-interrenal axis regulation. *J. Endocrinol.* **215**, 393–402.

Alderman, S. L., McGuire, A., Bernier, N. J. and Vijayan, M. M. (2012). Central and peripheral glucocorticoid receptors are involved in the plasma cortisol response to an acute stressor in rainbow trout. *Gen. Comp. Endocrinol.* **176**, 79–85.

Alsop, D. and Vijayan, M. M. (2008). Development of the corticosteroid stress axis and receptor expression in zebrafish. *Am. J. Physiol. Regul. Integr. Comp. Physiol.* **294**, R711–R719.

Alsop, D. and Vijayan, M. M. (2009). Molecular programming of the corticosteroid stress axis during zebrafish development. *Comp. Biochem. Physiol. A Mol. Integr. Physiol.* **153**, 49–54.

Alsop, D., Ings, J. S. and Vijayan, M. M. (2009). Adrenocorticotropic hormone suppresses gonadotropin-stimulated estradiol release from zebrafish ovarian follicles. *PLoS One* **4**, e6463.

Aluru, N. and Vijayan, M. M. (2006). Aryl hydrocarbon receptor activation impairs cortisol response to stress in rainbow trout by disrupting the rate-limiting steps in steroidogenesis. *Endocrinology* **147**, 1895–1903.

Aluru, N. and Vijayan, M. M. (2007). Hepatic transcriptome response to glucocorticoid receptor activation in rainbow trout. *Physiol. Genomics* **31**, 483–491.

Aluru, N. and Vijayan, M. M. (2008). Molecular characterization, tissue-specific expression, and regulation of melanocortin 2 receptor in rainbow trout. *Endocrinology* **149**, 4577–4588.

Aluru, N. and Vijayan, M. M. (2009). Stress transcriptomics in fish: a role for genomic cortisol signaling. *Gen. Comp. Endocrinol.* **164**, 142–150.

Aluru, N., Renaud, R., Leatherland, J. F. and Vijayan, M. M. (2005). Ah receptor-mediated impairment of interrenal steroidogenesis involves StAR protein and P450scc gene attenuation in rainbow trout. *Toxicol. Sci.* **84**, 260–269.

Aluru, N., Karchner, S. I., Franks, D. G., Nacci, D., Champlin, D. and Hahn, M. E. (2015a). Targeted mutagenesis of aryl hydrocarbon receptor 2a and 2b genes in Atlantic killifish (*Fundulus heteroclitus*). *Aquat. Toxicol.* **158**, 192–201.

Aluru, N., Kuo, E., Helfrich, L. W., Karchner, S. I., Linney, E. A., Pais, J. E., et al. (2015b). Developmental exposure to 2,3,7,8-tetrachlorodibenzo-p-dioxin alters DNA methyltransferase (dnmt) expression in zebrafish (*Danio rerio*). *Toxicol. Appl. Pharmacol.* **284**, 142–151.

Ando, H., Hasegawa, M., Ando, J. and Urano, A. (1999). Expression of salmon corticotrophin releasing hormone precursor gene in the preoptic nucleus in stressed rainbow trout. *Gen. Comp. Endocrinol.* **113**, 87–95.

Ansai, S., Ochiai, H., Kanie, Y., Kamei, Y., Gou, Y., Kitano, T., Yamamoto, T. and Kinoshita, M. (2012). Targeted disruption of exogenous EGFP gene in medaka using zinc-finger nucleases. *Dev. Growth Differ.* **54**, 546–556.

Ansai, S., Sakuma, T., Yamamoto, T., Ariga, H., Uemura, N., Takahashi, R., et al. (2013). Efficient targeted mutagenesis in medaka using custom-designed transcription activator-like effector nucleases. *Genetics* **193**, 739–749.

Ansai, S., Inohaya, K., Yoshiura, Y., Schartl, M., Uemura, N., Takahashi, R., et al. (2014). Design, evaluation, and screening methods for efficient targeted mutagenesis with transcription activator-like effector nucleases in medaka. *Dev. Growth Differ.* **56**, 98–107.

Argenton, F., Ramoz, N., Charlet, N., Bernardini, S., Colombo, L., Bortolussi, M., et al. (1996). Mechanisms of transcriptional activation of the promoter of the rainbow trout prolactin gene by GHF1/Pit1 and glucocorticoid the pituitary-restricted POU domain transcription factor GHF1/Pit1 is required for the expression of the growth hormone (GH). *Biochem. Biophys. Res. Commun.* **66**, 57–66.

Arnold-Reed, D. E. and Balment, R. J. (1994). Peptide hormones influence in vitro interrenal secretion of cortisol in the trout, *Oncorhynchus mykiss*. *Gen. Comp. Endocrinol.* **96**, 85–91.

Arterbery, A. S., Deitcher, D. L. and Bass, A. H. (2010). Corticosteroid receptor expression in a teleost fish that displays alternative male reproductive tactics. *Gen. Comp. Endocrinol.* **165**, 83–90.

Arukwe, A. (2008). Steroidogenic acute regulatory (StAR) protein and cholesterol side-chain cleavage (P450scc)-regulated steroidogenesis as an organ-specific molecular and cellular target for endocrine disrupting chemicals in fish. *Cell Biol. Toxicol.* **24**, 527–540.

Aruna, A., Nagarajan, G. and Chang, C. F. (2012). Involvement of corticotrophin-releasing hormone and corticosteroid receptors in the brain-pituitary-gill of tilapia during the course of seawater acclimation. *J. Neuroendocrinol.* **24**, 818–830.

Auer, T. O., Duroure, K., De Cian, A., Concordet, J. and Del Bene, F. (2014). Highly efficient CRISPR/Cas9-mediated knock-in in zebrafish by homology-independent DNA repair. *Genome Res.* **24** (1), 142–153.

Baker, B. I., Bird, D. J. and Buckingham, J. C. (1996). In the trout, CRH and AVT synergize to stimulate ACTH release. *Regul. Pept.* **67**, 207–210.

Barton, B. A. (2002). Stress in fishes: a diversity of responses with particular reference to changes in circulating corticosteroids. *Integr. Comp. Biol.* **42**, 517–525.

Barton, B. A. and Iwama, G. K. (1991). Physiological changes in fish from stress in aquaculture with emphasis on the response and effects of corticosteroids. *Annu. Rev. Fish Dis.* **1**, 3–26.

Basu, N., Nakano, T., Grau, E. G. and Iwama, G. K. (2001). The effects of cortisol on heat shock protein 70 levels in two fish species. *Gen. Comp. Endocrinol.* **124**, 97–105.

Basu, N., Todgham, A. E., Ackerman, P. A., Bibeau, M. R., Nakano, K., Schulte, P. M., et al. (2002). Heat shock protein genes and their functional significance in fish. *Gene* **295**, 173–183.

Benato, F., Colletti, E., Skobo, T., Moro, E., Colombo, L., Argenton, F., et al. (2014). A living biosensor model to dynamically trace glucocorticoid transcriptional activity during development and adult life in zebrafish. *Mol. Cell. Endocrinol.* **392** (1–2), 60–72.

Berezikov, E. (2011). Evolution of microRNA diversity and regulation in animals. *Nat. Rev. Genet.* **12**, 846–860.

Bernier, N. J. (2006). The corticotropin-releasing factor system as a mediator of the appetite-suppressing effects of stress in fish. *Gen. Comp. Endocrinol.* **146**, 45–55.

Bernier, N. J. and Peter, R. E. (2001). The hypothalamic-pituitary-interrenal axis and the control of food intake in teleost fish. *Comp. Biochem. Physiol. B Biochem. Mol. Biol.* **129**, 639–644.

Bernier, N. J., Bedard, N. and Peter, R. E. (2004). Effects of cortisol on food intake, growth, and forebrain neuropeptide Y and corticotropin-releasing factor gene expression in goldfish. *Gen. Comp. Endocrinol.* **135**, 230–240.

Bernier, N. J., Alderman, S. L. and Bristow, E. N. (2008). Heads or tails? Stressor-specific expression of corticotropin-releasing factor and urotensin I in the preoptic area and caudal neurosecretory system of rainbow trout. *J. Endocrinol.* **196**, 637–648.

Bernier, N. J., Gorissen, M. and Flik, G. (2012). Differential effects of chronic hypoxia and feed restriction on the expression of leptin and its receptor, food intake regulation and the endocrine stress response in common carp. *J. Exp. Biol.* **215**, 2273–2282.

Best, C., Melnyk-Lamont, N., Gesto, M. and Vijayan, M. M. (2014). Environmental levels of the antidepressant venlafaxine impact the metabolic capacity of rainbow trout. *Aquat. Toxicol.* **155**, 190–198.

Biga, P. R., Cain, K. D., Hardy, R. W., Schelling, G. T., Overturf, K., Roberts, S. B., et al. (2004). Growth hormone differentially regulates muscle myostatin1 and -2 and increases circulating cortisol in rainbow trout (*Oncorhynchus mykiss*). *Gen. Comp. Endocrinol.* **138**, 32–41.

Boone, A. N. and Vijayan, M. M. (2002a). Glucocorticoid-mediated attenuation of the Hsp70 response in trout hepatocytes involves the proteasome. *Am. J. Physiol. Regul. Integr. Comp. Physiol.* **283**, R680–R687.

Boone, A. N. and Vijayan, M. M. (2002b). Constitutive heat shock protein 70 expression in rainbow trout hepatocytes: effect of heat shock and heavy metal exposure. *Comp. Biochem. Physiol. C Toxiocol. Pharmacol.* **132** (2), 223–233.

Borski, R. J. (2000). Nongenomic membrane actions of glucocorticoids in vertebrates. *Trends Endocrinol. Metab.* **11**, 427–436.

Bouw, E., Huisman, M., Neggers, S. J. C. M. M., Themmen, A. P. N., van der Lely, A. J. and Delhanty, P. J. D. (2014). Development of potent selective competitive-antagonists of the melanocortin type 2 receptor. *Mol. Cell. Endocrinol.* **394**, 99–104.

Bradford, C. S., Fitzpatrick, M. S. and Schreck, C. B. (1992). Evidence for ultra-short-loop feedback in ACTH-induced interrenal steroidogenesis in coho salmon: acute self-suppression of cortisol secretion in vitro. *Gen. Comp. Endocrinol.* **87**, 292–299.

Bury, N. R. and Sturm, A. (2007). Evolution of the corticosteroid receptor signalling pathway in fish. *Gen. Comp. Endocrinol.* **153**, 47–56.

Bury, N. R., Sturm, A., Le, R. P., Lethimonier, C., Ducouret, B., Guiguen, Y., et al. (2003). Evidence for two distinct functional glucocorticoid receptors in teleost fish. *J. Mol. Endocrinol.* **31**, 141–156.

Cairns, M. T., Johnson, M. C., Talbot, A. T., Pemmasani, J. K., McNeill, R. E., Houeix, B., et al. (2008). A cDNA microarray assessment of gene expression in the liver of rainbow trout (*Oncorhynchus mykiss*) in response to a handling and confinement stressor. *Comp. Biochem. Physiol. Part D Genomics Proteomics* **3**, 51–66.

Calderwood, S. K., Mambula, S. S. and Gray, P. J. (2007). Extracellular heat shock proteins in cell signaling and immunity. *Ann. N. Y. Acad. Sci.* **39**, 28–39.

Calduch-Giner, J. A., Echasseriau, Y., Crespo, D., Baron, D., Planas, J. V., Prunet, P., et al. (2014). Transcriptional assessment by microarray analysis and large-scale meta-analysis of the metabolic capacity of cardiac and skeletal muscle tissues to cope with reduced nutrient availability in gilthead sea bream (*Sparus aurata* L.). *Mar. Biotechnol.* **16**, 423–435.

Callinan, P. A. (2006). The emerging science of epigenomics. *Hum. Mol. Genet.* **15**, R95–R101.

Cedar, H. and Bergman, Y. (2012). Programming of DNA methylation patterns. *Annu. Rev. Biochem.* **81**, 97–117.

Caseras, A., Metón, I., Fernández, F. and Baanante, I. V. (2000). Glucokinase gene expression is nutritionally regulated in liver of gilthead sea bream (*Sparus aurata*). *Biochim. Biophys. Acta Gene. Struct. Expr.* **1493**, 135–141.

Castillo, J., Castellana, B., Acerete, L., Planas, J. V., Goetz, F. W., Mackenzie, S., et al. (2008). Stress-induced regulation of steroidogenic acute regulatory protein expression in head kidney of Gilthead seabream (*Sparus aurata*). *J. Endocrinol.* **196**, 313–322.

Castillo, J., Teles, M., Mackenzie, S. and Tort, L. (2009). Stress-related hormones modulate cytokine expression in the head kidney of gilthead seabream (*Sparus aurata*). *Fish Shellfish Immunol.* **27**, 493–499.

Cerdá-Reverter, J. M., Agulleiro, M. J., Cortés, R., Sánchez, E., Guillot, R., Leal, E., et al. (2013). Involvement of melanocortin receptor accessory proteins (MRAPs) in the function of melanocortin receptors. *Gen. Comp. Endocrinol.* **188**, 133–136.

Chadzinska, M., Tertil, E., Kepka, M., Hermsen, T., Scheer, M. and Lidy Verburg-van Kemenade, B. M. (2012). Adrenergic regulation of the innate immune response in common carp (*Cyprinus carpio* L.). *Dev. Comp. Immunol.* **36**, 306–316.

Chandrasekar, G., Lauter, G. and Hauptmann, G. (2007). Distribution of corticotropin-releasing hormone in the developing zebrafish brain. *J. Comp. Neurol.* **505**, 337–351.

Chatzopoulou, A., Roy, U., Meijer, A. H., Alia, A., Spaink, H. P. and Schaaf, M. J. M. (2015). Transcriptional and metabolic effects of glucocorticoid receptor α and β signaling in zebrafish. *Endocrinology* **156**, 1757–1769.

Clark, B. J. and Cochrum, R. K. (2007). The steroidogenic acute regulatory protein as a target of endocrine disruption in male reproduction. *Drug Metab. Rev.* **39**, 353–370.

Collingwood, T. N., Urnov, F. D. and Wolffe, A. P. (1999). Nuclear receptors: coactivators, corepressors and chromatin remodeling in the control of transcription. *J. Mol. Endocrinol.* **23**, 255–275.

Colombe, L., Fostier, A., Bury, N., Pakdel, F. and Guiguen, Y. (2000). A mineralocorticoid-like receptor in the rainbow trout, *Oncorhynchus mykiss*: cloning and characterization of its steroid binding domain. *Steroids* **65**, 319–328.

Corrales, J., Fang, X., Thornton, C., Mei, W., Barbazuk, W. B., Duke, M., et al. (2014). Effects on specific promoter DNA methylation in zebrafish embryos and larvae following benzo[a] pyrene exposure. *Comp. Biochem. Physiol. Part C* **163**, 37–46.

Craig, P. M., Al-Timimi, H. and Bernier, N. J. (2005). Differential increase in forebrain and caudal neurosecretory system corticotropin-releasing factor and urotensin I gene expression associated with seawater transfer in rainbow trout. *Endocrinology* **146**, 3851–3860.

Cruz, S. A., Lin, C.-H., Chao, P. L. and Hwang, P. P. (2013). Glucocorticoid receptor, but not mineralocorticoid receptor, mediates cortisol regulation of epidermal ionocyte development and ion transport in zebrafish (*Danio rerio*). *PLoS One* **8**, e77997.

Deane, E. E. and Woo, N. Y. S. (2011). Advances and perspectives on the regulation and expression of piscine heat shock proteins. *Rev. Fish Biol. Fish.* **21**, 153–185.

Deane, E. E., Kelly, S. P., Lo, C. K. M. and Woo, N. Y. S. (1999). Effects of GH, prolactin and cortisol on hepatic heat shock protein 70 expression in a marine teleost *Sparus sarba*. *J. Endocrinol.* **161**, 413–421.

De Boeck, G., De Wachter, B., Vlaeminck, A. and Blust, R. (2003). Effect of cortisol treatment and/or sublethal copper exposure on copper uptake and heat shock protein levels in common carp, *Cyprinus carpio*. *Environ. Toxicol. Chem.* **22** (5), 1122–1126.

Dickmeis, T. (2009). Glucocorticoids and the circadian clock. *J. Endocrinol.* **200**, 3–22.

Dickmeis, T. and Foulkes, N. S. (2011). Glucocorticoids and circadian clock control of cell proliferation: At the interface between three dynamic systems. *Mol. Cell. Endocrinol.* **331**, 11–22.

Dickmeis, T., Lahiri, K., Nica, G., Vallone, D., Santoriello, C., Neumann, C. J., et al. (2007). Glucocorticoids play a key role in circadian cell cycle rhythms. *PLoS Biol.* **5**, e78.

Dindia, L., Murray, J., Faught, E., Davis, T. L., Leonenko, Z. and Vijayan, M. M. (2012). Novel nongenomic signaling by glucocorticoid may involve changes to liver membrane order in rainbow trout. *PLoS One* **7**, e46859.

Dindia, L., Faught, E., Leonenko, Z., Thomas, R. and Vijayan, M. M. (2013). Rapid cortisol signaling in response to acute stress involves changes in plasma membrane order in rainbow trout liver. *Am. J. Physiol. Endocrinol. Metab.* **304**, E1157–E1166.

Dong, Z., Ge, J., Li, K., Xu, Z., Liang, D., Li, J., et al. (2011). Heritable targeted inactivation of myostatin gene in yellow catfish (*Pelteobagrus fulvidraco*) using engineered zinc finger nucleases. *PLoS One* **6**, e28897.

Dores, R. M. (2013). Observations on the evolution of the melanocortin receptor gene family: distinctive features of the melanocortin-2 receptor. *Front. Neurosci.* **7**, 1–12.

Dores, R. M. and Baron, A. J. (2011). Evolution of POMC: origin, phylogeny, posttranslational processing, and the melanocortins. *Ann. N. Y. Acad. Sci.* **1220**, 34–48.

Dores, R. M. and Garcia, Y. (2015). Views on the co-evolution of the melanocortin-2 receptor, MRAPs, and the hypothalamus/pituitary/adrenal-interrenal axis. *Mol. Cell. Endocrinol.* **408**, 12–22.

Dores, R. M., Londraville, R. L., Prokop, J., Davis, P., Dewey, N. and Lesinski, N. (2014). Molecular evolution of GPCRs: melanocortin/melanocortin receptors. *J. Mol. Endocrinol.* **52**, T29–T42.

Dores, R. M., Liang, L., Hollmann, R. E., Sandhu, N. and Vijayan, M. M. (2016). Identifying the activation motif in the *N*-terminal of rainbow trout and zebrafish melanocortin-2 receptor accessory protein 1 (MRAP1) orthologs. *Gen. Comp. Endocrinol.* ⟨http://dx.doi.org/10.1016/j.ygcen.2015.12.031⟩.

Doyon, C., Trudeau, V. L. and Moon, T. W. (2005). Stress elevates corticotropin-releasing factor (CRF) and CRF-binding protein mRNA levels in rainbow trout (*Oncorhynchus mykiss*). *J. Endocrinol.* **186**, 123–130.

Doyon, C., Leclair, J., Trudeau, V. L. and Moon, T. W. (2006). Corticotropin-releasing factor and neuropeptide Y mRNA levels are modified by glucocorticoids in rainbow trout, *Oncorhynchus mykiss*. *Gen. Comp. Endocrinol.* **146**, 126–135.

Du, F., Xu, G., Nie, Z., Xu, P. and Gu, R. (2014). Transcriptome analysis gene expression in the liver of Coilia nasus during the stress response. *BMC Genomics* **15**, 558.

Ducouret, B., Tujague, M., Ashraf, J., Mouchel, N., Servel, N., Valotaire, Y., et al. (1995). Cloning of a teleost fish glucocorticoid receptor shows that it contains a deoxyribonucleic acid- binding domain different from that of mammals. *Endocrinology* **136**, 3774–3783.

Edvardsen, R. B., Leininger, S., Kleppe, L., Skaftnesmo, K. O. and Wargelius, A. (2014). Targeted mutagenesis in Atlantic salmon (*Salmo salar* L.) using the CRISPR/Cas9 system induces complete knockout individuals in the F0 generation. *PLoS One* **9**, e108622.

Enes, P., Panserat, S., Kaushik, S. and Oliva-Teles, A. (2009). Nutritional regulation of hepatic glucose metabolism in fish. *Fish Physiol. Biochem.* **35**, 519–539.

Eriksen, M. S., Bakken, M., Espmark, A., Braastad, B. O. and Salte, R. (2006). Prespawning stress in farmed Atlantic salmon *Salmo salar*: maternal cortisol exposure and hyperthermia during embryonic development affect offspring survival, growth and incidence of malformations. *J. Fish Biol.* **69**, 114–129.

Esbaugh, A. J. and Walsh, P. J. (2009). Identification of two glucocorticoid response elements in the promoter region of the ubiquitous isoform of glutamine synthetase in gulf toadfish, *Opsanus beta*. *Am. J. Physiol. Regul. Integr. Comp. Physiol.* **297**, R1075–R1081.

Fabbri, E. and Moon, T. W. (2015). Adrenergic signaling in teleost fish liver, a challenging path. *Comp. Biochem. Physiol. B Biochem. Mol. Biol.* http://dx.doi.org/10.1016/j.cbpb.2015.10.002. [Epub ahead of print].

Fabbri, E., Capuzzo, A. and Moon, T. W. (1998). The role of circulating catecholamines in the regulation of fish metabolism: an overview. *Comp. Biochem. Physiol. Part C* **120**, 177–192.

Fang, X., Corrales, J., Thornton, C., Scheffler, B. E. and Willett, K. L. (2013a). Global and gene specific DNA methylation changes during zebrafish development. *Comp. Biochem. Physiol. Part B Biochem. Mol. Biol.* **166**, 99–108.

Fang, X., Thornton, C., Scheffler, B. E. and Willett, K. L. (2013b). Benzo[a]pyrene decreases global and gene specific DNA methylation during zebrafish development. *Environ. Toxicol. Pharmacol.* **36**, 40–50.

Faught, E., Best, C. and Vijayan, M. M. (2016). Maternal stress-associated cortisol stimulation may protect embryos from cortisol excess in zebrafish. *R. Soc. Open Sci.* **3**, 160032.

Filby, A. L. and Tyler, C. R. (2007). Cloning and characterization of cDNAs for hormones and/or receptors of growth hormone, insulin-like growth factor I, thyroid hormone, and corticosteroid and the gender, tissue, and developmental specific expression of their mRNA transcripts in fathead minnow (*Pimephales promelas*). *Gen. Comp. Endocrinol.* **150**, 151–163.

Flik, G., Klaren, P. H. M., Van den Burg, E. H., Metz, J. R. and Huising, M. O. (2006). CRF and stress in fish. *Gen. Comp. Endocrinol.* **146**, 36–44.

Frolow, J. and Milligan, C. L. (2004). Hormonal regulation of glycogen metabolism in white muscle slices from rainbow trout (*Oncorhynchus mykiss* Walbaum). *Am. J. Physiol. Integr. Comp. Physiol.* **287**, 1344–1353.

Fryer, J., Leders, K. and Rivier, J. (1983). Urotensin I, a CRF-like neuropeptide stimulates ACTH release from the teleost pituitary. *Endrocrinology* **113**, 2308–2310.

Fuks, F. (2005). DNA methylation and histone modifications: teaming up to silence genes. *Curr. Opin. Genet. Dev.* **15**, 490–495.

Funder, J. W. (1997). Glucocorticoid and mineralocorticoid receptors: biology and clinical relevance. *Annu. Rev. Med.* **48**, 231–240.

Fuzzen, M. L. M., Van Der Kraak, G. and Bernier, N. J. (2010). Stirring up new ideas about the regulation of the hypothalamic-pituitary-interrenal axis in zebrafish (*Danio rerio*). *Zebrafish* **7**, 349–358.

Galt, N. J., Froehlich, J. M., Remily, E. A., Romero, S. R. and Biga, P. R. (2014). The effects of exogenous cortisol on myostatin transcription in rainbow trout, *Oncorhynchus mykiss*. *Comp. Biochem. Physiol. A Mol. Integr. Physiol.* **175**, 57–63.

Geislin, M. and Auperin, B. (2004). Relationship between changes in mRNAs of the genes encoding steroidogenic acute regulatory protein and P450 cholesterol side chain cleavage in head kidney and plasma levels of cortisol in response to different kinds of acute stress in the rainbow trout (*Oncorhynchus mykiss*). *Gen. Comp. Endocrinol.* **135**, 70–80.

Gesto, M., Soengas, J. L., Rodríguez-Illamola, A. and Míguez, J. M. (2014). Arginine vasotocin treatment induces a stress response and exerts a potent anorexigenic effect in rainbow trout, *Oncorhynchus mykiss*. *J. Neuroendocrinol.* **26**, 89–99.

Ghisleni, G., Capiotti, K. M., Da Silva, R. S., Oses, J. P., Piato, Â. L., Soares, V., et al. (2012). The role of CRH in behavioral responses to acute restraint stress in zebrafish. *Prog. Neuropsychopharmacol. Biol. Psychiatry* **36**, 176–182.

Geising, E. R., Suski, C. D., Warner, R. E. and Bell, A. M. (2011). Female sticklebacks transfer information via eggs: effects of maternal experience with predators on offspring. *Proc. Biol. Sci.* **278** (1712), 1753–1759.

Gravel, A. and Vijayan, M. M. (2006). Salicylate disrupts interrenal steroidogenesis and brain glucocorticoid receptor expression in rainbow trout. *Toxicol. Sci.* **93**, 41–49.

Greenwood, A. K., Butler, P. C., White, R. B., DeMarco, U., Pearce, D. and Fernald, R. D. (2003). Multiple corticosteroid receptors in a teleost fish: distinct sequences, expression patterns, and transcriptional activities. *Endocrinology* **144**, 4226–4236.

Griffiths, B. B., Schoonheim, P. J., Ziv, L., Voelker, L., Baier, H. and Gahtan, E. (2012). A zebrafish model of glucocorticoid resistance shows serotonergic modulation of the stress response. *Front Behav. Neurosci.* **6**, 1–10.

Hagen, I. J., Kusakabe, M. and Young, G. (2006). Effects of ACTH and cAMP on steroidogenic acute regulatory protein and P450 11β-hydroxylase messenger RNAs in rainbow trout interrenal cells: relationship with in vitro cortisol production. *Gen. Comp. Endocrinol.* **145**, 254–262.

Herzog, W., Zeng, X., Lele, Z., Sonntag, C., Ting, J. W., Chang, C. Y., et al. (2003). Adenohypophysis formation in the zebrafish and its dependence on sonic hedgehog. *Dev. Biol.* **254**, 36–49.

Hontela, A. and Vijayan, M. M. (2009). Adrenocortical toxicology in fishes. In: *Adrenal Toxicology, Target Organ Toxicology Series* (eds. P. W. Harvey, D. J. Everett and C. J. Springall), pp. 233–256. London: Informa Healthcare USA, Inc.

Hori, T. S., Gamperl, A., Booman, M., Nash, G. W. and Rise, M. L. (2012). A moderate increase in ambient temperature modulates the Atlantic cod (*Gadus morhua*) spleen transcriptome response to intraperitoneal viral mimic injection. *BMC Genomics* **13**, 431.

Huang, Y., Harrison, M. R., Osorio, A., Kim, J., Baugh, A., Duan, C., et al. (2013). Igf signaling is required for cardiomyocyte proliferation during zebrafish heart development and regeneration. *PLoS One* **8**, e67266.

Huising, M. O., Metz, J. R., van Schooten, C., Taverne-Thiele, A. J., Hermsen, T., Verburg-van Kemenade, B. M. L., et al. (2004). Structural characterisation of a cyprinid (*Cyprinus carpio* L.) CRH, CRH-BP and CRH-R1, and the role of these proteins in the acute stress response. *J. Mol. Endocrinol.* **32**, 627–648.

Huising, M. O., van der Aa, L. M., Metz, J. R., de Fátima Mazon, A., Kemenade, B. M. L. V. and Flik, G. (2007). Corticotropin-releasing factor (CRF) and CRF-binding protein expression in and release from the head kidney of common carp: evolutionary conservation of the adrenal CRF system. *J. Endocrinol.* **193**, 349–357.

Hwang, W. Y., Fu, Y., Reyon, D., Maeder, M. L., Tsai, S. Q., Sander, J. D., et al. (2013). Efficient genome editing in zebrafish using a CRISPR-Cas system. *Nat. Biotechnol.* **31**, 227–229.

Ings, J. S., Servos, M. R. and Vijayan, M. M. (2011a). Exposure to municipal wastewater effluent impacts stress performance in rainbow trout. *Aquat. Toxicol.* **103**, 85–91.

Ings, J. S., Servos, M. R. and Vijayan, M. M. (2011b). Hepatic transcriptomics and protein expression in rainbow trout exposed to municipal wastewater effluent. *Environ. Sci. Technol.* **45**, 2368–2376.

Ings, J. S., George, N., Peter, M. C. S., Servos, M. R. and Vijayan, M. M. (2012). Venlafaxine and atenolol disrupt epinephrine-stimulated glucose production in rainbow trout hepatocytes. *Aquat. Toxicol.* **106–107**, 48–55.

Ishikawa, T., Okada, T., Ishikawa-Fujiwara, T., Todo, T., Kamei, Y., Shigenobu, S., et al. (2013). ATF6α/β-mediated adjustment of ER chaperone levels is essential for development of the notochord in medaka fish-supplemental material. *Mol. Biol. Cell* **24**, 1387–1395.

Iwama, G. K. (2004). Are hsps suitable for indicating stressed states in fish? *J. Exp. Biol.* **207**, 15–19.

Iwama, G. K., Thomas, P. T., Forsyth, R. B. and Vijayan, M. M. (1998). Heat shock protein expression in fish. *Rev. Fish Biol. Fish.* **8**, 35–56.

Iwama, G. K., Vijayan, M. M., Forsyth, R. B. and Ackerman, P. A. (1999). Heat shock proteins and physiological stress in fish. *Integr. Comp. Biol.* **39**, 901–909.

Iwama, G. K., Afonso, L. O. B. and Vijayan, M. M. (2006). Stress in fishes. In *The Physiology of Fishes* (eds. D. H. Evans and J. B. Claiborne), pp. 319–342. Boca Raton, FL: CRC Press.

Janzen, W. J., Duncan, C. A. and Riley, L. G. (2012). Cortisol treatment reduces ghrelin signaling and food intake in tilapia, *Oreochromis mossambicus*. *Domest. Anim. Endocrinol.* **43**, 251–259.

Jao, L.-E., Wente, S. R. and Chen, W. (2013). Efficient multiplex biallelic zebrafish genome editing using a CRISPR nuclease system. *Proc. Natl. Acad. Sci.* **110**, 13904–13909.

Jeffrey, J. D., Gollock, M. J. and Gilmour, K. M. (2014). Social stress modulates the cortisol response to an acute stressor in rainbow trout (*Oncorhynchus mykiss*). *Gen. Comp. Endocrinol.* **196**, 8–16.

Johansen, I. B., Sandvik, G. K., Nilsson, G. E., Bakken, M. and Overli, O. (2011). Cortisol receptor expression differs in the brains of rainbow trout selected for divergent cortisol responses. *Comp. Biochem. Physiol. Part D Genomics Proteomics* **6**, 126–132.

Jorgensen, E. H., Bernier, N. J., Maule, A. G. and Vijayan, M. M. (2015). Effect of long-term fasting and a subsequent meal on mRNA abundances of hypothalamic appetite regulators, central and peripheral leptin expression and plasma leptin levels in rainbow trout. *Peptides* ⟨ http://dx.doi.org/10.1016/j.peptides.2015.08.010 ⟩.

Jurczyk, A., Roy, N., Bajwa, R., Gut, P., Lipson, K., Yang, C., et al. (2011). Dynamic glucoregulation and mammalian-like responses to metabolic and developmental disruption in zebrafish. *Gen. Comp. Endocrinol.* **170**, 334–345.

Kelly, S. P. and Chasiotis, H. (2011). Glucocorticoid and mineralocorticoid receptors regulate paracellular permeability in a primary cultured gill epithelium. *J. Exp. Biol.* **214**, 2308–2318.

Kelsall, C. J. and Balment, R. J. (1998). Native urotensins influence cortisol secretion and plasma cortisol concentration in the euryhaline flounder, *Platichthys flesus*. *Gen. Comp. Endocrinol.* **112**, 210–219.

Kiilerich, P., Kristiansen, K. and Madsen, S. S. (2007). Cortisol regulation of ion transporter mRNA in Atlantic salmon gill and the effect of salinity on the signaling pathway. *J. Endocrinol.* **194**, 417–427.

Kiilerich, P., Triqueneaux, G., Christensen, N. M., Trayer, V., Terrien, X., Lombès, M., et al. (2015). Interaction between the trout mineralocorticoid and glucocorticoid receptors in vitro. *J. Mol. Endocrinol.* **55**, 55–68.

Kim, H. and Kim, J. S. (2014). A guide to genome engineering with programmable nucleases. *Nat. Rev. Genet.* **15**, 321–334.

Kleppe, L., Karlsen, O., Edvardsen, R. B., Norberg, B., Andersson, E., Taranger, G. L., et al. (2013). Cortisol treatment of prespawning female cod affects cytogenesis related factors in eggs and embryos. *Gen. Comp. Endocrinol.* **189**, 84–95.

Kolmakov, N. N., Kube, M., Reinhardt, R. and Canario, A. V. (2008). Analysis of the goldfish Carassius auratus olfactory epithelium transcriptome reveals the presence of numerous non-olfactory GPCR and putative receptors for progestin pheromones. *BMC Genomics* **9**, 429.

Krasnov, A., Koskinen, H., Pehkonen, P., Rexroad, C. E., Afanasyev, S. and Mölsä, H. (2005). Gene expression in the brain and kidney of rainbow trout in response to handling stress. *BMC Genomics* **6**, 3.

Krasnov, A., Skugor, S., Todorcevic, M., Glover, K. A. and Nilsen, F. (2012). Gene expression in Atlantic salmon skin in response to infection with the parasitic copepod *Lepeophtheirus salmonis*, cortisol implant, and their combination. *BMC Genomics* **13**, 130.

Kregel, K. C. (2002). Heat shock proteins: modifying factors in physiological stress responses and acquired thermotolerance. *J. Appl. Physiol.* **92**, 2177–2186.

Krug, R. G., Poshusta, T. L., Skuster, K. J., Berg, M. R., Gardner, S. L. and Clark, K. J. (2014). A transgenic zebrafish model for monitoring glucocorticoid receptor activity. *Genes Brain Behav.* 478–487.

Kukreti, H., Amuthavalli, K., Harikumar, A., Sathiyamoorthy, S., Feng, P. Z., Anantharaj, R., et al. (2013). Muscle-specific MicroRNA1 (miR1) targets heat shock protein 70 (HSP70) during dexamethasone-mediated atrophy. *J. Biol. Chem.* **288**, 6663–6678.

Kure, E. H., Sæbø, M., Stangeland, A. M., Hamfjord, J., Hytterød, S., Heggenes, J., et al. (2013). Molecular responses to toxicological stressors: profiling microRNAs in wild Atlantic salmon (*Salmo salar*) exposed to acidic aluminum-rich water. *Aquat. Toxicol.* **138–139**, 98–104.

Kusakabe, M., Nakamura, I. and Young, G. (2003). 11B-Hydroxysteroid dehydrogenase complementary deoxyribonucleic acid in rainbow trout: cloning, sites of expression, and seasonal changes in gonads. *Endocrinology* **144**, 2534–2545.

Kusakabe, M., Zuccarelli, M. D., Nakamura, I. and Young, G. (2009). Steroidogenic acute regulatory protein in white sturgeon (*Acipenser transmontanus*): cDNA cloning, sites of expression and transcript abundance in corticosteroidogenic tissue after an acute stressor. *Gen. Comp. Endocrinol.* **162**, 233–240.

Lacroix, A. and Hontela, A. (2006). Role of calcium channels in cadmium-induced disruption of cortisol synthesis in rainbow trout (*Oncorhynchus mykiss*). *Comp. Biochem. Physiol. C Toxicol. Pharmacol.* **144**, 141–147.

Laird, P. W. and Jaenisch, R. (1996). The role of DNA methylation in cancer genetics and epigenetics. *Annu. Rev. Genet.* **30**, 441–464.

Laiz-Carrión, R., Fuentes, J., Redruello, B., Guzmán, J. M., Martín del Río, M. P., Power, D., et al. (2009). Expression of pituitary prolactin, growth hormone and somatolactin is modified in response to different stressors (salinity, crowding and food-deprivation) in gilthead sea bream *Sparus auratus*. *Gen. Comp. Endocrinol.* **162**, 293–300.

Leung, L. Y. and Woo, N. Y. (2010). Effects of growth hormone, insulin-like growth factor 1, triiodothyronine, thyroxine, and the cortisol on gene expression of carbohydrate metabolic enzymes in sea bream hepatocytes. *Comp. Biochem. Physiol. A Mol. Integr. Physiol.* **157**, 272–282.

Leung, L. Y., Kwong, A. K., Man, A. K. and Woo, N. Y. (2008). Direct actions of cortisol, thyroxine and growth hormone on IGF-1 mRNA expression in sea bream hepatocytes. *Comp. Biochem. Physiol. A Mol. Integr. Physiol.* **151**, 705–710.

Lewis, J. M., Hori, T. S., Rise, M. L., Walsh, P. J. and Currie, S. (2010). Transcriptome responses to heat stress in the nucleated red blood cells of the rainbow trout (*Oncorhynchus mykiss*). *Physiol. Genomics* **42**, 361–373.

Li, C. C., Maloney, C. A., Cropley, J. E. and Suter, C. M. (2010). Epigenetic programming by maternal nutrition: shaping future generations. *Epigenomics* **2**, 539–549.

Li, M., Christie, H. L. and Leatherland, J. F. (2012). The in vitro metabolism of cortisol by ovarian follicles of rainbow trout (*Oncorhynchus mykiss*): comparison with ovulated oocytes and pre-hatch embryos. *Reproduction* **144**, 713–722.

Li, M. H., Yang, H. H., Li, M. R., Sun, Y. L., Jiang, X. L., Xie, Q. P., et al. (2013). Antagonistic roles of Dmrt1 and Foxl2 in sex differentiation via estrogen production in tilapia as demonstrated by TALENs. *Endocrinology* **154**, 4814–4825.

Li, M., Yang, H., Zhao, J., Fang, L., Shi, H., Li, M., et al. (2014). Efficient and heritable gene targeting in tilapia by CRISPR/Cas9. *Genetics* **197**, 591–599.

Liang, B., Wei, D.-L., Cheng, Y.-N., Yuan, H.-J., Lin, J., Cui, X.-Z., et al. (2013). Restraint stress impairs oocyte developmental potential: role of CRH-Induced apoptosis of ovarian cells. *Biol. Reprod.* **89**, 1–12.

Liang, L., Schmid, K., Sandhu, N., Angleson, J. K., Vijayan, M. M. and Dores, R. M. (2015). Structure/function studies on the activation of the rainbow trout melanocortin-2 receptor. *Gen. Comp. Endocrinol.* **210**, 145–151.

Liu, N. A., Huang, H., Yang, Z., Herzog, W., Hammerschmidt, M., Lin, S., et al. (2003). Pituitary corticotroph ontogeny and regulation in transgenic zebrafish. *Mol. Endocrinol.* **17**, 959–966.

Liu, Y., Yuan, C., Chen, S., Zheng, Y., Zhang, Y., Gao, J., et al. (2014). Global and cyp19a1a gene specific DNA methylation in gonads of adult rare minnow *Gobiocypris rarus* under bisphenol A exposure. *Aquat. Toxicol.* **156**, 10–16.

Loosli, F., Staub, W., Finger-Baier, K. C., Ober, E. A., Verkade, H., Wittbrodt, J., et al. (2003). Loss of eyes in zebrafish caused by mutation of chokh/rx3. *EMBO Rep.* **4**, 894–899.

Lopez-Patino, M., Gesto, M., Conde-Siera, M., Soengas, J. L. and Miguez, J. M. (2014). Stress inhibition of melatonin synthesis in the pineal organ of rainbow trout (*Oncorhynchus mykiss*) is mediated by cortisol. *J. Exp. Biol.* **217**, 1407–1416.

Lorenzi, V., Mehinto, A. C., Denslow, N. D. and Schlenk, D. (2012). Effects of exposure to the β-blocker propranolol on the reproductive behavior and gene expression of the fathead minnow, *Pimephales promelas*. *Aquat. Toxicol.* **116–117**, 8–15.

Ma, P., Liu, Y., Reddy, K. P., Chan, W. K. and Lam, T. J. (2004). Characterization of the seabass pancreatic α-amylase gene and promoter. *Gen. Comp. Endocrinol.* **137**, 78–88.

MacKenzie, S., Iliev, D., Liarte, C., Koskinen, H., Planas, J. V., Goetz, F. W., et al. (2006a). Transcriptional analysis of LPS-stimulated activation of trout (*Oncorhynchus mykiss*) monocyte/macrophage cells in primary culture treated with cortisol. *Mol. Immunol.* **43**, 1340–1348.

MacKenzie, S., Montserrat, N., Mas, M., Acerete, L., Tort, L., Krasnov, A., et al. (2006b). Bacterial lipopolysaccharide induces apoptosis in the trout ovary. *Reprod. Biol. Endocrinol.* **4**, 46.

Madan, A. P. and DeFranco, D. B. (1993). Bidirectional transport of glucocorticoid receptors across the nuclear envelope. *Proc. Natl. Acad. Sci. U.S.A.* **90**, 3588–3592.

Madison, B. M., Tavakoli, S., Kramer, S. and Bernier, N. J. (2015). Chronic cortisol and the regulation of food intake and the encordine growth axis in rainbow trout. *J. Endocrinol.* **226**, 103–119.

Manna, P. R., Eubank, D. W., Lalli, E., Sassone-Corsi, P. and Stocco, D. M. (2003). Transcriptional regulation of the mouse steroidogenic acute regulatory protein gene by the cAMP response-element binding protein and steroidogenic factor 1. *J. Mol. Endocrinol.* **30**, 381–397.

Manuel, R., Gorissen, M., Zethof, J., Ebbesson, L. O. E., van de Vis, H., Flik, G., et al. (2014). Unpredictable chronic stress decreases inhibitory avoidance learning in Tuebingen long-fin zebrafish: stronger effects in the resting phase than in the active phase. *J. Exp. Biol.* **217**, 3919–3928.

Manuel, R., Zethof, J., Flik, G. and van den Bos, R. (2015). Providing a food reward reduces inhibitory avoidance learning in zebrafish. *Behav. Process.* **120**, 69–72.

Mattick, J. S. (2006). Non-coding RNA. *Hum. Mol. Genet.* **15**, R17–R29.

Mccormick, M. I. (1999). Experimental test of the effect of maternal hormones on larval quality of a coral reef fish. *Oecologia* **118**, 412–422.

McCormick, S. D., Regish, A., O'Dea, M. F. and Shrimpton, J. M. (2008). Are we missing a mineralocorticoid in teleost fish? Effects of cortisol, deoxycorticosterone and aldosterone on osmoregulation, gill Na+,K+ -ATPase activity and isoform mRNA levels in Atlantic salmon. *Gen. Comp. Endocrinol.* **157**, 35–40.

McDonald, M. D., Vulesevic, B., Perry, S. F. and Walsh, P. J. (2009). Urea transporter and glutamine synthetase regulation and localization in gulf toadfish gill. *J. Exp. Biol.* **212**, 704–712.

McGowan, P. O., Sasaki, A., D'Alessio, A. C., Dymov, S., Labonté, B., Szyf, M., et al. (2009). Epigenetic regulation of the glucocorticoid receptor in human brain associates with childhood abuse. *Nat. Neurosci.* **12**, 342–348.

Medeiros, L. R., Cartolano, M. C. and McDonald, M. D. (2014). Crowding stress inhibits serotonin 1A receptor-mediated increases in corticotropin-releasing factor mRNA expression and adrenocorticotropin hormone secretion in the Gulf toadfish. *J. Comp. Physiol. B Biochem. Syst. Environ. Physiol.* **184**, 259–271.

Melnyk-Lamont, N., Best, C., Gesto, M. and Vijayan, M. M. (2014). The antidepressant venlafaxine disrupts brain monoamine levels and neuroendocrine responses to stress in rainbow trout. *Environ. Sci. Technol.* **48**, 13434–13442.

Mennigen, J. A., Lado, W. E., Zamora, J. M., Duarte-Guterman, P., Langlois, V. S., Metcalfe, C. D., et al. (2010). Waterborne fluoxetine disrupts the reproductive axis in sexually mature male goldfish, *Carassius auratus.* *Aquat. Toxicol.* **100** (4), 354–364.

Milla, S., Terrien, X., Sturm, A., Ibrahim, F., Giton, F., Fiet, J., et al. (2008). Plasma 11-deoxycorticosterone (DOC) and mineralocorticoid receptor testicular expression during rainbow trout *Oncorhynchus mykiss* spermiation: implication with 17a,20B-dihydroxyprogesterone on the milt fluidity? *Reprod. Biol. Endocrinol.* **6**, 19.

Milligan, C. L. (1997). The role of cortisol in amino acid mobilization and metabolism following exhaustive exercise in rainbow trout (*Oncorhynchus mykiss* Walbaum). *Fish Physiol. Biochem.* **16**, 119–128.

Milligan, C. L. (2003). A regulatory role for cortisol in muscle glycogen metabolism in rainbow trout *Oncorhynchus mykiss* Walbaum. *J. Exp. Biol.* **206**, 3167–3173.

Mommsen, T. P., Vijayan, M. M. and Moon, T. W. (1999). Cortisol in teleosts: dynamics, mechanisms of action, and metabolic regulation. *Rev. Fish Biol. Fish.* **9**, 211–268.

Momoda, T. S., Schwindt, A. R., Feist, G. W., Gerwick, L., Bayne, C. J. and Schreck, C. B. (2007). Gene expression in the liver of rainbow trout, *Oncorhynchus mykiss*, during the stress response. *Comp. Biochem. Physiol. Part D Genomics Proteomics* **2**, 303–315.

Morishima, Y., Murphy, P. J. M., Li, D.-P., Sanchez, E. R. and Pratt, W. B. (2000). Stepwise assembly of a glucocorticoid receptor · hsp90 heterocomplex resolves two sequential ATP-dependent events involving first Hsp70 and then Hsp90 in opening of the steroid binding pocket. *J. Biol. Chem.* **275**, 18054–18060.

Mowbray, C., Hammerschmidt, M. and Whitfield, T. T. (2001). Expression of BMP signalling pathway members in the developing zebrafish inner ear and lateral line. *Mech. Dev.* **108**, 179–184.

Muto, A., Orger, M. B., Wehman, A. M., Smear, M. C., Kay, J. N., Page-McCaw, P. S., et al. (2005). Forward genetic analysis of visual behavior in zebrafish. *PLoS Genet.* **1**, e66.

Muto, A., Taylor, M. R., Suzawa, M., Korenbrot, J. I. and Baier, H. (2013). Glucocorticoid receptor activity regulates light adaptation in the zebrafish retina. *Front Neural. Circuits* **7**, 145.

Nakano, T., Afonso, L. O. B., Beckman, B. R., Iwama, G. K. and Devlin, R. H. (2013). Acute physiological stress down-regulates mRNA expressions of growth-related genes in coho salmon. *PLoS One* **8**, e71421.

Neave, B., Holder, N. and Patient, R. (1997). A graded response to BMP-4 spatially coordinates patterning of the mesoderm and ectoderm in the zebrafish. *Mech. Dev.* **62**, 183–195.

Nesan, D. and Vijayan, M. M. (2012). Embryo exposure to elevated cortisol level leads to cardiac performance dysfunction in zebrafish. *Mol. Cell. Endocrinol.* **363**, 85–91.

Nesan, D. and Vijayan, M. M. (2013a). Role of glucocorticoid in developmental programming: evidence from zebrafish. *Gen. Comp. Endocrinol.* **181**, 35–44.

Nesan, D. and Vijayan, M. M. (2013b). The transcriptomics of glucocorticoid receptor signaling in developing zebrafish. *PLoS One* **8**, e80726.

Nesan, D. and Vijayan, M. M. (2016). Maternal cortisol mediates hypothalamus-pituitary-interrenal axis development in zebrafish. *Sci. Rep.* **6**, 22582.

Nesan, D., Kamkar, M., Burrows, J., Scott, I. C., Marsden, M. and Vijayan, M. M. (2012). Glucocorticoid receptor signaling is essential for mesoderm formation and muscle development in zebrafish. *Endocrinol.* **153**, 1288–1300.

Oleksiak, M. F., Karchner, S. I., Jenny, M. J., Franks, D. G., Welch, D. B. M. and Hahn, M. E. (2011). Transcriptomic assessment of resistance to effects of an aryl hydrocarbon receptor (AHR) agonist in embryos of Atlantic killifish (Fundulus heteroclitus) from a marine Superfund site. *BMC Genomics* **12**, 263.

Olsson, P. E., Kling, P., Erkell, L. J. and Kille, P. (1995). Structural and functional analysis of the rainbow trout (*Oncorhyncus mykiss*) metallothionein-A gene. *Eur. J. Biochem.* **230**, 344–349.

Ortega, V. A., Lovejoy, D. A. and Bernier, N. J. (2013). Appetite-suppressing effects and interactions of centrally administered corticotropin-releasing factor, urotensin I and serotonin in rainbow trout (*Oncorhynchus mykiss*). *Front Neurosci.* **7**, 196.

Pankhurst, N. W. (2016). Reproduction and Development. In *Fish Physiology-Biology of Stress in Fish*, Vol. 35 (eds. C. B. Schreck, L. Tort, A. P. Farrell and C. J. Brauner), San Diego, CA: Academic Press.

Pepels, P. P. L. M., Meek, J., Wendelaar Bonga, S. E. and Balm, P. H. M. (2002). Distribution and quantification of the corticotropn-releasing hormone in the brain of the teleost fish *Oreochromis mossambicus* (tilapia). *J. Comp. Neurol.* **453**, 247–268.

Pérez-Jiménez, A., Guedes, M. J., Morales, A. E. and Oliva-Teles, A. (2007). Metabolic responses to short starvation and refeeding in *Dicentrarchus labrax*. Effect of dietary composition. *Aquaculture* **265**, 325–335.

Philip, A. M. and Vijayan, M. M. (2015). Stress-immune-growth interactions: cortisol modulates suppressors of cytokine signaling and JAK/STAT pathway in rainbow trout liver. *PLoS One* **10**, e0129299.

Philip, A. M., Daniel Kim, S. and Vijayan, M. M. (2012). Cortisol modulates the expression of cytokines and suppressors of cytokine signaling (SOCS) in rainbow trout hepatocytes. *Dev. Comp. Immunol.* **38**, 360–367.

Pierce, A. L., Dickey, J. T., Felli, L., Swanson, P. and Dickhoff, W. W. (2010). Metabolic hormones regulate basal and growth hormone-dependent igf 2 mRNA level in primary cultured coho salmon hepatocytes: effects of insulin, glucagon, dexamethasone, and triiodothyronine. *J. Endocrinol.* **204**, 331–339.

Pierce, A. L., Breves, J. P., Moriyama, S., Hirano, T. and Grau, G. E. (2011). Differential regulation of Igf1 and Igf2 mRNA levels in tilapia hepatocytes: effects of insulin and cortisol on GH sensitivity. *J. Endocrinol.* **211**, 187–200.

Pierron, F., Baillon, L., Sow, M., Gotreau, S. and Gonzalez, P. (2014a). Effect of low-dose cadmium exposure on DNA methylation in the endangered European eel. *Environ. Sci. Technol.* **48**, 797–803.

Pierron, F., Bureau du Colombier, S., Moffett, A., Caron, A., Peluhet, L., Daffe, G., et al. (2014b). Abnormal ovarian DNA methylation programming during gonad maturation in wild contaminated fish. *Environ. Sci. Technol.* **48**, 11688–11695.

Pikulkaew, S., Benato, F., Celeghin, A., Zucal, C., Skobo, T., Colombo, L., et al. (2011). The knockdown of maternal glucocorticoid receptor mRNA alters embryo development in zebrafish. *Dev. Dyn.* **240**, 874–889.

Pippal, J. B., Cheung, C. M. I., Yao, Y.-Z., Brennan, F. E. and Fuller, P. J. (2011). Characterization of the zebrafish (*Danio rerio*) mineralocorticoid receptor. *Mol. Cell. Endocrinol.* **332**, 58–66.

Pockley, A. G., Muthana, M. and Calderwood, S. K. (2008). The dual immunoregulatory roles of stress proteins. *Trends Biochem. Sci.* **33**, 71–79.

Poulain, M., Fürthauer, M., Thisse, B., Thisse, C. and Lepage, T. (2006). Zebrafish endoderm formation is regulated by combinatorial Nodal, FGF and BMP signalling. *Development* **133**, 2189–2200.

Pratt, W. B., Morishima, Y., Murphy, M. and Harrell, M. (2006). Chaperoning of glucocorticoid receptors. *Handb. Exp. Pharmacol.* 111–138.

Prunet, P., Sturm, A. and Milla, S. (2006). Multiple corticosteroid receptors in fish: from old ideas to new concepts. *Gen. Comp. Endocrinol.* **147**, 17–23.

Reid, S. D., Moon, T. W. and Perry, S. F. (1992). Rainbow trout hepatocyte adrenoceptors, catecholamine responsiveness and effects of cortisol. *Am. Physiol. Soc.* **262**, 794–799.

Reid, S. G., Bernier, N. J. and Perry, S. F. (1998). The adrenergic stress response in fish: control of catecholamine storage and release. *Comp. Biochem. Physiol. Part C* **120**, 1–27.

Reik, W., Santos, F. and Dean, W. (2003). Mammalian epigenomics: reprogramming the genome for development and therapy. *Theriogenology* **59**, 21–32.

Reindl, K. M. and Sheridan, M. A. (2012). Peripheral regulation of the growth hormone-insulin-like growth factor system in fish and other vertebrates. *Comp. Biochem. Physiol. A Mol. Integr. Physiol.* **163**, 231–245.

Reinick, C. L., Liang, L., Angleson, J. K. and Dores, R. M. (2012). Identification of an MRAP-independent melanocortin-2 receptor: functional expression of the cartilaginous fish, *Callorhinchus milii*, melanocortin-2 receptor in CHO cells. *Endocrinology* **153**, 4757–4765.

Rise, M. L., Jones, S. R. M., Brown, G. D., von Schalburg, K. R., Davidson, W. S. and Koop, B. F. (2004). Microarray analyses identify molecular biomarkers of Atlantic salmon macrophage and hematopoietic kidney response to Piscirickettsia salmonis infection. *Physiol. Genomics* **20**, 21–35.

Richter, K., Haslbeck, M. and Buchner, J. (2010). The heat shock response: life on the verge of death. *Mol. Cell* **40**, 253–266.

Rodela, T. M., Esbaugh, A. J., McDonald, M. D., Gilmour, K. M. and Walsh, P. J. (2011). Evidence for transcriptional regulation of the urea transporter in the gill of the Gulf toadfish, *Opsanus beta*. *Comp. Biochem. Physiol. B Biochem. Mol. Biol.* **160**, 72–80.

Rodgers, B. D., Weber, G. M., Kelley, K. M. and Levine, M. A. (2003). Prolonged fasting and cortisol reduce myostatin mRNA levels in tilapia larvae; short-term fasting elevates. *Am. J. Physiol. Regul. Integr. Comp. Physiol.* **284**, R1277–R1286.

Ronco, A. M., Llaguno, E., Epuñan, M. J. and Llanos, M. N. (2010). Effect of cadmium on cortisol production and 11β-hydroxysteroid dehydrogenase 2 expression by cultured human choriocarcinoma cells (JEG-3). *Toxicol. Vitro* **24**, 1532–1537.

Roy, B. and Jacobson, A. (2013). The intimate relationships of mRNA decay and translation. *Trends Genet* **29**, 691–699.

Salem, M., Xiao, C., Womack, J., Rexroad, C. E. and Yao, J. (2010). A microRNA repertoire for functional genome research in rainbow trout (Oncorhynchus mykiss). *Mar. Biotechnol.* **12**, 410–429.

Salem, M., Paneru, B., Al-Tobasei, R., Abdouni, F., Thorgaard, G. H., Rexroad, C. E., et al. (2015). Transcriptome assembly, gene annotation and tissue gene expression atlas of the rainbow trout. *PLoS One* **10**, 1–27.

Sánchez, C. C., Weber, G. M., Gao, G., Cleveland, B. M., Yao, J. and Rexroad, C. E. (2011). Generation of a reference transcriptome for evaluating rainbow trout responses to various stressors. *BMC Genomics* **12**, 626.

Sandhu, N. and Vijayan, M. M. (2011). Cadmium-mediated disruption of cortisol biosynthesis involves suppression of corticosteroidogenic genes in rainbow trout. *Aquat. Toxicol.* **103**, 92–100.

Sandhu, N., McGeer, J. C. and Vijayan, M. M. (2014). Exposure to environmental levels of waterborne cadmium impacts corticosteroidogenic and metabolic capacities, and compromises secondary stressor performance in rainbow trout. *Aquat. Toxicol.* **146**, 20–27.

Sarkar, A., Ray, D., Shrivastava, A. N. and Sarker, S. (2006). Molecular biomarkers: their significance and application in marine pollution monitoring. *Ecotoxicology* **15**, 333–340.

Sathiyaa, R. and Vijayan, M. M. (2003). Autoregulation of glucocorticoid receptor by cortisol in rainbow trout hepatocytes. *Am. J. Physiol. Cell Physiol.* **284**, 1508–1515.

Sathiyaa, R., Campbell, T. and Vijayan, M. M. (2001). Cortisol modulates HSP90 mRNA expression in primary cultures of trout hepatocytes. *Comp. Biochem. Physiol. B Comp. Biochem.* **129**, 679–685.

Schaaf, M. J. M., Champagne, D., van Laanen, I. H. C., van Wijk, D. C. W. A., Meijer, A. H. and Meijer, O. C. (2008). Discovery of a functional glucocorticoid receptor B-isoform in zebrafish. *Endocrinology* **149**, 1591–1599.

Schaaf, M. J. M., Chatzopoulou, A. and Spaink, H. P. (2009). The zebrafish as a model system for glucocorticoid receptor research. *Comp. Biochem. Physiol. Part A* **153**, 75–82.

Schoneveld, O. J. L. M., Gaemers, I. C. and Lamers, W. H. (2004). Mechanisms of glucocorticoid signalling. *Biochim. Biophys. Acta Gene Struct. Expr.* **1680**, 114–128.

Schulte, P. M., Glemet, H. C., Fiebig, A. A. and Powers, D. A. (2000). Adaptive variation in lactate dehydrogenase-B gene expression: role of a stress-responsive regulatory element. *Proc. Natl. Acad. Sci.* **97**, 6597–6602.

Sexton, T. and Cavalli, G. (2015). The role of chromosome domains in shaping the functional genome. *Cell* **160**, 1049–1059.

Shelley, L. K., Ross, P. S., Miller, K. M., Kaukinen, K. H. and Kennedy, C. J. (2012). Toxicity of atrazine and nonylphenol in juvenile rainbow trout (*Oncorhynchus mykiss*): effects on general health, disease susceptibility and gene expression. *Aquat. Toxicol.* **124–125**, 217–226.

Shi, Y., Mosser, D. D. and Morimoto, R. I. (1998). Molecular chaperones as HSF1-specific transcriptional repressors. *Genes Dev.* **12**, 654–666.

Skinner, M. K. (2014). Environmental stress and epigenetic transgenerational inheritance. *BMC Med.* **12**, 153.

Small, B. C., Murdock, C. A., Waldbieser, G. C. and Peterson, B. C. (2006). Reduction in channel catfish hepatic growth hormone receptor expression in response to food deprivation and exogenous cortisol. *Domest. Anim. Endocrinol.* **31**, 340–356.

Soengas, J. L., Polakof, S., Chen, X., Sangiao-Alvarellos, S. and Moon, T. W. (2006). Glucokinase and hexokinase expression and activitis in rainbow trout tissues: changes with food depravation and refeeding. *Am. J. Physiol.* **291**, R810–R821.

Stolte, E. H., van Kemenade, B. M. L. V., Savelkoul, H. F. J. and Flik, G. (2006). Evolution of glucocorticoid receptors with different glucocorticoid sensitivity. *J. Endocrinol.* **190**, 17–28.

Stolte, E. H., Nabuurs, S. B., Bury, N. R., Sturm, A., Flik, G., Savelkoul, H. F. J., et al. (2008). Stress and innate immunity in carp: corticosteroid receptors and pro-inflammatory cytokines. *Mol. Immunol.* **46**, 70–79.

Storey, K. B. (1997). Organic solutes in freezing tolerance. *Comp. Biochem. Physiol. Part A Physiol.* **117**, 319–326.

Stratholt, M. L., Donaldson, E. M. and Liley, N. R. (1997). Stress induced female coho reflected in appear elevation of plasma cortisol in adult salmon (*Oncorhynchus kisutch*), is egg cortisol content, but does not to affect early development. *Aquaculture* **158**, 141–153.

Takahashi, A. and Kawauchi, H. (2006). Evolution of melanocortin systems in fish. *Gen. Comp. Endocrinol.* **148**, 85–94.

Takahashi, H. and Sakamoto, T. (2013). The role of "mineralocorticoids" in teleost fish: relative importance of glucocorticoid signaling in the osmoregulation and "central" actions of mineralocorticoid receptor. *Gen. Comp. Endocrinol.* **181**, 223–228.

Takei, Y., and Hwang, P.-P. (2016). Homeostatic Responses to Osmotic Stress. In *Fish Physiology-Biology of Stress in Fish*, Vol. 35 (eds. C.B. Schreck, L. Tort, A.P. Farrell, and C.J. Brauner), San Diego, CA: Academic Press.

Teles, M., Boltaña, S., Reyes-López, F., Santos, M. A., Mackenzie, S. and Tort, L. (2013). Effects of chronic cortisol administration on global expression of GR and the liver transcriptome in *Sparus aurata*. *Mar. Biotechnol. (NY)* **15**, 104–114.

Terova, G., Gornati, R., Rimoldi, S., Bernadini, G. and Saroglia, M. (2005). Quantification of a glucocorticoid receptor in sea bass (*Dicentrarchus labrax* L.) reared at high stocking density. *Gene* **357**, 144–151.

Thiel, G., Al Sarraj, J. and Stefano, L. (2005). cAMP response element binding protein (CREB) activates transcription via two distinct genetic elements of the human glucose-6-phosphatase gene. *BMC Mol. Biol.* **6**, 1–4.

Thomas, M. A. and Klaper, R. D. (2012). Psychoactive pharmaceuticals induce fish gene expression profiles associated with human idiopathic autism. *PLoS One* **7**, e32917.

Tipsmark, C. K., Jørgensen, C., Brande-Lavridsen, N., Engelund, M., Olesen, J. H. and Madsen, S. S. (2009). Effects of cortisol, growth hormone and prolactin on gill claudin expression in Atlantic salmon. *Gen. Comp. Endocrinol.* **163**, 270–277.

Tiso, N., Filippi, A., Pauls, S., Bortolussi, M. and Argenton, F. (2002). BMP signalling regulates anteroposterior endoderm patterning in zebrafish. *Mech. Dev.* **118**, 29–37.

To, T. T., Hahner, S., Nica, G., Rohr, K. B., Hammerschmidt, M., Winkler, C., et al. (2007). Pituitary–interrenal interaction in zebrafish interrenal organ development. *Mol. Endocrinol.* **21**, 472–485.

Tokarz, J., Mindnich, R., Norton, W., Möller, G., Hrabé de Angelis, M., et al. (2012). Discovery of a novel enzyme mediating glucocorticoid catabolism in fish: 20beta-hydroxysteroid dehydrogenase type 2. *Mol. Cell. Endocrinol.* **349**, 202–213.

Tokarz, J., Norton, W., Möller, G., Hrabé de Angelis, M. and Adamski, J. (2013). Zebrafish 20β-hydroxysteroid dehydrogenase type 2 is important for glucocorticoid catabolism in stress response. *PLoS One* **8**, e54851.

Tort, L. (2011). Stress and immune modulation in fish. *Dev. Comp. Immunol.* **35**, 1366–1375.

Turecki, G. and Meaney, M. (2014). Effects of the social environment and stress on glucocorticoid receptor gene methylation: a systematic review. *Biol. Psychiatry* 1–10.

Venkatesh, S. and Workman, J. L. (2015). Histone exchange, chromatin structure and the regulation of transcription. *Nat. Rev. Mol. Cell Biol.* **16**, 178–189.

Villeneuve, D. L., Knoebl, I., Larkin, P., Miracle, A. L., Carter, B. J., Denslow, N. D., et al. (2008). Altered gene expression in the brain and liver of female fathead minnows Pimephales promelas Rafinesque exposed to fadrozole. *J. Fish Biol.* **72**, 2281–2340.

Vijayan, M. M., Raptis, S. and Sathiyaa, R. (2003). Cortisol treatment affects glucocorticoid receptor and glucocorticoid-responsive genes in the liver of rainbow trout. *Gen. Comp. Endocrinol.* **132**, 256–263.

Vijayan, M. M., Prunet, P. and Boone, A. N. (2005). Xenobiotic impact on corticosteroid signalling. In *Biochemical and Molecular Biology of Fishes, Vol 6. Environmental Toxicology* (eds. T. W. Moon and T. P. Mommsen), pp. 365–394. Amsterdam: Elsevier.

Vijayan, M. M., Aluru, N. and Leatherland, J. F. (2010). Stress response and the role of cortisol. In *Fish Diseases and Disorders. Vol 2: Non-Infectious Disorders* (eds. J. F. Leatherland and P. Woo), pp. 182–201. Oxfordshire: CABI.

Wang, R.-L., Bencic, D. C., Garcia-Reyero, N., Perkins, E. J., Villeneuve, D. L., Ankley, G. T., et al. (2014). Natural Variation in Fish Transcriptomes: Comparative Analysis of the Fathead Minnow (Pimephales promelas) and Zebrafish (Danio rerio). *PLoS One* **9**, e114178.

Wang, T. and Hong, Y. (2014). Direct gene disruption by TALENs in medaka embryos. *Gene* **543**, 28–33.

Weaver, I. C. G., Cervoni, N., Champagne, F. A. D.', Alessio, A. C., Sharma, S., Seckl, J. R., et al. (2004). Epigenetic programming by maternal behavior. *Nat. Neurosci.* **7**, 847–854.

Wendelaar Bonga, S. E. (1997). The stress response in fish. *Physiol. Rev.* **77**, 591–625.

Whitehead, A., Roach, J. L., Zhang, S. and Galvez, F. (2011). Genomic mechanisms of evolved physiological plasticity in killifish distributed along an environmental salinity gradient. *Proc. Natl. Acad. Sci.* **108**, 6193–6198.

Williams, T. D., Lee, J. S., Sheader, D. L. and Chipman, J. K. (2000). The cytochrome P450 1A gene (CYP1A) from European flounder (*Platichthys flesus*), analysis of regulatory regions and development of a dual luciferase reporter gene system. *Mar. Environ. Res.* **50**, 1–6.

Wilson, K. S., Matrone, G., Livingstone, D. E. W., Al-Dujaili, E. A. S., Mullins, J. J., Tucker, C. S., et al. (2013). Physiological roles of glucocorticoids during early embryonic development of the zebrafish (*Danio rerio*). *J. Physiol.* **591**, 6209–6220.

Wilson, K. S., Baily, J., Tucker, C. S., Matrone, G., Vass, S., Moran, C., et al. (2015). Early-life perturbations in glucocorticoid activity impacts on the structure, function and molecular composition of the adult zebrafish (*Danio rerio*) heart. *Mol. Cell. Endocrinol.* **414**, 120–131.

Wiseman, S., Osachoff, H., Bassett, E., Malhotra, J., Bruno, J., VanAggelen, G., et al. (2007). Gene expression pattern in the liver during recovery from an acute stressor in rainbow trout. *Comp. Biochem. Physiol. Part D Genomics Proteomics* **2**, 234–244.

Wunderink, Y. S., Martínez-Rodríguez, G., Yúfera, M., Martín Montero, I., Flik, G., Mancera, J. M., et al. (2012). Food deprivation induces chronic stress and affects thyroid hormone metabolism in Senegalese sole (*Solea senegalensis*) post-larvae. *Comp. Biochem. Physiol. Part A Mol. Integr. Physiol.* **162**, 317–322.

Xia, J. H., Liu, P., Liu, F., Lin, G., Sun, F., Tu, R., et al. (2013). Analysis of stress-responsive transcriptome in the intestine of Asian seabass (*Lates calcarifer*) using RNA-Seq. *DNA Res.* **2**, 449–460.

Yada, T. and Tort, L. (2016). Stress and Disease Resistance: Immune System and Immunoendocrine Interactions. In *Fish Physiology-Biology of Stress in Fish*, Vol. 35 (eds. C. B. Schreck, L. Tort, A. P. Farrell and C. J. Brauner), San Diego, CA: Academic Press.

Yan, H., Simola, D. F., Bonasio, R., Liebig, J., Berger, S. L. and Reinberg, D. (2014). Eusocial insects as emerging models for behavioural epigenetics. *Nat. Rev. Genet.* **15**, 677–688.

Yang, B.-Y., Chan, K.-M., Lin, C.-M. and Chen, T. T. (1997). Characterization of rainbow trout (*Oncorhynchus mykiss*) growth hormone 1 gene and the promoter region of growth hormone 2 hene. *Arch. Biochem. Biophys.* **340**, 359–368.

Yano, A., Nicol, B., Jouanno, E. and Guiguen, Y. (2014). Heritable targeted inactivation of the rainbow trout (*Oncorhynchus mykiss*) master sex-determining gene using zinc-finger nucleases. *Mar. Biotechnol.* **16**, 243–250.

Yeh, C.-M. (2015). The basal NPO crh fluctuation is sustained under compromised glucocorticoid signaling in diurnal zebrafish. *Front Neurosci.* 〈doi: 10.3389/fnins.2015.00436〉.

Zannas, A. S. and West, A. E. (2015). Epigenetics and the regulation of stress vulnerability and reslience. *Neuroscience* **264**, 157–170.

Zhang, T. Y., Labonté, B., Wen, X. L., Turecki, G. and Meaney, M. J. (2013). Epigenetic mechanisms for the early environmental regulation of hippocampal glucocorticoid receptor gene expression in rodents and humans. *Neuropsychopharmacology* **38**, 111–123.

Zheng, J., Xiao, X., Zhang, Q. and Yu, M. (2014). DNA methylation: the pivotal interaction between early-life nutrition and glucose metabolism in later life. *Br. J. Nutr.* **112**, 1850–1857.

Ziv, L., Muto, A., Schoonheim, P. J., Meijsing, S. H., Strasser, D., Ingraham, H. A., et al. (2013). An affective disorder in zebrafish with mutation of the glucocorticoid receptor. *Mol. Psychiatry* **18**, 681–691.

5

STRESS AND GROWTH

BASTIEN SADOUL

MATHILAKATH M. VIJAYAN

1. Introduction
2. A Conceptual Framework for Growth
   2.1. The Dynamic Energy Budget (DEB) Model for Growth
   2.2. Myocyte Growth
3. Stress Effect on Energy Available for Growth
   3.1. Food Intake
   3.2. Energy Substrate Absorption at the Gut
   3.3. Energy Demand for Maintenance
4. Stress Effects on Promoters of Muscle Formation
   4.1. Stress Effects on Myogenesis
   4.2. Stress Regulation of the GH–IGF Axis
5. Conclusion and Knowledge Gaps

In fish, muscle contributes over half the body mass and thus changes in the size of this organ system is considered to be of primary importance to growth. Muscle growth is the final consequence of a complex set of processes starting with the absorption of nutrients from the environment to their allocation for increases in myocyte number and size. Stress affects these processes, including energy utilization, absorption, and allocation, resulting in reduced muscle growth. Situations encountered in the wild such as predator avoidance or deteriorated water quality, or in captive situations such as handling, crowding, sorting, and grading may be stressful to the animal, entailing reallocation of energy away from growth to cope with stress. Here we highlight the potential role of cortisol, the principal glucocorticoid in teleosts, in regulating processes leading to muscle growth suppression. Stress-mediated elevation in plasma cortisol level affects energy intake, absorption, and utilization, including protein turnover and modulation of protective proteins expression, all leading to a decrease in

Biology of Stress in Fish: Volume 35
FISH PHYSIOLOGY

energy available for muscle growth. Cortisol also modulates muscle growth regulators, including growth factors and transcription factors, causing growth suppression. Overall, there is a paucity of information on the mechanisms leading to stress effects on growth, but the available literature suggests a key role for stressor-mediated elevation in circulating cortisol levels and modulating muscle protein accretion in fish.

1. INTRODUCTION

Growth is a highly heritable trait in many fish species, with selection for faster growth rate allowing for increased production (Dunham, 2011). In aquaculture, faster growth of fish is one of the most important selection traits (Gjedrem, 2005). However, the use of isogenic fish lines, in which all individuals have the same genome, has illustrated the importance of environment and life history in influencing the growth trajectory (Dupont-Nivet et al., 2009). Stress is one of the most important physiological factors affecting growth, and several biotic and abiotic stressors, including routine hatchery practices like handling and sorting, poor water quality, and crowding suppress growth (McCormick et al., 1998; DiBattista et al., 2006). Coping with stress alters the energy status, as the animal activates a complex suite of energy-consuming pathways to restore homeostasis and preserve functional integrity (Barton, 2002). As the available energy for growth at a given moment is limited (Guderley and Pörtner, 2010), coping with stress may partition energy substrate away from growth, leading to decreased fish production. The acute increase in energy demand in response to stressor exposure is mediated by the activation of the stress-responsive hormonal pathways, including the hypothalamic–pituitary–interrenal (HPI) axis, leading to cortisol production. This in turn allows energy substrate mobilization and reallocation to regain homeostasis (Faught et al., 2016; Chapter 4 in this volume). Consequently, the stress and growth axes are intricately linked and this chapter highlights our current understanding of stress-mediated growth suppression in teleosts, with an emphasis on the role of cortisol in mediating these effects on muscle. We have organized the text into three main sections: the first emphasizes growth in terms of resource allocation using a modeling approach, and the second and third sections describe the potential molecular mechanisms by which stress and/or cortisol affects energy allocation and modulates growth promoters, respectively, thereby affecting muscle growth. Major knowledge gaps and future challenges are also identified.

2. A CONCEPTUAL FRAMEWORK FOR GROWTH

Here we focus on somatic growth, while gametic growth is discussed in Pankhurst (2016; Chapter 8 in this volume). Growth in fish, considered as the increase in length and weight, is a complex process, influenced by foraging, nutrient assimilation, energy substrate allocations, and utilization. In fish, muscle contributes over half the body mass and thus changes in the size of this organ system is of primary importance to growth (Mommsen and Moon, 2001). Unlike other animal models, fish show indeterminate growth and this involves a combination of both new muscle fiber formation (hyperplasia) and an increase in size of existing muscle fibers (hypertrophy) (Mommsen, 2001). While in most other vertebrate groups muscle hyperplasia is limited to early development, in teleosts it can contribute to over 50% of growth in adult fish (Mommsen, 2001). Therefore, from a growth standpoint, the structural changes might be best reflected in muscle protein accretion. The majority of protein synthesis in myocytes takes place in myofibrils and involves the myosin heavy chain (MHC) proteins. Consequently, MHC abundance has been proposed as a marker of structural growth along with the supporting structures, including bone and cartilage (Mommsen, 2001). Excellent reviews are available on the mechanisms of muscle development, structure, and physiology in the context of growth in fish (Mommsen, 2001; Mommsen and Moon, 2001; Rescan, 2005; Johnston et al., 2011). Here we focus on skeletal muscle growth in the context of structural protein deposition with the view that stress impinges on the overall energy resource allocation.

2.1. The Dynamic Energy Budget (DEB) Model for Growth

Many models have been developed to describe growth, aiming to predict fish weight at given feeding rations (Dumas et al., 2010; Seginer, 2016). Among others, bioenergetic models based on thermodynamic laws allow quantitative assessment of energy partitioning between different processes during animal growth, including maintenance, growth, and reproduction. These predictive models fit well with fish growth curves in many species (Warren and Gerald, 1967; Kitchell et al., 1977; Pauly, 1981; Essington et al., 2001). More recently the dynamic energy budget (DEB) model has gained increased acceptance for growth predictions (Kooijman, 2000, 2010; Nisbet et al., 2000) because it covers the full life cycle of the organism and takes into account the effect of environmental variables (Sousa et al., 2010). This model has been convincingly applied to many vertebrate species, including teleosts (Table 5.1).

Table 5.1
Dynamic energy budget (DEB) model applied to fish species

Species	Scientific name	References
European plaice	*Pleuronectes platessa*	van der Veer et al. (2001)
Delta smelt	*Hypomesus transpacificus*	Fujiwara et al. (2005)
Flounder	*Platichthys flesus*	van der Veer et al. (2001), Freitas et al. (2010)
Dab	*Limanda limanda*	van der Veer et al. (2001), Freitas et al. (2010)
Sole	*Solea solea*	van der Veer et al. (2001), Freitas et al. (2010), Eichinger et al. (2010)
European hake	*Merluccius merluccius*	Bodiguel et al. (2009)
Fathead minnows	*Pimephales promelas*	Jager et al. (2009)
Common goby	*Pomatoschistus microps*	Freitas et al. (2010)
Sand goby	*Pomatoschistus minutus*	Freitas et al. (2010)
Eelpout	*Zoarces viviparous*	Freitas et al. (2010)
Bull-rout	*Myoxocephalus scorpius*	Freitas et al. (2010)
Seabass	*Dicentrarchus labrax*	Freitas et al. (2010)
Atlantic cod	*Gadus morhua*	Freitas et al. (2010), Klok et al. (2014)
Atlantic herring	*Clupea harengus*	Freitas et al. (2010)
Sprat	*Sprattus sprattus*	Freitas et al. (2010)
European anchovy	*Engraulis encrasicolus*	Pecquerie et al. (2009), Freitas et al. (2010), Pethybridge et al. (2013)
Icelandic capelin	*Mallotus villosus*	Einarsson et al. (2011)
Zebrafish	*Danio rerio*	Augustine et al. (2011, 2012)
Pacific bluefin tuna	*Thunnus orientalis*	Jusup et al. (2011)
Pink salmon	*Oncorhynchus gorbuscha*	Pecquerie et al. (2011)
Sockeye salmon	*Oncorhynchus nerka*	Pecquerie et al. (2011)
Coho salmon	*Oncorhynchus kisutch*	Pecquerie et al. (2011)
Chum salmon	*Oncorhynchus keta*	Pecquerie et al. (2011)
Chinook salmon	*Oncorhynchus tshawytscha*	Pecquerie et al. (2011)
White seabream	*Diplodus sargus*	Serpa et al. (2013)
Gilthead seabream	*Sparus aurata*	Serpa et al. (2013)
Central mudminnow	*Umbra limi*	Filgueira et al. (2016)

The DEB model describes the rates at which the animal assimilates and allocates energy for maintenance, growth, and reproduction, described as energy pathways (Fig. 5.1). The model assumes that energy reserves are constantly supplied by assimilated nutrients, with a maximum assimilation rate proportional to the body surface area (Kooijman, 2010). Assimilation is the result of food availability, ingestion (depending on appetite and activity of the animal), and the capacity for nutrient absorption in the intestine. Food availability is dictated by extrinsic factors and includes the amount of food in the environment, and the direct competition for food related to conspecific density and social status. In contrast, ingestion and nutrient absorption capacity are dictated by intrinsic factors, such as appetite

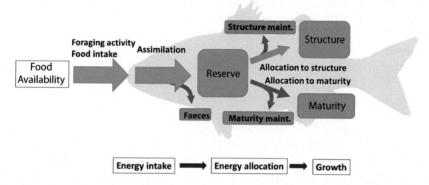

Figure 5.1. Energy fluxes in fish based on the Dynamic Energy Budget theory. Energy is obtained from the environment through foraging, food intake, and assimilation and transferred into the reserve compartment. Energy is then allocated to structure or maturity with a fraction allocated to their maintenance (maint.). In the proposed model growth is mostly related to the energy fraction allocated to structure and, therefore, positively correlated with the blue energy path (blue arrows and compartments). On the contrary, increased energy allocated to orange compartments (through orange arrows) is directly counterproductive for growth. The figure is based on the dynamic energy budget model proposed by Kooijman (2010).

and activity of the animal, and regulated by nervous and endocrine systems. Energy is continuously allocated to two main compartments, either building structures or maturity (described later), with a fixed fraction allocated to maintenance of each compartment (Fig. 5.1). The structures consist of all the constituents of the animal that form the body, including muscles and skeleton (Kooijman, 2010), while the maturity is described as the "complexity" of the animal and includes, for example, the brain, immune system, and gonads. Although the latter comprise only a small fraction of the mass of the animal, they consume a great deal of energy. For instance, while human brain accounts for only 2% of the body mass, it accounts for 20% of the total energy consumed (Mink et al., 1981). This is also true in goldfish (*Carassius auratus*) where brain is 1% of the total mass of the animal, yet consumes greater than 7% of the total energy budget (Mink et al., 1981). Similarly, immune system activation is also very energy demanding, despite contributing very little to the mass of the animal (Wolowczuk et al., 2008; Rauw, 2012). Consequently, an increase in the energy allocation to maturity that is essential for proper development of the animal will impact overall growth by reducing the resource allocation for structure.

The DEB model also assumes that a fraction of the allocated energy for structure and maturity is used for their maintenance (such as osmoregulation, respiration, and protein turnover) and, importantly, this resource allocation always takes priority (Kooijman, 2010). This is illustrated in feed-restricted fish, where growth and sexual maturation are greatly impaired (Bromley et al., 2000; Zhu et al., 2015), suggesting that under a low energy intake scenario, little to no energy is allocated to structure or maturity. The DEB model provides a promising tool to predict growth through an understanding of the energy allocated for structure, maturity, and maintenance not only under routine growth conditions, but also in a variable (stressful) environment in fish (Jager and Selck, 2011; Augustine et al., 2012). One of the most important aspects of growth in fish in terms of this review is the capacity of the animal to transfer energy from the environment to the formation of muscle structures. As stress increases energy consumption, the DEB model may provide a conceptual framework to understand where and how stressors may impact energy allocation, thereby affecting growth.

2.2. Myocyte Growth

Hyperplasia occurs through a complex set of processes involving specification, proliferation, and differentiation of precursor cells into myocytes, and fusion with one another creating multinuclear cells (Bentzinger et al., 2012). In parallel, hypertrophy occurs essentially through increased protein accretion, regulated by the balance between proteosynthesis and proteolysis. We are only beginning to understand the mechanisms involved in muscle hyperplasia and hypertrophy in fish.

2.2.1. MYOBLAST DIFFERENTIATION AND FUSION

Contrary to mammals, hyperplastic muscle growth in fish occurs throughout the whole life cycle (Mommsen, 2001). Muscle hyperplasia consists of the creation of new muscle fibers and is stimulated by myogenic regulatory factors (MRFs), including myogenic differentiation protein (MYOD), MYF5, myogenin, and MRF4. In mammals, the MRF genes *MYOD* and *MYF5* are required for specification of stem cells to myogenic progenitors (Weintraub et al., 1991; Rudnicki et al., 1993) and double *MYOD/MYF5* knockout mice lack the skeletal muscle completely (Rudnicki et al., 1993). During skeletal myogenesis, *MYF5* was found to be the first to be expressed, followed by *MYOD*. Later in muscle development, myogenin and MRF4 regulate the differentiation from myoblast to myocyte (Sumariwalla and Klein, 2001), followed by fusion, which is the last step in myogenesis. These four MRFs have been identified in fish species (Rescan et al., 1995; Rescan and Gauvry, 1996; Tan and Du, 2002; Galloway et al., 2006; García de la serrana et al., 2014), and their expression patterns are similar to those seen during mammalian myogenesis (Vélez et al., 2015). The fusion of myocytes, a fundamental step in muscle formation and development, is also a highly regulated process (Kim et al., 2015). Recent studies suggest a key role for nephrin, KIRREL3L, myomaker, and the complex JAMB/JAMC proteins in regulating myocyte fusion in zebrafish (*Danio rerio*) (Srinivas et al., 2007; Sohn et al., 2009; Powell and Wright, 2011; Landemaine et al., 2014). However, there is a scarcity of information on the role of stress and/or cortisol stimulation in affecting myocyte growth and differentiation in fish.

2.2.2. MUSCLE PROTEOLYSIS

Tissue proteins are constantly synthesized and degraded, and the balance between the two processes determines muscle growth (Jagoe et al., 2002; Lai et al., 2004). Proteolysis in fish myocytes seems to follow the general mechanisms observed in mammals, with the most important roles played by four highly conserved systems, including the ubiquitin-proteasome system (UPS), the autophagic-lysosomal system (ALS), the calpain-calpastatin system, and the apoptosis-protease system (Salem et al., 2006; Seiliez et al., 2008a, 2010; Salmerón et al., 2013). Among them, the UPS and ALS have garnered the most attention, accounting for about 55% of the total protein degradation in serum-deprived fish myocytes (Seiliez et al., 2014). Studies have shown that these two systems are modulated by environmental factors, including nutrient availability and growth factors, suggesting a key role in muscle growth and maintenance (Cleveland and Weber, 2010; Seiliez et al., 2012a).

UPS is essential for the degradation of damaged or short-lived proteins (Attaix et al., 2005) and involves two main steps consisting of (1) tagging the substrate through polyubiquitination and (2) its degradation by a proteasome complex. The first step permits the UPS to be selective to targeted proteins and is performed by a multistep pathway involving ubiquitin ligases, including ATROGIN-1 and muscle RING finger protein-1 (MURF1), two proteins expressed in fish tissues (Seiliez et al., 2008b; Wang et al., 2011). However, apart from their expression, limited information is available on the underlying molecular mechanisms of the UPS functioning in fish.

Unlike UPS, the ALS is a complex, mostly nonspecific system, degrading portions of cytoplasm and playing a primary role in the recycling of large-size proteins and organelles not targeted by the UPS (Sandri, 2010; Noda and Inagaki, 2015). First, a double lipid membrane is generated and expanded around intracellular cargo, forming an autophagosome vesicle, sequestering the substrates from the cytoplasm. The molecular mechanisms involved were thoroughly studied in yeast, *Saccharomyces cerevisiae*, demonstrating the central role of autophagy-related proteins (ATGs) for the formation of the vesicle. Initiation of the autophagosome formation starts with the assembly of the ATG1 complex, creating the phagophore assembly site or preautophagosomal structure (Suzuki et al., 2001). The elongation of the phagophore is then driven by the association of the enzyme Vps34, also called class III phosphatidylinositol 3-kinase (PI3K), with ATG6 (or BECN1 in mammals), forming a complex required for autophagy in yeast and mammals (Morris et al., 2015). The complex ATG2-ATG18/WIPI is recruited and necessary for autophagosome nucleation (Velikkakath et al., 2012). During expansion of autophagosome, ATG8 is conjugated to phosphatidylethanolamine by the ATG12/ATG5 complex, enabling ATG8 to anchor in the autophagosome membrane and control the expansion (Xie et al., 2008). Mammals express several ATG8 orthologs, with the most important ones being LC3 and GABARAP (Shpilka et al., 2011). All these ATGs are highly conserved in eukaryotes, and have recently been characterized in teleosts (Table 5.2).

Although the role of ALS in muscle growth regulation in fish is far from clear, the highly conserved nature of the key proteins suggests that regulatory mechanisms may be similar to those seen in mammals. Once the autophagosome is complete, it is transported along the microtubules toward the lysosome, where they fuse to form the autolysosome, leading to the enzymatic degradation of biological material inside the vesicle. Although little is known about the mechanisms involved in this fusion and degradation, the SNARE proteins, RAB7, and the HOPS complex (Gutierrez et al., 2004; Fader et al., 2009; McEwan et al., 2015), the GABARAP, and the enzyme phosphatidylinositol 4-kinase α (PI4Kα) are thought to play an essential role in the mammalian cell systems (Wang et al., 2015). Altogether, suppression of the protein degradation pathways may be a key aspect of muscle growth, while the stress-mediated activation of the protein breakdown pathways may be a mechanism for growth suppression, but this remains to be tested in fish.

2.2.3. REGULATION OF MUSCLE PROTEIN DYNAMICS

Although the molecular mechanisms involved in the regulation of protein synthesis and proteolysis are described in mammals, it is far from clear in fish. Studies have demonstrated that nutrients and growth factors are important regulators of growth in fish and this involves modulating protein dynamics (Fig. 5.2). Specifically, transforming growth factors-beta (TGFs-beta), insulin, insulin-like growth factors (IGFs), their binding proteins (IGFBPs), and growth hormone (GH) are key players in the regulation of myogenesis in fish (Mommsen and Moon, 2001; Reindl and Sheridan, 2012).

In mammals, myostatin, a protein belonging to the TGF-beta super family, is a strong inhibitor of muscle growth, with mutation in the gene

Table 5.2
Autophagy-related proteins (ATGs) examined in fish and their ortholog in yeast

Yeast protein	Fish-related proteins	Species	Tissue	References
ATG8	LC3B	Rainbow trout	Myoblasts primary culture and white muscle	Seiliez et al. (2010)
	LC3B	Zebrafish	Embryonic cells	Yabu et al. (2012)
	LC3	Zebrafish	Embryos	He et al. (2009)
	GABARAPL1	Rainbow trout	Myoblasts primary culture and white muscle	Seiliez et al. (2010)
	GABARAP	Zebrafish	Embryos	He et al. (2009)
ATG12	ATG12L	Rainbow trout	Myoblasts primary culture and white muscle	Seiliez et al. (2010)
	ATG12	Zebrafish	Embryos	Hu et al. (2011)
ATG4	ATG4B	Rainbow trout	Myoblasts primary culture and white muscle	Seiliez et al. (2010)
ATG5	ATG5	Zebrafish	Embryos	Hu et al. (2011)
	ATG5	Red cusk-eel	Myoblasts primary culture and white muscle	Aedo et al. (2015)
ATG7	ATG7	Zebrafish	Embryos	Hu et al. (2011)
ATG6	POBECLIN1	Olive flounder	Gill	Kong et al. (2011)
	BECLIN1	Zebrafish	Embryos	He et al. (2009)
	BECLIN1	Common carp	Kidney	Gao et al. (2014)
ATG1	ULK1A	Zebrafish	Embryos	He et al. (2009)
	ULK1B	Zebrafish	Embryos	He et al. (2009)
ATG9	ATG9A	Zebrafish	Embryos	He et al. (2009)
	ATG9B	Zebrafish	Embryos	He et al. (2009)
ATG16	ATG16L1	Red cusk-eel	Myoblasts primary culture and white muscle	Aedo et al. (2015)

leading to hypertrophy in several species (Kambadur et al., 1997; McPherron et al., 1997; Mosher et al., 2007). Teleosts have up to four homologs of myostatin (Rescan et al., 2001; Roberts and Goetz, 2001; Rodgers et al., 2007). Myostatin function in fish is similar to that in mammals, as knockdown of this protein led to a dramatic increase in zebrafish body mass (Acosta et al., 2005; Lee et al., 2009). Also, reducing myostatin protein expression did not result in weight increase, but led to muscle hyperplasia in zebrafish and medaka (*Oryzias latipes*) (Xu et al., 2003; Sawatari et al., 2010). Similarly, overexpression of follistatin, an antagonist of myostatin, enhanced hyperplasia in transgenic rainbow trout (*Oncorhynchus mykiss*) (Medeiros et al., 2009), supporting a key role of myostatin in muscle growth regulation. The growth suppression seen with myostatin involves the inhibition of cell proliferation (Seiliez et al., 2012b) and the stimulation of protein degradation in fish (Seiliez et al., 2013b). As in mammals, the mode of action of myostatin in fish is by binding to the activin type receptors, and activating the intracellular signaling cascade SMAD, by phosphorylating SMAD3 (Seiliez et al., 2012b; Fuentes et al., 2013a) and SMAD2 (Seiliez et al., 2013b), leading to inhibition of AKT/mTOR pathways (Seiliez et al., 2013b).

The AKT/mTOR pathway plays a central role in the regulation of the protein synthesis/proteolytic balance in mammals (Laplante and Sabatini, 2009), and this was also the case in fish and essential for muscle growth (Seiliez et al., 2008a, 2012a). Indeed, upon phosphorylation, the mechanistic target of rapamycin (mTOR) protein activates protein synthesis and inhibits autophagy leading to an increased accretion of proteins in muscle tissue (Fig. 5.2). Recent studies suggest a role for energy substrates in modulating mTOR activity in fish muscle cells. For instance, in cultured muscle cells of rainbow trout and seabream (*Sparus aurata* L.), medium deprivation of amino acids (AAs) attenuates mTOR activity (Seiliez et al., 2012a; Vélez et al., 2014), leading to an increase of LC3B2/LC3B1 ratio, a commonly used marker of autophagy (Seiliez et al., 2012a; Yabu et al., 2012), while AA supplementation reduces the expression of autophagy-related genes through a mTOR-independent pathway (Seiliez et al., 2012a). However, unlike mammals, glucose failed to show any effect on mTOR and its downstream pathway in trout and seabream myocytes in vitro (Belghit et al., 2013; Vélez et al., 2014); however, glucose showed a mTOR-independent autophagy activation in these cells (Belghit et al., 2013). These results suggest that plasma nutrient levels may also be a key regulator of the protein balance in muscle cells (Fig. 5.2).

Muscle protein synthesis and proteolysis are also regulated by the phosphoinositide 3-kinase/AKT (PI3K/AKT) signaling pathway (Rommel et al., 2001). This pathway inhibits expression of genes related to proteolysis and increases expression of MRFs genes involved in hyperplasia. Activation of the PI3K/AKT pathway leads to the phosphorylation and inactivation of fork-head box O (FOXO) transcription factors, reducing the expression of ubiquitin ligases (Mammucari et al., 2008). As in mammals, the PI3K/AKT pathway may also act by phosphorylating mTOR in fish (Fuentes et al., 2013b). For instance, the PI3K/AKT pathway can be modulated by refeeding in rainbow trout (Seiliez et al., 2008b, 2013a; Belghit et al., 2013) or by changing diet composition (Seiliez et al., 2011a). The phosphorylation of FOXO through the PI3K/AKT pathway has also been observed in rainbow trout with an associated downregulation of ATROGIN-1 and MURF1 (Cleveland and Weber, 2010; Seiliez et al., 2010). As IGF-1 and insulin are strong activators of the PI3K/AKT pathway in fish

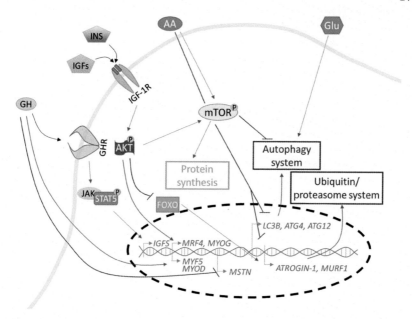

Figure 5.2. Regulations of myogenic genes and proteosynthesis/proteolysis balance in fish. Amino acids (AAs) downregulate the expression of authophagy-related genes and increase the activity of the mechanistic targets of rapamycin (mTOR), leading to a decrease in autophagy activation. In parallel, insulin (INS) and insulin-like growth factors (IGFs) can bind to the IGF-1 receptor (IGF1R), phosphorylating AKT leading to increased mTOR activity. Phosphorylated, AKT may also inhibit FOXO leading to a downregulation of genes involved in the ubiquitin/proteasome system. Growth hormone (GH) binds to GHR and activates the JAK/STAT5 pathway, leading to the upregulation of IGF-1 expression. The regulated genes are myogenic differentiation protein (*MYOD*), insulin-like growth factors (*IGFs*), myogenic regulator factor 4 (*MRF4*), myogenin (*MYOG*), myogenic factor 5 (*MYF5*), myostatin (*MSTN*), microtubule-associated protein *LC3B*, autophagy-related genes 4 and 12 (*ATG4* and *ATG12*), ATROGIN-1 and muscle RING-finger protein-1 (*MURF1*).

(Castillo et al., 2006; Montserrat et al., 2007; Cleveland and Weber, 2010; Seiliez et al., 2010; Vélez et al., 2014), the growth-promoting effects of these hormones are essentially due to enhanced protein synthesis, as well as inhibition of proteolysis (McCormick et al., 1992; Wood et al., 2005; Reindl and Sheridan, 2012; Fig. 5.2).

In addition to insulin and IGF-1, GH may also play a role in muscle growth either directly or indirectly by upregulating IGF-1 expression. GH has been shown to increase the expression of MRFs (MYOD2, MRF5) and IGF-1 in cultured myocytes of seabream (Fuentes et al., 2013b). The upregulation of IGF-1 expression by GH is the consequence of the activation of the JAK2-STAT5 signaling pathway in response to

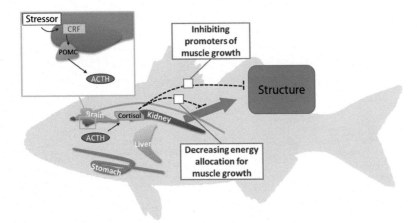

Figure 5.3. Proposed model for stress effect on growth. In response to stressor exposure, the hypothalamic–pituitary–interrenal (HPI) axis is activated leading to the production of corticotropin-releasing factor (CRF) in the hypothalamus, which stimulates the synthesis of proopiomelanocortin (POMC) in the anterior pituitary. POMC is posttranslationally modified to the bioactive adrenocorticotropic hormone (ACTH), which stimulates the release of cortisol from the interrenal tissue located in the head kidney. Cortisol affects muscle structure by (1) decreasing energy allocated for muscle growth and (2) inhibiting the promoters of muscle growth.

GH receptor (GHR) activation (Fuentes et al., 2011). GH treatment was also recently shown to have a negative effect on the expression of myostatin genes, *MSTN1A* and *MSTN1B* in trout myocytes (Seiliez et al., 2011b), suggesting other control factors that may modulate GH signaling on muscle growth. Stress, and the associated activation of the HPI axis leading to elevated cortisol in circulation, interferes with many of these mechanisms, thereby affecting growth. We propose that stress effects on growth suppression in fish (muscle structure) involves a role for the HPI axis in (1) reducing energy allocation for growth and (2) the downregulation of promoters of muscle formation (Fig. 5.3).

3. STRESS EFFECT ON ENERGY AVAILABLE FOR GROWTH

Stress and the associated activation of the HPI axis leading to cortisol secretion have been extensively covered by Faught and colleagues (2016; Chapter 4 in this volume). Briefly, stressor-induced activation of the HPI axis includes the secretion of corticotropin-releasing factor (CRF) from the hypothalamus, which stimulates the synthesis of proopiomelanocortin

(POMC) and the secretion of adrenocorticotropic hormone (ACTH) by posttranslational modification from the anterior pituitary. ACTH in turn stimulates the biosynthesis of cortisol in the steroidogenic cells located predominantly in the head kidney region in teleosts (Faught et al., 2016; Chapter 4 in this volume). Here we focus on the role of cortisol in affecting muscle growth during stress in fish.

Growth of muscle structure is dictated to a large extent by the energy substrate reaching the final structure compartment. This allocation is dependent on the overall nutrients entering the fish, and also on the energy needed for other bodily functions, including maintenance, maturation, and reproduction as described by the DEB model. Stressor exposure, and the associated elevation in circulating cortisol levels, may modify these energy fluxes via various mechanisms, including reduced feed intake, limited food absorption at the gut, and an increase in energy allocated for maintenance processes, all leading to the decrease in energy allocation for structure (Fig. 5.4).

3.1. Food Intake

One of the fundamental effects of stress in fish is the loss of appetite. For instance, food consumption is reduced by 62% after 17 days stressor exposure in Atlantic salmon (*Salmo salar*) (McCormick et al., 1998) and by almost 50% in rainbow trout facing a hypoxia challenge (Bernier and Craig, 2005). The decrease in food intake in response to stress lowers the amount of energy entering the reserve compartment of the fish. The limited quantity of available energy is allocated to maintenance, which has priority over other functions, leading to a decrease in growth (Fig. 5.4 (1)). Several studies have tried to address the mechanisms behind the reduced feed intake, including the contribution of neuropeptides and gut peptides, but the mechanisms are far from clear.

Food intake is regulated at the level of the hypothalamus by the orexigenic and the anorexigenic centers, respectively, stimulating and inhibiting appetite and feeding behaviors (Volkoff et al., 2005). Some of the key players in the brain include the neuropeptide Y (NPY), an orexigen, and the two HPI axis molecules CRF and POMC, which are anorexigens in fish. Intracerebroventricular injection of NPY has been shown to increase food consumption in goldfish, catfish (*Ictalurus punctatus*), and rainbow trout (López-Patiño et al., 1999; Silverstein and Plisetskaya, 2000; Aldegunde and Mancebo, 2006). Conversely, injection of NPY antagonist led to decreased food intake (Aldegunde and Mancebo, 2006). Studies in mammals have shown that the production of NPY is under the control of leptin (Wang et al., 1997). In fish, leptin was recently cloned and shown to

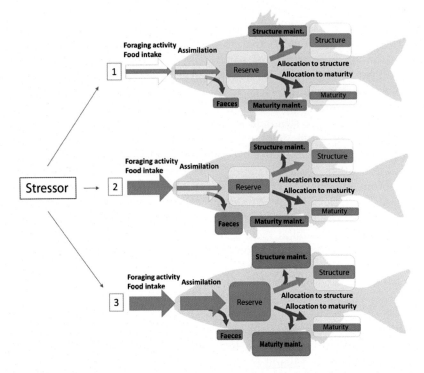

Figure 5.4. Proposed model for stress effect on overall resource allocation toward structure. The three main effects of a stressor leading to a decrease in energy available for growth include the following. (1) Reduction in energy availability due to a decrease in foraging activity; this is indicated by narrow arrows for food intake and assimilation and reduced overall energy reserve (small blue box), leading to reduced energy allocation for maturity (smaller orange box) and structure (smaller blue box). (2) Reduction in energy availability because of a decrease in food assimilation; this is indicated by a normal blue arrow for food intake but a narrow blue arrow for assimilation, leading to increased loss of energy in faeces (larger orange box); this also reduced overall energy reserve (small blue box), leading to reduced energy allocation for maturity (smaller orange box) and structure (smaller blue box). (3) Increase in the energy demand for maintenance; this is indicated by normal blue arrows for food intake and assimilation, leading to an overall normal energy reserve (blue box), but increased allocation for maintenance (larger orange boxes) limits energy allocation for maturity (smaller orange box) and structure (smaller blue box).

be mainly produced in the liver (Kurokawa et al., 2005; Huising et al., 2006). However, the regulation of leptin and its mode of action are far from clear in fish (Copeland et al., 2011; Gorissen and Flik, 2014). Indeed, administration of homologous leptin by central or peritoneal injection has a strong anorexic effect on food intake (Murashita et al., 2008; Li et al., 2010; Gong et al., 2015). The effect of leptin on feed intake (Fig. 5.5) may involve

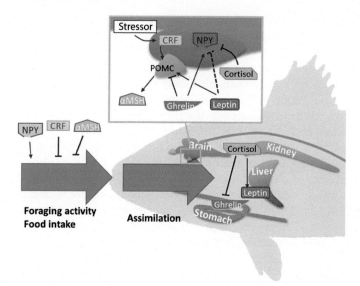

Figure 5.5. The role of stress and/or cortisol action on appetite regulation in fish. The neuropeptide Y (NPY) is a strong activator of food intake. Hypothalamic corticotropin-releasing factor (CRF) and alpha-melanocyte-stimulating hormone (αMSH), derived from POMC from the pituitary, are anorexigenic molecules released in response to stressor-mediated HPI axis activation. Cortisol produced in response to HPI activity downregulates the production of ghrelin, a orexigenic hormone inhibiting POMC but activating NPY secretion. Cortisol also stimulates the production of leptin in the liver. Leptin is considered anorexigenic and stimulates the HPI axis and may also downregulate NPY action.

downregulation of NPY, as well as upregulation of POMC in fish (Murashita et al., 2008, 2011; Li et al., 2010). The POMC stimulation by leptin may involve the activation of PI3K/AKT pathway in fish (Gong et al., 2015). Interestingly, a recent in vitro study has also demonstrated that leptin interacts with the GH–IGF axis, by stimulating hepatocyte gene expression of IGF-1 and GHRs (Won et al., 2016).

As in mammals, leptin has been identified as a regulator of the stress axis in fish (Roubos et al., 2012; Gorissen and Flik, 2014). A recent study showed that chronic cortisol administration increases liver leptin gene expression in rainbow trout (Madison et al., 2015). This effect was also observed in vitro in primary culture of hepatocytes, and blocked by the addition of the GR inhibitor RU486, supporting a role for cortisol-mediated GR signaling (Madison et al., 2015). Also, chronic stress by pathogen infection, which reduced feed intake and upregulated hypothalamic *POMC* transcript levels, elevated plasma leptin levels in fish (MacDonald et al., 2014).

Ghrelin synthesized in fish stomach and gut (Kaiya et al., 2003) is a strong orexigenic hormone in fish (Unniappan et al., 2004; Shepherd et al., 2007; Penney and Volkoff, 2014; Tinoco et al., 2014). This orexigenic effect is consistent with mammalian studies, and is mediated through the stimulation of NPY expression (Fig. 5.5) in the hypothalamus of goldfish (Miura et al., 2006) and rainbow trout (Velasco et al., 2016). Central injection of ghrelin also decreases the gene expression of POMC in rainbow trout (Velasco et al., 2016; Fig. 5.5). However, the effect of ghrelin on food intake and NPY expression are not consistent in fish. Studies showed that ghrelin could also have an inhibitory effect on food intake (Jönsson et al., 2010; Schroeter et al., 2015), and this is thought to be mediated indirectly due to an increase in CRF expression in rainbow trout (Jönsson et al., 2010; Jönsson, 2013). Confinement stress, accompanied by elevated plasma cortisol level, had either no effect or a suppressing effect on plasma ghrelin levels in juvenile salmonids, but always an inhibitory effect on food intake (Pankhurst et al., 2008a,b). A recent study injecting cortisol in tilapia (*Oreochromis mossambicus*) showed a decrease in plasma levels of ghrelin and this was associated with a downregulation of this gene expression in the stomach (Janzen et al., 2012). Overall, stress modulates plasma ghrelin levels and this may be mediated by cortisol, but the molecular mechanisms linking that to cessation in feeding is far from clear in fish.

Several anorexigenic peptides have been identified in fish (Volkoff et al., 2005), including key hormones of the HPI axis. Indeed, injection of CRF strongly decreases food intake compared to sham-injected fish (Bernier and Peter, 2001). The injection of a CRF-receptor antagonist was shown to reduce the food intake-suppressing effect of a hypoxia challenge, suggesting the central role of CRF in food intake regulation during stressful conditions (Bernier and Craig, 2005). Additionally, α-MSH, a product from the cleavage of the POMC protein, is an anorexigen in fish (Cerdá-Reverter et al., 2003). Consequently, stress and the activation of the HPI axis will reduce feed intake by directly producing anorexigenic peptides or indirectly through the action of cortisol (Fig. 5.5), and this will restrict the energy available for growth (Fig. 5.4 (1)).

3.2. Energy Substrate Absorption at the Gut

Once food is ingested, the gastrointestinal tract has to digest and absorb nutrients in order to provide usable energy to the reserve compartment. The digestive tract structure and functional characteristics are highly modified by environmental variables and internal factors, including reproduction in fish (Habibi and Ince, 1984; Pankhurst and Sorensen, 1984; Madsen et al., 2015). Apart from the decreased food intake, stress is also known to inhibit digestive functions by decreasing the capacity of the fish to assimilate nutrients (Fig. 5.4 (2)). Analyses of gut structure and microbial population show that an acute 15 min stress can transiently affect intestinal ultrastructure and functions in salmonids (Olsen et al., 2002, 2005). Stress also inhibits acid secretion in the stomach of the Nile tilapia (*Oreochromis niloticus*) despite a normal food intake, lowering digestion efficiency (Moriarty, 1973). During salinity acclimation, cortisol increases the anterior intestinal epithelial turnover (Takahashi et al., 2006b), necessary to cope with the salinity challenge. Cortisol levels have also been shown to modify intestinal fluid transport (Veillette et al., 1995). Similarly, the presence of glucocorticoid receptor (GR) transcripts in fish intestine suggests a direct effect of cortisol (Takahashi et al., 2006a). This was supported by the effect of chronic cortisol feeding on intestinal structure and the associated decreased digestive efficiency in trout (Barton et al., 1987). These results support a possible link between the cortisol stress axis activation and impaired nutrient absorption during stress in fish. Unfortunately, as often, the molecular mechanism(s) leading to cortisol effect on gut function is still lacking in fish. Consequently, stress can affect the structure and function of the digestive system in fish, leading to a decrease in the assimilation of energy from food (Fig. 5.4 (2)).

3.3. Energy Demand for Maintenance

The reestablishment of homeostasis in response to stressor exposure involves the activation of energy-demanding processes, including ion regulation, metabolic substrate availability, and transport and synthesis of stress-coping proteins. All these activities translate to increased energy allocation for structure and maturity maintenances in the DEB model (Fig. 5.4 (3)). These costs restrict production of structures, while increasing the chance of coping and surviving the stressor.

The increased cost for maintenance in stressful situations is difficult to measure empirically, but modifications of respiration and ventilation, cardiovascular activity, or more generally, metabolic rates are good indicators. Different types of stressors, including social stressors, increase oxygen consumption in fish (Barton and Schreck, 1987; Sloman et al., 2000; Sadoul et al., 2015a), and this coincides with hyperactivity of the cardiorespiratory systems (Laitinen and Valtonen, 1994). Similarly, increase in metabolic rate was also seen with exogenous cortisol administration suggesting that stress-induced elevation in plasma cortisol levels may be playing a role in increasing the metabolic rate of the animal (Morgan and Iwama, 1996). In support of this notion, studies have shown that stress and/or cortisol stimulation increases the metabolic scope of fish (Mommsen et al., 1999;

Aluru and Vijayan, 2007, 2009). One of the key metabolic responses to cortisol stimulation is the elevation in liver metabolic capacity, including enhanced AA catabolism and channeling of C3 substrates for oxidation and gluconeogenesis (Mommsen et al., 1999; Aluru and Vijayan, 2007, 2009), responses that were abolished in the presence of the GR antagonist mifepristone (Aluru and Vijayan, 2007). These metabolic changes, and the associated increase in energy demand, are essential for the animal to metabolically cope with the stress. However, this acute increase in energy demand to restore homeostasis may be at the cost of a reduction in energy availability for growth (Fig. 5.4 (3)).

At the cellular level, increased maintenance cost is observed through the activation of a variety of responses ranging from survival pathways to the initiation of cell death and replacement. All these mechanisms are energy-consuming since they involve the synthesis of new proteins, the replacement of cellular constituents, or at the tissue level, the replacement of cells (Fulda et al., 2010). Heat shock proteins (HSPs) are a family of proteins that are absolutely important for normal cell function under stressed (Schreck and Tort, 2016; Chapter 1 in this volume) and unstressed conditions and this has been reviewed extensively in fish (Iwama et al., 1998, 2004; Deane and Woo, 2010). The HSPs exert protective roles in cells, including chaperoning function, maintaining protein homeostasis in response to cellular stress, and as an essential component of the stress steroid signaling (Pratt, 1993; Vijayan et al., 2005; Deane and Woo, 2010). The synthesis of HSPs, especially during times of stress, are energetically expensive yet essential and may occur at the expense of other energy-demanding pathways. For instance, heat shock-induced upregulation of the 70 kDa (kilodaltons) family of HSPs in trout hepatocytes corresponded with a reduced capacity for cortisol-mediated gluconeogenesis (Boone et al., 2002), an energy-demanding pathway consuming at least 10 molecules of adenosine triphosphate per mole of glucose produced. Consequently, stressor-mediated changes in protein homeostasis, essential for protection, appear energetically expensive and occur at the expense of other energy-demanding pathways, including growth.

Cellular protein homeostasis is also affected by increased protein catabolism. Stress and/or cortisol stimulation have been associated with the upregulation of genes encoding for proteins involved in catabolic pathways, such as cathepsin D, other cathepsins, and aminotransferases (Mommsen et al., 1999; Aluru and Vijayan, 2007, 2009). In the presence of a nutrient stressor, fish cells sped up protein catabolism through an increase in the activity of the autophagy and UPSs in response to a nutrient stressor (Seiliez et al., 2008b; Yabu et al., 2012; Belghit et al., 2013, 2014). Similarly, a handling stress was found to increase the production of autophagy-related genes (*ATG9*, *ATG5*, *ATG16l*, *ATG4*, *ATG7*, and *LC3Bi*) in the red cusk-eel (*Genypterus chilensis*;

Aedo et al., 2015). Cultured myocytes from the same species showed upregulation of autophagy-related genes with cortisol, and this was associated with higher levels of ubiquitinated proteins (Aedo et al., 2015). Therefore, proteolytic processes are upregulated in response to cortisol treatment, and this led to a decrease in myocyte size (Aedo et al., 2015). Interestingly, these effects were inhibited in cells pretreated with RU486, suggesting a direct genomic role for cortisol in the regulation of proteolysis (Aedo et al., 2015). Additionally, in response to the intensity and duration of the stressor, pathways leading to cell death, including apoptosis, are activated. In fish, these effects were mostly observed in response to toxicant exposure (Krumschnabel et al., 2005; Wu et al., 2015; Morcillo et al., 2016), but recent transcriptome analysis also showed increased expression of apoptosis-related genes due to a handling stress (Du et al., 2014). These studies support the notion that stress-induced increase in cortisol levels enhance myocyte protein degradation and/or cell death. Together, from the individual level to the cellular level, a number of maintenance mechanisms are activated in order to cope with the stressor, and this may be mediated by cortisol signaling. As these pathway activations are energy demanding, the reallocation of resources for maintenance decreases the energy allocation to growth (Fig. 5.4 (3)).

4. STRESS EFFECTS ON PROMOTERS OF MUSCLE FORMATION

Surprisingly, very few studies have addressed the effect of stress and/or cortisol stimulation on muscle function in fish. While most studies have focused on liver as metabolic target for stress adaptation and cortisol action, the few studies that have examined muscle have been mostly descriptive (Milligan, 1997; Mommsen et al., 1999; Milligan et al., 2000). Based on the responses seen in the liver, possible effects on muscle function, especially as a target for proteolysis due to cortisol stimulation during stress, have been proposed (Mommsen et al., 1999; Aluru and Vijayan, 2009). Based on the preceding sections, it is a safe assumption that stress and/or cortisol action will affect muscle function in multiple ways, including protein synthesis and protein breakdown pathways, as well as muscle formation by decreasing the expression of myogenic genes and reducing the action of the GH–IGF axis. Understanding the underlying molecular mechanisms will be essential in establishing a cause–effect relationship between stress and growth.

4.1. Stress Effects on Myogenesis

The molecular mechanism of cortisol action has been described in detail in this book (Faught et al., 2016; Chapter 4 in this volume). Suffice it to say that the genomic action of cortisol includes binding to GR and activating target genes by interacting with the glucocorticoid response element (GRE). Cortisol reduces the expression of *myogenin*, *MYOD*, and *MYF5* in mammalian models, suggesting a role for the stress steroid in down-regulating myogenesis (Pandurangan et al., 2014). In fish, little is known about the effect of stress on MRFs. Recently, a handling stress decreased the expression of MYOD1 in red cusk-eel, while myocytes incubated with cortisol in vitro did not show any change in MYOD transcript levels (Aedo et al., 2015). The analysis of the promoter regions of *MSTN* genes has revealed the presence of a GRE in several fish species (Garikipati et al., 2006; Funkenstein et al., 2009; De Santis and Jerry, 2011), suggesting a possible direct regulation of cortisol on muscle growth in fish. To this end, knockdown of GR reduces *MSTN* expression in zebrafish (Nesan et al., 2012). However, the effect of stressors or cortisol administration on *MSTN* levels is not consistent among studies or species (Table 5.3). For instance, despite the lack of GRE in rainbow trout *MSTN* genes (Garikipati et al., 2006, 2007), a recent study showed increased *MSTN1B* transcript levels in response to cortisol treatment in trout myoblasts in vitro (Galt et al., 2014), suggesting a possible GR-independent cortisol effect on *MSTN* expression. It remains to be seen if cortisol stimulation of *MSTN* expression involves mineralocorticoid receptor activation and/or nongenomic signaling in fish (for a review of cortisol signaling see Faught et al., 2016; Chapter 4 in this volume).

Overall, it appears that the differences seen with cortisol effects on muscle growth regulation may be related to several factors, including species differences, developmental stage, and the nutrient status of the animal, and this needs to be well characterized in fish. In addition to effect of cortisol on muscle function, it is likely that this steroid may be modulating the levels of nutrients, growth factors, and hormones involved in muscle growth regulation, leading to an indirect effect on growth (Mommsen, 2001).

4.2. Stress Regulation of the GH–IGF Axis

GH is mainly secreted by the pituitary gland and under the control of hypothalamic peptides, neurotransmitters, and hormones (Björnsson et al., 2002; Wong et al., 2006). In vertebrates, including fish, GH secretion is under a complex suite of control mechanisms involving multiple GH release factors, including dopamine, ghrelin, cholecystokinin, and gonadotropin-releasing hormone, but also several somatostatins (SSTs) and other inhibitors, like serotonin or noradrenaline (Wong et al., 1993; Kaiya et al., 2003; Gahete et al., 2009; Sheridan and Hagemeister, 2010). In mammals, GH secretion has also been demonstrated to be under

Table 5.3

The effect of either a stressor or glucocorticoid treatment on myostatin (*MSTN1* or *MSTN2*) expression (downregulation (−), upregulation (+), or no significant difference (NS)) in fish

Species	Stressor or glucocorticoid treatment	Tissue	Myostatin expression levels compared to control		References
			MSTN 1	MSTN 2	
Zebrafish	Grown in high stocking density	Muscle	−		Vianello et al. (2003)
Tilapia	3 days fasting	Whole body	−		
	6 days fasting	Whole body	−		
	3 days fasting (experiment 2)	Whole body	+		Rodgers et al. (2003)
	9 days fasting (experiment 2)	Whole body	−		
	3 or 6 h immersion in cortisol	Whole body	−		
	28 days fasting	White muscle	NS		
Catfish	1 day fasting	White muscle	−		
	14 and 28 days fasting	White muscle	NS		
	1 day cold temperature	White muscle	−		Weber and Bosworth (2005)
	14 and 28 days cold temperature	White muscle	+		
Catfish	12 h after dexamethasone injection	White muscle	−		
	24 h after dexamethasone injection	White muscle	NS		Weber et al. (2005)
Rainbow trout	30 days of starvation	White muscle	−	NS	Johansen and Overturf (2006)
Zebrafish	3 days high stocking density	Muscle	NS	NS	
		Brain	NS	NS	Helterline et al. (2007)
		Spleen	+	+	
Barramundi	30 days of fasting	White muscle	+	NS	
		Liver	+	+	De Santis and Jerry (2011)
		Brain	−	NS	
		Gill	−	+	
Rainbow trout	12 or 24 h after cortisol injection	Red muscle	NS	+ (*2A*)	
		White muscle	NS	NS	Galt et al. (2014)
	24 h of cortisol incubation	Myoblast cell culture	+ (*1B*)	NS	

the control of glucocorticoids (Mazziotti and Giustina, 2013), but direct evidence for similar mechanisms in fish is lacking. Acute stress modulates circulating levels of GH in fish, suggesting an interaction between the stress axis and the GH axis in fish (Deane and Woo, 2008). A confinement stress, a sudden temperature change, or a bacterial infection decreased GH levels or GH mRNA levels in fish (Pickering et al., 1991; Auperin et al., 1997; Rotllant et al., 2000, 2001; Deane and Woo, 2006). Whether these responses are linked to an increased production of SST, a decreased production of GH releasing hormone (GHRH), or a different pathway altogether remains to be investigated in fish. Interestingly, the subordinate cichlid fish *Haplochromis burtoni*, which has elevated cortisol levels (Fox et al., 1997), has larger somatostatin-containing neurons (Hofmann and Fernald, 2000), arguing in favor of a positive regulation of cortisol on SST secretion (Fig. 5.6). To our knowledge, very little is known about the effect of stress on GHRH in fish. Therefore, while there is evidence that stress affects the GH–IGF axis, and this may involve cortisol stimulation, the mechanisms involved need to be elucidated in fish.

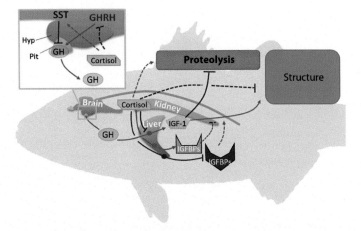

Figure 5.6. Cortisol effect on the GH–IGF axis in fish. Somatostatin (SST) and growth hormone-releasing hormone (GHRH), respectively, inhibit and upregulate the expression of growth hormone (GH) in the pituitary (Pit). GH is then released in the circulatory system and activates insulin-like growth factor (IGF) production mainly in the liver. IGFs downregulate proteolysis and enhance structure production. IGF binding proteins (IGFBPs) modulate IGFs activity, with small IGFBPs (orange) inhibiting IGF actions and big IGFBPs (black) increasing IGFs activity. Cortisol interacts with the GH–IGF axis by inhibiting GHRH secretion and increasing SST production. Cortisol also inhibits IGF-1 expression and modulates IGFBPs expression (inhibiting the secretion of big IGFBPs, but promoting small IGFBPs secretion) in the liver. Known links are illustrated using solid lines, whereas proposed regulations are illustrated using dashed lines.

Following synthesis and storage, GH is released into the plasma exerting its actions throughout the whole body by interacting with the GH receptors, predominantly in the liver and muscle of fish (Jiao et al., 2006). After binding to GH receptors, GH activates the production of IGF-1, mainly in hepatocytes and to a limited extent in myocytes, as evidenced by increased transcript abundance, as well as by circulating level of this growth factor (Reindl and Sheridan, 2012). Elevated levels of plasma IGF-1 increases muscle growth as described in Section 2.2.3. Several studies have demonstrated that glucocorticoids can affect IGF-sensitivity to GH, and IGF-1 production. For instance, intraperitoneal injection of cortisol decreases the expression of IGF-1 in the liver and the levels in the plasma in tilapia without impacting plasma GH levels (Kajimura et al., 2003). Similarly, dexamethasone or cortisol inhibited the GH-mediated upregulation of IGF-1 in hepatocytes from salmon (Pierce et al., 2005) and trout (Philip and Vijayan, 2015). This decreased IGF-sensitivity to GH may be related to the reduced hepatic GH-receptor expression in cortisol supplemented fish (Small et al., 2006). In vitro, cortisol increases the expression of suppressor of cytokine signaling (SOCS), SOCS-1, and SOCS-2 (Philip and Vijayan, 2015), two genes known for their inhibitory action on the JAK/STAT pathway, the main GH signaling pathway leading to the production of IGF-1 (Fig. 5.2). In isolated hepatocytes, cortisol also decreases IGF-1 expression directly, suggesting that cortisol may also regulate IGF-1 production independent of GH in fish hepatocytes (Leung et al., 2008). Together, these results underscore a key role for cortisol in affecting muscle growth in fish, either directly or indirectly by regulating GH–IGF axis during stress (Fig. 5.6).

The activity of IGFs on muscle function is dependent on high-affinity IGFBPs. In mammals, six IGFBPs of different sizes were discovered and shown to modulate the distribution and stability of IGFs, as well as their access to the IGF receptors (Rosenzweig, 2004). Depending on experimental conditions, IGFBPs either potentiate or inhibit the action of IGFs, and the production of IGFBPs are modulated by glucocorticoids (Conover et al., 1993; Okazaki et al., 1994; Gabbitas and Canalis, 1996; Pereira et al., 1999). For instance, highly phosphorylated or in excess amount, IGFBP-1 inhibits IGFs actions (Cox et al., 1994; Yu et al., 1998), whereas at low concentrations or dephosphorylated, IGFBP-1 increases IGFs activity (Yu et al., 1998). Studies in mammals also demonstrated that cortisol infusion increases the production of IGFBP-1 in vivo (Conover et al., 1993), and the expression of IGFBP-1 mRNA in human osteoblast-like cells in vitro (Okazaki et al., 1994). Similarly, IGFBP-3 plays a central role in the modulation of IGFs actions, since it transports IGFs to the target tissues and increases its half-life from minutes up to 12 h. In vitro, dexamethasone decreases IGFBP-3 expression in rats (Villafuerte et al., 1995). Altogether, IGFBPs may be playing a key role in IGF action and availability (Fig. 5.6), and may be a target for cortisol action during stress, but few studies have examined this in teleosts.

In fish, due to the 3R genome duplication and the 4R in salmonids, studies reported, respectively, 9–11 and 19 different IGFBPs (Daza et al., 2011; Macqueen et al., 2013). In tilapia, the injection of cortisol showed a quick (within 2 h) increase in the plasma of four IGFBPs and a decrease in two IGFBPs (Kajimura et al., 2003). The response was dependent on the size of the protein with IGFBPs of 24–32 kDa showing an increase, while the ≥ 40 kDa IGFBPs showed a decrease (Kajimura et al., 2003). Similarly, a confinement stress reduced the plasma levels of an IGFBP (33 kDa) equivalent to the mammalian-IGFBP-3, whereas two other smaller IGFBPs (24 and 28 kDa) were increased in sunshine bass, a hybrid between white bass, *Morone chrysops*, and striped bass, *Morone saxatilis* (Davis and Peterson, 2006). However, increase of plasma cortisol levels did not affect the levels of IGFBPs in the same study. It is not known whether the higher levels of big IGFBPs in fast-growing fish, and low levels of small IGFBPs in slow-growing fish are linked to stress and/or cortisol stimulation (Shimizu et al., 1999; Peterson and Small, 2004). We conclude the GH–IGF axis is regulated by stress in fish, with cortisol impacting the GH secretion, the IGF-sensitivity to GH, and the production of IGFBPs, all of which will contribute to the decreased energy demand for growth.

5. CONCLUSION AND KNOWLEDGE GAPS

Stress has a broad and ubiquitous impact on numerous steps during the allocation of energy to growth. The HPI axis activation and the associated increase in cortisol levels interact tightly with growth-related physiological mechanisms, at various levels of organization of the fish. The activation of stress response mechanisms is highly energy demanding, and occurs at the cost of other energy-demanding pathways including growth. Growth described as protein accretion in muscle cells is the result of the positive balance between structural protein synthesis and structural protein break-down. Stress reduces growth by shifting this balance toward structural protein breakdown (Fig. 5.7), through the activation of maintenance processes such as proteolysis and stress-coping mechanisms, and reducing the overall available energy by decreasing food intake and food absorption. Also, stress affects growth by directly inhibiting the molecular mechanisms of myogenesis and the hormonal regulation of growth, facilitating the

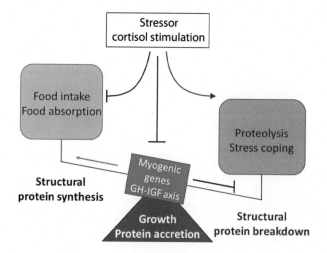

Figure 5.7. Proposed protein balance model for cortisol effect on muscle growth. Muscle growth, which is essentially structural protein accretion, is a fine balance between protein synthesis and protein breakdown pathways. We propose that stress and cortisol stimulation shift the balance toward increased protein breakdown leading to reduced growth. Increased protein synthesis and reduced breakdown lead to positive protein accretion. This occurs when more energy is allocated to muscle growth in response to increased food intake, food assimilation, along with increased stimulation of growth-promoting factors, including the GH–IGF axis. On the other hand, less protein accretion occurs when the balance shifts toward increased breakdown, as seen with stress (cortisol stimulation) when muscle proteolysis is enhanced, along with the activation of stress-coping mechanisms. Cortisol inhibits the mechanisms leading to an increase of energy allocation to growth and promotes the mechanisms leading to a decrease in energy available for growth.

shift from structural protein synthesis to structural protein catabolism (Fig. 5.7).

These mechanisms reducing muscle growth are supposed to help the animal to cope and survive the stressor. In the context of aquaculture, where growth is the primary phenotype of interest, activation of stress response mechanisms should be avoided as much as possible. This can be achieved by tight water quality management, limiting or removing pathogens, and reducing physical stressors, including handling and grading, but this may be a less profitable option (Bostock et al., 2010). A more sustainable solution may be to include robustness traits in breeding programs in order to select animals capable of being resistant to nonlife-threatening challenges (Sadoul et al., 2015a,b). However, in order to provide relevant biomarkers of robustness, underlying mechanisms of the effect of stress on growth needs to be ascertained.

As is clear from this chapter large gaps exist in our knowledge about stress effect on muscle growth. Specifically, we lack mechanistic insights on how stress and/or cortisol stimulation affects pathways leading to altered muscle protein dynamics. Some key knowledge gaps are identified here:

- What is the concentration of cortisol, and/or the mechanism(s) of action of this hormone, that tips the balance from eustress, facilitating anabolism and growth, to distress, which is catabolic and depresses growth?
- What are the molecular mechanisms involved in stress and/or cortisol effect on muscle protein synthesis and proteolysis?
- Does cortisol have a direct effect on the autophagy and/or ubiquitin/proteasome systems?
- How does cortisol play a role in energy substrate repartitioning during stress in fish? Are SOCS key regulators of the energy allocation for growth?
- Does stress affect the gut microbiome, leading to changes in gut function and nutrient absorption?
- What is the mechanism by which cortisol affects appetite regulation in fish?
- Does cortisol play a role in modulating gut structure and gut peptide regulation in fish?

REFERENCES

Acosta, J., Carpio, Y., Borroto, I., González, O. and Estrada, M. P. (2005). Myostatin gene silenced by RNAi show a zebrafish giant phenotype. *J. Biotechnol.* **119**, 324–331.

Aedo, J. E., Maldonado, J., Aballai, V., Estrada, J. M., Bastias-Molina, M., Meneses, C., et al. (2015). mRNA-seq reveals skeletal muscle atrophy in response to handling stress in a marine teleost, the red cusk-eel (*Genypterus chilensis*). *BMC Genomics* **16**, 1024.

Aldegunde, M. and Mancebo, M. (2006). Effects of neuropeptide Y on food intake and brain biogenic amines in the rainbow trout (*Oncorhynchus mykiss*). *Peptides* **27**, 719–727.

Aluru, N. and Vijayan, M. M. (2007). Hepatic transcriptome response to glucocorticoid receptor activation in rainbow trout. *Physiol. Genomics* **31**, 483–491.

Aluru, N. and Vijayan, M. M. (2009). Stress transcriptomics in fish: a role for genomic cortisol signaling. *Gen. Comp. Endocrinol.* **164**, 142–150.

Attaix, D., Ventadour, S., Codran, A., Béchet, D., Taillandier, D. and Combaret, L. (2005). The ubiquitin–proteasome system and skeletal muscle wasting. *Essays. Biochem.* **41**, 173–186.

Augustine, S., Gagnaire, B., Floriani, M., Adam-Guillermin, C. and Kooijman, S. A. L. M. (2011). Developmental energetics of zebrafish, *Danio rerio. Comp. Biochem. Physiol. A Mol. Integr. Physiol.* **159**, 275–283.

Augustine, S., Gagnaire, B., Adam-Guillermin, C. and Kooijman, S. A. L. M. (2012). Effects of uranium on the metabolism of zebrafish, *Danio rerio. Aquat. Toxicol.* **118–119**, 9–26.

Auperin, B., Baroiller, J. F., Ricordel, M. J., Fostier, A. and Prunet, P. (1997). Effect of confinement stress on circulating levels of growth hormone and two prolactins in freshwater-adapted tilapia (*Oreochromis niloticus*). *Gen. Comp. Endocrinol.* **108**, 35–44.

Barton, B. A. (2002). Stress in fishes: a diversity of responses with particular reference to changes in circulating corticosteroids. *Integr. Comp. Biol.* **42**, 517–525.

Barton, B. A. and Schreck, C. B. (1987). Metabolic cost of acute physical stress in juvenile steelhead. *Trans. Am. Fish. Soc.* **116**, 257–263.

Barton, B. A., Schreck, C. B. and Barton, L. D. (1987). Effects of chronic cortisol administration and daily acute stress on growth, physiological conditions, and stress responses in juvenile rainbow trout. *Dis. Aquat. Organ.* **2**, 173–185.

Belghit, I., Panserat, S., Sadoul, B., Dias, K., Skiba-Cassy, S. and Seiliez, I. (2013). Macronutrient composition of the diet affects the feeding-mediated down regulation of autophagy in muscle of rainbow trout (*O. mykiss*). *PLoS ONE.* **8**, e74308.

Belghit, I., Skiba-Cassy, S., Geurden, I., Dias, K., Surget, A., Kaushik, S., et al. (2014). Dietary methionine availability affects the main factors involved in muscle protein turnover in rainbow trout (*Oncorhynchus mykiss*). *Br. J. Nutr.* **112**, 493–503.

Bentzinger, C. F., Wang, Y. X. and Rudnicki, M. A. (2012). Building muscle: molecular regulation of myogenesis. *Cold Spring Harb. Perspect. Biol.* **4**, a008342.

Bernier, N. J. and Peter, R. E. (2001). Appetite-suppressing effects of urotensin I and corticotropin-releasing hormone in goldfish (*Carassius auratus*). *Neuroendocrinology* **73**, 248–260.

Bernier, N. J. and Craig, P. M. (2005). CRF-related peptides contribute to stress response and regulation of appetite in hypoxic rainbow trout. *Am. J. Physiol.– Regul. Integr. Comp. Physiol.* **289**, R982–R990.

Björnsson, B. T., Johansson, V., Benedet, S., Einarsdottir, I. E., Hildahl, J., Agustsson, T., et al. (2002). Growth hormone endocrinology of salmonids: regulatory mechanisms and mode of action. *Fish Physiol. Biochem.* **27**, 227–242.

Bodiguel, X., Maury, O., Mellon-Duval, C., Roupsard, F., Le Guellec, A.-M. and Loizeau, V. (2009). A dynamic and mechanistic model of PCB bioaccumulation in the European hake (*Merluccius merluccius*). *J. Sea Res.* **62**, 124–134.

Boone, A. N., Ducouret, B. and Vijayan, M. M. (2002). Glucocorticoid-induced glucose release is abolished in trout hepatocytes with elevated hsp70 content. *J. Endocrinol.* **172**, R1–R5.

Bostock, J., McAndrew, B., Richards, R., Jauncey, K., Telfer, T., Lorenzen, K., et al. (2010). Aquaculture: global status and trends. *Philos. Trans. R. Soc. Lond. B Biol. Sci.* **365**, 2897–2912.

Bromley, P. J., Ravier, C. and Witthames, P. R. (2000). The influence of feeding regime on sexual maturation, fecundity and atresia in first-time spawning turbot. *J. Fish. Biol.* **56**, 264–278.

Castillo, J., Ammendrup-Johnsen, I., Codina, M., Navarro, I. and Gutiérrez, J. (2006). IGF-I and insulin receptor signal transduction in trout muscle cells. *Am. J. Physiol.– Regul. Integr. Comp. Physiol.* **290**, R1683–R1690.

Cerdá-Reverter, J. M., Schiöth, H. B. and Peter, R. E. (2003). The central melanocortin system regulates food intake in goldfish. *Regul. Pept.* **115**, 101–113.

Cleveland, B. M. and Weber, G. M. (2010). Effects of insulin-like growth factor-I, insulin, and leucine on protein turnover and ubiquitin ligase expression in rainbow trout primary myocytes. *Am. J. Physiol.– Regul. Integr. Comp. Physiol.* **298**, R341–R350.

Conover, C. A., Divertie, G. D. and Lee, P. D. (1993). Cortisol increases plasma insulin-like growth factor binding protein-1 in humans. *Acta Endocrinol. (Copenh).* **128**, 140–143.

Copeland, D. L., Duff, R. J., Liu, Q., Prokop, J. and Londraville, R. L. (2011). Leptin in teleost fishes: an argument for comparative study. *Front. Physiol.* **2**, 26.

Cox, G. N., McDermott, M. J., Merkel, E., Stroh, C. A., Ko, S. C., Squires, C. H., et al. (1994). Recombinant human insulin-like growth factor (IGF)-binding protein-1 inhibits somatic growth stimulated by IGF-I and growth hormone in hypophysectomized rats. *Endocrinology* **135**, 1913–1920.

Davis, K. B. and Peterson, B. C. (2006). The effect of temperature, stress, and cortisol on plasma IGF-I and IGFBPs in sunshine bass. *Gen. Comp. Endocrinol.* **149**, 219–225.

Daza, D. O., Sundström, G., Bergqvist, C. A., Duan, C. and Larhammar, D. (2011). Evolution of the insulin-like growth factor binding protein (IGFBP) family. *Endocrinology* **152**, 2278–2289.

Deane, E. E. and Woo, N. Y. S. (2006). Molecular cloning of growth hormone from silver sea bream: effects of abiotic and biotic stress on transcriptional and translational expression. *Biochem. Biophys. Res. Commun.* **342**, 1077–1082.

Deane, E. E. and Woo, N. Y. S. (2008). Modulation of fish growth hormone levels by salinity, temperature, pollutants and aquaculture related stress: a review. *Rev. Fish Biol. Fish.* **19**, 97–120.

Deane, E. E. and Woo, N. Y. S. (2010). Advances and perspectives on the regulation and expression of piscine heat shock proteins. *Rev. Fish Biol. Fish.* **21**, 153–185.

De Santis, C. and Jerry, D. R. (2011). Differential tissue-regulation of myostatin genes in the teleost fish lates calcarifer in response to fasting. Evidence for functional differentiation. *Mol. Cell. Endocrinol.* **335**, 158–165.

DiBattista, J. D., Levesque, H. M., Moon, T. W. and Gilmour, K. M. (2006). Growth depression in socially subordinate rainbow trout *Oncorhynchus mykiss*: more than a fasting effect. *Physiol. Biochem. Zool.* **79**, 675–687.

Du, F., Xu, G., Nie, Z., Xu, P. and Gu, R. (2014). Transcriptome analysis gene expression in the liver of *Coilia nasus* during the stress response. *BMC Genomics* **15**, 558.

Dumas, A., France, J. and Bureau, D. (2010). Modelling growth and body composition in fish nutrition: where have we been and where are we going? *Aquacul. Res.* **41**, 161–181.

Dunham, R. A. (2011). *Aquaculture and fisheries biotechnology: genetic approaches* (second ed.). Wallingford, Oxfordshire, UK: Cabi.

Dupont-Nivet, M., Médale, F., Leonard, J., Le Guillou, S., Tiquet, F., Quillet, E., et al. (2009). Evidence of genotype–diet interactions in the response of rainbow trout (*Oncorhynchus mykiss*) clones to a diet with or without fishmeal at early growth. *Aquaculture* **295**, 15–21.

Eichinger, M., Loizeau, V., Roupsard, F., Le Guellec, A. M. and Bacher, C. (2010). Modelling growth and bioaccumulation of polychlorinated biphenyls in common sole (*Solea solea*). *J. Sea Res.* **64**, 373–385.

Einarsson, B., Birnir, B. and Sigurðsson, S. (2011). A dynamic energy budget (DEB) model for the energy usage and reproduction of the icelandic capelin (*Mallotus villosus*). *J. Theor. Biol.* **281**, 1–8.

Essington, T. E., Kitchell, J. F. and Walters, C. J. (2001). The von Bertalanffy growth function, bioenergetics, and the consumption rates of fish. *Can. J. Fish. Aquat. Sci.* **58**, 2129–2138.

Fader, C. M., Sánchez, D. G., Mestre, M. B. and Colombo, M. I. (2009). TI-VAMP/VAMP7 and VAMP3/cellubrevin: two v-SNARE proteins involved in specific steps of the autophagy/multivesicular body pathways. *Biochim. Biophys. Acta* **1793**, 1901–1916.

Faught, E., Aluru, N. and Vijayan, M. M. (2016). The Molecular Stress Response. In *Fish Physiology - Biology of Stress in Fish*, Vol. 35 (eds. C. B. Schreck, L. Tort, A. P. Farrell and C. J. Brauner), San Diego, CA: Academic Press.

Filgueira, R., Chapman, J. M., Suski, C. D. and Cooke, S. J. (2016). The influence of watershed land use cover on stream fish diversity and size-at-age of a generalist fish. *Ecol. Indic.* **60**, 248–257.

Fox, H. E., White, S. A., Kao, M. H. F. and Fernald, R. D. (1997). Stress and dominance in a social fish. *J. Neurosci.* **17**, 6463–6469.

Freitas, V., Cardoso, J. F. M. F., Lika, K., Peck, M. A., Campos, J., Kooijman, S. A. L. M., et al. (2010). Temperature tolerance and energetics: a dynamic energy budget-based comparison of North Atlantic marine species. *Philos. Trans. R. Soc. Lond. B Biol. Sci.* **365**, 3553–3565.

Fuentes, E. N., Einarsdottir, I. E., Valdes, J. A., Alvarez, M., Molina, A. and Björnsson, B. T. (2011). Inherent growth hormone resistance in the skeletal muscle of the fine flounder is modulated by nutritional status and is characterized by high contents of truncated GHR, impairment in the JAK2/STAT5 signaling pathway, and low IGF-I expression. *Endocrinology* **153**, 283–294.

Fuentes, E. N., Pino, K., Navarro, C., Delgado, I., Valdés, J. A. and Molina, A. (2013a). Transient inactivation of myostatin induces muscle hypertrophy and overcompensatory growth in zebrafish via inactivation of the SMAD signaling pathway. *J. Biotechnol.* **168**, 295–302.

Fuentes, E. N., Safian, D., Einarsdottir, I. E., Valdés, J. A., Elorza, A. A., Molina, A., et al. (2013b). Nutritional status modulates plasma leptin, AMPK and TOR activation, and mitochondrial biogenesis: implications for cell metabolism and growth in skeletal muscle of the fine flounder. *Gen. Comp. Endocrinol.* **186**, 172–180.

Fujiwara, M., Kendall, B. E., Nisbet, R. M. and Bennett, W. A. (2005). Analysis of size trajectory data using an energetic-based growth model. *Ecology* **86**, 1441–1451.

Fulda, S., Gorman, A. M., Hori, O. and Samali, A. (2010). Cellular stress responses: cell survival and cell death. *Int. J. Cell Biol.* **2010**, 214074.

Funkenstein, B., Balas, V., Rebhan, Y. and Pliatner, A. (2009). Characterization and functional analysis of the 5′ flanking region of *Sparus aurata* myostatin-1 gene. *Comp. Biochem. Physiol. A Mol. Integr. Physiol.* **153**, 55–62.

Gabbitas, B. and Canalis, E. (1996). Cortisol enhances the transcription of insulin-like growth factor-binding protein-6 in cultured osteoblasts. *Endocrinology* **137**, 1687–1692.

Gahete, M. D., Durán-Prado, M., Luque, R. M., Martínez-Fuentes, A. J., Quintero, A., Gutiérrez-Pascual, E., et al. (2009). Understanding the multifactorial control of growth hormone release by somatotropes. *Ann. Y. Acad. Sci.* **1163**, 137–153.

Galloway, T. F., Bardal, T., Kvam, S. N., Dahle, S. W., Nesse, G., Randøl, M., et al. (2006). Somite formation and expression of MyoD, myogenin and myosin in Atlantic halibut (*Hippoglossus hippoglossus* L.) embryos incubated at different temperatures: transient asymmetric expression of MyoD. *J. Exp. Biol.* **209**, 2432–2441.

Galt, N. J., Froehlich, J. M., Remily, E. A., Romero, S. R. and Biga, P. R. (2014). The effects of exogenous cortisol on myostatin transcription in rainbow trout, *Oncorhynchus mykiss*. *Comp. Biochem. Physiol. A Mol. Integr. Physiol.* **175**, 57–63.

Gao, D., Xu, Z., Kuang, X., Qiao, P., Liu, S., Zhang, L., et al. (2014). Molecular characterization and expression analysis of the autophagic gene Beclin 1 from the purse red common carp (*Cyprinus carpio*) exposed to cadmium. *Comp. Biochem. Physiol. C Toxicol. Pharmacol.* **160**, 15–22.

García de la serrana, D., Codina, M., Capilla, E., Jiménez-Amilburu, V., Navarro, I., Du, S.-J., et al. (2014). Characterisation and expression of myogenesis regulatory factors during in vitro myoblast development and in vivo fasting in the gilthead sea bream (*Sparus aurata*). *Comp. Biochem. Physiol. Part A Mol. Integr. Physiol.* **167**, 90–99.

Garikipati, D. K., Gahr, S. A. and Rodgers, B. D. (2006). Identification, characterization, and quantitative expression analysis of rainbow trout myostatin-1a and myostatin-1b genes. *J. Endocrinol.* **190**, 879–888.

Garikipati, D. K., Gahr, S. A., Roalson, E. H. and Rodgers, B. D. (2007). Characterization of rainbow trout myostatin-2 genes (rtMSTN-2a and -2b): genomic organization, differential expression, and pseudogenization. *Endocrinology* **148**, 2106–2115.

Gjedrem, T. (ed.). (2005). *Selection and breeding programs in aquaculture*. p. 364. Dordrecht, The Netherlands: Springer.

Gong, N., Jönsson, E. and Bjornsson, B. T. (2015). Acute anorexigenic action of leptin in rainbow trout is mediated by the hypothalamic Pi3k pathway. *J. Mol. Endocrinol.* 15–0279.

Gorissen, M. and Flik, G. (2014). Leptin in teleostean fish, towards the origins of leptin physiology. *J. Chem. Neuroanat.* **61–62**, 200–206.

Guderley, H. and Pörtner, H. O. (2010). Metabolic power budgeting and adaptive strategies in zoology: examples from scallops and fish. *Can. J. Zool.* **88**, 753–763.

Gutierrez, M. G., Munafó, D. B., Berón, W. and Colombo, M. I. (2004). Rab7 is required for the normal progression of the autophagic pathway in mammalian cells. *J. Cell. Sci.* **117**, 2687–2697.

Habibi, H. R. and Ince, B. W. (1984). A study of androgen-stimulated l-leucine transport by the intestine of rainbow trout (*Salmo gairdneri* Richardson) in vitro. *Comp. Biochem. Physiol. Part A Physiol.* **79**, 143–149.

He, C., Bartholomew, C. R., Zhou, W. and Klionsky, D. J. (2009). Assaying autophagic activity in transgenic GFP-Lc3 and GFP-Gabarap zebrafish embryos. *Autophagy* **5**, 520–526.

Helterline, D. L. I., Garikipati, D., Stenkamp, D. L. and Rodgers, B. D. (2007). Embryonic and tissue-specific regulation of myostatin-1 and -2 gene expression in zebrafish. *Gen. Comp. Endocrinol.* **151**, 90–97.

Hofmann, H. A. and Fernald, R. D. (2000). Social status controls somatostatin neuron size and growth. *J. Neurosci.* **20**, 4740–4744.

Hu, Z., Zhang, J. and Zhang, Q. (2011). Expression pattern and functions of autophagy-related gene atg5 in zebrafish organogenesis. *Autophagy* **7**, 1514–1527.

Huising, M. O., Geven, E. J. W., Kruiswijk, C. P., Nabuurs, S. B., Stolte, E. H., Spanings, F. A. T., et al. (2006). Increased leptin expression in common carp (*Cyprinus carpio*) after food intake but not after fasting or feeding to satiation. *Endocrinology* **147**, 5786–5797.

Iwama, G. K., Thomas, P. T., Forsyth, R. B. and Vijayan, M. M. (1998). Heat shock protein expression in fish. *Rev. Fish Biol. Fish.* **8**, 35–56.

Iwama, G. K., Afonso, L. O. B., Todgham, A., Ackerman, P. and Nakano, K. (2004). Are hsps suitable for indicating stressed states in fish? *J. Exp. Biol.* **207**, 15–19.

Jager, T. and Selck, H. (2011). Interpreting toxicity data in a DEB framework: a case study for nonylphenol in the marine polychaete *Capitella teleta*. *J. Sea Res.* **66**, 456–462.

Jager, T., Vandenbrouck, T., Baas, J., Coen, W. M. D. and Kooijman, S. A. L. M. (2009). A biology-based approach for mixture toxicity of multiple endpoints over the life cycle. *Ecotoxicology* **19**, 351–361.

Jagoe, R. T., Lecker, S. H., Gomes, M. and Goldberg, A. L. (2002). Patterns of gene expression in atrophying skeletal muscles: response to food deprivation. *FASEB J.* **16**, 1697–1712.

Janzen, W. J., Duncan, C. A. and Riley, L. G. (2012). Cortisol treatment reduces ghrelin signaling and food intake in tilapia, *Oreochromis mossambicus*. *Domest. Anim. Endocrinol.* **43**, 251–259.

Jiao, B., Huang, X., Chan, C. B., Zhang, L., Wang, D. and Cheng, C. H. K. (2006). The co-existence of two growth hormone receptors in teleost fish and their differential signal transduction, tissue distribution and hormonal regulation of expression in seabream. *J. Mol. Endocrinol.* **36**, 23–40.

Johansen, K. A. and Overturf, K. (2006). Alterations in expression of genes associated with muscle metabolism and growth during nutritional restriction and refeeding in rainbow trout. *Comp. Biochem. Physiol. Part B Biochem. Mol. Biol.* **144**, 119–127.

Johnston, I. A., Bower, N. I. and Macqueen, D. J. (2011). Growth and the regulation of myotomal muscle mass in teleost fish. *J. Exp. Biol.* **214**, 1617–1628.

Jönsson, E. (2013). The role of ghrelin in energy balance regulation in fish. *Gen. Comp. Endocrinol.* **187**, 79–85.

Jönsson, E., Kaiya, H. and Björnsson, B. T. (2010). Ghrelin decreases food intake in juvenile rainbow trout (*Oncorhynchus mykiss*) through the central anorexigenic corticotropin-releasing factor system. *Gen. Comp. Endocrinol.* **166**, 39–46.

Jusup, M., Klanjscek, T., Matsuda, H. and Kooijman, S. A. L. M. (2011). A full lifecycle bioenergetic model for bluefin tuna. *PLoS ONE* **6**, e21903.

Kaiya, H., Kojima, M., Hosoda, H., Moriyama, S., Takahashi, A., Kawauchi, H., et al. (2003). Peptide purification, complementary deoxyribonucleic acid (DNA) and genomic DNA cloning, and functional characterization of ghrelin in rainbow trout. *Endocrinology* **144**, 5215–5226.

Kajimura, S., Hirano, T., Visitacion, N., Moriyama, S., Aida, K. and Grau, E. G. (2003). Dual mode of cortisol action on GH/IGF-I/IGF binding proteins in the tilapia, *Oreochromis mossambicus*. *J. Endocrinol.* **178**, 91–99.

Kambadur, R., Sharma, M., Smith, T. P. L. and Bass, J. J. (1997). Mutations in myostatin (GDF8) in double-muscled Belgian blue and piedmontese cattle. *Genome Res.* **7**, 910–915.

Kim, J. H., Jin, P., Duan, R. and Chen, E. H. (2015). Mechanisms of myoblast fusion during muscle development. *Curr. Opin. Genet. Dev.* **32**, 162–170.

Kitchell, J. F., Stewart, D. J. and Weininger, D. (1977). Applications of a bioenergetics model to yellow perch (*Perca flavescens*) and walleye (*Stizostedion vitreum vitreum*). *J. Fish. Res. Board Can.* **34**, 1922–1935.

Klok, C., Nordtug, T. and Tamis, J. E. (2014). Estimating the impact of petroleum substances on survival in early life stages of cod (*Gadus morhua*) using the dynamic energy budget theory. *Mar. Environ. Res.* **101**, 60–68.

Kong, H. J., Moon, J.-Y., Nam, B.-H., Kim, Y.-O., Kim, W.-J., Lee, J.-H., et al. (2011). Molecular characterization of the autophagy-related gene beclin-1 from the olive flounder (*Paralichthys olivaceus*). *Fish. Shellfish. Immunol.* **31**, 189–195.

Kooijman, S. A. L. M. (2000). *Dynamic energy and mass budgets in biological systems*. Cambridge, United Kingdom: Cambridge University Press.

Kooijman, S. A. L. M. (2010). *Dynamic energy budget theory for metabolic organisation*. Cambridge, United Kingdom: Cambridge University Press.

Krumschnabel, G., Manzl, C., Berger, C. and Hofer, B. (2005). Oxidative stress, mitochondrial permeability transition, and cell death in Cu-exposed trout hepatocytes. *Toxicol. Appl. Pharmacol.* **209**, 62–73.

Kurokawa, T., Uji, S. and Suzuki, T. (2005). Identification of cDNA coding for a homologue to mammalian leptin from pufferfish, *Takifugu rubripes*. *Peptides* **26**, 745–750.

Lai, K.-M. V., Gonzalez, M., Poueymirou, W. T., Kline, W. O., Na, E., Zlotchenko, E., et al. (2004). Conditional activation of AKT in adult skeletal muscle induces rapid hypertrophy. *Mol. Cell. Biol.* **24**, 9295–9304.

Laitinen, M. and Valtonen, T. (1994). Cardiovascular, ventilatory and total activity responses of brown trout to handling stress. *J. Fish. Biol.* **45**, 933–942.

Landemaine, A., Rescan, P.-Y. and Gabillard, J.-C. (2014). Myomaker mediates fusion of fast myocytes in zebrafish embryos. *Biochem. Biophys. Res. Commun.* **451**, 480–484.

Laplante, M. and Sabatini, D. M. (2009). mTOR signaling at a glance. *J. Cell. Sci.* **122**, 3589–3594.

Lee, C.-Y., Hu, S.-Y., Gong, H.-Y., Chen, M. H.-C., Lu, J.-K. and Wu, J.-L. (2009). Suppression of myostatin with vector-based RNA interference causes a double-muscle effect in transgenic zebrafish. *Biochem. Biophys. Res. Commun.* **387**, 766–771.

Leung, L. Y., Kwong, A. K. Y., Man, A. K. Y. and Woo, N. Y. S. (2008). Direct actions of cortisol, thyroxine and growth hormone on IGF-I mRNA expression in sea bream hepatocytes. *Comp. Biochem. Physiol. A Mol. Integr. Physiol.* **151**, 705–710.

Li, G.-G., Liang, X.-F., Xie, Q., Li, G., Yu, Y. and Lai, K. (2010). Gene structure, recombinant expression and functional characterization of grass carp leptin. *Gen. Comp. Endocrinol.* **166**, 117–127.

López-Patiño, M. A., Guijarro, A. I., Isorna, E., Delgado, M. J., Alonso-Bedate, M. and de Pedro, N. (1999). Neuropeptide Y has a stimulatory action on feeding behavior in goldfish (*Carassius auratus*). *Eur. J. Pharmacol.* **377**, 147–153.

MacDonald, L. E., Alderman, S. L., Kramer, S., Woo, P. T. K. and Bernier, N. J. (2014). Hypoxemia-induced leptin secretion: a mechanism for the control of food intake in diseased fish. *J. Endocrinol.* **221**, 441–455.

Macqueen, D. J., Garcia de la Serrana, D. and Johnston, I. A. (2013). Evolution of ancient functions in the vertebrate insulin-like growth factor system uncovered by study of duplicated salmonid fish genomes. *Mol. Biol. Evol.* **30**, 1060–1076.

Madison, B. N., Tavakoli, S., Kramer, S. and Bernier, N. J. (2015). Chronic cortisol and the regulation of food intake and the endocrine growth axis in rainbow trout. *J. Endocrinol.* **226**, 103–119.

Madsen, S. S., Weber, C., Nielsen, A. M., Mohiseni, M., Bosssus, M. C., Tipsmark, C. K., et al. (2015). Sexual maturation and changes in water and salt transport components in the kidney and intestine of three-spined stickleback (*Gasterosteus aculeatus* L.). *Comp. Biochem. Physiol. A Mol. Integr. Physiol.* **188**, 107–119.

Mammucari, C., Schiaffino, S. and Sandri, M. (2008). Downstream of AKT: FoxO3 and mTOR in the regulation of autophagy in skeletal muscle. *Autophagy* **4**, 524–526.

Mazziotti, G. and Giustina, A. (2013). Glucocorticoids and the regulation of growth hormone secretion. *Nat. Rev. Endocrinol.* **9**, 265–276.

McCormick, S. D., Kelley, K. M., Young, G., Nishioka, R. S. and Bern, H. A. (1992). Stimulation of coho salmon growth by insulin-like growth factor I. *Gen. Comp. Endocrinol.* **86**, 398–406.

McCormick, S. D., Shrimpton, J. M., Carey, J. B., O'Dea, M. F., Sloan, K. E., Moriyama, S., et al. (1998). Repeated acute stress reduces growth rate of Atlantic salmon parr and alters plasma levels of growth hormone, insulin-like growth factor I and cortisol. *Aquaculture* **168**, 221–235.

McEwan, D. G., Popovic, D., Gubas, A., Terawaki, S., Suzuki, H., Stadel, D., et al. (2015). PLEKHM1 regulates autophagosome-lysosome fusion through HOPS complex and LC3/ GABARAP proteins. *Mol. Cell.* **57**, 39–54.

McPherron, A. C., Lawler, A. M. and Lee, S.-J. (1997). Regulation of skeletal muscle mass in mice by a new TGF-p superfamily member. *Nature* **387**, 83–90.

Medeiros, E. F., Phelps, M. P., Fuentes, F. D. and Bradley, T. M. (2009). Overexpression of follistatin in trout stimulates increased muscling. *Am. J. Physiol.– Regul. Integr. Comp. Physiol.* **297**, R235–R242.

Milligan, C. L. (1997). The role of cortisol in amino acid mobilization and metabolism following exhaustive exercise in rainbow trout (*Oncorhynchus mykiss* Walbaum). *Fish Physiol. Biochem.* **16**, 119–128.

Milligan, C. L., Hooke, G. B. and Johnson, C. (2000). Sustained swimming at low velocity following a bout of exhaustive exercise enhances metabolic recovery in rainbow trout. *J. Exp. Biol.* **203**, 921–926.

Mink, J. W., Blumenschine, R. J. and Adams, D. B. (1981). Ratio of central nervous system to body metabolism in vertebrates: its constancy and functional basis. *Am. J. Physiol.* **241**, R203–R212.

Miura, T., Maruyama, K., Shimakura, S.-I., Kaiya, H., Uchiyama, M., Kangawa, K., et al. (2006). Neuropeptide Y mediates ghrelin-induced feeding in the goldfish, *Carassius auratus*. *Neurosci. Lett.* **407**, 279–283.

Mommsen, T. P. (2001). Paradigms of growth in fish. *Comp. Biochem. Physiol. B: Biochem. Mol. Biol.* **129**, 207–219.

Mommsen, T. P. and Moon, T. W. (2001). Hormonal Regulation of Muscle Growth. In *Fish Physiology - Muscle Development and Growth*, Vol. 18 (ed. I. E. Johnston), London: Academic Press.

Mommsen, T. P., Vijayan, M. M. and Moon, T. W. (1999). Cortisol in teleosts: dynamics, mechanisms of action, and metabolic regulation. *Rev. Fish Biol. Fish.* **9**, 211–268.

Montserrat, N., Sánchez-Gurmaches, J., Garcia de la Serrana, D., Navarro, M. I. and Gutiérrez, J. (2007). IGF-I binding and receptor signal transduction in primary cell culture of muscle cells of gilthead sea bream: changes throughout in vitro development. *Cell Tissue Res.* **330**, 503–513.

Morcillo, P., Esteban, M.Á. and Cuesta, A. (2016). Heavy metals produce toxicity, oxidative stress and apoptosis in the marine teleost fish SAF-1 cell line. *Chemosphere* **144**, 225–233.

Morgan, J. D. and Iwama, G. K. (1996). Cortisol-induced changes in oxygen consumption and ionic regulation in coastal cutthroat trout (*Oncorhynchus clarki* clarki) parr. *Fish Physiol. Biochem.* **15**, 385–394.

Moriarty, D. J. W. (1973). The physiology of digestion of blue-green algae in the cichlid fish, *Tilapia nilotica*. *J. Zool.* **171**, 25–39.

Morris, D. H., Yip, C. K., Shi, Y., Chait, B. T. and Wang, Q. J. (2015). Beclin 1-vps34 complex architecture: understanding the nuts and bolts of therapeutic targets. *Front. Biol.* **10**, 398–426.

Mosher, D. S., Quignon, P., Bustamante, C. D., Sutter, N. B., Mellersh, C. S., Parker, H. G., et al. (2007). A mutation in the myostatin gene increases muscle mass and enhances racing performance in heterozygote dogs. *PLoS Genet.* **3**, e79.

Murashita, K., Uji, S., Yamamoto, T., Rønnestad, I. and Kurokawa, T. (2008). Production of recombinant leptin and its effects on food intake in rainbow trout (*Oncorhynchus mykiss*). *Comp. Biochem. Physiol. Part B Biochem. Mol. Biol.* **150**, 377–384.

Murashita, K., Jordal, A.-E. O., Nilsen, T. O., Stefansson, S. O., Kurokawa, T., Björnsson, B. T., et al. (2011). Leptin reduces Atlantic salmon growth through the central pro-opiomelanocortin pathway. *Comp. Biochem. Physiol. A Mol. Integr. Physiol.* **158**, 79–86.

Nesan, D., Kamkar, M., Burrows, J., Scott, I. C., Marsden, M. and Vijayan, M. M. (2012). Glucocorticoid receptor signaling is essential for mesoderm formation and muscle development in zebrafish. *Endocrinology* **153**, 1288–1300.

Nisbet, R. M., Muller, E. B., Lika, K. and Kooijman, S. A. L. M. (2000). From molecules to ecosystems through dynamic energy budget models. *J. Anim. Ecol.* **69**, 913–926.

Noda, N. N. and Inagaki, F. (2015). Mechanisms of autophagy. *Ann. Rev. Biophys.* **44**, 101–122.

Okazaki, R., Riggs, B. L. and Conover, C. A. (1994). Glucocorticoid regulation of insulin-like growth factor-binding protein expression in normal human osteoblast-like cells. *Endocrinology* **134**, 126–132.

Olsen, R. E., Sundell, K., Hansen, T., Hemre, G. I., Myklebust, R., Mayhew, T. M., et al. (2002). Acute stress alters the intestinal lining of Atlantic salmon, *Salmo salar* L.: an electron microscopical study. *Fish Physiol. Biochem.* **26**, 211–221.

Olsen, R. E., Sundell, K., Mayhew, T. M., Myklebust, R. and Ringo, E. (2005). Acute stress alters intestinal function of rainbow trout, *Oncorhynchus mykiss* (Walbaum). *Aquaculture* **250**, 480–495.

Pandurangan, M., Moorthy, H., Sambandam, R., Jeyaraman, V., Irisappan, G. and Kothandam, R. (2014). Effects of stress hormone cortisol on the mRNA expression of myogenenin, MyoD, Myf5, PAX3 and PAX7. *Cytotechnology* **66**, 839–844.

Pankhurst, N. W. (2016). Reproduction and Development. In *Fish Physiology - Biology of Stress in Fish*, Vol. 35 (eds. C. B. Schreck, L. Tort, A. P. Farrell and C. J. Brauner), San Diego, CA: Academic Press.

Pankhurst, N. W. and Sorensen, P. W. (1984). Degeneration of the alimentary tract in sexually maturing European *Anguilla anguilla* (L.) and American eels *Anguilla rostrata* (LeSueur). *Can. J. Zool.* **62**, 1143–1149.

Pankhurst, N. W., King, H. R. and Ludke, S. L. (2008a). Relationship between stress, feeding and plasma ghrelin levels in rainbow trout, *Oncorhynchus mykiss*. *Marine Freshwater Behav. Physiol.* **41**, 53–64.

Pankhurst, N. W., Ludke, S. L., King, H. R. and Peter, R. E. (2008b). The relationship between acute stress, food intake, endocrine status and life history stage in juvenile farmed Atlantic salmon, *Salmo salar*. *Aquaculture* **275**, 311–318.

Pauly, D. (1981). The relationship between gill surface area and growth performance in fishes: a generalization of von Bertalanffy's theory of growth. *Berichte des Deutschen Wissenschaftlichen Kommission fur Meeresforschung* **28**, 251–282.

Pecquerie, L., Petitgas, P. and Kooijman, S. A. L. M. (2009). Modeling fish growth and reproduction in the context of the dynamic energy budget theory to predict environmental impact on anchovy spawning duration. *J. Sea Res.* **62**, 93–105.

Pecquerie, L., Johnson, L. R., Kooijman, S. A. L. M. and Nisbet, R. M. (2011). Analyzing variations in life-history traits of pacific salmon in the context of dynamic energy budget (DEB) theory. *J. Sea Res.* **66**, 424–433.

Penney, C. C. and Volkoff, H. (2014). Peripheral injections of cholecystokinin, apelin, ghrelin and orexin in cavefish (Astyanax fasciatus mexicanus): effects on feeding and on the brain expression levels of tyrosine hydroxylase, mechanistic target of rapamycin and appetite-related hormones. *Gen. Comp. Endocrinol.* **196**, 34–40.

Pereira, R. C., Blanquaert, F. and Canalis, E. (1999). Cortisol enhances the expression of mac25/insulin-like growth factor-binding protein-related protein-1 in cultured osteoblasts. *Endocrinology* **140**, 228–232.

Peterson, B. C. and Small, B. C. (2004). Effects of fasting on circulating IGF-binding proteins, glucose, and cortisol in channel catfish (*Ictalurus punctatus*). *Domest. Anim. Endocrinol.* **26**, 231–240.

Pethybridge, H., Roos, D., Loizeau, V., Pecquerie, L. and Bacher, C. (2013). Responses of european anchovy vital rates and population growth to environmental fluctuations: an individual-based modeling approach. *Ecol. Model.* **250**, 370–383.

Philip, A. M. and Vijayan, M. M. (2015). Stress-immune-growth interactions: cortisol modulates suppressors of cytokine signaling and JAK/STAT pathway in rainbow trout liver. *PLoS ONE* **10**, e0129299.

Pickering, A. D., Pottinger, T. G., Sumpter, J. P., Carragher, J. F. and Le Bail, P. Y. (1991). Effects of acute and chronic stress on the levels of circulating growth hormone in the rainbow trout, *Oncorhynchus mykiss*. *Gen. Comp. Endocrinol.* **83**, 86–93.

Pierce, A. L., Fukada, H. and Dickhoff, W. W. (2005). Metabolic hormones modulate the effect of growth hormone (GH) on insulin-like growth factor-I (IGF-I) mRNA level in primary culture of salmon hepatocytes. *J. Endocrinol.* **184**, 341–349.

Powell, G. T. and Wright, G. J. (2011). Jamb and Jamc are essential for vertebrate myocyte fusion. *PLoS Biol.* **9**, e1001216.

Pratt, W. B. (1993). The role of heat shock proteins in regulating the function, folding, and trafficking of the glucocorticoid receptor. *J. Biol. Chem.* **268**, 21455–21458.

Rauw, W. M. (2012). Immune response from a resource allocation perspective. *Front. Genetics* **3**, 267.

Reindl, K. M. and Sheridan, M. A. (2012). Peripheral regulation of the growth hormone-insulin-like growth factor system in fish and other vertebrates. *Comp. Biochem. Physiol. A Mol. Integr. Physiol.* **163**, 231–245.

Rescan, P. Y. (2005). Muscle growth patterns and regulation during fish ontogeny. *Gen. Comp. Endocrinol.* **142**, 111–116.

Rescan, P.-Y. and Gauvry, L. (1996). Genome of the rainbow trout (*Oncorhynchus mykiss*) encodes two distinct muscle regulatory factors with homology to MyoD. *Comp. Biochem. Physiol. B Biochem. Mol. Biol.* **113**, 711–715.

Rescan, P.-Y., Gauvry, L. and Paboeuf, G. (1995). A gene with homology to myogenin is expressed in developing myotomal musculature of the rainbow trout and in vitro during the conversion of myosatellite cells to myotubes. *FEBS Lett.* **362**, 89–92.

Rescan, P.-Y., Jutel, I. and Rallière, C. (2001). Two myostatin genes are differentially expressed in myotomal muscles of the trout (*Oncorhynchus mykiss*). *J. Exp. Biol.* **204**, 3523–3529.

Roberts, S. B. and Goetz, F. W. (2001). Differential skeletal muscle expression of myostatin across teleost species, and the isolation of multiple myostatin isoforms. *FEBS Lett.* **491**, 212–216.

Rodgers, B. D., Weber, G. M., Kelley, K. M. and Levine, M. A. (2003). Prolonged fasting and cortisol reduce myostatin mRNA levels in tilapia larvae; short-term fasting elevates. *Am. J. Physiol.– Regul. Integr. Comp. Physiol.* **284**, R1277–R1286.

Rodgers, B. D., Roalson, E. H., Weber, G. M., Roberts, S. B. and Goetz, F. W. (2007). A proposed nomenclature consensus for the myostatin gene family. *Am. J. Physiol.– Endocrinol. Metabol.* **292**, E371–E372.

Rommel, C., Bodine, S. C., Clarke, B. A., Rossman, R., Nunez, L., Stitt, T. N., et al. (2001). Mediation of IGF-1-induced skeletal myotube hypertrophy by PI(3)K/AKT/mTOR and PI (3)K/AKT/GSK3 pathways. *Nat. Cell. Biol.* **3**, 1009–1013.

Rosenzweig, S. A. (2004). What's new in the IGF-binding proteins? *Growth Hormone IGF Res.* **14**, 329–336.

Rotllant, J., Balm, P. H. M., Wendelaar-Bonga, S. E., Pérez-Sánchez, J. and Tort, L. (2000). A drop in ambient temperature results in a transient reduction of interrenal ACTH responsiveness in the gilthead sea bream (*Sparus aurata*, L.). *Fish Physiol. Biochem.* **23**, 265–273.

Rotllant, J., Balm, P. H. M., Pérez-Sánchez, J., Wendelaar-Bonga, S. E. and Tort, L. (2001). Pituitary and interrenal function in gilthead sea bream (*Sparus aurata* L., Teleostei) after handling and confinement stress. *Gen. Comp. Endocrinol.* **121**, 333–342.

Roubos, E. W., Dahmen, M., Kozicz, T. and Xu, L. (2012). Leptin and the hypothalamo-pituitary–adrenal stress axis. *Gen. Comp. Endocrinol.* **177**, 28–36.

Rudnicki, M. A., Schnegelsberg, P. N. J., Stead, R. H., Braun, T., Arnold, H.-H. and Jaenisch, R. (1993). MyoD or Myf-5 is required for the formation of skeletal muscle. *Cell* **75**, 1351–1359.

Sadoul, B., Leguen, I., Colson, V., Friggens, N. C. and Prunet, P. (2015a). A multivariate analysis using physiology and behavior to characterize robustness in two isogenic lines of rainbow trout exposed to a confinement stress. *Physiol. Behav.* **140**, 139–147.

Sadoul, B., Martin, O., Prunet, P. and Friggens, N. C. (2015b). On the use of a simple physical system analogy to study robustness features in animal sciences. *PLoS ONE* **10**, e0137333.

Salem, M., Kenney, P. B., Rexroad, C. E. and Yao, J. (2006). Molecular characterization of muscle atrophy and proteolysis associated with spawning in rainbow trout. *Comp. Biochem. Physiol. Part D Genomics Proteomics* **1**, 227–237.

Salmerón, C., García de la serrana, D., Jiménez-Amilburu, V, Fontanillas, R., Navarro, I, Johnston, I. A., Gutiérrez, J. and Capilla, E. (2013). Characterisation and expression of calpain family members in relation to nutritional status, diet composition and flesh texture in gilthead sea bream (*Sparus aurata*). *PLoS ONE* **8**, e75349.

Sandri, M. (2010). Autophagy in skeletal muscle. *FEBS Lett.* **584**, 1411–1416.

Sawatari, E., Seki, R., Adachi, T., Hashimoto, H., Uji, S., Wakamatsu, Y., et al. (2010). Overexpression of the dominant-negative form of myostatin results in doubling of muscle-fiber number in transgenic medaka (*Oryzias latipes*). *Comp. Biochem. Physiol. A Mol. Integr. Physiol.* **155**, 183–189.

Schreck, C. B. and Tort, L. (2016). The Concept of Stress in Fish. In *Fish Physiology - Biology of Stress in Fish*, Vol. 35 (eds. C. B. Schreck, L. Tort, A. P. Farrell and C. J. Brauner), San Diego, CA: Academic Press..

Schroeter, J. C., Fenn, C. M. and Small, B. C. (2015). Elucidating the roles of gut neuropeptides on channel catfish feed intake, glycemia, and hypothalamic NPY and POMC expression. *Comp. Biochem. Physiol. A Mol. Integr. Physiol.* **188**, 168–174.

Seginer, I. (2016). Growth models of gilthead sea bream (*Sparus aurata* L.) for aquaculture: a review. *Aquacul. Eng.* **70**, 15–32.

Seiliez, I, Gabillard, J.-C., Skiba-Cassy, S., Garcia-Serrana, D., Gutiérrez, J., Kaushik, S., Panserat, S. and Tesseraud, S. (2008a). An in vivo and in vitro assessment of TOR signaling cascade in rainbow trout (*Oncorhynchus mykiss*). *Am. J. Physiol. Regul. Integr. Comp. Physiol.* **295**, R329–R335.

Seiliez, I, Panserat, S., Skiba-Cassy, S., Fricot, A., Vachot, C., Kaushik, S. and Tesseraud, S. (2008b). Feeding status regulates the polyubiquitination step of the ubiquitin-proteasome-dependent proteolysis in rainbow trout (*Oncorhynchus mykiss*) muscle. *J. Nutr.* **138**, 487–491.

Seiliez, I., Gutierrez, J., Salmerón, C., Skiba-Cassy, S., Chauvin, C., Dias, K., et al. (2010). An in vivo and in vitro assessment of autophagy-related gene expression in muscle of rainbow trout (*Oncorhynchus mykiss*). *Comp. Biochem. Physiol. B Biochem. Mol. Biol.* **157**, 258–266.

Seiliez, I., Panserat, S., Lansard, M., Polakof, S., Plagnes-Juan, E., Surget, A., et al. (2011a). Dietary carbohydrate-to-protein ratio affects TOR signaling and metabolism-related gene expression in the liver and muscle of rainbow trout after a single meal. *Am. J. Physiol.– Regul. Integr. Comp. Physiol.* **300**, R733–R743.

Seiliez, I, Sabin, N. and Gabillard, J.-C. (2011b). FoxO1 is not a key transcription factor in the regulation of myostatin (mstn-1a and mstn-1b) gene expression in trout myotubes. *Am. J. Physiol.– Regul. Integr. Comp. Physiol.* **301**, R97–R104.

Seiliez, I., Gabillard, J.-C., Riflade, M., Sadoul, B., Dias, K., Avérous, J., et al. (2012a). Amino acids downregulate the expression of several autophagy-related genes in rainbow trout myoblasts. *Autophagy* **8**, 364–375.

Seiliez, I., Sabin, N. and Gabillard, J.-C. (2012b). Myostatin inhibits proliferation but not differentiation of trout myoblasts. *Mol. Cell. Endocrinol.* **351**, 220–226.

Seiliez, I., Médale, F., Aguirre, P., Larquier, M., Lanneretonne, L., Alami-Durante, H., et al. (2013a). Postprandial regulation of growth- and metabolism-related factors in zebrafish. *Zebrafish* **10**, 237–248.

Seiliez, I., Taty Taty, G. C., Bugeon, J., Dias, K., Sabin, N. and Gabillard, J.-C. (2013b). Myostatin induces atrophy of trout myotubes through inhibiting the TORC1 signaling and promoting ubiquitin–proteasome and autophagy-lysosome degradative pathways. *Gen. Comp. Endocrinol.* **186**, 9–15.

Seiliez, I., Dias, K. and Cleveland, B. M. (2014). Contribution of the autophagy-lysosomal and ubiquitin-proteasomal proteolytic systems to total proteolysis in rainbow trout (*Oncorhynchus mykiss*) myotubes. *Am. J. Physiol.– Regul. Integr. Comp. Physiol.* **307**, R1330–R1337.

Serpa, D., Ferreira, P. P., Ferreira, H., da Fonseca, L. C., Dinis, M. T. and Duarte, P. (2013). Modelling the growth of white seabream (*Diplodus sargus*) and gilthead seabream (*Sparus aurata*) in semi-intensive earth production ponds using the dynamic energy budget approach. *J. Sea Res.* **76**, 135–145.

Shepherd, B. S., Johnson, J. K., Silverstein, J. T., Parhar, I. S., Vijayan, M. M., McGuire, A., et al. (2007). Endocrine and orexigenic actions of growth hormone secretagogues in rainbow trout (*Oncorhynchus mykiss*). *Comp. Biochem. Physiol. A Mol. Integr. Physiol.* **146**, 390–399.

Sheridan, M. A. and Hagemeister, A. L. (2010). Somatostatin and somatostatin receptors in fish growth. *Gen. Comp. Endocrinol.* **167**, 360–365.

Shimizu, M., Swanson, P. and Dickhoff, W. W. (1999). Free and protein-bound insulin-like growth factor-I (IGF-I) and IGF-binding proteins in plasma of coho salmon, *Oncorhynchus kisutch*. *Gen. Comp. Endocrinol.* **115**, 398–405.

Shpilka, T., Weidberg, H., Pietrokovski, S. and Elazar, Z. (2011). Atg8: an autophagy-related ubiquitin-like protein family. *Genome. Biol.* **12**, 226.

Silverstein, J. T. and Plisetskaya, E. M. (2000). The effects of NPY and insulin on food intake regulation in fish. *Am. Zool.* **40**, 296–308.

Sloman, K. A., Motherwell, G., O'Connor, K. I. and Taylor, A. C. (2000). The effect of social stress on the Standard Metabolic Rate (SMR) of brown trout, *Salmo trutta*. *Fish Physiol. Biochem.* **23**, 49–53.

Small, B. C., Murdock, C. A., Waldbieser, G. C. and Peterson, B. C. (2006). Reduction in channel catfish hepatic growth hormone receptor expression in response to food deprivation and exogenous cortisol. *Domest. Anim. Endocrinol.* **31**, 340–356.

Sohn, R. L., Huang, P., Kawahara, G., Mitchell, M., Guyon, J., Kalluri, R., et al. (2009). A role for nephrin, a renal protein, in vertebrate skeletal muscle cell fusion. *Proc. Natl. Acad. Sci.* **106**, 9274–9279.

Sousa, T., Domingos, T., Poggiale, J.-C. and Kooijman, S. A. L. M. (2010). Dynamic energy budget theory restores coherence in biology. *Philos. Trans. R. Soc. B Biol. Sci.* **365**, 3413–3428.

Srinivas, B. P., Woo, J., Leong, W. Y. and Roy, S. (2007). A conserved molecular pathway mediates myoblast fusion in insects and vertebrates. *Nat. Genet.* **39**, 781–786.

Sumariwalla, V. M. and Klein, W. H. (2001). Similar myogenic functions for myogenin and MRF4 but not MyoD in differentiated murine embryonic stem cells. *Genesis* **30**, 239–249.

Suzuki, K., Kirisako, T., Kamada, Y., Mizushima, N., Noda, T. and Ohsumi, Y. (2001). The pre-autophagosomal structure organized by concerted functions of APG genes is essential for autophagosome formation. *EMBO J.* **20**, 5971–5981.

Takahashi, H., Sakamoto, T., Hyodo, S., Shepherd, B. S., Kaneko, T. and Grau, E. G. (2006a). Expression of glucocorticoid receptor in the intestine of a euryhaline teleost, the Mozambique tilapia (*Oreochromis mossambicus*): effect of seawater exposure and cortisol treatment. *Life Sci.* **78**, 2329–2335.

Takahashi, H., Takahashi, A. and Sakamoto, T. (2006b). In vivo effects of thyroid hormone, corticosteroids and prolactin on cell proliferation and apoptosis in the anterior intestine of the euryhaline mudskipper (*Periophthalmus modestus*). *Life Sci.* **79**, 1873–1880.

Tan, X. and Du, S. (2002). Differential expression of two MyoD genes in fast and slow muscles of gilthead seabream (*Sparus aurata*). *Dev. Genes. Evol.* **212**, 207–217.

Tinoco, A. B., Näslund, J., Delgado, M. J., de Pedro, N., Johnsson, J. I. and Jönsson, E. (2014). Ghrelin increases food intake, swimming activity and growth in juvenile brown trout (*Salmo trutta*). *Physiol. Behav.* **124**, 15–22.

Unniappan, S., Canosa, L. F. and Peter, R. E. (2004). Orexigenic actions of ghrelin in goldfish: feeding-induced changes in brain and gut mRNA expression and serum levels, and responses to central and peripheral injections. *Neuroendocrinology* **79**, 100–108.

van der Veer, H. W., Kooijman, S. A. L. M. and van der Meer, J. (2001). Intra- and interspecies comparison of energy flow in North Atlantic flatfish species by means of dynamic energy budgets. *J. Sea Res.* **45**, 303–320.

Veillette, P. A., Sundell, K. and Specker, J. L. (1995). Cortisol mediates the increase in intestinal fluid absorption in Atlantic salmon during parr-smolt transformation. *Gen. Comp. Endocrinol.* **97**, 250–258.

Velasco, C., Librán-Pérez, M., Otero-Rodiño, C., López-Patiño, M. A., Míguez, J. M., Cerdá-Reverter, J. M., et al. (2016). Ghrelin modulates hypothalamic fatty acid-sensing and control of food intake in rainbow trout. *J. Endocrinol.* **228**, 25–37.

Vélez, E. J., Lutfi, E., Jiménez-Amilburu, V., Riera-Codina, M., Capilla, E., Navarro, I., et al. (2014). IGF-I and amino acids effects through TOR signaling on proliferation and differentiation of gilthead sea bream cultured myocytes. *Gen. Comp. Endocrinol.* **205**, 296–304.

Vélez, E. J., Lutfi, E., Azizi, S., Montserrat, N., Riera-Codina, M., Capilla, E., et al. (2015). Contribution of in vitro myocytes studies to understanding fish muscle physiology. *Comp. Biochem. Physiol. B Biochem. Mol. Biol.* Available at: ⟨http://www.sciencedirect.com/science/article/pii/S1096495915002158⟩.

Velikkakath, A. K. G., Nishimura, T., Oita, E., Ishihara, N. and Mizushima, N. (2012). Mammalian Atg2 proteins are essential for autophagosome formation and important for regulation of size and distribution of lipid droplets. *Mol. Biol. Cell.* **23**, 896–909.

Vianello, S., Brazzoduro, L., Valle, L. D., Belvedere, P. and Colombo, L. (2003). Myostatin expression during development and chronic stress in zebrafish (*Danio rerio*). *J. Endocrinol.* **176**, 47–59.

Vijayan, M. M., Prunet, P. and Boone, A. N. (2005). Xenobiotic impact on corticosteroid signaling. In *Biochemistry and Molecular Biology of Fishes.* (eds. T. W. Moon and T. P. Mommsen), pp. 365–394. Environmental Toxicology. Amsterdam, The Netherlands: Elsevier.

Villafuerte, B. C., Koop, B. L., Pao, C. I. and Phillips, L. S. (1995). Glucocorticoid regulation of insulin-like growth factor-binding protein-3. *Endocrinology* **136**, 1928–1933.

Volkoff, H., Canosa, L. F., Unniappan, S., Cerdá-Reverter, J. M., Bernier, N. J., Kelly, S. P., et al. (2005). Neuropeptides and the control of food intake in fish. *Gen. Comp. Endocrinol.* **142**, 3–19.

Wang, Q., Bing, C., Al-Barazanji, K., Mossakowaska, D. E., Wang, X. M., McBay, D. L., et al. (1997). Interactions between leptin and hypothalamic neuropeptide Y neurons in the control of food intake and energy homeostasis in the rat. *Diabetes* **46**, 335–341.

Wang, J., Salem, M., Qi, N., Kenney, P. B., Rexroad, C. E., III and Yao, J. (2011). Molecular characterization of the MuRF genes in rainbow trout: potential role in muscle degradation. *Comp. Biochem. Physiol. B Biochem. Mol. Biol.* **158**, 208–215.

Wang, H., Sun, H.-Q., Zhu, X., Zhang, L., Albanesi, J., Levine, B., et al. (2015). GABARAPs regulate PI4P-dependent autophagosome: lysosome fusion. *Proc. Natl. Acad. Sci.* **112**, 7015–7020.

Warren C. E. and Gerald, D. E. (1967). Laboratory studies on the feeding, bioenergetics and growth of fish. Pacific Cooperative Water Pollution and Fisheries Research Laboratories, Agricultural Experiment Station, Oregon State University.

Weber, T. E. and Bosworth, B. G. (2005). Effects of 28 day exposure to cold temperature or feed restriction on growth, body composition, and expression of genes related to muscle growth and metabolism in channel catfish. *Aquaculture* **246**, 483–492.

Weber, T. E., Small, B. C. and Bosworth, B. G. (2005). Lipopolysaccharide regulates myostatin and MyoD independently of an increase in plasma cortisol in channel catfish (*Ictalurus punctatus*). *Domest. Anim. Endocrinol.* **28**, 64–73.

Weintraub, H., Davis, R., Tapscott, S., Thayer, M., Krause, M., Benezra, R., et al. (1991). The myoD gene family: nodal point during specification of the muscle cell lineage. *Science* **251**, 761–766.

Wolowczuk, I., Verwaerde, C., Viltart, O., Delanoye, A., Delacre, M., Pot, B., et al. (2008). Feeding our immune system: impact on metabolism. *Clin. Dev. Immunol.* **2008**, 639803.

Won, E. T., Douros, J. D., Hurt, D. A. and Borski, R. J. (2016). Leptin stimulates hepatic growth hormone receptor and insulin-like growth factor gene expression in a teleost fish, the hybrid striped bass. *Gen. Comp. Endocrinol.* **229**, 84–91.

Wong, A. O. L., Chang, J. P. and Peter, R. E. (1993). Dopamine functions as a growth hormone-releasing factor in the goldfish, *Carassius auratus*. *Fish Physiol. Biochem.* **11**, 77–84.

Wong, A. O. L., Zhou, H., Jiang, Y. and Ko, W. K. W. (2006). Feedback regulation of growth hormone synthesis and secretion in fish and the emerging concept of intrapituitary feedback loop. *Comp. Biochem. Physiol. A Mol. Integr. Physiol.* **144**, 284–305.

Wood, A. W., Duan, C. and Bern, H. A. (2005). Insulin-like growth factor signaling in fish. *Int. Rev. Cytol.* **243**, 215–285.

Wu, S., Ji, G., Liu, J., Zhang, S., Gong, Y. and Shi, L. (2015). TBBPA induces developmental toxicity, oxidative stress, and apoptosis in embryos and zebrafish larvae (*Danio rerio*). *Environ. Toxicol.* Available at: ⟨http://onlinelibrary.wiley.com/doi/10.1002/tox.22131/abstract⟩.

Xie, Z., Nair, U. and Klionsky, D. J. (2008). Atg8 controls phagophore expansion during autophagosome formation. *Mol. Biol. Cell.* **19**, 3290–3298.

Xu, C., Wu, G., Zohar, Y. and Du, S.-J. (2003). Analysis of myostatin gene structure, expression and function in zebrafish. *J. Exp. Biol.* **206**, 4067–4079.

Yabu, T., Imamura, S., Mizusawa, N., Touhata, K. and Yamashita, M. (2012). Induction of autophagy by amino acid starvation in fish cells. *Marine Biotechnol.* **14**, 491–501.

Yu, J., Iwashita, M., Kudo, Y. and Takeda, Y. (1998). Phosphorylated insulin-like growth factor (IGF)-binding protein-1 (IGFBP-1) inhibits while non-phosphorylated IGFBP-1 stimulates IGF-I-induced amino acid uptake by cultured trophoblast cells. *Growth Horm. IGF Res.* **8**, 65–70.

Zhu, Z., Zeng, X., Lin, X., Xu, Z. and Sun, J. (2015). Effects of ration levels on growth and reproduction from larvae to first-time spawning in the female *Gambusia affinis*. *Int. J. Mol. Sci.* **16**, 5604–5617.

HOMEOSTATIC RESPONSES TO OSMOTIC STRESS

YOSHIO TAKEI

PUNG-PUNG HWANG

The blood of fish is rarely similar to the water in which they reside, and thus they are constantly exposed to some level of osmotic stress. In addition, physiological stressors and associated responses may have profound hydromineral balance consequences (Schreck and Tort, 2016; Chapter 1 in this volume). The osmotic gradient between the fish and its environment is perceived by osmosensors and transduced through the intracellular signaling pathway to effectors in the osmoregulatory organs and initiates both general and osmospecific adaptive responses over different time scales ranging from the immediate posttranslational modification to the longer-term

Biology of Stress in Fish: Volume 35
FISH PHYSIOLOGY

transcriptional regulation of genes. The former includes alterations in the activity of transport proteins and osmolyte-producing enzymes, cytoskeletal organization, vesicular trafficking, and metabolism. Recent studies using genomewide, molecular physiological approaches, together with emerging model species, help identify new transcription factors and effector molecules that play critical roles in adaptation to osmotic stress. In this chapter, we will review recent data on the osmotic stress response in fishes at different levels of biological organization (from gene to whole organism) and at different time scales (from acute to chronic response) to highlight what is known and to suggest areas for further study.

1. INTRODUCTION

Fishes actively regulate the osmolality and ion levels of body fluids different from those of their environment either in freshwater or in seawater (Edwards and Marshall, 2012). Both osmolality and ion levels may be influenced by a range of stressors (Schreck and Tort, 2016; Chapter 1 in this volume) (Winberg et al., 2016; Chapter 2 in this volume). To maintain body fluid homeostasis, the sympathetic nervous system and the endocrine system (hypothalamic–pituitary–interrenal axis) are activated as a primary response to the stressors, which results in catecholamine release from the chromaffin cells and cortisol release from the steroidogenic cells of interrenal gland (Wendelaar Bonga, 1997) (Gorissen and Flik, 2016; Chapter 3 in this volume). These hormones act on the gills to regulate the activity of ion transport proteins within ionocytes and/or alter relative circulation to the respiratory arterio-arteriolar shunt and the osmoregulatory arterio-venous shunt (Olson, 2002), and these can significantly affect ion and water balance (Redding et al., 1984; Takei and Loretz, 2006). Cortisol also promotes longer-term adaptation in both freshwater and seawater by reorganizing the osmoregulatory organs despite the need to transport ions and water in opposite directions in these two water types (Wendelaar Bonga, 1997). In addition, changes in plasma glucose induced by these hormones alter the activity of ATP-driven ion pumps in the gills and other osmoregulatory organs (Bickler and Buck, 2007) (see Sadoul and Vijayan, 2016; Chapter 5 in this volume; and Section 5 of this chapter).

As mentioned by Schreck and Tort (2016; Chapter 1 in this volume), this book aims to provide a holistic review of physiological responses to various stressors and does not focus on any particular stressor. In this regard, McDonald and Milligan (1997) reviewed the effect of various stressors on the primary stress response of hydromineral balance in salmonids in freshwater. Other papers have considered effects of general (nonosmotic) stressors on (1) growth rate and plasma osmolality in the red porgy fry, *Pagrus pagrus* (Vargau-Chacoff et al., 2011); (2) plasma ion concentrations in the sea bass, *Dicentrarchus labrax* (Sinha et al., 2015); (3) mobilization of oxygen supply to various osmoregulatory tissues for hydromineral and urea homeostasis in elasmobranchs (Skomal and Mandelman, 2012); and (4) tissue water permeability and Na^+ and K^+ losses in freshwater stingray, *Potamotrygon* cf. *histrix* (Brinn et al., 2012).

While the studies on the effect of general stressors on osmoregulation are scanty, there recently have been rapid and profound advances in knowledge concerning the osmotic stress responses in a range of animals including fishes. It seems that the mechanisms to maintain water and ion balance are similar among different stressors in terms of associated tissues and molecules. In this chapter, therefore, we mainly focus on how fishes maintain body fluid homeostasis in response to osmotic stresses. The signaling cascade from osmosensing to cell volume regulation has been studied extensively in nonfish model species and the basic mechanisms appear to be quite similar to those of fishes. Thus, we also review some of the literature on nonfish models to provide future directions for research in fishes.

1.1. Osmoregulation in Fishes

Here we briefly describe the basic principles of osmoregulation in fishes to better understand the effects of stressors. We also briefly discuss acid/base balance because it is intimately associated with ionocyte function and the stress response can involve a change in pH. This information is also useful when considering future research needs. Detailed information on this topic is available in McCormick et al. (2013). Maintenance of body fluid balance is a complex process that involves various osmoregulatory organs, principally the gills, digestive tract, and kidney (Fig. 6.1). Among them, the gills are the primary organ that directly senses external osmotic changes and actively takes up or excretes monovalent ions (eg, Na^+, K^+, and Cl^-), depending upon the environmental water type, to maintain plasma osmolality within a narrow range. The brain is not categorized as an osmoregulatory organ but is important for osmoregulation as it regulates drinking. In marine teleosts, ingested seawater is absorbed by the intestine together with Na^+ and Cl^- (Fig. 6.1), but the intestine plays a minor role in osmoregulation in freshwater fishes. The kidney is a critical organ for divalent ion (Mg^{2+}, Ca^{2+}, and SO_4^{2-}) secretion in marine teleosts (Beyenbach, 2004), while it actively secretes a large amount of dilute urine to counter copious water influx across the body surfaces in freshwater fishes.

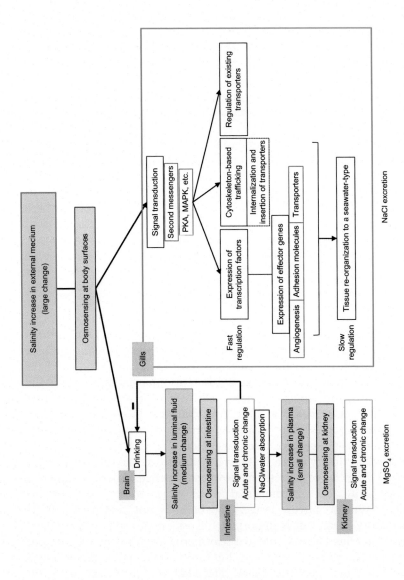

Figure 6.1. The sequential events that occur in response to osmotic stress (exemplified by a seawater challenge here) over time in different osmoregulatory tissues of fish. MAPK, mitogen-activated protein kinase; PKA, protein kinase A.

Epithelial cells that constitute osmoregulatory organs are exposed to different osmotic stresses, depending upon whether they are facing the apical or basolateral side; this varies further with tissue type. For example, the apical membrane of gill epithelial cells directly contacts external freshwater or seawater, while that of the intestinal cells is faced with varying concentrations of ions modified from external media. In seawater fishes, ingested seawater is processed gradually along the digestive tracts, with a decreasing concentration of monovalent ions and an increasing concentration of divalent ions progressing down the gut (Fig. 6.1). Renal tubular cells interface with filtered plasma, and the change in tubular fluid osmolality is slow and minor after osmotic stress (Fig. 6.1). Thus the fish kidney is an osmotically stress-free tissue compared with other osmoregulatory tissues of fishes and with the renal medullary cells of mammals that are exposed to high concentrations of urea and other osmolytes (Beuchat, 1996). In contrast to the apical membrane, the basolateral membrane of epithelial cells is bathed in interstitial fluid that changes slightly after osmotic stress. It is apparent that the severity of the stress is greatest at the gills, followed by the intestine and then kidney (Fig. 6.1).

The apical and basolateral membranes possess different ion-transporting proteins to absorb or secrete ions for homeostasis of extracellular fluids. For example, gill ionocytes in freshwater fish take up NaCl from the ion-deficient environment by different cell types using Na^+-Cl^- cotransporter (NCC), Na^+/H^+ exchanger (NHE), or other Na^+-coupled transporters on the apical side energized by Na^+/K^+-ATPase (NKA) on the basolateral side (Hwang and Lee, 2007; Hwang et al., 2011). Ionocytes of marine fish, however, excrete NaCl principally by joint actions of apical cystic fibrosis transport regulator (CFTR) Cl^- channel and basolateral Na^+-K^+-$2Cl^-$ cotransporter type 1 (NKCC1) and NKA (Evans et al., 2005; Hwang et al., 2011). Such transporter activity is modified after osmotic stress in all epithelial cells in the osmoregulatory tissues.

The response to osmotic stress can be divided temporarily into two phases; the acute phase and the chronic phase (Fig. 6.2). For instance, gill epithelial cells reverse ion regulation from uptake to excretion upon transfer from freshwater to seawater. This immediate switching may occur by removing the freshwater-type transporter/channels and recruiting the seawater-type from the intracellular pool to the plasma membrane (Fig. 6.1). It is also possible that the signal transduction from the osmosensor regulates various kinase/phosphatase systems to inactivate the freshwater-type and activate the seawater-type on the plasma membrane. However, little is yet known about such acute-phase regulation in fishes. In the chronic phase, osmoregulatory organs are reorganized as a new tissue with opposite functions after osmotic stress by apoptosis of old cell types and differentiation of progenitor cells to new cell types

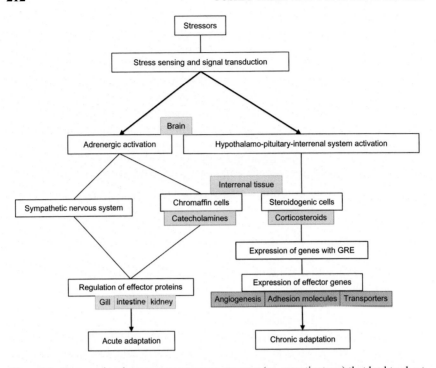

Figure 6.2. Hormonal and nervous responses to stressors (eg, osmotic stress) that lead to short-term and long-term acclimation to a new environment in fish. Other hormones involved in osmotic stress are introduced in the text. Detailed description of the cascade in response to general stress can be found in Schreck and Tort (2016; Chapter 1 in this volume). GRE, glucocorticoid responsive element.

(Fig. 6.2). For instance, 1 week following seawater transfer, the epithelial cells of the digestive tract become thinner and transparent, while the surface becomes reddish due to the development of blood vessels (Yuge et al., 2003). These changes apparently help active absorption of water and ions, which is an essential function of the marine teleost intestine. Together with the extensive studies on gill ionocytes (Evans et al., 2005; Hwang et al., 2011; Hiroi and McCormick, 2012; Hwang and Lin, 2014; Takei et al., 2014), such chronic changes are well characterized at the cellular and molecular levels compared with the acute changes.

1.2. pH Regulation in Freshwater and Seawater

Fishes have to maintain body fluid homeostasis under the stress of environmental (eg, due to ocean acidification or pollutions) or internal metabolism-derived pH fluctuations because CO_3^- and H^+ regulation are closely linked to those of other ions. Fishes, similar to mammals, maintain acid–base homeostasis by three compensatory mechanisms: (1) instantaneous physicochemical buffering with bicarbonate or nonbicarbonate buffers, (2) respiratory adjustments of the "open" CO_2-HCO_3^- buffer system, and (3) net transport of acid–base relevant molecules between animal and environment (Evans et al., 2005). Fishes, however, predominately adopt the third mechanism to achieve acid–base regulation because of respiratory limits (lower O_2 solubility and higher viscosity of water than air) in water-breathing fish compared to air-breathing mammals (Claiborne et al., 2002; Evans et al., 2005). The gills are generally thought to take over the majority of acid and base movements (90% or higher), compared to the kidney and intestine (Claiborne et al., 2002). It is generally accepted that seawater fishes excrete metabolic acids and bases through an NHE and HCO_3^-/Cl^- exchange (anion exchanger, AE) in the apical membrane of gill cells down the ionic gradients due to the high external Na^+ and Cl^- concentration in seawater. Distinct types of ionocytes for acid and base secretion have been identified in the gills of marine and euryhaline fishes. The acid-secreting ionocyte is characterized by apical NHE2/3 and/or basolateral NKA, and the base-secreting ionocyte is characterized by apical pendrin (SLC26A4) and basolateral H^+-ATPase (HA) for body fluid acid–base homeostasis (see Evans et al., 2005; Hwang et al., 2011; Hwang and Lin, 2014). The NHE-expressing ionocytes and their expression and function are well documented. AE (pendrin)-expressing ionocytes have been studied mostly in elasmobranchs while the information in teleosts is sparse.

Compared to seawater fishes, transport mechanisms of H^+ and HCO_3^- in the gills of freshwater fishes appear to be more complex and diverse, probably because of thermodynamic issues; ambient low Na^+, Cl^-, and H^+ gradients are unfavorable for operation of the related transporters. The operation of electroneutral NHE3 or other isoforms has been questioned, particularly regarding environments with low Na^+ or pH (Avella and Bornancin, 1989; Parks et al., 2008). The recent identification of ammonia-transporting Rh proteins has helped in our understanding of mechanisms involved (Nakada et al., 2007a,b). A subsequent molecular physiological study proposed a model in which apical Rhcg1 mediates the deprotonation of intracellular NH_4^+ and acid trapping at the apical membrane and the resultant H^+ gradient drives NHE3 for achieving the functions of H^+ secretion and Na^+ uptake (Wu et al., 2010). On the other hand, in a few species such as zebrafish, apical HA was found, instead of NHE, to play a predominant role in acid secretion and Rhcg1-mediated NH_4^+ excretion (Guh et al. 2015; Lin et al., 2006; Shih et al., 2008, 2012). To achieve transepithelial H^+ excretion, carbonic anhydrases (cytosolic CA2-like and membrane-form CA15) and basolateral AE1 (but not Na^+-HCO_3^- cotransporter, NBC) have to collaborate with apical NHE or HA in the ionocytes (Gilmour, 2012; Hwang and Chou, 2013; Lin et al., 2008; Hsu et al., 2014; Lee et al., 2011).

1.3. Cell Volume Regulation

Cell volume regulation in response to osmotic stress has been investigated intensively in isolated vertebrate cells such as erythrocytes and in unicellular organisms such as yeast. The basic mechanisms for osmosensing and the signal transduction system appear to be common between vertebrates and yeast, but the accumulated knowledge is much more extensive in the latter. Thus such nonfish data are introduced where appropriate for future research in fish cells.

After exposure to hyperosmotic stress, the cells shrink, which causes macromolecular crowding on the plasma membrane, an increase in intracellular ion concentration, and mechanical stimuli to the cytoskeleton. These changes may work together to sense the volume change and initiate distal signaling events that ultimately result in cell swelling and volume recovery. For instance, transient receptor potential vallinoid-type (TRPV) channel is a possible sensor molecule that changes intracellular Ca^{2+} levels to modulate a number of kinase/phosphatase activities, resulting in modification of the phosphorylation state of target transport proteins. The activated proteins may then alter intracellular osmolality to restore cell volume. It is also possible that the cytoskeleton plays a role for recruitment of transport proteins in the cytoplasmic vesicles to the plasma membrane via modification of contractile proteins (Fletcher and Mullins, 2010). A few transport proteins have been implicated in the ion influx following hyperosmotic stress; NKCC1 (SLC12A2) takes up four osmolytes into the cell at a time, while NHE1 (SLC9A1) and AE (probably SLC4 or SLC26 family) work together to take up Na^+ and Cl^- in exchange for H^+ and HCO_3^-, respectively. The discharged H^+ and HCO_3^- produce H_2O and CO_2 to decrease the medium osmolality. The NaCl influx is energized by the NKA that maintains low cytoplasmic Na^+ concentration. The use of either transport system is dependent on the cell type and mode of shrinkage. Various kinases have been suggested to phosphorylate the transporters and regulate their activity, including oxidative stress-response kinase (OSR1)/ Ste20p-related proline alanine-rich kinase (SPAK) complex, with no lysine kinase, and myosin light-chain kinase (see Hoffmann et al., 2009). Other possible transport systems include the organic osmolytes transporter (eg, taurine transporter) and hypertonicity-induced cation channels.

Following exposure to hypoosmotic stress, cell volume increases and subsequent volume recovery may be due to activated swelling-responsive or stretch-activated ion efflux via ion channels for Cl^- and K^+ secretion (Hoffmann et al., 2009). The possible Cl^- channels are members of the ClC or anoctamin families, but at present are not fully elucidated. A swelling-activated K^+ efflux has been demonstrated in various cell types, and several different K^+ channels have been implicated in the efflux depending on the cell types. The candidate channels include the members of two-pore-domain K^+ channel family or leak channels such as TASK2/KCNK5, TREK1/KCNK2, and TRAAK/KCNK4.

In the killifish, *Fundulus heteroclitus*, osmotic stress has been shown to modulate NaCl secretion from the seawater-type ionocytes (Marshall, 2011). In case of hypotonic shock, the ionocyte decreased Cl^- secretion by inactivation of basolateral NKCC1. The inactivation is caused by the dephosphorylation of focal adhesion kinase, which is connected with the osmosensor candidate, integrin β1 (Marshall et al., 2008). Using various kinase inhibitors, possible involvement of distal kinases in the signaling cascade has been suggested in fish. The kinases include p38 mitogen-activated protein kinase (MAPK), OSR1, SPAK that have been suggested in yeast. The effect of hypotonic shock on other ionocyte transporters (CFTR, K^+ channel and organic osmolytes transporters) have also been examined in the killifish (Duranton et al., 2000; Avella et al., 2009; Marshall et al., 2009), while mechanisms of the regulatory volume changes have been investigated in the intestinal cells of the killifish (Marshall et al., 2002) and eels (Lionetto et al., 2005). However, a lot remains to be investigated in terms of cell volume regulation in fishes.

2. RESPONSES TO HYPEROSMOTIC STRESS

The hyperosmotic marine environment is a stressor to iono/osmoregu-lators (Schreck and Tort, 2016; Chapter 1 in this volume), and fishes respond physiologically by actively retaining water and excreting ions (McCormick et al., 2013). Stress hormones have been shown to play critical roles in maintaining internal homeostasis (Gorissen and Flik, 2016; Chapter 3 in this volume). Since seawater contains high concentrations of divalent ions as well as monovalent ions, the homeostatic responses are discussed separately in the two ion types (Fig. 6.1).

2.1. Monovalent Ions as Stressors

Na^+ concentration is higher than that of Cl^- in plasma of marine teleosts (c. 170 mM Na^+ vs 130 mM Cl^-, respectively), but the relationship is reversed in seawater (450 mM Na^+ vs 520 mM Cl^-, respectively). This implies that Cl^- ions are subjected to more active regulation than Na^+ ions. The sensor to switch from the freshwater to the seawater form of ionoregulation is generally assumed to be Cl^- ions, not Na^+ ions, in fishes. The immediate drinking that occurs in eels after seawater transfer is triggered by Cl^- but not Na^+, evidenced by the fact that exposure to $MgCl_2$ and choline-Cl solutions initiate drinking while $NaNO_3$ and Na_2SO_4 solutions do not (Hirano, 1974). Also, Cl^- concentration in the luminal fluid of the intestine is responsible for inhibition of excess drinking in seawater eels as revealed by in situ perfusion experiments (Ando and Nagashima, 1996). Therefore, the Cl^- sensors likely exist on the body surface and the luminal surface of the intestine for regulation of drinking. Since the intestinal lumen is continuous with the external environment, the sensor(s) detect changes in external Cl^- ions after drinking seawater. Clearly, the Cl^- sensing mechanism in fish is an area worthy of further study.

Ingested seawater that enters the esophagus is processed gradually along the digestive tract and up to 80% is absorbed by the time it reaches the distal intestine (Takei and Loretz, 2011; Grosell, 2011). This ability for absorption in marine teleosts is surprising as terrestrial animals lose water when they drink seawater. Water absorption is achieved by efficient NaCl absorption by NKCC2, NCC, and NHE+AE, which accelerate parallel absorption of water via aquaporins (AQPs) (Grosell, 2011; Madsen et al., 2014; Ando and Takei, 2014). Likewise, profound HCO_3^- secretion into the intestinal lumen precipitates $Ca/MgCO_3$ (magnesian calcite), which decreases luminal fluid osmolality and enhances further water absorption (Wilson et al., 2009; Grosell et al., 2009). Thus, the luminal side of intestinal epithelial cells is exposed to high divalent ions (Mg^{2+}, 200 mM; Ca^{2+}, 30 mM; SO_4^{2-}, 100 mM) and low monovalent ions (Na^+, 4 mM; Cl^-, 30 mM), except for HCO_3^- (120 mM) in seawater eels (Tsukada and Takei, 2006). Because of the high concentration of HCO_3^-, the pH of the luminal fluid reaches >9.0 (Grosell et al., 2009). Thus, epithelial cells of the intestine are exposed to a unique osmotic stress on the luminal side. The molecular transition of intestinal epithelial cells after seawater transfer have been well characterized and reviewed extensively elsewhere (Grosell, 2011).

Na^+ and Cl^- enter the body passively across the concentration gradient via the gills and intestine when fishes are in seawater. When euryhaline fishes move from freshwater to seawater, a large amount of NaCl should enter the body immediately, because the regulatory system in the gills and intestine is directed to ion absorption when they are in freshwater. It seems that the hyperosmotic stress detected by the osmosensor immediately inhibits the absorptive system and activates the excretory system at epithelial cells of osmoregulatory organs (Fig. 6.1). Then, the osmoregulatory organs are reorganized by hormones through apoptosis of absorptive cells and newly differentiated excretory cells (Takei and McCormick, 2013).

Molecular and cellular mechanisms for ion secretion are described later in some detail using gill ionocytes as an example. Ionocytes of marine teleosts are mitochondrion-rich cells specialized for NaCl secretion, and have abundant NKA coupled with K^+ channels (Kir5.1) and NKCC1 on the basolateral membrane and CFTR on the apical membrane (Suzuki et al., 1999; Evans et al., 2005; Marshall, 2011; Hwang and Lin, 2014). Resting potential of the ionocytes should be extremely low as active NKA mobilizes $3Na^+$ out of the cell in exchange of 2 K^+ and excess K^+ leaves the cell through Kir5.1 according to the concentration gradient. In addition, cytosolic Cl^- concentration may be unusually high because of active 2 Cl^- influx by NKCC1. Low intracellular potential and high Cl^- concentration expel Cl^- through the apical CFTR into seawater, even though seawater contains > 500 mM Cl^-.

2.2. Divalent Ions as Stressors

The ratio of concentration of divalent ions in the plasma relative to seawater is generally smaller than that of monovalent ions (50 mM vs 1.6 mM for Mg^{2+}, 10 mM vs 1.2 mM for Ca^{2+}, and 35 mM vs 0.8 mM for SO_4^{2-}, respectively) than for monovalent ions (450 mM vs 170 mM for Na^+ and 520 mM vs 120 mM for Cl^-, respectively) in seawater-acclimated eels (Watanabe and Takei, 2012a). Such a profound concentration gradient of divalent ions between the body and the environment likely serves as stressors for fishes; hence seawater fishes prevent these divalent ions from entering the body. In the case of SO_4^{2-}, however, environmental SO_4^{2-} enters the body at a rate of 1.55 μmol/kg per h in seawater eels, of which 85% is across the body surfaces, mostly via the gills, and 15% is from the digestive tract, as shown by the difference in $^{35}SO_4^{2-}$ influx in esophagus-ligated eels. The intestine was thought to scarcely absorb divalent ions and thus an increase in SO_4^{2-} concentration of luminal fluid was used as a marker for water absorption, but significant amounts of SO_4^{2-} appear to be absorbed by the gut to help water absorption. The significant influx of divalent ions can be inferred by their high concentrations in the urine (116 mM Mg^{2+} and 45 mM SO_4^{2-}) in seawater eels (Watanabe and Takei, 2012a). Thus, the major site of excretion of divalent ions is the kidney (Beyenbach, 2004; Kurita et al., 2008; Watanabe and Takei, 2011), which differs from the gills where monovalent ions are excreted (Fig. 6.1). SO_4^{2-} is taken up at 1.76 μmol/kg per h from freshwater containing 0.3 mM of SO_4^{2-} against a concentration gradient, while all influx can be explained by passive diffusion in seawater (Watanabe and Takei, 2012a). Thus, the fish serves as an excellent model for Mg^{2+} and SO_4^{2-} regulation in vertebrates.

When eels are in freshwater, almost all SO_4^{2-} in the glomerular filtrate is reabsorbed by the proximal tubules using apical SLC13A1 and basolateral SLC26A1 (Kurita et al., 2008), and the apical SLC13A1 is replaced by SLC26A6 for excretion after seawater transfer (Watanabe and Takei, 2011). It is interesting to note that the switching from absorption to secretion at the proximal tubules is not triggered by SO_4^{2-} but Cl^- in seawater, as $MgCl_2$ in the environmental water was effective but $MgSO_4$ was ineffective for switching (Watanabe and Takei, 2012b). Again, Cl^- ions are responsible for salinity stress sensing by the kidney of teleost fishes.

2.3. Involvement of Hormones

Typical stress hormones such as cortisol and epinephrine influence water and ion balance in seawater (Gorissen and Flik, 2016; Chapter 3 in this volume). In addition, a number of fast-acting hormones, such as angiotensin II (Ang II) and atrial natriuretic peptide, stimulate the production of stress hormones in fishes (McCormick, 2001; Takei and Loretz, 2006; Takei and McCormick, 2013). In addition, the secretion of slow-acting hormones, such as cortisol and growth hormone (GH), are stimulated by the fast-acting hormones. The slow-acting hormones act chronically to upregulate gene transcription associated with long-term seawater acclimation. It is known that cortisol and GH act in concert to stimulate differentiation of seawater-type ionocytes in the gills in teleosts (Madsen et al., 2009; Armesto et al., 2014) and reorganization of the digestive epithelia in seawater in the mudskipper, *Periophthalmus modestus* (Takahashi et al., 2006). The role of peripheral GH/IGF-I system in fish physiology including osmoregulation has been reviewed (Reindl and Sheridan, 2012). Other stress-related osmoregulatory hormones include thyroid hormones (Peter, 2011) and leptin (Baltzegar et al., 2014). The signaling system and target osmoregulatory genes are the next step of research for the understanding of stress hormone action.

3. RESPONSES TO HYPOOSMOTIC STRESS

Compared with hyperosmotic seawater, hypoosmotic freshwater has lower buffering capacity and much more diversity in terms of ionic composition and pH levels that impact fish body fluid, ionic, and acid–base homeostasis. Here, we focus on the effect of different levels of Na^+, Cl^-, Ca^{2+}, and low pH, given that these parameters have been the most studied at the molecular and cellular level.

3.1. Ionic Compositions as Stressors

The freshwater environment with low ion levels (Na^+, Cl^-, or Ca^{2+}) relative to fish body fluids generally results in passive ion losses to the environment that must be compensated by active uptake to maintain ionic homeostasis (Mcdonald and Rogano, 1986; Perry and Laurent, 1989; Chou et al., 2002). Artificial freshwater with lower levels of Na^+, Cl^-, or Ca^{2+} compared to those in natural freshwater did not appear to significantly affect passive ion effluxes but was associated with reduced uptake of respective ions in the acute phase immediately following dilute water transfer of rainbow trout or tilapia (Perry and Laurent, 1989; Chou et al., 2002). The disturbed ion balance due to severe ionic losses was attenuated or restored by upregulation of uptake pathways and/or simultaneous suppression of passive effluxes, depending upon the stressors and fish species (Perry and Laurent, 1989; Hwang et al., 1996; Chou et al., 2002; Chen et al., 2003).

The absolute values of certain ions, such as Ca^{2+}, may influence the regulation of other ions. In rainbow trout, acute exposure to low-Ca^{2+} freshwater resulted in increased passive Na^+ efflux with no significant effect on Na^+ influx (McWilliams, 1982), while chronic high-Ca^{2+} stress for several days caused a net Na^+ loss (McDonald et al., 1980). The possible mechanisms involved have been investigated through further manipulation of water Ca^{2+} levels. Addition of 10 mM $CaCl_2$ to normal freshwater initially caused a decrease in the density of ionocytes, and thereafter a compensatory increase of ionocytes with a concomitant enhancement of Na^+ influx in the gills of rainbow trout (Avella et al., 1987). Stress of high ambient Ca^{2+} (with constant NaCl) for several weeks reveals similar effects of low NaCl, resulting in stimulated function/activity of ion uptake and/or ionocytes (Laurent et al., 1985; Chang et al., 2001).

Compensatory regulation of ion transport functions at the level of ionocytes is critical and important for fish to cope with the stress of ionic deficiency in freshwater. Stimulation of Cl^- uptake function resulted from the increase of activated NCC-expressing tilapia ionocytes according to sequential in vivo observations by immunocytochemistry and a noninvasive scanning ion-selective electrode technology (SIET). These appear to be derived from modification of the expression and function of transporters in the preexisting ionocytes but without changes in cell proliferation or differentiation during acute low-Cl^- stress (Chang et al., 2003; Lin et al., 2004; Horng et al., 2009a). Following the compensatory or acclamatory processes of the chronic phase of osmotic stress, regulatory events are generally involved in proliferation and/or differentiation of the related ionocytes (Hwang and Lin, 2014; Guh et al., 2015).

Molecular evidence including the identification and functional analyses of specific transporters for the transport of respective ions, have recently become available to shed insight into the compensatory mechanisms associated with exposure to hypoosmotic stress. The expression of NHE3 (tilapia and zebrafish), Rhcg1, AE1b, and membrane-form CA15a (zebrafish) in the gills or yolk sac epithelia of embryos were stimulated during a chronic low-Na^+ stress regime (Yan et al., 2007; Lin et al., 2008; Inokuchi et al., 2009; Lee et al., 2011; Shih et al., 2012). Similarly, the expressions of NCC and SCL26-type AE in zebrafish and that of epithelial calcium channel (ECaC) in trout and zebrafish gills were stimulated by low-Cl^- and low-Ca^{2+} stresses, respectively (Bayaa et al., 2009; Liao et al., 2007; Pan et al., 2005; Perry et al., 2009; Shahsavarani et al., 2006; Wang et al., 2009). The recent findings obtained in those model species should be extended to other species in the future.

3.2. Low pH as Stressors

Acute stress of environmental acidification impairs body fluid acid–base balance and other ionic/osmotic homeostasis (see later), and the responses in fishes are initiated within minutes to hours (Evans et al., 2005). Given the recent progress in molecular and cellular physiological approaches, the responses and related compensation in both acute and chronic phases were explored in more detail in some species. Using SIET (as described by Lin et al., 2006), H^+ secretion from the apical membrane of H^+-ATPase-rich (HR) ionocytes (the main cells responsible for acid secretion) in zebrafish embryonic skin was found to be stimulated as quickly as several hours after low-pH (pH 4) stress compared to control zebrafish (Furukawa et al., 2015). This enhanced acid secretion probably involves posttranslational regulation, but neither transcriptional nor translational regulation, of HA since its mRNA expression did not change until 4 days after pH 4 stress (Chang et al., 2009). During the chronic phase of pH 4 stress, on the other hand, the compensatory acid secretion function in HR ionocytes appears to be due to cytogenic and genomic modifications. These are achieved not only by increased number of HR ionocytes but also by enhancing the acid secretion function at a single cell based on the detailed SIET and immunocytochemical analyses (Horng et al., 2009b). Moreover, this increase in HR cells originates from the differentiation of glial cell missing 2 (GCM2)-labeled ionocyte precursor cells and the proliferation of epithelial p63-labeled stem cells (Chang et al., 2009; Horng et al., 2009b). These cytogenic and genomic events for the compensatory regulation to cope with low pH stress were also reported in other species (Hwang et al., 2011; Hwang and Lin, 2014).

Acidic stress causes impacts on numerous physiological mechanisms other than acid–base regulation. Among them, the impacts on body fluid Na^+ homeostasis have been the most explored. As described in a recent review by Kwong et al. (2014), ambient acidity generally results in net Na^+ loss, which is characterized by increased passive Na^+ efflux and simultaneously suppressed Na^+ influx. A significant increase in passive Na^+ efflux under acidic stress is probably associated with augmented paracellular permeability, which is mediated by tight junction proteins (Kwong et al., 2014). As such, freshwater species exhibit a significant whole-body Na^+ loss that result from increased passive Na^+ efflux and impaired Na^+ uptake in the acute phase of acidic stress. During the subsequent chronic phase, the compensatory regulation in body fluid Na^+ homeostasis appears to rely on the enhanced Na^+ uptake but not on any changes in Na^+ efflux in zebrafish (Kwong et al., 2014). This compensation of Na^+ uptake in zebrafish mainly originates from the functional stimulation of NCC (but not NHE3b) and NCC-expressing ionocytes through upregulation of cell differentiation (Chang et al., 2013), because NHE3b expression was suppressed by ambient acidity due to the unfavorable thermodynamics (Yan et al., 2007).

Ambient acidity also affects the regulation of Cl^- and Ca^{2+}. The impaired body fluid Cl^- balance under acidic stress was ascribed to a decrease in blood HCO_3^- that is coupled with Cl^- uptake via pendrin (Kwong et al., 2014). As for the compensatory regulation of Ca^{2+} uptake mechanism during acidic stress in zebrafish, pH 4 stress initially suppresses Ca^{2+} uptake function (and thus decreased whole-body Ca^{2+} content) (Horng et al., 2009b; Kumai et al., 2015), and the Ca^{2+} uptake mechanism is upregulated within several days after exposure to such stress (Horng et al., 2009b; Kumai et al., 2015).

3.3. Involvement of Hormones

Osmotic stresses cause both the general responses and the impacts on body fluid ionic/acid–base homeostasis in fishes, and the hormonal actions associated with the general responses have been described in Chapters 2, 3 and 4 earlier in this volume. Hormones are known to regulate the iono- and osmoregulatory mechanisms for body fluid ionic and acid–base homeostasis in fish as in mammals, and these hormonal actions are critical for compensation against disturbed homeostasis due to osmotic or ionic stress. Evans et al. (2005) have comprehensively reviewed the regulatory actions of many hormones on fish iono-/osmoregulation in freshwater; however, most of the proposed actions regarding the detailed pathways and precise target ion transport functions, ion transporters, and ionocytes (subtypes, cell stages) remain unclear. Precisely uncovering these pathways only became feasible recently, when specific subtypes of ionocytes and their respective ion transporters were identified in such model species as zebrafish, medaka, and tilapia. Furthermore, gene knockdown/overexpression of homologous/ endogenous levels of hormones and/or receptors, when combined with more traditional pharmacological methods, has led to more detailed understanding of the functional role of hormone signaling (Hwang and Chou, 2013; Guh et al., 2015). We summarize the emerging associated models of ionocyte function in Fig. 6.3, enabling us to better understand the hormonal actions that regulate ionic and acid–base transport for recovery of the disturbed body fluid homeostasis due to osmotic stress.

Cortisol, vitamin D (VitD) and parathyroid hormone (PTH) were reported to exert positive actions on fish Ca^{2+} uptake (Evans et al., 2005). Recent studies on zebrafish demonstrated that cortisol-glucocorticoid receptor (cortisol-GR), VitD-receptor (VDRa), and PTH1exert their actions on Ca^{2+} uptake through the regulation of the expression of apical ECaC but not that of the two basolateral transporters, PMCA2 and NCX1b (Lin et al., 2011, 2012, 2014). This supports the hypothesis that ECaC, but not PMCA2 and NCX1b, is the target gene for Ca^{2+} regulation during exposure to low-Ca^{2+} freshwater (Hwang and Lin, 2014) as in mammals (Hoenderop et al., 2005). Hypocalcemic hormones, like stanniocalcin (STC) and calcitonin (CT), are other important components in regulation of body fluid Ca^{2+} homeostasis under osmotic or ionic stress in fish, as concluded from studies employing traditional injection of hormones and/or surgical removal of the hormone-synthesizing gland. The expression of STC-1 and CT is upregulated under ambient high-Ca^{2+} stress (Tseng et al., 2009a; Lafont et al., 2011; Lin et al., 2014). Similar to the hypercalcemic hormones described earlier, STC-1 and CT also target ECaC in their control actions.

Cortisol regulates hydromineral balance under both hyperosmotic and hypoosmotic stresses (McCormick, 2001; Evans et al., 2005; Takei et al., 2014). Cortisol-GR-mediated activation of Na^+ uptake in zebrafish was suggested to contribute to the tolerance of acidic stress (Kumai et al., 2012a). On the other hand, prolactin (PRL) has long been shown to be an freshwater-adaption hormone; however, the precise target transporters are still not well understood (Breves et al., 2014). In tilapia and zebrafish, the action of PRL on transcriptional/translational regulation of NCC, but not that of NHE3 or ECaC, was clearly demonstrated (Breves et al., 2013). The role of the renin-angiotensin system was suggested to be involved in the control of body fluid volume and blood pressure in fish (Takei and Tsuchida, 2000), and recently was shown to stimulate not only Na^+, but also Cl^- handling in zebrafish (Kumai et al., 2014). Catecholamines released from either nerve terminals or chromaffin cells have long been proposed to exert their actions through α- and

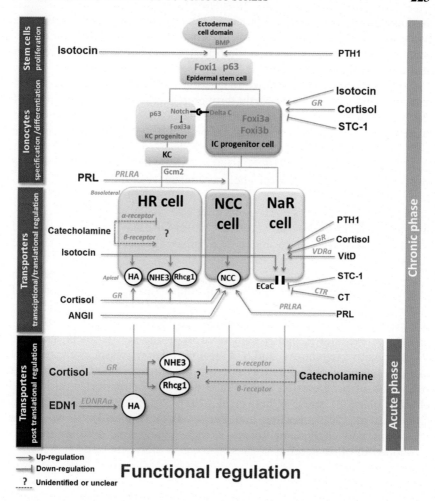

Figure 6.3. Hormone actions on the various aspects of ion regulatory processes in zebrafish under hypoosmotic stress. Different hormones differentially control ionocyte number (cell proliferation, specification, and differentiation), transporter abundance (mRNA and/or protein expression), and activity (posttranslational regulation). For ionocyte specification and differentiation, refer to Hwang and Chou (2013). CA, carbonic anhydrase; ClC, chloride channel; CT, calcitonin; CTR, calcitonin receptor; ECaC, epithelial calcium channel; EDN1, endothelin 1; EDNRAa, endothelin receptor type Aa; GCM2, glial cell missing 2; GR, glucocorticoid receptor; HA, H⁺-ATPase; HR, H⁺-ATPasc-rich; NaR, Na⁺/K⁺-ATPase-rich; NBC, Na⁺-HCO₃⁻ cotransporter; NCC, Na⁺-Cl⁻ cotransporter; NHE, Na⁺/H⁺ exchanger; NKA, Na⁺/K⁺-ATPase; PRL, prolactin; PRLRA, prolactin receptor type A; PTH, parathyroid hormone; RAS, renin-angiotensin system; Rhcg, Rh family C glycoprotein; STC, stanniocalcin; VDRa, vitamin D receptor type a; VitD, vitamin D.

β-adrenergic receptors on fish hyperosmotic regulation in freshwater (see Evans et al., 2005). In a recent study on zebrafish, β-adrenergic receptor was demonstrated to contribute to the compensatory increase in Na^+ uptake upon acidic stress (Kumai et al., 2012b).

SIET analyses demonstrated the action of cortisol on zebrafish H^+ secretion, and this is at least partly through regulating the differentiation of HR ionocytes based on the stimulatory effects of exogenous cortisol on the expression of FOXI3a (a transcriptional factor controlling ionocyte differentiation) and the number of HR cells (Cruz et al., 2013a,b; Guh et al., 2015). Endothelin 1 (EDN1) and one of its receptors, EDNRAa, are also involved in zebrafish in the mechanism regarding acid secretion, particularly under acute acidic stress, and this appears to be mediated by the posttranslational process of HA soon after acid challenge (Guh et al., 2014). As such, EDN1-EDNRAa and the cortisol-GR may regulate fish acid secretion during the acute and chronic phase, respectively, of an acidic stress.

Regulation of body fluid ionic and acid–base balance to cope with hypoosmotic stress is achieved not only through adjustments of transporter activities but also through cytogenetic effects on ionocytes. Regulation at the ionocyte level is important during the chronic phase of osmotic stress, and is tightly controlled by hormones. Isotocin, cortisol, PTH1, positive regulators, and STC-1, a negative regulator, were recently identified as control factors for zebrafish ion transport functions through regulation of proliferation and/or differentiation of ionocytes during the chronic phase of stress (Chou et al., 2011, 2015; Cruz et al., 2013a,b; Trayer et al., 2013; Kwong and Perry, 2015) (Fig. 6.3). It would be important and challenging to see if these cytogenetic actions are also developed in other species.

4. STRESS SENSING TO HOMEOSTASIS

Osmosensors detect salinity changes in media to initiate immediate, anticipatory response to ameliorate drastic future changes in salts and water balance, as exemplified by the Cl^- response for initiation of drinking that occurs within a minute after seawater transfer in eels (Hirano, 1974). However, there seem to be multiple sensors for osmotic stress, detecting changes in cell volume (see Section 1.3) or in specific ion concentrations in extracellular fluids, which initiate various physiological and behavioral responses to cope with the stress. Here, new findings in the field of osmosensing will be summarized, from the sensor molecules to the nongenomic and genomic responses, which leads to body fluid homeostasis (Fig. 6.1). An emphasis is placed on the recent identification of osmosensitive transcription factors and their target genes as revealed by a transcriptomic approach.

4.1. Osmosensors

Although osmosensing mechanisms have been reviewed recently in a volume of this series (Kültz, 2013), little is known about osmosensors in fish. It is likely that similar molecules may act as osmosensors throughout vertebrates, so we will introduce what is known about osmosensors in nonfish species for the future study in fishes. Salt-sensing mechanisms have been vigorously investigated in mammals in relation to the salt-sensitive hypertension (Orlov and Mongin, 2007), revealing the presence of at least four distinct sensing mechanisms: (1) osmolality of the cerebrospinal fluid (CSF) for vasopressin release by TRPV family members (Clapham et al., 2001); (2) osmolality by glial cells via yet unidentified volume-regulated anion channel (Bres et al., 2000); (3) Na^+ concentration in the CSF for drinking and sodium appetite by an atypical voltage-gated Na^+ channel, Nax (Noda, 2007); and (4) Cl^- in the tubular fluid at the macula densa of the kidney for tubulo-glomerular feedback and renin release by NKCC2 (Bell et al., 2003). The calcium sensing receptor (CaSR) is also suggested as a salinity sensor (Loretz, 2008). Changes in ambient osmolality alter cell volume, which is also detected by changes in the tension of the cytoskeleton.

TRPV is a group of mechano-, thermo-, and chemosensitive nonselective cation channels, of which TRPV1 and TRPV4 have been suggested to be osmosensors that respond to changes in internal (plasma) osmolality. The response is triggered by altered plasma membrane stretch, and the effectors are various transport proteins, osmoregulatory hormone release, and expression of various genes involved in osmoregulation (Orlov and Mongin, 2007; Sladek and Johnson, 2013). TRPV1 that lacks the N-terminal region is localized in the vasopressin-producing magnocellular neurons and in the sensory circumventricular organs such as subfornical organ (SFO) and organum vasculosum of the lamina terminalis (OVLT) of mammals. The TRPV1 appears to act as a sensor to hyperosmotic stress for vasopressin secretion from neurohypophysis and for regulation of drinking. In fact, *trpv1* knockout mice have a phenotype of reduced vasopressin and drinking response to intracerebroventricular (icv) administration of hypertonic NaCl solutions and to water deprivation compared with control littermates (Sharif Naeini et al., 2006). The *trpv4* gene is also expressed in these brain sites and its knockout impaired normal vasopressin secretion and drinking in mice, resulting in hypovolemia. The knockout mice also exhibited reduced vasopressin and drinking responses to hyperosmotic challenges (Liedtke and Friedman, 2003). Since TRPV4 alone is suggested to be a hypoosmotic sensor, TRPV1 and TRPV4 may form a heterodimer to detect hyperosmotic stress in intact animals.

Molecular evolution of the TRPV family members has been examined in silico and two TRPV1s and one TRPV4 have been identified in the pufferfish, *Takifugu rubripes* (Saito and Shingai, 2006). The presence of TRPV1 has been demonstrated on the sperm membrane of a teleost, *Labeo rohita* (Majhi et al., 2013), and on the neuronal cell membrane of zebrafish embryo (Gau et al., 2013). However, both studies examined its function as a thermosensor, but its function as an osmosensor needs to be investigated in fishes. TRPV4 has been shown to play an important role in hypoosmosis-induced prolactin secretion in Mozambique tilapia pituitary cells (Seale et al., 2012). As the *trpv* genes exist ubiquitously in all animal species thus far examined, their function as an osmosensor is an intriguing target of research.

A possible Na^+ sensor, Nax, has been identified in mammals (Noda, 2007). Nax monitors Na^+ concentrations in the extracellular fluids, and its increase caused by water deprivation suppresses sodium appetite via the SFO or OVLT. In $Nax^{-/-}$ mice, however, water deprivation induced greater neuronal activities and failed to suppress sodium appetite (Watanabe et al., 2000). As Nax immunoreactivity was localized on glial processes, there must be an inhibitory signal transduction from the glial cells to the neurons. In addition, the presence of epithelial Na^+ channel has been demonstrated in the magnocellular vasopressin neurons, which suggests its role as a Na^+ sensor (Teruyama et al., 2012). Although the Nax ortholog has not been identified in fishes, its function as a Na^+ sensor is an interesting topic in fishes. However, all experiments mentioned earlier examined the effect of hypertonic NaCl or water deprivation, both of which increased not only Na^+ but also Cl^- in the extracellular fluids. Therefore, it is also possible that a Cl^- sensor exists in these circumventricular organs, similar to the suggestion that they reside on the body surfaces of fishes (see Section 2.1).

It has been shown that the macula densa of the thick ascending limb of Henle's loop (TAHL) senses Cl^-, not Na^+, in the luminal fluid and regulates glomerular filtration rate and renin release in mammals (Kotchen et al., 1978; Wilcox, 1983). The sensor for Cl^- is suggested to be NKCC2, which is localized abundantly on the apical membrane of the epithelial cells of TAHL (Bell et al., 2003; Edwards et al., 2014). The increased cell volume caused by the influx of four osmolytes via NKCC2 induces the release of ATP and other vasoconstrictors and decreases the blood flow through the glomerulus and release of renin (Boudreault and Grygorczyk, 2002). The role of NKCC2 as a Cl^- sensor is an interesting topic in teleosts as their intestine has apical NKCC2 similar to the TAHL (Grosell, 2011). As perfusion of intestinal lumen with Cl^- containing solution inhibits drinking as mentioned earlier, it is possible that NKCC2 also senses Cl^- in fishes.

CaSR is a member of the family C, G protein-coupled receptors that includes chemoreceptors such as pheromone receptors and odorant receptors. After binding of inorganic divalent cations such as Ca^{2+} and Mg^{2+} in the low millimolar range, CaSR activates the intracellular signaling cascade utilizing phospholipase C and/or MAPK. As CaSR is localized at the body surface such as the gills and skin, and as seawater contains high concentrations of Ca^{2+} (10 mM) and Mg^{2+} (50 mM), it is reasonable to speculate that extracellular CaSR serves as an osmosensor via divalent cation sensing in fishes (Nearing et al., 2002; Loretz, 2008). In fact, low medium salinity significantly increased the affinity of CaSR to Ca^+ in Mozambique tilapia (Loretz et al., 2004) and shark, *Squalus acanthias* (Fellner and Parker, 2004). The affinity of CaSR to Mg^{2+} is threefold lower than that to Ca^{2+}, while seawater contains a fivefold higher concentration of Mg^{2+}. Thus CaSR may sense the change in Mg^{2+} as well as Ca^{2+} concentration in seawater.

4.2. Signal Transduction from Sensors

Osmotic stress detected by the osmosensors triggers a series of events via an intracellular signaling cascade to maintain body fluid homeostasis. While little is known about this in fishes, intensive research has been performed in isolated cells including yeast (Lópetz-Maury et al., 2008; de Nadal et al., 2011) and similar mechanisms may apply to fish cells in vivo.

When cells are placed in hypertonic NaCl solution, they lose water rapidly but restore it gradually. In order to counteract the shrinkage, the cells accumulate small organic molecules that do not interfere with cytosolic enzyme activities (Yancy et al., 1982) to balance the osmotic pressure with the media. Regulation of osmolyte production by *myo*-inositol biosynthetic pathway has recently been revealed in the eel and tilapia (Kalujnala et al., 2013; Sacchi et al., 2013). Coincidently, cellular functions are maintained by phosphorylation and dephosphorylation of various cytoplasmic proteins to immediately cope with the rapid changes such as cytoskeletal reorganization, vesicle trafficking, metabolic adjustment, and other mechanisms (de Nadal et al., 2011). In addition, transcriptional regulation of osmoregulatory genes occurs in the chronic response to osmotic stress, the signaling cascade of which is extensively clarified in yeast (Saito and Takahashi, 2004). After sensing increased osmolality, stress-activated protein kinase (SAPK) pathways are activated for signal transduction. The major molecule responsible for the pathway is p38-related MAPK, which is activated by phosphorylation through sequential actions of kinases (Capaldi et al., 2008), resulting in the expression of osmoregulatory genes by phosphorylation of transcription factors (Alepuz et al., 2001).

Similar responses to osmotic stress may be functioning in the medullary cells of the mammalian kidney (Sheikh-Hamad and Gustin, 2004) and probably in fish gill cells. The key player in the SAPK pathways is also p38 MAPK, which not only activates osmoregulatory gene expression but also modulates mRNA stability and translation (Cuadrado and Nebreda, 2010). In fish gills that sense osmotic stress directly, at least two routes may operate to cope with the stress (Fig. 6.1). The most rapid mechanism may utilize the cytoskeletal network; osmotic stress induces mechanical or morphological changes in ion-transporting cells, which close or expand the apical surface that contacts the external media. This recruits new transport proteins from the intracellular store by exocytosis, and removes old transport proteins by endocytosis into the cytoplasm as demonstrated in mammals by the vasopressin effect on aquaporin 2 (AQP2) water channel and insulin effect on glucose transporter (GLUT) 4 (Leto and Saltiel, 2012; Möller and Fenton, 2012). Such trafficking of transport protein-tagged vesicles is mediated by the function of the cytoskeleton whose activity is also regulated by kinases. However, vesicle trafficking has not been demonstrated in fishes, partly because vasotocin is not a tubular antidiuretic hormone and insulin is not a hypoglycemic hormone in fishes. The role of MAPKs and other kinases in the osmoregulatory epithelial cells is also a next target of research in fishes.

In addition to the actions via cytoskeleton, the signal from the osmosensor initiates a sequence of gene expression starting from the expression of immediate early genes, most likely osmosensitive transcription factors that govern the adaptation to hyperosmotic stresses as demonstrated by the tonicity enhancer-binding protein gene for adaptation of mammals to terrestrial environment (Lam et al., 2004). The gene was also called an osmotic response element-binding protein or nuclear factor of activated T cells 5. Consensus sequence of ORE has been elucidated (Ferraris et al., 1999) and found to exist in the promoter regions of the AQP2 and urea transporter-A1 genes in mammals. Data are emerging on the immediate early genes that respond to osmotic stress in fishes that may regulate the expression of genes involved in osmoregulation.

4.3. Targets of Intracellular Signaling Cascade

The activity of the target effector is principally regulated by the two mechanisms: cytoskeleton-based mechanisms that occur most quickly and transcriptional regulation by osmosensitive transcription factors. All cells are equipped with a cytoskeleton, an interconnected network of filamentous polymers, and regulatory proteins (Fletcher and Mullins, 2010). Accumulated evidence suggests that the cytoskeleton plays important roles in cell functions including volume sensing and transport of intracellular cargo. In the skin of mudskipper, NaCl-secreting ionocytes are buried under the surrounding cells, and only its apical pit is exposed to the external media through the cleft. After transfer of fish to hypoosmotic media, the cleft disappeared by the surface expansion of the surrounding cells, which may cover the ionocytes to stop ion excretion in seawater (Sakamoto et al., 2000). An in vitro study using an Ussing chamber, the opercular membrane of killifish increased transepithelial resistance immediately after exposure to freshwater, probably due to the closure of the cleft by surrounding cells (Daborn et al., 2001). The closure is due to the dynamic morphological changes in pavement cells that surround the cleft because it is inhibited by cytochalasin D.

When euryhaline fishes move from freshwater to seawater, on the other hand, ion-absorbing ionocytes must reverse their function to excretion, and this switching must be immediate to maintain internal ion balance within a narrow range. The easiest and quickest way is to replace a suite of transporters for ion absorption by those for excretion, which can be achieved by internalizing the absorptive transporters on the plasma membrane (endocytosis) and recruiting the excretory transporters in the cytosolic vesicles to the plasma membrane (exocytosis) via intracellular trafficking. The regulation of trafficking may be achieved by phosphorylation of contractile proteins in the cytoskeleton (Valenti et al., 2005), but nothing is thus far known about the membrane trafficking in fishes.

Transcriptional regulation of immediate early genes occurs within a few hours after osmotic stress in osmoregulatory organs of fishes. Such immediate genes are most likely transcription factors that regulate the expression of osmoregulatory effector genes. The candidate gene reported for the first time in fishes was osmotic transcription factor-1 (OSTF-1), which is an ortholog of TSC22 domain family protein 3 (TSC22D3) identified in humans as a leucine zipper protein that functions as a transcription regulator (Fiol and Kültz, 2004). The expression of these genes was upregulated in tilapia gills 2 h after osmotic stress. The OSTF-1 has attracted attention as a key gene for seawater acclimation as its expression was also upregulated in the gills of other euryhaline species such as eels and medaka after seawater challenges (Tse et al., 2008). However, in vivo knockdown of the OSTF-1 gene had no effect on seawater acclimation in medaka (Tse et al., 2011), suggesting a limited function of OSTF-1 in seawater acclimation. Another study by differential display detected an interesting glycine-rich RNA binding protein in Atlantic salmon, *Salmo salar*, similar to the cold inducible RNA binding protein of mammals, whose gene expression is upregulated in the gill lamella

48 h after hyperosmotic stress (Pan et al., 2004). The gene expression was not upregulated after cold stress, and the upregulation was not detected in the liver, kidney, and heart after hyperosmotic stress.

The osmotic stress responsive genes have been identified by a microarray-based method during 12 h after freshwater-seawater and seawater-freshwater transfer in the gills of goby, *Gillichthyes mirabilis* (Evans and Somero, 2008, 2009). Functional classification of the differentially expressed genes based on a gene ontology database showed that they are classified as cell signaling, energy production, transcription regulation, extracellular matrix, cytoskeleton, protein trafficking, ion homeostasis, and organic osmolytes, among others. Among the annotated genes, FK506-binding protein 51 (FKBP-51) and translationally controlled tumor protein are candidates for the regulation of cortisol actions and NKA activity that are both important for osmotic stress adaptation. Hormone-related genes such as insulin, prolactin, and GH (IGF-1 and somatostatin) responded to osmotic stress. Transcriptomic analyses were also performed in the gills and liver of homing sockeye salmon, *Oncorhynchus nerka*, from both freshwater and seawater, but differentially expressed genes were few (Evans et al., 2011). They suggested that the transcriptional changes of osmoregulatory genes are largely anticipatory, as suggested by the disappearance of seawater-type ionocytes and appearance of the freshwater-type in homing chum salmon, *Oncorhynchus keta*, before entering freshwater from seawater (Uchida et al., 1997). Microarray-based transcriptomic comparison was also performed in the gills of euryhaline killifish, *Lucania parva*, in freshwater and seawater populations and in the stenohaline freshwater species, *Lucania goodie*; many differentially expressed genes related to ion transport and cell adhesion were detected (Kozak et al., 2013). In addition, a large-scale proteomic analysis in whitefish, *Coregonus lavaretus*, larvae spawned in different osmotic environments identified cytokines and Ca^+ and Na^+ transporters as osmosensitive proteins (Papakostas et al., 2012).

Transcriptome analyses of gills and digestive tracts of euryhaline tilapia acclimated in freshwater, seawater, or double-strength seawater was conducted (Li et al., 2014); typical transporters known for NaCl excretion in the gills (NKAα1 and α3, CFTR, NKCC1a) and for NaCl absorption in the intestine (NKAα1 and α3, NKCC2) were upregulated in hyperosmotic media. In addition, NKCC1b and CFTR genes were upregulated in the tilapia intestine, suggesting the presence of secretory-type epithelial cells in the teleost intestine. Transcriptome comparison was performed on the intestine of euryhaline Mozambique tilapia and stenohaline Nile tilapia (Ronkin et al., 2015), revealing that angiotensin converting enzyme (AQP8), SLC34A2 (Na^+-phosphate cotransporter), SLC43 A2 (Na^+-independent amino acid transporter) were upregulated in Mozambique tilapia but downregulated in Nile tilapia after hyperosmotic stress. These genes may be involved in the high salinity tolerance of Mozambique tilapia to seawater. In medaka, *Oryzias latipes*, the expression of 19 genes was significantly upregulated in a few hours after hyperosmotic stress (Wong et al., 2014). However, only 5 of the 19 specifically responded to osmotic stress and 14 others also responded to freshwater-freshwater transfer, suggesting that they responded also to general (handling) stress. The stress-specific genes include CCAAT/enhancer-binding proteins (CEBP) β and δ, DNA repair and recombination protein RAD54-like 2 (RAD54L2), hypoxia-inducible factor 3 α (HIF3A), and LIM domain-binding protein 1 (LDB1). CEBP binding sites have been identified in the promoter region of the angiotensinogen gene and transporter genes such as NKCC2 and CFTR. The function of other genes in the osmotic stress response in fish remain to be explored. Among the genes that responded to both osmotic and handling stress reported osmosensitive transcription factors such as serum/glucocorticoid regulated kinase 1 (SKG1) and TSC22D3 (OSTF-1) were included. They are in the protein interaction network of other osmoregulatory genes, suggesting their involvement in osmoregulation (Wong et al., 2014).

5. ENERGY METABOLISM IN RESPONSE TO OSMOTIC STRESS

Energy supply is a rate-limiting factor for energy-consuming physiological processes such as responses to osmotic stress. Activating and newly synthesizing transporters, enzymes, or other related proteins require extra energy for fish to achieve necessary compensatory and regulatory processes to cope with osmotic stress. This may affect the development, growth, and reproduction of fish (Sadoul and Vijayan, 2016; Chapter 5 in this volume) for a thorough discussion of the metabolism during stress in fish.

5.1. Oxygen Consumption

A traditional way to estimate the energy cost associated with osmotic stress is to measure the changes in the entire organismal oxygen consumption rate of fish after treatments with different salinities. Measuring oxygen consumption rate has been routinely conducted in the studies on fish osmoregulation (see Boeuf and Payan, 2001; Ern et al., 2014; Morgan and Iwama, 1991). Growth rate declined following the increasing salinity with concomitantly changing oxygen consumption in salmonids. However, comparing different species showed inconsistent patterns of the relation between oxygen consumption and environmental salinity (Morgan and Iwama, 1991). The natural history of a species could determine the type of metabolic response to salinity changes; that is, the lowest metabolic rates are associated with the environment in which a species is most commonly found. On the other hand, a recent review covering a greater number of species concluded no clear trend of lowest oxygen uptake at either normal lifecycle salinity or isosmotic salinity (Ern et al., 2014). Boeuf and Payan (2001) concluded that many fish species spent $20 \sim >50\%$ of the total energy budget to cope with osmotic stress. As Morgan and Iwama (1991) suggested, quantification of the energy cost related to osmotic stress by determining entire organismal oxygen consumption is likely confounded by other metabolic processes that also respond to the stress. To overcome such problems inherent in studying whole fish, metabolic rates in isolated gills (or other organs) were used to estimate the energy cost for specific osmoregulatory organs (see Tseng and Hwang, 2008). The ouabain (an NKA inhibitor)-sensitive or bafilomycin (an HA inhibitor)-sensitive oxygen consumption in isolated trout gills are similar to the ATP or oxygen consumption by ionic fluxes (Kirschner, 1995; Morgan and Iwama, 1991). The oxygen consumption by the gill cells appears to be relatively small ($<4 \sim 20\%$) compared to that of the whole fish (McCormick et al., 1989; Morgan and Iwama, 1991).

Some fish species survive well in natural hypersaline waters; how the energy metabolism in those species operates to allow the fish to cope with this harsh environment is an important topic in fish osmoregulation. In a recent review, Gonzalez (2012) proposed that a general strategy appears to be to suppress branchial water permeability and increase gill NKA expression/activity and the density/size of ionocytes without a clear trend toward increased metabolic rate. The drop or lack of change in metabolic rate in hypersaline waters relative to that in seawater has been ascribed to compromised oxygen uptake due to decreased branchial oxygen permeability, lowering routine activity, or other factors; however, there are no conclusive answers (Gonzalez, 2012). As such, oxygen consumption does provide important and basic information related to osmotic stress. However, interpreting these data in relation to the energetic cost of coping with osmotic stress needs to be tempered with considerations of the species, life stages, and acclimation state as well as the physiological parameters measured, and most importantly, the limitations of the methodologies employed.

5.2. Metabolism Modifications

In addition to assessing the metabolic cost, the role of different metabolic fuels and metabolic reorganization of tissues is also important for understanding how fish cope with an osmotic stress. The oxygen-to-nitrogen quotient (O:N) is derived from the ratio of oxygen consumption rate and ammonia excretion rate and is reduced when substrate use is switched from lipid to protein. In fat snook (*Centropomus parallelus*) there is a trend of replacing proteins with lipid (or carbohydrates) as the energy substrates following transfer from freshwater to seawater over an approximately 15–30 day period (da Silva Rocha et al., 2005). A similar trend of decreasing proteins as an energy substrate was also observed in common snook, *Centropomus undecimalis*, following freshwater to seawater transfer over an approximately 1–2 week period (Gracia-Lopez et al., 2006). These findings indicate that there is a metabolic shift between different intermediary metabolic pathways in fish coping with osmotic stress, suggesting a strategy of the retention of cytosolic free amino acids likely associated with cell volume regulation in higher salinity (da Silva Rocha et al., 2005; Gracia-Lopez et al., 2006); this is supported by the fact that there is an increase in tissue amino acid levels in response to hyperosmotic stress in many species (Assem and Hanke, 1983; Bystriansky et al., 2007; Chang et al., 2007a; Fiess et al., 2007; Tok et al., 2009).

Direct investigation of energy substrates and related enzymes also support the notion of a metabolic reorganization or shifting that accompanies coping with osmotic stress. The swamp eel (*Monopterus albus*) showed increased levels of free amino acids (particularly glutamine) and glutamine synthetase expression in the muscle and liver and a concomitant decline of ammonia excretion after transfer from freshwater to seawater. This suggests a decrease in amino acid catabolism and increase in synthesis of certain nonessential amino acids to achieve the cell volume regulation upon osmotic stress (Tok et al., 2009). The metabolic reorganization due to osmotic stress appears to change temporally. Glycogenolysis may provide a glucose source during acute osmotic challenge from freshwater to seawater while gluconeogenesis from amino acids appears important as well (Baltzegar et al., 2014). Liver glycogen content becomes stable within 24 h of osmotic stress (Chang et al., 2007b) and thereafter gluconeogenesis is stimulated, followed by fatty acid synthesis (Baltzegar et al., 2014). The plasma levels of glucose and amino acids and the liver transaminase activity are stimulated after 96 h (Bystriansky et al., 2007) of salinity challenges in several species including tilapia, Arctic char, and sea bream.

The metabolic reorganization due to osmotic stress also has tissue-specific patterns. The liver experiences stimulated glycogenolysis and decreased glycolytic potential, while the gills and brain have enhanced glycolysis in sea bream transferred from freshwater to seawater (Sangiao-Alvarellos et al., 2003). These findings suggest that the liver is the major organ that supplies glucose through glycogenolysis to the gills and brain for driving ion transport mechanisms or other cellular processes to cope with osmotic stress.

Information on the lipid metabolism during osmotic stress is fragmentary and partially conflicting. The activities of carnitine palmitoyl transferase, 3-hydroxyacyl CoA dehydrogenase (HOAD), and malic enzyme in the gills, liver, and red muscle of Arctic char did not change after 96-h hyperosmotic challenge. These effects could be explained by the constant levels of plasma nonesterified fatty acids (Bystriansky et al., 2007). However, plasma triglyceride level was increased in sea bream during the first 96 h after transfer to higher salinity (Sangiao-Alvarellos et al., 2003). During the 2 months before Arctic char migrate to sea, there is an initial increase of glucose-6-phosphate dehydrogenase that provided reducing equivalents for lipogenesis and a subsequent stimulation of HOAD activity in the liver. This indicates the occurrence of an increased capacity for hepatic lipid catabolism at the time of seawater entry (Aas-Hansen et al., 2005). We speculate that this is to moderate the osmotic stress that would be experienced in the pending transfer to a hyperosmotic environment.

Gluconeogenesis is important in energy metabolism during osmotic stress as discussed earlier, and it is similarly important for acute acid stress. Phosphoenolpyruvate carboxykinase 1 (PCK1)-mediated gluconeogenesis is inducible by acid exposure in zebrafish for the fueling of a higher rate of acid secretion via HA in HR ionocytes as judged from experiments employing PCK1 knockdown, glucose supplement, and SIET (Furukawa et al., 2015). Furthermore, glutamine and glutamate appear to be the major source for replenishment of Krebs cycle intermediates, which are subtracted by PCK1 activity. In a recent study (Zikos et al., 2014), acclimation of tilapia to a hyperosmotic environment resulted in enhanced expression of NKCC, cytochrome c oxidase subunit IV (a key enzyme in the mitochondria electron transport chain), glycogen phosphorylase (GP), and peroxisome proliferator-activated receptor γ coactivator 1α (PGC-1α, a mitochondrial biogenesis regulator) in the gills. In addition, the standard metabolic rate becomes elevated. All of these data suggest the involvement of mitochondrial biogenesis in the higher energy cost for coping with hypoosmotic stress (Zikos et al., 2014). As such, recent studies have afforded deeper insight into the sophisticated pathway of energy metabolism associated with ionic and acid–base regulation at the cellular or subcellular level, which awaits further exploration in other species.

5.3. Metabolites Transport

The liver, through glycogenolysis, is the major organ that exports glucose to other organs (including the osmoregulatory organs) to meet various energy-consuming physiological and cellular demands. The brain and muscles also contain a substantial amount of glycogen reserves, although considerably less than the liver. In the mammalian brain, glycogen is mainly stored in astrocytes and astroglial cells and never in neurons (Pfeiffer-Guglielmi et al., 2003), and glycogen is degraded to lactate and shuttled from astrocytes to high-energy-consuming neurons, particularly during energy deprivation (Brown et al., 2003; Pfeiffer-Guglielmi et al., 2007). This same pattern was recently found in the gills of tilapia and zebrafish (Hwang et al., 2011) (Fig. 6.4). In tilapia, a novel type of gill cell, glycogen-rich (GLR) cell, was identified as an energy depository for supplying emergent energy to gill ionocytes under acute salinity stress (Tseng et al., 2007). GLR cells are adjacent to ionocytes, rich in glycogen, and both express GP (a gill-specific form) and glycogen synthase genes

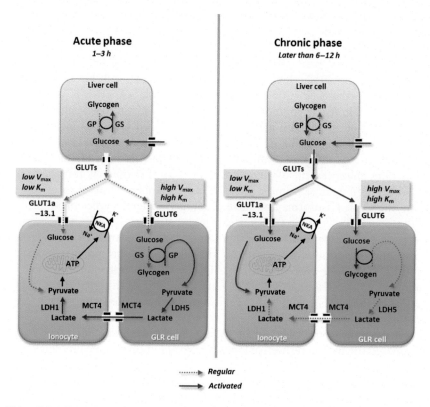

Figure 6.4. Model of the carbohydrate energy supply of ionocytes during osmotic stress. For details, refer to the text (Section 5.3). Blue dashed line and red line indicate regular state and activated state, respectively. GLUT, glucose transporter; GLR, glycogen-rich; GP, glycogen phosphorylase; GS, glycogen synthase; LDH, lactate dehydrogenase; MCT, monocarboxylate transporter.

(Chang et al., 2007b; Tseng et al., 2007). Glycogenolysis in gill GR cells is activated within $1 \sim 3$ h after seawater challenge to provide immediate energy for neighboring ionocytes, presumably used to upregulate ion secretion capacity. Subsequent energy supply ($6 \sim 12$ h seawater) via glycogenolysis likely comes from the liver (Chang et al., 2007b; Tseng et al., 2007). As such, the liver is the central carbohydrate reserve for the entire body while the gills appear to serve as a local energy source during acute salinity challenge (Fig. 6.4). This spatial and temporal relationship in energy translocation between GLR cells and ionocytes and liver cells and ionocytes is analogous to that between the astrocytes and neurons and the liver cells and neurons in mammals.

Tilapia gill cells were found to oxidize more glucose and lactate than other metabolites (alanine or oleate) (Perry and Walsh, 1989). How these energy substrates are transported to fuel ionocytes to cope with osmotic stress is another important issue (Tseng et al., 2009b). Many GLUTs have been identified in fish, and more than one are generally expressed in osmoregulatory organs (gills, kidney, intestine, etc.) (Hall et al., 2014; Polakof et al., 2012; Tseng et al., 2009b; Tseng and Hwang, 2008). Surprisingly, there is little reference to the role of GLUT in energy translocation in fish encountering osmotic stress (Balmaceda-Aguilera et al., 2012; Tseng et al., 2009b). mRNA expression of GLUT1 was increased in both the gills and brain of gilthead sea bream, *Sparus aurata*, during acclimation to different salinities (Balmaceda-Aguilera et al., 2012), suggesting enhancement of glucose transport to fuel the gills for operating ion regulation against osmotic stress. Since the gills generally express more than one GLUT isoforms and have developed different types of ionocytes with respective sets of ion transporters and functions, it is difficult to describe the precise molecular and cellular mechanisms of energy translocation during acclimation to osmotic stress without identification of the specific isoform(s) of GLUT in each type of gill cells. Of 18 GLUT isoforms identified in zebrafish, GLUT-1a and GLUT-13.1 are expressed by different types of gill and skin ionocytes, while GLUT6 is specifically expressed in GLR cells (Tseng et al., 2009b).

Glucose transport function of these GLUTs were demonstrated to be stimulated for fueling respective ionocyte type(s) upon specific ion-deficiency stresses (Tseng et al., 2009b, 2011). Further transport kinetic analyses reinforced the physiological importance of the proposed spatial relation (see earlier) between gill ionocytes and GLR cells for energy supply during osmotic stress. In zebrafish, GLUT-1a (Km=3.4 mM) and GLUT-13.1 (1.5 mM) enables ionocytes to effectively compete for hexose absorption to carry out ionic regulation under a situation of low blood glucose due to extensive energy consumption under osmotic stress (Tseng et al., 2011).

On the contrary, GLUT6 (Km=17.5 mM) allows GLR cells to absorb the excess glucose from the blood for energy deposition during a postprandial time period. This process appears to be of critical importance when acute osmotic stress has created the need for ion transport regulation and emergent energy demands in fish gills. Studies in more species are needed to support this notion.

6. CONCLUSIONS AND PERSPECTIVES

In this chapter, we attempted to review the most recent progress in how fishes maintain water and electrolyte balance in response to various stressors at different tissues and at different time scales. We mostly confined the topic to the osmotic stress as this area is most thoroughly elucidated and the basic mechanisms for ion and water regulation are similar among different stressors. The data are rapidly accumulating on how fish sense the stress, transmit the information via intracellular signaling pathways, and finally cope with the stress by various genomic and nongenomic mechanisms. The subjects that need immediate attention for future studies are discussed in each section, but important ones are:

1. Volume regulation of epithelial cells that directly contact environmental water, such as the skin and gills, needs to be elucidated with reference to the mechanisms suggested in yeast.
2. Regulatory mechanisms of divalent ions, particularly Mg^{2+} and SO_4^{2-}, in response to hyperosmotic stress are as important as those of monovalent ions (Na^+ and Cl^-).
3. The mechanism of pendrin-mediated base secretion is less understood compared with NHE-mediated acid secretion, which needs to be studied in teleost species.
4. The apical H^+-ATPase-mediated acid secretion mechanism was studied in zebrafish, which needs to be confirmed in other freshwater species because of diverse mechanisms among teleost species (Takei et al., 2014).
5. The compensatory responses of the transporters and ionocyte subtypes during hyperosmotic or acidic stress should be further investigated in species other than the few model species (zebrafish, tilapia, and trout).
6. The cytogenetic actions of isotocin, cortisol, PTH1, and STC-1 on the acclimation to osmotic stress shown in zebrafish should be examined in other species.
7. Studies on osmosensors and the signaling pathway therefrom are emerging in fishes, but further analyses are needed based on the data obtained in mammals and yeast.

8. Environmental osmotic stress appears to be sensed primarily by Cl^- sensors located on the epithelial cells of buccal and alimentary cavities, and molecular identity of the sensor needs to be elucidated using euryhaline fishes.
9. The recent progress in the detailed pathway of energy metabolism associated with osmotic stress in zebrafish awaits further exploration in other species.
10. The novel notion of the spatial and temporal relationship in energy translocation between the gill cells and liver cells during salinity stress needs more studies in species other than tilapia to support the notion.

REFERENCES

Aas-Hansen, O., Vijayan, M. M., Johnsen, H. K., Cameron, C. and Jorgensen, E. H. (2005). Resmoltification in wild, anadromous Arctic char (*Salvelinus alpinus*): a survey of osmoregulatory, metabolic, and endocrine changes preceding annual seawater migration. *Can. J. Fish. Aquat. Sci.* **62**, 195–204.

Alepuz, P. M., Jovanovic, A., Reiser, V. and Ammerer, G. (2001). Stress-induced MAP kinase Hog 1 is part of transcription activation complexes. *Mol. Cell.* **7**, 767–777.

Ando, M. and Nagashima, K. (1996). Intestinal Na^+ and Cl^- levels control drinking behavior in the seawater-adapted eel, *Anguilla japonica. J. Exp. Biol.* **199**, 711–716.

Ando, M. and Takei, Y. (2014). Intestinal absorption of salts and water. In *Eel Physiology* (eds. F. Trischitta, Y. Takei and P. Sébert), pp. 160–177. Boca Raton: CRC Press.

Armesto, P., Campinho, M. A., Rodriguez-Rua, A., Cousin, X., Power, D. M., Manchado, M., et al. (2014). Molecular characterization and transcriptional regulation of the Na^+/K^+ ATPase α subunit isoforms during development and salinity challenge in a teleost fish, the Senegalese sole (*Solea senegalensis*). *Comp. Biochem. Physiol.* **175B**, 23–38.

Assem, H. and Hanke, W. (1983). The significance of the amino-acids during osmotic adjustment in teleost fish. I. Changes in the euryhaline *Sarotherodon mossambicus. Comp. Biochem. Physiol.* **74A**, 531–536.

Avella, M. and Bornancin, M. (1989). A new analysis of ammonia and sodium transport through the gills of the freshwater rainbow trout (*Salmo gairdneri*). *J. Exp. Biol.* **142**, 155–175.

Avella, M., Masoni, A., Bornancin, M. and Mayergostan, N. (1987). Gill morphology and sodium influx in the rainbow-trout (*Salmo gairdneri*) acclimated to artificial freshwater environments. *J. Exp. Zool.* **241**, 159–169.

Avella, M., Ducoudret, O., Pisani, D. F. and Poujeol, P. (2009). Swelling-activated transport of taurine in cultured gill cells of sea bass: physiological adaptation and pavement cell plasticity. *Am. J. Physiol.* **296**, R1149–R1160.

Balmaceda-Aguilera, C., Martos-Sitcha, J. A., Mancera, J. M. and Martinez-Rodriguez, G. (2012). Cloning and expression pattern of facilitative glucose transporter 1 (GLUT1) in gilthead sea bream *Sparus aurata* in response to salinity acclimation. *Comp. Biochem. Physiol.* **163A**, 38–46.

Baltzegar, D. A., Reading, B. J., Douros, J. D. and Borski, R. J. (2014). Role for leptin in promoting glucose mobilization during acute hyperosmotic stress in teleost fishes. *J. Endocrinol.* **220**, 61–72.

Bayaa, M. B., Vulesevic, A., Esbaugh, M., Braun, M. E., Ekker, M. and Perry, S. F. (2009). The involvement of SLC26 anion transporters in chloride uptake in zebrafish (*Danio rerio*) larvae. *J. Exp. Biol.* **212**, 3283–3295.

Bell, P. D., Lapointe, J. Y. and Peti-Peterdi, J. (2003). Macula densa cell signaling. *Annu. Rev. Physiol.* **65**, 481–500.

Beuchat, C. A. (1996). Structure and concentrating ability of the mammalian kidney: correlations with habitat. *Am. J. Physiol.* **271**, R157–R179.

Beyenbach, K. W. (2004). Kidney sans glomeruli. *Am. J. Physiol.* **286**, F811–F827.

Bickler, P. E. and Buck, L. T. (2007). Hypoxia tolerance in reptiles, amphibians, and fishes: life with variable oxygen availability. *Annu. Rev. Physiol.* **69**, 145–170.

Boeuf, G. and Payan, P. (2001). How should salinity influence fish growth? *Comp. Biochem. Physiol.* **130C**, 411–423.

Boudreault, F. and Grygorczyk, R. (2002). Cell swelling-induced ATP release and gadolinium-sensitive channels. *Am. J. Physiol.* **282**, C216–C226.

Bres, V., Hurbin, A., Duboid, A., Orcel, H., Moos, F. C., Rabie, A., et al. (2000). Pharmacological characterization of voltage-sensitive, taurine permeable anion channels in rat supraoptic glial cells. *Br. J. Pharmacol.* **130**, 1976–1982.

Breves, J. P., Serizier, S. B., Goffin, V., McCormick, S. D. and Karlstrom, R. O. (2013). Prolactin regulates transcription of the ion uptake Na^+/Cl^- cotransporter (ncc) gene in zebrafish gill. *Mol. Cell. Endocrinol.* **369**, 98–106.

Breves, J. P., McCormick, S. D. and Karlstrom, R. O. (2014). Prolactin and teleost ionocytes: new insights into cellular and molecular targets of prolactin in vertebrate epithelia. *Gen. Comp. Endocrinol.* **203**, 21–28.

Brinn, R. P., Marcon, J. I., McComb, D. M., Gomes, L. C., Abreu, J. S. and Baldisseroto, B. (2012). Stress responses of the endemic freshwater cururu stingrey (*Potamotrygon* cf. *histrix*) during trasnportation in the Amazon region of the Rio Negro. *Comp. Biochem. Physiol.* **162A**, 139–145.

Brown, A. M., Tekkok, S. B. and Ransom, B. R. (2003). Glycogen regulation and functional role in mouse white matter. *J. Physiol.* **549**, 501–512.

Bystriansky, J. S., Frick, N. T. and Ballantyne, J. S. (2007). Intermediary metabolism of Arctic char *Salvelinus alpinus* during short-term salinity exposure. *J. Exp. Biol.* **210**, 1971–1985.

Capaldi, A. P., Kaplan, T., Liu, Y., Habib, N., Regev, A., Friedman, N., et al. (2008). Structure and function of a transcriptional network activated by the MAPK Hog 1. *Nat. Genet.* **40**, 1300–1306.

Chang, I. C., Lee, T. H., Yang, C. H., Wei, Y. Y., Chou, F. I. and Hwang, P. P. (2001). Morphology and function of gill mitochondria-rich cells in fish acclimated to different environments. *Physiol. Biochem. Zool.* **74**, 111–119.

Chang, I. C., Wei, Y. Y., Chou, F. I. and Hwang, P. P. (2003). Stimulation of Cl^- uptake and morphological changes in gill mitochondria-rich cells in freshwater tilapia (*Oreochromis mossambicus*). *Physiol. Biochem. Zool.* **76**, 544–552.

Chang, E. W., Loong, A. M., Wong, W. P., Chew, S. F., Wilson, J. M. and Ip, Y. K. (2007a). Changes in tissue free amino acid contents, branchial Na^+/K^+-ATPase activity and bimodal breathing pattern in the freshwater climbing perch, *Anabas testudineus* (Bloch), during seawater acclimation. *J. Exp. Zool.* **307**, 708–723.

Chang, J. C., Wu, S. M., Tseng, Y. C., Lee, Y. C., Baba, O. and Hwang, P. P. (2007b). Regulation of glycogen metabolism in gills and liver of the euryhaline tilapia (*Oreochromis mossambicus*) during acclimation to seawater. *J. Exp. Biol.* **210**, 3494–3504.

Chang, W. J., Horng, J. L., Yan, J. J., Hsiao, C. D. and Hwang, P. P. (2009). The transcription factor, glial cell missing 2, is involved in differentiation and functional regulation of H^+-ATPase-rich cells in zebrafish (*Danio rerio*). *Am. J. Physiol.* **296**, R1192–1201.

Chang, W. J., Wang, Y. F., Hu, H. J., Wang, J. H., Lee, T. H. and Hwang, P. P. (2013). Compensatory regulation of Na$^+$ absorption by Na$^+$/H$^+$ exchanger and Na$^+$-Cl$^-$ cotransporter in zebrafish (*Danio rerio*). *Front. Zool.* **10**, 46.

Chen, Y. Y., Lu, F. I. and Hwang, P. P. (2003). Comparisons of calcium regulation in fish larvae. *J. Exp. Zool.* **295**, 127–135.

Chou, M. Y., Yang, C. H., Lu, F. I., Lin, H. C. and Hwang, P. P. (2002). Modulation of calcium balance in tilapia larvae (*Oreochromis mossambicus*) acclimated to low-calcium environments. *J. Comp. Physiol. B* **172**, 109–114.

Chou, M. Y., Hung, J. C., Wu, L. C., Hwang, S. P. and Hwang, P. P. (2011). Isotocin controls ion regulation through regulating ionocyte progenitor differentiation and proliferation. *Cell. Mol. Life Sci.* **68**, 2797–2809.

Chou, M. Y., Lin, C. H., Chao, P. L., Hung, J. C., Cruz, S. A. and Hwang, P. P. (2015). Stanniocalcin-1 controls ion regulation functions of ion-transporting epithelium other than calcium balance. *Int. J. Biol. Sci.* **11**, 122–132.

Claiborne, J. B., Edwards, S. L. and Morrison-Shetlar, A. I. (2002). Acid-base regulation in fishes: cellular and molecular mechanisms. *J. Exp. Zool.* **293**, 302–319.

Clapham, D. E., Runnels, L. W. and Strubing, C. (2001). The TRP ion channel family. *Nat. Rev. Neurosci.* **2**, 387–396.

Cruz, S. A., Chao, P. L. and Hwang, P. P. (2013a). Cortisol promotes differentiation of epidermal ionocytes through Foxi3 transcription factors in zebrafish (*Danio rerio*). *Comp. Biochem. Physiol.* **164A**, 249–257.

Cruz, S. A., Lin, C. H., Chao, P. L. and Hwang, P. P. (2013b). Glucocorticoid receptor, but not mineralocorticoid receptor, mediates cortisol regulation of epidermal ionocyte development and ion transport in zebrafish (*Danio rerio*). *PLoS One* **8**, e77997.

Cuadrado, A. and Nebreda, A. R. (2010). Mehanisms and functions of p38 MAPK signalling. *Biochem. J.* **429**, 405–417.

Daborn, K., Cozzi, R. R. F. and Marshall, W. S. (2001). Dynamics of pavement cell-chloride cell interactions during abrupt salinity change in *Fundulus heteroclitus*. *J. Exp. Biol.* **204**, 1889–1899.

da Silva Rocha, A. J., Gomes, V., Van Ngan, P., de Arruda Campos Rocha Passos, M. J. and Rios Furia, R. (2005). Metabolic demand and growth of juveniles of *Centropomus parallelus* as function of salinity. *J. Exp. Mar. Biol. Ecol.* **316**, 157–165.

de Nadal, E., Ammerer, G. and Posas, F. (2011). Controlling gene expression in response to stress. *Nat. Rev.* **12**, 833–845.

Duranton, C., Mikulovic, E., Tauc, M., Avella, M. and Poujeol, P. (2000). Potassium channel in primary culture of seawater fish gill cells II. Channel activation by hypotonic shock. *Am. J. Physiol.* **279**, R1659–R1670.

Edwards, A., Castrop, H., Laghmani, K., Vallon, V. and Layton, A. T. (2014). Effects of NKCC2 isoform regulation on NaCl transport in thick ascending limb and macula densa: a modeling study. *Am. J. Physiol.* **307**, F137–F146.

Edwards, S. L. and Marshall, W. S. (2012). Principles and Patterns of Osmoregulation and Euryhalinity in Fishes. In *Fish Physiology - Euryhaline Fishes*, Vol. 32 (eds. S. C. McCormick, A. P. Farrell and C. J. Brauner), pp. 1–44. London: Academic Press.

Ern, R., Huong, D. T., Cong, N. V., Bayley, M. and Wang, T. (2014). Effect of salinity on oxygen consumption in fishes: a review. *J. Fish. Biol.* **84**, 1210–1220.

Evans, D. H., Piermarini, P. M. and Choe, K. P. (2005). The multifunctional fish gill: dominant site of gas exchange, osmoregulation, acid-base regulation, and excretion of nitrogenous waste. *Physiol. Rev.* **85**, 97–177.

Evans, T. G. and Somero, G. N. (2008). A macroarray-based transcriptomic time-course of hyper- and hypo-osmotic stress signaling events in the euryhaline fish, *Gillichthys mirabilis*: osmosensors to effectors. *J. Exp. Biol.* **211**, 3636–3649.

Evans, T. G. and Somero, G. N. (2009). Protein-protein interactions enable rapid adaptive response to osmotic stress in fish gills. *Commun. Integr. Biol.* **2**, 94–96.

Evans, T. G., Hammil, E., Kaukinen, K., Schulze, A. D., Patterson, D. A., English, K. K., et al. (2011). Transcriptomics of environmental acclimatization and survival in wild adult Pacific sockeye salmon (*Oncorhynchus nerka*) during spawning migration. *Mol. Ecol.* **20**, 4472–4489.

Fellner, S. K. and Parker, L. (2004). Ionic strength and the polyvalent cation receptor of shark rectal gland and artery. *J. Exp. Zool.* **301A**, 235–239.

Ferraris, J. D., Williams, C. K., Ohtaka, A. and García-Pérez, A. (1999). Functional consensus for mammalian osmotic response elements. *Am. J. Physiol.* **276**, C667–673.

Fiess, J. C., Kunkel-Patterson, A., Mathias, L., Riley, L. G., Yancey, P. H., Hirano, T., et al. (2007). Effects of environmental salinity and temperature on osmoregulatory ability, organic osmolytes, and plasma hormone profiles in the Mozambique tilapia (*Oreochromis mossambicus*). *Comp. Biochem. Physiol.* **146A**, 252–264.

Fiol, D. F. and Kültz, D. (2004). Rapid hyperosmotic coinduction of two tilapia (*Oreochromis mossambicus*) transcription factors in gill cells. *Proc. Natl. Acad. Sci. U. S. A.* **102**, 927–932.

Fletcher, D. A. and Mullins, R. D. (2010). Cell mechanics and the cytoskeleton. *Nature* **463**, 485–492.

Furukawa, F., Tseng, Y. C., Liu, S. T., Chou, Y. L., Lin, C. C., Sung, P. H., et al. (2015). Induction of phosphoenolpyruvate carboxykinase (PEPCK) during acute acidosis and its role in acid secretion by V-ATPase-expressing ionocytes. *Int. J. Biol. Sci.* **11**, 712–725.

Gau, P., Poon, J., Ufret-Vincenty, C., Snelson, C. D., Gordon, S. E., Raible, D. W., et al. (2013). The zebrafish ortholog of TRPV1 is required for heat-induced locomotion. *J. Neurosci.* **33**, 5249–5260.

Gilmour, K. M. (2012). New insights into the many functions of carbonic anhydrase in fish gills. *Respir. Physiol. Neurobiol.* **184**, 223–230.

Gonzalez, R. J. (2012). The physiology of hyper-salinity tolerance in teleost fish: a review. *J. Comp. Physiol. B* **182**, 321–329.

Gorissen, M. and Flik, G. (2016). Endocrinology of the Stress Response in Fish. In *Fish Physiology - Biology of Stress in Fish*, Vol. 35 (eds. C. B. Schreck, L. Tort, A. P. Farrell and C. J. Brauner), San Diego, CA: Academic Press.

Gracia-Lopez, V., Rosas-Vazquez, C. and Brito-Perez, R. (2006). Effects of salinity on physiological conditions in juvenile common snook *Centropomus undecimalis*. *Comp. Biochem. Physiol.* **145A**, 340–345.

Grosell, M. (2011). The role of the gastrointestinal tract in salt and water balance. In *Fish Physiology - The Multifunctional Gut of Fish*, Vol. 30 (eds. M. Grosell, A. P. Farrell and C. J. Brauner), pp. 135–164. San Diego: Academic Press.

Grosell, M., Mager, E. M., Williams, C. and Taylor, J. R. (2009). High rates of HCO_3^- secretion and Cl^- absorption against adverse gradients in the marine teleost intestine: the involvement of an electrogenic anion exchanger and H^+-pump metabolon?. *J. Exp. Biol.* **212**, 1684–1696.

Guh, Y. J., Tseng, Y. C., Yang, C. Y. and Hwang, P. P. (2014). Endothelin-1 regulates H^+-ATPase-dependent transepithelial H^+ secretion in zebrafish. *Endocrinology* **155**, 1728–1737.

YOSHIO TAKEI AND PUNG-PUNG HWANG

Guh, Y. J., Lin, C. H. and Hwang, P. P. (2015). Osmoregulation in zebrafish: ion transport mechanisms and functional regulation. *EXCLI J.* **14**, 627–659.

Hall, J. R., Clow, K. A., Short, C. E. and Driedzic, W. R. (2014). Transcript levels of class I GLUTs within individual tissues and the direct relationship between GLUT1 expression and glucose metabolism in Atlantic cod (*Gadus morhua*). *J. Comp. Physiol. B* **184**, 483–496.

Hirano, T. (1974). Some factors regulating water intake in the eel, *Anguilla japonica*. *J. Exp. Biol.* **61**, 171–178.

Hiroi, J. and McCormick, S. D. (2012). New insights into gill ionocyte and ion transporter function in euryhaline and diadromous fish. *Respir. Physiol. Neurobiol.* **184**, 257–268.

Hoenderop, J. G., Nilius, B. and Bindels, R. J. (2005). Calcium absorption across epithelia. *Physiol. Rev.* **85**, 373–422.

Hoffmann, E. K., Lambert, I. H. and Pedersen, S. F. (2009). Physiology of cell volume regulation in vertebrates. *Physiol. Rev.* **89**, 193–277.

Horng, J. L., Hwang, P. P., Shih, T. H., Wen, Z. H., Lin, C. S. and Lin, L. Y. (2009a). Chloride transport in mitochondrion-rich cells of euryhaline tilapia (*Oreochromis mossambicus*) larvae. *Am. J. Physiol.* **297**, C845–854.

Horng, J. L., Lin, L. Y. and Hwang, P. P. (2009b). Functional regulation of H$^+$-ATPase-rich cells in zebrafish embryos acclimated to an acidic environment. *Am. J. Physiol.* **296**, C682–692.

Hsu, H. H., Lin, L. Y., Tseng, Y. C., Horng, J. L. and Hwang, P. P. (2014). A new model for fish ion regulation: identification of ionocytes in freshwater- and seawater-acclimated medaka (*Oryzias latipes*). *Cell. Tissue. Res.* **357**, 225–243.

Hwang, P. P. and Chou, M. Y. (2013). Zebrafish as an animal model to study ion homeostasis. *Pflugers. Arch.* **465**, 1233–1247.

Hwang, P. P. and Lee, T. H. (2007). New insights into fish ion regulation and mitochondrion-rich cells. *Comp. Biochem. Physiol.* **148A**, 479–497.

Hwang, P. P. and Lin, L. Y. (2014). Gill ion transport, acid-base regulation and nitrogen excretion. In *The Physiology of Fishes* (eds. D. Evans, J. B. Claiborne and S. Currie), pp. 205–233. Boca Raton: CRC Press.

Hwang, P. P., Tung, Y. C. and Chang, M. H. (1996). Effect of environmental calcium levels on calcium uptake in tilapia larvae (*Oreochromis mossambicus*). *Fish Physiol. Biochem.* **15**, 363–370.

Hwang, P. P., Lee, T. H. and Lin, L. Y. (2011). Ion regulation in fish gills: recent progress in the cellular and molecular mechanisms. *Am. J. Physiol.* **301**, R28–47.

Inokuchi, M., Hiroi, J., Watanabe, S., Hwang, P. P. and Kaneko, T. (2009). Morphological and functional classification of ion-absorbing mitochondria-rich cells in the gills of Mozambique tilapia. *J. Exp. Biol.* **212**, 1003–1010.

Kalujnala, S., Gellatly, S. A., Hazon, N., Villasenor, A., Yancy, P. H. and Cramb, G. (2013). Seawater acclimation and inositol monophosphatase isoform expression in the Enropean eel (*Anguilla anguilla*) and Nile tilapia (*Oreochromis niloticus*). *Am. J. Physiol.* **305**, R369–R384.

Kirschner, L. B. (1995). Energetics of osmoregulation in fresh-water vertebrates. *J. Exp. Zool.* **271**, 243–252.

Kotchen, T. A., Galla, J. H. and Luke, R. G. (1978). Contribution of chloride to the inhibition of plasma renin by sodium chloride in rat. *Kidney Int.* **13**, 201–207.

Kozak, G. M., Brennan, R. S., Berdan, E. L., Fuller, R. C. and Whitehead, A. (2014). Functional and population genomic divergence within and between two species of killifish adapted to different osmotic niches. *Evolution* **68** (1), 63–80.

Kültz, D. (2012). Osmosensing. In *Fish Physiology - Euryhaline Fishes*, Vol. 32 (eds. S. C. McCormick, A. P. Farrell and C. J. Brauner), pp. 45–68. London: Academic Press.

Kumai, Y., Nesan, D., Vijayan, M. M. and Perry, S. F. (2012a). Cortisol regulates Na$^+$ uptake in zebrafish, *Danio rerio*, larvae via the glucocorticoid receptor. *Mol. Cell. Endocrinol.* **364**, 113–125.

Kumai, Y., Ward, M. A. and Perry, S. F. (2012b). beta-Adrenergic regulation of Na$^+$ uptake by larval zebrafish *Danio rerio* in acidic and ion-poor environments. *Am. J. Physiol.* **303**, R1031–1041.

Kumai, Y., Bernier, N. J. and Perry, S. F. (2014). Angiotensin-II promotes Na$^+$ uptake in larval zebrafish, *Danio rerio*, in acidic and ion-poor water. *J. Endocrinol.* **220**, 195–205.

Kumai, Y., Kwong, R. W. and Perry, S. F. (2015). A role for transcription factor glial cell missing 2 in Ca^{2+} homeostasis in zebrafish, *Danio rerio. Pflugers Arch.* **467**, 753–765.

Kurita, Y., Nakada, T., Kato, A., Doi, H., Mistry, A. C., Chang, M. H., et al. (2008). Identification of intestinal bicarbonate transporters involved in formation of carbonate precipitates to stimulate water absorption in marine teleost fish. *Am. J. Physiol.* **294**, R1402–R1412.

Kwong, R. W. and Perry, S. F. (2015). An essential role for parathyroid hormone in gill formation and differentiation of ion-transporting cells in developing zebrafish. *Endocrinology* en20141968.

Kwong, R. W., Kumai, Y. and Perry, S. F. (2014). The physiology of fish at low pH: the zebrafish as a model system. *J. Exp. Biol.* **217**, 651–662.

Lafont, A. G., Wang, Y. F., Chen, G. D., Liao, B. K., Tseng, Y. C., Huang, C. J., et al. (2011). Involvement of calcitonin and its receptor in the control of calcium-regulating genes and calcium homeostasis in zebrafish (*Danio rerio*). *J. Bone Miner. Res.* **26**, 1072–1083.

Lam, A. K. M., Ko, B. C. B., Tam, S., Morris, R., Yang, J. Y., Chung, S. K., et al. (2004). Osmotic response element-binding protein (OREBP) is an essential regulator of the urine concentrating mechanism. *J. Biol. Chem.* **279**, 48048–48054.

Laurent, P., Hobe, H. and Dunelerb, S. (1985). The role of environmental sodium-chloride relative to calcium in gill morphology of fresh-water salmonid fish. *Cell Tissue Res.* **240**, 675–692.

Lee, Y. C., Yan, J. J., Cruz, S. A., Horng, J. L. and Hwang, P. P. (2011). Anion exchanger 1b, but not sodium-bicarbonate cotransporter 1b, plays a role in transport functions of zebrafish H$^+$-ATPase-rich cells. *Am. J. Physiol.* **300**, C295–307.

Leto, D. and Saltiel, A. R. (2012). Regulation of glucose transport by insulin: traffic control of GLUT4. *Nat. Rev.* **13**, 383–396.

Li, Z., Lui, E. Y., Wilson, J. M., Ip, Y. K., Lin, Q., Lam, T. J., et al. (2014). Expression of key ion transporters in the gill and esophageal-gastrointestinal tract of euryhaline Mozambique tilapia, *Oreochromis mossambicus*, acclimated to fresh water, seawater and hypersaline water. *PLoS One* **9**, e87591.

Liao, B. K., Deng, A. N., Chen, S. C., Chou, M. Y. and Hwang, P. P. (2007). Expression and water calcium dependence of calcium transporter isoforms in zebrafish gill mitochondrion-rich cells. *BMC Genomics* **8**, 354.

Liedtke, W. and Friedman, J. M. (2003). Abnormal osmotic regulation in *trpv4*$^{-/-}$ mice. *Proc. Natl. Acad. Sci. U. S. A.* **100**, 13698–13703.

Lin, C. H., Huang, C. L., Yang, C. H., Lee, T. H. and Hwang, P. P. (2004). Time-course changes in the expression of Na, K-ATPase and the morphometry of mitochondrion-rich cells in gills of euryhaline tilapia (*Oreochromis mossambicus*) during freshwater acclimation. *J. Exp. Zool.* **301**, 85–96.

Lin, C. H., Tsai, I. L., Su, C. H., Tseng, D. Y. and Hwang, P. P. (2011). Reverse effect of mammalian hypocalcemic cortisol in fish: cortisol stimulates Ca^{2+} uptake via glucocorticoid receptor-mediated vitamin D$_3$ metabolism. *PLoS One* **6**, e23689.

Lin, C. H., Su, C. H., Tseng, D. Y., Ding, F. C. and Hwang, P. P. (2012). Action of vitamin D and the receptor, VDRα, in calcium handling in zebrafish (*Danio rerio*). *PLoS One* **7**, e45650.

Lin, C. H., Su, C. H. and Hwang, P. P. (2014). Calcium-sensing receptor mediates Ca^{2+} homeostasis by modulating expression of PTH and stanniocalcin. *Endocrinology* **155**, 56–67.

Lin, L. Y., Horng, J. L., Kunkel, J. G. and Hwang, P. P. (2006). Proton pump-rich cell secretes acid in skin of zebrafish larvae. *Am. J. Physiol.* **290**, C371–378.

Lin, T. Y., Liao, B. K., Horng, J. L., Yan, J. J., Hsiao, C. D. and Hwang, P. P. (2008). Carbonic anhydrase 2-like a and 15a are involved in acid-base regulation and Na^+ uptake in zebrafish H^+-ATPase-rich cells. *Am. J. Physiol.* **294**, C1250–1260.

Lionetto, M. G., Giordano, M. E., de Nuccio, F., Nicolardi, G., Hoffmann, E. K. and Schettino, T. (2005). Hypotonicity induced K^+ and anion conductive pathways activation in eel intestinal epithelium. *J. Exp. Biol.* **208**, 749–760.

Lópetz-Maury, L., Marguerat, S. and Bähler, J. (2008). Tuning gene expression to changing environments: from rapid responses to evolutionary adaptation. *Nat. Rev.* **9**, 583–593.

Loretz, C. A. (2008). Extracellular calcium-sensing receptors in fish. *Comp. Biochem. Physiol.* **149A**, 225–245.

Loretz, C. A., Pollina, C., Hyodo, S., Takei, Y., Chang, W. and Shoback, D. (2004). cDNA cloning and functional expression of a Ca^{2+}-sensing receptor with truncated carboxyterminal tail from the Mozambique tilapia (*Oreochromis mossambicus*). *J. Biol. Chem.* **279**, 53288–53297.

Madsen, S. S., Kiilerich, P. and Tipsmark, C. K. (2009). Multiplicity of expression of Na^+,K^+– ATPase α-subunit isoforms in the gill of Atlantic salmon (*Salmo salar*): cellular localisation and absolute quantification in response to salinity change. *J. Exp. Biol.* **212**, 78–88.

Madsen, S. S., Bujak, J. and Tipsmark, C. K. (2014). Aquaporin expression in the Japanese medaka (*Oryzias latipes*) in freshwater and seawater: challenging the paradigm of intestinal water transport?. *J. Exp. Biol.* **217**, 3108–3121.

Majhi, R. K., Kumar, A., Yadav, M., Swain, N., Kumari, S., Saha, A., et al. (2013). Thermosensitive ion channel TRPV1 is endogenously expressed in the sperm of a fresh water teleost fish (*Labeo rohita*) and regulates sperm motility. *Channels* **7**, 483–492.

Marshall, W. S. (2011). Mechanosensitive signalling in fish gill and other ion transporting epithelia. *Acta Physiol.* **202**, 487–499.

Marshall, W. S., Howard, J. A., Cozzi, R. R. F. and Lynch, E. M. (2002). NaCl and fluid secretion by the intestine of the teleost *Fundulus heteroclitus*: involvement of CFTR. *J. Exp. Biol.* **205**, 745–758.

Marshall, W. S., Katoh, F., Main, H. P., Sers, N. and Cozzi, R. R. F. (2008). Focal adhesion kinase and β1 integrin regulation of Na^+, K^+, $2Cl^-$ cotransporter in osmosensing ion transporting cells of killifish, *Fundulus heteroclitus*. *Comp. Biochem. Physiol.* **150A**, 288–300.

Marshall, W. S., Watters, K. D., Hovdestad, L. R., Cozzi, R. R. F. and Katoh, F. (2009). CFTR Cl^- channel functional regulation by phosphorylation of focal adhesion kinase at tyrosine 407 in osmosensitive ion transporting mitochondria rich cells of euryhaline killifish. *J. Exp. Biol.* **212**, 2365–2377.

McCormick, S. D. (2001). Endocrine control of osmoregulation in teleost fish. *Am. Zool.* **41**.

McCormick, S. D., Moyes, C. D. and Ballantyne, J. S. (1989). Influence of salinity on the energetics of gill and kidney of Atlantic salmon (*Salmo salar*). *Fish Physiol. Biochem.* **6**, 243–254.

McCormick, S. C., Farrell, A. P. and Brauner, C. J. (2012). *Fish Physiology - Euryhaline Fishes*, Vol. 32. London: Academic Press.

McDonald, D. G. and Rogano, M. S. (1986). Ion regulation by the rainbow-trout, *Salmo gairdneri*, in ion-poor water. *Physiol. Zool.* **59**, 318–331.

McDonald, D. G., Hobe, H. and Wood, C. M. (1980). The influence of calcium on the physiological responses of the rainbow trout, *Salmo gairdneri*, to low environmental pH. *J. Exp. Biol.* **88**, 109–131.

McDonald, G. and Milligan, L. (1997). Ionic, osmotic and acid-base regulation in stress. In *Fish Stress and Health in Aquaculture* (eds. G. K. Iwama, A. D. Pickering, J. P. Sumpter and C. B. Schreck), pp. 119–144. Cambridge: Cambridge University Press.

McWilliams, P. G. (1982). The effects of calcium on sodium fluxes in the brown trout, *Salmo trutta*, in neutral and acid water. *J. Exp. Biol.* **96**, 439–442.

Möller, H. B. and Fenton, R. A. (2012). Cell biology of vasopressin-regulated aquaporin-2 trafficking. *Pflugers. Arch.* **464**, 133–144.

Morgan, J. D. and Iwama, G. K. (1991). Effects of salinity on growth, metabolism, and ion regulation in juvenile rainbow and steelhead trout (*Oncorhynchus mykiss*) and fall chinook salmon (*Oncorhynchus tshawytscha*). *Can. J. Fish. Aquat. Sci.* **48**, 2083–2094.

Nakada, T., Hoshijima, K., Esaki, M., Nagayoshi, S., Kawakami, K. and Hirose, S. (2007a). Localization of ammonia transporter Rhcg1 in mitochondrion rich cells of yolk sac, gill, and kidney of zebrafish and its ionic strength-dependent expression. *Am. J. Physiol.* **293**, R1743–1753.

Nakada, T., Westhoff, C. M., Kato, A. and Hirose, S. (2007b). Ammonia secretion from fish gill depends on a set of Rh glycoproteins. *FASEB J.* **21**, 1067–1074.

Nearing, J., Betka, M., Quinn, S., Hentschel, H., Elger, M., Baum, M., et al. (2002). Polyvalent cation receptor proteins (CaRs) are salinity sensor in fish. *Proc. Natl. Acad. Sci. U. S. A.* **99**, 9231–9236.

Noda, M. (2007). Hydromineral neuroendocrinology: mechanism of sensing sodium levels in the mammalian brain. *Exp. Physiol.* **92**, 513–522.

Olson, K. R. (2002). Vascular anatomy of the fish gill. *J. Exp. Zool.* **293**, 214–231.

Orlov, S. N. and Mongin, A. A. (2007). Salt sensing mechanisms in blood pressure regulation and hypertension. *Am. J. Physiol.* **293**, H2039–H2053.

Pan, F., Zarate, J., Choudhury, A., Rupprecht, R. and Bradley, T. M. (2004). Osmotic stress of salmon stimulates upregulation of a cold inducible RNA binding protein (CIRP) similar to that of mammals and amphibians. *Biochemie* **86**, 451–461.

Pan, T. C., Liao, B. K., Huang, C. J., Lin, L. Y. and Hwang, P. P. (2005). Epithelial Ca^{2+} channel expression and Ca^{2+} uptake in developing zebrafish. *Am. J. Physiol.* **289**, R1202–1211.

Papakostas, S., Vasemagi, A., Vaha, J.-P., Himberg, M., Peil, L. and Primmer, C. R. (2012). A proteomics approach reveals divergent molcular responses to salinity in populations of European whitefish (*Coregonus lavaretus*). *Mol. Ecol.* **21**, 3516–3530.

Parks, S. K., Tresguerres, M. and Goss, G. G. (2008). Theoretical considerations underlying Na^{+} uptake mechanisms in freshwater fishes. *Comp. Biochem. Physiol.* **148C**, 411–418.

Perry, S. F. and Laurent, P. (1989). Adaptational responses of rainbow-trout to lowered external NaCl concentration - contribution of the branchial chloride cell. *J. Exp. Biol.* **147**, 147–168.

Perry, S. F., Vulesevic, B. and Bayaa, M. (2009). Evidence that SLC26 anion transporters mediate branchial chloride uptake in adult zebrafish (*Danio rerio*). *Am. J. Physiol.* **297**, R988–997.

Perry, S. F. and Walsh, P. J. (1989). Metabolism of isolated fish gill cells: contribution of epithelial chloride cells. *J. Exp. Biol* **144**, 507–520.

Peter, M. C. S. (2011). The role of thyroid hormones in stress response of fish. *Gen. Comp. Endocrinol.* **172**, 198–210.

Pfeiffer-Guglielmi, B., Fleckenstein, B., Jung, G. and Hamprecht, B. (2003). Immunocyto-chemical localization of glycogen phosphorylase isozymes in rat nervous tissues by using isozyme-specific antibodies. *J. Neurochem.* **85**, 73–81.

Pfeiffer-Guglielmi, B., Francke, M., Reichenbach, A. and Hamprecht, B. (2007). Glycogen phosphorylase isozymes and energy metabolism in the rat peripheral nervous system-an immunocytochemical study. *Brain Res.* **1136**, 20–27.

Polakof, S., Panserat, S., Soengas, J. L. and Moon, T. W. (2012). Glucose metabolism in fish: a review. *J. Comp. Physiol. B* **182**, 1015–1045.

Redding, J. M., Schreck, C. B., Barks, E. K. and Ewing, R. D. (1984). Cortisol and its effects on plasma thyroid hormone and electrolyte concentrations in fresh water and during seawater acclimation in yearing coho salmon, *Oncorhynchus kisutch. Gen. Comp. Endocrinol.* **56**, 146–155.

Reindl, K. M. and Sheridan, M. A. (2012). Peripheral regulation of the growth hormone-insulin-like growth factor system in fish and other vertebrates. *Comp. Biochem. Physiol.* **163A**, 231–245.

Ronkin, D., Seroussi, E., Nitzan, T., Dron-Faigenboim, A. and Cnaani, A. (2015). Intestinal transcriptome analysis revealed differential salinity adaptation between two tilapiiine species. *Comp. Biochem. Physiol.* **13D**, 35–43.

Sacchi, R., Li, J., Villarreal, F., Gardell, A. M. and Kultz, D. (2013). Salinity-induced regulation of the myo-inositol biosynthetic pathway in tilapia gill epithelium. *J. Exp. Biol.* **216**, 4626–4638.

Sadoul, B. and Vijayan, M. M. (2016). Stress and Growth. In *Fish Physiology - Biology of Stress in Fish*, Vol. 35 (eds. C. B. Schreck, L. Tort, A. P. Farrell and C. J. Brauner), San Diego, CA: Academic Press.

Saito, H. and Takahashi, K. (2004). Regulation of the osmoregulatory HOC MAPK cascade in yeast. *J. Biochem.* **136**, 267–272.

Saito, S. and Shingai, R. (2006). Evolution of thermoTRP ion channel homologs in vertebrates. *Physiol. Genomics* **27**, 219–230.

Sakamoto, T., Yokota, S. and Ando, M. (2000). Rapid morphological oscillation of mitochondrion-rich cell in estuarine mudskipper following salinity changes. *J. Exp. Zool.* **286**, 666–669.

Sangiao-Alvarellos, S., Laiz-Carrion, R., Guzman, J. M., Martin del Rio, M. P., Miguez, J. M., Mancera, J. M., et al. (2003). Acclimation of *S. aurata* to various salinities alters energy metabolism of osmoregulatory and nonosmoregulatory organs. *Am. J. Physiol.* **285**, R897–907.

Schreck, C. B. and Tort, L. (2016). The Concept of Stress in Fish. In *Fish Physiology - Biology of Stress in Fish*, Vol. 35 (eds. C. B. Schreck, L. Tort, A. P. Farrell and C. J. Brauner), San Diego, CA: Academic Press.

Seale, A. P., Watanabe, S. and Grau, E. G. (2012). Osmoreception: perspectives on signal transduction and environmental modulation. *Gen. Comp. Endocrinol.* **176**, 354–360.

Shahsavarani, A., McNeill, B., Galvez, F., Wood, C. M., Goss, G. G., Hwang, P. P., et al. (2006). Characterization of a branchial epithelial calcium channel (ECaC) in freshwater rainbow trout (*Oncorhynchus mykiss*). *J. Exp. Biol.* **209**, 1928–1943.

Sharif Naeini, R., Witty, M. F., Seguera, P. and Bourque, C. W. (2006). An *N*-terminal variant of Trpv1 channel is required for osmosensory transduction. *Nat. Neurosci.* **9**, 93–98.

Sheikh-Hamad, D. and Gustin, M. C. (2004). MAP kinases and the adaptive response to hypertonicity: functional preservation from yeast to mammals. *Am. J. Physiol.* **287**, F1102–F1110.

Shih, T. H., Horng, J. L., Hwang, P. P. and Lin, L. Y. (2008). Ammonia excretion by the skin of zebrafish (*Danio rerio*) larvae. *Am. J. Physiol.* **295**, C1625–1632.

Shih, T. H., Horng, J. L., Liu, S. T., Hwang, P. P. and Lin, L. Y. (2012). Rhcg1 and NHE3b are involved in ammonium-dependent sodium uptake by zebrafish larvae acclimated to low-sodium water. *Am. J. Physiol.* **302**, R84–93.

Sinha, A. K., Rasoloniriana, R., Dasan, A. F., Pipralia, N., Blust, R. and De Boeck, G. (2015). Interactive effect of high environmental ammonia and nutritional status on ecophysiological performance of European sea bass (*Dicentrarchus labrax*) acclimated to reduced seawater salinities. *Aquat. Toxicol.* **160**, 39–56.

Skomal, G. B. and Mandelman, J. W. (2012). The physiological response to anthropogenic stressors in marine elasmobranch fishes: a review with a focus on the secondary response. *Comp. Biochem. Physiol.* **162A**, 146–155.

Sladek, C. D. and Johnson, A. K. (2013). Integration of thermal and osmotic regulation of water homeostasis: the role of TRPV channels. *Am. J. Physiol.* **305**, R669–R678.

Suzuki, Y., Itakura, M., Kashiwagi, M., Nakamura, N., Matsuki, T., Sakuta, H., et al. (1999). Identification by differential display of a hypertonicity-inducible inward rectifier potassium channel highly expressed in chloride cells. *J. Biol. Chem.* **274**, 11376–11382.

Takahashi, H., Sakamoto, T. and Narita, K. (2006). Cell proliferation and apoptosis in the anterior intestine of an amphibious, euryhaline mudskipper (*Periophthalmus modestus*). *J. Comp. Physiol.* **176B**, 463–468.

Takei, Y. and Loretz, C. A. (2006). Endocrinology. In *The Physiology of Fishes* (eds. D. H. Evans and J. B. Claiborne), third ed., pp. 271–318. Boca Raton: CRC Press.

Takei, Y. and Loretz, C. A. (2010). The gastrointestinal tract as an endocrine/neuroendocrine/paracrine organ: organization, chemical messengers and physiological targets. In *Fish Physiology - The Multifunctional Gut of Fish*, Vol. 30 (eds. M. Grosell, A. P. Farrell and C. J. Brauner), pp. 261–317. London: Academic Press.

Takei, Y. and McCormick, S. D. (2012). Hormonal control of fish euryhalinity. In *Fish Physiology - Euryhaline Fishes*, vol. 32 (eds. S. D. McCormick, A. P. Farrell and C. J. Brauner), pp. 69–123. San Diego: Academic Press.

Takei, Y. and Tsuchida, T. (2000). Role of the renin-angiotensin system in drinking of seawater-adapted eels *Anguilla japonica*: a reevaluation. *Am. J. Physiol.* **279**, R1105–R1111.

Takei, Y., Hiroi, J., Takahashi, H. and Sakamoto, T. (2014). Diverse mechanisms for body fluid regulation in teleost fishes. *Am. J. Physiol.* **307**, R778–R792.

Teruyama, R., Sakuraba, M., Wilson, L. L., Wandrey, L. E. and Armstrong, W. E. (2012). Epithelial Na^+ sodium channels in magnocellular cells of the rat supraoptic and paraventricular nuclei. *Am. J. Physiol.* **302**, E273–285.

Tok, C. Y., Chew, S. F., Peh, W. Y., Loong, A. M., Wong, W. P. and Ip, Y. K. (2009). Glutamine accumulation and up-regulation of glutamine synthetase activity in the swamp eel, *Monopterus albus* (Zuiew), exposed to brackish water. *J. Exp. Biol.* **212**, 1248–1258.

Trayer, V., Hwang, P. P., Prunet, P. and Thermes, V. (2013). Assessment of the role of cortisol and corticosteroid receptors in epidermal ionocyte development in the medaka (*Oryzias latipes*) embryos. *Gen. Comp. Endocrinol.* **194**, 152–161.

Tse, K. F. W., Chow, S. C. and Wong, C. K. C. (2008). The cloning of eel osmotic stress transcription factor and the regulation of its expression in primary gill cell culture. *J. Exp. Biol.* **211**, 1964–1968.

Tse, K. F. W., Lai, K. P. and Takei, Y. (2011). Medaka osmotic stress transcription factor 1b (Ostf1b/TSC22D3-2) triggers hyperosmotic responses of different ion transporters in medaka gill and human embryonic kidney cells via the JNK signaling pathway. *Int. J. Biochem. Cell. Biol.* **43**, 1764–1775.

Tseng, Y. C. and Hwang, P. P. (2008). Some insights into energy metabolism for osmoregulation in fish. *Comp. Biochem. Physiol.* **148C**, 419–429.

Tseng, Y. C., Huang, C. J., Chang, J. C., Teng, W. Y., Baba, O., Fann, M. J., et al. (2007). Glycogen phosphorylase in glycogen-rich cells is involved in the energy supply for ion regulation in fish gill epithelia. *Am. J. Physiol.* **293**, R482–491.

Tseng, D. Y., Chou, M. Y., Tseng, Y. C., Hsiao, C. D., Huang, C. J., Kaneko, T., et al. (2009a). Effects of stanniocalcin 1 on calcium uptake in zebrafish (*Danio rerio*) embryo. *Am. J. Physiol.* **296**, R549–557.

Tseng, Y. C., Chen, R. D., Lee, J. R., Liu, S. T., Lee, S. J. and Hwang, P. P. (2009b). Specific expression and regulation of glucose transporters in zebrafish ionocytes. *Am. J. Physiol.* **297**, R275–290.

Tseng, Y. C., Lee, J. R., Lee, S. J. and Hwang, P. P. (2011). Functional analysis of the glucose transporters-1a, -6, and -13.1 expressed by zebrafish epithelial cells. *Am. J. Physiol.* **300**, R321–329.

Tsukada, T. and Takei, Y. (2006). Integrative approach to osmoregulatory action of atrial natriuretic peptide in seawater eels. *Gen. Comp. Endocrinol.* **147**, 31–38.

Uchida, K., Kaneko, T., Yamaguchi, A., Ogasawara, T. and Hirano, T. (1997). Reduced hypoosmoregulatory ability and alteration in fill chloride cell distribution in mature chum salmon (*Oncorhynchus keta*) migrated upstream for spawining. *Marine Biol.* **129**, 247–253.

Valenti, G., Procino, G., Tamma, G., Carmosino, M. and Svelto, M. (2005). Minireview: aquaporin 2 trafficking. *Endocrinology* **146**, 5063–5070.

Vargau-Chacoff, L., Calvo, A., Ruiz-Jarabo, I., Villarroel, F., Munoz, J. L., Tinoco, A. B., et al. (2011). Growth performance, osmoregulatory and metabolic modifications in red porgy fry, *Pagrus pagrus*, under different environmental salinities and stocking densities. *Aquac. Res.* **42**, 1269–1278.

Wang, Y. F., Tseng, Y. C., Yan, J. J., Hiroi, J. and Hwang, P. P. (2009). Role of SLC12A10.2, a Na-Cl cotransporter-like protein, in a Cl uptake mechanism in zebrafish (*Danio rerio*). *Am. J. Physiol.* **296**, R1650–1660.

Watanabe, E., Fujikawa, A., Matsunaga, H., Yasoshima, Y., Sako, N., Yamamoto, T., et al. (2000). Nav2/NaG channel is involved in control of salt-intake behavior in the CNS. *J. Neurosci.* **20**, 7743–7751.

Watanabe, T. and Takei, Y. (2011). Molecular physiology and functional morphology of sulfate excretion by the kidney of seawater-adapted eels. *J. Exp. Biol.* **214**, 1783–1790.

Watanabe, T. and Takei, Y. (2012a). Vigorous SO_4^{2-} influx via the gills is balanced by enhanced SO_4^{2-} excretion by the kidney in eels after seawater adaptation. *J. Exp. Biol.* **215**, 1775–1781.

Watanabe, T. and Takei, Y. (2012b). Environmental factors responsible for switching on the SO_4^{2-} excretory system in the kidney of seawater eels. *Am. J. Physiol.* **301**, R402–R411.

Wendelaar Bonga, S. E. (1997). The stress response in fish. *Physiol. Rev.* **77**, 591–625.

Wilcox, C. S. (1983). Regulation of renal blood flow by plasma chloride. *J. Clin. Invest.* **71**, 726–735.

Wilson, R. W., Millero, F. J., Taylor, J. R., Walsch, P. J., Christensen, V., Jennings, S., et al. (2009). Contribution of fish to marine inorganic carbon cycle. *Science* **323**, 359–362.

Winberg, S., Höglund, E. and Øverli, Ø. (2016). Variation in the Neuroendocrine Stress Response. In *Fish Physiology - Biology of Stress in Fish*, Vol. 35 (eds. C. B. Schreck, L. Tort, A. P. Farrell and C. J. Brauner), San Diego, CA: Academic Press.

Wong, M. K. S., Ozaki, H., Suzuki, Y., Iwasaki, W. and Takei, Y. (2014). Discovery of osmotic sensitive transcription factors in fish intestine via a transcriptomic approach. *BMC Genomics* **15**, 1134.

Wu, S. C., Horng, J. L., Liu, S. T., Hwang, P. P., Wen, Z. H., Lin, C. S., et al. (2010). Ammonium-dependent sodium uptake in mitochondrion-rich cells of medaka (*Oryzias latipes*) larvae. *Am. J. Physiol.* **298**, C237–250.

Yan, J. J., Chou, M. Y., Kaneko, T. and Hwang, P. P. (2007). Gene expression of Na^+/H^+ exchanger in zebrafish H^+-ATPase-rich cells during acclimation to low-Na^+ and acidic environments. *Am. J. Physiol.* **293**, C1814–1823.

Yancy, P. H., Clark, M. E., Hand, S. C., Bowlus, R. D. and Somero, G. N. (1982). Living with water stress: evolution of osmolytes systems. *Science* **217**, 1214–1222.

Yuge, S., Inoue, K., Hyodo, S. and Takei, Y. (2003). A novel guanylin family (guanylin, uroguanylin and renoguanylin) in eels: possible osmoregulatory hormones in intestine and kidney. *J. Biol. Chem.* **278**, 22726–22733.

Zikos, A., Seale, A. P., Lerner, D. T., Grau, E. G. and Korsmeyer, K. E. (2014). Effects of salinity on metabolic rate and branchial expression of genes involved in ion transport and metabolism in Mozambique tilapia (*Oreochromis mossambicus*). *Comp. Biochem. Physiol.* **178A**, 121–131.

7

THE STRESS AND STRESS MITIGATION EFFECTS OF EXERCISE: CARDIOVASCULAR, METABOLIC, AND SKELETAL MUSCLE ADJUSTMENTS

KENNETH J. RODNICK

JOSEP V. PLANAS

Fish use swimming as their mode of locomotion and many species swim constantly to engage in feeding, migratory, reproductive, and predator avoidance behaviors. The vastly diverse lifestyles among fish species in different aquatic environments (eg, pelagic, benthic, anadromous) are reflected by extremely different capacities for swimming activity. Swimming, by way of increased activity of skeletal muscle and the cardiorespiratory systems, demands an increased production of metabolic energy to maintain homeostasis during and after exercise. Irrespective of swimming activity, a consistent feature of a stress response is the stimulation of the cardiovascular system and oxygen transfer and uptake to tissues. Fish must prioritize oxygen delivery in response to elevated metabolic states during stress and/or physical exercise. The provision and use of energy are fundamental

Biology of Stress in Fish: Volume 35
FISH PHYSIOLOGY

determinants of physiological performance and swimming exercise can be viewed as both a potential physiological stressor and a stress-reducing mechanism. Swimming is generally classified as either burst or sustained according to intensity and duration. Burst swimming can impose significant stress upon many physiological systems, causing disturbances in metabolic, acid–base, osmotic, and electrolyte balance. In contrast, sustained swimming for extended periods does not lead to significant changes in circulating cortisol and catecholamines and yet can induce positive physiological responses and improved resistance to subsequent stressors. Specifically, sustained swimming in active species increases growth, improves aerobic performance, reduces cortisol levels, promotes schooling, and reduces aggressive interactions. In this light, restrictions in the natural swimming behavior imposed by aquaculture or a research setting may deprive fish of the physiologically beneficial effects of swimming and, consequently, may be stressful to fish. In this chapter we approach the topic of swimming as a potential stressor as well as a stress-reducing behavior that contributes to homeostasis and diverse phenotypic adaptations.

1. INTRODUCTION

Stress can be defined as a condition that disrupts organismal homeostasis by external or internal stimuli, referred to as stressors (Schreck, 2010) (Schreck and Tort, 2016; Chapter 1 in this volume). Fish are exposed to diverse stressors in the natural environment, as well as in aquaculture and the laboratory setting. The stress response in fishes is a normal response to cope with stressors and promote homeostatic recovery. This response begins with a rapid release of catecholamines and cortisol into the circulation and can change an array of biochemical activities that alters circulating substrates and energy metabolism. Acute stress has pronounced effects on hydromineral balance, respiration, and cardiovascular function. We could argue that a major outcome of the stress response is the hormone-mediated mobilization of energy stores and neurohumoral-enhanced delivery of oxygen to energy-demanding tissues. Individual responses to threatened homeostasis can be behavioral as well as physiological and the rate of recovery to pre-stress levels may affect life processes such as growth and reproduction, and even determine whether survival occurs. Increased energy costs associated with stress and changes in energy budgets can ultimately compromise one or more physiological functions. Fish experience elevated metabolic rates during swimming activity and it is important to consider the interactions between exercise, at various intensities and durations, and stress-induced changes in performance. While numerous studies measure the swimming limits or capabilities of fish, fewer studies define mechanisms by which lower intensity exercise induces physiological adaptations and identify exercise as a potential stress-reducing behavior.

Exercise, by way of increased activity of skeletal muscle and the cardiorespiratory systems, demands an increased production of metabolic energy to maintain homeostasis during and after exercise. Unlike terrestrial animals, many fish are constantly swimming for purposes of feeding, migrating, and avoiding predation. The capacity for swimming activity is extremely variable between species and swimming is generally classified as either burst or sustained according to intensity and duration. At one extreme, studies of exhaustive exercise and recovery have been extremely valuable because of the physiological knowledge attained and because of the large number of stressful activities that induce exhaustive exercise and an extended recovery period lasting many hours. For example, catch and release fishing, electrofishing for biological surveys, fish transportation and release, and a variety of handling stresses are associated with aquaculture practices and stimulate high levels of swimming activity. The rapid rise in skeletal muscle and blood lactate, and corresponding reductions in blood oxygen content and pH are biochemical markers of acute and severe stress responses (Wendelaar Bonga, 1997). The stress associated with exhaustive exercise can even exceed homeostatic mechanisms and cause fish death several hours postexercise (Black, 1958; Wood et al., 1983; van Ginneken et al., 2008). Assuming that a stress response does not exceed homeostatic capabilities, some studies suggest that active swimming during recovery can accelerate the recovery of plasma cortisol levels, acid–base balance, hydromineral balance, and/or energy metabolism following crowding (Veiseth et al., 2006) or exhaustive exercise (Pagnotta et al., 1994; Milligan, 1996). However, this is not a universal finding (Meyers and Cook, 1996; Powell and Nowak, 2003; Kieffer et al., 2011), and whether swimming affects the rate of recovery may depend upon the stressor, the species investigated, and the swimming intensity during recovery.

Generally speaking, chronic stress is considered to be detrimental to fish (Schreck and Tort, 2016; Chapter 1 in this volume). In contrast to burst swimming or exhaustive exercise for just seconds to minutes, sustained swimming under controlled conditions for hours at intensities lower than critical swimming speed (U_{crit}) does not elicit peaks in circulating stress hormones (Hughes et al., 1988; Wood, 1991; Perry and Bernier, 1999) or stimulate anaerobic energy metabolism in skeletal muscle. For example, adult sockeye salmon (*Oncorhynchus nerka*) can swim at 75% of U_{crit} for >4 h in a swim tunnel at 15°C without increasing plasma lactate levels (Steinhausen et al., 2008). As salmonids approach U_{crit}, they change their swimming gait, recruit glycolytic, fast-twitch white muscles, and increase blood lactate and decrease blood pH (Brauner et al., 2000). If swimming exercise persists at low to moderate intensities for extended periods, below the neuroendocrine threshold for a stress response, adaptations occur through a process of hormesis and lead to positive physiological responses and resistance to subsequent stressors. Thus, depending on the species and the type and duration of exercise, exercise can be considered a significant stressor and a stress-modifying factor. In fact, there is a substantial body of literature reporting on the positive, physiological effects of sustained swimming in a variety of fish species (Davison, 1997; Palstra and Planas, 2013; see Section 3). Sustained swimming at moderate intensities has been reported to reduce plasma cortisol levels (Boesgaard et al., 1993; Postlethwaite and McDonald, 1995), and improve performance and the health and welfare status of fish in the face of stressors (Jorgensen and Jobling, 1993; Ward and Hilwig, 2004; Veiseth et al., 2006; Castro et al., 2011; Larsen et al., 2012; McKenzie et al., 2012). Conversely, an underlying assumption is that restriction of the ability of fish to swim and the ensuing reduction in the potential physiologically beneficial effects of swimming, as it commonly occurs in aquaculture, should be stressful to fish.

Our goal is to review the physiological responses of cardiorespiratory and skeletal muscle systems to acute exercise and relate those activities to a generalized stress response in fishes. We will also highlight the recent findings that long-term exercise can serve as a valuable stimulus for favorable physiological adaptation in fishes, including a reduced response to stressors. We concentrate our efforts on peer-reviewed literature spanning over 60 years and refer, when possible, to other chapters in the current volume and previous reviews. The majority of studies on the effects of swimming have focused on blood (for practical reasons), skeletal muscle (because this tissue defines performance and is affected by fatigue), and the cardiorespiratory system. It is important to mention that most work on physiological stress responses and exercise has targeted only a small number of fish species—mostly teleosts and freshwater salmonids—in the laboratory setting. As a result, readers should be cautious because the reported findings and trends may not extend to all members of this large and diverse group of vertebrates, in their natural environment.

2. PHYSIOLOGICAL DEMANDS OF SWIMMING EXERCISE AND THE STRESS CONTINUUM

2.1. Introduction

The goal of this section is to provide a working definition of exercise and extend out to the associated physiological disturbances/responses. Much of our understanding about swimming performance in fish has resulted from experiments conducted under laboratory conditions. High intensity and exhaustive exercise involving anaerobic white muscle lasting just 20 s can impose significant stress upon many physiological systems. These include major disturbances in metabolic, acid–base, osmotic, and electrolyte balance (Wood, 1991; Kieffer, 2000). At the other extreme of performance, "sustained swimming" occurs at lower speeds for extended periods (>200 min) and is powered predominantly by oxidative red muscle and aerobic metabolism. However, based upon initial hormonal and metabolic changes, the effects at the onset of even moderate and low intensity exercise resemble those of acute stress (Nielsen et al., 1994). Thus, whether or not sustained swimming, per se, becomes a stressor will depend upon the neuroendocrine response and timing. It is now appreciated that exercise in fishes, similar to mammals, promote physiological adaptation and affects both swim performance and responses to a variety of stressors.

2.2. Neuroendocrine Aspects of Stress and Exercise

Gorissen and Flik review the endocrine stress axis and control of the stress response (Gorissen and Flik, 2016; Chapter 3 in this volume). A fundamental role of stress hormones is to make energy available for systems involved in fight, flight (swimming activity), or coping. Briefly, blood levels of catecholamines epinephrine and norepinephrine increase during or immediately following a variety of physical and environmental stressors. These stressors include external hypoxia, hypercapnia, air exposure, metabolic acidosis, exposure to soft water, and exhaustive exercise (reviewed in Randall and Perry, 1992). The rapid release of circulating stress hormones during exercise is highly dependent on intensity and duration, and plasma catecholamine levels remain essentially unchanged during sustained, lower intensity exercise (Hughes et al., 1988; Wood, 1991; Perry and Bernier, 1999). The highest concentrations of epinephrine and norepinephrine in blood are observed immediately after exhaustive exercise, with epinephrine concentrations exceeding norepinephrine by a factor of 2–3 (Milligan, 1996). There is a rapid recovery of catecholamines to resting levels within approximately 2 h of exercise cessation. As mentioned earlier, it may be possible to use active swimming to ameliorate circulating cortisol levels and enhance recovery after exhaustive swimming (Pagnotta et al., 1994; Milligan, 1996).

The stress response can also be mediated directly by adrenergic innervation to select organs such as the heart (although not in hagfish and elasmobranchs) and vasculature in the gill and systemic circulation. Overall, catecholamines, directly or indirectly, lead to an increase in energy metabolism, oxygen supply, and the provision of energy substrates and oxygen to a variety of tissues (Wendelaar Bonga, 1997; Pankhurst, 2011). Additional actions of catecholamines include increasing plasma glucose levels (Wright et al., 1989), changing gill diffusive capacity, increasing cardiac output and gill ventilation, increasing blood oxygen capacity by stimulating erythrocyte release from the spleen, and elevating erythrocyte pH (see later and Fig. 7.1). Catecholamines may also indirectly increase ion exchange across the gill by increasing cardiac output and therefore the surface area of perfused gill lamellae in contact with the aquatic environment.

Cortisol affects a wide variety of physiological functions, including energy metabolism, amino acid metabolism, ion regulation, and immune functions. Major targets of cortisol action in teleosts include the liver and skeletal muscle. Increases in plasma cortisol increase protein turnover, ammonia output, gluconeogenesis, and lipolysis (Mommsen et al., 1999). Cortisol also has important actions at the gills for osmoregulation by increasing Na^+/K^+ ATPase activity and Na^+ transport (McCormick, 1995). Overall, cortisol facilitates the mobilization of energy substrates and restoration of hydromineral balance during a stress response. Very few studies have examined the actions of cortisol during acute exercise; however, unlike the serious deleterious effects that chronic plasma cortisol can have on fish feeding behavior, growth, physical condition, and survival, aerobic swimming performance appears to be unaffected by chronic stress (Gregory and Wood, 1999). It also appears that the elevation of cortisol after exercise may inhibit glycogen synthesis in skeletal muscle and prolong the recovery process (Milligan, 2003).

2.3. Energy Metabolism During Stress and Exercise

In this section, an attempt will be made to compare and link the metabolic responses during stress and exercise. Sadoul and Vijayan (2016; Chapter 5 in this volume) cover the effects of stress on metabolic rate, energy homeostasis, and energy substrate partitioning to a variety of physiological stressors. The term "metabolism" reflects the summation of all chemical reactions in an organism. These include anabolic and catabolic processes. The most important cellular processes are the anaerobic and aerobic formation of adenosine triphosphate (ATP) and its hydrolysis to adenosine diphosphate (ADP), inorganic phosphate (Pi), and H^+. Stress is an energy-demanding process and energy supply is a limiting factor for all physiological activities. Stressors can promote the rapid mobilization and utilization of energy reserves. The fact that stressors can also modify oxygen consumption, hydromineral and acid–base balance, cardiovascular function, and the activity of skeletal muscle has a direct impact on the scope for physical activity and physiological performance.

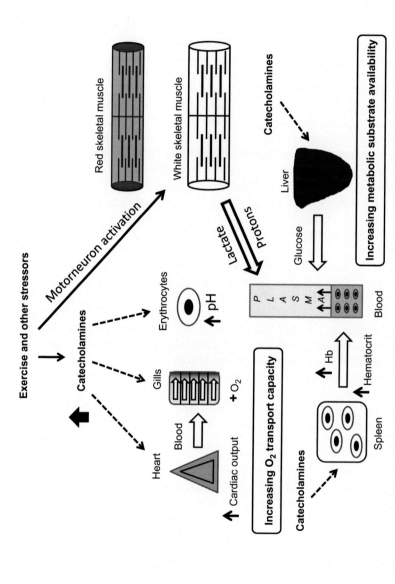

Figure 7.1. Circulatory, respiratory, and metabolic responses to catecholamines and muscle activation during exercise and other stressors in fish. The scheme highlights adrenergic effects on the heart, gills, erythrocytes, and spleen to increase blood O_2 transport function during stress, exercise, and postexercise recovery. The liver increases blood glucose concentrations by glycogenolysis. During exhaustive exercise, energy stores in anaerobic white muscle may be depleted with lactate and protons being released into the blood, causing acidosis.

2.3.1. Exercise Effects on Water and Ion Homeostasis

Takei and Hwang (2016; Chapter 6 in this volume), provides a detailed perspective on osmoregulation, hydromineral balance, and physiological responses to osmotic stresses. It is well established that stress in fish has a major impact on water and ion homeostasis and results in additional energy needs for ion transport mechanisms (Wendelaar Bonga, 1997; Tseng and Hwang, 2008). Depending on salinity and experimental methodology, estimates suggest that the gill accounts for <4% (Morgan and Iwama, 1999), 7% (Mommsen, 1984), and <20% (McCormick et al., 1989) of a fish's total oxygen consumption, and the increased oxygen demand during stress compromises osmotic and ionic regulation (Wendelaar Bonga, 1997). The energy required to maintain hydromineral balance in the face of swimming at different intensities for various durations remains to be defined.

The respiratory-osmoregulatory compromise between gas exchange and osmoregulatory burden is well recognized (Wood and Randall, 1973; Nilsson and Sundin, 1998) and illustrated by acute stress and elevated levels of stress hormones. Epinephrine facilitates environmental oxygen uptake by (1) increasing gill perfusion by way of elevated cardiac output and ventral aortic pressure; (2) decreasing vascular resistance of the gills; (3) increasing permeability of the tight junctions between branchial epithelial cells; and (4) recruiting more brachial lamellae and therefore increasing effective respiratory surface area for diffusive exchange of gases, water, and ions. As a result, increased passive loss of ions and water uptake are associated with higher rates of oxygen consumption and elevated concentration of circulating catecholamines. Both acute stress and any change in swimming activity can affect the passive ion fluxes across the gill. However, there is some uncertainty about the hormonal control of ion regulatory responses. Although the osmoregulatory responses to confinement stress and aerobic exercise are similar, the circulating hormone profiles are different (Postlethwaite and McDonald, 1995).

2.3.2. Muscle Activity and Locomotion

Comparative studies of physiological responses to exercise have been invaluable for an understanding of locomotory muscle activity, neuroendocrine responsiveness, metabolic compensation, and cardiorespiratory abilities to cope with a stressful situation. Locomotion is energetically costly in fishes, and the majority of this energy need comes from the contracting skeletal musculature (Gerry and Ellerby, 2014). Not surprisingly, all vertebrate animals raise their metabolism during exercise and other stressful activities. The cost of swimming is directly related to intensity and duration of contractile activity of muscle tissue and may be fueled by both aerobic and anaerobic metabolism (see later).

When a fish moves through water, drag is exerted against the body. Drag is proportional to the square of the velocity of the fish pushing against the water. As a result, oxygen consumption increases exponentially with increasing velocity (Brett, 1964) and the maximal capacity to consume and transport oxygen (respiratory metabolism) limits swimming velocity (Bennett, 1991). Stressors can initiate behavioral change and disrupt physiological performance. For example, under high stocking densities (Anras and Lagardère, 2004) and during transportation (Chandroo et al., 2005), cultured rainbow trout demonstrate elevated swimming activity. However, even beyond a 48 h resting period, 50 min in a commercial shipping tank reduced swimming performance (U_{crit} and endurance) relative to nontransported controls (Chandroo et al., 2005). Conversely, although chronic elevation of cortisol had negative effects on feeding behavior and growth, elevated cortisol, by itself, does not impact swimming performance of rainbow trout (Gregory and Wood, 1999) and common carp (*Cyprinus carpio*; Liew et al., 2013). Other fish behaviors affected by stressors that likely impact energy metabolism include avoidance, chemoreception, feeding, predator evasion, and thermoregulation (Schreck et al., 1997). If juvenile rainbow trout experience exhaustive exercise, swimming performance is impaired for at least 6 h and maladaptive swimming behaviors would increase the risk of predation and impair foraging ability (Lee-Jenkins et al., 2007). Together, evidence to date shows that stressors can impact swimming performance of fishes and swimming to exhaustion affects a variety of other behaviors.

An appreciation of myotomal muscle characteristics is fundamental to the current understanding of the energetic demands and limitations of swimming activity. Skeletal muscle of fish has been categorized into two major types, white and red fibers, which are spatially distinct with different metabolic and contractile characteristics (Driedzic and Hochachka, 1978). White muscle has limited vascularity, few mitochondria, little myoglobin, and relies predominantly on anaerobic metabolism for energy production. White muscle constitutes the majority (80–95%) of the swimming musculature and provides power for high-intensity, short-term burst activity. The two muscle fiber types have different substrate preference and metabolic strategies for producing ATP (see later). Based on the accumulation of lactate in the white muscle, it also appears that there is an increasing recruitment of white muscle fibers as fish transition from low to high swimming speeds (Johnston and Goldspink, 1973). Superficial red muscle fibers have an extensive vascular supply, a high content of mitochondria and myoglobin, and much higher rates of oxygen consumption than white muscle fibers. Red-oxidative muscle powers slow swimming activity over extended periods using aerobic energy metabolism and both red- and white-glycolytic muscle are used at the highest swimming velocities (Rome et al., 1993; Johnson et al., 1994; Richards et al., 2002c).

While numerous studies have used enforced exercise at high intensity and exhaustion to define physiological capacity and recovery (reviewed in Wood et al., 1983; Milligan, 1996; Kieffer, 2000), this activity is probably rarely experienced under natural conditions. Exceptions, due to human intervention, include hook and line angling capture and netting fish for sport or commercial purposes. Acute stressors such as capture, handling, transport, forced exercise, hypoxia, osmotic and temperature shocks, crowding, or exposure to water pollutants have been shown to increase muscle and plasma lactate levels, decrease blood pH, and increase oxygen utilization (reviewed in Wendelaar Bonga, 1997). Free-swimming fish typically choose lower intensity speeds that can be sustained for prolonged periods (Krohn and Boisclair, 1994). Comprehensive perspectives on substrate utilization by fish during exercise have been provided by Driedzic and Hochachka (1978), Moyes and West (1995), and Richards et al. (2002c).

2.3.3. ENERGY PRODUCTION AND UTILIZATION

A hallmark of the physiological stress response is an increase in whole-animal metabolic rate and the provision of circulating substrates to fuel cellular needs to reestablish homeostasis (Wendelaar Bonga, 1997) (Sadoul and Vijayan, 2016; Chapter 5 in this volume). Fry introduced the concept of "scope for metabolic activity," which is defined as the difference between maximum metabolic rate and standard metabolic rate. This concept is limited to aerobic metabolism and yet it highlights energy constraints for physiological activities and behavioral responses to various environments (Fry, 1947). Cells have multiple biochemical options and energy substrates for regenerating ATP, for purposes for biosynthesis, ion pumping, and mechanical work. Catecholamines stimulate glycogenolysis and anaerobic glycolysis, whereas cortisol stimulates both anaerobic and aerobic metabolism, nitrogen excretion, and gluconeogenesis. Metabolic rates in all animals are limited and stressors could lead to a shortage of energy and define whether an animal can tolerate a particular stressor. The energetic costs and capacity to perform multiple tasks concurrently can be summarized mathematically by a bioenergetics equation (Schreck and Tort, 2016; Chapter 1 in this volume). Clearly, maintaining a positive energy balance is a fundamental requirement to accommodate stressors and maintain a scope for activities like swimming (Schreck, 2010). However, diverse environmental and anthropogenic stressors can ultimately decrease the ability of fish to perform through physiological limitations and behavioral impairments, potentially leading to mortality (Portz et al., 2006).

2.3.3.1. Phosphagens. An elevated metabolic rate during and after swimming requires a concurrent match between energy demand and supply to maintain homeostasis. While ATP is an immediate source of energy for muscle contraction and other cellular functions, endogenous ATP supplies are limited and would be quickly exhausted during high intensity muscle contractions. Instead, intracellular creatine phosphate (CrP) and the enzyme creatine kinase can rapidly resynthesize ATP from ADP and H^+, and help maintain cellular levels of ATP in the absence of O_2. This mechanism (ie, phosphagen mobilization) is of major importance in select tissues (striated and smooth muscle plus nervous tissue). The phosphorylation of ADP by CrP is particularly important at the onset of high intensity exercise in white muscle to prevent large changes in ATP concentrations, before the rate of glycolysis has been increased (Newsholme and Leech, 1983; Hochachka and Somero, 2002; Richards et al., 2002a). White muscle in fishes has very high "resting" concentrations of CrP and ATP compared with other tissues and recovery of CP in white muscle is quite rapid, usually within 1 h postexhaustive exercise (van Ginneken et al., 2008). CrP therefore plays a prominent role in maintaining cellular energy status, especially during high-intensity, nonsustainable swimming activity (eg, 90% of U_{crit}). Depending on the tissue, specific ATP demands of a stressor or stressors, and intrinsic levels of the intracellular creatine, the CrP system can serve as a protective mechanism to prevent critical reductions in free energy.

2.3.3.2. Glycolysis and Gluconeogenesis. Glucose is a fundamental energy substrate for all cells. Certain fish tissues (eg, brain, gills, heart, and blood cells) rely primarily on glucose for energy metabolism (Polakof et al., 2011). During stress, elevations in plasma cortisol and catecholamines stimulate the mobilization of glycogen stores from the liver, thereby raising circulating glucose levels and increasing substrate supply to meet higher energy demands (see Sadoul and Vijayan (2016; Chapter 5 in this volume), for a more comprehensive review of effects of stress on tissue-specific energy metabolism and homeostasis). It is noteworthy that blood glucose shows larger changes in some fish (rainbow trout) compared with mammals when fish are subjected to stressors such as hypoxia and crowded rearing conditions (Polakof et al., 2011).

Glycolysis is important because it represents the first pathway of cellular respiration of glucose. The energy required for white muscle contraction during strenuous exercise is provided mainly by the anaerobic conversion of local glycogen stores to lactate (Richards et al., 2002a). The conversion of glycogen, rather than glucose, to lactate has a 50% energetic advantage (3 vs 2 ATP) per glucose residue converted to lactate (Newsholme and Leech, 1983). Because glycolysis is a very inefficient means of producing ATP compared with cellular respiration, it has to occur at a high rate during metabolic challenges such as strenuous exercise, and it can lead to the accumulation of lactate and protons. Lactate, in turn, can serve as an energy source for oxidative tissues and gluconeogenic substrate (Suarez and Mommsen, 1987; Gladden, 2004). After exhausting swimming, lactate is retained within white muscle and this process prolongs recovery from maximal exercise (VanRaaij et al., 1996).

Glycolytic activity within cells is regulated by stress hormones, ions, adenylates, and key enzymes. Catecholamines are well known to affect local and systemic carbohydrate metabolism in fishes (Fabbri et al., 1998). For example, glycogen phosphorylase (GP), the enzyme that catalyzes the flux-generating step for glycogen conversion to lactate in white skeletal muscle, is activated by epinephrine through β-adrenoreceptors and second messenger cyclic adenosine monophosphate (cAMP). The rise in sarcoplasmic Ca^{2+} concentration that occurs during the contraction process also activates GP, leading to intracellular glycogenolysis and regeneration of ATP. During exercise, the cellular accumulation of Pi, free ADP, free AMP, inosine monophosphate (IMP), and NH_3 can also activate glycolysis for ATP production in white muscle of rainbow trout (Richards et al., 2002a). During the first 10 s of burst activity, high rates of ATP turnover are supported primarily by CrP hydrolysis and glycolysis (Richards et al., 2002a). In addition, exhaustive exercise increases plasma NH_3 dramatically (Mommsen and Hochachka, 1988) and this metabolite may help stimulate glycolytic production of ATP.

2.3.3.3. Oxidative Phosphorylation. A third option for regenerating cellular ATP requires O_2 and a series of complex pathways—the tricarboxylic (or citric) acid cycle, electron transfer, and oxidative phosphorylation—contained within mitochondria. The convective transport of O_2 to metabolically active cells, especially contracting skeletal muscle during swimming, depends on a coordinated interplay involving cardiovascular and respiratory responses. Whereas the use of phosphagens (ATP and CrP) and anaerobic glycolysis for short-term energy production relies predominantly on intracellular fuels, sustained oxidative phosphorylation during exercise or stress can also use circulating fuel supplies. As a result, the cardiovascular system serves a vital role to provide circulating fuels to energy-requiring cells during both exercise and stress.

Organs and tissues that demand a continuous supply of ATP for biosynthesis, ion transport, and/or contractile activity will exhibit high rates of oxygen consumption and large cellular complement of mitochondria (Hochachka, 1994). Given that mitochondria function is critical for cellular energy homeostasis and oxygen consumption increases during stress and sustained exercise, oxidative capacity is an important determinant of an organism's maximum metabolic rate and therefore metabolic scope. Although lipids would not appear to be an important fuel for anaerobic white muscle during exhaustive exercise and recovery, evidence suggests that multiple species of fishes activate oxidative phosphorylation and lipid catabolism for ATP synthesis in this tissue during postexercise recovery (Richards et al., 2002b).

Very little is known about the direct effects of catecholamines and cortisol on mitochondria and oxidative metabolism in fishes. Elevated cortisol due to implants does not impair aerobic swimming performance (Gregory and Wood, 1999), but it does increase whole animal O_2 consumption after handling stress (Davis and Schreck, 1997) and after cortisol administration only (Morgan and Iwama, 1996; De Boeck et al., 2001). Although not confirmed in fish, recent studies demonstrate that mitochondria contain receptors for glucocorticoids and catecholamines and can therefore both sense and respond to cellular stress (Manoli et al., 2007; Picard et al., 2014). Among many of the mitochondrial functions and dysfunctions elicited by glucocorticoids, these hormones can increase mitochondrial biogenesis and therefore increase the capacity of mitochondria to generate ATP for elevated energy demands during stress and recovery after stress.

2.3.3.4. Measurements of Oxygen Consumption.   Changes in metabolic rate can be estimated by indirect calorimetry using oxygen consumption and a respirometer. Rates of whole-animal aerobic metabolism are highly variable in fishes and sensitive to many factors such as handling, time since feeding, and temperature (Fry, 1971). Numerous studies highlight the stimulation of energy metabolism and oxygen uptake during acute stress and exercise in fishes. Acute physical stress doubles oxygen consumption in juvenile rainbow trout, and elevated oxygen consumption and plasma cortisol levels were positively correlated (Barton and Schreck, 1987). These authors estimated that up to 25% of the scope for activity might be utilized to support mild and brief disturbances. Exhaustive exercise lasting just 5 min results in a 2- to 2.5-fold increase in oxygen consumption postexercise; near depletion of whole-body glycogen, ATP, and CP stores; and a fivefold increase in whole-body lactate (Scarabello et al., 1991).

Measurements of whole-animal oxygen consumption cannot delineate the energy costs to a particular organ or physiological process. Although increased metabolic rates (O_2 consumption) are observed during exposure to many different stressors (Barton and Schreck, 1987), very little is known about the distribution of metabolic energy during stress. Likely candidates for elevated energy requirements during stress include osmoregulation, protein synthesis, and contractile activity of skeletal and cardiac muscle. Furthermore, increases in energy demands associated with stress do not end with the termination of a specific stressor or stressors. Brief handling plus air exposure stress, in particular, can elevate oxygen consumption for 1 h (Davis and Schreck, 1997) to 3–4 days (Korovin et al., 1982), depending on the species, life stage, and severity of the stressor. The elevated oxygen consumption following exercise, termed excess post-exercise oxygen consumption (EPOC), can be used to assess metabolic recovery time and the nonaerobic energy cost of exercise in fishes (Lee et al., 2003). Following exhaustive exercise, EPOC and circulating lactate can persist for 6 h in rainbow trout whereas recovery of CP, ATP, and glycogen stores took 5 min, ~1 h, and at least 6 h, respectively (Scarabello et al., 1991). Elevated blood lactate also persists in young salmonids after other stressors like acute handling (2–8 h elevation of lactate, depending upon nutritional status; Vijayan and Moon, 1992) and fish loading and transport (>48 h elevation of lactate for transport lasting 0.5 to 4.5 h; Iversen et al., 1998).

2.4. Cardiovascular and Respiratory Adjustments to Stress and Exercise

A key component of an effective stress response is the stimulation of oxygen transfer and uptake to tissues (reviewed in Wendelaar Bonga, 1997). Fish must prioritize oxygen delivery and adjust cardiorespiratory activities in response to elevated metabolic states (Hicks and Bennett, 2004), such as physical exercise and/or stress. During a stress response, oxygen and energy substrates may be reallocated from one or several energy-demanding functions to support the most immediate needs for survival. This response involves both circulating and neurally released catecholamines and well-documented changes in cardiorespiratory function (see review by Randall and Perry, 1992). However, our current understanding of nongenomic effects of cortisol on cardiorespiratory functions in fishes is extremely limited.

The Fick Principle states that oxygen consumption (MO_2, of an organism or organ) is equal to the product of blood flow or the cardiac output of blood and oxygen extraction from the blood (Fick, 1870). Thus, the ways that whole animal oxygen consumption can increase are increasing cardiac output, raising arterial oxygen content (CaO_2), and/or reducing venous oxygen content (CvO_2). Cardiac output is determined by the stroke volume and heart rate. Stress, spontaneous activity, and blood loss all tend to increase resting cardiac output (reviewed by Farrell and Jones, 1992). CaO_2–CvO_2 defines the arteriovenous oxygen content difference. The ability to increase CaO_2–CvO_2 depends primarily on the arterial O_2 content and concentration of hemoglobin (Hb), although a reduction in CvO_2 is a normal outcome during sustained swimming activity (Kiceniuk and Jones, 1977). The concentration of Hb in the blood is dependent upon the total circulating mass of Hb and plasma volume. The importance of cardiac output and blood oxygen content during stress relates directly to aerobic energy metabolism and the use of increasing amounts of oxygen for energy production to support cellular homeostasis and cope with stressors.

The capabilities of the O_2 transport system have been determined during steady state exercise up to critical swimming speeds (U_{crit}), which defines aerobic capacity. Depending on fish size and water temperature, O_2 delivery and tissue consumption can increase 5–15 times the basal level (standard metabolic rate) in salmonids (Brett, 1972). At U_{crit}, MO_2 was increased above resting levels by 7.5 times in rainbow trout (Kiceniuk and Jones, 1977). For salmonids at U_{crit}, arterial blood remains fully saturated with O_2 and venous blood has been maximally depleted of O_2 by active tissues (Farrell, 2007). Integrated responses to sustained exercise include (1) increasing ventilation of the gills; (2) corresponding increases in blood flow perfusing gills due to increasing cardiac output and ventral aortic blood pressure; and (3) a redistribution of blood away from inactive tissues to the contracting axial musculature. Cardiac output increases with increasing swimming speed to maximum values close to U_{crit}. Consistent with the Fick Principle, increased O_2 delivery was accomplished by increasing cardiac output through higher heart rate and stroke volume, plus increased extraction of O_2 in the blood (lowering CvO_2). Ultimately there is a physiological limit to increased O_2 delivery in fish, and compromises in gill ventilation, cardiovascular function, or blood characteristics may lead to stress and impaired functions (see next).

2.4.1. Cardiovascular Adjustments

Modulation of autonomic nervous output to the heart and vasculature promote circulatory homeostasis and increase delivery of oxygen to energy-demanding tissues. Similar to other vertebrates, the dual autonomic innervation pattern of the teleost heart exits and has been reviewed extensively by Nilsson (1983) and Santer (1985). Adrenergic innervation of the heart is present in most teleosts studied, but not in cyclostomes, elasmobranchs, and pleuronectids. The teleost heart receives adrenergic fibers releasing both epinephrine and norepinephrine. Depending on the species, these catecholamines bind to adrenoreceptors (α or β) on the myocardium and pacemaker tissue, and may exert positive chronotropic effects via β-adrenoreceptors (Nilsson, 1983), or negative chronotropism via α-adrenoreceptors (Tirri and Ripatti, 1982). Vagal (cholinergic) fibers also innervate the fish heart, except the hagfish, and, together with adrenergic fibers, define heart rate. Catecholamines increase cardiac sensitivity to preload, thus promoting an increase in stroke volume (Farrell et al., 1986). Adrenergic nerves innervate systemic blood vessels of most fishes and catecholamines usually increase blood pressure and induce vasoconstriction (via α-adrenoreceptors) in areas like the gut under stressful conditions (Axelsson et al., 1989). Catecholamines and cortisol released during stress also decrease viscosity of deoxygenated blood and therefore the resistance of erythrocytes to blood flow (Sorensen and Weber, 1995).

The role of circulating catecholamines in regulating blood flow during low intensity swimming may be insignificant because circulating catecholamine levels do not generally change during mild or moderate stress, and therefore do not promote cardiovascular responses. Only during acute stress is there a pronounced release of catecholamines into the blood (Perry and Bernier, 1999). The tachycardia during aerobic exercise may be determined by the reduction in vagal (cholinergic) input to the heart rather than additional adrenergic stimulation (Farrell and Jones, 1992). Conversely, burst swimming often results in decreases in heart rate, cardiac output, and blood pressure during activity, followed by elevated cardiovascular function during recovery (Farrell, 1982). Exhaustive exercise reduces plasma volume due to fluid shifts between muscle and blood and to a general diuresis. This hemoconcentration, together with splenic contraction, probably contribute to an increase in arterial blood O_2 content (reviewed in Randall and Perry, 1992).

Potential stressors such as elevated water temperature and decreased O_2 levels elicit somewhat different cardiovascular responses than exercise. Acute temperature elevation increases heart rate and cardiac output, but not stroke volume, due to direct stimulatory effects on cardiac pacemaker cells (Harper et al., 1995). However, this response, and cardiac pumping capacity, is limited and may define the upper limits of thermal tolerance. There are also indications that the adrenergic component is important for thermal tolerance and cardiac performance at temperatures approaching the upper lethal limit in fish (Ekström et al., 2014). Moderate to severe hypoxia, on the other hand, results in a reduced heart rate due to increased cholinergic tone to the heart, and a compensatory increase in stroke volume to maintain cardiac output in several species (Farrell and Jones, 1992). Hypoxia also increases vascular resistance to the gill, although the combination of bradycardia and arterial hypertension does not elevate arterial O_2 levels and enhance gas transfer in rainbow trout (Perry and Desforges, 2006).

Consistent with studies in mammals and amphibians, cortisol can rapidly modulate cardiac function in fishes through nongenomic mechanisms. More specifically, physiological levels of cortisol rapidly (within 3–10 min) increase myocardial contractility under oxygenated conditions, in vitro, in rainbow trout (Farrar and Rodnick, 2004). The positive effects of cortisol are additive to epinephrine and likely involve mechanisms such as polyamine and nitric oxide signaling cascades. Additional studies are needed to define the cardiovascular responsiveness to cortisol in fishes and appreciate whether stress-induced cortisol levels help maintain elevated levels of cardiac function and possibly whole-animal performance.

2.4.2. RESPIRATORY ADJUSTMENTS

Fish increase the flow of water over their gills (hyperventilate) as well as elevate cardiac output—and therefore gill perfusion with blood—for enhanced gas exchange during periods of mild and moderate stress. During sustained exercise, the increase in gill ventilation is due to small increases in breathing rate and much larger changes in stroke volume (Jones and Randall, 1978). At burst speeds, however, gill ventilation may cease altogether (Satchell, 1991). Hyperventilation may enhance the transfer of O_2 across the gill by reducing the boundary layer at the surface of lamellae. While the role of circulating catecholamines for regulating ventilation is unclear and still is a matter of debate, the regulation of gill ventilation involves external and internal O_2 tensions. Moreover, a lowering of O_2 content in blood during stress appears to be a proximate stimulus for the release of catecholamines in teleosts (Randall and Perry, 1992). However, the value of stimulating ventilation during stress may extend beyond the maintenance of normal blood O_2 levels. For example, fish hyperventilate for extended periods during recovery from exhaustive exercise, even though arterial blood O_2 levels are normal (Wood, 1991). Proposed benefits for increasing ventilation beyond the metabolic requirements for O_2 include acid-base homeostasis, CO_2 excretion, and ammonia excretion (Zhang and Wood, 2009). After exhaustive exercise, and throughout recovery, intracellular and extracellular ammonia increases (Wang et al., 1994).

Given that all cardiac ventricular output must pass through the gills, gill blood flow is increased in direct proportion to exercise-induced increases in cardiac output. Catecholamines elevate O_2 uptake at the gill indirectly by raising cardiac output and dorsal aortic blood pressure, thereby increasing the number of lamellae being perfused with blood, and providing a more even distribution of blood throughout each lamellae (Randall and Perry, 1992). Increasing pulse pressure due to increased stroke volume will perfuse more lamellae, especially at the distal end of the filaments (Farrell, 1980; Farrell et al., 1980). Stress-induced catecholamines also affect blood O_2 capacity through β-adrenergic activation of erythrocyte Na^+/H^+ exchange, and the subsequent elevation of intracellular pH to counteract the reduction in Hb-O_2 affinity resulting from metabolic acidosis (reviewed by Nikinmaa, 1992). During severe exercise, catecholamines induce splenic contraction by way of α-adrenoreceptors, resulting in the release of erythrocytes into the circulation, and therefore higher hematocrit, CaO_2, and elevated concentration of Hb in the circulation (Nilsson, 1983). A general scheme of some cardiovascular, respiratory, and metabolic responses to exercise is shown in Fig. 7.1.

2.5. Limits of Swimming Exercise and Stress

Stressed fish can reach a stage of exhaustion, whereby homeostatic mechanisms cannot compensate for the stressful challenge. One important example of this, which has received little attention, is the significant mortality of fish that occurs during an unsuccessful recovery after exercise to exhaustion in the laboratory (Wood et al., 1983; van Ginneken et al., 2008). Mortality also occurs in wild anadromous salmonids swimming anaerobically during their freshwater spawning migration (Burnett et al., 2014). As mentioned previously, even if fish survive, exhausted fish have impaired swimming abilities for extended periods following exercise (Lee-Jenkins et al., 2007). Proposed causes for the postexercise mortality include high rates of energy consumption due to excess stress hormones, depletion of energy stores, reduced phosphorylation potential, and severe acidosis (Wood et al., 1983; van Ginneken et al., 2008). Burst swimming also disrupts osmotic and electrolyte balance (reviewed in Wood, 1991; Milligan, 1996; Kieffer, 2000). Physiological responses to these challenges would involve multiple tissues, require considerable energy production during recovery, and yet may not achieve homeostasis. Finally, blood acidosis, hyperkalemia, and hypoxia following exhaustive exercise can limit cardiac pumping activity (Driedzic and Gesser, 1994) and therefore reduce O_2 delivery to tissues during recovery. Clearly, there will be a point where cardiac output cannot support additional metabolic demands. Although adrenergic stimulation attenuates the cardioplegic effects of acidosis, hyperkalemia, and hypoxia, the beneficial effects of epinephrine decrease at high temperature (Farrell et al., 1996). Thus, depending on environmental conditions, the maintenance of O_2 transport and restoration of energy reserves, acid–base, and ion homeostasis may ultimately set the physiological limits for a successful recovery after high intensity swimming and other severe stressors associated with fish culture procedures and recreational fisheries.

3. PHYSIOLOGICAL ADAPTATIONS TO SWIMMING AND RELEVANCE TO STRESS

3.1. Introduction

As stated previously, swimming is an integral aspect of the life history of teleost fish. Swimming can also be considered a physiologically important behavior because it is intimately linked to the ability of fish to fully display many of the behaviors that are essential for their survival: reproductive (eg, migration, courtship), feeding, predator avoidance, environmental (eg, temperature, salinity, light, depth) preference, social behaviors, and so on. Depending on the intensity and duration, swimming can be viewed as a physical stressor capable of eliciting a significant stress response. Consequently, the swimming capacity, performance, and activities in fish are important determinant of fitness.

A majority of the swimming behaviors exhibited by wild fish, including feeding and reproductive migrations, foraging activities, and exploratory activities, can be classified as sustained swimming with limited evidence of a stress response. This type of swimming, unlike high intensity burst swimming, which can cause the rapid release of circulating stress hormones, defines swimming behaviors that can be maintained for a long period of time (ie, hours, days, months) before the fish suffers from exhaustion (Beamish, 1979) and exceeds homeostatic capabilities. Sustained swimming is aerobic; that is, it is fueled by the oxidation of nutrients that provide the necessary ATP for continued skeletal and cardiac muscle contractile activity. This also is in contrast to burst (ie, escape reactions, predation) or burst-and-glide swimming behaviors that are primarily supported by CrP and anaerobic glycolysis. In view of the aerobic component of sustained swimming, its energetic cost can be estimated by the amount of oxygen consumed (Tudorache et al., 2013). Sustained swimming represents swimming at a speed that can be maintained at a low metabolic (ie, aerobic) cost. Experimentally, the speed at which fish can maintain sustained swimming is that at which the energy spent on swimming a certain distance (ie, cost of transport) is the lowest and therefore most efficient. This swimming speed is referred to as the optimal swimming speed (U_{opt}) and can vary across species and across developmental stages within the same species. Interestingly, fish subjected to an aerobic swimming regime at U_{opt} show a clear adaptive response. For example, fish demonstrate increased growth rates, increased food conversion efficiency, increased hypertrophy and vascularization of skeletal muscle, and increased metabolic utilization of dietary nutrients.

Aerobic swimming regimes at U_{opt} may therefore be beneficial from a economic, physiological, and possibly from a behavioral perspective, and their potential for improving production, health, and welfare of fish in aquaculture is significant (Davison, 1997; Palstra and Planas, 2011). This section reviews available knowledge on the diverse effects of aerobic swimming training on the cardiovascular system; on feeding, digestion, and metabolism; on skeletal muscle growth; and finally, on the health and welfare of fish. A major goal of this section is to highlight phenotypic adaptations to swim training that are pertinent to stress responses and may provide mechanisms for enhanced performance capacity and stress reduction.

3.2. Effects of Swim Training on the Cardiovascular System

The plasticity of striated muscle is well established in vertebrates and exercise training can improve the convective O_2 transport system in both mammals and fishes. As a result, changes in the morphology and performance of the cardiovascular system occur in select species of fish and may improve swimming performance and alter subsequent responses to stressors. Exercise training in adult salmonid fishes induces small but significant increases in cardiac size, maximum cardiac output and whole animal O_2 consumption, hematocrit, arterial O_2 content, and tissue O_2 extraction while swimming (reviewed in Gamperl and Farrell, 2004). Swim training also causes increases in axial muscle capillarity (vascularization) in cyprinids (Sänger and Pötscher, 2000), rainbow trout (Davie et al., 1986), and zebrafish (*Danio rerio*) (Pelster et al., 2003), thereby reducing diffusion distance between erythrocytes and nearby myocytes. These adaptations can occur early in development and may improve tolerance to environmental stressors. For instance, endurance-trained zebrafish larvae consume less O_2 during swimming than untrained larvae, and trained larvae are more tolerant of hypoxia (Bagatto et al., 2001). Exercise training clearly has important effects on the heart at the cellular and molecular levels that underlie the hypertrophic and angiogenic response to increased cardiac activity (Ellison et al., 2012). However, despite evidence supporting a swimming-induced increase in cardiac growth (reviewed in Takle and Castro, 2013), there is very little evidence regarding the possible factors responsible for this effect. In a recent study, Castro et al. (2013a) reported on the stimulatory effects of swim training in Atlantic salmon (*Salmo salar*) on the expression of ventricular cardiomyocyte proliferation and hypertrophic markers (eg, PCNA, MEF2C, and ACTA1) as well as markers of angiogenesis (ie, VEGF, VEGF-R2, and erythropoietin). These results suggest that swim training in salmonids may induce hypertrophy and hyperplasia in the heart through the activation of a specific set of genes known in mammals to be involved in the regulation of cardiac growth. Cardiac growth, in turn, can potentially increase cardiac output and facilitate greater delivery of O_2 and nutrients during stressful periods. Based upon studies showing a positive correlation between myocardial enlargement and poststressor cortisol levels in salmonid fishes (Johansen et al., 2011), it is also possible that cortisol may be facilitating myocardial gene expression and remodeling during swim training. Ultimately, the identification of molecular targets and pathways for exercise-induced adaptations such as cardiac enlargement and increased vascularization in fishes will help improve our understanding of cardiovascular plasticity and beneficial responses to stressors.

3.3. Effects of Swim Training on Feeding and Energy Metabolism

Swimming is an energetically demanding activity and, consequently, it has a profound impact on nutritional and metabolic processes in fish. Swim training at sustained swimming has been reported to stimulate food intake and, most importantly, to stimulate food conversion efficiency in a number of fish species (reviewed in Davison, 1997; Magnoni et al., 2013b), contributing to its growth-promoting effects. Stress, on the other hand, is well known to reduce somatic growth in the aquaculture setting (Sadoul and Vijayan, 2016; Chapter 5 in this volume). Skeletal muscle is one of the most important tissues from a metabolic point of view in fish due to its large contribution to whole body mass and its activity during swimming. Previous reviews of studies performed predominantly using salmonid species (Magnoni et al., 2013b,c) indicate that sustained swimming promotes nutrient turnover by stimulating the entry and utilization of lipids and carbohydrates as fuel and also the retention of protein in skeletal muscle that is invested into growth and, particularly, into skeletal muscle mass. Recent studies have extended these observations on the stimulation of energy substrate utilization in skeletal muscle by sustained swimming to nonsalmonid species, such as sea bream (Martin-Perez et al., 2012; Felip et al., 2013), common carp (He et al., 2013), and qingbo (*Spinibarbus sinensis*) (Zhao et al., 2012). Several studies investigating the effects of sustained swimming on the transcriptome and proteome of skeletal muscle have begun to decipher the molecular basis of the metabolic adaptive changes in skeletal muscle in fish subjected to sustained swimming. Currently, there is evidence, both at the protein and at the gene expression levels in several fish species, that suggests that the increase in carbohydrate metabolism in skeletal muscle in response to swimming involves the stimulation of the GLUT4-mediated entry of glucose and its intracellular phosphorylation and metabolism through increased glycolysis (Martin-Perez et al., 2012; Magnoni et al., 2013a, 2014; Palstra et al., 2014). As a result, swimming-induced contractile activity has hypoglycemic effects (Felip et al.,

2012) that appear to be the result of increased GLUT4-dependent glucose uptake in skeletal muscle cells, as shown by the direct stimulation of the transcription of the GLUT4 gene (Marín-Juez et al., 2013) and of glucose uptake by electrical pulse stimulation in skeletal muscle cells in vitro (Magnoni et al., 2014). Interestingly, sustained swimming has been shown to improve the utilization and efficiency of diets containing high levels of carbohydrates, such as those used in aquaculture, and to improve glycemic control (Felip et al., 2012; Martin-Perez et al., 2012). For this reason, sustained swimming can be viewed as a metabolic promoter that can optimize the use of available diets for fish. Furthermore, in line with the known predominant role of lipids as fuel in skeletal muscle to support sustained swimming (Magnoni and Weber, 2007), there is evidence for increased expression of genes involved in lipoprotein catabolism (eg, lipoprotein lipase), fatty acid uptake (eg, CD36), and mitochondrial fatty acid transport (CPT1) to fuel beta-oxidation in skeletal muscle of exercised fish (Magnoni et al., 2013a, 2014; Palstra et al., 2014). These observations, coupled with the stimulation by swimming of the activity of fatty acid catabolic enzymes (eg, 3-OH Acyl CoA dehydrogenase) in skeletal muscle (Johnston and Moon, 1980a, b), strongly suggest that sustained swimming may stimulate lipid metabolism through β-oxidation and enhance the aerobic capacity of skeletal muscle. Whether the enhanced biochemical capability for aerobic metabolism in muscle of exercise-trained fish, per se, are adaptive by modifying individual stress and coping responses remains to be confirmed.

As stated earlier, an important metabolic consequence of sustained swimming in skeletal muscle is the increase in protein metabolism, resulting in net protein deposition due to increased protein synthesis resulting in a protein-sparing effect that preserves muscle proteins for contraction (Felip et al., 2012; Martin-Perez et al., 2012). Proteomic and transcriptomic studies on exercised rainbow trout, sea bream, and zebrafish have demonstrated that sustained swimming increases the expression of protein and gene families that participate in protein synthesis in skeletal muscle (Martin-Perez et al., 2012; Magnoni et al., 2013a; Palstra et al., 2013, 2014). As in mammals, the canonical signaling pathway IGF-1/PI3K/mTOR is one of the likely mechanisms responsible for the increase in protein synthesis by swimming in fish since it is increased by swimming in white muscle of zebrafish. This signaling pathway has also been associated with the increase of skeletal muscle mass by hypertrophy in exercised zebrafish (Palstra et al., 2014). Interestingly, transcriptomic studies of the exercised skeletal muscle in fish show that the increase in protein synthesis induced by swimming occurs concomitantly with an increase in protein degradation. Specifically, swimming elevates the expression of genes that participate in the ubiquitin-proteasome system, atrophy regulators, and proteolytic systems (eg, calpains, cathepsins)

(Magnoni et al., 2013a; Palstra et al., 2013, 2014) that are likely essential for tissue remodeling under growth conditions. Therefore, these observations provide convincing evidence that the high level of skeletal muscle protein turnover that is induced by sustained swimming that results in a net gain in protein deposition is the basis for swimming-induced growth in fish (see Section 3.4). By promoting higher protein deposition and skeletal muscle mass, swimming may also increase a key energy reserve for certain energy-demanding behaviors (ie, migration and sexual maturation), therefore increasing overall fitness of fish.

3.3.1. ENERGY SENSING MECHANISMS IN SKELETAL MUSCLE IN SWIMMING FISH: AMPK

Sadoul and Vijayan (2016; Chapter 5 in this volume), provides a valuable overview of energy sensors and mechanisms responsible for energy sensing during stress. Here we highlight a particular heterotrimeric enzyme, Adenosine monophosphate-dependent protein kinase (AMPK), which serves as both an energy sensor and metabolic master switch to maintain cellular energy homeostasis (Hardie and Sakamoto, 2006). The activation of AMPK occurs by reductions in cellular energy status (ie, increases in the AMP:ATP ratio), and therefore can be elicited by stressors that accelerate ATP use (eg, exercise, elevated temperature, and osmotic imbalances) or limit ATP synthesis (such as hypoxia) (Weber, 2011). AMPK serves as a glycogen sensor (McBride et al., 2009) and may further enhance the provision of energy during stressful events by inhibiting anabolic, energy-demanding processes, such as glycogen synthesis, gluconeogenesis, lipid synthesis, and protein synthesis (Weber, 2011). In skeletal muscle, contractile activity of muscle fibers is fueled by energy derived from ATP hydrolysis. One of the most important consequences of AMPK activation is to stimulate metabolic pathways that will regenerate ATP such as the uptake and oxidation of free fatty acids and glucose in skeletal muscle (Richter and Ruderman, 2009). In fish, few studies have addressed whether AMPK is a potential intracellular metabolic mediator of the effects of swimming-induced contractile activity of skeletal muscle fibers. Magnoni et al. (2012) reported on the identification and characterization of the heterotetrameric nature of AMPK in trout skeletal muscle cells and its activation by synthetic AMPK agonists such as AICAR and metformin. Furthermore, pharmaco-logical activation of AMPK resulted in increased GLUT4-mediated glucose uptake and utilization by trout skeletal muscle cells (Magnoni et al., 2012). Subsequently, it was demonstrated that AMPK activity in white and in red trout muscle fibers increased in response to sustained swimming and that AMPK activity was induced by electrical stimulation in cultured trout myotubes (Magnoni et al., 2014). This study also showed that AMPK activity is required for the stimulatory effects of electrical stimulation on glucose uptake and utilization. Interestingly, pharmacological activation of AMPK in trout muscle cells in culture increased the expression of the transcriptional regulator PGC-1α (Magnoni et al., 2012), known in mammals to induce mitochondrial biogenesis and stimulate oxidative metabolism, and therefore, ATP production (Lin et al., 2005). These studies clearly suggest that AMPK is an important factor in fish skeletal muscle that translates the activation of muscle fiber contraction into a metabolic adaptive response for energy homeostasis and sustained function. Thus, the importance of AMPK should also extend to stress responses that involve fish swimming activity and recovery, plus any stressor that challenges muscle energy production and performance.

In addition to turning on ATP regenerating mechanisms, activated AMPK turns off ATP consuming (ie, anabolic) pathways such as protein synthesis, mostly to preserve energy for skeletal muscle contraction (Hardie and Sakamoto, 2006). However, one of the problems with this model is that the reported increase in protein synthesis and mass of skeletal muscle in response to swimming (Felip et al., 2012) is difficult to reconcile with the increase in AMPK activity in skeletal muscle (Magnoni et al., 2014). How AMPK activation stimulates catabolic pathways to regenerate ATP while protein synthesis is increased under exercise conditions is a question that, unfortunately, has not been resolved yet, even in mammals.

3.4. Effects of Swim Training on Skeletal Muscle Growth

Swimming is both a function of the contractile capabilities of skeletal muscle and a physiological stimulus that exerts profound changes in skeletal muscle (Fig. 7.3). Vertebrate skeletal muscle is greatly influenced by its physiological environment and with increased usage it can change its contractile activity, excitability, metabolic activity, and mass (Egan and Zierath, 2013). In consequence, skeletal muscle is considered a tissue with unique plasticity and adaptability that may augment swimming capabilities and modulate behavioral responses to stressors. As indicated earlier (Section 2.3), it is widely believed that sustained swimming in fish is accomplished primarily by the contractile activity of the slow, or red muscle fibers, whereas burst swimming is accomplished primarily by the contractile activity of fast, or white muscle fibers. However, it is difficult to reconcile how sustained swimming, such as that performed during long reproductive or feeding migrations (eg, anguillids, salmonids, thunniforms) can be accomplished solely by the contractile activity of red muscle fibers. The notion that white muscle fibers are almost exclusively involved in high-intensity swimming (eliciting a pronounced stress response) has been challenged recently (Videler, 2011). In fact, there is evidence that white muscle fibers can also be recruited to contract during lower intensity sustained swimming (Moyes and West, 1995; Johnston, 1999; Coughlin, 2002).

3.4.1. AEROBIC SWIMMING AND ITS REGULATION OF SKELETAL MUSCLE MASS

Somatic growth is an integrated response that is regulated by multiple variables: feeding, gastrointestinal digestion and absorption, circulating and local hormones, and stress (Sadoul and Vijayan, 2016; Chapter 5 in this volume). Sustained swimming has been shown to promote growth in a number of fish species, particularly when fish are forced to swim at or close to their U_{opt} (Davison and Herbert, 2013). This is particularly true of species whose sustained swimming speed in their natural environment is actually close to the experimentally derived U_{opt}, such as most salmonid species. Swimming also promotes growth in several nonsalmonid species, including the striped bass (*Morone saxatilis*) (Young and Cech, 1993), matrinxa (*Brycon amazonicus*) (Arbelaez-Rojas and Moraes, 2009, 2010), yellowtail kingfish (*Seriola lalandi*) (Yogata and Oku, 2000; Brown et al., 2011), sea bream (Ibarz et al., 2011), qingbo (Li et al., 2010), pacu (*Piaractus mesopotamicus*) (Da Silva Nunes et al., 2013), and zebrafish (Pelster et al., 2003; van der Meulen et al., 2006; Palstra et al., 2010). In those fish species in which it has been investigated, swimming-induced growth is associated with hypertrophy of white and red skeletal muscle fibers (Johnston and Moon, 1980a, b; Davison, 1997; Bugeon et al., 2003), supporting the notion that white muscle fibers may indeed play an important role in sustained swimming. At a cellular level, hypertrophic changes in white muscle fibers due to swimming-induced activity include increases in morphometric parameters such as fiber cross-sectional area and fiber perimeter, resulting in a shift toward a higher abundance of larger fibers (Johnston and Moon, 1980a, b; Ibarz et al., 2011; Palstra et al., 2014) that will eventually lead to increased muscle mass. Therefore, it is believed that white fiber hypertrophy, given that white skeletal muscle constitutes more than 50% of the body mass in fish, represents one of the mechanisms underlying the growth-promoting effects of swimming in fish.

The mechanisms responsible for the stimulation of hypertrophic growth in white muscle fibers by swimming in fish have started to be deciphered with the use of transcriptomic approaches. Recent studies point out that skeletal muscle fiber hypertrophy in response to sustained swimming can be explained by the profound transcriptomic changes that characterizes the working skeletal muscle, as reported for adult rainbow trout and zebrafish (Magnoni et al., 2013a; Palstra et al., 2013, 2014). At a molecular level, swimming-induced contractile activity results in an upregulation of the expression of genes coding for structural and regulatory contractile elements as well as components of the extracellular matrix (Magnoni et al., 2013a; Palstra et al., 2014). Specifically, swimming results in the increased expression of genes involved in the activation of neuromuscular communication, excitation-contraction coupling, sarcomere contraction, cytoskeletal transmission of sarcomeric contractile force to the sarcolemma, and force transmission and muscle structure maintenance by the extracellular matrix (Magnoni et al., 2013a; Palstra et al., 2013, 2014). These studies suggest that sustained swimming promotes myofibrillogenesis in white muscle fibers and strongly support that hypertrophic muscle growth in fish subjected to sustained swimming is due, in great part, to increased accretion of myofibrillar proteins (Johnston et al., 2011). Support for this idea is derived from the observed increase in protein deposition in the trout white muscle in response to swimming (Felip et al., 2012; Section 3.3). Transcriptomic analyses of white skeletal muscle from exercised fish have also provided a first evidence of regulatory factors potentially involved in translating the contractile signal into an activated transcriptional program that leads to increased myofibrillogenesis. In this regard, the zebrafish has been particularly useful for the identification of factors that are potential regulators of skeletal muscle mass in fish and that are transcriptionally activated in response to swimming-induced contractile activity (Palstra et al., 2010, 2014). These studies indicate that hypertrophic growth occurs in white skeletal muscle in response to contractile activity induced by sustained swimming in teleost fish and that some of the muscle-derived factors potentially responsible for the stimulation of hypertrophic growth have been identified. Importantly, these factors could serve as potential markers of white skeletal muscle growth and function that could also be valuable indicators of muscle responses to stress.

To date, no conclusive stimulatory effects of swimming on skeletal muscle hyperplasia have been published. Several studies have reported lack of changes in fiber density or number in white muscle in response to swimming (Ibarz et al., 2011; Rasmussen et al., 2011; Palstra et al., 2014). However, there are also reports that suggest that swimming may stimulate the proliferation of skeletal muscle progenitor cells in zebrafish (van der Meulen et al., 2006; Palstra et al., 2014), cells that have been shown to retain their proliferative capacity in adulthood (Zhang and Anderson, 2014). Regardless of the mechanism of exercise-induced muscle growth, expansion of the contractile machinery could potentially enhance swimming capabilities of fish and have direct impacts on behaviors like feeding, territoriality, predator avoidance, migration, and reproduction.

3.4.2. ACQUISITION OF AN AEROBIC PHENOTYPE IN SKELETAL MUSCLE BY SWIM TRAINING

Physiological and biochemical adaptations that involve intracellular organelles, membrane channels, and expansion of the vascular network in skeletal muscle may provide helpful mechanisms to respond to stressors and enhance performance. Swim training protocols applying a swimming speed at, or close to, U_{opt} potentiate aerobic metabolism in skeletal muscle. This is particularly evident in red muscle, a tissue with high basal mitochondrial oxidative enzyme activities. Some studies have also shown that sustained swimming increases the activity and mRNA expression levels of citrate synthase (Johnston and Moon, 1980a, b; van der Meulen et al., 2006; LeMoine et al., 2010; Magnoni et al., 2013a) as well as mitochondrial density (Pelster et al., 2003). Interestingly, the oxidative capacity of white muscle also appears to be increased by sustained swimming, as shown by the increased expression of mitochondrial oxidative markers and myogenin (Johnston and Moon, 1980a, b; McClelland et al., 2006; van der Meulen et al., 2006; LeMoine et al., 2010; Magnoni et al., 2013a). Furthermore, the increased aerobic phenotype of red and white muscles in response to sustained swimming is accompanied by an increase in tissue vascularization (see next).

3.4.3. INCREASED AEROBIC PHENOTYPE IS ASSOCIATED WITH IMPROVED PERFORMANCE AFTER SWIM TRAINING

A relevant question that arises in view of the profound effects of swimming-induced contractile activity on skeletal muscle is whether swim training affects swimming performance. Several studies have addressed this question and have reported on a significant improvement of swimming performance in a variety of fish species subjected to swim training protocols applying sustained swimming speeds (Pearson et al., 1990; Davison, 1997; Holk and Lykkeboe, 1998; McDonald et al., 1998; Liu et al., 2009; Li et al., 2010; Anttila et al., 2011; Zhao et al., 2012). In Atlantic salmon, brown trout (*Salmo trutta*), and whitefish (*Coregonus lavaretus*), improved swimming performance was associated with increased density of dihydropyridine and ryanodine receptors, known to play important roles in the release of intracellular calcium in the excitation-contraction coupling mechanism, in both white and red muscle (Anttila et al., 2008, 2011). Furthermore, increased swimming performance has also been associated with an increase in the oxidative capacity of white and red muscle, as determined by the abundance and activity increases in mitochondrial enzymes (Farrell et al., 1991; Anttila et al., 2006, 2011; McClelland et al., 2006) as well as by the upregulation of oxidative pathways at the transcriptional level (McClelland et al., 2006; LeMoine et al., 2010; Magnoni et al., 2013a; Palstra et al., 2013, 2014). Importantly, swim training has been shown to increase capillarization in white and red muscles (Davie et al., 1986; Pelster et al., 2003; Ibarz et al., 2011; Palstra et al., 2014) that may result in improved gas and metabolite exchange between the blood and skeletal muscle. The aforementioned observations lead to the conclusion that swim training at sustained swimming speeds may induce adaptive responses in working muscle that increase aerobic capacity and diffusive exchange. Therefore, the corresponding improvement of swimming performance by swim training is likely the result of improved contractile activity of skeletal muscle, due in part to an increased generation of ATP via oxidative phosphorylation and enhanced cycling of intracellular calcium. Although not tested directly, the muscular adaptations following swim training could help determine the likelihood or magnitude of a stress response, or the energetics and recovery associated with a stress response.

3.5. Effects of Swim Training on Stress: Behavior, Health, and Welfare

One of the inherent problems of confining fish in captivity and stocking them in high densities (ie, aquaculture) is the disruption of their natural swimming behavior. Fish, under current aquaculture conditions, cannot display the same level of swimming activity than their wild counterparts; therefore, fish are raised under more sedentary conditions than in their natural environment. Alterations of swimming behavior imposed by confinement may represent deviations from natural behaviors, and therefore, could be considered a source of stress. Confinement and the reduction in physical activity associated with swimming may also deprive fish of the beneficial physiological effects of swimming, thereby constituting an additional source of stress. Although the practical implications of confinement-induced disruptions in swimming behavior are very important, the relationship between swimming activity and behavior, health, and welfare in fish and their implications in the stress response has received relatively little attention to date. However, a few studies have set out to test the hypothesis whether restoration of swimming behavior by swim training (ie, forced swimming) can improve the health and welfare status of fish.

3.5.1. EFFECTS OF SWIM TRAINING ON BEHAVIOR AND STRESS

Increased swimming activity in salmonid fish in response to increased water currents has been shown to alter the behavior of fish in captivity primarily by inducing schooling behavior, as reported in Arctic char and rainbow trout (Grünbaum et al., 2008; Larsen et al., 2012). The rheotactic response of fish to a moderate water current reduces spontaneous activity and promotes schooling. In addition to water currents, moving light stimuli are able to induce an optomotor response in fish, which has been shown to increase schooling behavior and swimming activity in caged Atlantic salmon (Herbert et al., 2011). The applications of inducing the optomotor response to manipulate the swimming behavior of fish in large sea cages, both by altering swimming speed and depth within the cages, are of potential value for large-scale production of actively swimming fish.

In addition to being metabolically more efficient (Herskin and Steffensen, 1998), swim training-induced schooling reduces aggressive behaviors in fish (Christiansen et al., 1989; Adams et al., 1995; Davison, 1997; Brannas, 2009), probably due to the energetic cost of swimming. Swim training disrupts hierarchical individual relationships that develop in low water flow conditions and reduces the incidence of aggressive interactions, particularly during feeding sessions (Brannas, 2009). Forced swimming and the induction of schooling in fish have also been shown to reduce stress, which is likely a consequence of decreased aggressive interactions. Swim training reduces the blood levels of cortisol (Woodward and Smith, 1985; Boesgaard et al., 1993; Postlethwaite and McDonald, 1995; Herbert et al., 2011; Arbelaez-Rojas et al., 2013) and accelerates the return of poststress cortisol levels to basal, unstressed levels (Veiseth et al., 2006). Thus, sustained swimming may reduce stress in fish and, at the same time, reduce the recovery time from stressors, thereby improving swimming and growth performance (Milligan et al., 2000; Veiseth et al., 2006; McKenzie et al., 2012) (Figs. 7.2 and 7.3). However, other studies have not shown a clear effect of swimming during metabolic recovery from an acute stressor (Kieffer et al., 2011). Although a stress-reducing effect of swimming is very plausible, differences in swim training protocols, species, nutritional status, and developmental stages may contribute to some of the conflicting reports in the literature.

3.5.2. EFFECTS ON SWIM TRAINING ON DISEASE RESISTANCE

Exercise in humans is known to confer beneficial effects for the prevention of a number of diseases, most importantly those of cardiovascular and metabolic origin. In fish, there have been efforts to examine the potential beneficial effects of swimming regarding their response to pathogens, particularly in view of the relatively high prevalence of diseases in fish under aquaculture practices. It is hypothesized that fish that are not being forced to swim at their sustained speed, together with the high densities employed in aquaculture, may have a higher susceptibility to diseases. This idea was supported by the observation that Atlantic salmon raised under sustained swimming conditions experienced a lower incidence

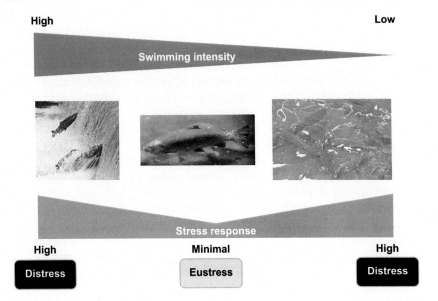

Figure 7.2. Conceptual model depicting the relationship between swimming intensity and stress responses under different environmental conditions. In salmonid fishes, stressful responses are realized at both high and low swimming intensities. Intermediate or optimal swimming speeds limit the stress response and may lead to several adaptations if swimming persists over extended periods.

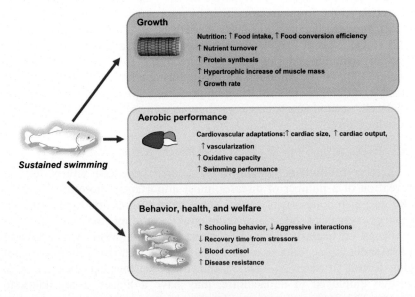

Figure 7.3. Summary of positive effects of sustained swimming on physiological and behavioral variables leading to enhanced performance and improved resistance to subsequent stressors.

of mortalities than resting fish (Totland et al., 1987). Along this line of thought, other researchers hypothesized that swimming under aerobic, sustained conditions would improve pathogen resistance. Two recent studies using Atlantic salmon have tested this hypothesis. In the first study, fish were subjected to either an interval or a continuous aerobic swim training protocol for 6 weeks and, after a 6-week detraining period, fish were challenged with the infectious pancreatic necrosis virus (IPNV). The results obtained showed that interval aerobic training improved resistance to IPNV over the untrained fish, although not significantly (Castro et al., 2011). In the second study, Atlantic salmon smolts were again subjected to a continuous training period (at three different swimming intensities) or an interval training period (at two different durations) for a total of 10 weeks prior to a challenge with IPNV (Castro et al., 2013b). In this study fish subjected to the medium intensity during swim training displayed the highest level of resistance to IPNV over any other group, including the unchallenged fish, but no statistically significant differences were detected. However, when fish were initially screened according to swimming performance and good versus poor swimmers were separated and subsequently challenged with IPNV, good swimmers showed significantly higher resistance to IPNV over poor swimmers (Castro et al., 2013b). The results of these studies show that swim training under sustained conditions tended to improve resistance to a lethal viral infection. Although no significant effects were observed, these results should stimulate more research in the use of swim training protocols to improve resistance to various diseases and define specific mechanisms of stress reduction.

Interestingly, more significant effects of swim training were observed when fish were separated according to their inherent swimming performance, suggesting that better swimmers may be more resistant to diseases. From a practical point of view, it may be possible to use swimming performance screens as a measure of robustness in the face of stressors to immune function in aquaculture. One particular aspect that deserves attention is whether swimming can modulate the immune system to improve immunocompetence and immunoresponsiveness in fish. To date, the possible activation of the innate or adaptive branches of the immune system by swimming in fish and improved resistance to a variety of biological and other stressors has not been investigated. Furthermore, there is no information on how the contractile muscle could play a part, if any, in potentiating the immune system in fish subjected to sustained swimming conditions. However, similar to exercised mammals (Wiendl et al., 2005), swimming-induced contractile activity results in the altered expression of genes coding for immune-related molecules in white and red skeletal muscle in fish, including members of the antigen-presenting machinery, complement factors, immunoglobulins, cytokines, chemokines, and their receptors (Magnoni et al., 2013a; Palstra et al., 2013, 2014). The clear immunological capabilities of the exercised skeletal muscle in fish provide an interesting possible scenario in which to further evaluate resistance to pathogens and other diverse stressors in trained fish or in fish with better swimming performance. In summary, chronic swimming activity can modulate the physiological response of the immune system to potential biological stressors and may ultimately define the tolerance capacity of fish when they are exposed to specific pathogens. Whether the reduction in circulating cortisol and lack of immunosuppression in exercised fish serve as relevant mediators for maintaining immunological capabilities requires confirmation.

3.5.3. Effects on Swim Training on Stress and Fish Welfare

As mentioned by Schreck and Tort (2016; Chapter 1 in this volume), the ambient environment—wild or artificial—is quite variable and stressful, with direct implications for the physiological response to stressors. The reported effects of swim training on the stimulation of swimming behavior, reduction of blood cortisol levels, improvement of recovery from acute stress, and potential improvement of resistance to pathogenic infections strongly suggest that swim training may reduce stress responses and improve fish welfare (Fig. 7.3). In a recent review, Huntingford and Kadri (2013) postulate that sustained swimming, in addition to improving the physiological status of fish, may improve behavioral processes related to stress by increasing schooling behavior, decreasing aggressive interactions, and also by promoting neurogenesis and the related learning capacity. Although improvements in neurogenesis and learning capacity by swimming in fish still need to be demonstrated, swimming has been shown to increase brain dopaminergic activity (Ortiz and Lutz, 1995). Other studies have also shown an association between increased dopaminergic activity and swimming activity (Johansson et al., 2004). Overall, the physiological and behavioral adaptations induced by sustained swimming appear to comply with a functional definition of welfare and stress reduction. However, the accurate determination of the sustained, optimal speed and its physiological and behavioral benefits for each species and developmental stage is a prerequisite for the meaningful application of swim training for reducing social stress, increasing growth, and improving disease resistance in aquaculture species.

3.6. Summary, Future Perspectives, and Key Unknowns

Although there is a significant amount of information on the physiological effects of swimming activity, no attempts have been made to establish a relationship between the variability of individual stress responses

(Winberg et al., 2016; Chapter 2 in this volume) and the physiological responses to induced swimming. Assuming a continuum of individual response to stressors, with proactive individuals in one extreme and reactive individuals in the other extreme of the spectrum, a relevant question is whether there is a swimming coping style based on swimming performance. Several studies suggest that there may be intrinsic individual differences in swimming performance in a fish population that allows researchers to distinguish poor versus good swimmers (Claireaux et al., 2005; Castro et al., 2013b). For example, Atlantic salmon that are classified as good swimmers, according to their performance in a swim test, have been reported to show higher resistance to viral infections than poor swimmers (Castro et al., 2013b), suggesting a possible link between individual swimming performance and immune capabilities. Further research is clearly needed to assess whether swim training at sustained speeds can promote the health status of fish and stress resistance through activation of the immune system.

There are additional benefits of sustained swimming in fish, most notably due to the increase in growth and health and welfare status and also to the decrease in aggressive interactions and in the stress response. As a result, it is clear that the implementation of sustained swimming practices would be beneficial for the cultivation of fish (ie, aquaculture). One of the most important limitations today for implementing swimming in aquaculture is in fact a technical one: how do we make fish swim at the desired speed in an industrial setting? Future research in this area is likely to establish the necessary conditions for the aquaculture industry to produce healthy and robust fish.

Finally, a potential area of interest is related to the effects of swimming on behavior and the neural processes responsible for memory acquisition and retrieval, and spatial and social learning. Stress is well known to impair learning and alter behavioral strategies (Noakes and Jones, 2016; Chapter 9 in this volume). Recent studies in mammals have linked exercise and cognitive function through the exercise-induced production of a muscle-derived factor (ie, irisin), which may act in the brain increasing expression of brain-derived neurotrophic factor (BDNF) (Wrann et al., 2013). Interestingly, the known antidepressant effects of exercise in mammals have been associated with the ability of contractile skeletal muscle to alter tryptophan metabolism in this tissue, resulting in an improvement of cognition and mood in exercised animals (Agudelo et al., 2014). These studies illustrate the chemical communication between skeletal muscle and brain in mammals, and provide important, unanswered questions for fish physiologists. Does swimming causes similar effects in the fish brain? Regarding the relationship between swimming, animal welfare, and stress reduction, does swimming provides a positive and rewarding experience for fish? Comparative studies that combine molecular, physiological, and behavioral measurements to sustained swimming and various stressors will ultimately increase our mechanistic understanding of swimming and its connections to stress responses of fish populations in the wild or in captivity.

REFERENCES

Adams, C. E., Huntingford, F. A., Krpal, J., Jobling, M. and Burnett, S. J. (1995). Exercise, agonistic behaviour and food acquisition in Arctic charr *Salvelinus alpinus*. *Environ. Biol. Fishes* **43**, 213–218.

Agudelo, L. Z., Femenia, T., Orhan, F., Porsmyr-Palmertz, M., Goiny, M., Martinez-Redondo, V., et al. (2014). Skeletal muscle PGC-1 alpha1 modulates kynurenine metabolism and mediates resilience to stress-induced depression. *Cell* **159**, 33–45.

Anras, M.-L. B. and Lagardère, J. P. (2004). Measuring cultured fish swimming behaviour: first results on rainbow trout using acoustic telemetry in tanks. *Aquaculture* **240**, 175–186.

Anttila, K., Järvilehto, M. and Mänttäri, S. (2008). The swimming performance of brown trout and whitefish: the effects of exercise on Ca^{2+} handling and oxidative capacity of swimming muscles. *J. Comp. Physiol. B* **178**, 465–475.

Anttila, K., Jokikokko, E., Erkinaro, J., Jaervilehto, M. and Maenttaeri, S. (2011). Effects of training on functional variables of muscles in reared Atlantic salmon *Salmo salar* smolts: connection to downstream migration pattern. *J. Fish. Biol.* **78**, 552–566.

Anttila, K., Mänttäri, S. and Järvilehto, M. (2006). Effects of different training protocols on Ca^{2+} handling and oxidative capacity in skeletal muscle of Atlantic salmon (*Salmo salar* L.). *J. Exp. Biol.* **209**, 2971–2978.

Arbelaez-Rojas, G. A., Hackbarth, A., Inoue, L. A. K. A., Dias de Moraes, F. and Moraes, G. (2013). Sustained swimming mitigates stress in juvenile fish *Brycon amazonicus*. *J. Appl. Aquac.* **25**, 271–281.

Arbelaez-Rojas, G. A. and Moraes, G. (2009). Sustained swimming and stocking density iteration in the performance and body composition of matrinxa *Brycon amazonicus* juveniles. *Cienc. Rural* **39**, 201–208.

Arbelaez-Rojas, G. A. and Moraes, G. (2010). Optimization of sustaining swimming speed of matrinxa *Brycon amazonicus*: performance and adaptive aspects. *Sci. Agr.* **67**, 253–258.

Axelsson, M., Driedzic, W. R., Farrell, A. P. and Nilsson, S. (1989). Regulation of cardiac-output and gut blood-flow in the sea raven, *Hemitripterus americanus*. *Fish Physiol. Biochem.* **6**, 315–326.

Bagatto, B., Pelster, B. and Burggren, W. W. (2001). Growth and metabolism of larval zebrafish: effects of swim training. *J. Exp. Biol.* **204**, 4335–4343.

Barton, B. A. and Schreck, C. B. (1987). Metabolic cost of acute physical stress in juvenile steelhead. *Trans. Am. Fish. Soc.* **116**, 257–263.

Beamish, F. W. H. (1979). Swimming capacity. In *Fish Physioogy - Locomotion*, Vol. 7 (eds. W. S. Hoar and D. J. Randall), pp. 101–187. New York, NY: Academic Press.

Bennett, A. F. (1991). The evolution of activity capacity. *J. Exp. Biol.* **160**, 1–23.

Black, E. C. (1958). Hyperactivity as a lethal factor in fish. *J. Fish. Res. Bd. Can.* **15**, 573–586.

Boesgaard, L., Nielsen, E. N. and Rosenkilde, P. (1993). Moderate exercise decreases plasma cortisol levels in Atlantic salmon (*Salmo salar*). *Comp. Biochem. Physiol. A* **106**, 641–643.

Brannas, E. (2009). The effect of moderate exercise on growth and aggression depending on social rank in groups of Arctic charr (*Salvelinus alpinus* L.). *Appl. Anim. Behav. Sci.* **119**, 115–119.

Brauner, C. J., Thorarensen, H., Gallaugher, P., Farrell, A. P. and Randall, D. J. (2000). The interactions between O_2 and CO_2 exchange in rainbow trout during graded sustained exercise. *Resp. Physiol.* **119**, 83–96.

Brett, J. R. (1964). The respiratory metabolism and swimming performance of young sockeye salmon. *J. Fish. Res. Bd. Can.* **21**, 1183–1226.

Brett, J. R. (1972). Metabolic demand for oxygen in fish, particularly salmonids, and a comparison with other vertebrates. *Resp. Physiol.* **14**, 151–170.

Brown, E. J., Bruce, M., Pether, S. and Herbert, N. A. (2011). Do swimming fish always grow fast? Investigating the magnitude and physiological basis of exercise-induced growth in juvenile New Zealand yellowtail kingfish, *Seriola lalandi. Fish Physiol. Biochem.* **37**, 327–336.

Bugeon, J., Lefevre, F. and Fauconneau, B. (2003). Fillet texture and muscle structure in brown trout (*Salmo trutta*). *Aquac. Res.* **34**, 1287–1295.

Burnett, N. J., Hinch, S. G., Braun, D. C., Casselman, M. T., Middleton, C. T., Wilson, S. M., et al. (2014). Burst swimming in areas of high flow: delayed consequences of anaerobiosis in wild adult sockeye salmon. *Physiol. Biochem. Zool.* **87**, 587–598.

Castro, V., Grisdale-Helland, B., Helland, S. J., Kristensen, T., Joergensen, S. M., Helgerud, J., et al. (2011). Aerobic training stimulates growth and promotes disease resistance in Atlantic salmon (*Salmo salar*). *Comp. Biochem. Physiol. A* **160**, 278–290.

Castro, V., Grisdale-Helland, B., Helland, S. J., Torgersen, J., Kristensen, T., Claireaux, G., et al. (2013a). Cardiac molecular-acclimation mechanisms in response to swimming-induced exercise in Atlantic salmon. *PLoS One* **8**, e55056.

Castro, V., Grisdale-Helland, B., Jorgensen, S. M., Helgerud, J., Claireaux, G., Farrell, A. P., et al. (2013b). Disease resistance is related to inherent swimming performance in Atlantic salmon. *BMC Physiol.* **13**, 1.

Chandroo, K. P., Cooke, S. J., McKinley, R. S. and Moccia, R. D. (2005). Use of electromyogram telemetry to assess the behavioural and energetic responses of rainbow trout, *Oncorhynchus mykiss* (Walbaum) to transportation stress. *Aquac. Res.* **36**, 1226–1238.

Christiansen, J. S., Ringo, E. and Jobling, M. (1989). Effects of sustained exercise on growth and body composition of first-feeding fry of Arctic charr, *Salvelinus alpinus* L. *Aquaculture* **79**, 329–335.

Claireaux, G., McKenzie, D. J., Genge, A. G., Chatelier, A., Aubin, J. and Farrell, A. P. (2005). Linking swimming performance, cardiac pumping ability and cardiac anatomy in rainbow trout. *J. Exp. Biol.* **208**, 1775–1784.

Coughlin, D. J. (2002). Aerobic muscle function during steady swimming in fish. *Fish Fisher.* **3**, 63–78.

Da Silva Nunes, C., Moraes, G., Fabrizzi, F., Hackbarth, A. and Arbelaez-Rojas, G. A. (2013). Growth and hematology of pacu subjected to sustained swimming and fed different protein levels. *Pesq. Agropec. Bras.* **48**, 645–650.

Davie, P. S., Wells, R. M. and Tetens, V. (1986). Effects of sustained swimming on rainbow trout muscle structure, blood oxygen transport, and lactate dehydrogenase isozymes: evidence for increased aerobic capacity of white muscle. *J. Exp. Zool.* **237**, 159–171.

Davis, L. E. and Schreck, C. B. (1997). The energetic response to handling stress in juvenile coho salmon. *Trans. Am. Fish. Soc.* **126**, 248–258.

Davison, W. (1997). The effects of exercise training on teleost fish, a review of recent literature. *Comp. Biochem. Physiol. A* **117**, 67–75.

Davison, W. and Herbert, N. A. (2013). Swimming-enhanced growth. In *Swimming Physiology of Fish* (eds. A. P. Palstra and J. V. Planas), pp. 177–202. Heidelberg: Springer.

De Boeck, G., Alsop, D. and Wood, C. (2001). Cortisol effects on aerobic and anaerobic metabolism, nitrogen excretion, and whole-body composition in juvenile rainbow trout. *Physiol. Biochem. Zool.* **74**, 858–868.

Driedzic, W. R. and Gesser, H. (1994). Energy metabolism and contractility in ectothermic vertebrate hearts - hypoxia, acidosis, and low temperature. *Physiol. Rev.* **74**, 221–258.

Driedzic, W. R. and Hochachka, P. W. (1978). Energy metabolism during exercise. In *Fish Physiology - Locomotion*, Vol. 7 (eds. W. S. Hoar and D. J. Randall), pp. 503–543. New York, NY: Academic Press.

Egan, B. and Zierath, J. R. (2013). Exercise metabolism and the molecular regulation of skeletal muscle adaptation. *Cell Metab.* **17**, 162–184.

Ekström, A., Jutfelt, F. and Sandblom, E. (2014). Effects of autonomic blockade on acute thermal tolerance and cardioventilatory performance in rainbow trout, *Oncorhynchus mykiss*. *J. Therm. Biol.* **44**, 47–54.

Ellison, G. M., Waring, C. D., Vicinanza, C. and Torella, D. (2012). Physiological cardiac remodelling in response to endurance exercise training: cellular and molecular mechanisms. *Heart* **98**, 5–10.

Fabbri, E., Capuzzo, A. and Moon, T. W. (1998). The role of circulating catecholamines in the regulation of fish metabolism: an overview. *Comp. Biochem. Physiol. C* **120**, 177–192.

Farrar, R. S. and Rodnick, K. J. (2004). Sex-dependent effects of gonadal steroids and cortisol on cardiac contractility in rainbow trout. *J. Exp. Biol.* **207**, 2083–2093.

Farrell, A. P. (1980). Vascular pathways in the gill of ling cod, *Ophiodon elongatus*. *Can. J. Zool.* **58**, 796–806.

Farrell, A. P. (1982). Cardiovascular changes in the unanaesthetized lingcod (*Ophiodon elongatus*) during short-term progressive hypoxia and spontaneous activity. *Can. J. Zool.* **60**, 933–941.

Farrell, A. P. (2007). Cardiorespiratory performance during prolonged swimming tests with salmonids: a perspective on temperature effects and potential analytical pitfalls. *Philos. T. Roy. Soc. B* **362**, 2017–2030.

Farrell, A. P., Gamperl, A. K., Hicks, J. M. T., Shiels, H. A. and Jain, K. E. (1996). Maximum cardiac performance of rainbow trout (*Oncorhynchus mykiss*) at temperatures approaching their upper lethal limit. *J. Exp. Biol.* **199**, 663–672.

Farrell, A. P., Johansen, J. A. and Suarez, R. K. (1991). Effects of exercise-training on cardiac performance and muscle enzymes in rainbow trout, *Oncorhynchus mykiss*. *Fish Physiol. Biochem.* **9**, 303–312.

Farrell, A. P. and Jones, D. R. (1992). The heart. In *Fish Physiology - The Cardiovascular System*, Vol. 12A (eds. W. S. Hoar, D. J. Randall and A. P. Farrell), pp. 1–88. San Diego, CA: Academic Press.

Farrell, A. P., Macleod, K. R. and Chancey, B. (1986). Intrinsic mechanical-properties of the perfused rainbow-trout heart and the effects of catecholamines and extracellular calcium under control and acidotic conditions. *J. Exp. Biol.* **125**, 319–345.

Farrell, A. P., Sobin, S. S., Randall, D. J. and Crosby, S. (1980). Intralamellar blood-flow patterns in fish gills. *Am. J. Physiol.* **239**, R428–R436.

Felip, O., Blasco, J., Ibarz, A., Martin-Perez, M. and Fernandez-Borras, J. (2013). Beneficial effects of sustained activity on the use of dietary protein and carbohydrate traced with stable isotopes ^{15}N and ^{13}C in gilthead sea bream (*Sparus aurata*). *J. Comp. Physiol. B* **183**, 223–234.

Felip, O., Ibarz, A., Fernández-Borrás, J., Beltrán, M., Martín-Pérez, M., Planas, J. V., et al. (2012). Tracing metabolic routes of dietary carbohydrate and protein in rainbow trout (*Oncorhynchus mykiss*) using stable isotopes ([^{13}C]starch and [^{15}N]protein): effects of gelatinisation of starches and sustained swimming. *Br. J. Nutr.* **107**, 834–844.

Fick, A. (1870). Über die Messung den Blutquantums in der Herzventrikeln. *Sitzungb. Phys. Med. Ges. Würzburg* **16**.

Fry, F. E. J. (1947). Effects of environment on animal activity. *Univ. Toronto Stud. Biol. Ser.* **55**, 1–62.

Fry, F. E. J. (1971). The effect of environmental factors on the physiology of fish. In *Fish Physiology - Environmental Relations and Behavior*, Vol. 6 (eds. W. S. Hoar and D. J. Randall), pp. 1–98. New York, NY: Academic Press.

Gamperl, A. K. and Farrell, A. P. (2004). Cardiac plasticity in fishes: environmental influences and interspecific differences. *J. Exp. Biol.* **207**, 2539–2550.

Gerry, S. P. and Ellerby, D. J. (2014). Resolving shifting patterns of muscle energy use in swimming fish. *PLoS One* **9**, e106030.

Gladden, L. B. (2004). Lactate metabolism: a new paradigm for the third millennium. *J. Physiol. Lond.* **558**, 5–30.

Gorissen, M. and Flik, G. (2016). Endocrinology of the Stress Response in Fish. In *Fish Physiology - Biology of Stress in Fish*, Vol. 35 (eds. C. B. Schreck, L. Tort, A. P. Farrell and C. J. Brauner), San Diego, CA: Academic Press.

Gregory, T. R. and Wood, C. M. (1999). The effects of chronic plasma cortisol elevation on the feeding behaviour, growth, competitive ability, and swimming performance of juvenile rainbow trout. *Physiol. Biochem. Zool.* **72**, 286–295.

Grünbaum, T., Cloutier, R. and Le François, N. R. (2008). Positive effects of exposure to increased water velocity on growth of newly hatched Arctic charr, *Salvelinus alpinus* L. *Aquac. Res.* **39**, 106–110.

Hardie, D. and Sakamoto, K. (2006). AMPK: a key sensor of fuel and energy status in skeletal muscle. *Physiology* **21**, 48–60.

Harper, A. A., Newton, I. P. and Watt, P. W. (1995). The effect of temperature on spontaneous action-potential discharge of the isolated sinus venosus from winter and summer plaice (*Pleuronectes platessa*). *J. Exp. Biol.* **198**, 137–140.

He, W., Xia, W., Cao, Z.-D. and Fu, S.-J. (2013). The effect of prolonged exercise training on swimming performance and the underlying biochemical mechanisms in juvenile common carp (*Cyprinus carpio*). *Comp. Biochem. Physiol. A* **166**, 308–315.

Herbert, N. A., Kadri, S. and Huntingford, F. A. (2011). A moving light stimulus elicits a sustained swimming response in farmed Atlantic salmon, *Salmo salar* L. *Fish Physiol. Biochem.* **37**, 317–325.

Herskin, J. and Steffensen, J. F. (1998). Energy savings in sea bass swimming in a school: measurements of tail beat frequency and oxygen consumption at different swimming speeds. *J. Fish. Biol.* **53**, 366–376.

Hicks, J. W. and Bennett, A. F. (2004). Eat and run: prioritization of oxygen delivery during elevated metabolic states. *Resp. Physiol. Neurobiol.* **144**, 215–224.

Hochachka, P. W. (1994). *Muscles as Molecular and Metabolic Machines*. Boca Raton, FL: CRC Press.

Hochachka, P. W. and Somero, G. N. (2002). Biochemical Adaptation. Mechanisms and Process in Physiological Evolution. New York, NY: Oxford University Press.

Holk, K. and Lykkeboe, G. (1998). The impact of endurance training on arterial plasma K$^+$ levels and swimming performance of rainbow trout. *J. Exp. Biol.* **201**, 1373–1380.

Hughes, G. M., Le-Bras-Pennec, Y. and Pennec, J.-P. (1988). Relationships between swimming speed, oxygen consumption, plasma catecholamines and heart performance in rainbow trout (S. gairdneri R.). Exp. Biol. **48**, 45–49.

Huntingford, F. A. and Kadri, S. (2013). Exercise, stress and welfare. In Swimming Physiology of Fish (eds. A. P. Palstra and J. V. Planas), pp. 161–176. Heidelberg: Springer.

Ibarz, A., Felip, O., Fernández-Borràs, J., Martín-Pérez, M., Blasco, J. and Torrella, J. (2011). Sustained swimming improves muscle growth and cellularity in gilthead sea bream. J. Comp. Physiol. B **181**, 209–217.

Iversen, M., Finstad, B. and Nilssen, K. J. (1998). Recovery from loading and transport stress in Atlantic salmon (Salmo salar L.) smolts. Aquaculture **168**, 387–394.

Johansen, I. B., Lunde, I. G., Rosjo, H., Chrisyensen, G., Nilsson, G. E., Bakken, M., et al. (2011). Cortisol response to stress is associated with myocardial remodeling in salmonid fishes. J. Exp. Biol. **214**, 1313–1321.

Johansson, V., Winberg, S., Jonsson, E., Hall, D. and Bjornsson, B. T. (2004). Peripherally administered growth hormone increases brain dopaminergic activity and swimming in rainbow trout. Horm. Behav. **46**, 436–443.

Johnson, T. P., Syme, D. A., Jayne, B. C., Lauder, G. V. and Bennett, A. F. (1994). Modeling red muscle power output during steady and unsteady swimming in largemouth bass. Am. J. Physiol. **267**, R481–R488.

Johnston, I. A. (1999). Muscle development and growth: potential implications for flesh quality in fish. Aquaculture **177**, 99–115.

Johnston, I. A., Bower, N. I. and Macqueen, D. J. (2011). Growth and the regulation of myotomal muscle mass in teleost fish. J. Exp. Biol. **214**, 1617–1628.

Johnston, I. A. and Goldspink, G. (1973). Study of swimming performance of crucian carp Carassius carassius (L.) in relation to effects of exercise and recovery on biochemical changes in myotomal muscles and liver. J. Fish. Biol. **5**, 249–260.

Johnston, I. A. and Moon, T. W. (1980a). Endurance exercise training in the fast and slow muscles of a teleost fish (Pollachius virens). J. Comp. Physiol. B **135**, 147–156.

Johnston, I. A. and Moon, T. W. (1980b). Exercise training in skeletal muscle of brook trout (Salvelinus fontinalis). J. Exp. Biol. **87**, 177–194.

Jones, D. R. and Randall, D. J. (1978). The respiratory and circulatory systems during exercise. In Fish Physiology - Bioenergetics and Growth, Vol. 8 (eds. W. S. Hoar and D. J. Randall), pp. 425–501. New York, NY: Academic Press.

Jorgensen, E. H. and Jobling, M. (1993). The effect of exercise on growth, food utilization and osmoregulatory capacity of juvenile Atlantic salmon, Salmo salar. Aquaculture **116**, 233–246.

Kiceniuk, J. W. and Jones, D. R. (1977). Oxygen transport system in trout (Salmo gairdneri) during sustained exercise. J. Exp. Biol. **69**, 247–260.

Kieffer, J. D. (2000). Limits to exhaustive exercise in fish. Comp. Biochem. Physiol. A **126**, 161–179.

Kieffer, J. D., Kassie, R. S. and Taylor, S. G. (2011). The effects of low-speed swimming following exhaustive exercise on metabolic recovery and swimming performance in brook trout (Salvelinus fontinalis). Physiol. Biochem. Zool. **84**, 385–393.

Korovin, V. A., Zybin, A. S. and Legomin, V. B. (1982). Response of pond fish young to stress factors caused by transfers in geo thermal fish farming. Vopr. Ikhtiol. **22**, 280–284.

Krohn, M. M. and Boisclair, D. (1994). Use of a stereo video system to estimate the energy expenditure of free-swimming fish. Can. J. Fish. Aquat. Sci. **51**, 1119–1127.

Larsen, B. K., Skov, P. V., McKenzie, D. J. and Jokumsen, A. (2012). The effects of stocking density and low level sustained exercise on the energetic efficiency of rainbow trout (Oncorhynchus mykiss) reared at 19°C. Aquaculture **324**, 226–233.

Lee, C. G., Farrell, A. P., Lotto, A., Hinch, S. G. and Healey, M. C. (2003). Excess post-exercise oxygen consumption in adult sockeye (*Oncorhynchus nerka*) and coho (*O. kisutch*) salmon following critical speed swimming. *J. Exp. Biol.* **206**, 3253–3260.

Lee-Jenkins, S. S. Y., Binder, T. R., Karch, A. P. and McDonald, D. G. (2007). The recovery of locomotory activity following exhaustive exercise in juvenile rainbow trout (*Oncorhynchus mykiss*). *Physiol. Biochem. Zool.* **80**, 88–98.

LeMoine, C. M. R., Craig, P. M., Dhekney, K., Kim, J. J. and McClelland, G. B. (2010). Temporal and spatial patterns of gene expression in skeletal muscles in response to swim training in adult zebrafish (*Danio rerio*). *J. Comp. Physiol. B* **180**, 151–160.

Li, X.-M., Cao, Z.-D., Peng, J.-L. and Fu, S.-J. (2010). The effect of exercise training on the metabolic interaction between digestion and locomotion in juvenile darkbarbel catfish (*Peltebagrus vachelli*). *Comp. Biochem. Physiol. A* **156**, 67–73.

Liew, H. J., Chiarella, D., Pelle, A., Faggio, C., Blust, R. and De Boeck, G. (2013). Cortisol emphasizes the metabolic strategies employed by common carp, *Cyprinus carpio* at different feeding and swimming regimes. *Comp. Biochem. Physiol. A* **166**, 449–464.

Lin, J., Handschin, C. and Spiegelman, B. M. (2005). Metabolic control through the PGC-1 family of transcription coactivators. *Cell Metab.* **1**, 361–370.

Liu, Y., Cao, Z. D., Fu, S.-J., Peng, J.-L. and Wang, Y. X. (2009). The effect of exhaustive chasing training and detraining on swimming performance in juvenile darkbarbel catfish (*Peltebagrus vachelli*). *J. Comp. Physiol. B* **179**, 847–855.

Magnoni, L. J., Crespo, D., Ibarz, A., Blasco, J., Fernández-Borràs, J. and Planas, J. V. (2013a). Effects of sustained swimming on the red and white muscle transcriptome of rainbow trout (*Oncorhynchus mykiss*) fed a carbohydrate-rich diet. *Comp. Biochem. Physiol. A* **166**, 1–12.

Magnoni, L. J., Felip, O., Blasco, J. and Planas, J. V. (2013b). Metabolic fuel utilization during swimming: optimizing nutritional requirements for enhanced performance. In *Swimming Physiology of Fish* (eds. A. P. Palstra and J. V. Planas), pp. 203–235. Heidelberg: Springer.

Magnoni, L. J., Marquez-Ruiz, P., Palstra, A. P. and Planas, J. V. (2013c). Physiological consequences of swimming-induced activity in trout. In *Trout: From Physiology to Conservation* (eds. S. Polakof and T. W. Moon), pp. 321–350. New York, NY: Nova Science Publishers.

Magnoni, L. J., Palstra, A. P. and Planas, J. V. (2014). Fueling the engine: induction of AMP-activated protein kinase in trout skeletal muscle by swimming. *J. Exp. Biol.* **217**, 1649–1652.

Magnoni, L. J., Vraskou, Y., Palstra, A. P. and Planas, J. V. (2012). AMP-activated protein kinase plays an important evolutionary conserved role in the regulation of glucose metabolism in fish skeletal muscle cells. *PLoS One* **7**, e31219.

Magnoni, L. and Weber, J. M. (2007). Endurance swimming activates trout lipoprotein lipase: plasma lipids as a fuel for muscle. *J. Exp. Biol.* **210**, 4016–4023.

Manoli, I., Alesci, S., Blackman, M. R., Su, Y. A., Rennert, O. M. and Chrousos, G. P. (2007). Mitochondria as key components of the stress response. *Trends. Endocrinol. Metab.* **18**, 190–198.

Marín-Juez, R., Díaz, M., Morata, J. and Planas, J. V. (2013). Mechanisms regulating GLUT4 transcription in skeletal muscle cells are highly conserved across vertebrates. *PLoS One* **8**, e80628.

Martin-Perez, M., Fernandez-Borras, J., Ibarz, A., Millan-Cubillo, A., Felip, O., de Oliveira, E., et al. (2012). New insights into fish swimming: a proteomic and isotopic approach in gilthead sea bream. *J. Proteome. Res.* **11**, 3533–3547.

McBride, A., Ghilagaber, S., Nikolaev, A. and Hardie, D. G. (2009). The glycogen-binding domain on the AMPK beta subunit allows the kinase to act as a glycogen sensor. *Cell Metab.* **9**, 23–34.

McClelland, G. B., Craig, P. M., Dhekney, K. and Dipardo, S. (2006). Temperature- and exercise-induced gene expression and metabolic enzyme changes in skeletal muscle of adult zebrafish (*Danio rerio*). *J. Physiol.* **577**, 739–751.

McCormick, S. D. (1995). Hormonal control of gill Na$^+$, K$^+$-ATPase and chloride cell function. In *Cellular and Molecular Approaches to Fish Ionic Regulation* (eds. C. M. Wood and T. J. Shuttleworth), pp. 285–315. San Diego, CA: Academic Press.

McCormick, S. D., Moyes, C. D. and Ballantyne, J. S. (1989). Influence of salinity on the energetics of gill and kidney of Atlantic salmon (*Salmo salar*). *Fish. Physiol. Biochem.* **6**, 243–254.

McDonald, D. G., Milligan, C. L., McFarlane, W. J., Croke, S., Currie, S., Hooke, B., et al. (1998). Condition and performance of juvenile Atlantic salmon (*Salmo salar*): effects of rearing practices on hatchery fish and comparison with wild fish. *Can. J. Fish. Aquat. Sci.* **55**, 1208–1219.

McKenzie, D. J., Hoglund, E., Dupont-Prinet, A., Larsen, B. K., Skov, P. V., Pedersen, P. B., et al. (2012). Effects of stocking density and sustained aerobic exercise on growth, energetics and welfare of rainbow trout. *Aquaculture* **338**, 216–222.

Meyer, W. F. and Cook, P. A. (1996). An assessment of the use of low-level aerobic swimming in promoting recovery from handling stress in rainbow trout. *Aquac. Int.* **4**, 169–171.

Milligan, C. L. (1996). Metabolic recovery from exhaustive exercise in rainbow trout. *Comp. Biochem. Physiol. A* **113**, 51–60.

Milligan, C. L. (2003). A regulatory role for cortisol in muscle glycogen metabolism in rainbow trout *Oncorhynchus mykiss* Walbaum. *J. Exp. Biol.* **206**, 3167–3173.

Milligan, C. L., Hooke, G. B. and Johnson, C. (2000). Sustained swimming at low velocity following a bout of exhaustive exercise enhances metabolic recovery in rainbow trout. *J. Exp. Biol.* **203**, 921–926.

Mommsen, T. P. (1984). Metabolism of the fish gill. In *Fish Physiology - Gills: Ion and Water Transfer*, Vol. 10B (eds. W. S. Hoar and D. J. Randall), pp. 203–238. San Diego, CA: Academic Press.

Mommsen, T. P. and Hochachka, P. W. (1988). The purine nucleotide cycle as 2 temporally separated metabolic units - a study on trout muscle. *Metabolism* **37**, 552–556.

Mommsen, T. P., Vijayan, M. M. and Moon, T. W. (1999). Cortisol in teleosts: dynamics, mechanisms of action, and metabolic regulation. *Rev. Fish Biol. Fish.* **9**, 211–268.

Morgan, J. D. and Iwama, G. K. (1996). Cortisol-induced changes in oxygen consumption and ionic regulation in coastal cutthroat trout (*Oncorhynchus clarki clarki*) parr. *Fish Physiol. Biochem.* **15**, 385–394.

Morgan, J. D. and Iwama, G. K. (1999). Energy cost of NaCl transport in isolated gills of cutthroat trout. *Am. J. Physiol.* **277**, R631–R639.

Moyes, C. D. and West, T. G. (1995). Exercise metabolism of fish. In *Biochemistry and Molecular Biology of Fishes*, vol. 4 (eds. P. W. Hochachka and T. P. Mommsen), pp. 368–392. Amsterdam: Elseiver Science.

Newsholme, E. A. and Leech, A. R. (1983). *Biochemistry for the Medical Sciences*. Chichester: Wiley and Sons.

Nielsen, M. E., Boesgaard, L., Sweeting, R. M., McKeown, B. A. and Rosenkilde, P. (1994). Plasma levels of lactate, potassium, glucose, cortisol, growth hormone and triiodo-L-thyronine in rainbow-trout (*Oncorhynchus mykiss*) during exercise at various levels for 24 h. *Can. J. Zool.* **72**, 1643–1647.

Nikinmaa, M. (1992). How does environmental pollution affect red cell function in fish. *Aquat. Toxicol.* **22**, 227–238.

Nilsson, S. (1983). *Autonomic Nerve Function in Vertebrates*. Berlin: Springer Verlag.

Nilsson, S. and Sundin, L. (1998). Gill blood flow control. *Comp. Biochem. Physiol. A* **119**, 137–147.

Noakes, D. L. G. and Jones, K. M. M. (2016). Cognition, Learning, and Behavior. In *Fish Physiology - Biology of Stress in Fish*, Vol. 35 (eds. C. B. Schreck, L. Tort, A. P. Farrell and C. J. Brauner), San Diego, CA: Academic Press.

Ortiz, M. and Lutz, P. L. (1995). Brain neurotransmitter changes associated with exercise and stress in a teleost fish (*Sciaenops ocellatus*). *J. Fish. Biol.* **46**, 551–562.

Pagnotta, A., Brooks, L. and Milligan, L. (1994). The potential regulatory roles of cortisol in recovery from exhaustive exercise in rainbow trout. *Can. J. Zool.* **72**, 2136–2146.

Palstra, A. P., Beltran, S., Burgerhout, E., Brittijn, S. A., Magnoni, L. J., Henkel, C. V., et al. (2013). Deep RNA sequencing of the skeletal muscle transcriptome in swimming fish. *PLoS One* **8**, e53171.

Palstra, A. P. and Planas, J. V. (2011). Fish under exercise. *Fish Physiol. Biochem.* **37**, 259–272.

Palstra, A. P. and Planas, J. V. (2013). Swimming Physiology of Fish: Towards Using Exercise to Farm a Fit Fish in Sustainable Aquaculture. Heidelberg: Springer.

Palstra, A. P., Rovira, M., Rizo-Roca, D., Torrella, J. R., Spaink, H. P. and Planas, J. V. (2014). Swimming-induced exercise promotes hypertrophy and vascularization of fast skeletal muscle fibres and activation of myogenic and angiogenic transcriptional programs in adult zebrafish. *BMC Genomics* **15**, 1136.

Palstra, A. P., Tudorache, C., Rovira, M., Brittijn, S. A., Burgerhout, E., Van den Thillart, G. E., et al. (2010). Establishing zebrafish as a novel exercise model: swimming economy, swimming-enhanced growth and muscle growth marker gene expression. *PLoS One* **5**, e14483.

Pankhurst, N. W. (2011). The endocrinology of stress in fish: an environmental perspective. *Gen. Comp. Endocrinol.* **170**, 265–275.

Pearson, M. P., Spriet, L. L. and Stevens, E. D. (1990). Effect of sprint training on swim performance and white muscle metabolism during exercise and recovery in rainbow trout (*Salmo gairdneri*). *J. Exp. Biol.* **149**, 45–60.

Pelster, B., Sänger, A. M., Siegele, M. and Schwerte, T. (2003). Influence of swim training on cardiac activity, tissue capillarization, and mitochondrial density in muscle tissue of zebrafish larvae. *Am. J. Physiol.* **285**, R339–R347.

Perry, S. F. and Bernier, N. J. (1999). The acute humoral adrenergic stress response in fish: facts and fiction. *Aquaculture* **177**, 285–295.

Perry, S. F. and Desforges, P. R. (2006). Does bradycardia or hypertension enhance gas transfer in rainbow trout (*Oncorhynchus mykiss*)?. *Comp. Biochem. Physiol. A* **144**, 163–172.

Picard, M., Juster, R. P. and McEwen, B. S. (2014). Mitochondrial allostatic load puts the 'gluc' back in glucocorticoids. *Nat. Rev. Endocrinol.* **10**, 303–310.

Polakof, S. M., Mommsen, T. P. and Soengas, J. L. (2011). Glucosensing and homeostasis: from fish to mammals. *Comp. Biochem. Physiol. B* **160**, 123–149.

Portz, D. E., Woodley, C. M. and Cech, J. J., Jr. (2006). Stress-associated impacts of short-term holding on fishes. *Rev. Fish Biol. Fish.* **16**, 125–170.

Postlethwaite, E. K. and McDonald, D. G. (1995). Mechanisms of Na^+ and Cl^- regulation in freshwater-adapted rainbow trout (*Oncorhynchus mykiss*) during exercise and stress. *J. Exp. Biol.* **198**, 295–304.

Powell, M. D. and Nowak, B. F. (2003). Acid-base and respiratory effects of confinement in Atlantic salmon affected with amoebic gill disease. *J. Fish. Biol.* **62**, 51–63.

Randall, D. J. and Perry, S. F. (1992). Catecholamines. In *Fish Physiology - The Cardiovascular System*, Vol. 12B (eds. W. S. Hoar, D. J. Randall and A. P. Farrell), pp. 255–300. San Diego, CA: Academic Press.

Rasmussen, R. S., Heinrich, M. T., Hyldig, G., Jacobsen, C. and Jokumsen, A. (2011). Moderate exercise of rainbow trout induces only minor differences in fatty acid profile, texture, white muscle fibres and proximate chemical composition of fillets. *Aquaculture* **314**, 159–164.

Richards, J. G., Heigenhauser, G. J. F. and Wood, C. M. (2002a). Glycogen phosphorylase and pyruvate dehydrogenase transformation in white muscle of trout during high-intensity exercise. *Am. J. Physiol.* **282**, R828–R836.

Richards, J. G., Heigenhauser, G. J. F. and Wood, C. M. (2002b). Lipid oxidation fuels recovery from exhaustive exercise in white muscle of rainbow trout. *Am. J. Physiol.* **282**, R89–R99.

Richards, J. G., Mercado, A. J., Clayton, C. A., Heigenhauser, G. J. F. and Wood, C. M. (2002c). Substrate utilization during graded aerobic exercise in rainbow trout. *J. Exp. Biol.* **205**, 2067–2077.

Richter, E. A. and Ruderman, N. B. (2009). AMPK and the biochemistry of exercise: implications for human health and disease. *Biochem. J.* **418**, 261–275.

Rome, L. C., Swank, D. and Corda, D. (1993). How fish power swimming. *Science* **261**, 340–343.

Sadoul, B. and Vijayan, M. M. (2016). Stress and Growth. In *Fish Physiology - Biology of Stress in Fish*, Vol. 35 (eds. C. B. Schreck, L. Tort, A. P. Farrell and C. J. Brauner), San Diego, CA: Academic Press.

Sänger, A. M. and Pötscher, U. (2000). Endurance exercise training affects fast white axial muscle in the cyprinid species *Chalcalburnus chalcoides mento* (Agassiz, 1832), cyprinidae, teleostei. *Basic Appl. Myol.* **10**, 297–300.

Santer, R. M. (1985). Morphology and Innervation of the Fish Heart, *vi*. Berlin: Springer Verlag.

Satchell, G. H. (1991). Physiology and Form of Fish Circulation.. Cambridge: Cambridge University Press.

Scarabello, M., Heigenhauser, G. J. F. and Wood, C. M. (1991). The oxygen debt hypothesis in juvenile rainbow trout after exhaustive exercise. *Resp. Physiol.* **84**, 245–259.

Schreck, C. B. (2010). Stress and fish reproduction: the roles of allostasis and hormesis. *Gen. Comp. Endocrinol.* **165**, 549–556.

Schreck, C. B., Olla, B. L. and Davis, M. W. (1997). Behavior responses to stress. In Fish Stress and Health in Aquaculture *vol. Society of Experimental Biology Seminar Series 62* (eds. G. K. Iwama, A. D. Pickering, J. P. Sumpter and C. B. Schreck), pp. 145–170. Cambridge: Cambridge University Press.

Schreck, C. B. and Tort, L. (2016). The Concept of Stress in Fish. In *Fish Physiology - Biology of Stress in Fish*, Vol. 35 (eds. C. B. Schreck, L. Tort, A. P. Farrell and C. J. Brauner), San Diego, CA: Academic Press.

Sorensen, B. and Weber, R. E. (1995). Effects of oxygenation and the stress hormones adrenaline and cortisol on the viscosity of blood from the trout *Oncorhynchus mykiss*. *J. Exp. Biol.* **198**, 953–959.

Steinhausen, M. F., Sandblom, E., Eliason, E. J., Verhille, C. and Farrell, A. P. (2008). The effect of acute temperature increases on the cardiorespiratory performance of resting and swimming sockeye salmon (*Oncorhynchus nerka*). *J. Exp. Biol.* **211**, 3915–3926.

Suarez, R. K. and Mommsen, T. P. (1987). Gluconeogenesis in teleost fishes. *Can. J. Zool.* **65**, 1869–1882.

Takei, Y. and Hwang, P.-P. (2016). Homeostatic Responses to Osmotic Stress. In *Fish Physiology - Biology of Stress in Fish*, Vol. 35 (eds. C. B. Schreck, L. Tort, A. P. Farrell and C. J. Brauner), San Diego, CA: Academic Press.

Takle, H. and Castro, V. (2013). Molecular adaptive mechanisms in the cardiac muscle of exercised fish. In *Swimming Physiology of Fish* (eds. A. P. Palstra and J. V. Planas), pp. 257–274. Heidelberg: Springer.

Tirri, R. and Ripatti, P. (1982). Inhibitory adrenergic control of heart-rate of perch (*Perca fluviatilis*) in vitro. *Comp. Biochem. Physiol. C* **73**, 399–401.

Totland, G. K., Kryvi, H., Jødestøl, K. A., Christiansen, E. N., Tangeras, A. and Slinde, E. (1987). Growth and composition of the swimming muscle of adult Atlantic salmon (*Salmo salar* L.) during long-term sustained swimming. *Aquaculture* **66**, 299–313.

Tseng, Y. C. and Hwang, P. P. (2008). Some insights into energy metabolism for osmoregulation in fish. *Comp. Biochem. Physiol. C* **148**, 419–429.

Tudorache, C., De Boeck, G. and Claireaux, G. (2013). Forced and preferred swimming speeds of fish: a methodological approach. In *Swimming Physiology of Fish* (eds. A. P. Palstra and J. V. Planas), pp. 81–108. Heidelberg: Springer.

van der Meulen, T., Schipper, H., van den Boogaart, J. G. M., Huising, M. O., Kranenbarg, S. and van Leeuwen, J. L. (2006). Endurance exercise differentially stimulates heart and axial muscle development in zebrafish (*Danio rerio*). *Am. J. Physiol.* **291**, R1040–R1048.

van Ginneken, V., Coldenhoff, K., Boot, R., Hollander, J., Lefeber, F. and van den Thillart, G. (2008). Depletion of high energy phosphates implicates post-exercise mortality in carp and trout; an in vivo ^{31}P-NMR study. *Comp. Biochem. Physiol. A* **149**, 98–108.

VanRaaij, M. T. M., VandenThillart, G., Vianen, G. J., Pit, D. S. S., Balm, P. H. M. and Steffens, B. (1996). Substrate mobilization and hormonal changes in rainbow trout (*Oncorhynchus mykiss*, L.) and common carp (*Cyprinus carpio*, L.) during deep hypoxia and subsequent recovery. *J. Comp. Physiol. B* **166**, 443–452.

Veiseth, E., Fjaera, S. O., Bjerkeng, B. and Skjervold, P. O. (2006). Accelerated recovery of Atlantic salmon (*Salmo salar*) from effects of crowding by swimming. *Comp. Biochem. Physiol. B* **144**, 351–358.

Videler, J. J. (2011). An opinion paper: emphasis on white muscle development and growth to improve farmed fish flesh quality. *Fish Physiol. Biochem.* **37**, 337–343.

Vijayan, M. M. and Moon, T. W. (1992). Acute handling stress alters hepatic glycogen metabolism in food-deprived rainbow trout (*Oncorhynchus mykiss*). *Can. J. Fish. Aquat. Sci.* **49**, 2260–2266.

Wang, Y. X., Heigenhauser, G. J. F. and Wood, C. M. (1994). Integrated responses to exhaustive exercise and recovery in rainbow trout white muscle - acid-base, phosphogen, carbohydrate, lipid, ammonia, fluid volume and electrolyte metabolism. *J. Exp. Biol.* **195**, 227–258.

Ward, D. L. and Hilwig, K. D. (2004). Effects of holding environment and exercise conditioning on swimming performance of southwestern native fishes. *N. Am. J. Fish Manage.* **24**, 1083–1087.

Weber, J. M. (2011). Metabolic fuels: regulating fluxes to select mix. *J. Exp. Biol.* **214**, 286–294.

Wendelaar Bonga, S. E. (1997). The stress response in fish. *Physiol. Rev.* **77**, 591–625.

Wiendl, H., Hohlfeld, R. and Kieseier, B. (2005). Immunobiology of muscle: advances in understanding an immunological microenvironment. *Trends. Immunol.* **26**, 373–380.

Winberg, S., Höglund, E. and Øverli, Ø. (2016). Variation in the Neuroendocrine Stress Response. In *Fish Physiology - Biology of Stress in Fish*, Vol. 35 (eds. C. B. Schreck, L. Tort, A. P. Farrell and C. J. Brauner), San Diego, CA: Academic Press.

Wood, C. M. (1991). Acid-base and ion balance, metabolism, and their interactions, after exhaustive exercise in fish. *J. Exp. Biol.* **160**, 285–308.

Wood, C. M. and Randall, D. J. (1973). Influence of swimming activity on sodium balance in rainbow trout (*Salmo gairdneri*). *J. Comp. Physiol.* **82**, 207–233.

Wood, C. M., Turner, J. D. and Graham, M. S. (1983). Why do fish die after severe exercise? *J. Fish. Biol.* **22**, 189–201.

Woodward, J. J. and Smith, L. S. (1985). Exercise training and the stress response in rainbow trout, *Salmo gairdneri* Richardson. *J. Fish. Biol.* **26**, 435–447.

Wrann, C. D., White, J. P., Salogiannnis, J., Laznik-Bogoslavski, D., Wu, J., Ma, D., et al. (2013). Exercise induces hippocampal BDNF through a PGC-1alpha/FNDC5 pathway. *Cell Metab.* **18**, 649–659.

Wright, P. A., Perry, S. F. and Moon, T. W. (1989). Regulation of hepatic gluconeogenesis and glycogenolysis by catecholamines in rainbow trout during environmental hypoxia. *J. Exp. Biol.* **147**, 169–188.

Yogata, H. and Oku, H. (2000). The effects of swimming exercise on growth and whole-body protein and fat contents of fed and unfed fingerling yellowtail. *Fish. Sci.* **66**, 1100–1105.

Young, P. S. and Cech, J. J. (1993). Improved growth, swimming performance, and muscular development in exercised-conditioned young-of-the-year striped bass (*Morone saxatilis*). *Aquat. Sci.* **50**, 703–707.

Zhang, H. and Anderson, J. E. (2014). Satellite cell activation and populations on single muscle-fiber cultures from adult zebrafish (*Danio rerio*). *J. Exp. Biol.* **217**, 1910–1917.

Zhang, L. and Wood, C. M. (2009). Ammonia as a stimulant to ventilation in rainbow trout *Oncorhynchus mykiss*. *Resp. Physiol. Neurobiol.* **168**, 261–271.

Zhao, W.-W., Pang, X., Peng, J.-L., Cao, Z.-D. and Fu, S.-J. (2012). The effects of hypoxia acclimation, exercise training and fasting on swimming performance in juvenile qingbo (*Spinibarbus sinensis*). *Fish Physiol. Biochem.* **38**, 1367–1377.

8

REPRODUCTION AND DEVELOPMENT

N.W. PANKHURST

Stress has a consistent inhibitory effect on reproductive performance in fish of both sexes, but in a smaller subset of conditions can have stimulatory effects. Inhibitory effects include the suppression of ovarian and testicular development, inhibition of ovulation and spawning, and the production of smaller eggs and larvae. Long-term effects on progeny remain largely undescribed. Endocrine effects include the suppression of hypothalamic, pituitary, and gonadal steroid hormones, with the effects on the production of gonadal androgens and estrogens generally being most profound. Understanding the mechanisms by which stress interferes with reproduction is complicated by the fact that stress-modulated hormones can have systemic effects as well as direct effects on the reproductive endocrine system, and experimental paradigms often don't allow distinction between the two. With that caveat, there is evidence for inhibitory effects on reproduction from all

Biology of Stress in Fish: Volume 35
FISH PHYSIOLOGY

levels in the stress endocrine axis but strongest evidence is available for the role of corticosteroids, noting that the dominance of the literature by studies on the effects of cortisol is partly a reflection of the relative ease of measurement of steroid hormones. Proposed mechanisms of action include systemic metabolic effects, genomic glucocorticoid receptor-mediated effects and direct action through nongenomic processes that may include substrate competition for steroid-converting enzymes and binding proteins. The majority of stress studies have involved laboratory assessment of captive or cultured fish populations and there is much less information on the effects of stress among free-living, wild fishes. It is clear that reproductive processes can be maintained over a wide range of corticosteroid concentrations, but there is also increasing evidence that social control of reproduction may be mediated by stress processes. There remains scope for improved under-standing of stress-reproduction interactions at all levels of reproductive function.

1. INTRODUCTION

The understanding that physiological stress has a largely inhibitory effect on reproduction is well established in higher vertebrates and has the longest history of examination in mammals (Pottinger, 1999). There was initially less interest in and understanding of the process in lower vertebrates, particularly fish, with John Sumpter and colleagues writing in 1987, "Although there is some circumstantial evidence obtained from the study of both natural and laboratory populations of fish, there is very little direct evidence. Further, the mechanisms whereby stress might affect the reproductive axis of fish are unknown" (Sumpter et al., 1987). This was somewhat surprising given the parallel conversations on the consistent failure of many species to undergo sexual maturation in captivity or culture, and the increasing interest in endocrine manipulation as a management strategy (Fostier and Jalabert, 1982). The response was a pioneering set of studies on salmonids (Pickering et al., 1987; Sumpter et al., 1987; Carragher et al., 1989; Carragher and Sumpter, 1990; Pottinger and Pickering, 1990; Pottinger et al., 1991) that firmly established the link between stress and reproduction, began to explore the possible mechanisms involved and laid the foundation for a substantial body of research that would follow (reviewed in Pankhurst and Van Der Kraak, 1997; Schreck et al., 2001; Milla et al., 2009; Leatherland et al., 2010; Schreck, 2010; Fuzzen et al., 2011). These studies addressed a detailed assessment of the effects of stress across a broad range of taxa, extended our understanding of the mechanisms by which stress exerted its effects, and examined how to circumvent them in managed reproduction.

Recent advances in molecular techniques now offer the exciting opportunity to both reassess and extend existing paradigms of understanding (eg, Maruska, 2014), and also to explore the more complex interactions and relationships that undoubtedly exist at the transcriptomic level (Aluru and Vijayan, 2009). Much of the research effort to date has concentrated on the effects of stress in recently captive or cultured populations and with a few notable exceptions, left us with little understanding of the extent to which these considerations apply in natural populations (Pankhurst, 2011). For both wild and cultured populations there is also the still unanswered question as to the longer term effects of stress on reproductive fitness, and the extent to which these effects cross generations. Here again there is very little information for fishes, but some clear pointers from studies on higher vertebrates (Chand and Lovejoy, 2011; Lane et al., 2014) and the emerging understanding of the role of epigenesis in long-term individual and population effects of toxicants on fish (Bhandari et al., 2015; Kamstra et al., 2014). These questions set the agenda for the next phase of research into stress and reproduction in fishes and will be explored in the context of current understanding and knowledge gaps.

2. REGULATION OF REPRODUCTION

2.1. Patterns and Environmental Regulation of Reproduction

The patterns of reproduction in fishes are highly variable, and understanding of both their commonality and also their diversity is an essential prerequisite for the assessment of the effects of stress on reproduction. Patterns of gamete development, gonadal growth and development, and spawning are detailed in Pankhurst (1998a) but will be briefly summarized here. *Semelparous* species (which spawn only once, eg, anguillid eels and Pacific salmon) mature a single clutch of ovarian follicles through to ovulation into the oviduct (or body cavity in salmonids) in females, and males show progression to more or less complete maturation of all gametes to spermatozoa. This is described as *synchronous* gamete development. *Iteroparous* species spawn more than once during their lifetime and can be divided into those that mature a single clutch of gametes per spawning season (eg, other salmonids, which have *group synchronous* gamete development), and species that have multiple ovulatory and spawning events within a single season, described as having *asynchronous* gamete development. Asynchronous development is the most common teleost strategy and is characterized by the presence of multiple developing clutches of gametes present in the gonad at any one time. Following ovulation in females and spermiation (the production of hydrated milt) in males, gamete fertilization occurs via release of gametes to the water via the behavioral act of spawning (sometimes erroneously used to describe the process of ovulation), or for the smaller number of live-bearing species, copulation. Spawning patterns are also quite variable ranging from pelagic or demersal spawning in pairs or groups through to substrate spawning and fertilization in nest-protecting species.

Reproductive cycles are in turn environmentally synchronized and regulated, with broad seasonal phasing typically being entrained by photoperiod, and finer "within-season" tuning being strongly influenced by temperature and also depending on context, a range of other factors including rainfall, lunar phase, nutrition, and social status (Pankhurst and Porter, 2003). As in other vertebrates, sex in fishes is primarily determined genetically; however, unlike in other vertebrate systems where genetic mechanisms are highly conserved, fish show high variability of sex-determining genes (Devlin and Nagahama, 2002; Martinez et al., 2014). An additional complication is that many species also undergo sex inversion either during puberty or as reproductive adults, either beginning their lives as males and changing at some stage to females (*protandry*), beginning as females and changing to males (*protogyny*), or in a smaller number of species, undergoing bidirectional sex inversion (Frisch, 2004).

2.2. Endocrine Control of Reproduction

Environmental signals are transduced into an endocrine signal through activation of the hypothalamic–pituitary–gonadal (HPG) axis (Fig. 8.1), initially through synthesis and release of gonadotropin-releasing hormone (GnRH), a decapeptide, from the hypothalamus. All vertebrates including fish express two or more forms of GnRH, and to date, eight forms have been identified from among teleost fishes (Zohar et al., 2010). GnRHs are expressed in several tissues but the form produced in the hypothalamus (typically termed GnRH1) appears to have the primary role in activation of the HPG axis via synaptic stimulation of the gonadotropin-producing cells (gonadotrophs) of the pituitary. In some, but apparently not all species, pituitary gonadotrophs are also under the inhibitory action of dopamine (DA)-producing neurons, and the balance between the inhibitory DA tone and GnRH stimulation determines the rate and level of gonadotropin synthesis and release (Dufour et al., 2010). Further modulation of GnRH synthesis and release occurs through the more recently discovered

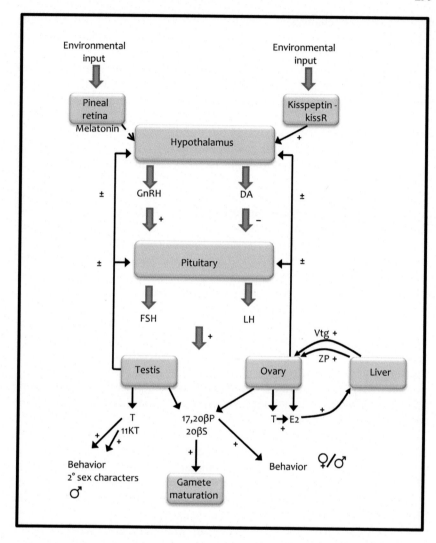

Figure 8.1. A summary of regulation of the hypothalamic–pituitary–gonadal (HPG) axis in fish. Abbreviations are given in the text. +, stimulatory effect; −, inhibitory effect; dashed line indicates probable but unconfirmed regulatory input.

kisspeptin-kissreceptor system, which is increasingly thought to be the mechanism whereby exogenous and endogenous environmental input exerts its effect on the reproductive system (Zohar et al., 2010; Shahjahan et al., 2013), although the mechanism of action remains undescribed. There is also inferred photoperiod modulation of the hypothalamo-pituitary axis by light-induced changes in melatonin secretion from the pineal gland (Migaud et al., 2010), and here also the mechanism of action is not yet understood.

GnRH stimulation of the pituitary results in the synthesis and release of the protein hormones, follicle stimulating hormone (FSH) and luteinizing hormone (LH) (Planas and Swanson, 2008; Levavi-Sivan et al., 2010) (these are synonyms for the hormones described in earlier studies on fish as GTH-I and GTH-II). FSH is involved in stimulating the early stages of gametogenesis, whereas LH modulates later stages of ovarian and testicular maturation, with both acting thought membrane-bound G protein-coupled receptors (Planas and Swanson, 2008). Ligand-receptor binding of both hormones activates adenylate cyclases with consequent increases in intracellular cyclic adenosine monophosphate (cAMP) concentrations and activation of a protein kinase-modulated cascade of steroid hormone synthesis and cleavage from the base molecule cholesterol (reviewed in Pankhurst, 2008). The first stage of this cascade is the delivery of cholesterol to cleavage sites on the inner mitochondrial membrane under the regulatory activity of steroid acute regulatory (StAR) protein, which appears to be one of the main rate-limiting steps in the cleavage pathway. Sequential enzymatically-regulated cleavage of carbon atoms from the steroid nucleus and the addition of active groups gives rise in cleavage order to 21 C progestins (and corticosteroids in interrenal tissues), and the shorter 19 C androgens and 18 C estrogens.

A bewilderingly wide range of intermediates and potentially active steroid products is produced by gonadal tissues, but the main active progestins are thought to be 17,20β-dihydroxy-4-pregnen-3-one (17,20βP) and 17,20β,21-trihydroxy-4-pregnen-3-one (20βS) in both sexes, the androgens testosterone (T) (both sexes) and 11-ketotestosterone (11KT) (principally males), and the estrogen 17β-estradiol (E2) (predominantly females) (Pankhurst, 2008). The progestins typically regulate gamete final maturation and are implicated in some aspects of reproductive behavior. Androgens, as well as being precursors for estrogen production in females, are responsible for testis development in males and development of secondary sexual characteristics, and are strongly implicated in territorial and aggressive behavior (Schulz et al., 2010; Munakata and Kobayashi, 2010). Estrogens are primarily responsible for the estrogen receptor (ER)-mediated production of the yolk precursor vitellogenin (VTG) by hepatocytes of the liver for subsequent uptake into growing oocytes during the process of vitellogenesis, and the production of egg shell, or zona pellucida (ZP), proteins (Pankhurst, 2008). Interestingly, estrogens are also implicated in regulation of gene expression in the testis and are thought to have a role in stimulating spermatogonial proliferation (Schulz et al., 2010). In sex-inverting protandrous species, male stages mainly produce 11KT, transitional phases have low or reducing levels of 11KT and increasing E2, and in protogynous species the pattern is reversed (Frisch, 2004). Steroids in turn can exert positive and negative feedback on higher levels in the HPG axis through classic (genomic) action via intracellular receptors (Pankhurst, 2008), and shorter term modulatory effects though membrane-bound steroid receptors (Thomas, 2003). The complexity of the HPG axis and its control provides a rich canvas for the potential action of stress on the reproductive system.

3. EFFECTS OF STRESS ON REPRODUCTION

3.1. Effects on Reproductive Performance

Among the first recognition of the effect of stress on reproduction were the observations of De Montalembert et al. (1978) reporting the development of ovarian atresia in captive pike (*Esox lucius*), and discussion of the effects of environmental stress on desert pupfish (*Cyprinodon nevadensis*) (Gerking, 1982). Subsequent studies showed a consistent inhibitory effect of artificially-generated stress (eg, capture, confinement, and husbandry) on reproductive performance, with the effects generally being more profound in nonsalmonid species, among species with asynchronous ovarian development, and in fish captured from wild populations.

Weekly exposure of rainbow trout (*Oncorhynchus mykiss*) to disturbance stress over a 9-month period resulted in delayed ovulation and reduced egg weight and volume in females, and reduced sperm density in males (Campbell et al., 1992). Shorter periods (2 weeks) of confinement stress produced similar effects in rainbow trout, but not brown trout (*Salmo trutta*) (Campbell et al., 1994). In contrast, exposure of rainbow trout to daily disturbance stress accelerated ovulation by an average of 2 weeks when the stress was applied throughout the period of reproductive development, or during final maturation; but exposure to stress during early vitellogenesis resulted in the production of smaller eggs (Contreras-Sánchez et al., 1998). Pottinger and Carrick (2000) reported increased egg mortality in high cortisol response strains of rainbow trout following 1 week of confinement, and Arctic char (*Salvelinus alpinus*) stressed by exposure to high sea louse densities had delayed ovulation, but no changes in egg quality or survival (Tveiten et al., 2010).

Exposure of nonsalmonids to stress appears to have more profound effects on reproductive performance. Australasian snapper (*Pagrus auratus*) held for up to 7 days after line capture from the wild showed cessation in ovulation from day 3 after capture (Carragher and Pankhurst, 1991) and ovarian atresia after 7 days of captivity (Cleary et al., 2000). Red gurnard (*Chelidonychthys kumu*) showed very high levels of ovarian atresia from 48 h onward after capture from the wild (Clearwater and Pankhurst, 1997) (Fig. 8.2); loss of larger vitellogenic follicles from the ovary occurred in striped trumpeter (*Latris lineata*) following daily sampling for 9 days (Morehead, 1998); and increased atresia and inhibited spawning was found in red belly tilapia (*Tilapia zillii*) stressed by crowding (Coward et al., 1998). The ovulatory response to analogs of GnRH, or human chorionic gonadotropin (hCG), was also suppressed following one day of confinement stress after capture in (*P. auratus*) (Cleary et al., 2002), black bream (*Acanthopagrus butcheri*) (Haddy and Pankhurst, 2000), and in stressed captive jundia (*Rhamdia quelen*) (Soso et al., 2008).

The still relatively limited data available show that stress impacts on reproductive performance and output are also carried through into effects on progeny. There was reduced survival to hatching, swim up, and 28 days posthatching in rainbow trout progeny from females exposed to stress (Campbell et al., 1992, 1994), but no effects on survival of up to 3 months

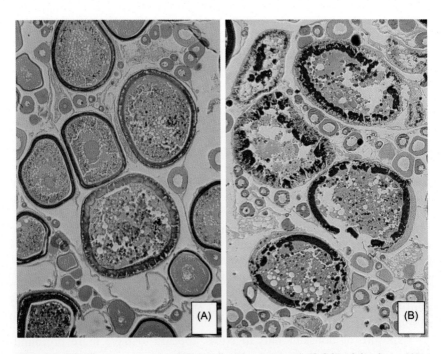

Figure 8.2. Micrographs of ovarian follicles of wild red gurnard *Chelidonichthys kumu* (A) at capture and (B) 96 h after capture and subsequent confinement, showing advanced atresia of vitellogenic follicles.

stress during vitellogenesis in another study with different stressors (Contreras-Sánchez et al., 1998). Stressed laboratory-acclimated populations of Atlantic cod (*Gadus morhua*) has similar spawning, egg production, and fertility to low-stress fish but showed a higher proportion of abnormally developing larvae (Morgan et al., 1999). Damselfish (*Pomacentrus amboinensis*) larvae from females stressed in the wild by increased conspecific and egg-predator density were smaller at hatching, and the same effect could be generated by treating spawning females with cortisol (McCormick, 1998, 1999, 2006). Larvae from females treated with cortisol also showed greater morphological variability. Female cichlids (*Neolamprologus pulcher*) showed increased intervals between spawning, and the production of smaller eggs when exposed to daily netting and chasing (Mileva et al., 2011), and eggs from Atlantic salmon (*Salmo salar*) treated 6 days before fertilization with exogenous cortisol showed higher mortality, smaller size at hatch, and a higher proportion of larval abnormality (Eriksen et al., 2006, 2007).

Essentially nothing is known about the effects of stress on the subsequent life history performance of progeny from stressed adults, nor is there any information on the possible intergenerational effects of stress. Auperin and Geslin (2008) did demonstrate that exposure of eggs and larvae of rainbow trout from 14–55 days postfertilization (hatching occurred at 32 days postfertilization) to stress dampened the subsequent cortisol increase in response to stress at 5 months of age, and that the same effect could be generated by treating eggs with exogenous cortisol. Presumably, this means that cortisol loading of eggs due to maternal stress may have the same effect. Outcomes from mammalian studies and multigenerational toxicology studies in fish also provide clues as to possible cross-generation effects of stress. There is now an increased understanding of the process of epigenesis where environmental effects mediate genetic expression through a number of processes including the methylation of nucleotide bases, primarily cytosines, and posttranslational modification of histones (Chand and Lovejoy, 2011; Siklenkia et al., 2015). This in turn interferes with the binding affinity of transcription factors in the promoter region of a gene. Methylation changes are quite stable and can persist across generations.

In mammals many epigenetic markers in oocytes are eliminated at fertilization, but at some loci the epigenetic changes are not cleared, allowing the possibility of the persistence across generations (Lane et al., 2014). In males, epigenetic changes include DNA methylation, but also changes to noncoding microRNAs involved in regulating gene transcription. Epigenetic changes in mammals have been found in response to exposure to toxins, endocrine disruptors, smoking and obesity in humans, and in rodents, exposure to stress (Lane et al., 2014). There is not yet any evidence for this effect being generated in fish by stress, but toxicant exposure does cause epigenetic changes in a large number of genes through DNA methylation, histone modification, and microRNA effects (Bhandari et al., 2015; Kamstra et al., 2014), and there is some evidence of transgenerational effects with exposure of fathead minnow (*Pimephales promelas*) to xenoestrogens producing reduced fecundity in F1 generation females (Staples et al., 2011). A reasonable assumption is that stress effects on adults could also be expressed in progeny and possibly subsequent generations, through stress-mediated epigenesis.

3.2. Effects of Stress on the Reproductive Endocrine System

Due to the ubiquity of structure and relative ease of measurement, most studies initially examined the effects of stress on steroid hormone synthesis, and for a smaller number of species for which assay systems were available, pituitary hormones. More recently, the proliferation of molecular techniques has greatly extended the range of possible enquiry. Early studies on male brown trout showed that acute handling stress resulted in elevation in plasma cortisol and adrenocorticotrophic hormone (ACTH), and depressions in plasma T and 11KT within 1–4 h of stress, and that the effects on steroids were maintained over longer periods of chronic stress (Pickering et al., 1987; Sumpter et al., 1987). In female brown, and male rainbow trout, 2 weeks of confinement elevated plasma cortisol, suppressed plasma T levels, had no effect on plasma E2 levels in either species, but depressed plasma VTG levels in female rainbow trout (Campbell et al., 1994). In contrast, Pottinger and Carrick (2000) reported declines in plasma E2, 2–24 h after imposition of confinement stress in female rainbow trout, and Pankhurst and Dedual (1994) found postcapture depression of plasma T and E2 in migrating wild rainbow trout held in stream cages for 24 h after capture, with associated increases in plasma cortisol. Cultured sockeye salmon (*Oncorhynchus nerka*) showed depression of T in both sexes, and 11KT in males within 15–30 min of net confinement (Kubokawa et al., 1999), and Arctic char showed depressed plasma T in both sexes, plasma 11KT in male, and plasma E2 in female fish exposed to high densities of sea louse parasites that elevated plasma cortisol (Tveiten et al., 2010).

Nonsalmonids show similar effects on gonadal steroids, with depressions of plasma T and 11KT in wild male spotted sea trout (*Cynoscion nebulosus*) by 60 min postcapture (Safford and Thomas, 1987); Australasian snapper confined after capture from the wild had elevated cortisol and depressed E2 1 and 6 h after capture, depressed T at 6 h, and elevated 17,20βP at both times (Carragher and Pankhurst, 1991); and wild black bream had elevated cortisol by 15 min postcapture, depressions in T by 30 min, E2 by 1 h and 11KT by 6 h, but initial elevations in plasma 17,20βP in both sexes

(Haddy and Pankhurst, 1999) (Fig. 8.3). Similarly, wild female red gurnard showed chronically elevated cortisol and depressed E2 and T from 24 h postcapture (Clearwater and Pankhurst, 1997), wild female striped trumpeter showed sustained elevation of cortisol and depression of T and E2 but no change in plasma 17,20βP (Morehead, 1998), and spiny damselfish (*Acanthochromis polyacanthus*) showed depressed T at 3 and 6 h postcapture and depressed E2 at 6 h in females, and depressed T and 11KT 6 h postcapture in males (Pankhurst, 2001). Roach (*Rutilus rutilus*) are characterized by often very high resting cortisol levels and extremely high elevations in cortisol in response to stress, but here also, there is depressed plasma E2 in response to both acute and chronic stress (Pottinger et al., 1999). Similarly, white suckers (*Catostomus commersoni*) show depressed plasma T and 11KT in males, and T and E2 in females following acute confinement stress (Van Der Kraak et al., 1992; Jardine et al., 1996). A different pattern was shown by cultured female jundia exposed to daily

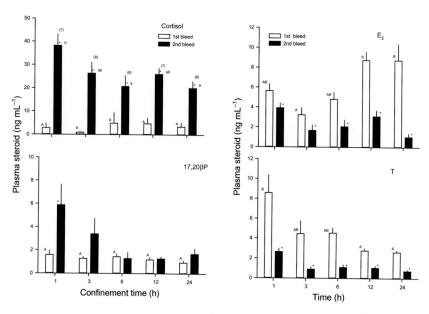

Figure 8.3. Effects of various periods of postcapture confinement of wild female black bream *Acanthopagrus butcheri* on plasma levels of cortisol, 17,20 βP, T and E₂. Values are mean + SE (n); upper- and lowercase letters indicate differences across sample times among first (at point of capture—clear bars) and second (after confinement—filled bars) bleeds, respectively; values that are significantly different between first and second bleeds at each confinement time are marked with an asterisk. Reproduced with permission from Haddy, J. A. and Pankhurst, N. W. (1999). Stress-induced changes in concentrations of plasma sex steroids in black bream. *J. Fish. Biol.* **55**, 1304–1316.

netting disturbance where there was no T suppression but reduced E2 levels (Soso et al., 2008). Male striped bass (*Morone saxatilis*) showed suppressed T and 11KT levels in response to repeated netting, but the magnitude of the response was linked to the individual stress response status with low cortisol responders showing more marked stress-suppression of plasma androgens (Castranova et al., 2005). Collectively, these observations suggest that it is likely that there will be individual variability in the response of the reproductive system to stress in all populations and species. There is also evidence that susceptibility to stress may change with reproductive stage. A field study of smallmouth bass (*Micropterus dolomieu*) showed poststress declines in plasma androgens in territorial males captured from nest sites, but the effect was ameliorated in males undertaking parental care (O'Connor et al., 2011). It is well established that stress responsiveness can vary with reproductive stage in other species (Pankhurst, 2011). What is less clear is whether this variability extends to the effects of stress on reproduction.

In higher vertebrates, stress inhibits both the synthesis and release of GnRH, suppression of LH and FSH release and the frequency of release pulses, and as in fish, suppression of the synthesis of gonadal steroids (Pottinger, 1999; Fuzzen et al., 2011). The effects of stress on higher levels of the HPG axis in fish are much less well understood. Early studies on the effects of stress on pituitary function were ambiguous. Acute stress resulted in increases in plasma LH in male brown trout (Pickering et al., 1987; Sumpter et al., 1987), and white suckers showed decreases in plasma LH in males but not females following serial handling and sampling (Van Der Kraak et al., 1992). In contrast, acute low water stress that elevated plasma cortisol and suppressed T had no effect on plasma LH in preovulatory rainbow trout (Pankhurst and Van Der Kraak, 2000). Gene expression of FSHβ and LHβ subunits were not different between wild and captive jack mackerel (*Trachurus japonicus*) of equivalent reproductive stage and there was also no effect on plasma E2 levels, despite significant suppression of GnRH1 in captive fish (Imanaga et al., 2014). Subordinate nonterritorial male African cichlids (*Astatotilapia* (=*Haplochromis*) *burtoni*) have lower pituitary LHβ and FSHβ mRNA, and lower plasma levels of FSH, LH, T, and 11KT than territorial males and the effect is interpreted as being a socially mediated stress response (Maruska, 2014). Nonterritorial males also have increased levels of corticotropin releasing factor receptor (CRFR1) mRNA and decreased GnRH1 mRNA in the preoptic area (POA) of the brain, and increased CRF and CRFR1 mRNA and decreased GnRH1-R1 mRNA in the pituitary. Earlier studies identified the same phenomenon in the form of smaller GnRH-producing neurons on the POA of nonterritorial males (Fox et al., 1997).

The identification of glucocorticoid receptor (GR) response elements on both gonadal aromatase (cyp19a1) and brain aromatase (cyp19b1) genes of a number of teleost species has led to speculation that corticosteroids may be implicated in gene regulation associated with sex inversion (Gardner et al., 2005). However, the evidence for stress-induced mechanisms playing a role in adult sex inversion is equivocal. Removal of a harem-protecting male blue banded goby (*Lythrypnus dalli*) results in rapid behavior changes in a dominant female and subsequent inversion to male sexual status (Black et al., 2011). Increased female aggression is associated with markedly reduced brain aromatase activity, but this follows rather than drives the behavior change. Dominance relationships can result in subordinate fish displaying elevated plasma cortisol levels but this is usually only during periods of dominance establishment (Sloman and Armstrong, 2002) and it seems unlikely that stress-related dominance effects could explain the social maintenance of female status in protogyny, given the typically inhibitory effect of stress on estrogen synthesis. Neither is the evidence for stress maintaining female status by suppressing androgen production, particularly strong. In an experiment to test whether substrate competition for the androgen and corticosteroid converting enzymes, 11β-hydroxylase (11βH) and 11β-hydroxysteroid dehydrogenase (11βHSD) could inhibit 11KT production, protogynous sandperch (*Parapercis cylindrica*) were treated with cortisol in silastic implants (Frisch et al., 2007). 11βH mediates conversion of androstenedione (A) and T to their respective 11β-hydroxylated derivatives, and 11βHSD is responsible both for the conversion of 11β-hydroxyA to 11-ketoandrostenedione (11KA) (an immediate precursor for 11KT synthesis), and the conversion of 11β-hydroxyT to 11KT. 11βH also mediates conversion of 11-deoxycortisol to cortisol, and 11βHSD mediates conversion of cortisol to the less active metabolite, cortisone (Perry and Grober, 2003). There was no effect on plasma T, 11KT after 21 days and hormone-treated fish underwent sex inversion at the same rate as controls. This is clearly an area where there is considerable scope for further study.

3.3. Thermal Stress: A Special Case?

It is now well established that elevated temperatures can exert major effects both on reproductive processes at the stage of sex determination in larvae and juveniles, and also on the processes of sexual maturation and spawning by adults, and there is increasing evidence that sex determination may be at least partially regulated by stress. In a number of species, sex determination is thermally labile, with higher temperatures typically driving sex determination toward the male phenotype. This has usually been interpreted as being a result of the demonstrated inhibitory effect of high temperature on aromatases and the consequent reduction in the conversion of androgens to estrogens (Devlin and Nagahama, 2002; Guiguen et al., 2010). Recent studies suggest that stress mechanisms may be involved in regulating this process. Larvae of the pejerrey (*Odontesthes bonariensis*) reared at 29°C were 100% male and had higher whole body concentrations of cortisol, T and 11KT than larvae held at feminizing temperatures (17°C) (Hattori et al., 2009). The masculinizing effects of elevated temperature could be mimicked by treatment with cortisol and the synthetic corticosteroid, dexamethasone. In similar experiments with medaka (*Oryzias latipes*), maintenance of genotypic females at 32–34°C resulted in the production of phenotypic males in association with elevated whole body cortisol and reduced FSH receptor (FSHR) mRNA expression (Hayashi et al., 2010). Here also, the effects could be reproduced by treatment with cortisol and ameliorated by the cortisol synthesis inhibitor metyrapone, leading to the conclusion that elevated temperature induced increased cortisol secretion which in turn suppressed germ cell proliferation and FSHR expression. A later study on medaka showed that both abnormally high temperature and cortisol inhibited the expression of cyp19a1 and gonadal soma-derived growth factor (Kitano et al., 2012). The effect could be reversed by cotreatment or subsequent treatment with E2, suggesting that the masculinizing effects were modulated by the inhibition of aromatase expression. Genotypic females of Japanese flounder (*Paralichthys olivaceus*) can be induced to develop either as phenotypic females or males by rearing at the normal- range temperatures of 18°C or 27°C, respectively, and fish at high temperature have higher whole body cortisol levels. In vivo, high temperature masculinization can be prevented by cotreatment with metyrapone, and cortisol suppresses cyp19a1 gene expression in vitro (Yamaguchi et al., 2010). In this study, there was no effect of cortisol on FSHR, but recombinant sea bream FSH abolished cortisol inhibition of cyp19a1 in vitro. There was also GR binding to the cAMP response element of the cyp19a1 promoter at 27°C, but not 18°C. An additional effect of stress is proposed by Fernandino et al. (2013), addressing the possibility that male phenotypes are androgen-driven rather than simply being a default condition of no estrogen. Under this scenario, the upregulating effect of stress-induced increases in cortisol on gonadal 11βHSD expression serves to both increase the synthesis of 11KT and to deactivate cortisol through conversion to cortisone. These authors suggest that none of these mechanisms are mutually exclusive and that a combination of effects could be in play.

The role of stress in modulating temperature effects on reproduction in adults is less clear. Elevated temperatures have a largely inhibitory effect on the reproductive process and these operate at multiple levels in the endocrine cascade (Pankhurst and Munday, 2011). Exposure to abnormally elevated temperature suppresses kisspeptin and kissR expression (Shahjahan et al., 2013); reduces expression of GnRH, GnRHR, LHβ (Okuzawa et al., 2003; Okuzawa and Gen, 2013), and FSHR (Soria et al., 2008); increases the dopaminergic inhibitory tone on pituitary release of LH (Gillet and Breton, 2009); inhibits steroid-converting enzymes, particularly aromatase (Lim et al., 2003; Anderson et al., 2012a); and reduces plasma steroids, and VTG and ZP gene expression (Tveiten and Johnsen, 2001; Soria et al., 2008; Pankhurst and King, 2010; Anderson et al., 2012b; Okuzawa and Gen, 2013). This is accompanied by retarded maturation, increased levels of ovarian atresia, and delayed ovulation—effects that look very similar to some of the suppressive effects of stress on reproduction. However, the limited evidence available suggests that corticosteroid-regulated mechanisms may not be involved in this process.

An extensive examination of the effects of moderately elevated (normal range) temperatures on cultured Atlantic salmon has shown consistent inhibitory effects on vitellogenesis and ovarian development in association with suppressed cyp19a1; VTG and ZP gene expression; reduced plasma levels of T, E2, and VTG; and delayed ovulation and production of 17,20βP in response to GnRH analogs (Pankhurst and King, 2010; Pankhurst et al., 2011; Anderson et al., 2012a). Similar effects are described for Arctic char with reduced follicular responsiveness to LH, reduced 17,20βP production, and reduced oocyte sensitivity to 17,20βP at higher temperatures (Gillet et al., 2011). In the Atlantic salmon experiments, there were no differences in plasma cortisol levels in response to thermal manipulation and these remained within the range of normal husbandry-related variations found in this domestic stock (Pankhurst et al., 2011). A stronger candidate for the thermal modulation of particularly cyp19a1 expression may be forkhead transcription factor (foxl2), which is involved in transcriptional regulation of cyp19a1 and is thermally impaired at elevated temperatures (Nakamoto et al., 2006; Wang et al., 2007; Yamaguchi et al., 2007). The presence of GR response elements on cyp19a1 genes (Gardner et al., 2005) does not preclude the action of cortisol in thermally stressed fish, but many of the inhibitory changes of elevated temperatures appear to occur within the range at which fish can maintain normal homeostasis and where thermal hypothalamic–pituitary–interrenal (HPI) activation is probably not occurring. In contrast, thermal shock associated with rapid reductions in temperature does activate the HPI axis. Exposure of juvenile male carp to a rapid temperature change from 25°C to 11°C resulted in elevated plasma cortisol levels and subsequent reduction in gonad size (Goos and Consten, 2002). It is possible that rapid and large upward shifts in temperature could elicit similar effects.

3.4. Effects of Hypoxia

An environmental factor that can elicit an endocrine stress response and that also has inhibitory effects on reproductive endocrine processes is exposure to hypoxia. Riverine and estuarine species appear to have evolved to cope with periods of naturally occurring hypoxia (Dabrowski et al., 2003; Landry et al., 2007; Thomas et al., 2007), but increasing temperatures across many aquatic habitats, and the increasing frequency of hypoxic conditions in marine coastal and surface waters as a result of nutrient loading, suggest that a much wider range of species is now at risk from hypoxic stress. Experimental studies all show that there is some degree of inhibition of reproduction in response to exposure to hypoxia. Pacu (*Piaractus brachypomus*) showed depressed plasma T and 11KT levels in males and T and E2 in females, and reduced larval survival and increased prevalence of larval deformities following 3 days exposure to hypoxic conditions (Dabrowski et al., 2003). Similarly, up to 12 weeks of hypoxia resulted in reduced gonadosomatic index (GSI) in both sexes of common carp (*Cyprinus carpio*), along with reductions in T and E2 in females; reduced T, spermatozoa, and sperm motility in males; and reduced egg fertility, hatching success, and larval survival (Wu et al., 2003). In a later study, hypoxic female carp had reduced plasma LH levels and delayed final maturation, although in this study, there were no effects on gonadal steroids (Wang et al., 2008). Gulf killifish (*Fundulus grandis*) exposed to hypoxia for 1 month showed reduced GSI in both sexes, lower plasma E2 in females and 11KT in males, fewer eggs, and later initiation of spawning (Landry et al., 2007).

A series of experiments exposing Atlantic croaker (*Micropogonias undulatus*) to extended periods (up to 10 weeks) of hypoxia resulted in reduced GSI, retarded oocyte development, reduced plasma E2 and VTG, and reduced ER gene expression in females; and reduced GSI, plasma 11KT levels, and proportion of spermatozoa in the testes of males (Thomas et al., 2006, 2007). Similar effects were found in fish sampled directly from hypoxic sites in estuaries. Laboratory-treated fish also showed reductions in plasma LH levels, and brain levels of GnRH mRNA and hypothalamic levels of serotonin (5HT) (which has a stimulatory effect on GnRH release in Atlantic croaker) (Thomas et al., 2007). In a later study on Atlantic croaker, 4 weeks of hypoxia resulted in reduced GSI, lower plasma 20βS levels, and less in vitro maturational responsiveness of oocytes to hCG and 20βS (Thomas and Rahman, 2009).

More recent experiments with zebrafish (*Danio rerio*) have shown that hypoxia affects gene expression at multiple levels (Lu et al., 2014). Three weeks of exposure to hypoxia resulted in reduced plasma E2 and T in females and males, respectively, but also falls in sGnRH, FSHβ, cyp19a1, pituitary, and gonadal ER and VTG gene expression in females, and LHβ, LHR, StAR, 20β-hydroxysteroid dehydrogenase (20βHSD—responsible for hyrdoxylatyion of maturational progestins), and 11βH gene expression in males. The conclusion was that these were all gene-specific effects rather than the result of general downregulation of gene expression. Fertilized eggs of genotypically female medaka (*O. latipes*) exposed to hypoxia for 12 days developed as phenotypic males along with downregulation of *vasa* gene expression (responsible for the proliferation of primordial germ cells and differentiation into ovary), and upregulation of the DNA-binding motif on the Y chromosome (DMY) gene responsible for testis development (Cheung et al., 2014), confirming that hypoxia is another environmental variable capable of affecting sex determination and that the effects appear to be gene-specific.

As with thermal effects, there is a question as to whether hypoxia is having direct effects on the endocrine cascade, or exercising at least some of its action through activation of the HPI axis. Acute severe hypoxia can induce increases in plasma cortisol levels in fish (Perry and Gilmour, 1999), but it is not clear whether this also occurs in response to less acute, or sustained hypoxia. As a further complication, cortisol responses to confinement stress are actually suppressed under conditions of poor water quality (including hypoxia) generated by holding fish at high densities (Pickering and Pottinger, 1987). Suppressive effects of hypoxia on reproduction were initially interpreted as being part of a stress response (Dabrowski et al., 2003), but the finding that hypoxia depresses 5HT levels in Atlantic croaker (where 5HT typically rises in response to stress; Thomas et al., 2007), and the demonstration of gene-specific effects (Cheung et al., 2014; Lu et al., 2014) suggest that the activation of the HPI axis is not requisite for the action of hypoxia on reproductive processes in fish.

3.5. Stimulatory Effects of Stress on Reproduction

While the majority of studies have reported inhibitory effects of stress on reproduction, there appear to be at least some conditions where corticosteroids may play permissive or stimulatory roles during reproduction, and most of these effects are seen in females during the periovulatory period (Milla et al., 2009). A number of studies have demonstrated the potentiating or direct stimulatory role of corticosteroids on in vitro oocyte maturation (Goetz, 1983; Greeley et al., 1986; Upadhyaya and Haider, 1986; Patino and Thomas, 1990), but with corticosteroids typically having much lower direct maturational potency than progestins, and in turn, 11-oxygenated corticosteroids being less effective than 11-deoxy corticosteroids. However, in other studies there is no potentiating effect of cortisol on in vitro oocyte maturation (Pankhurst, 1985). There is also evidence for the direct action of corticosteroids with 11-deoxycortisol having the highest maturational potency in vitro, and also being present in the plasma of fish undergoing final maturation in the orange roughy (*Hoplostethus atlanticus*) (Pankhurst, 1987) and the cyprinid (*Chalcalburnus tarichi*) (Ünal et al., 2008). *Chalcalburnus tarichi* also showed an increase in plasma levels of 11-deoxycortisol following treatment with hCG, and there is evidence for the synthesis of corticosteroids from radiolabeled steroid precursors by both ovarian (Kime et al., 1992) and testicular (Lee et al., 2000) tissue. It is still unclear whether corticosteroids have direct functional roles in gamete maturation, or show maturational activity as a result of having functional groups (eg, 21-hydroxylation) that mimic the action of hydroxylated progestins.

As noted in the earlier section, the typical effect of stress on maturational steroid levels in vivo is either maintenance of press-stress levels, or an increase in plasma concentrations. In salmonids at least, maturational steroids also appear to be able to stimulate the interrenal production of cortisol (Barry et al., 1997), and this is consistent with the periovulatory increases in corticosteroids reported for some species. Effects on males during the periovulatory period appear to be more uniformly negative, with little evidence that corticosteroids have a role in potentiating or supporting maturation of the testis (Milla et al., 2009).

4. MECHANISMS OF STRESS ACTION

The assessment of mechanisms by which stress impacts the reproductive endocrine axis is complicated by the fact that there is scope for both direct action of hormone products of the HPI axis on reproductive performance, and indirect effects arising from the regulatory action that the HPI axis has on behavior, metabolism, and growth (Leatherland et al., 2010). With that caveat, a reasonable working assumption was that as in mammals, there would in fish be effects at multiple levels of the endocrine cascade, and many of these would be mediated by the action of corticosteroids. GRs have been identified at multiple sites in the reproductive axis including GnRH neurons in the caudal telencephalon and POA of the brain, FSH- and LH-secreting cells of the pituitary (Teitsma et al., 1999), hepatocytes of the liver (Lethimonier et al., 2002), and ovarian and testicular tissues (Takeo et al., 1996; Milla et al., 2008). Borski et al. (2001) also reported nongenomic effects of cortisol on prolactin secreting cells in the pituitary of Mozambique tilapia, and mammalian studies have identified membrane-bound corticosteroid receptors mediating nongenomic effects in a wide range of tissues (Tasker et al., 2006). This resulted in an initial focus on the role of cortisol in modulating stress effects on reproduction in fish.

4.1. The Role of Cortisol: In Vivo Protocols

In vivo administration of cortisol generally mimics the inhibitory effects seen in response to exposure to stress. Eighteen days after implant with cortisol pellets, brown trout showed decreased ovary and testes weight, reduced T, and in females, E2 and VTG levels, and reduced pituitary LH content in both sexes. In contrast there was no effect on plasma 11KT levels in males or plasma LH in either sex. In male rainbow trout there was a fall in plasma LH, but no effect on other reproductive parameters, and immature female rainbow trout showed a fall in plasma VTG (Carragher et al., 1989). Mozambique tilapia (*Oreochromis mossambicus*) implanted with cortisol for 18 days similarly showed reduced body weight, GSI, plasma T and E2 in females, and plasma T in males. Females also showed increased incidence of follicular atresia (Foo and Lam, 1993a,b). Implants of cortisol in immature female rainbow trout resulted in reduced hepatic E2 binding sites after 4 weeks, but an increase in plasma E2 binding capacity, and this was interpreted as one of the mechanisms by which cortisol might suppress VTG production (Pottinger and Pickering, 1990). Similar effects were reported in a later study where cortisol implants in brown trout resulted in reduced plasma E2, VTG, and hepatic E2 binding capacity, and increased binding capacity of plasma E2 (Pottinger et al., 1991). Previtellogenic rainbow trout implanted with cortisol for 15 days showed reduced hepatic ER mRNA, and the autostimulation of ER mRNA by E2 was inhibited by in vitro treatment of hepatocytes with dexamethasone (Lethimonier et al., 2002). A reverse effect of stress on steroid binding protein (SBP) capacity was found in black bream, with confinement stress resulting in reduced SBP capacity in black bream 6 h poststress but having no effect on either binding capacity or affinity of SBP in rainbow trout (Hobby et al., 2000). In companion relative-binding studies, cortisol added at $100 \times$ the concentration of gonadal steroids effectively displaced E_2 from the binding protein, indicating that cortisol effects could arise from both reduced SBP capacity and the direct competitive binding effects of cortisol in species where normal gonadal steroid titers were low compared to stress-induced levels of cortisol (Hobby et al., 2000).

A series of experiments on juvenile male common carp (Consten et al., 2000, 2001a,b; Goos and Consten, 2002) where fish were fed cortisol-laced food for periods of up to 160 days resulted in reduced GnRH content in the brain; falls in pituitary FSH and LH mRNA, plasma T, 11KT, 11KA, and GSI; and retarded spermatogenesis. The effects took some time to materialize and were accompanied by inhibition of growth, again making it difficult to ascribe the effects of cortisol to direct action on the HPG axis. Restoration of 11KT levels by cotreatment of fish with 11KA did not offset the suppression of gonad growth, suggesting that the suppressive effects were not mediated by falls in plasma androgens (Goos and Consten, 2002). However, as noted previously for the effects of stress, not all cortisol effects are inhibitory. Four weeks of weekly cortisol injections resulted in upregulation of LHβ expression and increased pituitary LH in the European eel (*Anguilla anguilla*) (Huang et al., 1999).

In an attempt to avoid the ambiguity of effects of longer term cortisol treatment, Pankhurst and Van Der Kraak (2000) examined the effect of transitory elevation in plasma cortisol following cortisol injection of female rainbow trout. Treatment of fish in mid-vitellogenesis had no effect on plasma T or E2 levels, whereas preovulatory fish showed suppressed T at 1 and 3 h postinjection (pi), and suppressed E2 at 3 and 6 h pi. There was no effect on plasma LH, suggesting that the steroid-suppressing effects of cortisol can occur through direct action and accrue in the endocrine cascade, somewhere between LH-LHR binding and the synthesis of T. Short-term elevations of plasma cortisol to very high levels through injection of cortisol in an oil vehicle also had an effect on reproductive behavior in wild male largemouth bass (*Micropterus salmoides*), with an increased incidence of nest abandonment by territorial males following cortisol treatment. However, cortisol-treated fish also showed increased incidence of bacterial infection and the behavioral effect was interpreted as arising from a systemic effect of cortisol on metabolism and health status (O'Connor et al., 2009). This is supported by the absence of any change in nesting behavior of wild male spiny damselfish in response to underwater cortisol injection over short periods (24–48 h) (Pankhurst, 2001).

4.2. The Role of Cortisol: In Vitro Protocols

The interpretation of in vivo effects of cortisol administration assumes that they are through direct action. This assumption was supported by initial in vitro experiments that incubated ovarian follicles of rainbow trout with cortisol. Basal T and E2 secretion were suppressed in a dose-dependent manner at physiological doses of cortisol, with effects on T production being present at cortisol levels as low as 5 ng mL^{-1} (Carragher and Sumpter, 1990). Extensive replication of the same protocols in subsequent studies found a much less consistent effect, with suppression of basal E2 in only 4 out of 20 experiments at cortisol concentrations of 100–1000 ng mL^{-1}, and no effects of cortisol on either 17-hydroxyprogesterone (17 P)- or hCG-stimulated E2 production (Pankhurst, 1998b). Neither was there any consistent effect on the production of conjugated steroids nor absorption of steroid into oocytes as possible sinks for loss of free steroid from the incubation medium. The possibility that more subtle effects might be masked by the high levels of steroids typically produced by trout follicles in culture was investigated by applying the same procedures to nonsalmonids where in vitro steroid production is typically much lower. Cortisol at concentrations of up to $1000 \, \text{ng mL}^{-1}$ had no effect on basal T and E2 production, the conversion of the steroid precursor 25-hydroxycholesterol, or the steroidogenic response to hCG or carp LH in isolated ovarian follicles from goldfish (*Carassius auratus*), common carp, or Australasian snapper (Pankhurst et al., 1995). This led to the conclusion that the direct action of cortisol at the ovarian level was not consistent enough to explain the sometimes profound and rapid falls in plasma steroid levels occurring after the imposition of stress.

More recently, Alsop et al. (2009) also failed to find any effect of cortisol on hCG-stimulated E2 production by follicles of zebrafish. Further examination of rainbow trout follicles (Reddy et al., 1999) reported suppressive effects of cortisol at $100 \, \text{ng mL}^{-1}$ on both basal and LH-stimulated production of T and E2 in late vitellogenic, but not preovulatory follicles. The conclusion here was that the suppressive effect of cortisol is present but only when basal levels of T and E2 were high, although the reported levels were similar to, or lower than those found in the earlier study (Pankhurst, 1998b). A later study using mid-vitellogenic rainbow trout follicles did show depression of both basal, and cAMP-stimulated T and E2 secretion in incubations containing cortisol at $10 \, \text{ng mL}^{-1}$, but cortisol had no effect on the in vitro conversion of T to E2. Cortisol treatment also suppressed expression of StAR and P_{450} side chain cleavage enzyme (responsible for mediation of conversion of cholesterol to pregnenolone) mRNA levels (Barkataki et al., 2011). The conclusion was that the inhibitory effect of cortisol resulted from inhibition of both delivery of cholesterol to inner mitochondrial membranes, and its conversion to pregnenolone.

A similar effect appears to have been present in follicles harvested from stressed spiny damselfish 24 h postcapture where plasma steroid levels were depressed but follicles still converted 17 P to T and T to E2 but were unresponsive to hCG (Pankhurst, 2001). It remains unclear whether the inconsistency of the effects of in vitro exposure of ovarian follicles to cortisol is indicative of cortisol not having a strong inhibitory role at the ovarian level, or results from the inadequacy of in vitro systems for suitably modeling the dynamic secretion, transport, metabolism, and clearance processes that collectively determine circulating levels of plasma steroids. The fact that the effects of stress on plasma steroid levels can be very rapid suggests that if cortisol is the agent, then at least the early stages of this effect will need to occur through nongenomic mechanisms.

Effects on testicular tissue appear to be more consistent. In vitro production of 11KT from common carp testis stimulated with LH was lower in fish that had been treated for up to 160 days with cortisol via feed, but the effect was only present during early stages of pubertal development (Consten et al., 2000). Incubation of testis with high concentrations of cortisol (2000 ng mL^{-1}) resulted in decreased concentrations of 11KA (Consten et al., 2002). The mechanism was proposed to be through substrate competition with cortisol for 11βHSD, which as noted earlier mediates conversion of 11β-hydroxyandrostenedione to 11KA, and cortisol to cortisone. There is also in vitro evidence for direct action of cortisol at other sites in the reproductive axis. Lethimonier et al. (2002) reported the GR-mediated suppression of E2-induced upregulation of ER expression in incubations of rainbow trout hepatocytes. The mechanism involved cortisol inhibition of binding of a transcriptional factor C/EPBβ to the ER gene promoter region.

4.3. Effects of Other Stress Factors

While the majority of studies have examined the effect of cortisol, a smaller number of investigations have been targeted at other stress-related endocrine factors. Melanocortin2 receptors (MC2R) are strongly expressed in ovarian and testicular tissue of zebrafish, and the MC2R ligand ACTH inhibits the hCG-stimulated, but not basal secretion of E2 from isolated ovarian follicles (Alsop et al., 2009). The suppressive effect of ACTH could be abolished by cotreatment with forskolin, a cyclic AMP activator, and 8-bromoc AMP, indicating that the level of action was above adenylate cyclase activation. The suppressive effect also disappeared at longer incubation times, suggesting that the failure to detect any effect of ACTH on LH-stimulated E2 production by brown trout follicles (Sumpter et al., 1987), or basal T and E2 production by rainbow trout follicles (Fiztgibbon and Pankhurst, unpublished data) could arise either from the lack of a similar effect in salmonids, or the choice of incubation parameters used. There is also evidence for the role of other pituitary-derived factors in stress-suppression of reproduction. In Mozambique tilapia, a 22 days administration of β-endorphin reduced immunohistochemical staining of LHβ subunits in gonadotrophs in the *par distalis* of the pituitary (Ganesh and Chabbi, 2013). The effect could be reproduced by multiple daily applications of handling and netting stress, and was attenuated by cotreatment with the opioid-receptor antagonist naltrexone, indicating that the effect was opioid-receptor mediated.

CRF is also implicated in the mediation of stress-related effects but here also there is some uncertainty in terms of whether the effects are direct or mediated through hormones lower in the HPI endocrine axis. Daily intraperitoneal injection of Mozambique tilapia with CRF for 22 days resulted in increased plasma cortisol levels, reduction in the intensity of LHβ staining in the pituitary, reduced GSI, the loss of vitellogenic follicles in the ovary, and increased follicular atresia (Chabbi and Ganesh, 2013). Netting and handling stress produced similar effects, and these effects could be ameliorated by cotreatment with metyrapone, suggesting that the effects were cortisol-mediated rather than through the direct action of CRF. CRF does have direct effects on a range of behaviors including suppression of appetite and feeding, reduction in foraging, and increased locomotor activity (Bernier, 2006; Lowry and Moore, 2006), and some evidence exists that CRF may also have a role in modulating reproductive behavior. As noted earlier, nonterritorial African cichlids have elevated brain and pituitary CRFR; elevated pituitary CRF mRNA in association with smaller GnRH neurons, depressed GnRHR, FSHβ, and LHβ mRNA; and low plasma levels of LH, FSH, and androgens (Maruska, 2014). Changes in social status associated with nonterritorial fish gaining access to vacated territories result in rapid behavioral changes (within 10 min); increases in plasma LH, FSH, and androgens within 30 min; and longer term (5–7 days) increases in GnRH neuron size. The inference is that CRF has a key role in suppressing reproductive development in subordinate males. The short timeframes for change in response to social status suggest that either the effects are due to the direct action of CRF, or if they are cortisol-mediated then, here again, they must be occurring through nongenomic processes.

The most proximate component of the response to acute stress involves the neutrally mediated release of the catecholamines noradrenaline (NA) and adrenaline (AD) from the chromaffin tissue of the kidney to the circulation, with effects including increased hemoglobin affinity for oxygen, cardiac stimulation, increased arterial blood pressure, and the release of hepatic glycogen stores to increase plasma glucose levels (reviewed in Pankhurst, 2011). The evidence available suggests that acute elevations of plasma catecholamine levels are primarily stimulatory rather than inhibitory, aiding overall recovery from stress and also protecting organs such as the heart (Eliason et al., 2011). The same appears to be true for reproductive processes. Yu and Peter (1992) reported the stimulatory effect of NA on in vitro release of GnRH from slices of goldfish hypothalamus, and the effect could be reproduced by treatment with α_1-adrenergic receptor agonists and blocked by α_1 receptor antagonists. In mammals, NA typically has permissive or stimulatory effects on ovarian steroidogenesis (Aguado, 2002), and in vitro incubation of rainbow trout and spiny damselfish follicles with NA and α-adrenergic receptor agonists results in either no change or increases in basal E2 production (Fitzgibbon, Gonzalez-Reynoso, and Pankhurst, unpublished data). AD also stimulates follicle contraction and ovulation in brook trout (*Salvelinus fontinalis*) follicles, an effect blocked by α-adrenergic receptor antagonists (Goetz and Bradley, 1994). Catecholamines could also generate indirect inhibitory effects in species such as European sea bass (*Dicentrarchus labrax*), where both NA and AD stimulate corticosteroid production by interrenal tissue through β-adrenergic receptor stimulation (Rotllant et al., 2006).

A summary of the recorded effects of stress on reproductive performance and the possible mechanisms involved is presented in Fig. 8.4.

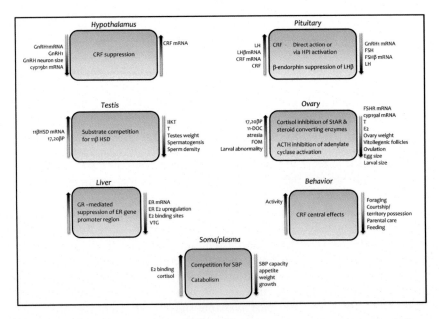

Figure 8.4. Summary of the effects of stress on components of the hypothalamic–pituitary–gonadal (HPG) axis. Upward arrows indicate stimulatory effects; downward arrows indicate inhibitory effects. Text boxes summarize possible mechanisms of action.

5. STRESS EFFECTS ON REPRODUCTION IN NATURAL ENVIRONMENTS

In considering the question of whether animals in the wild typically experience stress over normal ranges of activity and conditions, Pankhurst (2011) concluded that generally the answer is no. Our understanding of the physiology of stress tends to be biased by the domination of the literature by studies on captive and cultured fishes held under unnatural conditions, and with limited capacity to avoid or ameliorate stress through changing behavior. Further understanding of the processes in fish in the wild is limited by the small number of studies addressing the phenomenon in free-living fishes, and the fact that natural events that might be predicted to be stressful (eg, storms and floods) generally preclude both observation and sampling (Pankhurst, 2011). Field studies on birds show that both environmental harshness and the demands of reproduction are not stressful under conditions that are predictable (ie, animals are generally good at coping) (Wingfield, 1994). There is also some evidence from fish that predictable stressors produce less marked stress responses than unpredictable ones (Galhardo et al., 2011). Under conditions sufficiently challenging to elevate corticosteroids (for birds, typically storms or periods of high temperature), there is a shift in behavior from nesting, to foraging or refuge seeking, which serves to ameliorate the stress and also support recovery and resumption of reproductive behavior when conditions permit (Romero et al., 2000). The general view is that events such as migration and reproduction are demanding but not necessarily stressful, but as Schreck (2010) notes, if resistance to stressors is energetically costly, then this is likely to be reflected in reduced capacity to allocate energy resources in support of reproduction.

The limited field data available show that cortisol levels vary quite considerably among individuals in natural populations, and also that this does not seem to be correlated with any negative effect on reproduction. Plasma cortisol levels in territorial adult spiny damselfish captured and sampled underwater ranged from <1 and $42\,ng\,mL^{-1}$, but there was no relationship between cortisol plasma T and E2 levels in females, or T and 11KT in males, (and as noted earlier stress-suppression of sex steroids occurs in both sexes in the laboratory) (Pankhurst, 2001). Similarly, male bluegill sunfish (*Lepomis macrochirus*) engaged in parental care show normal cycles of plasma T and 11KT in association with spawning and egg protection even when cortisol levels are as high as $125\,ng\,mL^{-1}$ (Magee et al., 2006). As noted in Pankhurst (2011), the conclusion from these studies was that reproductive activity can be maintained over a range of corticosteroid states, and either that moderately elevated plasma cortisol levels in wild fish are not indicative of stress, or that the elevation is stress-related but not of sufficiently long duration to negatively affect reproduction.

Nutritional status clearly has a marked effect on the capacity of fish to respond to or manage the effects of stress in terms of maintaining the positive energy balance that enables the continued support or reproductive processes (Schreck, 2010). Evidence from natural populations of tropical damselfishes supports this view. Habitat variability in the form of coral cover and quality had no effect on plasma cortisol levels in either spiny damselfish or the staghorn damselfish (*Amblyglyphidodon curacao*), but ovarian condition was best in habitats of low fish density and consequent reduced competition for access to planktonic food resources (Pankhurst et al., 2008a,b). In contrast, poorest ovarian condition was found in habitats with high coral cover and fish density—the prettiest ones to human eyes were curiously the more demanding habitats for reproductive success. Bicolor damselfish (*Stegastes partitus*) did show differences in behavior in different habitats, with fish in coral rubble showing higher aggression, more frequent courtship and shelter-seeking behavior than fish on live coral, despite there being no differences in resting plasma cortisol levels, and increases in cortisol occurring with stress in fish from both habitats (Schrandt and Lema, 2011). There were, however, sex-related differences in stress-stimulated expression of CRF and urotensin I (UI) mRNA that were habitat-specific, but the significance of these differences is not yet understood.

An interesting subset of natural fish behavior, where stress modulates reproduction, exists among species where there is social regulation of reproductive status. This is strongly expressed in cichlids, where only a minority of the males in the population are the dominant, territorial sexually active fish. The majority of subordinate, nonterritorial fish are not sexually active and have low levels of reproductive endocrine activity in concert with elevated pituitary expression of CRF and CRFR1 mRNA (Maruska, 2014). The inference is that the effect in subordinate fish is a stress response to social dominance, but whether this is a classical stress response or an expression of the more general neuromodulatory role of CRF is still not clear. More recent studies suggest that the effect is not corticosteroid-mediated. Induction of social instability by shifting dominant male Mozambique tilapia among groups resulted in increases in plasma T and 11KT with the reestablishment of social dominance but no effects on plasma cortisol levels of either dominant or subordinate males (Almeida et al., 2014).

6. FUTURE DIRECTIONS

Despite the significant volume of work discussed here and the very substantial advances made in understanding since Sumpter and colleagues (Sumpter et al., 1987) reflected on how little we knew about the effects of stress and reproduction, there is still much to know. We have made little substantial progress on the understanding of the effects of stress on progeny beyond the earliest stages of development, and none on the effects on reproductive performance of progeny, or the multigenerational effects of stress. There is a generally consistent demonstration of the effects of stress at the gonadal level but still a very patchy understanding of the effects and interactions at higher levels in the HPG axis. Our understanding of the mechanisms by which these effects are generated is similarly incomplete. There is still ambiguity over the action of corticosteroids; there are certainly long-term systemic effects, possibly direct genomic effects over shorter timeframes, and possibly rapid proximate effects mediated by nongenomic mechanisms, but how and when these effects operate is poorly understood. There is also still very limited understanding of the role of CRF in the direct modulation of the HPG axis as well as its effects through the downstream agencies of ACTH, UI, β-endorphin, and corticosteroids. There is also no information on the possible effects of stress on the kisspeptin-kissR system, and still relatively modest understanding of the possible roles of catecholamines in regulating reproductive processes. The proliferation of molecular tools offers the opportunity for rapid expansion of inquiry into areas previously limited by access to suitable parameter measurement, but here also the process is unlikely to be straightforward. Chand and Lovejoy (2011) summarized the problem: "delineating molecular mechanisms by which this (the inhibition of reproductive function by activation of the stress response) occurs in vertebrates continues to present a major challenge." This is still the case.

It is also still unclear whether the widely demonstrated inhibitory effects of thermal stress and the equally profound effects of hypoxia are modulated through the HPI axis. This has significant implications in terms of how we might predict effects of climate and environmental change in natural populations, or manage the effects of thermal stress in cultured populations of fish. Amelioration strategies have tended to focus on endocrine manipulation of the HPG axis (Pankhurst and King, 2010). The possibility that stress may be the intermediary between thermal insult and reproductive inhibition suggests that there may be unexplored and more effective potential for protection from thermal damage by manipulation of the HPI axis. Finally, there is still the strong suspicion that our perceptions about the role of stress in natural habitats and populations are still overly influenced by what we have learned about stress processes from laboratory-based experiments. We still do not know whether (or when) natural conditions are stressful enough to trigger inhibitory effects on reproduction of the type seen in laboratory experiments, and what role stress plays in the normal control and organization of reproduction in free-living populations. Among fish where social control of reproduction in association with activation of the HPI axis has been demonstrated it seems possible that stress may initiate changes, but it does not seem as likely that it can maintain them. The sum effect of this is that there is no shortage of substrate for inquiry into the effects of stress on reproduction right from the benchtop to the natural environment but there is also a solid foundation on which to base that inquiry.

REFERENCES

Aguado, L. I. (2002). Role of the central and peripheral nervous system in the ovarian function. *Microscopy Res. Tech.* **59**, 462 473.

Almeida, O., Gonçalves-de-Freitas, E., Lopes, J. S. and Oliveira, R. F. (2014). Social instability promotes hormone-behavior associated patterns in a cichlid fish. *Horm. Behav.* **66**, 369–382.

Alsop, D., Ings, J. S. and Vijayan, M. M. (2009). Adrenocorticotropic hormone suppresses gonadotropin-stimulated estradiol release from zebrafish ovarian follicles. *PLoS ONE* **4**, e6463.

Aluru, N. and Vijayan, M. M. (2009). Stress transcriptomics in fish: a role for genomic cortisol signalling. *Gen. Comp. Endocrinol.* **164**, 142–150.

Anderson, K., Swanson, P., Pankhurst, N., King, H. and Elizur, A. (2012a). Effect of thermal challenge on plasma gonadotropin levels and ovarian steroidogenesis in female and repeat spawning Tasmanian Atlantic salmon (*Salmo salar*). *Aquaculture* **334**, 205–212.

Anderson, K., King, H., Pankhurst, N., Ruff, N., Pankhurst, P. and Elizur, A. (2012b). Effect of elevated temperature on estrogenic induction of vitellogenesis and zonagenesis in juvenile Atlantic salmon (*Salmo salar*). *Mar. Freshwat. Behav. Physiol.* **45**, 1–15.

Auperin, B. and Geslin, M. (2008). Plasma cortisol response to stress in juvenile rainbow trout is influenced by their life history during early development and by egg cortisol content. *Gen. Comp. Endocrinol.* **158**, 234–239.

Barkataki, S., Aluru, N., Li, M., Lin, L., Christie, H., Vijayan, M. M., et al. (2011). Cortisol inhibition of 17β-estradiol secretion by rainbow trout ovarian follicles involves modulation of *Star* and *P450scc* gene expression. *J. Aquac. Res. Devel.* **S2**, 001. ⟨http://dx.doi.org/10.4172/2155-9546.S2-001⟩.

Barry, T. P., Riebe, J. D., Parrish, J. J. and Malison, J. A. (1997). Effects of 17,20β-dihydroxy-4-pregnen-3-one on cortisol production by rainbow trout interrenal tissue in vitro. *Gen. Comp. Endocrinol.* **107**, 172–181.

Bernier, N. J. (2006). The corticotrophin-releasing factor system as a mediator of the appetite-suppressing effects of stress in fish. *Gen. Comp. Endocrinol.* **146**, 45–55.

Bhandari, R. K., Deem, S. L., Holliday, D. K., Jandegian, C. M., Kassotis, C. D., Nagel, S. C., et al. (2015). Effects of the environmental estrogenic contaminants bisphenol A and 17α-ethinyl estradiol on sexual development and adult behaviors in aquatic wildlife species. *Gen. Comp. Endocrinol.* **214**, 195–219.

Black, M. P., Balthazart, J., Baillien, M. and Grober, M. S. (2011). Rapid increase in aggressive behavior precedes the decrease in brain aromatase activity during socially mediated sex change in *Lythrypnus dalli*. *Gen. Comp. Endocrinol.* **170**, 119–124.

Borski, R. J., Hyde, G. N., Fruchtman, S. and Tsai, W. S. (2001). Cortisol suppresses prolactin release through a non-genomic mechanism involving interactions with the plasma membrane. *Comp. Biochem. Physiol.* **129B**, 533–541.

Campbell, P. M., Pottinger, T. G. and Sumpter, J. P. (1992). Stress reduces the quality of gametes produced by rainbow trout. *Biol. Reprod.* **47**, 1140–1150.

Campbell, P. M., Pottinger, T. G. and Sumpter, J. P. (1994). Preliminary evidence that chronic confinement stress reduces the quality of gametes produced by brown and rainbow trout. *Aquaculture* **120**, 151–169.

Carragher, J. F. and Pankhurst, N. W. (1991). Stress and reproduction in a commercially important marine fish, *Pagrus auratus* (Sparidae). In *Reproductive Physiology of Fish 1991*, 91 (eds. A. P. Scott, J. P. Sumpter, D. E. Kime and M. S. Rolfe), pp. 253–255. Sheffield: FishSymp.

Carragher, J. F. and Sumpter, J. P. (1990). The effect of cortisol on the secretion of sex steroids from cultured ovarian follicles of rainbow trout. *Gen. Comp. Endocrinol.* **77**, 403–407.

Carragher, J. F., Sumpter, J. P., Pottinger, T. G. and Pickering, A. D. (1989). The deleterious effects of cortisol implantation on reproductive function in two species of trout, *Salmo trutta* L. and *Salmo gairdneri* Richardson. *Gen. Comp. Endocrinol.* **76**, 310–321.

Castranova, D. A., King, W. and Woods, L. C. (2005). The effects of stress on androgen production, spermiation response and sperm quality in high and low cortisol responsive domesticated male striped bass. *Aquaculture* **246**, 413–422.

Chabbi, A. and Ganesh, C. B. (2013). Glucocorticoid synthesis inhibitor metyrapone blocks stress-induced suppression along luteinizing hormone secreting cells-ovary axis in the fish *Oreochromis mossambicus*. *J. Exp. Zool.* **321A**, 125–134.

Chand, D. and Lovejoy, D. A. (2011). Stress and reproduction: controversies and challenges. *Gen. Comp. Endocrinol.* **171**, 253–257.

Cheung, C. H. Y., Chiu, J. M. Y. and Wu, R. S. S. (2014). Hypoxia turns female medaka fish into phenotypic males. *Ecotoxicology* **23**, 1260–1269.

Clearwater, J. C. and Pankhurst, N. W. (1997). The response to capture and confinement stress of plasma cortisol, plasma sex steroids and vitellogenic oocytes in the marine teleost, red gurnard (*Chelidonichthys kumu*) (Triglidae). *J. Fish. Biol.* **50**, 429–441.

Cleary, J. J., Pankhurst, N. W. and Battaglene, S. C. (2000). The effect of capture and handling stress on plasma steroid levels and gonadal condition in wild and farmed snapper, *Pagrus auratus* (Sparidae). *J. World Aquacult. Soc.* **31**, 558–569.

Cleary, J. J., Battaglene, S. C. and Pankhurst, N. W. (2002). Capture and handling stress affects the endocrine and ovulatory response to exogenous hormone treatment in snapper, *Pagrus auratus* (Bloch & Schneider). *Aquacult. Res.* **33**, 1–10.

Consten, D., Lambert, J. G. D. and Goos, H. J. Th (2000). Inhibitory effects of cortisol on in vivo and in vitro androgen secretion in male common carp, *Cyprinus carpio*. In *Proceedings of the 6th International Symposium on the Reproductive Physiology of Fish* (eds. B. Norberg, O. S. Kjesbu, G. L. Taranger, E. Andersson and S. O. Stefansson), p. 192. Bergen: John Grieg AS.

Consten, D., Bogerd, J., Komen, J., Lambert, J. G. D. and Goos, H. J. Th (2001a). Long-term cortisol treatment inhibits pubertal development in male common carp, *Cyprinus carpio* L. *Biol. Reprod.* **64**, 1063–1071.

Consten, D., Lambert, J. G. D. and Goos, H. J. Th (2001b). Cortisol affects testicular development in male common carp, *Cyprinus carpio* L., but not via an effect on LH secretion. *Comp. Biochem. Physiol.* **129B**, 671–677.

Consten, D., Keuning, E. D., Terlou, M., Lambert, J. G. D. and Goos, H. J. Th (2002). Cortisol effects on the testicular androgen synthesizing capacity in common carp, *Cyprinus carpio* L. *Fish Physiol. Biochem.* **25**, 91–98.

Contreras-Sánchez, W. M., Schreck, C. B., Fitzpatrick, M. S. and Pereira, C. B. (1998). Effects of stress on the reproductive performance of rainbow trout (*Oncorhynchus mykiss*). *Biol. Reprod.* **58**, 439–447.

Coward, K., Bromage, N. R. and Little, D. C. (1998). Inhibition of spawning and associated suppression of sex steroid levels during confinement in the substrate-spawning *Tilapia zillii*. *J. Fish. Biol.* **52**, 152–165.

Dabrowski, K., Rinchard, J., Ottobre, J. S., Alcantara, F., Padilla, P., Ciereszko, A., et al. (2003). Effect of oxygen saturation in water on reproductive performance of pacu *Piaractus brachypomus*. *J. World Aquacult. Soc.* **34**, 441–449.

De Montalembert, G., Jalabert, B. and Bry, C. (1978). Precocious induction of maturation and ovulation in northern pike (*Esox lucius*). *Ann. Biol. Ann. Bioch. Biophys.* **18**, 969–975.

Devlin, R. H. and Nagahama, Y. (2002). Sex determination and sex differentiation in fish: an overview of genetic, physiological and environmental influences. *Aquaculture* **208**, 191–364.

Dufour, S., Sebert, M.-E., Weltzein, F.-A., Rousseau, K. and Pasqualini, C. (2010). Neuroendocrine control by dopamine of teleost reproduction. *J. Fish. Biol.* **76**, 129–160.

Eliason, E. J., Clark, T. D., Hague, M. J., Hanson, L. M., Gallagher, Z. S., Jeffries, K. M., et al. (2011). Differences in thermal tolerance among sockeye salmon populations. *Science* **332**, 109 112.

Eriksen, M. S., Bakken, M., Espmark, Å. and Braastad, B. O. (2006). Prespawning stress in farmed Atlantic salmon *Salmo salar*: maternal cortisol exposure and hyperthermia during embryonic development affect offspring survival, growth and incidence of malfunctions. *J. Fish. Biol.* **69**, 114–129.

Eriksen, M. S., Espmark, Å., Braastad, B. O., Salte, R. and Bakken, M. (2007). Long-term effects of maternal cortisol exposure and mild hyperthermia during embryogeny on survival, growth and morphological anomalies in farmed Atlantic salmon *Salmo salar* offspring. *J. Fish. Biol.* **70**, 462–473.

Fernandino, J. I., Hattori, R. S., Moreno Acosta, O. D., Strüssman, C. A. and Somoza, G. M. (2013). Environment stress-induced testis differentiation: androgen as a by-product of cortisol inactivation. *Gen. Comp. Endocrinol.* **192**, 36–44.

Foo, J. T. W. and Lam, T. J. (1993a). Retardation of ovarian growth and depression of serum steroid levels in the tilapia *Oreochromis mossambicus*, by cortisol implantation. *Aquaculture* **115**, 133–143.

Foo, J. T. W. and Lam, T. J. (1993b). Serum cortisol response to handling stress and the effect of cortisol implantation on testosterone level in the tilapia *Oreochromis mossambicus*. *Aquaculture* **115**, 145–158.

Fostier, A. and Jalabert, B. (1982). Physiological basis of practical means to induce ovulation in fish. In *Proceedings of the International Symposium on the Reproductive Physiology of Fish 1982* (eds. C. J. J. Richter and H. J. Th Goos), pp. 164–173. Wageningen: Pudoc.

Fox, H. E., White, S. A., Kao, M. H. F. and Fernald, R. D. (1997). Stress and dominance in a social fish. *J. Neurosci.* **17**, 6463–6469.

Frisch, A. (2004). Sex-change and gonadal steroids in sequentially-hermaphroditic teleost fish. *Rev. Fish Biol. Fish.* **14**, 481–489.

Frisch, A. J., Walker, S. P. W., McCormick, M. I. and Solomon-Lane, T. K. (2007). Regulation of protogynous sex change by competition between corticosteroids and androgens: an experimental test using sandperch, *Parapercis cylindrica*. *Horm. Behav.* **52**, 540–545.

Fuzzen, M. L. M., Bernier, N. J. and Van Der Kraak, G. (2011). Stress and reproduction. In *Hormones and Reproduction of Vertebrates. Volume 1: Fishes* (eds. D. O. Norris and K. H. Lopez), pp. 103–117. Amsterdam: Elsevier.

Galhardo, L., Vital, J. and Oliveira, R. F. (2011). The role of predictability in the stress response of a cichlid fish. *Physiol. Behav.* **102**, 367–372.

Ganesh, C. B. and Chabbi, A. (2013). Naltrexone attenuates stress-induced suppression of LH secretion in the pituitary gland in the cichlid fish *Oreochromis mossambicus*: evidence for the opioidergic mediation of reproductive stress response. *Fish Physiol. Biochem.* **39**, 627–636.

Gardner, L., Anderson, T., Place, A. R., Dixon, B. and Elizur, A. (2005). Sex change strategy and aromatase genes. *Steroid Biochem. Mol. Biol.* **94**, 395–404.

Gerking, S. D. (1982). The sensitivity of reproduction in fishes to stressful environmental conditions. In *Proceedings of the International Symposium on the Reproductive Physiology of Fish 1982* (eds. C. J. J. Richter and H. J. Th Goos), pp. 224–228. Wageningen: Pudoc.

Gillet, C. and Breton, B. (2009). LH secretion and ovulation following exposure of Arctic charr to different temperature and photoperiod regimes: responsiveness of females to a gonadotropin-releasing hormone analogue and a dopamine antagonist. *Gen. Comp. Endocrinol.* **162**, 210–218.

Gillet, C., Breton, B., Mikolajczyk, T., Bodinier, P. and Fostier, A. (2011). Disruption of the secretion and action of 17,20β-dihydroxy-4-pregnen-3-one in response to a rise in temperature in the Arctic charr, *Salvelinus alpinus*. Consequences on oocyte maturation and ovulation. *Gen. Comp. Endocrinol.* **172**, 392–399.

Goetz, F. W. (1983). Hormonal control of oocyte final maturation and ovulation in fishes. In *Fish Physiology - Reproduction: Behavior and Fertility Control*, Vol. 9B (eds. W. S. Hoar, D. J. Randall and E. M. Donaldson), pp. 117–170. New York, NY: Academic Press.

Goetz, F. W. and Bradley, J. A. (1994). Stimulation of in vitro ovulation and contraction of brook trout (*Salvelinus fontinalis*) follicles by adrenaline through α-adrenoreceptors. *J. Reprod. Fertil.* **100**, 381–385.

Goos, H. J. Th and Consten, D. (2002). Stress adaptation, cortisol and pubertal development in the male common carp, *Cyprinus carpio*. *Mol. Cell. Endocrinol.* **197**, 105–116.

Greeley, M. S., Calder, D. R., Taylor, M. H., Hols, H. and Wallace, R. A. (1986). Oocyte maturation in the mummichog (*Fundulus heteroclitus*): effects of steroids on germinal vesicle breakdown of intact follicles in vitro. *Gen. Comp. Endocrinol.* **62**, 281–289.

Guiguen, Y., Fostier, A., Piferrer, F. and Chang, C.-F. (2010). Ovarian aromatase and estrogens: a pivotal role for gonadal sex differentiation and sex change in fish. *Gen. Comp. Endocrinol.* **165**, 352–366.

Haddy, J. A. and Pankhurst, N. W. (1999). Stress-induced changes in concentrations of plasma sex steroids in black bream. *J. Fish. Biol.* **55**, 1304–1316.

Haddy, J. A. and Pankhurst, N. W. (2000). The efficacy of exogenous hormones in stimulating changes in plasma steroids and ovulation in wild black bream *Acanthopagrus butcheri* is improved by treatment at capture. *Aquaculture* **191**, 351–366.

Hattori, R. S., Fernandino, J. I., Kishii, A., Kimura, H., Kinno, T., Oura, M., et al. (2009). Cortisol-induced masculinization: does thermal stress affect gonadal fate in pejerrey, a teleost fish with temperature-dependent sex determination? *PLoS One* **4**, e6548.

Hayashi, Y., Kobira, H., Yamaguchi, T., Shiraishi, E., Yazawa, T., Hirai, T., et al. (2010). High temperature causes masculinization of genetically female medaka by elevation of cortisol. *Mol. Reprod. Dev.* **77**, 679–686.

Hobby, A. C., Pankhurst, N. W. and Haddy, J. A. (2000). The effect of short term confinement stress on binding characteristics of sex steroid binding protein (SBP) in female black bream (*Acanthopagrus butcheri*) and rainbow trout (*Oncorhynchus mykiss*). *Comp. Biochem. Physiol.* **125A**, 85–94.

Huang, Y.-S., Rousseau, K., Sbaiti, M., Le Belle, N., Schmitz, M. and Dufour, S. (1999). Cortisol selectively stimulates pituitary gonadotropin β-subunit in a primitive teleost, *Anguilla anguilla*. *Endocrinology* **140**, 1228–1235.

Imanaga, Y., Nyuji, M., Amano, M., Takahashi, A., Kitano, H., Yamaguchi, A., et al. (2014). Characterization of gonadotropin-releasing hormone and gonadotropin in jack mackerel (*Trachurus japonicus*): comparative gene expression analysis with respect to reproductive dysfunction in captive and wild fish. *Aquaculture* **428–429**, 226–235.

Jardine, J. J., Van Der Kraak, G. J. and Munkittrick, K. R. (1996). Capture and confinement stress in white sucker exposed to bleached kraft pulp mill effluent. *Ecotoxicol. Environ. Saf.* **33**, 287–298.

Kamstra, J. H., Aleström, P., Kooter, J. M. and Legler, J. (2014). Zebrafish as a model to study the role of DNA methylation in environmental toxicology. *Environ. Sci. Pollut. Res.* ⟨http://dx.doi.org/10.1007/s11356-014-3466-7⟩.

Kime, D. E., Scott, A. P. and Canario, A. V. M. (1992). In vitro biosynthesis of steroids, including 11-deoxycortisol and 5α-pregnane-3β,7α,20β-tetrol, by ovaries of the goldfish *Carassius auratus* during the stage of oocyte final maturation. *Gen. Comp. Endocrinol.* **87**, 375–384.

Kitano, T., Hayashi, Y., Shiraishi, E. and Kamei, Y. (2012). Estrogen rescues masculinization of genetically female medaka by exposure to cortisol or high temperature. *Mol. Reprod. Dev.* **79**, 719–726.

Kubokawa, K., Watanabe, T., Yoshioka, M. and Iwata, M. (1999). Effects of acute stress on plasma cortisol, sex steroid hormone and glucose levels in male and female sockeye salmon during the breeding season. *Aquaculture* **172**, 335–349.

Landry, C. A., Steele, S. L., Manning, S. and Cheek, A. O. (2007). Long term hypoxia suppresses reproductive capacity in the estuarine fish, *Fundulus grandis*. *Comp. Biochem. Physiol.* **148A**, 317–323.

Lane, M., Robker, R. L. and Robertson, S. A. (2014). Parenting from before conception. *Science* **345**, 756–760.

Leatherland, J. F., Li, M. and Barkataki, S. (2010). Stressors, glucocorticoids and ovarian function in teleosts. *J. Fish. Biol.* **76**, 86–111.

Lee, S. T. L., Lam, T. J. and Tan, C. H. (2000). Corticosteroid biosynthesis in vitro by testes of the grouper (*Epinephelus coioides*) after 17α-methyltestosterone-induced sex inversion. *J. Exp. Zool.* **287**, 453–457.

Lethimonier, C., Flouriot, G., Kah, O. and Ducouret, B. (2002). The glucocorticoid receptor represses the positive autoregulation of the trout estrogen receptor gene by preventing the enhancer effect of a C/EBPβ-like protein. *Endocrinology* **143**, 2961–2974.

Levavi-Sivan, B., Bogerd, J., Mañanós, E. L., Gómez, A. and Lareyre, J. J. (2010). Perspectives on fish gonadotropins and their receptors. *Gen. Comp. Endocrinol.* **165**, 412–437.

Lim, B.-S., Kagawa, H., Gen, K. and Okuzawa, K. (2003). Effects of water temperature on the gonadal development and expression of steroidogenic enzymes in the gonad of juvenile red seabream, Pagrus major. *Fish Physiol. Biochem.* **28**, 161–162.

Lowry, C. A. and Moore, F. L. (2006). Regulation of behavioural responses by corticotrophin releasing factor. *Gen. Comp. Endocrinol.* **146**, 19–27.

Lu, X., Yu, R. M. K., Murphy, M. B., Lau, K. and Wu, R. S. S. (2014). Hypoxia disrupts gene modulation along the brain-pituitary-gonad (BPG)-liver axis. *Ecotoxicol. Environ. Saf.* **102**, 70–78.

Magee, S. E., Neff, B. D. and Knapp, R. (2006). Plasma levels of androgen and cortisol in relation to breeding behaviour in parental male bluegill sunfish, *Lepomis macrochirus*. *Horm. Behav.* **49**, 598–609.

Martinez, P., Viñas, A. M., Sánchez, L., Díaz, N., Ribas, L. and Piferrer, F. (2014). Genetic architecture of sex determination in fish: applications to sex ratio control in aquaculture. *Front. Genet.* **5**, 340.

Maruska, K. P. (2014). Social regulation of reproduction in male cichlid fishes. *Gen. Comp. Endocrinol.* **207**, 2–12.

McCormick, M. I. (1998). Behaviorally induced maternal stress in a fish influences progeny quality by a hormonal mechanism. *Ecology* **79**, 1873–1883.

McCormick, M. I. (1999). Experimental test of the effect of maternal hormones on larval quality of a coral reef fish. *Oecologia* **118**, 412–422.

McCormick, M. I. (2006). Mothers matter: crowding leads to stressed mothers and smaller offspring in marine fish. *Ecology* **87**, 1104–1109.

Migaud, H., Davie, A. and Taylor, J. F. (2010). Current knowledge on the photoneuroendocrine regulation of reproduction in temperate fish species. *J. Fish. Biol.* **76**, 27–68.

Mileva, V. R., Gilmour, K. M. and Balshine, S. (2011). Effects of maternal stress on egg characteristics in a cooperatively breeding fish. *Comp. Biochem. Physiol.* **158A**, 22–29.

Milla, S., Terrien, X., Sturm, A., Ibrahim, F., Giton, F., Fiet, J., et al. (2008). Plasma 11-deoxycorticosterone (DOC) and mineralocorticoid receptor testicular expression during rainbow trout *Oncorhynchus mykiss* spermiation: implication with 17α,20β-dihydroxyprogesterone on the milt fluidity?. *Reprod. Biol. Endocrinol.* **6**, 19.

Milla, S., Wang, N., Mandiki, S. N. M. and Kestemont, P. (2009). Corticosteroids: friends or foes of teleost reproduction? *Comp. Biochem. Physiol.* **153A**, 242–251.

Morehead, D. T. (1998). Effect of capture, confinement and repeated sampling on plasma steroid concentrations and oocyte size in female striped trumpeter *Latris lineata* (Latrididae). *Mar. Freshwat. Res.* **49**, 373–377.

Morgan, M. J., Wilson, C. E. and Crim, L. W. (1999). The effect of stress on reproduction in Atlantic cod. *J. Fish. Biol.* **54**, 477–488.

Munakata, A. and Kobayashi, M. (2010). Endocrine control of sexual behaviour in teleost fish. *Gen. Comp. Endocrinol.* **165**, 456–468.

Nakamoto, M., Matsuda, M., Wang, D.-S., Nagahama, Y. and Shibata, N. (2006). Molecular cloning and analysis of gonadal expression of foxl2 in the medaka, *Oryzias latipes*. *Biochem. Biophys. Res. Comm.* **344**, 353–361.

O'Connor, C. M., Gilmour, K. M., Arlinghaus, R., Van Der Kraak, G. and Cooke, S. J. (2009). Stress and parental care in a wild teleost fish: insights from exogenous supraphysiological cortisol implants. *Physiol. Biochem. Zool.* **82**, 709–719.

O'Connor, C. M., Yick, C. Y., Gilmour, K. M., Van Der Kraak, G. and Cooke, S. J. (2011). The glucocorticoid stress response is attenuated but unrelated to reproductive investment during parental care in a teleost fish. *Gen. Comp. Endocrinol.* **170**, 215–221.

Okuzawa, K. and Gen, K. (2013). High water temperature impairs ovarian activity and gene expression in the brain-pituitary-gonadal axis in female red seabream during the spawning season. *Gen. Comp. Endocrinol.* **194**, 24–30.

Okuzawa, K., Kumakura, N., Gen, K., Yamaguchi, S., Lim, B.-S. and Kagawa, H. (2003). Effect of high water temperature on brain-pituitary-gonad axis of the red seabream during its spawning season. *Fish Physiol. Biochem.* **28**, 439–440.

Pankhurst, N. W. (1985). Final maturation and ovulation of oocytes of the goldeye, *Hiodon alosoides* (Rafinesque), in vitro. *Can. J. Zool.* **63**, 1003–1009.

Pankhurst, N. W. (1987). In vitro steroid production by ovarian follicles of orange roughy (*Hoplostethus atlanticus* Collett), from the continental slope off New Zealand. In *Reproductive Physiology of Fish 1987* (eds. D. R. Idler, L. W. Crim and J. M. Walsh), p. 266. St John's: Memorial University of Newfoundland.

Pankhurst, N. W. (1998a). Reproduction. In *Biology of Farmed Fish* (eds. K. D. Black and A. D. Pickering), pp. 1–26. Sheffield: Sheffield Academic Press.

Pankhurst, N. W. (1998b). Further evidence of the equivocal effects of cortisol on in vitro steroidogenesis by ovarian follicles of rainbow trout *Oncorhynchus mykiss*. *Fish Physiol. Biochem.* **19**, 315–323.

Pankhurst, N. W. (2001). Stress inhibition of reproductive endocrine processes in a natural population of the spiny damselfish *Acanthochromis polyacanthus*. *Mar. Freshwat. Res.* **52**, 753–761.

Pankhurst, N. W. (2008). Gonadal steroids: functions and patterns of change. In *Fish Reproduction* (eds. M. J. Rocha, A. Arukwe and B. G. Kapoor), pp. 67–111. Enfield, NH: Science Publishers.

Pankhurst, N. W. (2011). Stress in fish: an environmental perspective. *Gen. Comp. Endocrinol.* **170**, 265–275.

Pankhurst, N. W. and Dedual, M. (1994). Effects of capture and recovery on plasma levels of cortisol, lactate and gonadal steroids in a natural population of rainbow trout, *Oncorhynchus mykiss*. *J. Fish. Biol.* **45**, 1013–1025.

Pankhurst, N. W. and King, H. R. (2010). Temperature and salmonid reproduction: implications for aquaculture. *J. Fish. Biol.* **76**, 69–85.

Pankhurst, N. W. and Munday, P. L. (2011). Effects of climate change on fish reproduction and early life history stages. *Mar. Freshwat. Res.* **62**, 1015–1026.

Pankhurst, N. W. and Porter, M. J. R. (2003). Cold and dark or warm and light: variations on the theme of environmental control of reproduction. *Fish Physiol. Biochem.* **28**, 385–389.

Pankhurst, N. W. and Van Der Kraak, G. (1997). Effects of stress on growth and reproduction. In *Fish Stress and Health in Aquaculture* (eds. G. K. Iwama, A. D. Pickering, J. P. Sumpter and C. B. Schreck), pp. 73–93. Cambridge: Cambridge University Press.

Pankhurst, N. W. and Van Der Kraak, G. (2000). Evidence that acute stress inhibits ovarian steroidogenesis in rainbow trout in vivo, through the action of cortisol. *Gen. Comp. Endocrinol.* **117**, 225–237.

Pankhurst, N. W., Van Der Kraak, G. and Peter, R. E. (1995). Evidence that the inhibitory effects of stress on reproduction in teleost fish are not mediated by the action of cortisol on ovarian steroidogenesis. *Gen. Comp. Endocrinol.* **99**, 249–257.

Pankhurst, N. W., Fitzgibbon, Q. P., Pankhurst, P. M. and King, H. R. (2008a). Habitat-related variation in reproductive endocrine condition in the coral reef damselfish *Acanthochromis polyacanthus*. *Gen. Comp. Endocrinol.* **155**, 386–397.

Pankhurst, N. W., Fitzgibbon, Q., Pankhurst, P. and King, H. (2008b). Density effects on reproduction in natural populations of the staghorn damsel *Amblyglyphidodon curacao*. *CYBIUM* **32** (Suppl. 2), 297–299.

Pankhurst, N. W., King, H. R., Anderson, K., Elizur, A., Pankhurst, P. M. and Ruff, N. (2011). Thermal impairment of reproduction is differentially expressed in maiden and repeat spawning Atlantic salmon. *Aquaculture* **316**, 77–87.

Patino, R. and Thomas, P. (1990). Induction of maturation of Atlantic croaker oocytes by 17α,20β,21-trihydroxy-4-pregnen-3-one in vitro – consideration of some biological and environmental variables. *J. Exp. Zool.* **255**, 97–109.

Perry, A. N. and Grober, M. S. (2003). A model for the social control of sex change: interactions of behaviour, neuropeptides, glucocorticoids, and sex steroids. *Horm. Behav.* **43**, 31–38.

Perry, S. F. and Gilmour, K. M. (1999). Respiratory and cardiovascular systems during stress. In *Stress Physiology in Animals* (ed. P. H. M. Balm), pp. 52–107. Sheffield: Sheffield Academic Press.

Pickering, A. D. and Pottinger, T. G. (1987). Poor water quality suppresses the cortisol response of salmonid fish to handling and confinement. *J. Fish. Biol.* **30**, 363–374.

Pickering, A. D., Pottinger, T. G., Carragher, J. and Sumpter, J. P. (1987). The effects of acute and chronic stress on the levels of reproductive hormones in the plasma of mature male brown trout, *Salmo trutta* L. *Gen. Comp. Endocrinol.* **68**, 249–259.

Planas, J. V. and Swanson, P. (2008). Physiological function of gonadotropins in fish. In *Fish Reproduction* (eds. M. J. Rocha, A. Arukwe and B. G. Kapoor), pp. 37–66. Enfield, NH: Science Publishers.

Pottinger, T. G. (1999). The impact of stress on animal reproductive activities. In *Stress Physiology in Animals* (ed. P. H. M. Balm), pp. 130–177. Sheffield: Sheffield Academic Press.

Pottinger, T. G. and Carrick, T. R. (2000). Indicators of reproductive performance in rainbow trout *Oncorhynchus mykiss* (Walbaum) selected for high and low responsiveness to stress. *Aquacult. Res.* **31**, 367–375.

Pottinger, T. G. and Pickering, A. D. (1990). The effect of cortisol administration on hepatic and plasma estradiol-binding capacity in immature female rainbow trout (*Oncorhynchus mykiss*). *Gen. Comp. Endocrinol.* **80**, 264–273.

Pottinger, T. G., Campbell, P. M. and Sumpter, J. P. (1991). Stress-induced disruption of the salmonid liver-gonad axis. In *Reproductive Physiology of Fish 1991* (eds. A. P. Scott, J. P. Sumpter, D. E. Kime and M. S. Rolfe), pp. 114–116. Sheffield: FishSymp 91.

Pottinger, T. G., Yeomans, W. E. and Carrick, T. R. (1999). Plasma cortisol and 17β-oestradiol levels in roach exposed to acute and chronic stress. *J. Fish. Biol.* **54**, 525–532.

Reddy, P. K., Renaud, R. and Leatherland, J. F. (1999). Effects of cortisol and triiodo-L-thyronine on the steroidogenic capacity of rainbow trout ovarian follicles at two stages of oocyte maturation. *Fish Physiol. Biochem.* **21**, 129–140.

Romero, L. M., Reed, J. M. and Wingfield, J. C. (2000). Effects of weather on corticosterone responses in wild free-living passerine birds. *Gen. Comp. Endocrinol.* **118**, 113–122.

Rotllant, J., Ruane, N. M., Dinis, M. T., Canario, A. V. M. and Power, D. M. (2006). Intra-adrenal interactions in fish: catecholamine stimulated cortisol release in sea bass (*Dicentrarchus labrax* L.). *Comp. Biochem. Physiol.* **143A**, 375–381.

Safford, S. E. and Thomas, P. (1987). Effects of capture and handling on circulating levels of gonadal steroids and cortisol in the spotted seatrout, *Cynoscion nebulosus*. In *Reproductive Physiology of Fish 1987* (eds. D. R. Idler, L. W. Crim and J. M. Walsh), p. 312. St John's: Memorial University of Newfoundland.

Schrandt, M. N. and Lema, S. C. (2011). Habitat-associated intraspecific variation in behavior and stress responses in a demersal coral reef fish. *Mar. Ecol. Prog. Ser.* **443**, 153–166.

Schreck, C. B. (2010). Stress and fish reproduction: the roles of allostasis and hormesis. *Gen. Comp. Endocrinol.* **165**, 549–556.

Schreck, C. B., Conteras-Sánchez, W. and Fitzpatrick, M. S. (2001). Effects of stress on fish reproduction, gamete quality and progeny. *Aquaculture* **197**, 3–24.

Schulz, R. W., de Franca, L. R., Lareyre, J.-J., Le Gac, F., Chiarini-Garcia, H., Nobrega, R. H., et al. (2010). Spermatogenesis in fish. *Gen. Comp. Endocrinol.* **165**, 390–411.

Shahjahan, Md, Kitahashi, T., Ogawa, S. and Parhar, I. S. (2013). Temperature differentially regulates the two kisspeptin systems in the brain of zebrafish. *Gen. Comp. Endocrinol.* **193**, 78–85.

Siklenkia, K., et al. (2015). Disruption of histone methylation in developing sperm impairs offspring health transgenerationally. *Science* **350**, 651.

Sloman, K. A. and Armstrong, J. D. (2002). Physiological effects of dominance hierarchies: laboratory artefacts or natural phenomena? *J. Fish. Biol.* **61**, 1–23.

Soria, F. N., Strüssman, C. A. and Miranda, L. A. (2008). High water temperatures impair the reproductive ability of the pejerrey fish *Odontesthes bonariensis*: effects on the hypophyseal-gonadal axis. *Physiol. Biochem. Zool.* **81**, 898–905.

Soso, A. B., Barcellos, L. J. G., Ranzani-Paiva, M. J., Kreutz, L. C., Quevedo, R. M., Lima, M., et al. (2008). The effects of stressful broodstock handling on hormonal profiles and reproductive performance of *Rhamdia quelen* (Quoy & Gaimard) females. *J. World Aquacult. Soc.* **39**, 835–841.

Staples, C. A., Tilghman Hall, A., Friederich, U., Caspers, N. and Klecka, G. M. (2011). Early life-stage and multigeneration toxicity study with bisphenol A and fathead minnows (*Pimephales promelas*). *Ecotoxicol. Environ. Saf.* **74**, 1548–1557.

Sumpter, J. P., Carragher, J. F., Pottinger, T. G. and Pickering, A. D. (1987). Interaction of stress and reproduction in trout. In *Reproductive Physiology of Fish 1987* (eds. D. R. Idler, L. W. Crim and J. M. Walsh), pp. 299–302. St John's: Memorial University of Newfoundland.

Takeo, J., Hata, J. H., Segawa, C., Toyohara, H. and Yamashita, S. (1996). Fish glucocorticoid receptor with splicing variants in the DNA binding domain. *FEBS Lett.* **389**, 244–248.

Tasker, J. G., Di, S. and Malcher-Lopes, R. (2006). Minireview: rapid glucocorticoid signalling via membrane-associated receptors. *Endocrinology* **147**, 5549–5556.

Teitsma, C. A., Anglade, I., Lethimonier, C., Le Dréan, G., Saligaut, D., Ducouret, B., et al. (1999). Glucocorticoid receptor immunoreactivity in neurons and pituitary cells implicated in reproductive functions in rainbow trout: a double immunohistochemical study. *Biol. Reprod.* **60**, 642–650.

Thomas, P. (2003). Rapid, nongenomic steroid actions initiated at the cell surface: lessons from studies with fish. *Fish Physiol. Biochem.* **28**, 3–12.

Thomas, P. and Rahman, M. S. (2009). Chronic hypoxia impairs gamete maturation in Atlantic croaker induced by progestins through nongenomic mechanisms resulting in reduced reproductive success. *Environ. Sci. Technol.* **43**, 4175–4180.

Thomas, P., Rahman, M. S., Kummer, J. A. and Lawson, S. (2006). Reproductive endocrine dysfunction in Atlantic croaker exposed to hypoxia. *Mar. Env. Res.* **62**, S249–S252.

Thomas, P., Rahman, M. S., Khan, I. A. and Kummer, J. A. (2007). Widespread endocrine disruption and reproductive impairment in an estuarine fish population exposed to seasonal hypoxia. *Proc. R. Soc. B* **274**, 2693–2701.

Tveiten, H. and Johnsen, H. K. (2001). Thermal influences on temporal changes in plasma testosterone and oestradiol-17β concentrations during gonadal recrudescence in female common wolfish. *J. Fish. Biol.* **59**, 175–178.

Tveiten, H., Bjørn, P. A., Johnson, H. K., Finstad, B. and McKinley, R. S. (2010). Effects of the sea louse *Lepeophtheirus salmonis* on temporal changes in cortisol, sex steroids, growth and reproductive investment in Arctic charr *Salvelinus alpinus*. *J. Fish. Biol.* **76**, 2318–2341.

Ünal, G., Erdoğan, E., Oğuz, A. R., Kaptaner, B., Kankaya, E. and Elp, M. (2008). Determination of hormones inducing oocyte maturation in *Chalcalburnus tarichi* (Pallas, 1811). *Fish Physiol. Biochem.* **34**, 447–454.

Upadhyaya, N. and Haider, S. (1986). Germinal vesicle breakdown in oocytes of catfish, *Mystus vittatus* (Bloch): Relative in vitro effectiveness of estradiol-17β, androgens, corticosteroids, progesterone, and other pregnene derivatives. *Gen. Comp. Endocrinol.* **63**, 70–76.

Van Der Kraak, G. J., Munkittrick, K. R., McMaster, M. E., Portt, C. B. and Chang, J. P. (1992). Exposure to bleached kraft pulp mill effluent disrupts the pituitary-gonadal axis of white sucker at multiple sites. *Toxicol. Appl. Pharmacol.* **115**, 224–233.

Wang, D.-S., Kobayashi, T., Zhou, L.-Y., Paul-Prasanth, B., Ijiri, S., Sakai, F., et al. (2007). Foxl2 up-regulates aromatase gene transcription in a female-specific manner by binding to the promoter as well as interacting with Ad4 binding protein/steroidogenic factor 1. *Mol. Endocrinol.* **21**, 712–725.

Wang, S., Yuen, S. S. F., Randall, D. J., Hung, C. Y., Tsui, T. K. N., Poon, W. L., et al. (2008). Hypoxia inhibitis fish spawning via LH-dependent final oocyte maturation. *Comp. Biochem. Physiol.* **148C**, 363–369.

Wingfield, J. C. (1994). Modulation of the adrenocortical response to stress in birds. In *Perspectives in Comparative Endocrinology* (eds. K. G. Davey, R. E. Peter and S. S. Tobe), pp. 520–528. Ottawa: National Research Council of Canada.

Wu, R. S. S., Zhou, B. S., Randall, D. J., Woo, N. Y. S. and Lam, P. K. S. (2003). Aquatic hypoxia is an endocrine disruptor and impairs fish reproduction. *Environ. Sci. Technol.* **37**, 1137–1141.

Yamaguchi, T., Yamaguchi, S., Hirai, T. and Kitano, T. (2007). Follicle-stimulating hormone signalling and foxl2 are involved in transcriptional regulation of aromatase gene during gonadal sex differentiation in Japanese flounder, *Paralichthys olivaceus*. *Biochem. Biophys. Res. Comm.* **359**, 935–940.

Yamaguchi, T., Yoshinaga, N., Yazawa, T., Gen, K. and Kitano, T. (2010). Cortisol is involved in temperature-dependent sex determination in the Japanese flounder. *Endocrinology* **151**, 3900–3908.

Yu, K. L. and Peter, R. E. (1992). Adrenergic and dopaminergic regulation of gonadotropin-releasing hormone release from goldfish preoptic-anterior hypothalamus and pituitary in vitro. *Gen. Comp. Endocrinol.* **85**, 138–146.

Zohar, Y., Muñoz-Cueto, J. A., Elizur, A. and Kah, O. (2010). Neuroendocrinology of reproduction in teleost fish. *Gen. Comp. Endocrinol.* **165**, 438–455.

9

COGNITION, LEARNING, AND BEHAVIOR

DAVID L.G. NOAKES

KATHERINE M.M. JONES

1. How Stress Can Affect Behavior, and Vice Versa
2. Optimality, Preferences, and Decision-Making
3. Salmon as Model Species
4. Learning in Relation to Stress in Fishes
    4.1. Learning, Plasticity, and Problem Solving
5. Some Critical Knowledge Gaps

We review the correlations, the connections, and the cause–effect relationships between behavior and stress in fish species. We relate the physiological aspects of stress to studies of fish behavior built on a foundation of observational and experimental studies that have stressed ecological and evolutionary considerations. Theoretical models of fish behavior, including contributions from experimental and comparative psychology, help us to understand the ways in which physiology can influence or direct behavior, and conversely the physiological consequences of behavior. Productive areas of current research bringing together studies of physiology, stress, and behavior include subjects as diverse as foraging and feeding behavior, migration, learning, parental and social behavior, and life history patterns. Broader studies of additional model fish species provide dramatic increases in our understanding of both mechanisms at the level of molecular genetics and consequences at the level of evolutionary ecology. We consider the central role of the concept of optimality and how it links the physiological and behavioral aspects of stress. Optimality in terms of physiology is considered in terms of proximate cause-and-effect relationships. For behavior, considerations of optimality more often refer to ultimate, evolutionary consequences. We show how optimality can bring together proximate and ultimate considerations of stress. We propose possible future research directions that will continue to enhance our understanding of both proximate and ultimate aspects of behavior and stress in fishes.

Biology of Stress in Fish: Volume 35
FISH PHYSIOLOGY

1. HOW STRESS CAN AFFECT BEHAVIOR, AND VICE VERSA

Fishes[1] must deal with complex, constantly changing external environments. The physical parameters of some environments, perhaps the abyssal regions of the world's oceans, might vary much less over an annual cycle than do the shallow waters of a temperate pond in the northern hemisphere, or a tropical coral reef, or a freshwater stream (Fig. 9.1). Nonetheless, the biological environment will always vary, often unpredictably, in terms of presence of predators, availability of food, and proximity to conspecifics (eg, Jones, 2005, 2007). At the same time they must deal with a complex internal environment and regulate a myriad of physiological parameters within tolerance limits (Hoar, 1966). Fishes must deal with all aspects of their internal and external environments, survive, and reproduce, if possible (Dawkins, 1976). It is useful to consider the ways and means by which fishes respond to those external and internal environments as an operational definition of behavior.

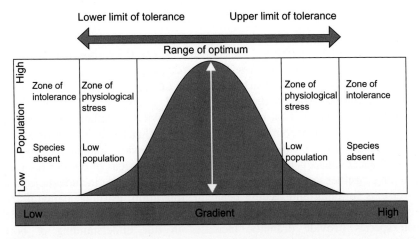

Figure 9.1. Hypothetical tolerance limits, optimum range, and zones of stress for a fish in relation to a given environmental gradient, temperature for example (modified from McFarland, D. (1999). *Animal Behaviour*. New York, NY: Prentice-Hall, with permission). Performance of individual fish, and hence population numbers, is minimal at tolerance limits and highest at the optimum value.

[1]We follow the convention of using the term "fish" to refer to one or more individuals of one species, and the term "fishes" to refer to one or more individuals of more than one species.

The behavior of fishes is a complex subject (Noakes and Baylis, 1990), but we take a predictive approach in relation to physiology and stress. Our approach to the study of behavior is based on the underlying hypothesis that the behavioral responses that we observe in fishes are adaptive, since they have been shaped by natural selection. This in keeping with the admonition that "nothing in biology makes sense except in the light of evolution" (Dobzhansky, 1973), and this basic hypothesis has also been shown to account for an enormous range of complexities of the behavior of individual animals (Hamilton, 1998; Ridley, 2003).

Any fish alive today must have descended from an unbroken ancestral line that survived and reproduced so we assume that the behavior we observe today must reflect that success (ie, behavior is adaptive). It is sometimes convenient shorthand to refer to this collectively as the life history strategy of a given species. Of course the sheer diversity of living fishes, together with the forms known only from the fossil record, demonstrates that there are many possible strategies (Dawkins, 2009).

It is conventional to think of animal behavior as being a consequence of physiology; for example, elevated levels of testosterone "cause" aggressive behavior (Fernald, 1976), elevated levels of thyroxine "cause" increased locomotor activity (Hoar, 1953). While we acknowledge this line of reasoning for causal mechanisms, in addition we will consider the perspective that behavior can influence physiology, and we will emphasize how behavior is a manifestation of what we typically consider the stress response.

We will review the relationships among stress, cognition, learning, and behavior in fishes. Our conclusion is that we gain a better understanding of these relationships if we place our considerations within the context of the adaptive nature of behavior (Spagnoli et al., 2016). Rather than attempting to catalog all the possible conditions that might cause stress, we review an extensive literature and consider the behavioral responses to stressors as adaptive. Furthermore, we conclude that because of the many conflicting demands placed upon any fish, and despite the significant adaptability of behavior, that any fish will probably very seldom be able to achieve a stress-free condition.

2. OPTIMALITY, PREFERENCES, AND DECISION-MAKING

There is an extensive literature, with a multitude of definitions of stress (Schreck and Tort, 2016; Chapter 1 in this volume). We will begin our consideration with a general operational definition of a stressor as anything that tends to disrupt the homeostasis of the internal physiological state of a fish (McFarland, 1999). That disruption can come about as a result of either internal or external stressors. Details of the physiological stress responses of fishes are considered elsewhere in this volume, so we will not elaborate on them here. However, we do consider the significant distinction between distress and eustress (see later) in a more comprehensive framework for behavioral alternatives shown by individuals. We find it useful to think of fishes in terms of their limits of tolerance in relation to external and internal fluctuations (actual or potential stressors). In the case of some external anthropogenic stressors the consequences can be lethal for some fishes (Paetzold et al., 2009).

Suppose we consider temperature as an example of a single physical factor. There is an extensive literature over many years of the physiological and behavioral responses of fishes to temperature (Coutant, 1977; Jobling, 1981). We have appreciated for many years that in part that is because we can readily measure temperature (Beamish, 1970), but more importantly because temperature has significant fundamental effects on the metabolism of any fish (Brett, 1971; Wood and MacDonald, 1997; Allan et al., 2015; Lawrence et al., 2015). We assume that the external temperature can range from the lower lethal to the upper lethal for a given fish. By definition the fish cannot survive at temperatures below or above the lower or upper lethal temperatures, respectively (Fig. 9.2). Within the tolerance range, between the lower and upper lethal extremes, an individual fish will have an optimal temperature (Brett, 1952) that may be influenced by its thermal history (acclimation or acclimatization). The temperature selected by a fish can be influenced by stress. Stressed zebrafish (*Danio rerio*) have been reported to respond with hyperthermia (Rey et al., 2015), and the growth responses of some species can be altered by stress (Eldridge et al., 2015).

The concept of optimality is one that has been widely invoked for many years, in considerations of both physiology and behavior. Optimality is central to our consideration of behavior and stress in fishes. We realize that in terms of physiology fishes can be either conformers (ie, their internal bodily condition follows the external condition) or regulators (ie, they maintain their internal condition relatively independent of the external conditions, within some limits) (Hoar, 1966). While some fishes can regulate their internal temperatures to some extent, this does not affect the general nature of our argument (Carey et al., 1971; Block and Finnerty, 1994; Wegner et al., 2015). In practice, the optimal temperature will almost certainly be a narrow range of temperatures (Jobling, 1981; McCauley and Casselman, 1981). At that optimal temperature the physiological performance of the fish will be maximal, by definition in terms of scope for activity (Fry, 1947; Beamish, 1970). At temperatures below or above the optimal range the fish will increasingly be in a range of physiological stress (Iwama et al., 2011). Those conditions have sometimes been described as aversive, in either physiological or behavioral terms, because they result in stress responses.

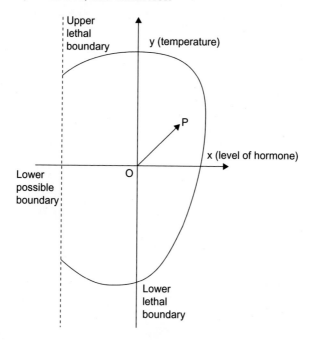

Figure 9.2. Physiological state (point p at the end of a vector) of a fish within the state space defined by the lethal boundaries for temperature (y axis) and hormone level (x axis). The value (O) represents the optimum value for both temperature and hormone level. The physiological state can be anywhere within the lethal boundaries for the two factors. Modified from McFarland, D. (1999). *Animal Behaviour*. New York, NY: Prentice-Hall, with permission.

The term aversive clearly implies that fish do (or should) attempt to avoid those conditions. As part of this perspective, the assumption has often been made that the preferred temperature of a fish (given access to a range of temperatures) will correspond to the optimum temperature (Hazel and Prosser, 1974; Beitinger and Fitzpatrick, 1979; McCauley and Casselman, 1981). If the temperature is only slightly below or above the optimum the fish can likely compensate by physiological responses (often described as homeostatic responses) (Hazel and Prosser, 1974; Beitinger and Fitzpatrick, 1979; Ott et al., 1980; Pankhurst, 1997, 2011). However, there will be limits to this physiological compensation (McFarland, 1999). Such responses might involve changes in the pattern of blood flow to different tissues, or changes in metabolic activity (Beitinger and Fitzpatrick, 1979).

Those physiological responses combined with behavioral responses or compensation will tend to bring the temperature of the fish back toward the optimum (Fig. 9.3). So, for example, the fish could move into a volume of water closer to the optimum value (behavioral thermoregulation, thermal

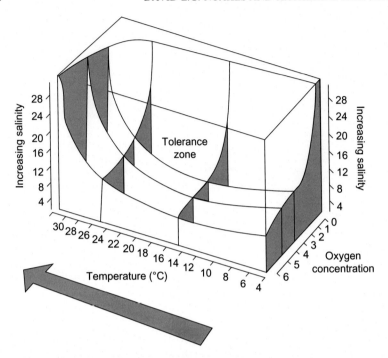

Figure 9.3. Physiological response surface of a fish to combinations of salinity, oxygen, and temperature (modified from McLeese (1956), with permission). The fish can tolerate the combinations of those environmental factors defined by the three-dimensional surface. Conditions outside that response surface are lethal to the fish.

preference) (Coutant, 1987). If we now consider two external environmental factors (temperature and pH, for example) we can begin to appreciate the complex situation facing any fish that has to maintain an optimal internal physiological state. A fish might be able to regulate its internal temperature near the optimum, but if environmental pH varies spatially and temporally then the fish will also have to respond to variation of that factor.

If we expand our consideration to include the host of possible factors or internal and external stimuli affecting a fish at any one time (eg, temperature, dissolved oxygen, pH, hormonal levels in the blood, presence of a predator, proximity to food, salinity, etc.), it is obvious that the individual will have a complex task to achieve anything close to its optimal internal physiological state (Takei and Hwang, 2016; Chapter 6 in this volume). For example, behavior could at once move an animal closer to the optimal condition for temperature while simultaneously move it farther away from the optimal condition for overhead cover. However, despite what naïve observers might claim, "doing nothing" is never an option for a fish.

It must always do "something," and it is the something that we recognize as behavior. What the fish does at any given moment can be considered as a consequence of behavioral decisions (Houston and McNamara, 1999) related to the state of the animal. Thus the behavior of the fish at any point in time will almost inevitably be a compromise with regard to conflicting demands to achieve optimal conditions (Sneddon et al., 2016; Chapter 12 in this volume). We know that it is possible to observe the behavioral responses of animals when they have the opportunity to alternate between mutually incompatible activities (Noakes, 1986a,b). Even more convincing estimates of the priorities that animals place on behavioral activities are based on measures of how much effort the animal will expend to engage in a specific activity (Dawkins, 2003).

Fish behavior will be constrained by what is recognized as the final common behavioral pathway (McFarland, 1999). With very few exceptions, a fish can only do one thing at a time. For example, a male stickleback might feed, but could not simultaneously defend a territory, or build a nest, or court a female (Noakes, 1986a,b). A fish could behave so as to reduce stress caused by one external (or internal) factor, but it probably could not simultaneously behave to reduce stress caused by any of numerous other internal or external factors that are also likely to cause stress at the moment.

We also note that the behavior of an individual at any point in time will necessarily be a consequence of a decision that will be affected by the age and developmental state of the animal (eg, immature juvenile vs sexually mature adult), and the interactions of both internal and external environmental conditions. The internal state of the animal can vary, depending upon previous experience (see our discussion of various kinds of learning), in addition to the developmental condition. A clear example is the downstream movement of juvenile salmonids (smolts), moving from freshwater to the marine environment (Seals Price, 2003). The movement of fishes between freshwater and saltwater environments is a very significant ecological, physiological, and behavioral transition (McCormick et al., 2013). There are many such fishes, and salmon (*Salmo* species; *Oncorhynchus* species) have been the subject of many studies of the movements of juvenile from fresh- to saltwater and the reverse movements of adults from the marine environment back to freshwater habitats for spawning (Boeuf, 1993; Romer et al., 2013; Thompson et al., 2015). The anadromous migrations of salmon are active, volitional behavioral decisions by individuals, based upon their developmental state, chronological age, and a combination of external environmental factors (time of year, water temperature, water flow) and internal conditions (sex, size, growth history) (Thompson et al., 2015).

There are many other examples of other fishes whose behavioral responses are a consequence of internal factors, including parental effects (Allan et al., 2014), and combinations of external environmental factors (Domenici et al., 2014; Donelson et al., 2014; Allan et al., 2015; Ferrari et al., 2015). Some of these are examples of what we might consider natural situations, while others are considered to predict what we might expect with changing ocean conditions (eg, acidification, warming). In all these cases the behavior that we observe will be an outcome of the same processes.

In extreme cases, an individual could behave so as to respond to certain stimuli in a very obvious, dramatic fashion. The classic example of this would be the fight-or-flight (freezing vs fleeing) situation postulated in many studies (Hertz et al., 1982; Jansen et al., 1995; Dawkins, 2012; Adamo, 2014; Baciadonna and McElligott, 2015; Domenici and Ruxton, 2015; Wiseman, 2015). Those behavioral alternatives would certainly be obvious, and might characterize situations where individuals either have no alternatives, or where the benefits (or costs) of alternatives are extreme (eg, defense of young, escape from a predator). We consider these as examples at one extreme of a continuum, as we will elaborate later.

Behavior is the significant immediate means by which a fish interacts with its environment. It is obvious that any fish alive today must have descended from a long ancestral line that successfully adapted to a long series of environmental conditions. The responses of a fish to internal and external environmental changes have therefore been shaped by natural selection over an evolutionary time span. There is a bewildering history of nomenclature associated with the potential responses of fishes to their environment (Reebs, 2001; Keenleyside, 1997). The entire suite of behavioral responses available to a fish can conveniently be thought of as the behavioral repertoire, sometimes described as the ethogram of the species (Manning and Dawkins, 2012).

Some have defined the ethogram as the innate, or genetically determined, or inborn set of behavioral actions available to the fish. While it is clear that natural selection must influence the behavior of a fish, just as it influences physiology or morphology, it is seriously misleading to think of behavior in such simplistic terms as innate versus learned. The fallacy of the innate versus learned dichotomy has long been recognized and laid to rest (Dawkins, 2004).

Perhaps the most important distinction to appreciate is that when we consider behavior we must think of the mechanisms that are responsible for differences in behavior. Is a difference in a particular behavior between two fish, say migrating to the ocean rather than remaining in freshwater, the result of a genetic difference between those individuals or a result of some environmental difference to which they had been exposed? In more realistic terms, those differences among individuals are likely to be the result of the interactions of both genetic and environmental differences (Manning and Dawkins, 2012). We expect that the typical behavior of a fish, when exposed to the usual range of conditions for that species, would be an expression of the adaptive responses of the ancestors to the forces of natural selection (Charnov, 1993; Houston and McNamara, 1999; Dawkins, 2009). By definition, the behavioral strategy specifies how the fish decides on behavior at each point in time (Houston and McNamara, 1999).

There has sometimes been a tendency to consider the responses of animals to what we define or recognize as stressful stimuli or situations as maladaptive (Iwama et al., 2011; Careau et al., 2014). There is an extensive literature on stressors and the stresses (and distress) they cause in fishes (Noakes and Leatherland, 1977; Young et al., 2010; Laursen et al., 2011; Careau et al., 2014). The relationship between physiological stress and behavior has been reviewed a number of times (Schreck et al., 1997; Yue et al., 2006). It is certainly true that prolonged or extreme conditions of stress will have negative consequences for a fish. Thus we might equally consider the behavioral responses to stress as adaptive, since they can reduce the stress to which the fish is exposed.

Certainly there is general acceptance that the consequences of stress are negative for fishes, whether domesticated species in the aquarium or wild species in their natural environment (Ejike and Schreck, 1980; Conte, 2004; Cachat et al., 2011; Galhardo et al., 2011; Cook et al., 2014; Monaghan and Spencer, 2014; Pavlidis, et al., 2015). But first and foremost, it is important to distinguish between stress and distress (Selye, 1975, 1976a,b,c; Szabo et al., 2012). More correctly, we distinguish between distress and eustress (Selye, 1978; Le Fevre et al., 2003; Hargrove et al., 2015). The distinction was captured best by Selye's statement, "Stress without distress" (Selye, 1976c). Distress produces negative consequences for the individual, and we consider that as the extreme situation in some of examples that follow. Eustress is positive, so far as the individual is concerned, for a great variety of reasons detailed by others in this volume. Eustress is the condition that we consider at greater length, since it is clear that individuals will always be subjected to potentially conflicting external and internal influences that will affect the behavior at any point in time (Selye, 1976c). There is no doubt that any situation of extreme, or prolonged, physical or physiological conditions will cause injury or ill health to a fish. For example, exposure to toxic chemicals can cause physiological distress in fishes (Paetzold et al., 2009; Grassie et al., 2013). We do not include those situations and their effects on fish within our concept of the adaptive nature of responses to stress. We can agree that distress is clearly negative for the fish, and should be avoided or minimized. It is not so clear that stress, short of distress, is necessarily negative for the fish, and there are theoretical reasons at least to predict that some stress may be unavoidable at any time. We will consider this possibility in some detail by referring to salmonids as model species.

3. SALMON AS MODEL SPECIES

We consider the behavior of salmonid fishes as a model for a number of reasons. There is a very extensive literature on the biology, behavior, ecology and evolution of salmonids (Magnan et al., 2002; Quinn, 2005; Piccolo et al., 2014; Woo and Noakes, 2014). That literature incorporates the full range of physiology and behavior that we want to consider within the natural environment of the species. We will generalize the life history of salmonids for the sake of convenience, but acknowledge that while this is convenient we recognize significant variations among genera, species, and local populations. Full disclosure, our generalized salmonid life history assumes the fish is anadromous, semelparous, and living within its native range in the northern hemisphere (Fraser et al., 2011).

There is probably no more contentious area of salmon biology than the consideration of the differences between hatchery and wild salmon and the causes of those differences (Fagerlund et al., 1981; Noakes and Corrarino, 2010, Noakes, 2014). There are numerous descriptive studies that document differences between wild and hatchery salmon, including some related to stressors and stress responses (Fenderson et al., 1968; Woodward and Strange, 1987; Ruzzante, 1994; Jonsson and Jonsson, 2006; Matsuda, 2013; Goetz et al., 2015). However, those differences appear to be quantitative rather than qualitative, so we will include examples of behavioral responses to stressors from studies of both wild and hatchery salmonids.

While there are undoubtedly many stressors in the life of salmonids, the extensive anadromous migrations of these fishes must rank high in that regard. Hormones typically thought to be associated with physiological stress in salmon can be involved in behavioral events during the normal life history of the species (Carruth et al., 2002). The natural environmental changes that migratory salmon encounter certainly include salinity, temperature, water currents, and presence of predators (Hodgson et al., 2006; Hayes et al., 2012; Romer et al., 2013; Kavanagh and Olson, 2014; Melnychuk et al., 2014). In addition, there are numerous examples of anthropogenic causes of stress during migration in fishes (Danley et al., 2002). Of course a dramatic feature of the life history of Pacific salmon (*Oncorhynchus* species) is their virtually suicidal semelparous behavior

(Quinn, 2005). Not surprisingly there is extensive literature on the stress responses of these salmonids (Contreras-Sanchez et al., 1998; Schreck, 2000; Iwama et al., 2011).

4. LEARNING IN RELATION TO STRESS IN FISHES

4.1. Learning, Plasticity, and Problem Solving

If changes (stressors) in the life of fishes were predictable over long periods, then evolutionary adaptations would result. If changes in the environment are predictable over the relatively short term during the life of an individual, the fish will respond in an adaptive manner that we recognize as learning (Hollis, 1982; Houston and McNamara, 1999; Braithwaite, 2010). Before we consider the effects of stress on learning in fishes we need a definition of learning that is general enough to be useful for a diversity of species across a wide variety of ecological and physiological situations. We use the definition of learning first proposed by W. E. Thorpe and other ethologists (Manning and Dawkins, 2012). We also note that other perspectives on behavior include different operational definitions of learning and other adaptive aspects of behavior (Houston and McNamara, 1999). By the definition we are using, learning is an adaptive change in the behavior of an individual that comes about as a result of experience. The first key point in this definition is that the change in behavior is adaptive, in other words the animal gains some benefit (short or long term) as a result of learning. In this case as a result of learning the fish would reduce the stress response. The second key point is that the change takes place within an individual (rather than a population, or over successive generations, for example). The third key point is that the change in behavior comes about as a result of experience (rather than simple growth or maturation, for example).

Learning is often categorized according to a relatively simple classification scheme that does not truly reflect the complexity of the phenomenon. We follow this simple scheme for the sake of convenience, and also it reflects the majority of published studies of learning responses of fishes. The terminology we use describes learning as one of the following: habituation, classical conditioning, instrumental conditioning, or imprinting.

Habituation is often described as the simplest form of learning. Habituation is conveniently defined as the decline in response of a fish to repeated presentations of a neutral stimulus that initially has evoked a response by the fish (Manning and Dawkins, 2012). The neutral stimulus can be virtually anything that the fish can detect from visual, olfactory, or tactile stimuli, for example. Habituation is clearly adaptive since the animal ceases to respond to stimuli that have no significant consequences (Nilsson et al., 2012). Initially the fish responds to the stimulus in question as a stressor (eg, a sudden change in some environmental feature such as a passing shadow, or a low frequency sound). The response of the fish could be either a physiological or a behavioral response. That might be an increase in heart rate, a surge in the release of a hormone, or movement away from the stimulus. However, because the stimulus has no relevance for the fish, it will cease to respond to repeated occurrence of the stimulus (stressor). Cases of habituation will be so frequent and so common that we might not even consider them as examples of learning. We think of the fish "getting used to" any of a whole series of sudden environmental stimuli or changes. Nonetheless, habituation will be extremely important for the stress responses of fishes, perhaps more than any other form of learning. Virtually everything and anything in the external environment that can be detected by a fish could be significant, either as a stressor or as an indicator of something that will become a stressor. But most of the things from the external environment have no significance for a fish, at least most of the time.

In essence, habituation is the means by which we and other animals can differentiate between meaningful and meaningless correlations. There *might* be a cause-and-effect relationship, but that is generally not the significant question. All we need to know for any given environmental signal or cue is whether it is correlated with a stressor for us. If there is a correlation, presumably the operational rule (strategy) is to respond appropriately by physiological or behavioral responses to minimize the effect of that stressor and move the physiological state of the individual back toward the optimum condition. If there is no correlation, then the operational rule is to ignore the signal or cue. The broader phenomenon is known as stimulus filtering, and it is recognized as the means by which animals selectively respond to relevant stimuli out of the very large amount of potential information that it could not possibly deal with. Given that a fish can detect something from the environment, habituation will be the means by which it comes to limit its responses only to the information that has relevance. The total lack of response of fish in public aquaria to the myriad human activities adjacent to their aquarium is an obvious example.

Classical conditioning, sometimes referred to as Pavlovian conditioning, is also clearly adaptive, but unlike habituation this involves the animal developing an association between two previously unrelated stimuli (Bitterman, 1964; Hollis et al., 1989). An initial stimulus or signal that has no relevance to the animal is followed by a consequence (in this case a stressor) that is significant for the animal (Sneddon et al., 2016; Chapter 12 in this volume). Pavlov rings a bell, a dog can hear that bell and might initially show some response to the sound (eg, turn toward the sound, increase in heart rate, move away from the sound), but would otherwise quickly habituate (stop responding) to repeated ringing of the bell. However, if ringing of the bell is followed by a relevant event (eg, meat powder in the mouth), the dog will quickly come to respond to the repeated ringing of the bell by anticipating delivery of the meat powder (salivation). Soon the dog comes to salivate in response to the sound of the bell ringing alone. A fish that detects an overhead shadow will respond by seeking cover, showing a change in heart rate, or some other response to that potential stressor. If repeated overhead shadows are linked with subsequent delivery of food in a hatchery, then the fish will quickly come to respond to the shadow by swimming up to the surface to feed. However, if the overhead shadow is followed by a predatory bird diving into the water to attack the fish (Penaluna et al., 2015), repeated presentations of an overhead shadow will result in the fish moving toward cover and showing a number of other responses anticipating the stress of a potential predatory attack. Several authors have suggested that this specific example might account for significant differences in the behavior of hatchery and wild salmon, to the detriment of hatchery salmon when released into open waters (Thorpe and Morgan, 1978; Pickering et al., 1987; Wiley et al., 1993; Olla et al., 1994; Berejikian et al., 2001; Brown et al., 2003; Malavasi et al., 2008).

Classical conditioning of heart rate with mild electric shock has been frequently used to determine sensory capabilities of fishes to visual, olfactory, and other external stimuli (Offitt, 1971; Beauchamp and Rowe, 1977; Beauchamp et al., 1979; Holland and Teeter, 1981; Hawryshyn and McFarland, 1987; Hawryshyn et al., 1989; Morin et al., 1989; Hawryshyn, 1992, 2000; Rodriguez et al., 2005). It is even possible to show classical conditioning of agonistic responses of some fishes (Thompson and Sturm, 1965).

Karen Hollis and her associates have demonstrated a number of elegant examples of classical conditioning in fishes that include anticipatory response for agonistic or courtship activities (Hollis, 1997, 1982; Hollis et al., 1989, 1995, 1997). Fish that have formed an association between some otherwise neutral stimulus (eg, light flashing) and subsequent appearance of a territorial intruder quickly come to respond to the light itself and initiate agonistic responses more quickly to an invading fish that do naïve individuals. As a result they are much more likely to become dominant in paired encounters with naïve individuals. There is no doubt that this kind of learning can also have very pervasive effects on fish behavior, especially responses to stressors. There is a somewhat subtle possibility that we also emphasize, which concerns the physiological or behavioral responses of fishes to repeated, predictable events, whether in captivity or in the wild.

The situation is described as a conditioned compensatory response (Siegel and Ramos, 2002). We know homeostasis as the complex of internal physiological responses that tend to restore the physiological state of the animal back to (or toward) the optimum value (=constant physiological state) (Gross, 1998; McEwen, 2005; Cooper, 2008). If an animal is exposed to an external stressor then homeostatic responses will tend to counteract the effect of the stressor as a compensatory response. For example, if the stressor tends to increase heart rate then the homeostatic response will be to reduce the heart rate. If that particular stressor is repeated, especially in a predictable manner, then the animal will come to anticipate the stressor in a physiological sense, through classical conditioning. With increasing repetitions of that stressor, the stress response of the animal will progressively decrease as a result of the anticipatory compensatory response (Siegel and Ramos, 2002). We could then have the paradoxical situation that repeated exposure, in a predictable fashion, to an anticipated stressor would ultimately produce no stress in the animal. An example might be the regular injection of a drug at the same time each day over some period of time. Initially we would measure a distinct stress response in the animal, but that would decline over time so that we might eventually measure no significant response. We might think of this as the animal developing a tolerance to repeated exposure to the stressor. But if that same stressor were to be presented in an unpredictable manner or at an unpredictable time we would expect the stress response of the animal to be as great as it was to the initial exposure. Obviously we would predict exactly the same pattern of conditioned compensatory responses by animals under natural circumstances. The key feature would again be the predictability of the stressor, with the consequent classical conditioning result (Fernö et al., 2011). Thus learning, at least in the form of classical conditioning, might significantly reduce stress in both short- and long-term situations.

Instrumental conditioning (sometimes referred to as operant conditioning) is a form of learning usually associated with the name Skinner, or Skinner box (Mackintosh, 1969; Fernö et al., 2011; Agrillo et al., 2012), although the experimental approach was first used by Thorndike as puzzle boxes (Burnham, 1972). There are numerous examples demonstrating the ability of fishes to learn to perform activities such as pushing on a lever, swimming through a tube, or crossing a light beam to trigger food rewards, control environmental temperature, or even provide access to conspecifics (Attia et al., 2012). The use of self-feeders operated by the fish has become standard practice for many aquaculture systems (Alanärä, 1992, 1996; Boujard et al., 1996; Millot et al., 2011; Patton and Braithwaite, 2015). These are all textbook examples of instrumental conditioning (Sneddon et al., 2016; Chapter 12 in this volume).

There have been a number of studies of learning in fishes, using model species (typically the goldfish, *Carassius auratus*). The intent of those studies has usually been to compare performance of the fish in particular learning situations, to compare with the bird, and the mammal, or other model species (Mackintosh, 1969; Fernö et al., 2011). While those studies have documented fish performance, they have seldom investigated the effects of stress or stressful circumstances on fish learning. There are even some studies comparing learning in different fish species in relation to their abilities to move about their habitats or direct their spatial movements (McAroe et al., 2015). While these comparisons may tell us something about learning in a comparative sense they are typically not directed to specific questions of learning relevant to the ecology of life history of fishes. However, it seems likely that instrumental conditioning could also lead to reduced stress in fishes, since it provides the animals an opportunity to bring about changes in their environment (McAroe et al., 2015).

Studies of the relationship between stress and learning have shown a number of possible causal linkages (Askey et al., 2006; Ferrari et al., 2012; Roche et al., 2012; Eaton et al., 2015; Mesquita et al., 2015). Not surprisingly, the effects can be immediate, but there is also evidence that effects can extend across generations. For example, first generation sticklebacks (*Gasterosteus aculeatus*) from maternal fish subjected to stress show reduced learning capacity, compared to individuals from maternal fish not subjected to such stress (Mommer and Bell, 2013). Similarly, the associative learning responses and spontaneous aggression of offspring of female guppies (*Poecilia reticulata*) subjected to mild stress are significantly disrupted by that treatment of their mothers (Eaton et al., 2015). The relationship between stress and learning can be complex, and in some cases counterintuitive. For example, in some cases stress can enhance learning (Larra et al., 2014), in others stress can detract from or block learning (Zoladz et al., 2011; Dominique and McGaugh, 2014). The physiological mechanisms are not known in most of those examples, but it is clear that the relationship between stress and learning is neither simple, nor uniform.

The stereotypical responses of some fishes to chemicals (Schreckstoff pheromones) released from injured conspecifics is known as the alarm reaction (Frisch, 1942; Smith, 1992). The alarm reaction is clearly a response to a stressful situation for the fish. That response has been a standardized test situation for the effects of stress on learning in fishes (Smith, 1992; Manek et al., 2012). Rearing or holding conditions can significantly affect the alarm responses of fishes, including cyprinids and salmonids. In some cases those changes in alarm responses are a consequence of learning by the fish, or at least environmental effects acting for a long period of time during early development (Berejikian et al., 1999; Manek et al., 2012).

There are some examples that provide a great deal of detailed information about learning in the life history of fishes. Studies of imprinting are best illustrated here. Imprinting is recognized by a kind of associative learning that is restricted to a brief period (the sensitive period) early in the life of a fish but that has delayed consequences much later in the life of that individual. Perhaps the best examples are the behavior of Atlantic (*Salmo salar*) and Pacific salmon (*Oncorhynchus* species). We know a great deal about learning in the context of the natural history of species from studies of the role of early learning in salmonid fishes (Suboski and Templeton, 1989). It is clear that early experience of chemical (probably olfactory) and geomagnetic cues can result in lasting changes in salmon behavior. Perhaps one of the best known examples of learning in the life of fishes comes from studies of olfactory imprinting and homing in salmon (Hasler, 1966; Hasler et al., 1978; Nordeng, 1977, 1983, 2009; Johnsen and Hasler, 1980; Nordeng and Bratland, 2006). While details continue to be investigated, this phenomenon is so widely accepted that it is the basis for standard hatchery practice in many areas (Dittman et al., 2015). "Smells are notorious triggers of memory" (Dawkins, 2013).

It is necessary to briefly summarize the life cycle of salmon to provide the context for this example. There are significant differences between Atlantic and Pacific salmon; the use of the common name is not always helpful. In fact some have stated that the imposition of the common name salmon on the Pacific species by early European explorers in North America has caused undue confusion (Dymond, 1932). Nonetheless, the important pattern of anadromous salmonid life history is that sexually mature adults return to spawn in their natal streams. The resulting offspring spend some time (months or years, perhaps) in the natal stream before they repeat the migration to the ocean where they feed, grow, and become sexually mature and migrate back to freshwater where they repeat the life cycle by spawning in the natal stream. The remarkably accurate return of salmon to their respective natal streams is a diagnostic feature of these fishes (Hansen and Quinn, 1998; Quinn, 2005; Waples and Hendry, 2008).

Some of the earliest studies involving transplants of salmon between different watersheds provided the first evidence that their precise homing behavior was a result of learning (Dittman and Quinn, 1996). Studies of the mechanisms have provided a wealth of information on the details of this phenomenon. The physiological correlates of olfactory imprinting are known in detail. Surges of thyroxine and cortisol early in the life of juveniles while in freshwater, and when they return as sexually mature adults, sensitize the olfactory receptors of the salmon to the chemicals characteristic of their home streams (Hasler, 1966; Hasler et al., 1978; Stabell, 1984; Carruth et al., 2002; Lema and Nevitt, 2004). Those chemicals may come from conspecifics, or from other sources in the stream, but the salmon form a strong association between the chemical cues and the location, and return the area where they reared as juveniles when they come to spawn in their turn (Brett and Groot, 1963; Hasler et al., 1978; Fisknes and Døving, 1982; Quinn and Hara, 1986; Brannon and Quinn, 1990; Dittman and Quinn, 1996; Ueda, 2014).

Some might argue that this early chemical imprinting in salmonids is not, strictly speaking, related to stress. The reasoning would be that olfactory imprinting occurs as part of the normal life history of these fishes, and stress should be thought of as something abnormal for the fish. We understand that perspective, but we consider the concept of stress to include any of a number of events or occasions during the normal life history of a fish (see other chapters in this volume). Thus we consider the elevated levels of thyroxine, cortisol, and perhaps other hormones around both the time of yolk absorption and downstream migration in young salmonids and subsequent upstream migration of adults to represent physiological stress to the fish (Schreck, 1990; Dittman and Quinn, 1996; Schreck et al., 1997; Congleton et al., 2000). As an aside, we know the remarkably specific memories that people can have associated with specific stressful instances in our lives. The extraordinarily detailed recollections of where we were, what we saw, and what we heard on Nov. 22, 1963, or Sep. 11, 2001, are examples only too familiar for some of us. Yet we cannot remember anything about Nov. 21, 1963 or Sep. 10, 2001.

The elevated levels of thyroxine in Pacific salmon sensitize the olfactory receptors for the initial learning and subsequent response to the chemical stimuli (Cooper and Scholz, 1976; Carruth et al., 2002; Lema and Nevitt, 2004). There will certainly be comparable cases for other hormones and other examples of learning in salmonids and other fishes. The elaborate parental behavior of cichlid fishes, with specific recognition of their own young and even different developmental stages, involves this kind of learning (Noakes, 1999; Galhardo et al., 2011; Morandini et al., 2014; Ramallo et al., 2014). There are almost certainly numerous times during the complex social interactions of cichlids that learning and associated stress are closely linked (Francis, 1988; Maruska et al., 2013). Learning, in the form of classical conditioning (Hollis et al., 1995; Hollis et al., 1997) or instrumental (operant) conditioning (Sneddon et al., 2016; Chapter 12 in this volume) can regulate both the behavior and the physiological state of fishes (Fernald, 2015). It is also possible for maternal stress to influence learning in their offspring in some fishes (Eaton et al., 2015).

There have also been important studies of learning within the broad category of behavioral ecology of fishes (Charnov, 1976; Dill, 1983; Drew and Gumm, 2015; Fürtbauer et al., 2015). Fishes can show considerable plasticity and adaptability in foraging, mate choice, habitat selection, and activities that relate directly to survival and life history (Charnov, 1976, 1978, 1993; Godin, 1999; Conrad et al., 2011; Davies et al., 2012; Mittelbach et al., 2014; Bond et al., 2015). Some authors have reported that this behavioral variability can be described and defined in terms of consistent behavioral phenotypes (McLaughlin et al., 1999; Pham et al., 2012; Oswald et al., 2013; Thörnqvist et al., 2015). These studies involve a considerable diversity of species, from salmonids to zebrafish (*Danio rerio*), under circumstances that range from landscape ecology to highly controlled and contrived laboratory testing situations (Grant and Noakes, 1987a,b; Dittman and Quinn, 1996; Brown and Laland, 2001; Brown et al., 2003; Askey et al., 2006; Oswald et al., 2013; Tudorache et al., 2015). These include some of the most sophisticated examples of learning known for fishes. We can safely assume that virtually all those examples will also include at least some degree of physiological stress (perhaps even distress) for the fishes.

Habitat and habitat selection are fundamental aspects of the behavior of fishes in natural environments (Nakano, 1994; Hughes, 2000; Brockmark et al., 2010; Jones et al., 2014). Responses of fishes to habitat features, and habitat preferences can be affected by the physiological condition of the fish. Of course behavioral responses to habitat features are among the most important and the most frequently studied for fishes in nature (Fausch and Northcote, 1992; Hughes, 2000; Hughes and Grand, 2000; Piccolo et al., 2007; Berejikian et al., 2013; Bunt et al., 2013; White and Brown, 2015). As mentioned, the behavior in relation to habitat features will include responses to cover, temperature, water quality, predators, and conspecifics, among others. Of these, responses to prey items have been most frequently studied and provide an extensive literature dealing with optimal foraging (Charnov, 1976; Green, 1987; Kamil et al., 1987; Stephens and Krebs, 1987). These are undoubtedly all examples of optimality and behavioral decisions that we have previously discussed.

The foraging behavior of fishes has been studied extensively, over many years, using a variety of technical and theoretical approaches. A basic technique that has been widely applied is to identify the stomach contents of individual fish in nature (Chiasson et al., 1995; Kawaguchi and Nakano, 2001; Nakano and Murakami, 2001). It is difficult to conclude anything about learning, or stress, from such data. More sophisticated techniques, such as analyses of stable isotopes will tell us more about the long-term feeding behavior of individuals. However, while these data will allow us to infer long-term feeding habits, they can tell us very little about stress or learning in relation to foraging and feeding behavior of wild fishes (Church et al., 2008; Gao et al., 2008). However, there is compelling evidence from studies of both captive and wild fishes to suggest that learning can have significant effects on selection of food items, foraging sites, and foraging activities (Dill, 1983; Grant and Noakes, 1987a,b; Nielsen, 1992; McLaughlin et al., 1999).

There is direct experimental evidence, as well as correlational data, to show that salmonids and other fishes can quickly adjust their foraging activities in relation to prey availability (Piccolo et al., 2014). Individuals can develop consistent prey preferences (Grant and Noakes, 1987a,b; McLaughlin et al., 1994), whether based on intrinsic features of the fish, or as a result of experience (ontogeny or learning). We know from detailed studies of feeding fish species in aquaculture that responses to feed can be affected by particle size, texture, movement, color, proximate composition, and manner of presentation to the fish (Jobling, 1985; Noakes and Grant, 1992; Brown et al., 2003; Fernö et al., 2011; Attia et al., 2012; Nilsson et al., 2012). In most, if not all, of these cases the fishes are also learning to respond to these features of food items and food delivery.

5. SOME CRITICAL KNOWLEDGE GAPS

It is important to differentiate between information and knowledge (or understanding). Syntheses, hypotheses, and principles are the objects of scientific research. Sometimes these are referred to as laws in the vernacular of popular culture. That is in large part a persistence of the 19th century perspective of natural laws governing our world. The important point is that such laws or rules give us the basis for real understanding, rather than the continued accumulation of simple facts or observations.

Based upon our review of the literature in this and related areas of fish behavior we suggest three important areas for analysis and information to provide critical information in the study of stress and fish behavior. All three are at the respective interfaces between active areas of productive research that hold considerable promise for our understanding of the mechanisms of stress and behavior.

The first of these is the interface between gene regulation and behavior. We believe this is a critical knowledge gap that can be a primary subject for research focus. There is a wealth of information from the very productive study of gene regulation (Brakefield, 2011; Bourret et al., 2013; Oswald et al., 2013; McKinney et al., 2015). The model fish system is of course the zebra danio (*Danio rerio*) but there is considerable potential for study of salmonids, especially related to the differences in life history in some species.

We are long past the time when we assumed that genes were either present or absent and that there were genes for behavior. Prime examples from salmonids come from studies of differences between hatchery and wild steelhead that appear to result from differences in gene activation attributed to domestication of fish in hatcheries (Araki et al., 2007, 2008; Christie et al., 2012). The approach of those studies will lead to an understanding of the causal relationships between genes and behavior. That will also provide us with the knowledge to manipulate gene activation through the external environment to produce fish with desired behavioral phenotypes. The production of fishes, especially salmonids, with specified behavioral phenotypes has long been the goal of those responsible for fisheries management, ranging from conservation and restoration to supplementation and support of harvest programs.

The second area that we propose for future studies is transgenerational effect of stress on behavior. These are sometimes referred to as nongenetic maternal effects or epigenetic effects (Leblanc et al., 2011). We could consider this area as a subset of gene regulation and behavior, for transgenerational effects will surely be a consequence of differences in gene regulation. However, we propose this as a specific need for future study because of the importance of understanding the role of stress in the life of fishes (Sloman, 2010). Evidence from detailed studies of the transgenerational effects of hatchery rearing on steelhead (*Oncorhynchus mykiss*) clearly shows the pattern we can predict (Araki et al., 2007, 2009; Christie et al., 2012, 2014). The challenge in this area of study will be to unravel and analyze the myriad of potential relationships between stress responses of the fish and the complex array of external and internal environmental factors that can be causally related to stress responses (Sneddon et al., 2016; Chapter 12 in this volume). Probably the search for common pathways linking stress and causal factors will be the most productive.

The third area that we propose is decision-making by fish (behavioral)— predictive models of behavior in relation to stress. This area of study presupposes an optimality model. As we have outlined in this chapter, our position is that the responses of fishes to stress can best be considered as adaptive. Their response will have been shaped by natural selection. However, that requires us to determine what is being optimized. Houston and McNamara (1999) have developed what must be one of the most comprehensive and elegant examples of this approach to understanding behavior. Perhaps their model will also apply to considerations that must include physiology, ecology, evolution, life history, sex differences, and their relationships to stress. This area of study will certainly be challenging, but in the end might be the most useful for a more comprehensive understanding of stress in the lives of fishes.

ACKNOWLEDGMENTS

We thank Carl Schreck and Lluis Tort for their invitation to contribute this chapter, and for their helpful comments and suggestions on various drafts. We offer our special thanks to Victoria Braithwaite for her insightful review and comments, and our gratitude to Lynn Bouvier for her invaluable assistance with innumerable details in revisions. We thank a number of colleagues, students, and associates for their shared thoughts, ideas, and research over a number of years. We have tried to acknowledge those by direct references to their published papers. We also thank all those institutions, agencies, and organizations that have supported our research activities and projects, and our current home departments and institutions for providing our intellectual homes to carry out our writing.

REFERENCES

Adamo, S. A. (2014). The effects of stress hormones on immune function may be vital for the adaptive reconfiguration of the immune system during fight-or-flight behavior. *Integr. Comp. Biol.* **54**, 419–426.

Agrillo, C., Petrazzini, M. E. M., Piffer, L., Dadda, M. and Bisazza, A. (2012). A new training procedure for studying discrimination learning in fish. *Behav. Brain Res.* **230**, 343–348.

Alanärä, A. (1992). Demand feeding as a self-regulating feeding system for rainbow trout (*Oncorhynchus mykiss*) in net-pens. *Aquaculture* **108**, 347–356.

Alanärä, A. (1996). The use of self-feeders in rainbow trout (*Oncorhynchus mykiss*) production. *Aquaculture* **145**, 1–20.

Allan, B. J., Miller, G. M., McCormick, M. I., Domenici, P. and Munday, P. L. (2014). Parental effects improve escape performance of juvenile reef fish in a high-CO_2 world. *Proc. R. Soc. Lond. B Biol. Sci.* **281**, 20132179.

Allan, B. J., Domenici, P., Munday, P. L. and McCormick, M. I. (2015). Feeling the heat: the effect of acute temperature changes on predator–prey interactions in coral reef fish. *Conserv. Physiol.* **3**, cov011.

Araki, H., Cooper, B. and Blouin, M. S. (2007). Genetic effects of captive breeding cause a rapid, cumulative fitness decline in the wild. *Science* **318**, 100–103.

Araki, H., Berejikian, B. A., Ford, M. J. and Blouin, M. S. (2008). Fitness of hatchery-reared salmonids in the wild. *Evol. Appl.* **1**, 342–355.

Araki, H., Cooper, B. and Blouin, M. S. (2009). Carry-over effect of captive breeding reduces reproductive fitness of wild-born descendants in the wild. *Biol. Lett.* **5**, 621–624.

Askey, P. J., Richards, S. A., Post, J. R. and Parkinson, E. A. (2006). Linking angling catch rates and fish learning under catch-and-release regulations. *N. Am. J. Fish. Manage.* **26**, 1020–1029.

Attia, J., Millot, S., Di-Poï, C., Bégout, M.-L., Noble, C., Sanchez-Vazquez, F. J., et al. (2012). Demand feeding and welfare in farmed fish. *Fish Physiol. Biochem.* **38**, 107–118.

Baciadonna, L. and McElligott, A. (2015). The use of judgement bias to assess welfare in farm livestock. *Anim. Welfare* **24**, 81–91.

Beamish, F. W. H. (1970). Influence of temperature and salinity acclimation on temperature preferenda of the euryhaline fish Tilapia nilotica. *J. Fish. Res. Board Can.* **27**, 1209–1214.

Beauchamp, R. D. and Rowe, J. S. (1977). Goldfish spectral sensitivity: a conditioned heart rate measure in restrained or curarized fish. *Vision. Res.* **17**, 617–624.

Beauchamp, R., Rowe, J. and O'Reilly, L. (1979). Goldfish spectral sensitivity: identification of the three cone mechanisms in heart-rate conditioned fish using colored adapting backgrounds. *Vision Res.* **19**, 1295–1302.

Beitinger, T. L. and Fitzpatrick, L. C. (1979). Physiological and ecological correlates of preferred temperature in fish. *Am. Zool.* **19**, 319–329.

Berejikian, B. A., Smith, R. J. F., Tezak, E. P., Schroder, S. L. and Knudsen, C. M. (1999). Chemical alarm signals and complex hatchery rearing habitats affect antipredator behavior and survival of Chinook salmon (*Oncorhynchus tshawytscha*) juveniles. *Can. J. Fish. Aquat. Sci.* **56**, 830–838.

Berejikian, B., Tezak, E., Riley, S. and LaRae, A. (2001). Competitive ability and social behaviour of juvenile steelhead reared in enriched and conventional hatchery tanks and a stream environment. *J. Fish. Biol.* **59**, 1600–1613.

Berejikian, B. A., Campbell, L. A. and Moore, M. E. (2013). Large-scale freshwater habitat features influence the degree of anadromy in eight Hood Canal *Oncorhynchus mykiss* populations. *Can. J. Fish. Aquat. Sci.* **70**, 756–765.

Bitterman, M. (1964). Classical conditioning in the goldfish as a function of the CS-UCS interval. *J. Comp. Physiol. Psychol.* **58**, 359.

Block, B. A. and Finnerty, J. R. (1994). Endothermy in fishes: a phylogenetic analysis of constraints, predispositions, and selection pressures. *Environ. Biol. Fish.* **40**, 283–302.

Boeuf, G. (1993). Salmonid smolting: a pre-adaptation to the oceanic environment. In *Fish ecophysiology* (eds. J. C. Rankin and F. B. Jensen), pp. 105–135. London: Chapman and Hall.

Bond, M. H., Miller, J. A. and Quinn, T. P. (2015). Beyond dichotomous life histories in partially migrating populations: cessation of anadromy in a long-lived fish. *Ecology* **96** (7), 1899–1910.

Boujard, T., Jourdan, M., Kentouri, M. and Divanach, P. (1996). Diel feeding activity and the effect of time-restricted self-feeding on growth and feed conversion in European sea bass. *Aquaculture* **139**, 117–127.

Bourret, V., Kent, M. P., Primmer, C. R., Vasemägi, A., Karlsson, S., Hindar, K., et al. (2013). SNP-array reveals genome-wide patterns of geographical and potential adaptive divergence across the natural range of Atlantic salmon (*Salmo salar*). *Mol. Ecol.* **22**, 532–551.

Braithwaite, V. (2010). *Do Fish Feel Pain?* Oxford: Oxford University Press.

Brakefield, P. M. (2011). Evo-devo and accounting for Darwin's endless forms. *Philos. Trans. R. Soc. B Biol. Sci.* **366**, 2069–2075.

Brannon, E. and Quinn, T. (1990). Field test of the pheromone hypothesis for homing by pacific salmon. *J. Chem. Ecol.* **16**, 603–609.

Brett, J. R. (1952). Temperature tolerance in young Pacific salmon, genus *Oncorhynchus*. *J. Fish. Res. Board Can.* **9**, 265–323.

Brett, J. R. (1971). Energetic responses of salmon to temperature. A study of some thermal relations in the physiology and freshwater ecology of sockeye salmon (*Oncorhynchus nerka*). *Am. Zool.* **11**, 99–113.

Brett, J. R. and Groot, C. (1963). Some aspects of olfactory and visual responses in Pacific salmon. *J. Fish. Res. Board Can.* **20**, 287–303.

Brockmark, S., Adriaenssens, B. and Johnsson, J. I. (2010). Less is more: density influences the development of behavioural life skills in trout. *Proc. R. Soc. B Biol. Sci.* **277**, 3035–3043.

Brown, C. and Laland, K. (2001). Social learning and life skills training for hatchery reared fish. *J. Fish. Biol.* **59**, 471–493.

Brown, C., Davidson, T. and Laland, K. (2003). Environmental enrichment and prior experience of live prey improve foraging behaviour in hatchery-reared Atlantic salmon. *J. Fish. Biol.* **63**, 187–196.

Bunt, C. M., Mandrak, N. E., Eddy, D. C., Choo-Wing, S. A., Heiman, T. G. and Taylor, E. (2013). Habitat utilization, movement and use of groundwater seepages by larval and juvenile Black Redhorse, Moxostoma duquesnei. *Environ. Biol. Fish.* **96**, 1281–1287.

Burnham, J. C. (1972). Thorndike's puzzle boxes. *J. Hist. Behav. Sci.* **8**, 159–167.

Cachat, J. M., Canavello, P. R., Elegante, M. F., Bartels, B. K., Elkhayat, S. I., Hart, P. C., et al. (2011). Modeling stress and anxiety in zebrafish. In *Zebrafish models in neurobehavioral research* (eds. A. V. Kalueff and J. M. Cachat), pp. 73–88. Dordrecht: Springer.

Careau, V., Buttemer, W. A. and Buchanan, K. L. (2014). Developmental stress can uncouple relationships between physiology and behaviour. *Biol. Lett.* **10**, 20140834.

Carey, F. G., Teal, J. M., Kanwisher, J. W., Lawson, K. D. and Beckett, J. S. (1971). Warm-bodied fish. *Am. Zool.* **11**, 137–143.

Carruth, L. L., Jones, R. E. and Norris, D. O. (2002). Cortisol and Pacific salmon: a new look at the role of stress hormones in olfaction and home-stream migration. *Integr. Comp. Biol.* **42**, 574–581.

Charnov, E. L. (1976). Optimal foraging, the marginal value theorem. *Theor. Popul. Biol.* **9**, 129–136.

Charnov, E. L. (1978). Evolution of eusocial behavior: offspring choice or parental parasitism? *J. Theor. Biol.* **75**, 451–465.

Charnov, E. L. (1993). *Life History Invariants*. Oxford, UK: Oxford University Press.

Chiasson, P., Beamish, F. W. H. and Noakes, D. L. G. (1995). Benthic invertebrates and stomach contents of juvenile sturgeon in the Moose river basin, Ontario. *Can. J. Fish. Aquat. Sci.* **54**, 2866–2871.

Christie, M. R., Marine, M. L., French, R. A. and Blouin, M. S. (2012). Genetic adaptation to captivity can occur in a single generation. *Proc. Natl. Acad. Sci.* **109**, 238–242.

Christie, M. R., French, R. A., Marine, M. L. and Blouin, M. S. (2014). How much does inbreeding contribute to the reduced fitness of hatchery-born steelhead (*Oncorhynchus mykiss*) in the wild? *J. Hered.* **105**, 111–119.

Church, M. R., Ebersole, J. L., Rensmeyer, K. M., Couture, R. B., Barrows, F. T. and Noakes, D. L. (2008). Mucus: a new tissue fraction for rapid determination of fish diet switching using stable isotope analysis. *Can. J. Fish. Aquat. Sci.* **66**, 1–5.

Congleton, J. L., LaVoie, W. J., Schreck, C. B. and Davis, L. E. (2000). Stress indices in migrating juvenile Chinook salmon and steelhead of wild and hatchery origin before and after barge transportation. *Trans. Am. Fish. Soc.* **129**, 946–961.

Conrad, J. L., Weinersmith, K. L., Brodin, T., Saltz, J. and Sih, A. (2011). Behavioural syndromes in fishes: a review with implications for ecology and fisheries management. *J. Fish. Biol.* **78**, 395–435.

Conte, F. (2004). Stress and the welfare of cultured fish. *Appl. Anim. Behav. Sci.* **86**, 205–223.

Contreras-Sanchez, W., Schreck, C., Fitzpatrick, M. and Pereira, C. (1998). Effects of stress on the reproductive performance of rainbow trout (*Oncorhynchus mykiss*). *Biol. Reprod.* **58**, 439–447.

Cook, K. V., Crossin, G. T., Patterson, D. A., Hinch, S. G., Gilmour, K. M. and Cooke, S. J. (2014). The stress response predicts migration failure but not migration rate in a semelparous fish. *Gen. Comp. Endocrinol.* **202**, 44–49.

Cooper, S. J. (2008). From claude bernard to walter cannon. Emergence of the concept of homeostasis. *Appetite* **51**, 419–427.

Cooper, J. C. and Scholz, A. T. (1976). Homing of artificially imprinted steelhead (rainbow) trout, *Salmo gairdneri*. *J. Fish. Res. Board Can.* **33**, 826–829.

Coutant, C. C. (1977). Compilation of temperature preference data. *J. Fish. Board Can.* **34**, 739–745.

Coutant, C. C. (1987). Thermal preference: when does an asset become a liability? *Environ. Biol. Fish.* **18**, 161–172.

Danley, M. L., Mayr, S. D., Young, P. S. and Cech, J. J. (2002). Swimming performance and physiological stress responses of splittail exposed to a fish screen. *N. Am. J. Fish. Manage.* **22**, 1241–1249.

Davies, N. B., Krebs, J. R. and West, S. A. (2012). *An Introduction to Behavioural Ecology.* Oxford: John Wiley & Sons.

Dawkins, R. (1976). *The Selfish Gene.* Oxford: Oxford University Press.

Dawkins, M. S. (2003). Behaviour as a tool in the assessment of animal welfare. *Zoology* **106**, 383–387.

Dawkins, R. (2004). Extended phenotype–but not too extended. A reply to Laland, Turner and Jablonka. *Biol. Philos.* **19**, 377–396.

Dawkins, R. (2009). *The Greatest Show on Earth. The Evidence for Evolution.* New York, NY: Simon & Schuster.

Dawkins, M. S. (2012). *Why Animals Matter: Animal Consciousness, Animal Welfare, and Human Well-being.* Oxford: Oxford University Press.

Dawkins, R. (2013). *An Appetie for Wonder.* New York, NY: Harper Collins.

Dill, L. M. (1983). Adaptive flexibility in the foraging behavior of fishes. *Can. J. Fish. Aquat. Sci.* **40**, 398–408.

Dittman, A. and Quinn, T. (1996). Homing in Pacific salmon: mechanisms and ecological basis. *J. Exp. Biol.* **199**, 83–91.

Dittman, A. H., Pearsons, T. N., May, D., Couture, R. B. and Noakes, D. L. (2015). Imprinting of hatchery-reared salmon to targeted spawning locations: a new embryonic imprinting paradigm for hatchery programs. *Fisheries* **40**, 114–123.

Dobzhansky, T. (1973). Nothing in biology makes sense except in the light of evolution. *Am. Biol. Teach.* **35**, 125–129.

Domenici, P. and Ruxton, G. D. (2015). *8 Prey behaviors during fleeing: escape trajectories, signaling, and sensory defenses. Escaping From Predators: An Integrative View of Escape Decisions.* Cambridge: Cambridge University Press.

Domenici, P., Allan, B. J., Watson, S.-A., McCormick, M. I. and Munday, P. L. (2014). Shifting from right to left: the combined effect of elevated CO_2 and temperature on behavioural lateralization in a coral reef fish. *PLoS One* **9** (1), e87969.

Dominique, J.-F. and McGaugh, J. L. (2014). Stress and the regulation of memory: from basic mechanisms to clinical implications. *Neurobiol. Learn. Mem.* **112**, 1.

Donelson, J. M., McCormick, M. I., Booth, D. J. and Munday, P. L. (2014). Reproductive acclimation to increased water temperature in a tropical reef fish. *PLoS One* **9** (5), e97223.

Drew, J. A. and Gumm, J. M. (2015). Learning and behavior in reef fish: fuel for microevolutionary change? *Ethology* **121**, 2–7.

Dymond, J. R. (1932). *The Trout and Other Game Fishes of British Columbia.* Ottawa, ON: Department of Fisheries.

Eaton, L., Edmonds, E., Henry, T., Snellgrove, D. and Sloman, K. (2015). Mild maternal stress disrupts associative learning and increases aggression in offspring. *Horm. Behav.* **71**, 10–15.

Ejike, C. and Schreck, C. B. (1980). Stress and social hierarchy rank in coho salmon. *Trans. Am. Fish. Soc.* **109**, 423–426.

Eldridge, W. H., Sweeney, B. W. and Law, J. M. (2015). Fish growth, physiological stress, and tissue condition in response to rate of temperature change during cool or warm diel thermal cycles. *Can. J. Fish. Aquat. Sci.* **72**, 1527–1537.

Fagerlund, U. H. M., McBride, J. R. and Stone, E. T. (1981). Stress-related effects of hatchery rearing density on coho salmon. *Trans. Am. Fish. Soc.* **110**, 644–649.

Fausch, K. D. and Northcote, T. G. (1992). Large woody debris and salmonid habitat in a small Coastal British Columbia stream. *Can. J. Fish. Aquat. Sci.* **49**, 682–693.

Fenderson, O. C., Everhart, W. H. and Muth, K. M. (1968). Comparative agonistic and feeding behavior of hatchery-reared and wild salmon in aquaria. *J. Fish. Res. Board Can.* **25**, 1–14.

Fernald, R. (1976). The effect of testosterone on the behavior and coloration of adult male cichlid fish (Haplochromis burtoni, Günther). *Horm. Res. Paediatr.* **7**, 172–178.

Fernald, R. D. (2015). Social behaviour: can it change the brain? *Anim. Behav.* **103**, 259–265.

Fernö, A., Huse, G., Jakobsen, P. J., Kristiansen, T. S. and Nilsson, J. (2011). *Fish behaviour, learning, aquaculture and fisheries. Fish Cognition and Behavior.* Oxford: Wiley-Blackwell.

Ferrari, M. C., Manassa, R. P., Dixson, D. L., Munday, P. L., McCormick, M. I., Meekan, M. G., et al. (2012). Effects of ocean acidification on learning in coral reef fishes. *PLoS One* **7**, e31478.

Ferrari, M., Munday, P. L., Rummer, J. L., McCormick, M. I., Corkill, K., Watson, S. A., et al. (2015). Interactive effects of ocean acidification and rising sea temperatures alter predation rate and predator selectivity in reef fish communities. *Glob. Change Biol.* **21**, 1848–1855.

Fisknes, B. and Døving, K. (1982). Olfactory sensitivity to group-specific substances in Atlantic salmon (*Salmo salar* L.). *J. Chem. Ecol.* **8**, 1083–1092.

Francis, R. C. (1988). Socially mediated variation in growth rate of the Midas cichlid: the primacy of early size differences. *Anim. Behav.* **36**, 1844–1845.

Fraser, D. J., Weir, L. K., Bernatchez, L., Hansen, M. M. and Taylor, E. B. (2011). Extent and scale of local adaptation in salmonid fishes: review and meta-analysis. *Heredity* **106**, 404–420.

Frisch, K. V. (1942). Über einen schreckstoff der cischhaut und seine biologische bedeutung. *Zeit. Vergl. Physiol.* **29**, 46–145.

Fry, F. E. J. (1947). Effects of the environment on animal activity. University of Toronto Studies, Biological Series No. 55. Publications of the Ontario Fisheries Research Laboratory No. 68, 62.

Fürtbauer, I., Pond, A., Heistermann, M. and King, A. J. (2015). Personality, plasticity and predation: linking endocrine and behavioural reaction norms in stickleback fish. *Funct. Ecol* **29**, 931–940.

Galhardo, L., Vital, J. and Oliveira, R. F. (2011). The role of predictability in the stress response of a cichlid fish. *Physiol. Behav.* **102**, 367–372.

Gao, Y., Bean, D. and Noakes, D. L. (2008). Stable isotope analyses of otoliths in identification of hatchery origin of Atlantic salmon (*Salmo salar*) in maine. *Environ. Biol. Fish.* **83**, 429–437.

Godin, J.-G. J. E. (1999). *Behavioural Ecology of Teleost Fishes.* New York, NY: Oxford University Press.

Goetz, F. A., Jeanes, E., Moore, M. E. and Quinn, T. P. (2015). Comparative migratory behavior and survival of wild and hatchery steelhead (*Oncorhynchus mykiss*) smolts in riverine, estuarine, and marine habitats of Puget Sound, Washington. *Environ. Biol. Fish.* **98**, 357–375.

Grant, J. W. A. and Noakes, D. L. G. (1987a). Movers and stayers: foraging tactics of young-of-the-year brook charr, *Salvelinus fontinalis. J. Anim. Ecol.* **56**, 1001–1013.

Grant, J. W. A. and Noakes, D. L. G. (1987b). A simple model of optimal territory size for drift-feeding fish. *Can. J. Zool.* **65**, 270–276.

Grassie, C., Braithwaite, V. A., Nilsson, J., Nilsen, T. O., Teien, H.-C., Handeland, S. O., et al. (2013). Aluminum exposure impacts brain plasticity and behavior in Atlantic salmon (*Salmo salar*). *J. Exp. Biol.* **216** (Pt 16), 3148–3155.

Green, R. F. (1987). Stochastic models of optimal foraging. In *Foraging behavior* (eds. A. C. Kamil, J. R. Krebs and H. R. Pulliam), pp. 273–302. Dordrecht: Springer.

Gross, C. G. (1998). Claude bernard and the constancy of the internal environment. *Neuroscientist* **4**, 380–385.

Hamilton, W. (1998). Narrow roads of gene land: The collected papers of W. D. Hamilton Volume 1: Evolution of social behaviour. Oxford: Oxford University Press.

Hansen, L. P. and Quinn, T. P. (1998). The marine phase of the Atlantic salmon (*Salmo salar*) life cycle, with comparisons to Pacific salmon. *Can. J. Fish. Aquat. Sci.* **55**, 104–118.

Hargrove, M. B., Becker, W. S. and Hargrove, D. F. (2015). The HRD eustress model generating positive stress with challenging work. *Hum. Resour. Dev. Rev.* **14**, 279–298.

Hasler, A. (1966). *Underwater Guideposts. Madison: University Wisconsin Press.*

Hasler, A. D., Scholz, A. T. and Horrall, R. M. (1978). Olfactory imprinting and homing in salmon. *Am. Sci.* **66**, 347–355.

Hawryshyn, C. W. (1992). Polarization vision in fish. *Am. Sci.* 164–175.

Hawryshyn, C. W. (2000). Ultraviolet polarization vision in fishes: possible mechanisms for coding e–vector. *Philos. Trans. R. Soc. Lond.* B Biol. Sci. **355**, 1187–1190.

Hawryshyn, C. W. and McFarland, W. N. (1987). Cone photoreceptor mechanisms and the detection of polarized light in fish. *J. Comp. Physiol.* A **160**, 459–465.

Hawryshyn, C. W., Arnold, M. G., Chaisson, D. J. and Martin, P. C. (1989). The ontogeny of ultraviolet photosensitivity in rainbow trout (*Salmo gairdneri*). *Vis. Neurosci.* **2**, 247–254.

Hayes, S. A., Hanson, C. V., Pearse, D. E., Bond, M. H., Garza, J. C. and MacFarlane, R. B. (2012). Should I stay or should I go? The influence of genetic origin on emigration behavior and physiology of resident and Anadromous juvenile *Oncorhynchus mykiss*. *N. Am. J. Fish. Manage.* **32**, 772–780.

Hazel, J. R. and Prosser, C. L. (1974). Molecular mechanisms of temperature compensation in poikilotherms. *Physiol. Rev.* **54**, 620–677.

Hertz, P. E., Huey, R. B. and Nevo, E. (1982). Fight versus flight: body temperature influences defensive responses of lizards. *Anim. Behav.* **30**, 676–679.

Hoar, W. S. (1953). Control and timing of fish migration. *Biol. Rev.* **28**, 437–452.

Hoar, W. S. (1966). *General and Comparative Physiology.* New York, NY: Prentice Hall.

Hodgson, S., Quinn, T. P., Hilborn, R. A. Y., Francis, R. C. and Rogers, D. E. (2006). Marine and freshwater climatic factors affecting interannual variation in the timing of return migration to fresh water of sockeye salmon (*Oncorhynchus nerka*). *Fish. Oceanogr.* **15**, 1–24.

Holland, K. N. and Teeter, J. H. (1981). Behavioral and cardiac reflex assays of the chemosensory acuity of channel catfish to amino acids. *Physiol. Behav.* **27**, 699–707.

Hollis, K. L. (1982). Pavlovian conditioning of signal-centered action patterns and autonomic behavior: a biological analysis of function. *Adv. Study Behav.* **12**, 1–64.

Hollis, K. L. (1997). Contemporary research on Pavlovian conditioning: a "new" functional analysis. *Am. Psychol.* **52**, 956.

Hollis, K. L., Cadieux, E. L. and Colbert, M. M. (1989). The biological function of Pavlovian conditioning: a mechanism for mating success in the blue gourami (*Trichogaster trichopterus*). *J. Comp. Psychol.* **103**, 115.

Hollis, K. L., Dumas, M. J., Singh, P. and Fackelman, P. (1995). Pavlovian conditioning of aggressive behavior in blue gourami fish (*Trichogaster trichopterus*): winners become winners and losers stay losers. *J. Comp. Psychol.* **109**, 123.

Hollis, K. L., Pharr, V. L., Dumas, M. J., Britton, G. B. and Field, J. (1997). Classical conditioning provides paternity advantage for territorial male blue gouramis (*Trichogaster trichopterus*). *J. Comp. Psychol.* **111**, 219.

Houston, A. I. and McNamara, J. M. (1999). *Models of Adaptive Behaviour.* Cambridge, UK: Cambridge University Press.

Hughes, N. F. (2000). Testing the ability of habitat selection theory to predict interannual movement patterns of a drift-feeding salmonid. *Ecol. Freshw. Fish* **9**, 4–8.

Hughes, N. and Grand, T. (2000). Physiological ecology meets the ideal-free distribution: predicting the distribution of size-structured fish populations across temperature gradients. *Environ. Biol. Fishes* **59**, 285–298.

Iwama, G. K., Pickering, A., Sumpter, J. and Schreck, C. (2011). *Fish Stress and Health in Aquaculture*. Cambridge: Cambridge University Press.

Jansen, A. S., Van Nguyen, X., Karpitskiy, V., Mettenleiter, T. C. and Loewy, A. D. (1995). Central command neurons of the sympathetic nervous system: basis of the fight-or-flight response. *Science* **270**, 644–646.

Jobling, M. (1981). Temperature tolerance and the final preferendum—rapid methods for the assessment of optimum growth temperatures. *J. Fish. Biol.* **19**, 439–455.

Jobling, M. (1985). Physiological and social constraints on growth of fish with special reference to Arctic charr, *Salvelinus alpinus* L. *Aquaculture* **44**, 83–90.

Johnsen, P. B. and Hasler, A. D. (1980). The use of chemical cues in the upstream migration of coho salmon, *Oncorhynchus kisutch* Walbaum. *J. Fish. Biol.* **17**, 67–73.

Jones, K. (2005). The effect of territorial damselfish (family Pomacentridae) on the space use and behaviour of the coral reef fish, *Halichoeres bivittatus* (Bloch, 1791)(family Labridae). *J. Exp. Mar. Biol. Ecol.* **324**, 99–111.

Jones, K. M. M. (2007). Distribution of behaviours and species interactions within home range contours in five Caribbean reef fish species (Family Labridae). *Environ. Biol. Fishes* **80**, 35–49.

Jones, K. M. M., McGrath, P. E. and Able, K. W. (2014). White perch morone americana (Gmelin, 1789) habitat choice and movements: comparisons between phragmites-invaded and spartina reference marsh creeks based on acoustic telemetry. *J. Exp. Mar. Biol. Ecol.* **455**, 14–21.

Jonsson, B. and Jonsson, N. (2006). Cultured Atlantic salmon in nature: a review of their ecology and interaction with wild fish. *ICES J. Marine Sci. J. du Conseil* **63**, 1162–1181.

Kamil, A. C., Krebs, J. R. and Pulliam, H. R. (1987). *Foraging Behavior*. New York, NY: Plenum Press.

Kavanagh, M. and Olson, D. E. (2014). The effects of rearing density on growth, fin erosion, survival, and migration behavior of hatchery winter steelhead. *N. Am. J. Aquac.* **76**, 323–332.

Kawaguchi, Y. and Nakano, S. (2001). Contribution of terrestrial invertebrates to the annual resource budget for salmonids in forest and grassland reaches of a headwater stream. *Freshw. Biol.* **46**, 303–316.

Keenleyside, M. H. (2012). *Diversity and Adaptation in Fish Behaviour*. Springer Science & Business Media.

Larra, M. F., Schulz, A., Schilling, T. M., de Sá, D. S. F., Best, D., Kozik, B., et al. (2014). Heart rate response to post-learning stress predicts memory consolidation. *Neurobiol. Learn. Mem.* **109**, 74–81.

Laursen, D. C., Olsén, H. L., de Lourdes Ruiz-Gomez, M., Winberg, S. and Höglund, E. (2011). Behavioural responses to hypoxia provide a non-invasive method for distinguishing between stress coping styles in fish. *Appl. Anim. Behav. Sci.* **132**, 211–216.

Lawrence, D. J., Beauchamp, D. A. and Olden, J. D. (2015). Life-stage-specific physiology defines invasion extent of a riverine fish. *J. Anim. Ecol.* **84**, 879–888.

Leblanc, C. A. L., Benhaïm, D., Hansen, B. R., Kristjánsson, B. K. and Skúlason, S. (2011). The importance of egg size and social effects for behaviour of Arctic charr juveniles. *Ethology* **117**, 664–674.

Le Fevre, M., Matheny, J. and Kolt, G. S. (2003). Eustress, distress, and interpretation in occupational stress. *J. Manage. Psychol.* **18**, 726–744.

Lema, S. C. and Nevitt, G. A. (2004). Evidence that thyroid hormone induces olfactory cellular proliferation in salmon during a sensitive period for imprinting. *J. Exp. Biol.* **207**, 3317–3327.

Mackintosh, N. (1969). Comparative studies of reversal and probability learning: rats, birds and fish. In *Animal Discrimination Learning* (eds. R. M. Gilbert and N. S. Sutherland), pp. 137–162. New York, London: Academic Press.

Magnan, P., Audet, C., Legault, M., Rodríguez, M. A. and Taylor, E. B. (2002). *Ecology, behaviour and conservation of the charrs, genus* Salvelinus. Dordrecht: Springer.

Malavasi, S., Georgalas, V., Mainardi, D. and Torricelli, P. (2008). Antipredator responses to overhead fright stimuli in hatchery–reared and wild European sea bass (*Dicentrarchus labrax* L.) juveniles. *Aquac. Res.* **39**, 276–282.

Manek, A. K., Ferrari, M. C., Sereda, J. M., Niyogi, S. and Chivers, D. P. (2012). The effects of ultraviolet radiation on a freshwater prey fish: physiological stress response, club cell investment, and alarm cue production. *Biol. J. Linn. Soc.* **105**, 832–841.

Manning, A. W. G. and Dawkins, M. S. (2012). *Introduction to Animal Behaviour.* Cambridge: Cambridge University Press.

Maruska, K. P., Zhang, A., Neboori, A. and Fernald, R. D. (2013). Social opportunity causes rapid transcriptional changes in the social behaviour network of the brain in an African cichlid fish. *J. Neuroendocrinol.* **25**, 145–157.

Matsuda, K. (2013). Regulation of feeding behavior and psychomotor activity by corticotropin-releasing hormone (CRH) in fish. *Front. Neurosci.* **7**.

McAroe, C. L., Craig, C. M. and Holland, R. A. (2015). Place versus response learning in fish: a comparison between species. *Anim. Cogn.* 1–9.

McCauley, R., and Casselman, J. (1981). The final preferendum as an index of the temperature for optimum growth in fish. In: *Proceedings of the World Symposium on Aquaculture in Heated Effluents and Recirculation Systems.* Vol. 2, pp. 81–93, Stavanger Norway.

McCormick, S. D., Farrell, A. P. and Brauner, C. J. (2013). *Fish Physiology: Euryhaline Fishes,* Vol. 32. Academic Press. London: Academic Press.

McEwen, B. S. (2005). Stressed or stressed out: what is the difference? *J. Psychiatry Neurosci.* **30**, 315.

McFarland, D. (1999). *Animal Behaviour.* New York, NY: Prentice - Hall.

McKinney, G. J., Hale, M. C., Goetz, G., Gribskov, M., Thrower, F. P. and Nichols, K. M. (2015). Ontogenetic changes in embryonic and brain gene expression in progeny produced from migratory and resident *Oncorhynchus mykiss.* Mol. Ecol. **24** (8), 1792–1809.

McLaughlin, R. L., Grant, J. W. A. and Kramer, D. L. (1994). Foraging movements in relation to morphology, water-column use, and diet for recently emerged brook trout (*Salvelinus fontinalis*) in still-water pools. *Can. J. Fish. Aquat. Sci.* **51**, 268–279.

McLaughlin, R. L., Ferguson, M. M. and Noakes, D. L. G. (1999). Adaptive peaks and alternative foraging tactics in brook charr: evidence of short-term divergent selection for sitting-and-waiting and actively searching. *Behav. Ecol. Sociobiol.* **45**, 386–395.

McLeese, D. W. (1956). Effects of temperature, salinity and oxygen on the survival of the American lobster. *J. Fish. Res. Bd. Can* 13, 247–272.

Melnychuk, M. C., Korman, J., Hausch, S., Welch, D. W., McCubbing, D. J., Walters, C. J., et al. (2014). Marine survival difference between wild and hatchery-reared steelhead trout determined during early downstream migration. *Can. J. Fish. Aquat. Sci.* **71**, 831–846.

Mesquita, F. O., Borcato, F. L. and Huntingford, F. A. (2015). Cue-based and algorithmic learning in common carp: a possible link to stress coping style. *Behav. Processes* **115**, 25–29.

Millot, S., Péan, S., Chatain, B. and Bégout, M.-L. (2011). Self-feeding behavior changes induced by a first and a second generation of domestication or selection for growth in the European sea bass, *Dicentrarchus labrax. Aquat. Living Resour.* **24**, 53–61.

Mittelbach, G. G., Ballew, N. G., Kjelvik, M. K. and Fraser, D. (2014). Fish behavioral types and their ecological consequences. *Can. J. Fish. Aquat. Sci.* **71**, 927–944.

Mommer, B. C. and Bell, A. M. (2013). A test of maternal programming of offspring stress response to predation risk in threespine sticklebacks. *Physiol. Behav.* **122**, 222–227.

Monaghan, P. and Spencer, K. A. (2014). Stress and life history. *Curr. Biol.* **24**, R408–R412.

Morandini, L., Honji, R. M., Ramallo, M. R., Moreira, R. G. and Pandolfi, M. (2014). The interrenal gland in males of the cichlid fish Cichlasoma dimerus: relationship with stress and the establishment of social hierarchies. *Gen. Comp. Endocrinol.* **195**, 88–98.

Morin, P.-P., Dodson, J. J. and Doré, F. Y. (1989). Thyroid activity concomitant with olfactory learning and heart rate changes in Atlantic salmon *Salmo salar*, during smoltification. *Can. J. Fish. Aquat. Sci.* **46**, 131–136.

Nakano, S. (1994). Variation in agonistic encounters in a dominance hierarchy of freely interacting red-spotted masu salmon (*Oncorhynchus masou ishikawai*). *Ecol. Freshw. Fish* **3**, 153–158.

Nakano, S. and Murakami, M. (2001). Reciprocal subsidies: dynamic interdependence between terrestrial and aquatic food webs. *Proc. Natl. Acad. Sci.* **98**, 166–170.

Nielsen, J. L. (1992). Microhabitat-specific foraging behavior, diet, and growth of juvenile coho salmon. *Trans. Am. Fish. Soc.* **121**, 617–634.

Nilsson, J., Stien, L. H., Fosseidengen, J. E., Olsen, R. E. and Kristiansen, T. S. (2012). From fright to anticipation: reward conditioning versus habituation to a moving dip net in farmed Atlantic cod (*Gadus morhua*). *Appl. Anim. Behav. Sci.* **138**, 118–124.

Noakes, D. L. (1986a). When to feed: decision making in sticklebacks, *Gasterosteus aculeatus*. In *Contemporary studies on fish feeding: the proceedings of GUTSHOP'84* (eds. C. A. Simenstad and G. M. Cailliet), pp. 95–104. Dordrecht: Springer.

Noakes, D. L. G. (1986b). Genetic basis of fish behaviour. In *The behaviour of teleost fishes* (ed. T. J. Pitcher), pp. 3–22. London: Croom Helm.

Noakes, D. L. G. (1999). Onogeny of behaviour. In *Behaviour of Cichlid Fishes* (ed. M. H. A. Keenleyside), London: Croom-Helm.

Noakes, D. L. G. (2014). Behavior and genetics of salmon. In *Salmon: Biology, Ecological Impacts and Economic Importance* (eds. P. T. Wood and D. J. Noakes), pp. 195–222. New York, NY: Nova. Nova Science Publishers, Orlando.

Noakes, D. L. G. and Baylis, J. R. (1990). Fish behavior. In *Methods in Fish Biology* (eds. C. B. Schreck and P. B. Moyle), pp. 553–585. Bethesda, MD: American Fisheries Society.

Noakes, D. L. G. and Corrarino, C. (2010). The oregon hatchery research center: an experimental laboratory in a natural setting. *World Aquac.* **41**, 33–37.

Noakes, D. L. and Grant, J. (1992). *Feeding and social behaviour of brook and lake charr. Feeding Behavior and Culture of Salmonid Fishes*. The Oregon Hatchery Research Center, Orlando: World Aquaculture Society.

Noakes, D. L. and Leatherland, J. F. (1977). Social dominance and interrenal cell activity in rainbow trout, *Salmo gairdneri* (Pisces, Salmonidae). *Environ. Biol. Fishes* **2**, 131–136.

Nordeng, H. (1977). A pheromone hypothesis for homeward migration in anadromous salmonids. *Oikos* **28**, 155–159.

Nordeng, H. (1983). Solution to the "Char Problem" based on Arctic char (*Salvelinus alpinus*) in Norway. *Can. J. Fish. Aquat. Sci.* **40**, 1372–1387.

Nordeng, H. (2009). Char ecology. Natal homing in sympatric populations of anadromous Arctic char *Salvelinus alpinus* (L.): roles of pheromone recognition. *Ecol. Freshw. Fish* **18**, 41–51.

Nordeng, H. and Bratland, P. (2006). Homing experiments with parr, smolt and residents of anadromous Arctic char *Salvelinus alpinus* and brown trout *Salmo trutta*: transplantation between neighbouring river systems. *Ecol. Freshw. Fish* **15**, 488–499.

Offitt, G. C. (1971). Response of the tautog (Tautoga onitis, teleost) to acoustic stimuli measured by classically conditioning the heart rate. *Cond. Reflex* **6**, 205–214.

Olla, B., Davis, M. and Ryer, C. (1994). Behavioural deficits in hatchery-reared fish: potential effects on survival following release. *Aquac. Fish. Manage.* **25** (Suppl. 1), 19–34.

Oswald, M. E., Singer, M. and Robison, B. D. (2013). The quantitative genetic architecture of the bold-shy continuum in zebrafish, *Danio rerio. PLoS One* **8**, e68828.

Ott, M. E., Heisler, N. and Ultsch, G. R. (1980). A re-evaluation of the relationship between temperature and the critical oxygen tension in freshwater fishes. *Comp. Biochem. Physiol. A Physiol.* **67**, 337–340.

Paetzold, S. C., Ross, N. W., Richards, R. C., Jones, M., Hellou, J. and Bard, S. M. (2009). Up-regulation of hepatic ABCC2, ABCG2, CYP1A1 and GST in multixenobiotic-resistant killifish (*Fundulus heteroclitus*) from the Sydney Tar Ponds, Nova Scotia, Canada. *Mar. Environ. Res.* **68**, 37–47.

Pankhurst, N. (1997). *Temperature effects on the reproductive performance of fishGlobal Warming: Implications for Freshwater and Marine Fish*, vol. 61. Cambridge: Cambridge University Press.

Pankhurst, N. W. (2011). The endocrinology of stress in fish: an environmental perspective. *Gen. Comp. Endocrinol.* **170**, 265–275.

Patton, B. W. and Braithwaite, V. A. (2015). Changing tides: ecological and historical perspectives on fish cognition. *Wiley Interdiscip. Rev. Cogn. Sci.* **6**, 159–176.

Pavlidis, M., Theodoridi, A. and Tsalafouta, A. (2015). Neuroendocrine regulation of the stress response in adult zebrafish, *Danio rerio. Prog. Neuropsychopharmacol. Biol. Psychiatry* **60**, 121–131.

Penaluna, B. E., Dunham, J. B. and Noakes, D. L. (2015). Instream cover and shade mediate avian predation on trout in semi-natural streams. *Ecol. Freshw. Fish.* 25, 405–411.

Pham, M., Raymond, J., Hester, J., Kyzar, E., Gaikwad, S., Bruce, I., et al. (2012). Assessing social behavior phenotypes in adult zebrafish: shoaling, social preference, and mirror biting tests. In *Zebrafish protocols for neurobehavioral research* (eds. A. V. Kalueff and A. M. Stewart), pp. 231–246. Dordrecht: Humana Press, Springer.

Piccolo, J. J., Hughes, N. F. and Bryant, M. D. (2007). The effects of water depth on prey detection and capture by juvenile coho salmon and steelhead. *Ecol. Freshw. Fish* **16**, 432–441.

Piccolo, J., Noakes, D. L. and Hayes, J. W. (2014). Preface to the special drift foraging issue of environmental biology of fishes. *Environ. Biol. Fishes* **97**, 449–451.

Pickering, A., Griffiths, R. and Pottinger, T. (1987). A comparison of the effects of overhead cover on the growth, survival and haematology of juvenile Atlantic salmon, *Salmo salar* L., brown trout, *Salmo trutta* L., and rainbow trout, *Salmo gairdneri* Richardson. *Aquaculture* **66**, 109–124.

Quinn, T. P. (2005). *The Behavior and Ecology of Pacific Salmon and Trout*. Bethesda, MD: American Fisheries Society.

Quinn, T. P. and Hara, T. J. (1986). Sibling recognition and olfactory sensitivity in juvenile coho salmon (*Oncorhynchus kisutch*). *Can. J. Zool.* **64**, 921–925.

Ramallo, M. R., Morandini, L., Alonso, F., Birba, A., Tubert, C., Fiszbein, A., et al. (2014). The endocrine regulation of cichlids social and reproductive behavior through the eyes of the chanchita, *Cichlasoma dimerus* (Percomorpha; Cichlidae). *J. Physiol. Paris* **108**, 194–202.

Reebs, S. G. (2001). *Fish Behaviour in the Aquarium and in the Wild*. Ithaca, NY: Cornell University Press.

Rey, S., Huntingford, F. A., Boltana, S., Vargas, R., Knowles, T. G. and Mackenzie, S. (2015). Fish can show emotional fever: stress-induced hyperthermia in zebrafish. *Proc. Biol. Sci.* **282**,pp. 20152266: The Royal Society

Ridley, M. (2003). *Evolution* (third ed.). New York, NY: Wiley.

Roche, D. P., McGhee, K. E. and Bell, A. M. (2012). Maternal predator-exposure has lifelong consequences for offspring learning in threespined sticklebacks. *Biol. Lett.* **8**, 932–935.

Rodriguez, F., Duran, E., Gomez, A., Ocana, F., Alvarez, E., Jimenez-Moya, F., et al. (2005). Cognitive and emotional functions of the teleost fish cerebellum. *Brain Res. Bull.* **66**, 365–370.

Romer, J. D., Leblanc, C. A., Clements, S., Ferguson, J. A., Kent, M. L., Noakes, D., et al. (2013). Survival and behavior of juvenile steelhead trout (*Oncorhynchus mykiss*) in two estuaries in Oregon, USA. *Environ. Biol. Fishes* **96**, 849–863.

Ruzzante, D. E. (1994). Domestication effects on aggressive and schooling behavior in fish. *Aquaculture* **120**, 1–24.

Schreck, C. B. (1990). Physiological, behavioral, and performance indicators of stress. *Am. Fish. Soc. Symp.* **8**, 29–37.

Schreck, C. (2000). Accumulation and long-term effects of stress in fish. *Biol. Anim. Stress* 147–158.

Schreck, C., Olla, B., Davis, M., Iwama, G., Pickering, A., Sumpter, J., et al. (1997). *Behavioral responses to stressFish Stress and Health in Aquaculture*, vol. 62. Cambridge: Cambridge University Press.

Schreck, C. B. and Tort, L. (2016). The Concept of Stress in Fish. In *Fish Physiology - Biology of Stress in Fish*, Vol. 35 (eds. C. B. Schreck, L. Tort, A. P. Farrell and C. J. Brauner), San Diego, CA: Academic Press.

Seals Price, C. C. B. S. (2003). Stress and saltwater entry behavior of juvenile chinook salmon (*Oncorhynchus tshawytscha*): conflict in physiological motivation. *Can. J. Fish. Aquat. Sci.* **60**, 910–918.

Selye, H. (1975). Confusion and controversy in the stress field. *J. Hum. Stress* **1**, 37–44.

Selye, H. (1976a). Forty years of stress research: principal remaining problems and misconceptions. *Can. Med. Assoc. J.* **115**, 53.

Selye, H. (1976b). The stress concept. *Can. Med. Assoc. J.* **115**, 718.

Selye, H. (1976c). *Stress without distress.* New York, NY: Signet.

Selye, H. (1978). On the real benefits of eustress. *Psychol. Today* **11**, 60–70.

Siegel, S. and Ramos, B. (2002). Applying laboratory research: drug anticipation and the treatment of drug addiction. *Exp. Clin. Psychopharmacol.* **10**, 162.

Sloman, K. (2010). Exposure of ova to cortisol pre-fertilisation affects subsequent behaviour and physiology of brown trout. *Horm. Behav.* **58**, 433–439.

Smith, R. J. F. (1992). Alarm signals in fishes. *Rev. Fish Biol. Fish.* **2**, 33–63.

Sneddon, L. U., Wolfenden, D. C. C. and Thomson, J. S. (2016). Stress Management and Welfare. In *Fish Physiology - Biology of Stress in Fish*, Vol. 35 (eds. C. B. Schreck, L. Tort, A. P. Farrell and C. J. Brauner), San Diego, CA: Academic Press.

Spagnoli, S., Kent, M. L. and Lawrence, C. (2016). Stress in laboratory fishes. In *Stress in Fishes* (ed. L. T. C. B. Schreck), New York, NY: Academic Press.

Stabell, O. B. (1984). Homing and olfaction in salmonids: a critical review with special reference to the Atlantic salmon. *Biol. Rev.* **59**, 333–388.

Stephens, D. W. and Krebs, J. R. (1987). *Forgaing Theory.* Princeton, NJ: Princeton University Press.

Suboski, M. D. and Templeton, J. J. (1989). Life skills training for hatchery fish: social learning and survival. *Fish. Res.* **7**, 343–352.

Szabo, S., Tache, Y. and Somogyi, A. (2012). The legacy of Hans Selye and the origins of stress research: a retrospective 75 years after his landmark brief "Letter" to the Editor# of nature. *Stress* **15**, 472–478.

Takei, Y. and Hwang, P.-P. (2016). Homeostatic Responses to Osmotic Stress. In *Fish Physiology - Biology of Stress in Fish*, Vol. 35 (eds. C. B. Schreck, L. Tort, A. P. Farrell and C. J. Brauner), San Diego, CA: Academic Press.

Thompson, T. and Sturm, T. (1965). Classical conditioning of aggressive display in Siamese fighting fish. *J. Exp. Anal. Behav.* **8**, 397–403.

Thompson, N. F., Leblanc, C. A., Romer, J. D., Schreck, C. B., Blouin, M. S. and Noakes, D. L. (2015). Sex–biased survivorship and differences in migration of wild steelhead (*Oncorhynchus mykiss*) smolts from two coastal Oregon rivers. *Ecol. Freshw. Fish*

Thörnqvist, P.-O., Höglund, E. and Winberg, S. (2015). Natural selection constrains personality and brain gene expression differences in Atlantic salmon (*Salmo salar*). *J. Exp. Biol.* **218**, 1077–1083.

Thorpe, J. E. and Morgan, R. I. G. (1978). Parental influence on growth rate, smolting rate and survival in hatchery reared juvenile Atlantic salmon, *Salmo salar. J. Fish. Biol.* **13**, 549–556.

Tudorache, C., ter Braake, A., Tromp, M., Slabbekoorn, H. and Schaaf, M. J. (2015). Behavioral and physiological indicators of stress coping styles in larval zebrafish. *Stress* **18**, 1–8.

Ueda, H. (2014). Homing ability and migration success in Pacific salmon: mechanistic insights from biotelemetry, endocrinology, and neurophysiology. *Mar. Ecol. Progr. Ser.* **496**, 219–232.

Waples, R. S. and Hendry, A. P. (2008). Evolutionary perspectives on salmonid conservation and management. *Evol. Appl.* **1**, 183–188.

Wegner, N. C., Snodgrass, O. E., Dewar, H. and Hyde, J. R. (2015). Whole-body endothermy in a mesopelagic fish, the opah, *Lampris guttatus. Science* **348**, 786–789.

White, G. and Brown, C. (2015). Microhabitat use affects goby (Gobiidae) cue choice in spatial learning task. *J. Fish. Biol.* **86** (4), 1305–1318.

Wiley, R. W., Whaley, R. A., Satake, J. B. and Fowden, M. (1993). An evaluation of the potential for training trout in hatcheries to increase poststocking survival in streams. *N. Am. J. Fish. Manage.* **13**, 171–177.

Wiseman, S. M. (2015). Soundscape response in animals. *J. Acoust. Soc. Am.* **138**, 1749-1749.

Woo, P. T. K. and Noakes, D. J. (2014). *Salmon: Biology, Ecological Impacts and Economic Importance*. New York, NY: Nova Science.

Wood, C. M. and MacDonald, D. G. (1997). *Global warming. Society for Experimental Biology Seminar Series (No. 61)*. Cambridge: Cambridge University Press.

Woodward, C. C. and Strange, R. J. (1987). Physiological stress responses in wild and hatchery-reared rainbow trout. *Trans. Am. Fish. Soc.* **116**, 574–579.

Young, P. S., Swanson, C. and Cech, J. J. (2010). Close encounters with a fish screen III: behavior, performance, physiological stress responses, and recovery of adult delta smelt exposed to two-vector flows near a fish screen. *Trans. Am. Fish. Soc.* **139**, 713–726.

Yue, S., Duncan, I. and Moccia, R. D. (2006). Do differences in conspecific body size influence social stress in domestic rainbow trout? *Environ. Biol. Fishes* **76**, 425–431.

Zoladz, P. R., Clark, B., Warnecke, A., Smith, L., Tabar, J. and Talbot, J. N. (2011). Pre-learning stress differentially affects long-term memory for emotional words, depending on temporal proximity to the learning experience. *Physiol. Behav.* **103**, 467–476.

10

STRESS AND DISEASE RESISTANCE: IMMUNE SYSTEM AND IMMUNOENDOCRINE INTERACTIONS

TAKASHI YADA

LLUIS TORT

1. Introduction
2. Effects of Stressors on the Immune Response
   2.1. Suppressive Versus Enhancing Effects
   2.2. Perception of Stress After Immune Stimulation: Systemic Versus Local Responses
   2.3. Stress and the Cellular and Humoral Immune Response
3. Organization of the Immune Response Following Stress: The Neuroimmunoendocrine Connection and the Role of the Head Kidney
4. Effects of Hormones on the Immune System
   4.1. Hypothalamic Hormones
   4.2. Pituitary Hormones
   4.3. Interrenal Hormones
   4.4. Receptor-Mediating Action of Cortisol in Fish Immunity During Stress Response
   4.5. Somatotropic Axis and Fish Immune System
5. Environmental Stressors and Fish Immunity
   5.1. Environmental Salinity
   5.2. Temperature and Seasonality
6. Future Directions

The endocrine-immune relationship of fish, particularly related to the stress response, is mediated by the close interaction of hormones and cytokines. In essence, stress can depress certain elements of the immune system and render fish vulnerable to infection and disease. This chapter summarizes the effects of stressors on disease resistance and the immune system and updates the knowledge on endocrine regulation of the immune system in fish, the effects at systemic and local levels, and the organization of the immune responses under stressed conditions, with special emphasis on the roles of hormones, their receptors, and system interactions. Basically, low levels of severity of stress (eustress) may lead to enhanced immune competence while greater severities

Biology of Stress in Fish: Volume 35
FISH PHYSIOLOGY

tend to be immunosuppressive. The immune response to stressors are mediated by the endocrine system at both central and peripheral levels.

1. INTRODUCTION

Diseases typically appear in fish, like in all animals, when they have previously been subjected to a stressful situation. All types of stressors may affect immunocompetence as long as they affect mechanisms involving energy supply or metabolic pathways related to key immune molecules. Stressors significantly activate hormones and have direct, receptor-mediated interactions with the immune system molecules, resulting in loss of immune competence and disease resistance. Stressors known to have such immune-suppressive effects include those associated with husbandry practices, as well as those resulting from aggression and behavioral challenges, environmental pollution, and dietary alterations. The allostatic load posed by a stressor (Schreck and Tort, 2016; Chapter 1 in this volume) can reduce the effective functioning of immune mechanisms, allowing pathogens to act with greater virulence. For example, it has been shown that acute temperature changes (from 27°C to either 19–23°C or 31–35°C) decreased the resistance to pathogens and reduced the immune response of tilapia (*Oreochromis mossambicus*) (Ndong et al., 2007). Variable water temperature increased susceptibility to pathogens such as *Nodavirus* and resulted in dramatic changes in plasma cortisol concentration, osmolality, IgM levels, and body mass in the sea bass (*Dicentrarchus labrax*) (Varsamos et al., 2006). Similarly, gilthead sea bream (*Sparus aurata*) showed high susceptibility to *Pseudomonas anguilliseptica* infection when experiencing winter syndrome and had high plasma cortisol levels and severe immune suppression (Tort et al., 1998). Behavioral challenges, such as hierarchy establishment that led to dominant aggressive fish, caused decreased disease resistance, including elimination of the existing beneficial adherent gut microbiota and protective mucus in Arctic char (*Salvelinus alpinus*) (Ringø et al., 2014). Pathogen susceptibility and mortality increased when catfish (*Ictalurus punctatus*) were challenged by low-water stress (Small and Bilodeau, 2005) or when subjected to a stressor or cortisol treatment after exposure to pathogenic protozoans (Davis et al., 2003).

This chapter reviews the effects of stressors on the immune response, including the central and local mechanisms for perception of stressors, the cellular and molecular responses generated, and the resulting enhancing and suppressing effects. We also emphasize the hormones as the mediators between the endocrine and immune systems, including hypothalamic, pituitary, interrenal, and somatotropic hormones. The effects of environmental stressors on fish immunity is also considered.

2. EFFECTS OF STRESSORS ON THE IMMUNE RESPONSE

A wide variety of immune changes that occur in stressed fish may be described as being suppressive to the immune processes (Fig. 10.1). Although some responses may enhance or activate immune responses, as discussed later, most of them are deleterious to immune competence. In particular, longer term stressors normally result in suppressive effects. Crowded sea bass have reduced immunocompetence, as indicated by the reduced rates of cytotoxicity and chemiluminescence activities (Vazzana et al., 2002). Similarly, high density rearing of sea bream (Montero et al., 1999; MacKenzie et al., 2006a,b) and red porgy (*Pagrus pagrus*) (Rotllant and Tort, 1997) reduced levels of complement proteins and elevated cortisol. Crowding also depressed complement proteins and phagocytic activities in head-kidney leukocytes, and activated migration of cells into the blood (Ortuño et al., 2001). Repeated chasing of sea bream also resulted in decreased complement activity, lysozyme levels, agglutination activity, and antibody titers after 3 days and beyond (Sunyer and Tort, 1995).

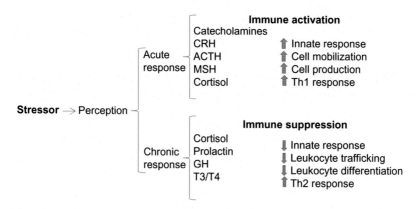

Figure 10.1. Relationship between stressors and immune activation and suppression. The diagram shows the main immune responses to stressors and the hormones involved. Acute responses are characterized by activation mechanisms, in particular, cell production and mobilization in the head kidney and proinflammatory responses synthesized by Th1 lymphocytes. Chronic responses are most often suppressive, leading to a decrease of innate response, decrease of leukocyte differentiation, and a shift to suppression of proinflammation mediated by Th2 lymphocytes. CRH, corticotropic releasing hormone; ACTH, adrenocorticotropic hormone; MSH, melanocyte stimulating hormone; GH, growth hormone; T3, triiodothyronine; T4, thyroxine; Th1, T helper 1 lymphocyte; Th2, T helper 2 lymphocyte. Upward arrows = stimulation. Downward arrows = suppression.

These types of responses are observed also at the genomic level, because immune-relevant genes can be significantly regulated in stressed fish. For example, rainbow trout (*Oncorhynchus mykiss*) subjected to stressors such as copper exposure or bacterial lipopolisaccharides (LPS) showed significant differences in the expression of the immune genes interleukin (IL)-1β, IL-6, and tumor necrosis factor-alpha (TNF-α) (Teles et al., 2011). Substitution of fish oil by vegetable oil in sea bream diets also induced changes in the expression of proinflammatory cytokine genes together with suppression of immune responses (Montero et al., 2003, 2010). Repeated handling stress in Atlantic cod (*Gadus morhua*) increased constitutive expression of IL-1β, followed by decreased leukocyte survival in Atlantic salmon (*Salmo salar*) exposed to *Aeromonas salmonicida* (Fast et al., 2008). Hypoxic Eurasian perch (*Perca fluviatilis*) reduced complement C3 expression and overexpressed transferrin (Douxfils et al., 2014).

2.1. Suppressive Versus Enhancing Effects

While stressors tend to induce an overall suppressive or adverse effect on the organism, some singular immune responses to stressors can also show enhancing, adaptive responses (see eustress as described in Schreck and Tort, 2016; Chapter 1 in this volume). Thus, a complex event such as the overall stress response will depend on the time period involved, the response itself, and the intensity of the stressor and its duration. While specific enhancing responses are sometimes present, most of the responses observed are suppressive (Miller et al., 1995; Dhabhar, 2002, 2008). Thus, altered trafficking of lymphocytes to the skin and other peripheral locations where the stress stimulus is acting may serve to enhance immune responses in those places (Dhabhar, 2000), but concomitantly the response may be delayed in other locations. Previous stress exposure influences the ability of what the immune system responds to and how it does so. Therefore, the efficacy of regulatory systems needed to overcome a stressor, including the activation of immune activity, can be affected (Dhabhar, 2002). Current understanding of these effects at the cellular and molecular levels is not sufficient to allow prediction of the effects of any particular stressor on a specific immune or inflammatory reaction (Pruett, 2003). More details on the consequences of the fish's previous experience to stressors, predictability, and variation of the response can be found in other chapters of this book (see Chapters 2, 3 and 9 in this volume).

The immune response to stressors can be characterized by two phases. In the first, the main reaction is activation of a number of mechanisms to respond directly to the challenge, stimulating receptors and triggering rapid responses. In particular, this activation phase is related to the production of acute phase-type proteins, the release of hormones and peptides that are temporarily stored or produced, and the mobilization of existing resources. Some activating responses include increased inflammatory markers, increased lysozyme activity, and increased complement C3 proteins as bacteriolytic mechanisms (Sunyer and Tort, 1995; Demers and Bayne, 1997), which involve enhancing adaptive responses. MacArthur et al. (1984) detected an increased number of myeloid-type leukocytes in an intraperitoneal lavage after a peritoneal injection of endotoxin and cortisol. Short-term handling stress also increased the number of glucocorticoid receptor (GR) sites per cell in coho salmon (*Oncorhynchus kisutch*) head kidney leukocytes (Maule and Schreck, 1991). Short-term crowding stress upregulated proinflammatory cytokines IL-1β and IL-8, as well as antibacterial genes and g-type lysozyme (Caipang et al., 2008).

If the stressor persists, the second phase will normally conduct to suppression. This suppression is caused by two main factors: the lack of resources (allostatic load) to support the mechanisms behind the maintenance of highly demanding processes like cell division and proliferation of protein synthesis, and the effect of relevant mediators that may induce downregulation or decrease of the potency of immune mechanisms. The suppressive effects are mostly associated with the release of final stress hormones of the hypothalamic–pituitary–interrenal (HPI) axis (in particular, cortisol); however, part of this response is also mediated by the regulatory hormones, such as catecholamines or corticotropin-releasing hormone (CRH). This dual response will depend on the time course of the response, the persistence of the stressor, and the previous experience, as said before.

Predictive indicators for activation or suppression have been difficult to identify due to, once again, the conditions of the stressor and time-course. While an acute stress given before an antigenic challenge significantly enhanced cell-mediated immunity and leukocyte redeployment in skin, suppression of the skin immune response was recorded when chronic stress exposure was given prior to sensitization (Dhabhar, 2008). This suggests bidirectional effects of stress on cutaneous cell-mediated immunity. Overall, it must be understood that at present, no single marker can be indicative of an unequivocal direction toward activation or suppression, or predictive of the severity of stress effects on the immune system.

2.2. Perception of Stress After Immune Stimulation: Systemic Versus Local Responses

Stressors activate the neuroendocrine response after being perceived by one or more of the sensory systems (ie, neural sensors, hormonal, or immune receptors). Subsequent activation of immune mechanisms is mediated by hormones, mainly cortisol, that have receptors in all cells; such activation can follow a large number of pathways (Fig. 10.2). Immune cells also have GRs and therefore direct responses to circulating cortisol (Gorissen and Flik, 2016; Chapter 3 in this volume). The effects of corticosteroids on the immune system are reviewed in Section 10.4.

However, this hormone-induced immune response is accompanied by a local response that is also relevant. External stimuli, such as physicochemical changes in the water, skin abrasion, and diet-induced gut alteration, among others, become stressors that are detectable at fish-specific interactive surfaces (more often gills, skin, or intestine). It has been shown that this local response involves mechanisms that are not only stressor-specific, but are also alarm-type response changes that include immune pathways. Thus, while the organs and systemic compartments (head kidney, thymus, and spleen) are responsible for leukocyte production, proliferation of T cells, and antigen capture, the local response involves innate and acquired immune agents both cellular and humoral, including factors such as lysozyme, lectins, proteases, or complement proteins that help to keep the tissues free from pathogens. Local responses are produced particularly in mucosal surfaces, without prior activation of a response in the central organs. This is a key point in the overall response to alarm situations, as both local and systemic responses will combine for an effective result.

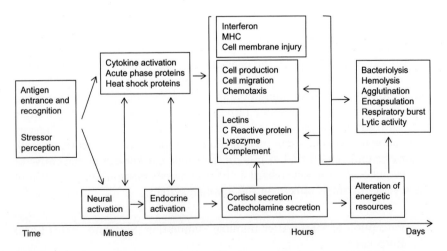

Figure 10.2. Time course for coactivation between immune and neuroendocrine responses to both stressors and antigenic agents. Arrows indicate interactions between factors.

While the systemic immune-endocrine response has been comparatively better studied, recently the importance of the local response, basically the response at the portals of entry such as gills, skin, or intestine and, in particular, the response at the mucosae, has been investigated. This immune response at the mucosal sites not only plays a primary role in the identification of pathogen danger, but it is also involved in a variety of immune mechanisms, phagocytosis, antibody response, and antimicrobial molecules, and, even more significant, in the identification of a wide range of other stressors (Salinas et al., 2011; Caipang et al., 2011; Esteban et al., 2012) (Fig. 10.3).

Mucosal surfaces acquire information continuously from the environment, detecting changes and processing this information so the fish can adapt, thereby maintaining homeostasis and thus survival. Functions of these mucosal surfaces are various, from interspecific or intraspecific communication, nutrient uptake, gas exchange, or microorganism detection (Salinas et al., 2011). As immunological sites, mucosae are capable of mounting a robust immune response after a pathogen challenge in all vertebrates and many invertebrates (Gomez et al., 2013; Zhang et al., 2014). Despite the evident morphological and physiological differences between species, all vertebrates possess these mucosal immune tissues or mucosa-associated lymphoid tissues (MALT) that control the immune response at mucosal sites. In teleosts, four MALTs have been described: nose-associated lymphoid tissue (NALT), skin-associated lymphoid tissue (SALT), gill-associated lymphoid tissue (GIALT), and gut-associated lymphoid tissue (GALT). These mucosal tissues not only develop a precise immune response, but also react to stressors even when the challenge is not specifically immune. For

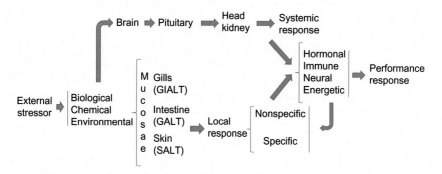

Figure 10.3. Pattern of mucosal and systemic responses to external challenges. Both local and systemic responses influence the overall performance and interactions occur between them. GIALT, gills-associated lymphoid tissue; GALT, gut-associated lymphoid tissue; SALT, skin-associated lymphoid tissue.

instance, genes related to antibacterial activity, antiviral response, cytokine production, stress response, and antiapoptotic activity were found in Atlantic cod skin, thus demonstrating that skin is an immunologically active site, especially on the ventral side of this fish (Caipang et al., 2011).

No specific studies have been published about the consequences of stress to immune cell populations in mucosal surfaces. It is known that cortisol reduces the expression of genes related with B cell activation in trout skin (Krasnov et al., 2012). Importantly, secretory IgA, a mucosal immunoglobulin in mammals that varies in mucosal surfaces depending on the nature of the stressor, is used as a marker for stress (Volkmann and Weekes, 2006). It is thus possible that sIgT/Z, as the mucosal specialized immunoglobulin in teleosts, could modify its secretion after stress. In general stressors are known to stimulate immune response processes regardless of whether or not the source is an immune challenge. In particular, it has been possible to define some characteristic traits of the stress reaction that may include the suppression of the immune system such as alteration of the response of thymus, spleen, or other lymphoid structures in the body, the change in the number and distribution of white blood cells in the body, or the appearance of bleeding or ulcers in the alimentary tract (Tort, 2011).

2.3. Stress and the Cellular and Humoral Immune Response

Stress induces changes in the number and the recruitment rate of immune cells that are directed to tissues or organs to increase effective immunopro- tection; this involves substantial differences in leukocyte distribution in the various body compartments. Acute stress stimulates the production of blood cells in the head kidney, including both red blood cells and leukocytes, which results in an increased number of circulating leukocytes that are distributed to target tissues. This is caused by the activation of the sympathetic nervous system and the release of catecholamines. The frequency of leukocyte cell types in the blood changes due to acute stress. The number and percentage of lymphocytes and monocytes is significantly reduced, while neutrophil numbers and percentage increases during stress (Wojtaszek et al., 2002). Acutely stressed animals showed higher numbers of activated macrophages in blood, as well as increased T-cell activation and enhanced recruitment of surveillance T cells to skin. Thus, acute stress significantly enhances the kinetics and magnitude of a leukocyte subpopula- tion's ability to infiltrate an immune activation site (Dhabhar, 2002). These findings show important interactive components of the innate (macrophages) and adaptive immunity (surveillance T cells) as important consequences of the stress-induced enhancement of a primary response, such as the catecholamine secretion. This trafficking of cells is performed in a chemoattractant-specific manner, in which tissue damage or antigen/pathogen-driven chemoattractants synergize with acute stress to further determine the specific leukocyte subpopulations that are recruited (Dhabhar, 2002). There is sparse information on similar effects due to chronic stressors in fish. Due to the continuous activation of the hypothalamic–pituitary–adrenal axis consequent to continuous stress, the initial mobilization and distribution of blood leukocyte subpopulations is followed by a general decrease in blood leukocyte numbers that are dependent on the stress intensity and duration. Leukocytes will mainly concentrate in affected organs and therefore both central organs and blood show a decrease in cell numbers (Dhabhar, 2002; Tort, 2011).

One of the key cell types of the immune response are the B-lymphocytes or B-cells. Teleost B-lymphocytes share both innate and adaptive immune characteristics. Innate features include natural antibody secretion, cytokine production, and phagocytic capabilities (Sunyer, 2013) that are relevant in terms of developing an alarm response. Teleosts express three different classes of immunoglobulins: IgM, IgT (also called IgZ in some species), and IgD. IgM is the most common immunoglobulin in serum and skin mucus and the key player in systemic immune responses, whereas IgT is the main immunoglobulin in the gut mucosal tissues (Salinas et al., 2011). Although mammalian and fish B-lymphocytes share most of these innate capabilities, their importance in organism defense may not be comparable in some cases. For instance, in mammals phagocytic B cells are confined to the peritoneal cavity and represent 10–15% of all B cells (Parra et al., 2012), whereas in teleost fish, phagocytic B cells are present in all systemic compartments (including blood, spleen, and head kidney) and represent 60% of all B-cells (Li et al., 2006; Zhang et al., 2010). Thus, the innate role of B-cells in fish seems to be much more important than in mammals, and their function bridging innate and adaptive responses might predominate in teleost fish (Zhu et al., 2014). Stress appears to reduce the number of circulating B-lymphocytes and decreases the antibody response after immunization in carp (Cyprinus carpio) (Stolte et al., 2009). Stress also induces a reduction in IgM levels and an increased susceptibility to Nodavirus in sea bass (Varsamos et al., 2006). Taken together, this is indicative of a suppressive effect of cortisol or other mediators of the stress response on the fish immune competence and particularly on B-cells.

Treatment of fish with cortisol produces a reduction in immunoglobulin secretion or in IgM mRNA; in addition, a decrease of the antibody-secreting cell number has also been described following cortisol treatment (Hou et al., 1999; Saha et al., 2004). Cortisol administration induces B-cell apoptosis and reduces proliferation in carp (Weyts et al., 1998b). Although in vivo and in vitro cortisol studies have been widely used as the model to understand the stress process in teleosts, other hormones could also be involved in the stress response. For instance, carp subjected to sudden temperature changes showed higher cortisol levels but no reduction of antibody production or changes in the antibody-secreting cell number (Saha et al., 2002), indicating that other components participate to activate the immune system under stressful situations. Therefore, although cortisol is the main hormone involved in the immune-related stress changes, immune regulation by the neuroendocrine system is more complex than just cortisol release and signaling.

Stressors also induce a number of changes on the humoral immune responses in fish, primarily characterized by the activation of numerous proteins related to nonspecific immune responses, and secondarily by a significantly large release of cytokines that mediate the immune response in cells and tissues. Many studies have demonstrated that stressors induce alteration of innate immune factors such as lysozyme, complement proteins, or antimicrobial peptides (Montero et al., 2009; Saurabh and Sahoo, 2008; Mauri et al., 2011; Costas et al., 2011; Teles et al., 2012). Cortisol is known to inhibit the release of proinflammatory cytokines during the resolution of inflammation. Activation of the stress hormones in mammals inhibits systemically T helper1(Th1)/proinflammatory responses but potentiates a Th2 shift by suppressing production of Th1 cytokines (TNF-α, interferon-γ, IL-2, IL-12) and inducing production of cytokines belonging to other Th subsets (IL-4, IL-10, IL-13, transforming growth factor (TGF)-β) (Elenkov and Chrousos, 1999). Alternatively, glucocorticoids modulate the adaptive immunity mediators, inhibiting the synthesis of IL-12 and IFN or upregulating IL-4, IL-10, and inducing the secretion of TGF-β (Borghetti et al., 2009).

There is evidence that several primary immune response mechanisms are also stimulated even when a pathogen is not present; that is, when nonantigenic stressors are applied. This can be clearly observed when analyzing the overall gene expression following a stressful episode in which immune response genes are also upregulated (Mackenzie et al., 2005; Teles et al., 2012; Xia et al., 2013; Zhu et al., 2013). Asian sea bass (*Lates calcarifer*) subjected to several stressors such as LPS treatment, *Vibrio harveyi* exposure, high salinity, or fasting, and sequenced for intestinal cDNA libraries, show an overall similar expression profile with 37 genes differentially expressed in response to all challenges, regardless of which stressor was involved. These data suggest that there is a global coordination and fine-tuning of gene regulation during these different challenges (Xia et al., 2013).

3. ORGANIZATION OF THE IMMUNE RESPONSE FOLLOWING STRESS: THE NEUROIMMUNOENDOCRINE CONNECTION AND THE ROLE OF THE HEAD KIDNEY

The neural, immune, and endocrine systems in vertebrates work together to process and respond to external or internal inputs, particularly under stressful situations when quick and/or powerful responses have to be organized. The nervous system is the main component of the regulatory network modulating the overall nonspecific immediate response and modification of hormonal activity. Stress hormones released during a stress episode may help to prepare the immune system for potential challenges (eg, infection, wound repair, tissue alteration), for which stress perception by the brain may serve as an early warning signal. This connection is present in all vertebrates, and molecules involved in stress, CRH, adrenocorticotrophic hormone (ACTH), glucocorticoids, biogenic amines, and cytokines are fundamentally similar and well preserved throughout evolution in different species and taxa (Ottaviani and Franceschi, 1996). Cytokines and neuropeptides have roles in both neuroendocrine and immune systems. This explains why stress is characterized by the involvement of immune and neuroendocrine cells and messengers at the systemic level. Thus, it seems clear that the multisystem interaction modulates and fine-tunes the first response to avoid excessive activation and to allow the organization of further mechanisms to adapt to a specific challenge. These interactive mechanisms will not only serve for a particular stress episode but also be useful for adapting and preparing the response for future challenges.

The head kidney plays a central role in organizing the stress response involving a close communication between the three regulatory systems within that organ (Verburg-Van Kemenade et al., 2009; Tort, 2011) (Gorissen and Flik, 2016; Chapter 3 in this volume). The function of the fish head kidney is unique in vertebrates because within this organ all three regulatory systems exert significant actions: catecholamines (adrenalin and noradrenalin) are released through activation of the chromaffin tissue by specific autonomic fibers; cortisol is secreted by the interrenal cells after the stimulation by ACTH; immune cells are produced and processed for maturation and early differentiation; and red blood cells are produced for transporting oxygen to the tissues. All these functions of the head kidney are very relevant when a stress response has to be organized: regulating metabolic and osmotic and ionic exchanges, modulating the immune response, and providing blood cells and oxygen for energetic resources. In this way red blood cells could also play a role in the overall immune response, being nucleated cells and therefore able to synthesize relevant immune molecules. In fact, trout erythrocytes express and regulate specific pattern recognition receptor mRNAs and are capable of detecting specific pathogen associated molecular patterns (PAMP), which is central to the innate immune response (Morera et al., 2011). In vitro challenge with diverse PAMPs led to de novo specific mRNA synthesis of immune receptors and response factors including IFN-α (Morera et al., 2011). Overall, the head kidney is an organ with a key role in organizing an integrated stress response involving several processes such as synthesis and mobilization of blood cells and release of hormones leading to metabolic changes. Moreover it is also the place where neuroimmunoendocrine interaction is most likely to occur, thus giving opportunity for cross-systems regulation.

Among the bidirectional relationships between these three regulatory systems, the endocrine-immune connection has been most extensively studied, in particular, the interaction between cortisol and immune responses. None-theless, it has been shown in mammals that interaction between endocrine and immune systems also occurs in adrenalectomized animals, thus suggesting that various neurotransmitters, neuropeptides, and the hypothalamic peptide CRH may also be involved (Leonard, 2005). Overall, any of the hormones assessed so far can have an influence on immune agents or mechanisms, though some of them play a principal role in the regulation of the immune response (Gorissen and Flik, 2016; Chapter 3 in this volume). Concerning allostatic load, the energetic costs of stress are highly demanding; it is clear that other hormone axes, such as the somatotropic axis involving growth hormone (GH) that is mostly devoted to metabolic support, will have an influence on the immune function. In addition to the increased energy requirements for the immune system, during the period of immune activation complex interactions between cytokines, glucocorticoids, and somatotropic hormones may result in altered metabolism and therefore decreased resources for growth. Moreover, food intake, energy balance, cell metabolism, and immune function are all interrelated and likely depend on central (brain, pituitary) and head-kidney interactions (Yada et al., 2002, 2005; Tort, 2011) (see next section).

4. EFFECTS OF HORMONES ON THE IMMUNE SYSTEM

Basically all the hormones discussed in this section will be related to the stress in some way, either directly as a part of the stress response or indirectly via altered or disrupted functions because of stress. Further, immunostimu-latory hormones may possibility modify stress-related immunosuppression. Multifunction of hormones is one of the characteristics of the fish endocrine system, and appears to complicate understanding of their roles in the stress response.

4.1. Hypothalamic Hormones

The importance of hypothalamic control of head-kidney functions during the stress response in fish is discussed thoroughly in other chapters of this volume (in particular, Gorissen and Flik, 2016; Chapter 3 in this volume). CRH is thought to be the most important secretagogue for corticosteroids as the initiator of the HPI axis in fish; hypothalamic CRH stimulates ACTH secretion, and cortisol secretion is controlled by circulating ACTH level. In addition to the role as a secretagogue in the HPI axis, several hypothalamic hormones are known to directly or indirectly affect the fish immune system.

Hypothalamic nerve endings form the neurohypophysis in the posterior pituitary and release neurohypophyseal hormones, such as melanin-concentrating hormone (MCH). The effect of MCH on the fish immune system has been investigated in relation to background color adaptation (Harris and Bird, 2000). An in vitro study using leukocytes isolated from head kidney of rainbow trout revealed that MCH directly stimulated phagocytosis (Harris and Bird, 1998). Gonadotropin-releasing hormone (GnRH) is primarily produced in the hypothalamus and stimulates gonadotropin release from the adenohypophysis of vertebrates (Van Der Kraak, 2009; Tsutsui et al., 2010; Gopurappilly et al., 2013). GnRH is one of the most important stimulators of gonadal steroids, involving immunosuppression during sexual maturation of fish. Roles of GnRH are not limited to the hypothalamic–pituitary–gonadal (HPG) axis, and are also thought to interact with endogenously produced growth factors and cytokines (Marchetti et al., 2001). Administration of GnRH increased proliferation, superoxide production during phagocytosis, and mRNA levels of a proinflammatory cytokine, TNF-α, in leukocytes of rainbow trout (Yada, 2012). Two other hypothalamic peptides have been found to regulate reproduction through the action of GnRH; kisspeptin stimulates but gonadotropin-inhibitory hormone (GnIH) inhibits gonadotropin secretion through the action of GnRH neurons (Tsutsui et al., 2010; Gopurappilly et al., 2013). These stimulating and inhibiting factors for the HPG axis could be responsible for immunomodulation during sexual maturation.

4.2. Pituitary Hormones

Pituitary hormones are important factors modulating fish immune functions. Early studies examining the effects of hypophysectomy on teleost species suggest the importance of hypophyseal hormones in maintaining an effective immune system (Slicher, 1961; Pickford et al., 1971; Yada and Nakanishi, 2002). Administration of GH enhances many aspects of specific and nonspecific defenses in fish immunity (Balm, 1997; Weyts et al., 1999; Yada, 2007; Tort, 2011). Calduch-Giner et al. (1995) revealed the presence of GH-binding sites in head-kidney lymphocytes of sea bream. Fish GH receptors were isolated and sequenced from several fish species (Calduch-Giner et al., 2000; Fukada et al., 2001; Lee et al., 2001; Shved et al., 2009, 2011). In contrast to the stimulatory actions just described, long-term GH treatment seems to suppress fish immune functions (Cuesta et al., 2005, 2006). Effects of transgenic GH on fish immune functions are equivocal, some indicating stimulation and others inhibition or no effect (Jhingan et al., 2003; Wang et al., 2006; Mori et al., 2007). Influences of GH on the regulation of the cytokine, TNF-α, production are also equivocal in the tilapia *Oreochromis niloticus* (Shved et al., 2011). The contradictory results may be related to differences in developmental and energetic conditions between different study systems.

Three hypophyseal peptides, GH, prolactin (PRL), and somatolactin (SL), are thought to be derived from a common ancestral molecule (Kaneko, 1996; Kawauchi et al., 2009). The effects of PRL on the fish immune system closely resembles those of GH, PRL-enhanced phagocytosis, and prolifera-tion of rainbow trout leukocytes (Sakai et al., 1996a,b; Yada et al., 2004). Expression of PRL receptor mRNA has been detected in kidney and intestine of several fish species (Sandra et al., 2000; Prunet et al., 2000; Tse et al., 2000; Higashimoto et al., 2001; Santos et al., 2001). PRL receptor expression was also observed in leukocytes isolated from peripheral blood and head kidney of Nile tilapia, indicating direct action of PRL on the fish immune system (Sandra et al., 2000). An increased level of PRL receptor mRNA was observed in head-kidney leukocytes of seawater-acclimated tilapia, coinciding with an enhanced respiratory burst (Yada et al., 2002). SL had no significant effect on fish immune function (Sakai et al., 1996a). Yet in a variant of rainbow trout (cobalt) that lacks most of pars intermedia, plasma cortisol levels and lysozyme activity are lower than in those in normal fish (Yada et al., 2006). SL-producing cells are distributed in neurohypophysial tissue in pars intermedia, and the cobalt variant had extremely low levels of circulating SL (Kaneko et al., 1993a,b). There is a possibility that SL is involved in the fish immune function indirectly through other immunomodulatory factors.

Besides their roles in the HPI axis, ACTH and other proopiomelano-cortin (POMC)-derived peptides have direct effects on fish immune functions as shown similarly in higher vertebrates. ACTH-receptor-like molecules expressed in thymus and spleen of sea bass (Mola et al., 2005). In vitro administration of ACTH, melanocyte stimulating hormone, and *N*-terminal peptides of POMC enhanced several innate functions in trout and carp (Bayne and Levy, 1991a,b; Harris and Bird, 1997, 1998, 2000;

Harris et al., 1998; Takahashi et al., 2000; Sakai et al., 2001). Furthermore, ACTH modulates gene expression of some proinflammatory cytokines in gilthead sea bream leukocytes, inhibits IL-1β, and stimulates IL-6, TNF-α, and TGF-β1 (Castillo et al., 2009). β-Endorphin is another POMC-derived peptide, which directly stimulates several immune functions in trout and carp (Watanuki et al., 1999, 2000; Takahashi et al., 2000). Opioid peptides modulate immune function in mammals, and the modulatory effect of opioids on inflammation in fish is supported by the localization of specific binding for the opioid receptor antagonist naloxone to head-kidney cells of goldfish (*Carassius auratus*) (Chadzinska et al., 1997).

4.3. Interrenal Hormones

The majority of studies on endocrine control of fish immune functions focus on suppression of the immune response by increased secretion of interrenal hormones in response to environmental stressors. However, interrenal hormones, such as adrenaline and cortisol, are both inhibitory and stimulatory for several sites of fish immune system.

4.3.1. IMMUNE MODULATION BY THE HYPOTHALAMIC–SYMPATHETIC–CHROMAFFIN CELL AXIS

The hypothalamic–sympathetic–chromaffin cell axis is an important regulatory system for mediating stress response brought by catecholamine release (Randall and Perry, 1992). The release of catecholamines from chromaffin cells is primarily regulated by preganglionic sympathetic nerves. Endocrine and nonendocrine factors, such as fluctuating plasma ion levels, are also involved in the regulation of catecholamine release (Randall and Perry, 1992; Wendelaar Bonga, 1997). Spontaneous regulation of the fish immune system by sympathetic nerves would seem to be inhibitory at least for specific immunity. Flory (1989) demonstrated that chemical sympathectomy of coho salmon resulted in an enhanced antibody response. Administrations of adrenaline and noradrenaline reduced phagocytosis in spotted murrel (*Channa punctatus*) (Roy and Rai, 2008). Furthermore, in vitro administration of adrenaline lowered mRNA levels of proinflammatory cytokines in gilthead sea bream leukocytes (Castillo et al., 2009).

The effects of catecholamines on fish immune functions have been examined using receptor agonists and antagonists; however, the results appeared to be paradoxical. Inhibition of the fish immune system by a β adrenergic agonist isoproterenol has been repeatedly observed (Flory, 1990; Bayne and Levy, 1991a,b; Flory and Bayne, 1991; Finkenbine et al., 1997). In contrast, an α1 adrenergic agonist, phenylephrine, enhanced antibody response and chemiluminescence in rainbow trout leukocytes (Flory, 1990;

Flory and Bayne, 1991). Stimulation of antibody response by α_2 agonist clonidine was blocked by an α_2 antagonist, yohimbine, but not by an α_1 antagonist, prazosin, suggesting the presence of two different types of receptors for adrenergic agents (Flory, 1990). Respiratory burst in phagocytic leukocytes from rainbow trout was also stimulated by phenylephrine in vitro (Bayne and Levy, 1991a,b). However in the same species the phagocytic index, estimated as the number of engulfed yeast cells by adherent leukocytes, was suppressed by administration of the same agonist phenylephrine (Narnaware et al., 1994; Narnaware and Baker, 1996).

Various adrenergic and cholinergic receptors have been shown in lymphoid organs of several fish species (Józefowski et al., 1995; Nickerson et al., 2003; Owen et al., 2007). Stress-induced changes in the kinetics of β adrenergic receptors were observed in rainbow trout erythrocytes, however changes in leukocytes were not well identified (Reid and Perry, 1991; Reid et al., 1993). In mammals, the receptor-specific signal transduction pathways associated with neurotransmitters and changes in gene expression following receptor stimulation have been described in detail (Sanders et al., 2001). Molecular studies on the functional expression of adrenergic receptors in the nervous system indicate the presence of both α_1 and α_2 receptors in fish (Svensson et al., 1993; Yasuoka et al., 1996). Functional characterization of adrenergic and cholinergic receptors is needed to clarify the roles of catecholamine in the fish immune function.

Neurotransmitters and neuropeptides often bind to G-protein-coupled receptors (GPCRs) that activate signaling pathways, such as protein kinases, which are signals triggered also by immune mediators in higher vertebrates. Adrenaline reduces the in vitro expression of the proinflammatory IL-1β cytokines but does not modify the expression of TNF-α, TGF-β, and IL-6 cytokines of cultured macrophages of sea bream (Castillo et al., 2009). Moreover, when adrenaline is administered in combination with LPS, the hormone lowers the LPS-induced expression of both IL-1β and TNF-α (Castillo et al., 2009).

4.3.2. Immune Modulation by Corticosteroids

Cortisol suppresses many aspects of the fish immune system, such as antibody production, leukocyte mitosis, and phagocytosis (Balm, 1997; Wendelaar Bonga, 1997; Weyts et al., 1999; Yada and Nakanishi, 2002; Tort, 2011). Proliferation of rainbow trout macrophages is regulated by cortisol (Pagniello et al., 2002). Conversely, a few studies indicate a significant enhancement of immune function by cortisol. White and Fletcher (1985) observed that cortisol administration to plaice (*Pleuronectes platessa*) stimulated the production of C-reactive protein, which is an acute phase protein involved in the inflammation process. Cortisol's ability to stimulate inflammation (and its other related parameters of immunity) are in agreement with cortisol's known roles in healing and tissue repair in higher vertebrates (Buckingham et al., 1996). Gene expression profiling using microarray analyses revealed that most genes related to inflammatory processes are reduced by cortisol treatment (MacKenzie et al., 2006a,b). This was shown using phagocytic leukocytes from rainbow trout; however, gene expression of several proinflammatory cytokines was also observed to be inhibited by cortisol in leukocytes from the head kidney of gilthead sea bream (Castillo et al., 2009).

Cortisol can have a biphasic effect on apoptosis of lymphocytes (Weyts et al., 1998a,b; Wojtaszek et al., 2002). Differential effects of cortisol on apoptosis and proliferation have shown in carp B cells isolated from head kidney, spleen, and peripheral blood (Verburg-Van Kemenade et al., 1999). These differential effects of cortisol between leukocyte subtypes seem to coincide with the fluctuations in B-cell to neutrophil ratios in stressed catfish (Ellsaesser and Clem, 1986; Ainsworth et al., 1991). Differential effects of cortisol on apoptosis and mitosis of leukocyte subtypes seem to be correlated with the changes in total leukocyte numbers. In mammals, catecholamines appear to act in concert with corticosteroids in the regulation of the leukocyte population by apoptotic mechanisms (Boomershine et al., 2001). Evidence suggests that corticosteroids do not suppress total fish immune function, but does regulate the redistribution of lymphoid cells into different locations in the organism during the stress response (Weyts et al., 1999; Dhabhar and McEwen, 2001; Schreck and Maule, 2001).

Glucocorticoids inhibit NF-kB signaling, leading to decreased production of proinflammatory cytokines such as IL-1, IL-6, and TNF (Sternberg, 2006). Cortisol inhibits fish inflammatory cytokine expression (Saeij et al., 2003). Cortisol and LPS synergistically stimulate the expression of IL-1 mRNA in head-kidney phagocytes (Engelsma et al., 2003). Cortisol suppresses cytokine expression (TNF, TGF-β, IL-6), and when cortisol and LPS are given together the cytokine induction is downregulated in sea bream. Moreover, cortisol may activate certain macrophage functions that lead to the resolution of inflammation (Castro et al., 2011).

4.4. Receptor-Mediating Action of Cortisol in Fish Immunity During Stress Response

Receptor-mediated mechanisms of hormonal action in the fish immune system were examined by the administration of receptor-specific agonists and antagonists, or by binding assays using labeled ligands. Distinct receptors for corticosteroids, GR-1, GR-2, and mineralocorticoid receptor

(MR), have been identified in rainbow trout (Ducouret et al., 1995; Colombe et al., 2000; Bury et al., 2003). Those distinct types of corticoid receptors have been identified in several fish species, implying the differential regulation of expression of the corticoid receptor gene between organs to express different functions (Prunet et al., 2006; Stolte et al., 2006, 2008a,b; McCormick et al., 2008; Flores et al., 2012).

4.4.1. In Vivo Studies

Radioreceptor assay has revealed the differences in specific binding of cortisol among leukocytes isolated from several lymphoid tissues of coho salmon (Maule and Schreck, 1990, 1991). Differences in sensitivity to cortisol between leukocyte subtypes may be related to the kinetic characteristics of the corticoid receptors. Low binding affinity for cortisol in gill cytosol would seem to be related to high plasma cortisol levels in chub (*Leuciscus cephalus*), which is assumedly a stress-resistant fish in comparison with rainbow trout (Pottinger et al., 2000). GRs have been cloned and sequenced in several fish species, and the distribution of mRNA was detected in lymphoid organs, especially the spleen (Ducouret et al., 1995; Takeo et al., 1996; Tagawa et al., 1997). However, there are few detailed studies on the regulation of gene expression of the GR in the fish immune system. In chum salmon (*Oncorhynchus keta*) the GR mRNA was detected in the gill and its level was influenced by environmental salinity (Uchida et al., 1998). These findings further supported the discovery that fish likely use the GR for the mediations of both dietary metabolism and osmoregulation (Ducouret et al., 1995). On the other hand, Colombe et al. (2000) cloned fish MR and showed a clear homology with the MR in higher vertebrates, but its recombinant protein from the fish sequence bound cortisol preferentially. There is a possibility that cortisol-induced immunomodulation in fish might be mediated by both GR and MR.

After an acute stress, a transient increase in corticoid receptor mRNA levels has been observed in the leukocytes of trout and carp, and it was followed by immunosuppression (Yada et al., 2007; Stolte et al., 2008a,b). An increased expression of corticoid receptor genes is observed in several tissues, including spleen, of gilthead sea bream after stimulation of immunity by LPS injection (Acerete et al., 2007). In rainbow trout, however, that transient increase in corticoid receptor mRNA levels is followed by a consistent decrease (Yada et al., 2007). Implant of slow-release cortisol into rainbow trout produces increases in mRNA levels of corticosteroid receptor expressed in head kidney, brain, gills, and gonads; and a mRNA reduction in the spleen, liver, and muscle (Teles et al., 2013).

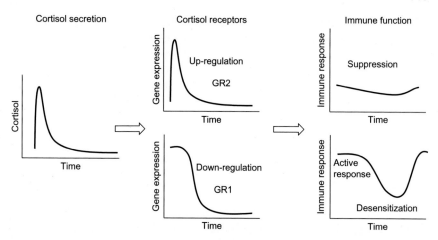

Figure 10.4. Conceptual representation of receptor-mediated cortisol effects on immune function. Cortisol secretion regulates specific GR receptors leading to suppression or desensitization of immune responses. GR, glucocorticoid receptor.

Corticosteroid receptors in the fish immune system have two patterns of expression following an elevation in secreted cortisol (Fig. 10.4). A rapid upregulation of corticoid receptors seems to be correlated with immunosuppression. Conversely, a chronic downregulation of corticoid receptor expression may result in a desensitization of immune function to cortisol stimuli (Yada et al., 2007; Stolte et al., 2008a,b; Teles et al., 2013). In saltwater-acclimated steelhead trout, leukocytes also had significant decreases in mRNA levels of corticoid receptor, when compared to those in freshwater-acclimated fish (Yada et al., 2008). The regulation of stress response in the fish immune system via corticoid receptors should be investigated during desensitization to stressors as well as an adaptation to environmental changes.

4.4.2. In vitro Studies

A binding study using peripheral blood leukocytes in carp showed the presence of a single class of cortisol-binding sites, and a corticoid receptor antagonist inhibited cortisol's effect on apoptosis reduction in neutrophils (Weyts et al., 1998a,c). Carp leukocytes activated by LPS are more sensitive to cortisol-induced apoptosis and suppression of proliferation than those from head kidney or spleen (Overberg-van Lemonade et al., 1999). However, cortisol had no significant effect on GR and MR mRNA levels in LPS-stimulated leukocytes from carp kidney (Stolte et al., 2008a,b).

Members of the heat shock family of proteins (HSPs) are ubiquitously expressed, conserved proteins that are classified by their molecular mass,

HSP30, HSP70, and HSP90. They are also known as key molecules for corticoid receptor-mediated intracellular signaling. The binding of the corticoid receptor with the 90 kDa HSP (HSP90) is known to increase its stability and promote the formation of homodimer for consensus sequences in the promoter region of the target genes (Buckingham et al., 1996; Pratt and Toft, 1997; Richter and Buchner, 2001). Furthermore, transcriptional regulations of HSP90 and some of cytokines are known to interact with each other (Auphan et al., 1995; Scheinman et al., 1995; Stephanou and Latchman, 1999). In fish, HSP expression can be regulated by the endocrine system. Heat shock-induced expression of HSP30, HSP70, and HSP90 in hepatocytes or the gill was suppressed by the introduction of cortisol, although there was no significant effect of cortisol on the basal HSP level without heat shock (Deane et al., 1999; Iwama et al., 1999; Ackerman et al., 2000; Basu et al., 2001; Sathiyaa et al., 2001). In vitro administration of adrenaline increased HSP70 levels in hepatocytes of rainbow trout, and the β blocker propranolol abolished the effect (Ackerman et al., 2000). In vivo administration of GH and PRL reduced HSP70 expression in the liver of sea bream (*Sparus sarba*) (Deane et al., 1999). Although the involvement of HSPs in fish immunosuppression is still unclear, HSPs mediate endocrine control of the stress response at the level of the corticoid receptor.

4.5. Somatotropic Axis and Fish Immune System

As in higher vertebrates, the growth-promoting action of GH in fish is mediated to a large extent by insulin-like growth factor (IGF)-I (Duan, 1998). In mammalian species, IGF-I is also known to be an important factor that stimulates many aspects of the immune system, including specific and nonspecific defenses (Clark, 1997; Dorshkind and Horseman, 2000; Venters et al., 2001). Coinciding with the reports in mammals, IGF-I possesses immunomodulatory functions also seen in fish (Yada, 2007, 2009).

Expression of the IGF-I gene has been detected in lymphoid tissues of several fish species (Duguay et al., 1992, 1996; Shamblott and Chen, 1993; Loffing-Cueni et al., 1998; Yada et al., 2002; Shved et al., 2009). Phagocytic leukocytes of tilapia expressed the IGF-I gene and secreted significant amounts of IGF-I in vitro; however, the addition of GH to the medium did not produce significant change in IGF-I secretion (Yada, 2007). This is consistent with findings from studies on human lymphocytes treated with GH (Clayton et al., 1994). GH treatment increased the number of IGF-I–producing leukocytes from rat spleen (Baxter et al., 1991). Mouse lymphoma cells transfected with GH-overexpressing vector also showed higher levels of immunoreactive IGF-I and IGF-I mRNA than the control cells (Weigent and Arnold, 2005). These seemingly contradictory findings could derive from differences in species or in the preparation of cells. In euryhaline fish, IGF-I is also an important regulator for hydromineral balance, and GH directly stimulates IGF-I gene expression in gills, a major osmoregulatory organ during seawater acclimation (Sakamoto and McCormick, 2006). Taken together, these findings indicate that the GH/IGF-I axis plays an important regulatory role in the teleost fish immune system. Further investigations are needed to clarify the role of GH/IGF-I axis in the fish immune system across different species under different environmental conditions.

Extrapituitary expression of the GH gene in fish leukocytes suggests autocrine or paracrine action of GH in the modulation of immune functions (Yada and Nakanishi, 2002; Yada et al., 2005). Ghrelin, a hunger signaling hormone, was originally discovered in the rat stomach as an endogenous ligand for the GH secretagogue-receptor (Kojima et al., 1999). Ghrelin is found among vertebrates, and its GH-secreting action is also shown in several species of fish (Unniappan and Peter, 2005; Kaiya et al., 2012). Administration of ghrelin increased superoxide production in rainbow trout phagocytic leukocytes, whereas GH immunoneutralization (by adding antisalmon GH serum into the medium) blocked the stimulatory effect of ghrelin (Yada et al., 2006). These results suggest that ghrelin stimulates fish leukocytes, in part, through GH secreted by leukocytes.

GnRH is known as a potent stimulator not only for gonadotropin, but also for GH secretion in several species of fish, including rainbow trout (Holloway and Leatherland, 1997; Melamed et al., 1998; Canosa et al., 2007; Gahete et al., 2009). As described earlier, GnRH stimulated several immune functions in rainbow trout (Yada, 2012). In contrast to the action of ghrelin, there was no significant effect of GnRH on GH mRNA levels expressed in leukocytes, and immunoneutralization of GH did not block the effect of GnRH, indicating that GnRH has direct action on fish immunity without mediation through paracrine expression of GH in fish leukocytes (Yada, 2012).

Other possible somatotropic factors affecting the fish immune function are thyroid hormones. However, only a few studies examined the effects of these hormones on fish immune functions. Hypothyroidism in mummichog (*Fundulus heteroclitus*) induced by radioactive iodine injection produced a significant decrease in the number of circulating leukocytes (Slicher, 1961). Ball and Hawkins (1976) observed that an administration of thyroxine or mammalian thyroid-stimulating hormone (TSH) restored the number of circulating leukocytes in hypophysectomized sailfin molly (*Poecilia latipinna*). Further studies are needed to clarify the role of thyroid hormone in the fish immune system considering the importance of the pituitary–thyroid axis in the regulation of both the immune system in higher vertebrates (Marsh and

Erf, 1996; Dorshkind and Horseman, 2000) and in early stages of ontogeny and following metamorphosis. For instance, while thyroid hormone stimulated the shift of erythrocyte population from larval to adult types during the metamorphosis of Japanese flounder (*Paralichthys olivaceus*) (Miwa and Inui, 1991), the effect of this hormone on leukocyte populations or lymphoid tissues has not been elucidated. Thyroid hormones are also known as a key factor controlling parr-smolt transformation of salmonid fishes followed by various morphological and physiological changes including immunity (McCormick, 1995), as described more fully next.

5. ENVIRONMENTAL STRESSORS AND FISH IMMUNITY

5.1. Environmental Salinity

The stress response to changes in environmental salinity have been considered by Takei and Hwang (2016; Chapter 6 in this volume). Here we consider how these stress responses influence fish immunity. Rainbow trout acutely exposed to seawater elevated plasma cortisol and decreased antibody, but prolonged seawater exposure did not affect immune function (Betoulle et al., 1995). Indeed, acclimation to seawater improved nonspecific immunity of rainbow trout (Yada et al., 2001, 2012). Brown trout (Salmo trutta) also had activation of plasma lysozyme and phagocytosis of head kidney leukocytes in seawater (Marc et al., 1995). In sea bream, environmental salinity did not affect phagocytosis (Narnaware et al., 1997, 1998). Mozambique tilapia acclimated to seawater had an increased respiratory burst following phagocytosis, while plasma lysozyme levels were unchanged (Yada et al., 2002). Conversely, tilapia showed increased levels of plasma lysozyme and immunoglobulin with high environmental salinities (Dominguez et al., 2004, 2005). We suggest that differences in the effects of water salinity on immune functions between species or experimental conditions may be related to the organism's adaptability to a hyperosmotic environment. Euryhaline fishes, those that tolerate stress induced by salinity change, appear to promote immune functions through changes in endocrine regulation while adapting to a hyperosmotic environment. However, there seems to be a species-specific difference in the response of each immune function to salinity changes among euryhaline fish, including salmonids, tilapia, and sea bream. In relation to fish osmoregulation, further studies comparing the effect of salinity on immune functions between fish species, including stenohaline species, is needed.

In some euryhaline fish, the activation of immune functions observed during seawater acclimation appears to be associated with the osmoregulatory action of GH. During acclimation of brown trout to seawater, positive relationships between the enhancement of immune functions and an elevation of plasma GH level have been observed (Marc et al., 1995). In addition to stimulating body growth and metabolism, GH is known to facilitate euryhaline teleosts' adaptation to a hyperosmotic environment; the secretion of endogenous GH is stimulated in response to environmental salinity (Sakamoto et al., 1993; Yada et al., 1994; McCormick, 1995; Björnsson, 1997; Shepherd et al., 1997). An increased secretion of GH in seawater-acclimated euryhaline fish appears to enhance not only osmor-egulation but also immune function. The inhibitory action on PRL was well known for fish osmoregulation in seawater (Sakamoto and McCormick, 2006). However, a stimulatory effect of PRL on immune function has been observed even after seawater acclimation accompanied with the expression of the receptor (Yada et al., 2002). Immunomodulatory actions of PRL would appear to be independent of its osmoregulatory action, and involve local expression of hormones in the fish immune system.

5.2. Temperature and Seasonality

One of the major environmental factors that can modulate fish immune function is water temperature (Fries, 1986; Bly and Clem, 1992; Manning and Nakanishi, 1996; Schreck, 1996; Bly et al., 1997; Le Morvan et al., 1998; Hernández and Tort, 2003). Temperatures below the range at which optimal immune responses occur, but still within the physiological range, suppress both cellular- and humoral-specific immune functions (reviewed by Manning and Nakanishi, 1996). Earlier studies suggested that helper T-cells (and not memory T-cells or B-cells) are sensitive to lower temperature based on the proliferative response of lymphocytes to mitogens, mixed leukocyte reactions, and antibody production (reviewed by Bly and Clem, 1992; Manning and Nakanishi, 1996; Le Morvan et al., 1998). The thermosensitiveness of helper T-cells was attributed to the characteristics of fatty acid and/or glucide composition of plasma membranes (Bly and Clem, 1992; Le Morvan et al., 1998). In contrast, nonspecific immunity (eg, phagocytosis and nonspecific cytotoxicity) tend to be more resistant to low temperature than specific immunity (Ainsworth et al., 1991; Dexiang and Ainsworth, 1991; Collazos et al., 1994; Kurata et al., 1995). Of course, appropriate water temperature produces higher levels of both specific and nonspecific immune functions than warmer or colder environments (Dominguez et al., 2004, 2005). It is well known that a difference of temperature affects immune function through other mechanisms, such as stress, in addition to the direct effect of lower temperature on immune functions. For nonspecific immune functions, temperature change rather than a constant low temperature would seem to act as a stressor (Elliott,

1981; Le Morvan et al., 1998). Cold stress appears to be mediated by the hypothalamic–pituitary–thyroid axis and affects the immune system in mammals (Davis, 1998). However, as described earlier, the action of thyroid hormones on the fish immune system is still unclear.

Sexual maturation is generally accompanied by signs of immunosuppression. In salmonid species, decreased bactericidal activity and increased frequency of infection have been reported during sexual maturation (Richards and Pickering, 1978; Pickering and Christie, 1980; Iida et al., 1989). Sexual maturation also coincides with lower plasma lysozyme activity, and decreased numbers of total leukocytes and antibody-producing cells in salmonids (Pickering and Pottinger, 1987; Maule et al., 1996). Plasma immunoglobulin levels showed a tendency to decrease during the period of reproduction in rainbow trout, goldfish, and rock fish (*Sebastiscus marmoratus*) (Nakanishi, 1986; Suzuki et al., 1996, 1997). An elevated level of circulating cortisol has also been observed during the reproductive period of many fish species, suggesting the mediation of immunosuppression by this stress hormone (see Schreck and Maule, 2001). Changes in the secretion of gonadotropin and sex steroids and in the responsiveness to those hormones are noteworthy endocrine events during sexual maturation among vertebrates (Bentley, 1998; Blázquez et al., 1998). The direct action of sex steroids on immune functions was also observed in fish, and there is a difference in the effects between androgen and estrogen. In higher vertebrates, differences in immune responses of females and males have been documented in detail (Chapman et al., 1996; Lin et al., 1996; Marchetti et al., 2001). Sexual differences in immune functions have also been reported in fish, implying the importance of sex steroids in the regulation of the immune system (Tatner, 1996; Yano, 1996). Stress-induced secretion of cortisol is known to affect the process of fish reproduction (Barton and Iwama, 1991; Pankhurst and Van Der Kraak, 1997; Wendelaar Bonga, 1997; Schreck and Maule, 2001). Changes in the fish immune response during the period of sexual maturation appear to stem from an interaction between cortisol and sex steroids.

The parr-smolt transformation, smolting, or smoltification in anadromous salmonids, involves morphological, behavioral, and physiological changes that prepare the freshwater juvenile "parr" for the migration to and life in the ocean (Hoar, 1988; Barron, 1986; Dickhoff et al., 1997). There is a general elevation in circulating hormones such as the thyroid hormones, cortisol and GHs (Hoar, 1988; Dickhoff et al., 1997), and the administration of those hormones to juveniles promoted the appearance of smolt characteristics (Björnsson, 1997; McCormick, 2013). Transient reductions in immune responses have been observed during the parr-smolt transformation of anadromous salmonids (Schreck, 1996). Differential patterns of plasma hormone levels during the parr-smolt transformation have been observed between species, and even within the same species under different conditions (Björnsson, 1997; Dickhoff et al., 1997). Olsen et al. (1993) revealed that plasma lysozyme activity did not show a significant change during the parr-smolt transformation of Atlantic salmon under a natural lighting regime. Environmental modulation of the parr-smolt transformation is well known, particularly when caused by the lighting regime (Hoar, 1988). In two strains of *O. mykiss*, rainbow trout and steelhead trout, there are seasonal variations of mRNA levels of corticoid receptors expressed in leukocytes and spleen that coincide with increased plasma levels of thyroid hormone and cortisol in spring (Yada et al., 2014). Lighting or day–night rhythm is thought to be one of the environmental modulators of the immune system (Esteban et al., 2006).

6. FUTURE DIRECTIONS

There are a plethora of important research topics regarding the effects of stressors on the immune system and the relationship between the other regulatory systems such as the neural and endocrine systems. One important area concerns the fact that stressors generally induce some responses that appear as immune responses, even when the stressor is not a pathogen or a "nonself" organism, for instance a chemical or physical stressor. Why and how these responses are activated is a key unknown. Another information gap of crucial importance is the immune response at the mucosal surfaces as portals of entry of the different stress stimuli, and the relationship between stress, mucosal responses, and local microbiota. Although this area is currently intensively investigated, it needs considerably more attention. Regarding interaction, there is also much work to be done in order to understand how hormones, cytokines, and neuropeptides relate to each other and whether the different tissues have receptors for these molecules to respond to stress challenges. In particular, the interaction between neuropeptides and/or catecholamines and the immune response needs elucidation, given that much more is known regarding other mediators such as the corticosteroids and their immune effects and the pathways involved. Concerning specific immune responses, how the fish are able generate a differential immune response depending on the behavioral characteristics or depending on the individual perception of the stimuli or coping strategy needs investigation. Moreover, there is a considerable lack of general knowledge regarding the fish immune system needed to develop solutions for protecting their health status and to resist pathogens, particularly when stressors are also present. In addition, it will be of interest to determine if there are phylogenetic differences in immune tactics employed under stressful conditions between different species, genera, and families of fish.

REFERENCES

Acerete, L., Balasch, J. C., Castellana, B., Redruello, B., Roher, N., Canario, A. V., et al. (2007). Cloning of the glucocorticoid receptor (GR) in gilthead seabream (*Sparus aurata*). Differential expression of GR and immune genes in gilthead seabream after an immune challenge. *Comp. Biochem. Physiol.* **148B**, 32–43.

Ackerman, P. A., Forsyth, R. B., Mazur, C. F. and Iwama, G. K. (2000). Stress hormones and the cellular stress response in salmonids. *Fish Physiol. Biochem.* **23**, 327–336.

Ainsworth, A. J., Dexiang, C., Waterstrat, P. R. and Greenway, T. (1991). Effect of temperature on the immune system of channel catfish (*Ictalurus punctatus*)-I. Leucocyte distribution and phagocyte function in the anterior kidney at 10°C. *Comp. Biochem. Physiol.* **100A**, 907–912.

Auphan, N., DiDonato, J. A., Rosette, C., Helmberg, A. and Karin, M. (1995). Immunosuppression by glucocorticoids: inhibition of NF-kB activity through induction of IlB synthesis. *Science* **270**, 286–290.

Ball, J. N. and Hawkins, E. F. (1976). Adrenocortical (interrenal) responses to hypophysectomy and adenohypophysial hormones in the teleost *Poecilia latipinna. Gen. Comp. Endocrinol.* **28**, 59–70.

Balm, P. H. M. (1997). Immune-endocrine interactions. In *Fish Stress and Health in Aquaculture* (eds. G. K. Iwama, A. D. Pickering, J. P. Sumpter and C. B. Schreck), pp. 195–221. Cambridge: Cambridge University Press.

Barron, M. G. (1986). Endocrine control of smoltification in anadromous salmonids. *J. Endocrinol.* **108**, 313–319.

Barton, B. A. and Iwama, G. K. (1991). Physiological changes in fish from stress in aquaculture with emphasis on the response and effects of corticosteroids. *Ann. Rev. Fish Dis.* **1**, 3–26.

Basu, N., Nakano, T., Grau, E. G. and Iwama, G. K. (2001). The effects of cortisol on heat shock protein 70 levels in two fish species. *Gen. Comp. Endocrinol.* **124**, 97–105.

Baxter, J. B., Blalock, J. E. and Weigent, D. A. (1991). Characterization of immunoreactive insulin-like growth factor-I from leukocytes and its regulation by growth hormone. *Endocrinology* **129**, 1727–1734.

Bayne, C. J. and Levy, S. (1991a). Modulation of the oxidative burst in trout myeloid cells by adrenocorticotropic hormone and catecholamines: mechanisms and action. *J. Leukoc. Biol.* **50**, 554–560.

Bayne, C. J. and Levy, S. (1991b). The respiratory burst of rainbow trout, *Oncorhynchus mykiss* (Walbaum), phagocytes is modulated by sympathetic neurotransmitters and the 'neuro' peptide ACTH. *J. Fish Biol.* **38**, 609–619.

Bentley, P. J. (1998). *Comparative Vertebrate Endocrinology* (third ed.). Cambridge: Cambridge University Press.

Betoulle, S., Troutaud, D., Khan, N. and Deschaux, R. (1995). Résponse anticorps, cortisolémie et prolactinémie chez la truite arc-en-ciel. *C. R. Acad. Sci. Paris* **318**, 677–681.

Björnsson, B. Th (1997). The biology of salmon growth hormone: from daylight to dominance. *Fish Physiol. Biochem.* **17**, 9–24.

Blázquez, M., Bosma, P. T., Fraser, E. J., Van Look, K. J. W. and Trudeau, V. L. (1998). Fish as models for the neuroendocrine regulation of reproduction and growth. *Comp. Biochem. Physiol.* **119C**, 345–364.

Bly, J. E. and Clem, L. W. (1992). Temperature and teleost immune functions. *Fish Shellfish Immunol.* **2**, 159–171.

Bly, J. E., Quiniou, S. M.-A. and Clem, L. W. (1997). Environmental effects of fish immune mechanisms. *Dev. Biol. Stand.* **90**, 33–43.

Boomershine, C. S., Wang, T. and Zwilling, B. S. (2001). Neuroendocrine regulation of macrophage and neutrophil function. In *Psychoneuroimmunology*, vol. 1 (eds. R. Ader, D. L. Felten and N. Cohen), pp. 289–300. San Diego: Academic Press.

Borghetti, P., Saleri, R., Mocchegiani, E., Corradi, A. and Martelli, P. (2009). Infection, immunity and the neuroendocrine response. *Vet. Immunol. Immunopathol.* **130**, 141–162.

Buckingham, J. C., Christian, H. C., Gillies, G. E., Philip, J. G. and Taylor, A. D. (1996). The hypothalamo-pituitary-adrenocortical immune axis. In *The Physiology of Immunity* (eds. J. A. Marsh and M. D. Kendall), pp. 331–354. Boca Raton: CRC Press.

Bury, N. R., Sturm, A., Le Rouzic, P., Lethimonier, C., Ducouret, B., Guiguen, Y., et al. (2003). Evidence for two distinct functional glucocorticoid receptors in teleost fish. *J. Mol. Endocrinol.* **31**, 141–156.

Caipang, C. M. A., Brinchmann, M. F. and Berg, I. (2008). Changes in selected stress and immune-related genes in Atlantic cod, *Gadus morhua*, following overcrowding. *Aquac. Res.* **39**, 1533–1540.

Caipang, C. M. A., Lazado, C. C., Brinchmann, M. F., Rombout, J. H. W. M. and Kiron, V. (2011). Differential expression of immune and stress genes in the skin of Atlantic cod (*Gadus morhua*). *Comp. Biochem. Physiol. D* **6**, 158–162.

Calduch-Giner, J. A., Sitjà-Bobadilla, A., Alvarez-Pellitero, P. and Pérez-Sánchez, J. (1995). Evidence for a direct action of GH on haemopoietic cells of a marine fish, the gilthead sea bream (*Sparus aurata*). *J. Endocrinol.* **146**, 459–467.

Calduch-Giner, J. A., Duval, H., Chesnel, F., Boeuf, G., Prérez-Sánchez, J. and Boujard, D. (2000). Fish growth hormone receptor: molecular characterization of two membrane-anchored forms. *Endocrinology* **142**, 3269–3273.

Canosa, L. F., Chang, J. P. and Peter, R. E. (2007). Neuroendocrine control of growth hormone in fish. *Gen. Comp. Endocrinol.* **151**, 1–26.

Castillo, J., Teles, M., Mackenzie, S. and Tort, L. (2009). Stress-related hormones modulate cytokine expression in the head kidney of gilthead seabream (*Sparus aurata*). *Fish Shellfish Immunol.* **27**, 493–499.

Castro, R., Zou, J., Secombes, C. J. and Martin, S. A. M. (2011). Cortisol modulates the induction of inflammatory gene expression in a rainbow trout macrophage cell line. *Fish Shellfish Immunol.* **30**, 215–223.

Chadzinska, M., Józefowski, S., Bigaj, J. and Plytycz, B. (1997). Morphine modulation of thioglycollate-elicited peritoneal inflammation in the goldfish, *Carassius auratus*. *Arch. Immunol. Ther. Exp.* **45**, 321–327.

Chapman, J. C., Despande, R. and Michael, S. D. (1996). Estrogen-mediated interactions between the immune and female reproductive systems. In *The Physiology of Immunity* (eds. J. A. Marsh and M. D. Kendall), pp. 239–261. Boca Raton: CRC Press.

Clark, R. (1997). The somatogenic hormones and insulin-like growth factor-1: stimulators of lymphopoiesis and immune function. *Endocr. Rev.* **18**, 157–179.

Clayton, P. E., Day, R. N., Silva, C. M., Hellmann, P., Day, H. K. and Thorner, M. O. (1994). Growth hormone induces tyrosine phosphorylation but does not alter insulin-like growth factor-I gene expression in human IM9 lymphocytes. *J. Mol. Endocrinol.* **13**, 127–136.

Collazos, M. E., Ortega, E. and Barriga, C. (1994). Effect of temperature on the immune system of a cyprinid fish (*Tinca tinca*, L.). Blood phagocyte function at low temperature. *Fish Shellfish Immunol.* **4**, 231–238.

Colombe, L., Fostier, A., Bury, N., Pakdel, F. and Guiguen, Y. (2000). A mineralcorticoid-like receptor in the rainbow trout, *Oncorhynchus mykiss*: cloning and characterization of its steroid binding domain. *Steroids* **65**, 319–328.

Costas, B., Conceição, L., Aragão, C., Martos, J. A., Ruiz-Jarabo, I., Mancera, J. M., et al. (2011). Physiological responses of Senegalese sole (Solea senegalensis Kaup, 1858) after stress challenge: effects on non-specific immune parameters, plasma free amino-acids. *Aquaculture* **316**, 68–76.

Cuesta, A., Laiz-Carrión, R., Del Río, M. P., Meseguer, J., Mancera, J. M. and Esteban, M. A. (2005). Salinity influences the humoral immune parameters of gilthead seabream (*Sparus aurata* L.). *Fish Shellfish Immunol.* **18**, 255–261.

Cuesta, A., Arjona, R. L.-C. F., del Río, M. P. M., Meseguer, J., Mancera, J. M. and Esteban, M.Á. (2006). Effect of PRL, GH and cortisol on the serum complement and IgM levels in gilthead seabream (*Sparus aurata* L.). *Fish Shellfish Immunol.* **20**, 427–432.

Davis, K. B., Griffin, B. R. and Gray, W. L. (2003). Effect of dietary cortisol on resistance of channel catfish to infection by *Ichthyophthirius multifiliis* and channel catfish virus disease. *Aquaculture* **218**, 121–130.

Davis, S. L. (1998). Environmental modulation of the immune system via the endocrine system. *Dom. Anim. Endocrinol.* **15**, 283–289.

Deane, E. E., Kelly, S. P., Lo, C. K. and Woo, N. Y. (1999). Effects of GH, prolactin and cortisol on hepatic heat shock protein 70 expression in a marine teleost *Sparus sarba*. *J. Endocrinol.* **161**, 413–421.

Demers, N. E. and Bayne, C. J. (1997). The immediate effects of stress on hormones and plasma lysozyme in rainbow trout. *Dev. Comp. Immunol.* **21**, 363–373.

Dexiang, C. and Ainsworth, A. J. (1991). Effect of temperature on the immune system of channel catfish (*Ictalurus punctatus*)-II. Adaptation of anterior kidney phagocytes to 10°C. *Comp. Biochem. Physiol.* **100A**, 913–918.

Dhabhar, F. S. (2000). Acute stress enhances while chronic stress suppresses skin immunity. The role of stress hormones and leukocyte trafficking. *Ann. N. Y. Acad. Sci.* **9178**, 76–93.

Dhabhar, F. S. and McEwen, B. S. (2001). Bidirectional effects of stress and glucocorticoid hormones on immune function: Possible explanations for paradoxical observations. In *Psychoneuroimmunology*, vol. 1 (eds. R. Ader, D. L. Felten and N. Cohen), pp. 301–338. San Diego: Academic Press.

Dhabhar, F. S. (2002). Stress-induced augmentation of immune function—the role of stress hormones, leukocyte trafficking, and cytokines. *Brain Behav. Immun.* **16**, 785–798.

Dhabhar, F. S. (2008). Enhancing versus suppressive effects of stress on immune function: implications for immunoprotection versus immunopathology. *Allergy Asthma Clin. Immunol.* 2–11.

Dickhoff, W. W., Beckman, B. R., Larsen, D. A., Duan, C. and Moriyama, S. (1997). The role of growth in endocrine regulation of salmon smoltification. *Fish Physiol. Biochem.* **17**, 231–236.

Dominguez, M., Takemura, A., Tsuchiya, M. and Nakamura, S. (2004). Impact of different environmental factors on the circulating immunoglobulin levels in the Nile tilapia, *Oreochromis niloticus*. *Aquaculture* **241**, 491–500.

Dominguez, M., Takemura, A. and Tsuchiya, M. (2005). Effects of changes in environmental factors on the non-specific immune response of Nile tilapia, *Oreochromis niloticus* L. *Aquac. Res.* **36**, 391–397.

Dorshkind, K. and Horseman, N. D. (2000). The roles of prolactin, growth hormone, insulin-like growth factor-I, and thyroid hormones in lymphocyte development and function: insights from genetic models of hormone and hormone receptor deficiency. *Endocr. Rev.* **21**, 292–312.

Douxfils, J., Lambert, S., Mathieu, C., Milla, S., Mandiki, S. N. M., Henrotte, E., et al. (2014). Influence of domestication process on immune response to repeated emersion stressors in Eurasian perch (*Perca fluviatilis*, L.). *Comp. Biochem. Physiol. A* **173**, 52–60.

Duan, C. (1998). Nutritional and developmental regulation of insulin-like growth factors in fish. *J. Nutr.* **128**, 306S–314S.

Ducouret, B., Tujague, M., Ashraf, J., Mouchel, N., Servel, N., Valotaire, Y., et al. (1995). Cloning of a teleost fish glucocorticoid receptor shows that it contains a deoxyribonucleic acid-binding domain different from that of mammals. *Endocrinology* **136**, 3774–3783.

Duguay, S. J., Park, L. K., Samadpour, M. and Dickhoff, W. W. (1992). Nucleotide sequence and tissue distribution of three insulin-like growth factor I prohormones in salmon. *Mol. Endocrinol.* **6**, 1202–1210.

Duguay, S. J., Lai-Zhang, J., Steiner, D. F., Funkenstein, B. and Chan, S. J. (1996). Developmental and tissue-regulated expression of IGF-I and IGF-II mRNAs in *Sparus aurata*. *J. Mol. Endocrinol.* **16**, 123–132.

Elenkov, I. J. and Chrousos, G. P. (1999). Stress hormones, Th1/Th2 patterns, pro/anti-inflammatory cytokines and susceptibility to disease. *Trends Endocrinol. Metab.* **10**, 359–368.

Elliott, J. M. (1981). Some aspects of thermal stress on freshwater teleosts. In *Stress and Fish* (ed. A. D. Pickering), pp. 209–245. London: Academic Press.

Ellsaesser, C. F. and Clem, L. W. (1986). Haematological and immunological changes in channel catfish stressed by handling and transport. *J. Fish Biol.* **28**, 511–521.

Engelsma, M. Y., Hougee, S., Nap, D., Hofenk, M., Rombout, J. H., van Muiswinkel, W. B., et al. (2003). Multiple acute temperature stress affects leucocyte populations and antibody responses in common carp, *Cyprinus carpio* L. *Fish Shellfish Immunol.* **15**, 397–410.

Esteban, M.Á., Cuesta, A., Rodríguez, A. and Meseguer, J. (2006). Effect of photoperiod on the fish innate immune system: a link between fish pineal gland and the immune system. *J. Pineal Res.* **41**, 261–266.

Fast, M. D., Hosoya, S., Johnson, S. C. and Afonso, L. O. (2008). Cortisol response and immune-related effects of Atlantic salmon (*Salmo salar* Linnaeus) subjected to short- and long-term stress. *Fish Shellfish Immunol.* **24**, 194–204.

Finkenbine, S. S., Gettys, T. W. and Brunett, K. G. (1997). Direct effects of catecholamines on T and B cell lines of the channel catfish, *Ictalurus punctatus*. *Dev. Comp. Immunol.* **21**, 155.

Flores, A.-M., Shrimpton, J. M., Patterson, D. A., Hills, J. A., Cooke, S. J., Yada, T., et al. (2012). Physiological and molecular endocrine changes in maturing wild sockeye salmon, *Oncorhynchus nerka*, during ocean and river migration. *J. Comp. Physiol. B* **182**, 77–90.

Flory, C. M. (1989). Automatic innervation of the spleen of the coho salmon, *Oncorhynchus kisutch:* a histochemical demonstration and preliminary assessment of its immunoregulatory role. *Brain Behav. Immun.* **3**, 331–344.

Flory, C. M. (1990). Phylogeny of neuroimmunoregulation: effects of adrenergic and cholinergic agents on the in vitro antibody response of the rainbow trout, *Oncorhynchus mykiss*. *Dev. Comp. Immunol.* **14**, 283–294.

Flory, C. M. and Bayne, C. J. (1991). The influence of adrenergic and cholinergic agents on the chemiluminescent and mitogenic responses of leukocytes from the rainbow trout, *Oncorhynchus mykiss*. *Dev. Comp. Immunol.* **15**, 135–142.

Fries, C. R. (1986). Effects of environmental stressors and immunosuppressants on immunity in *Fundulus heteroclitus*. *Am. Zool.* **26**, 271–282.

Fukada, H., Ozaki, Y., Adachi, S., Yamauchi, K., Hara, A. (2001). GeneBank accession number AB071216.

Gahete, M. D., Durán-Prado, M., Luque, R. M., Martínez-Fuentes, A. J., Quintero, A., Gutiérrez-Pascual, E., et al. (2009). Understanding the multifactorial control of growth hormone release by somatotropes. *Ann. N. Y. Acad. Sci.* **1163**, 137–153.

Gomez, D., Sunyer, J. O. and Salinas, I. (2013). The mucosal immune system of fish: the evolution of tolerating commensals while fighting pathogens. *Fish Shellfish Immunol.* **35**, 1729–1739.

Gopurappilly, R., Ogawa, S. and Parhar, I. S. (2013). Functional significance of GnRH and kisspeptin, and their cognate receptors in teleost reproduction. *Front. Endocrinol.* **4**, 24. ⟨http://dx.doi.org/10.3389/fendo.2013.00024⟩.

Gorissen, M. and Flik, G. (2016). Endocrinology of the Stress Response in Fish. In *Fish Physiology - Biology of Stress in Fish*, Vol. 35 (eds. C. B. Schreck, L. Tort, A. P. Farrell and C. J. Brauner), San Diego, CA: Academic Press.

Harris, J. and Bird, D. J. (1997). The effects of α-MSH and MCH on the proliferation of rainbow trout (*Oncorhynchus mykiss*) lymphocytes in vitro. In *Advances in Comparative Endocrinology* (eds. S. Kawashima and S. Kikuyama), pp. 1023–1026. Bologna: Monduzzi Editore.

Harris, J. and Bird, D. J. (1998). Alpha-melanocyte stimulating hormone (α-MSH) and melanin-concentrating hormone (MCH) stimulate phagocytosis by head kidney leucocytes of rainbow trout (*Oncorhynchus mykiss*) in vitro. *Fish Shellfish Immunol.* **8**, 631–638.

Harris, J. and Bird, D. J. (2000). Supernatants from leucocytes treated with melanin-concentrating hormone (MCH) and α-melanocyte stimulating hormone (α-MSH) have a stimulatory effect on rainbow trout (*Oncorhynchus mykiss*) phagocytes in vitro. *Vet. Immunol. Immunopathol.* **76**, 117–124.

Harris, J., Bird, D. J. and Yeatman, L. A. (1998). Melanin-concentrating hormone (MCH) stimulates the activity of rainbow trout (*Oncorhynchus mykiss*) head kidney phagocytes in vitro. *Fish Shellfish Immunol.* **8**, 639–642.

Hernández, A. and Tort, L. (2003). Annual variation of complement, lysozyme and haemagglutinin levels in serum of the gilthead sea bream *Sparus aurata*. *Fish Shellfish Immunol.* **15**, 479–481.

Higashimoto, Y., Nakao, N., Ohkubo, T., Tanaka, M. and Nakashima, K. (2001). Structure and tissue distribution of prolactin receptor mRNA in Japanese flounder (*Paralichthys olivaceus*): conserved and preferential expression in osmoregulatory organs. *Gen. Comp. Endocrinol.* **123**, 170–179.

Hoar, W. S. (1988). The physiology of smolting salmonids. In *Fish Physiology - The Physiology of Developing Fish: Viviparity and Posthatching Juveniles*, Vol. 11B (eds. W. S. Hoar and D. J. Randall), pp. 275–343. San Diego: Academic Press.

Holloway, A. C. and Leatherland, J. F. (1997). The effects of *N*-methyl-D,L-aspartate and gonadotropin-releasing hormone on in vitro growth hormone release in steroid-primed immature rainbow trout, *Oncorhynchus mykiss*. *Gen. Comp. Endocrinol.* **107**, 32–43.

Hou, Y., Suzuki, Y. and Aida, K. (1999). Effects of steroids on the antibody producing activity of lymphocytes in rainbow trout. *Fish Sci.* **65**, 850–855.

Iida, T., Takahashi, K. and Wakabayashi, H. (1989). Decrease in the bactericidal activity of normal serum during the spawning period of rainbow trout. *Bull. Jpn. Soc. Sci. Fish.* **55**, 463–465.

Iwama, G. K., Vijayan, M. M., Forsyth, R. B. and Ackenrian, P. A. (1999). Heat shock proteins and physiological stress in fish. *Am. Zool.* **39**, 901–909.

Jhingan, E., Devlin, R. H. and Iwama, G. K. (2003). Disease resistance, stress response and effects of triploidy in growth hormone transgenic coho salmon. *J. Fish Biol.* **63**, 806–823.

Józefowski, S., Gruca, P., Józkowicz, A. and Plytycz, B. (1995). Direct detection and modulatory effects of adrenergic and cholinergic receptors on goldfish leukocytes. *J. Mar. Biotechnol.* **3**, 171–173.

Kaiya, H., Hosoda, H., Kangawa, K. and Miyazato, M. (2012). Determination of nonmammalian ghrelin. In *Ghrelin* (eds. M. Kojima and K. Kangawa), pp. 75–87. San Diego: Academic Press.

Kaneko, T. (1996). Cell biology of somatolactin. *Int. Rev. Cytol.* **169**, 1–24.

Kaneko, T., Kakizawa, S. and Yada, T. (1993a). Pituitary of "cobalt" variant of the rainbow trout separated from the hypothalamus lacks most pars intermedial and neurohypophysial tissue. *Gen. Comp. Endocrinol.* **92**, 31–40.

Kaneko, T., Kakizawa, S., Yada, T. and Hirano, T. (1993b). Gene expression and intercellular localization of somatolactin in the pituitary of rainbow trout. *Cell Tissue Res.* **272**, 11–16.

Kawauchi, H., Sower, S. A. and Moriyama, S. (2009). The neuroendocrine regulation of prolactin and somatolactin secretion in fish. In *Fish Neuroendocrinology* (eds. N. J. Bernier, G. van der Kraak, A. P. Farrell and C. J. Brauner), pp. 197–234. San Diego: Elsevier Academic Press.

Kojima, M., Hosoda, H., Date, Y., Nakazato, M., Matsuo, H. and Kangawa, K. (1999). Ghrelin is a growth-hormone-releasing acylated peptide from stomach. *Nature* **402**, 656–660.

Krasnov, A., Skugor, S., Todorcevic, M., Glover, K. A. and Nilsen, F. (2012). Gene expression in Atlantic salmon skin in response to infection with the parasitic copepod *Lepeophtheirus salmonis*, cortisol implant, and their combination. *BMC Genomics* **13**, 130. ⟨http://dx.doi.org/10.1186/1471-2164-13-130⟩.

Kurata, O., Okamoto, N., Suzumura, E., Sano, N. and Ikeda, Y. (1995). Accommodation of carp natural killer-like cells to environmental temperature. *Aquaculture* **129**, 421–424.

Le Morvan, C., Troutaud, D. and Deschaux, P. (1998). Differential effects of temperature on specific and nonspecific immune defences in fish. *J. Exp. Biol.* **201**, 165–168.

Lee, L. T. O., Nong, G., Chan, Y. H., Tse, D. L. Y. and Cheng, C. H. K. (2001). Molecular cloning of a teleost growth hormone receptor and its functional interaction with human growth hormone. *Gene* **270**, 121–129.

Leonard, B. E. (2005). The HPA and immune axes in stress: the involvement of the serotonergic system. *Eur. Psychiatry* **20** (Suppl. 3S), 302–306.

Li, J., Barreda, D. R., Zhang, Y. A., Boshra, H., Gelman, A. E., Lapatra, S., et al. (2006). B lymphocytes from early vertebrates have potent phagocytic and microbicidal abilities. *Nat. Immunol.* **7**, 1116–1124.

Lin, T.-M., Lustig, R. H. and Chang, C. (1996). The role of androgens-androgen receptor in immune system activity. In *The Physiology of Immunity* (eds. J. A. Marsh and M. D. Kendall), pp. 263–276. Boca Raton: CRC Press.

Loffing-Cueni, D., Schmid, A. C., Graf, H. and Reinecke, M. (1998). IGF-I in the bony fish *Cottus scorpius*: cDNA, expression and differential localization in brain and islets. *Mol. Cell. Endocrinol.* **141**, 187–194.

MacArthur, J. I., Fletcher, T. C., Pirie, B. J. S., Davidson, R. J. L. and Thomson, A. W. (1984). Peritoneal inflammatory cells in plaice, *Pleuronectes platessa* L.: effects of stress and endotoxin. *J. Fish Biol.* **25**, 69–81.

MacKenzie, S., Iliev, D., Liarte, C., Koskinen, H., Planas, J. V., Goetz, F. W., et al. (2006a). Transcriptional analysis of LPS-stimulated activation of trout (*Oncorhynchus mykiss*) monocyte/macrophage cells in primary culture treated with cortisol. *Mol. Immunol.* **43**, 1340–1348.

MacKenzie, S., Liarte, D. I., Koskinen, H., Planas, J. V., Goetz, F. W., Mölsä, H., et al. (2006b). Transcriptional analysis of LPS-stimulated activation of trout (*Oncorhynchus mykiss*) monocyte/macrophage cells in primary culture treated with cortisol. *Mol. Immunol.* **43**, 1340–1348.

Manning, M. J. and Nakanishi, T. (1996). The specific immune system: cellular defenses. In *The Fish Immune System: Organism, Pathogen, and Environment* (eds. G. Iwama and T. Nakanishi), pp. 159–205. San Diego: Academic Press.

Marc, A. M., Quentel, C., Severe, A., Le Bail, P. Y. and Boeuf, G. (1995). Changes in some endocrinological and non-specific immunological parameters during seawater exposure in the brown trout. *J. Fish Biol.* **46**, 1065–1081.

Marchetti, B., Morale, M. C., Gallo, F., Lomeo, E., Testa, N., Tirolo, C., et al. (2001). The hypothalmo-pituitary-gonadal axis and the immune system. In *Psychoneuroimmunology*, vol. 1 (ed. R. Ader), pp. 363–389. San Diego: Elsevier Academic Press.

Marsh, J. A. and Erf, G. F. (1996). Interactions between the thyroid and the immune system. In *The Physiology of Immunity* (eds. J. A. Marsh and M. D. Kendall), pp. 211–235. Boca Raton: CRC Press.

Maule, A. G. and Schreck, C. B. (1990). Glucocorticoid receptors in leukocytes and gill of juvenile coho salmon (*Oncorhynchus kisutch*). *Gen. Comp. Endocrinol.* **77**, 448–455.

Maule, A. G. and Schreck, C. B. (1991). Stress and cortisol treatment changed affinity and number of glucocorticoid receptors in leukocytes and gill of coho salmon. *Gen. Comp. Endocrinol.* **84**, 83–93.

Maule, A. G., Schrock, R., Slater, C., Fitzpatrick, M. S. and Schreck, C. B. (1996). Immune and endocrine responses of adult chinook salmon during freshwater migration and sexual maturation. *Fish Shellfish Immunol.* **6**, 221–233.

Mauri, I., Romero, A., Acerete, L., MacKenzie, S., Roher, N., Callol, A., et al. (2011). Changes in complement responses in Gilthead seabream (*Sparus aurata*) and European seabass (*Dicentrarchus labrax*) under crowding stress, plus viral and bacterial challenges. *Fish Shellfish Immunol.* **30**, 182–188.

McCormick, S. D. (1995). Hormonal control of gill Na$^+$, K$^+$-ATPase and chloride cell function. In *Cellular and Molecular Approaches to Fish Ionic Regulation* (eds. C. M. Wood and T. J. Shuttleworth), pp. 285–315. San Diego: Academic Press.

McCormick, S. D. (2013). Smolt physiology and endocrinology. In *Fish Physiology Vol. 32 Euryhaline Fishes* (eds. S. D. McCormick, A. P. Farrell and C. J. Brauner), pp. 200–251. Amsterdam: Academic Press.

McCormick, S. D., Regish, A., O'Dea, M. F. and Shrimpton, J. M. (2008). Are we missing a mineralocorticoid in teleost fish? Effects of cortisol, deoxycorticosterone and aldosterone on osmoregulation, gill Na$^+$,K$^+$-ATPase activity and isoform mRNA levels in Atlantic salmon. *Gen. Comp. Endocrinol.* **157**, 35–40.

Melamed, P., Rosenfeld, H., Elizur, A. and Yaron, Z. (1998). Endocrine regulation of gonadotropin and growth hormone gene transcription in fish. *Comp. Biochem. Physiol.* **119C**, 325–338.

Miwa, S. and Inui, Y. (1991). Thyroid hormone stimulates the shift of erythrocyte populations during metamorphosis of the flounder. *J. Exp. Zool.* **259**, 222–228.

Mola, L., Gambarelli, A., Pederzoli, A. and Ottaviani, E. (2005). ACTH response to LPS in the first stages of development of the fish *Dicentrarchus labrax* L. *Gen. Comp. Endocrinol.* **143**, 99–103.

Montero, D., Marrero, M., Izquierdo, M. S., Robaina, L., Vergara, J. M. and Tort, L. (1999). Effect of Vitamin E and C and dietary supplementation on some immune parameters of gilthead seabream (*Sparus aurata*) juveniles subjected to crowding stress. *Aquaculture* **171**, 269–278.

Montero, D., Kalinowski, T., Obach, A., Robaina, L., Tort, L., Caballero, M. J., et al. (2003). Vegetable lipid sources for gilthead seabream (*Sparus aurata*): effects on fish health. *Aquaculture* **225**, 353–370.

Montero, D., Lalumera, G., Izquierdo, M. S., Caballero, M. J., Saroglia, M. and Tort, L. (2009). Establishment of dominance relationships in gilthead sea bream (*Sparus aurata*) juveniles during feeding: effects on feeding behaviour, feed utilization, and fish health. *J. Fish Biol.* **74**, 1–16.

Montero, D., Mathlouthi, F., Tort, L., Afonso, J. M. and Torrecillas, S. (2010). Replacement of dietary fish oil by vegetable oils affects humoral immunity and expression of pro-inflammatory cytokines genes in gilthead sea bream *Sparus aurata*. *Fish Shellfish Immunol.* **29**, 1073–1081.

Morera, D., Roher, N., Ribas, L., Balasch, J. C., Doñate, C., Callol, A., et al. (2011). RNA-Seq reveals an integrated immune response in nucleated erythrocytes. *PloS One* **6** (10), e26998. ⟨http://dx.doi.org/10.1371/journal.pone.0026998⟩.

Mori, T., Hiraka, I., Kurata, Y., Kawachi, H., Mano, N., Devlin, R. H., et al. (2007). Changes in hepatic gene expression related to innate immunity, growth and iron metabolism in GH-transgenic amago salmon (*Oncorhynchus masou*) by cDNA subtraction and microarray analysis, and serum lysozyme activity. *Gen. Comp. Endocrinol.* **151**, 42–54.

Nakanishi, T. (1986). Seasonal changes in the humoral immune response and the lymphoid tissues of the marine teleost, *Sebastiscus marmoratus*. *Vet. Immunol. Immunopathol.* **12**, 213–221.

Narnaware, Y. K. and Baker, B. I. (1996). Evidence that cortisol may protect against the immediate effects of stress on circulating leukocytes in the trout. *Gen. Comp. Endocrinol.* **103**, 359–366.

Narnaware, Y. K., Baker, B. I. and Tomlinson, M. G. (1994). The effect of various stresses, corticosteroids and adrenergic agents on phagocytosis in the rainbow trout *Oncorhynchus mykiss*. *Fish Physiol. Biochem.* **13**, 31–40.

Narnaware, Y. K., Kelly, S. P. and Woo, N. Y. S. (1997). Effect of injected growth hormone on phagocytosis in silver sea bream (*Sparus sarba*) adapted to hyper- and hypo-osmotic salinities. *Fish Shellfish Immunol.* **7**, 515–517.

Narnaware, Y. K., Kelly, S. P. and Woo, N. Y. S. (1998). Stimulation of macrophage phagocytosis and lymphocyte count by exogenous prolactin administration in silver sea bream (*Sparus sarba*) adapted to hyper- and hypo-osmotic salinities. *Vet. Immunol. Immunopathol.* **61**, 387–391.

Ndong, D., Chen, Y. Y., Lin, Y. H., Vaseeharan, B. and Chen, J. C. (2007). The immune response of tilapia *Oreochromis mossambicus* and its susceptibility to *Streptococcus iniae* under stress in low and high temperatures. *Fish Shellfish Immunol.* **22**, 686–694.

Nickerson, J. G., Dugan, S. G., Drouin, G., Perry, S. F. and Moon, T. W. (2003). Activity of the unique-adrenergic Na^+/H^+ exchanger in trout erythrocytes is controlled by a novel β_3-AR subtype. *Am. J. Physiol. Regul. Integr. Comp. Physiol.* **285**, R526–R535.

Olsen, Y. A., Reitan, F. J. and Roed, K. H. (1993). Gill Na^+,K^+-ATPase activity, plasma cortisol level, and non-specific immune response in Atlantic salmon (*Salmo salar*) during parr-smolt transformation. *J. Fish Biol.* **43**, 559–573.

Ortuño, J., Esteban, M. A. and Meseguer, J. (2001). Effects of short-term crowding stress on the gilthead seabream (*Sparus aurata* L.) innate immune response. *Fish Shellfish Immunol.* **11**, 187–197.

Ottaviani, E. and Franceschi, C. (1996). The neuroimmunology of stress from invertebrates to man. *Prog. Neurobiol.* **48**, 421–440.

Owen, S. F., Giltrow, E., Huggett, D. B., Hutchinson, T. H., Saye, J.-A., Winter, M. J., et al. (2007). Comparative physiology, pharmacology and toxicology of β-blockers: mammals versus fish. *Aquat. Toxicol.* **82**, 145–162.

Pagniello, K. B., Bols, N. C. and Lee, L. E. (2002). Effect of corticosteroids on viability and proliferation of the rainbow trout monocyte/macrophage cell line, RTS11. *Fish Shellfish Immunol.* **13**, 199–214.

Pankhurst, N. W. and Van Der Kraak, G. (1997). Effects of stress on reproduction and growth of fish. In *Fish Stress and Health in Aquaculture* (eds. G. K. Iwama, A. D. Pickering, J. P. Sumpter and C. B. Schreck), pp. 73–93. Cambridge: Cambridge University Press.

Parra, D., Rieger, A. M., Li, J., Zhang, Y. A., Randall, L. M., Hunter, C. A., et al. (2012). Pivotal advance: peritoneal cavity B-1 B cells have phagocytic and microbicidal capacities and present phagocytosed antigen to CD4+ T cells. *J. Leukoc. Biol.* **91**, 525–536.

Pickering, A. D. and Christie, P. (1980). Sexual differences in the incidence and severity of ectoparasitic infestation of the brown trout, *Salmo trutta* L. *J. Fish Biol.* **16**, 669–683.

Pickering, A. D. and Pottinger, T. G. (1987). Lymphocytopenia and interrenal activity during sexual maturation in the brown trout, *Salmo trutta* L. *J. Fish Biol.* **30**, 41–50.

Pickford, G. E., Srivastave, A. K., Slicher, A. M. and Pang, P. K. T. (1971). The stress response in the abundance of circulating leucocytes in the killifish, *Fundulus heteroclitus*. I. The cold-shock sequence and the effects of hypophysectomy. *J. Exp. Zool.* **177**, 89–96.

Pottinger, T. G., Carrick, T. R., Appleby, A. and Yeomans, W. E. (2000). High blood cortisol levels and low cortisol receptor affinity: is the chub, *Leuciscus cephalus*, a cortisol-resistant teleost? *Gen. Comp. Endocrinol.* **120**, 108–117.

Pratt, W. B. and Toft, D. O. (1997). Steroid receptor interactions with heat shock protein and immunophilin chaperones. *Endocr. Rev.* **18**, 306–360.

Pruett, S. B. (2003). Stress and the immune system. *Pathophysiology* **9**, 133–153.

Prunet, P., Sandra, O., Le Rouzic, P., Marchand, O. and Laudet, V. (2000). Molecular characterization of the prolactin receptor in two fish species, tilapia *Oreochromis niloticus* and rainbow trout, *Oncorhynchus mykiss*: a comparative approach. *Can. J. Physiol. Pharmacol.* **78**, 1086–1096.

Prunet, P., Sturm, A. and Milla, S. (2006). Multiple corticosteroid receptors in fish: from old ideas to new concepts. *Gen. Comp. Endocrinol.* **147**, 17–23.

Randall, D. J. and Perry, S. F. (1992). Catecholamines. In *Fish Physiology - The Cardiovascular System*, Vol. 12B (eds. W. S. Hoar, D. J. Randall and A. P. Farrell), pp. 255–300. San Diego: Academic Press.

Reid, S. D. and Perry, S. F. (1991). The effects and physiological consequences of raised levels of cortisol on rainbow trout (*Oncorhynchus mykiss*) erythrocyte β-adrenoreceptors. *J. Exp. Biol.* **158**, 217–240.

Reid, S. D., Lebras, Y. and Perry, S. F. (1993). The in vitro effect of hypoxia on the trout erythrocyte β-adrenergic signal transduction system. *J. Exp. Biol.* **176**, 103–116.

Richards, R. H. and Pickering, A. D. (1978). Frequency and distribution patterns of *Saprolegnia* infection in wild and hatchery-reared brown trout *Salmo trutta* L. and char *Salvelinus alpinus* (L.). *J. Fish Dis.* **1**, 69–82.

Richter, K. and Buchner, J. (2001). HSP90: chaperoning signal transduction. *J. Cell. Physiol.* **188**, 281–290.

Ringø, E., Zhou, Z., He, S. and Erik, R. (2014). Effect of stress on intestinal microbiota of Arctic charr, Atlantic salmon, rainbow trout and Atlantic cod: a review. *Afr. J. Microbiol. Res.* **8**, 609–618.

Rotllant, J. and Tort, L. (1997). Cortisol and glucose responses after acute stress by net handling in the sparid red porgy previously subjected to crowding stress. *J. Fish Biol.* **51**, 21–28.

Roy, B. and Rai, U. (2008). Role of adrenoceptor-coupled second messenger system in sympatho-adrenomedullary modulation of splenic macrophage functions in live fish *Channa punctatus*. *Gen. Comp. Endocrinol.* **155**, 298–306.

Saeij, J. P., Verburg-van Kemenade, L. B., van Muiswinkel, W. B. and Wiegertjes, G. F. (2003). Daily handling stress reduces resistance of carp to *Trypanoplasma borreli*: in vitro modulatory effects of cortisol on leukocyte function and apoptosis. *Dev. Comp. Immunol.* **27**, 233–245.

Saha, N. R., Usami, T. and Suzuki, Y. (2002). Seasonal changes in the immunr activities of common carp (*Cyprinus carpio*). *Fish Physiol. Biochem.* **26**, 379–387.

Saha, N. R., Usami, T. and Suzuki, Y. (2004). In vitro effects of steroid hormones on IgM-secreting cells and IgM secretion in common carp (*Cyprinus carpio*). *Fish Shellfish Immunol.* **17**, 149–158.

Sakai, M., Kobayashi, M. and Kawauchi, H. (1996a). In vitro activation of fish phagocytic cells by GH, PRL and somatolactin. *J. Endocrinol.* **151**, 113–118.

Sakai, M., Kobayashi, M. and Kawauchi, H. (1996b). Mitogenic effect of growth hormone and prolactin on chum salmon *Oncorhynchus keta* leukocytes in vitro. *Vet. Immunol. Immunopathol.* **53**, 185–189.

Sakai, M., Yamaguchi, T., Wananuki, H., Yasuda, A. and Takahashi, A. (2001). Modulation of fish phagocytic cells by *N*-terminal peptides of proopiomelanocortin (NPP). *J. Exp. Zool.* **290**, 341–346.

Sakamoto, T. and McCormick, S. D. (2006). Prolactin and growth hormone in fish osmoregulation. *Gen. Comp. Endocrinol.* **147**, 24–30.

Sakamoto, T., McCormick, S. D. and Hirano, T. (1993). Osmoregulatory actions of growth hormone and its role of action in salmonids: a review. *Fish Physiol. Biochem.* **11**, 155–164.

Salinas, I., Zhang, Y. A. and Sunyer, J. O. (2011). Mucosal immunoglobulins and B cells of teleost fish. *Dev. Comp. Immunol.* **35**, 1346–1365.

Sanders, V. M., Kasprowicz, D. J., Kohm, A. P. and Swanson, M. A. (2001). Neurotransmitter receptors on lymphocytes and lymphoid cells. In *Psychoneuroimmunology*, vol. 1 (eds. R. Ader, D. L. Felten and N. Cohen), pp. 161–196. San Diego: Academic Press.

Sandra, O., Le Rouzic, P., Cauty, C., Edery, M. and Prunet, P. (2000). Expression of the prolactin receptor (tiPRL-R) gene in tilapia *Oreochromis niloticus*: tissue distribution and cellular localization in osmoregulatory organs. *J. Mol. Endocrinol.* **24**, 215–224.

Santos, C. R. A., Ingleton, P. M., Cavaco, J. E. B., Kelly, P. A., Edery, M. and Power, D. M. (2001). Cloning, characterization, and tissue distribution of prolactin receptor in the sea bream (*Sparus aurata*). *Gen. Comp. Endocrinol.* **121**, 32–47.

Sathiyaa, R., Campbell, T. and Vijayan, M. M. (2001). Cortisol modulates HSP90 mRNA expression in primary cultures of trout hepatocytes. *Comp. Biochem. Physiol.* **129B**, 679–685.

Saurabh, S. and Sahoo, P. K. (2008). Lysozyme: an important defence molecule of fish innate immune system. *Aquac. Res.* **39**, 223–229.

Scheinman, R. I., Cogwell, P. C., Lofquist, A. K. and Baldwin, A. S., Jr. (1995). Role of transcriptional activation of IkBα in mediation of immunosuppression by glucocorticoids. *Science* **270**, 283–286.

Schreck, C. B. (1996). Immunomodulation: endogenous factors. In *Fish Physiology-The Fish Immune System: Organism, Pathogen, and Environment*, Vol. 15 (eds. G. Iwama and T. Nakanishi), pp. 311–337. New York, NY: Academic Press.

Schreck, C. B. and Maule, A. G. (2001). Are the endocrine and immune systems really the same thing? In *Perspective in Comparative Endocrinology: Unity and Diversity* (eds. H. J. Th Goos, R. K. Rastogi, H. Vaudry and R. Pierantoni), pp. 351–357. Bologna: Monduzzi Editore.

Schreck, C. B. and Tort, L. (2016). The Concept of Stress in Fish. In *Fish Physiology - Biology of Stress in Fish*, Vol. 35 (eds. C. B. Schreck, L. Tort, A. P. Farrell and C. J. Brauner), San Diego, CA: Academic Press.

Shamblott, M. J. and Chen, T. T. (1993). Age-related and tissue-specific levels of five forms of insulin-like growth factor mRNA in a teleost. *Mol. Mar. Biol. Biotechnol.* **2**, 351–361.

Shepherd, B. S., Ron, B., Burch, A., Sparks, R., Richman, N. H., III, Shimoda, S. K., et al. (1997). Effects of salinity, dietary level of protein and 17α-methyltestosterone on growth hormone (GH) and prolactin (tPRL$_{177}$ and tPRL$_{188}$) levels in the tilapia, *Oreochromis mossambicus*. *Fish Physiol. Biochem.* **17**, 279–288.

Shved, N., Berishvili, G., Häusermann, E., D'Cotta, H., Baroiller, J.-F. and Eppler, E. (2009). Challenge with 17α-ethinylestradiol (EE2) during early development persistently impairs growth, differentiation, and local expression of IGF-I and IGF-II in immune organs of tilapia. *Fish Shellfish Immunol.* **26**, 524–530.

Shved, N., Berishvili, G., Mazel, P., Baroiller, J.-F. and Eppler, E. (2011). Growth hormone (GH) treatment acts on the endocrine and autocrine/paracrine GH/IGF-axis and on TNF-α expression in bony fish pituitary and immune organs. *Fish Shellfish Immunol.* **31**, 944–952.

Slicher, A. M. (1961). Endocrinological and hematological studies in *Fundulus heteroclitus* (Linn.). *Bull. Bingham Ocean Coll.* **17**, 3–55.

Small, B. C. and Bilodeau, A. L. (2005). Effects of cortisol and stress on channel catfish (*Ictalurus punctatus*) pathogen susceptibility and lysozyme activity following exposure to *Edwardsiella ictaluri*. *Gen. Comp. Endocrinol.* **142**, 256–262.

Stephanou, A. and Latchman, D. S. (1999). Transcriptional regulation of the heat shock protein genes by STAT family transcription factors. *Gene Expr.* **7**, 311–319.

Sternberg, E. M. (2006). Neural regulation of innate immunity: a coordinated nonspecific host response to pathogens. *Nat. Rev. Immunol.* **6**, 318–328.

Stolte, E. H., Verburg van Kemenade, B. M. L., Savelkoul, H. F. J. and Flik, G. (2006). Evolution of glucocorticoid receptors with different glucocorticoid sensitivity. *J. Endocrinol.* **190**, 17–28.

Stolte, E. H., de Mazon, A. F., Leon-Koosterziel, K. M., Jęsiak, M., Bury, N. R., Strum, A., et al. (2008a). Corticosteroid receptors involved in stress regulation in common carp, *Cyprinus carpio*. *J. Endocrinol.* **198**, 403–417.

Stolte, E. H., Nabuurs, S. B., Bury, N. R., Sturm, A., Flik, G., Savelkoul, H. F., et al. (2008b). Stress and innate immunity in carp: corticosteroid receptors and pro-inflammatory cytokines. *Mol. Immunol.* **46**, 70–79.

Sunyer, J. O. (2013). Fishing for mammalian paradigms in the teleost immune system. *Nat. Immunol.* **14**, 320–326.

Sunyer, J. O. and Tort, L. (1995). Natural hemolytic and bactericidal activities of sea bream *Sparus aurata* serum are effected by the alternative complement pathway. *Vet. Immunol. Immunopathol.* **45**, 333–345.

Suzuki, Y., Orito, M., Iigo, M., Kezuka, H., Kobayashi, M. and Aida, K. (1996). Seasonal changes in blood IgM levels in goldfish, with special reference to water temperature and gonadal maturation. *Fish Sci.* **62**, 754–759.

Suzuki, Y., Otaka, T., Sato, S., Hou, Y. Y. and Aida, K. (1997). Reproduction related immunoglobulin changes in rainbow trout. *Fish Physiol. Biochem.* **17**, 415–421.

Svensson, S. P., Bailey, T. J., Pepperl, D. J., Grundström, N., Ala-Uotila, S., Scheinin, M., et al. (1993). Cloning and expression of a fish β$_2$-adrenoceptor. *Br. J. Pharmacol.* **110**, 54–60.

Tagawa, M., Hagiwara, H., Takemura, A., Hirose, S. and Hirano, T. (1997). Partial cloning of the hormone-binding domain of the cortisol receptor in tilapia, *Oreochromis mossambicus*, and changes in the mRNA level during early embryonic development. *Gen. Comp. Endocrinol.* **108**.

Takahashi, A., Takasaka, T., Yasuda, A., Amemiya, Y., Sakai, M. and Kawauchi, H. (2000). Identification of carp proopiomelanocortin-related peptides and their effects on phagocytes. *Fish Shellfish Immunol.* **10**, 273–284.

Takei, Y. and Hwang, P.-P. (2016). Homeostatic Responses to Osmotic Stress. In *Fish Physiology - Biology of Stress in Fish*, Vol. 35 (eds. C. B. Schreck, L. Tort, A. P. Farrell and C. J. Brauner), San Diego, CA: Academic Press.

Takeo, J., Hara, S., Segawa, C., Toyokawa, H. and Yamashita, S. (1996). Fish glucocorticoid receptor with splicing variants in the DNA binding domain. *FEBS Lett.* **389**, 244–248.

Tatner, M. F. (1996). Natural changes in the immune system of fish. In *The Fish Immune System: Organism, Pathogen, and Environment* (eds. G. Iwama and T. Nakanishi), pp. 255–287. San Diego: Academic Press.

Teles, M., Mackenzie, S., Boltaña, S., Callol, A. and Tort, L. (2011). Gene expression and TNF-alpha secretion profile in rainbow trout macrophages following exposures to copper and bacterial lipopolysaccharide. *Fish Shellfish Immunol.* **30**, 340–346.

Teles, M., Tridico, R., Callol, A., Fierro-Castro, C. and Tort, L. (2013). Differential expression of the corticosteroid receptors GR1, GR2 and MR in rainbow trout organs with slow release cortisol implants. *Comp. Biochem. Physiol.* **164A**, 506–511.

Tort, L. (2011). Stress and immune modulation in fish. *Dev. Comp. Immunol.* **35**, 1366–1375.

Tort, L., Padros, F., Rotllant, J. and Crespo, S. (1998). Winter syndrome in the gilthead sea bream *Sparus aurata*. Immunological and histopathological features. *Fish Shellfish Immunol.* **8**, 37–47.

Tse, D. L. Y., Chow, B. K. C., Chan, C. B., Lee, L. T. O. and Cheng, C. H. K. (2000). Molecular cloning and expression of a prolactin receptor in goldfish (*Carassius auratus*). *Life Sci.* **66**, 593–605.

Tsutsui, K., Bentley, G. E., Kriegsfeld, L. J., Osugi, T., Seong, J. Y. and Vaudry, H. (2010). Discovery and evolutionary history of gonadotrophin-inhibitory hormone and kisspeptin: new key neuropeptides controlling reproduction. *J. Neuroendocrinol.* **22**, 716–727.

Uchida, K., Kaneko, T., Tagawa, M. and Hirano, T. (1998). Localization of cortisol receptor in branchial chloride cells in chum salmon fry. *Gen. Comp. Endocrinol.* **109**, 175–185.

Unniappan, S. and Peter, R. E. (2005). Structure, distribution and physiological functions of ghrelin in fish. *Comp. Biochem. Physiol.* **140A**, 396–408.

Van Der Kraak, G. (2009). The GnRH system and the neuroendocrine regulation of reproduction. In *Fish Neuroendocrinology* (eds. N. J. Bernier, G. Van Der Kraak, A. P. Farrell and C. J. Brauner), pp. 113–149. London: Academic Press.

Varsamos, S., Flik, G., Pepin, J. F., Bonga, S. E. W. and Breuil, G. (2006). Husbandry stress during early life stages affects the stress response and health status of juvenile sea bass, *Dicentrarchus labrax*. *Fish Shellfish Immunol.* **20**, 83–96.

Vazzana, M., Cammarata, M., Cooper, E. L. and Parrinello, N. (2002). Confinement stress in sea bass (*Dicentrarchus labrax*) depresses peritoneal leukocyte cytotoxicity. *Aquaculture.* **210**, 231–243.

Venters, H. K., Dantzer, R., Freund, G. G., Broussard, S. R. and Kelley, K. W. (2001). Growth hormone and insulin-like growth factor as cytokines in the immune system. In *Psychoneuroimmunology*, vol 1 (eds. R. Ader, D. L. Felten and N. Cohen), pp. 339–362. San Diego: Academic Press.

Verburg-Van Kemenade, B. M. L., Nowak, B., Engelsma, M. Y. and Weyts, F. A. A. (1999). Differential effects of cortisol on apoptosis and proliferation of carp B-lymphocytes from head kidney, spleen and blood. *Fish Shellfish Immunol.* **9**, 405–415.

Verburg-Van Kemenade, B. M. L., Stolte, E. H., Metz, J. R. and Chadzinska, M. (2009). Neuroendocrine- immune interactions in teleost fish. *Fish Neuroendocrinology* 313–364.

Volkmann, E. R. and Weekes, N. Y. (2006). Basal SIgA and cortisol levels predict stress-related health outcomes. *Stress Health* **22**, 11–23.

Wang, W.-B., Wang, Y.-P., Hu, W., Li, A.-H., Cai, T.-Z., Zhu, Z.-Y., et al. (2006). Effects of the "all-fish" growth hormone transgene expression on non-specific immune functions of common carp, *Cyprinus carpio* L. *Aquaculture* **259**, 81–87.

Watanuki, H., Gushiken, Y., Takahashi, A., Yasuda, A. and Sakai, M. (2000). In vitro modulation of fish phagocytic cells by β-endorphin. *Fish Shellfish Immunol.* **10**, 203–212.

Watanuki, N., Takahashi, A., Yasuda, A. and Sakai, M. (1999). Kidney leucocytes of rainbow trout, *Oncorhynchus mykiss*, are activated by intraperitoneal injection of β-endorphin. *Vet. Immunol. Immunopathol.* **71**, 89–97.

Weigent, D. A. and Arnold, R. E. (2005). Expression of insulin-like growth factor-1 and insulin-like growth factor-1 receptors in EL4 lymphoma cells overexpressing growth hormone. *Cell Immunol.* **234**, 54–66.

Wendelaar Bonga, S. E. (1997). The stress response in fish. *Physiol. Rev.* **77**, 591–625.

Weyts, F. A. A., Flik, G. and Verburg-Van Kemenade, B. M. L. (1998a). Cortisol inhibits apoptosis in carp neutrophilic granulocytes. *Dev. Comp. Immunol.* **22**, 563–572.

Weyts, F. A. A., Flik, G., Rombout, J. H. W. M. and Verburg-Van Kemenade, B. M. L. (1998b). Cortisol induces apoptosis in activated B cells, not in other lymphoid cells of common carp, *Cyprinus carpio*. *Dev. Comp. Immunol.* **22**, 551–562.

Weyts, F. A. A., Verburg-Van Kemenade, B. M. L. and Flik, G. (1998c). Characterization of glucocorticoid receptors in peripheral blood leukocytes of carp, *Cyprinus carpio* L. *Gen. Comp. Endocrinol.* **111**, 1–8.

Weyts, F. A. A., Cohen, N., Flik, G. and Verburg-Van Kemenade, B. M. L. (1999). Interactions between the immune system and the hypothalamo-pituitary-interrenal axis in fish. *Fish Shellfish Immunol.* **9**, 1–20.

White, A. and Fletcher, T. C. (1985). The influence of hormones and inflammatory agents on C-reactive protein, cortisol and alanine aminotransferase in the place (*Pleuronectes platessa* L.). *Comp. Biochem. Physiol.* **80C**, 99–104.

Wojtaszek, J., Dziewulska-Szwajkowska, D., Lozińska-Gabska, M., Adamowicz, A. and Dzugaj, A. (2002). Hematological effects of high dose of cortisol on the carp (*Cyprinus carpio* L.): cortisol effect on the carp blood. *Gen. Comp. Endocrinol.* **125**, 176–183.

Yada, T. (2007). Growth hormone and fish immune system. *Gen. Comp. Endocrinol.* **152**, 353–358.

Yada, T. (2009). Effects of insulin-like growth factor-I on non-specific immune functions in rainbow trout. *Zool. Sci.* **26**, 338–343.

Yada, T. (2012). Effect of gonadotropin-releasing hormone on phagocytic leucocytes of rainbow trout. *Comp. Biochem. Physiol.* **155C**, 375–380.

Yada, T. and Nakanishi, T. (2002). Interaction between endocrine and immune systems in fish. *Int. Rev. Cytol.* **220**, 35–92.

Yada, T., Hirano, T. and Grau, E. G. (1994). Changes in plasma levels of the two prolactins and growth hormone during adaptation to different salinities in the euryhaline tilapia, *Oreochromis mossambicus. Gen. Comp. Endocrinol.* **93**, 214–223.

Yada, T., Azuma, T. and Takagi, Y. (2001). Stimulation of non-specific immune functions in seawater-adapted rainbow trout, *Oncorhynchus mykiss*, with reference to the role of growth hormone. *Comp. Biochem. Physiol.* **129B**, 695–701.

Yada, T., Uchida, K., Kajimura, S., Azuma, T., Hirano, T. and Grau, E. G. (2002). Immunomodulatory effects of prolactin and growth hormone in the tilapia, *Oreochromis mossambicus. J. Endocrinol.* **173**, 483–492.

Yada, T., Misumi, I., Muto, K., Azuma, T. and Schreck, C. B. (2004). Effects of prolactin and growth hormone on proliferation and survival of cultured trout leucocytes. *Gen. Comp. Endocrinol.* **136**, 298–306.

Yada, T., Muto, K., Azuma, T., Hyodo, S. and Schreck, C. B. (2005). Cortisol stimulates growth hormone gene expression in rainbow trout leucocytes in vitro. *Gen. Comp. Endocrinol.* **142**, 248–255.

Yada, T., Muto, K., Azuma, T., Fukamachi, S., Kaneko, T. and Hirano, T. (2006). Effects of acid water exposure on plasma cortisol, ion balance, and immune functions in the "cobalt" variant of rainbow trout. *Zool. Sci.* **23**, 707–713.

Yada, T., Azuma, T., Hyodo, S., Hirano, T., Grau, E. G. and Schreck, C. B. (2007). Differential expression of corticosteroid receptor genes in trout immune system in response to acute stress. *Can. J. Fish. Aquat. Sci.* **64**, 1382–1389.

Yada, T., Hyodo, S. and Schreck, C. B. (2008). Effects of seawater acclimation on mRNA levels of corticosteroid receptor genes in osmoregulatory and immune systems in trout. *Gen. Comp. Endocrinol.* **156**, 622–627.

Yada, T., McCormick, S. D. and Hyodo, S. (2012). Effects of environmental salinity, biopsy, and GH and IGF-I administration on the expression of immune and osmoregulatory genes in the gills of Atlantic salmon (*Salmo salar*). *Aquaculture* **362–363**, 177–183.

Yada, T., Miyamoto, K., Miura, G. and Munakata, A. (2014). Seasonal changes in gene expression of corticoid receptors in anadromous and non-anadromous strains of rainbow trout *Oncorhynchus mykiss*. *J. Fish Biol.* **85**, 1263–1278.

Yano, T. (1996). The nonspecific immune system: humoral defense. In *The Fish Immune System: Organism, Pathogen, and Environment* (eds. G. Iwama and T. Nakanishi), pp. 105–157. San Diego: Academic Press.

Yasuoka, A., Abe, K., Arai, S. and Emori, Y. (1996). Molecular cloning and functional expression of the 1α-adrenoceptor of medaka fish, *Oryzias latipes*. *Eur. J. Biochem.* **235**, 501–507.

Zhang, T., Qiu, L., Sun, Z., Wang, L., Zhou, Z., Liu, R., et al. (2014). The specifically enhanced cellular immune responses in Pacific oyster (*Crassostrea gigas*) against secondary challenge with *Vibrio splendidus*. *Dev. Comp. Immunol.* **45**, 141–150.

Zhang, Y. A., Salinas, I., Li, J., Parra, D., Bjork, S., Xu, Z., et al. (2010). IgT, a primitive immunoglobulin class specialized in mucosal immunity. *Nat. Immunol.* **11**, 827–835.

Zhu, L. Y., Lin, A. F., Shao, T., Nie, L., Dong, W. R., Xiang, L. X., et al. (2014). B cells in teleost fish act as pivotal initiating APCs in priming adaptive immunity: an evolutionary perspective on the origin of the B-1 cell subset and B7 molecules. *J. Immunol.* **192**, 2699–2714.

11

STRESS INDICATORS IN FISH

NATALIE M. SOPINKA
MICHAEL R. DONALDSON
CONSTANCE M. O'CONNOR
CORY D. SUSKI
STEVEN J. COOKE

A fish is chased with a net in an aquarium before being captured, scooped out of the water, and placed in a nearby testing arena. Is it stressed? How can we tell? Are our indicators reliable? Quantification of stress in fish has evolved from the initial development of radioimmunoassays to measure cortisol in plasma to the rapidly expanding suite of genome-based assays. Indicators range from the intracellular to whole-animal level. Expression of heat shock proteins (HSPs) and activity of metabolic enzymes can be paired with straightforward observations of reflexes and survival. Both traditional and emerging indicators have advantages and disadvantages, and their use is

Biology of Stress in Fish: Volume 35
FISH PHYSIOLOGY

tissue- and context-specific. Ecological, biological, and methodological factors must be considered when selecting, measuring, and interpreting stress indicators. Inter- and intraspecific, sex, life stage, and temporal differences in physiological responses to stressors can confound confirmation of a stressed state. Despite numerous types of indicators, our understanding of how absolute levels of indicators relate to stressor severity and recovery to date remains limited. How accurately indicators characterize stress in wild populations naturally exposed to stressors is still an evolving discussion. The integration of research disciplines and involvement of stakeholders and user groups will aid in filling these knowledge gaps, as well as the translation of individual-level indicators to population- and ecosystem-level processes.

1. WHY DO WE MEASURE STRESS?

As explained by Schreck and Tort (2016; Chapter 1 in this volume), stress is an inherent component of the life of all vertebrates, including fishes. Measures of stress inform us how effectively a fish resists death and resets homeostatic norms when faced with noxious stimuli. This information is then translated into evolutionary and ecological theory to understand how animals are adapted to, or able to adapt to, future stressors. From a fundamental perspective, measuring stress contributes to our knowledge of carryover effects (eg, O'Connor et al., 2010), parental effects (eg, Sopinka et al., 2014), personality (eg, Aubin-Horth et al., 2012), and life history variation (eg, Pottinger and Carrick, 2001; Ricklefs and Wikelski, 2002). Without classifying, quantifying, refining, and interpreting indicators of stress, the significance and implications of a fish's response to external and internal environmental change would be unclear (Schulte, 2014). For example, without a clear stress indicator, can an animal's response be defined as stress? Can undisturbed animals be characterized as adequate experimental controls? Further, could an environmental change be classified as a stressor? Stress transcends levels of biological organization; measuring stress also serves to link organismal responses to population-level processes (Calow and Forbes, 1998; Fefferman and Romero, 2013) and ecosystem health (Dale and Beyeler, 2001).

From an applied perspective, measuring stress is necessary to determine how the health, performance, and welfare of fishes are being influenced by interactions with humans. For example, assessing stress in captive populations of fish (eg, hatcheries, farms, aquariums) is often done with a goal of reducing stress to maximize growth and survival. In fact, the empirical study of stress in fishes is rooted, and continues to be prolific, in aquaculture studies that examine how handling, rearing, and transport (Barton et al., 1980; Portz et al., 2006), as well as anesthesia (Iwama et al., 1989; Trushenski et al., 2012), affect captive broodstock health and production efficiency. Foundational mechanistic work quantifying stress was, and continues to be, performed using domesticated species (eg, rainbow trout, *Oncorhynchus mykiss*; tilapia, *Oreochromis* spp.). Stress indicators are now used as objective indices of the welfare status of fish (Iwama, 2007), and to understand the impacts of recreational (eg, Morrissey et al., 2005; Landsman et al., 2011) and commercial fisheries (Marçalo et al., 2006; Raby et al., 2015), as well as laboratory animal husbandry (Brydges et al., 2009; Eaton et al., 2015) on fish performance. There is also increasing interest in optimizing the use of stress indicators to quantify the condition of wild populations (Madliger and Love, 2014). Combining basic and applied motivations to measure stress is especially valuable. Linking evolutionary and ecological underpinnings of stress with measures of stress relevant to industry and conservation practitioners can guide management strategies that effectively take into account fish biology, and facets of human livelihood and culture.

Here, we provide an overview of stress indicators ranging from the cellular to whole-animal level (Section 3). We then outline important considerations when measuring and interpreting these indicators (Section 4), discuss the extension of individual-level indicators to population- and ecosystem-level processes (Section 5), and conclude with avenues of research and novel indicators that warrant further investigation (Section 6). We acknowledge the pioneering syntheses on this topic in S. Marshall Adams' book, *Biological Indicators of Stress in Aquatic Ecosystems* (Adams, 2002), and especially the chapter on physiological and condition-related indicators of stress in fish by Barton et al. (2002). We encourage readers to consult these works for additional insight on stress indicators in fishes.

2. QUANTIFYING STRESS

Approaches to quantifying stress in fish are varied depending on the targeted indicators (Figs. 1 and 2, Section 3). A main distinction in quantification is whether the stress response itself is being measured (ie, the primary response/HPI axis activity) (Gorissen and Flik, 2016; Chapter 3 in this volume), or other physiological, behavioral, or life-history traits (ie, secondary and tertiary measures; Mazeaud et al., 1977; see Section 3) predicted to change in concert with, or as a result of, stimulation of the HPI axis. Assessing the magnitude of the response that an individual mounts when exposed to an acute or chronic stressor typically utilizes repeated sampling over a period of time. Establishing as accurately as possible baseline/resting levels of the desired indicators is necessary and serves as the first time-point in a series of measurements. Changes in key primary parameters (eg, catecholamines, cortisol) over time are measured to characterize the stress response (see Section 3.2). When chronic stress is quantified, resting levels of primary indicators are measured upon cessation of the stressor exposure (vs levels induced by an acute stressor following the chronic stressor exposure). These established methods for quantifying the stress response, though largely replicated and consistent, represent only one piece of the whole-animal response to a stressor (see Section 3.3).

Experimental assessment of the stress response is not without its challenges. Inability to repeatedly sample individuals due to body size constraints (ie, fish that are extremely large or extremely small), unintentional omission of quantifying peak levels of the indicator based on predetermined sampling intervals, and capture and handling during tissue sampling initiating a secondary mounting of the stress response (Baker and Vynne, 2014), can all limit statistical power and comprise interpretation of data. Habituation during chronic stressor exposures can also skew interpretation (see Section 4.4). Ultimately without measurements of fitness, or proxies thereof, such mechanistic studies are limited in their extension to population-level trends (Calow and Forbes, 1998; see Section 5). Mechanistic studies are, however, essential in building fundamental knowledge of stress, motivating and guiding design of research which focuses on physiological and behavioral processes interconnected with stress, and streamlining future validation of stress (ie, targeting a single poststressor time-point to confirm mounting of the stress response in study animals).

Based on established relationships between stressor exposure, HPI activity, and a range of whole-animal responses (reviewed in Wendelaar-Bonga, 1997; Iwama et al., 1997; Mommsen et al., 1999; Schreck, 2010; Barton, 2002), quantification of stress can refer to quantification of primary, secondary, and tertiary responses that occur following HPI axis activation (eg, gene expression, immune function, metabolism, growth, reproduction, performance, behavior; see Section 3). Quantification of responses typically requires measurement of pre- and poststressor levels of the indicator. Often, the latter response is measured once following stressor exposure. Differences observed pre- and poststressor are treated as an indication of stress. This sampling regime provides a snapshot of how the trait has changed in the stressor-exposed animal but does not guarantee that maximum response or response recovery is captured. Generally, when responses are quantified, a primary indicator directly related to HPI axis function (eg, plasma cortisol) is also measured prestressor exposure, and at a single poststressor time point. The time-points for primary, secondary, and tertiary stress indicators ideally should be based on preliminary time course sampling using the study species and specific stressor (eg, Pickering et al., 1982; Donaldson et al., 2014). A number of confounds can arise when the time-point at which the indicator will be sampled is chosen arbitrarily. For example, two fish can have the same elevated stress indicator at the sampling time-point, but one individual on a trajectory to death and the other individual on a trajectory to recovery. Further, depending on whether stressors are continuous or sequential (see Section 4.4) it is possible that indicator levels will oscillate throughout an exposure (Schreck and Tort, 2016; Chapter 1 in this volume, Figure 1.5). Thus, making definitive mechanistic connections between the stressor and any secondary/tertiary measures is not always straightforward. Overall, however, our understanding of how the HPI axis and organismic performance varies throughout a stressful event (including recovery) is limited but is explored conceptually by Schreck and Tort (2016; Chapter 1 in this volume, Figures 1.5 and 1.6).

Studies linking the quantification of the stress response with fitness traits have the potential to provide insight into the physiological basis of life history. For example, relationships between stressor-induced plasma glucocorticoids and fitness (reviewed in Breuner et al., 2008) have been detected in both birds (eg, MacDougall-Shackleton et al., 2009) and reptiles (eg, Romero and Wikelski, 2001), with data emerging now in fish (Cook et al., 2014). This integrative approach of quantifying the stress response itself (or a component thereof) in combination with downstream changes in fitness facilitates collaboration between physiologists and ecologists, and advances interdisciplinary fields such as ecophysiology or conservation physiology (Wikelski and Cooke, 2006; Cooke et al., 2013; Boonstra, 2013a).

3. SPECIFIC MEASURES OF FISH STRESS

It is important to note that although quantification of stress is extensively studied and numerous indicators exist that identify a stressed fish (Tables 11.1–11.3), our grasp of what absolute levels of stress indicators mean is rudimentary, especially in nonexperimental contexts (eg, How do you know a fish in the wild is stressed or not?). Elevated levels of an indicator can signal a stressed fish, but lower levels of an indicator do not

Table 11.1
Cellular and molecular stress indicators

Indicator	Sampling and analytical considerations	Strengths of indicator	Weaknesses of indicator
Oxidative stress • Metabolic pathways produce reactive oxygen species (ROS) as a natural by-product (Costantini, 2008) • ROS are damaging to biological molecules, especially lipids, proteins, RNA, and DNA • Antioxidants prevent the damage caused by ROS either by preventing the formation of ROS, or by removing ROS • Oxidative stress occurs when ROS production overwhelms the counterbalancing capacities of antioxidants, and damage occurs to biological molecules (Lesser, 2006)	• Since oxidative stress can occur from either the overproduction of ROS, or from insufficient antioxidants, oxidative stress can be quantified by (1) measuring ROS; (2) measuring antioxidant levels; or (3) measuring damage to biomolecules • There are multiple markers available for measuring ROS, antioxidants, and biomolecule damage (Lesser, 2006) • ROS tend to be unstable, and so measurement of antioxidants or biomolecule damage are more common • Markers are measured through a variety of colorimetric assays depending on what is being measured, and different markers can be measured more easily in plasma, serum, urine, tissue homogenates, or cell cultures (Valavandis et al., 2006)	• Oxidative stress is an inevitable by-product of metabolism, and can therefore be taken as a cost of life • Oxidative stress has been measured in an ecological context as a cost of reproduction (eg, Alonso-Alverz et al., 2004), a cost of immune responses (eg, Torres and Velando, 2007), or a cost of strenuous energy expenditure such as migration (eg, Rankin and Burchsted, 1992) • Oxidative stress is also the result of exposure to challenging environments, such as areas that are heavily contaminated (eg, Bacanskas et al., 2004)	• Since oxidative stress arises from complex processes, results can be difficult to interpret • Measuring oxidative stress typically requires specialized equipment and can be relatively expensive, although commercial kits are becoming more widely available

Telomere length

- Environmental stressors can cause oxidative stress, which if not counteracted can cause telomere shortening, accelerating cellular (and possibly organismal) senescence (Monaghan et al., 2009)
- Environmental stressors such as psychological stress (Epel et al., 2004) or elevated reproductive effort (Kotrschal et al., 2007) have been linked to telomere shortening

- To determine relative telomere length, quantitative PCR can be used to measure the factor by which the DNA sample differed from a reference DNA sample in its ratio of telomere repeat copy number to single copy gene copy number (Cawthon, 2002)

- Potentially powerful indicator to bridge the gap between environmental stressors, oxidative stress, and organismal senescence

- The links between environmental stressors, cellular stress, telomere shortening, and organismal senescence are still largely untested (Monaghan and Haussmann, 2006). More research is needed to understand the tertiary outcomes associated with telomere shortening in order to effectively use telomere length as an indicator of stress

Heat shock proteins (HSPs)

- HSPs, under the control of Heat Shock Factor 1 (HSF1), indicate a cellular stress response and HSP expression increases to maintain cellular homeostasis (Iwama et al., 2004)
- Most HSPs are molecular chaperones which fold, repair, and catabolize proteins (Moseley, 1997)

- HSP expression can be determined using quantitative real-time PCR (qRT-PCR), requiring isolation of genomic DNA, total RNA extraction, and reverse transcriptase PCR amplification (eg, Fangue et al., 2006)
- Hsp70 has previously been quantified by ELISA, BIAcore, and bead-based assays for use by FACS; BIAcore and FACS are more sensitive and require less sample than ELISA
- HSP expression levels are context-dependent, meaning that the establishment of baseline expression levels is required (Tomanek and Somero, 1999)

- HSPs are sensitive to a range of stressors (eg, rapid temperature changes, salinity challenges, handling; Palmisano et al., 2000; Donaldson et al., 2008)
- Widely studied, well-understood function

- The expression of HSPs is context-dependent since they are sensitive to the magnitude and duration of the stressor (Iwama et al., 2004) as well as acclimation to previous stressors (Somero and Hoffman, 1996)

(Continued)

Table 11.1 (Continued)

Indicator	Sampling and analytical considerations	Strengths of indicator	Weaknesses of indicator
Immediate early genes and transcription factors • Immediate early genes (IEGs) are induced within minutes of the cellular stress response (Hughes and Dragunow, 1995) • Commonly measured IEGs include transcription factors c-fos, fosB, c-jun, JUNB, c-myc, egr-1 (Inuzuka et al., 1999) • Transcription factors, such as hypoxia-inducible factor 1 (HIF-1a) and NUPR1 are commonly activated during stress (Semenza, 1998; Momoda et al., 2007)	• IEG expression can be determined using qRT-PCR or microarray approaches • Can be measured in multiple tissues (eg, heart, liver, gill) • Nonlethal gill biopsies can be used, which facilitate integrative studies (eg, physiological telemetry; Miller et al., 2011)	• Sensitive indicators of stress and recovery (Momoda et al. 2007; Donaldson et al., 2014)	• Most studies have focused on mammals but interpreting IEG activation across a broader range of animal taxa would help identify upstream regulators of transcriptional stress responses (Kassahn et al., 2009) • Studies conducted on fish to date often focus on different timecourses (Krasnov et al., 2005), species, tissues (Kassahn et al., 2009), techniques (Prunet et al., 2008), and genes of interest • Future studies required to understand functional roles and downstream effects across species
Intracellular enzymes • Intracellular enzymes such as alanine transaminase (ALT), aspartate transaminase (AST), lactate dehydrogenase (LDH), or creatine kinase (CK), are released by cell damage or death • The presence of these enzymes in the plasma is therefore a useful indicator that tissue damage has occurred (Henry, 1996) • Many enzymes are tissue specific, and can therefore provide information about the type of tissue damage that has occurred (Wagner and Congleton, 2004)	• Indicators of tissue damage can be measured through colorimetric assays in plasma • Commercial labs also routinely measure enzymes in plasma using autoanalyzers for a fee	• The presence of intracellular enzymes in the plasma is a good indicator of injury • The availability of commercial labs measuring indicators of tissue damage relatively inexpensive and easy to measure in plasma without investing in specialized equipment	• These indicators are often not elevated unless physical injury has occurred, and so are less useful as an indicator of stressors that do not include tissue damage (Wagner and Congleton, 2004)

Table 11.2

Primary and secondary physiological stress indicators

Indicator	Sampling and analytical considerations	Strengths of indicator	Weaknesses of indicator
Catecholamines			
• When an individual is faced with a challenge, the physiological response is first an immediate release of the catecholamine hormones epinephrine and norepinephrine from the chromaffin cells (Reid et al., 1998; Gallo and Civinini, 2003) • The release of epinephrine and norepinephrine is associated with the classic fight-or-flight stress response	• Catecholamines are typically measured in plasma using chromatography with electrochemical detection (Woodward, 1982)	• Catecholamines are responsive to a variety of stressors (Reid et al., 1998; Pottinger, 2008), and the measurement of catecholamines therefore provides very accurate information about the response to acute stressors at fine timescales	• Catecholamines respond extremely rapidly (ie, within seconds) to capture and handling, and so it is difficult to quantify catecholamine levels without specialized equipment and animals that are held in the laboratory
Steroid hormones			
• Following the release of catecholamines, the stress response is characterized by activation of the hypothalamic–pituitary–interrenal (HPI) axis. HPI axis activation involves a complex set of interactions that culminate in the production and release of glucocorticoid steroid hormones (Mommsen et al., 1999; Barton, 2002; Pottinger, 2008)	• Steroid hormones are typically measured by either radio-immunoassay or enzyme-linked immunoassay in plasma or tissue homogenates (Pottinger, 2008; Sheriff et al., 2011) • Glucocorticoids can also be measured in urine and feces, and so can be extracted and measured in water samples (Ellis et al., 2004; Pottinger, 2008)	• Poststress glucocorticoid levels can provide information about how individuals are affected by specific stimuli (eg, capture and handling stress, different holding conditions, acute exposures; Sapolsky et al., 2000; Barton, 2002) • Baseline glucocorticoid levels can provide information about whether animals are experiencing chronic	• Circulating glucocorticoids respond rapidly (ie, often within 3–5 min) to capture and handling (Romero and Reed, 2005), and so it is often difficult to obtain baseline levels in wild animals • The relationships between both baseline (Bonier et al., 2009) and stress-induced (Breuner et al., 2008) glucocorticoid levels and future performance and survival

(*Continued*)

Table 11.2 (Continued)

Indicator	Sampling and analytical considerations	Strengths of indicator	Weaknesses of indicator
• Measurements of circulating glucocorticoids are therefore an indicator of whether an individual is experiencing a stressor • Corticotropin releasing hormone (CRH) and adrenocorticotropic hormone (ACTH), the intermediate hormones involved in HPI axis activation, are also commonly measured	• Since glucocorticoids change rapidly to challenges, glucocorticoid measurements are often taken both before and after exposure to stressors to obtain a measure of stress responsiveness; both baseline and poststress levels, or stress responsiveness, provide information about the state of an individual (Breuner et al., 2008; Bonier et al., 2009)	environmental stressors, and in some cases can be predictive of future performance and survival (Bonier et al., 2009) • Stress responsiveness in some cases can be predictive of future performance and survival (Breuner et al., 2008) • Circulating glucocorticoid levels can also be linked to life history traits and trade-offs (Wingfield et al., 1998; Ricklefs and Wikelski, 2002)	are context- and species-specific, and results can be difficult to interpret
Metabolites • Once glucocorticoids are produced and circulating, they are associated with a suite of secondary responses that help the animal survive and recover from the challenge (Sapolsky et al., 2000), including the mobilization of stored glucose (Barton, 2002)	• Glucose and lactate can be measured through colorimetric assays in plasma or tissue homogenates • Lactate and glucose can also can be measured in whole blood using portable meters designed for diabetic patients or for athletic training (Wells and	• Metabolites are very useful for assessing the acute response to specific stressors (Barton et al., 2002), particularly exercise stressors in the case of lactate (Wood et al., 1983) • The readily available portable meters make both lactate and glucose inexpensive and easy to measure, using very	• Since both glucose and lactate are affected by general metabolic processes outside of stress responses, baseline results can be difficult to interpret, and these indicators are most useful as measures of acute responses to specific stressors

• Anaerobic stressors (eg, exhaustive exercise) generate anaerobic metabolites, such as lactate (Wood et al., 1983)	Pankhurst, 1999; Beecham et al., 2006; Stoot et al., 2014)	small blood samples with no specialized equipment, which makes them good parameters to measure under field conditions when assessing acute stressors (Wells and Pankhurst, 1999; Beecham et al., 2006)
Osmolality and ion concentrations • When fish experience an acute stressor, the rise in adrenaline causes vasoconstriction and increased cardiac output (Mazeaud and Mazeaud, 1981), which in turn increases gill diffusing capacity due to increased perfusion of the lamallae (Randall and Perry, 1992) • This increased capacity for diffusion causes an increase in ion transfer at the gills, and subsequent changes in plasma osmolality, particularly in circulating concentrations of Na^+ and Cl^- (see McDonald and Milligan, 1997, for review)	• Plasma osmolality is measured using an osmometer • Ions can be measured in plasma using spectrophotometry • There are commercially available meters that will measure some of the common ions in plasma (eg, Na^+) • Commercial labs also routinely measure ions in plasma using autoanalyzers for a fee	• Changes in overall osmolality or ion balance are good indicators of acute stress • The availability of commercial labs makes measuring ions relatively inexpensive and easy to measure in plasma without investing in specialized equipment	• These indicators are useful indicators of acute stress, but are often difficult to interpret in the context of chronic exposures because they are context-specific and influenced by multiple internal and external factors (McDonald and Milligan, 1997)

(Continued)

Table 11.2 (Continued)

Indicator	Sampling and analytical considerations	Strengths of indicator	Weaknesses of indicator
Nutritional indicators • Nutritional indicators in the plasma (eg, total protein, total cholesterol, triglycerides) provide information about current levels of mobilized energy stores that are available to fuel activities (Wagner and Congleton, 2004; Congleton and Wagner, 2006)	• Nutritional indicators in the plasma are typically measured through colorimetric assays • Commercial labs also routinely measure nutritional indicators in plasma using autoanalyzers for a fee	• Nutritional indicators in the plasma can provide information about the recent feeding history of fishes (Congleton and Wagner, 2006) • The availability of commercial labs makes measuring nutritional indicators relatively inexpensive and easy to measure in plasma without investing in specialized equipment	• Nutritional indicators do not show consistent responses to fasting and stressors, and results are species- and context-specific. Nutritional indicators can in some cases be difficult to interpret as indicators of general health (Wagner and Congleton, 2004; Congleton and Wagner, 2006; O'Connor et al., 2011)
Bioenergetics • Energy is the currency of life, so understanding its allocation to bodily processes can serve as a sensitive indicator to organismal stress (Beyers et al., 2002) • Energy stores and lipid content are linked to survival, reproduction, and life history strategies (Henderson and Tocher, 1987; Adams, 1999)	• Glycogen, a long-term energy reserve, is typically measured in tissue homogenates using hydrolysis and enzymatic assays • Phosphocreatine (PCr) and adenosine triphosphate (ATP) from tissue homogenates are typically measured through colorimetric assays • Proximate body composition (ie, proportion of the body that is lipid, protein, water, organic ash) and analysis can be performed to determine how energy is allocated among various compartments	• The availability of commercial labs makes measuring nutritional indicators relatively inexpensive and easy to measure in plasma without investing in specialized equipment • Provides long-term indication of organismal status • Information can be incorporated into bioenergetics models (particularly when combined with information on metabolic rates) to put in a broader context (Beyers et al., 2002)	• Since some indicators of energetic stores change very rapidly in response to acute exercise stress (eg, PCr or ATP); it is important to know the recent history of the animal in order to interpret results • Some measures such as proximate body composition analysis require relatively large quantities of tissue so tend to be lethal • Energy varies inherently among fish of different sex and size so need to control for these factors

- Ω bomb calorimetry provides information on overall tissue energy density
- Can also assess lipid constituents (eg, cholesterol, fatty acids, triglycerides), although these are rarely used as stress indicators
- Commercial kits are available for measuring energetic stores in tissue homogenates

- Can use nonlethal electronic devices (eg, handheld microwave energy meter, Crossin and Hinch, 2005; bioelectrical impedance analysis, Kushner, 1992)

- Nonlethal sampling tools require calibration
- Generally unresponsive to acute stressors (Schreck, 2000)

Leukocytes

- Leukocytes, or white blood cells (WBCs), are a collection of cells in the blood that serve an important role in immune defense and inflammation
- There are 5 types of WBCs in most vertebrates: basophils, eosinophils, lymphocytes, monocytes, and neutrophils (heterophils in birds and reptiles)
- The relative proportions of each WBC type are influenced by stressors (Dhabhar et al., 1996), and thus provide a useful measure of animal health and exposure to stress

- Leukocyte profiles are typically obtained by light microscope examination of 100 leukocytes in a stained blood smear (Davis et al., 2008)

- Leukocyte profiles are predictive of future performance and viability, such as susceptibility to infection (Al-Murrani et al., 2002), growth rates (Moreno et al., 2002), and survival (Lobato et al., 2005; Kilgas et al., 2006)
- Blood smears are relatively inexpensive and easy to obtain from captured wild animals
- Leukocytes respond relatively slowly to capture and handling (ie, within hours or days; Davis et al., 2008), and so leukocyte profiles provide a convenient measure of baseline stress levels in wild animals

- Leukocyte profiles are influenced by disease and infection, as well as stress, and so it can be difficult to interpret changes in leukocyte profiles (Davis et al., 2008)

(Continued)

Table 11.2 (Continued)

Indicator	Sampling and analytical considerations	Strengths of indicator	Weaknesses of indicator
• The most common measure is the ratio of neutrophils or heterophils to lymphocytes (N:L or H:L ratio) (Davis et al., 2008)		• The leukocyte response is conserved across taxonomic groups, and so results obtained from one taxa should be widely applicable (Davis et al., 2008)	
Hematocrit • Erythrocytes, or red blood cells (RBCs), are the oxygen-transporting cells within the blood • Hematocrit is the volume percentage of RBCs in the blood	• Hematocrit is most commonly measured by packed cell volume (PCV), which is obtained by centrifuging a whole blood sample within a capillary tube, which separates the blood into layers • The volume of packed red blood cells divided by the total volume of the blood sample gives the PCV	• Hematocrit typically increases with exposure to stressors, is relatively inexpensive and simple to measure, and requires no specialized assays	• Hematocrit can in some cases increase or decrease in response to challenges, depending on the specific challenge, and results can be difficult to interpret

Table 11.3

Whole-organism stress indicators

Indicator	Sampling and analytical considerations	Strengths of indicator	Weaknesses of indicator
Reflexes • Simple reflex indicators such as ability to flip upright • Becoming increasing popular to characterize neurological responses of fish to external stimuli or functions of the autonomic nervous system (Davis, 2010)	• Reflexes can be assessed individually (as present or absent) or as a composite to derive a score (Davis, 2010) • Need to validate reflexes for each species but some common ones relevant to most species are righting-reflex (ie, roll fish on back and see if it regains upright orientation after 3 s) and tail-grab reflex (ie, grab the tail and see if fish attempts to burst away) • Does not require any specialized equipment and provides an immediate (<20 s) measure of fish vitality	• Predictive of mortality in a number of fish species in the lab (Davis, 2010) and field (Raby et al., 2012) • Rapid, simple, and inexpensive to evaluate reflexes without observer bias • Can train stakeholders (eg, anglers, fisheries observers), as it does not require any equipment or scientific skills • Developed given inability of traditional physiological measures to predict mortality (Davis et al., 2001) • Not dependent on fish size, motivation states, or acclimation (Davis, 2010)	• Need to validate reflexes for all species as not all fish have the same reflexes • Relatively new approach so relatively few published examples • Exact mechanisms by which reflexes predict mortality are unclear • Improper reflex choice or interpretation can be ambiguous

(Continued)

Table 11.3 (Continued)

Indicator	Sampling and analytical considerations	Strengths of indicator	Weaknesses of indicator
Behavior • Locomotion, feeding, social interactions, predator-prey dynamics, habitat selection and other aspects of behavior • Behavior is an ecologically relevant indicator that requires integration of various physiological systems (Schreck et al., 1997)	• Can be measured in the lab and field • Growing number of tools for remotely studying the behavior of free-swimming fish using biotelemetry and biologgers (eg, electromyogram [EMG] telemetry; Cooke et al., 2004) • Action cameras provide opportunity to study fish behavior in water (Struthers et al., 2015)	• Ability to integrate with other measures including those more directly linked to organismal physiology (Scott and Sloman, 2004) given that physiology and behavior are inherently linked (Cooke et al., 2014) • Ecologically relevant given that many behaviors related directly to food acquisition and predator avoidance (Schreck et al., 1997) • Many behavioral endpoints are inexpensive (ie, no assay costs) but require specialized equipment that can be expensive or require technical expertise	• Difficult to identify specific mechanisms as behavior depends on capacity, motivation, sensory acuity, and responsiveness (Schreck et al., 1997) • Necessary to control for experimental artifacts and observer influence • Can be subjective (rather than objective) if endpoints are not clear
Swimming performance • Examples include speed, intensity, duration of swimming • Swimming requires the integration of numerous biological systems and is thus regarded as a sensitive integrator of stress and whole-organism status (Hammer, 1995)	• Most often quantified in laboratory environments with the use of swim tunnel/flume/annular respirometer (Ellerby and Herskin, 2013) or drag-strip (Nelson et al., 2002), although some swim tunnels are mobile	• Can be combined with other indicators such as metabolic rate (oxygen consumption) if swimming conducted in a respirometer (Farrell et al., 2003) • Swimming performance considered to have strong	• To some extent swimming performance is a reflection of fish motivation (eg, maturation, appropriate environmental cues), which can be independent of stressed state

	ecological relevance (Plaut, 2001) • and can be used in field settings (Farrell et al., 2003) • Multiple forms of swimming (eg, burst, critical swimming speed, endurance) can be measured (see Beamish, 1978) • Repeat swimming performance approaches account for interindividual variation and are useful for evaluating performance impairments from stressors (Jain et al., 1998)	• Initial costs associated with purchasing equipment can be relatively high • Swimming performance can be influenced by energetic state (fed vs fasted) (Gingerich et al., 2010)
Metabolic rate • Rate of oxygen consumption as food converted to energy • Indicates the minimum metabolic rate required to maintain life • There is likely a context-dependent link between metabolic rate with fitness, growth, or survival (Burton et al., 2011)	• Samples typically collected on resting, postabsorbtive animals, free of stimuli, in isolation, in a laboratory setting (Nelson and Chabot, 2011) • Can define standard metabolic rate, maximum metabolic rate, or aerobic scope (difference between maximum and standard metabolic rate) • Elevation in metabolic rate can be interpreted as stressed state relative to controls (Barton and Schreck, 1987) • Can incorporate data into bioenergetics models to make inferences about several different processes	• Requires specialized equipment and standardized procedures to generate data (ie, animal must be postabsorbtive, isolated from external stimuli) • Factors such as handling stress, individual variation, social status, acclimation time, and nutritional status can all influence results (Sloman et al., 2000; Nelson and Chabot, 2011) • Data are strongly influenced by size, making comparisons across size classes challenging • Techniques are highly variable across research groups, making intraspecific comparisons challenging (Nelson and Chabot, 2011)

(Continued)

Table 11.3 (Continued)

Indicator	Sampling and analytical considerations	Strengths of indicator	Weaknesses of indicator
Ventilation rate • Most fish move water over their gills through active ventilation, which involves opening and closing of the opercula, and can be used as a proxy for respiration (Barreto and Volpato, 2004)	• Can be counted with a stopwatch from direct observation or video (White et al., 2008) • Can be measured remotely by use of bioelectric sensors placed in water near fish (Altimiras and Larsen, 2000) • Possible to telemeter opercular activity (either radio or acoustic) to estimate ventilation rate (Oswald, 1978)	• Ability to differentiate between different stressors (Barreto and Volpato, 2006) • Relatively simple indicator that can be measured noninvasively and with little expense	• Ventilation rate does not appear to reflect the severity of a stressor in all species (Barreto and Volpato, 2004) • Rate alone might not be sufficient, and may be necessary to also quantify amplitude
Cardiac activity • The heart is essential for circulation and life-support such that cardiac activity (eg, heart rate, stroke volume, cardiac output) are relevant indicators of whole-organism stress (Farrell, 1991)	• Heart rate/ECG is relatively simple to measure, but, given that some fish are volume modulators, measuring cardiac output is often better (Farrell, 1991) • Doppler and transsonic cuffs can be used to measure blood flow in ventral aorta (Farrell, 1991)	• Robust indicator of stress in many species • Used to document responses to different husbandry conditions (Rabben and Furevik, 1993), environmental conditions (Claireaux et al., 1995; Lefrançois et al., 1998), and human-induced disturbances (Anderson et al., 1998)	• Baseline values of cardiac activity can be variable, making it difficult to identify when fish have recovered following a stressor • Although a large number of tools are available for measuring cardiac activity, they are all rather technically challenging

	Heart rate can be measured remotely in free-swimming fish in the field (Priede, 1983; Armstrong, 1998) or in the lab using ECG biologgers (Raby et al., 2015) or bioelectric sensors placed in water near the fish (Altimiras and Larsen, 2000)	• Tend to be very responsive (eg, rapid change in the face of a stressor)	

Growth and Life History

• Growth and reproduction occur only after the energetic demands of other processes are met • Reduced growth rate can result from stress • Several hard structures in fish (eg, scales, otoliths, bone) deposit growth rings to allow accurate age determination	• Can use a range of hard structures, some of which can be collected nonlethally (eg, scales) • Can generate proxy for reproduction with various gonadal indices	• Individual size correlates positively with fecundity in females, and with reproductive output in males of some species (Suski and Philipp, 2004) • Reduced growth rate is a well established indicator of stress (Pankhurst and Van der Kraak, 1997)	• May need lethal samples to generate data • May need to sample populations over long time scales to discern trends in growth, by which point it may be too late to stop or alter trajectories • Size/age alone might not be sufficient to discern population-level trends, and demographic data (ie, fecundity, survival) may be required • Need to validate growth across age classes (Beamish and McFarlane, 1983)

Condition indices

• Condition indices of stress include length–weight relationships, organosomatic	• Approaches range from being relatively noninvasive (eg, simple measurements on live	• Simple and inexpensive (Bolger and Connolly, 1989) • Good indicator of the aggregate condition of the fish (eg, can	• Not overly sensitive to short-term stressors (Bolger and Connolly, 1989)

(Continued)

Table 11.3 (Continued)

Indicator	Sampling and analytical considerations	Strengths of indicator	Weaknesses of indicator
indices, and necropsy (Barton et al., 2002)	fish) to lethal (eg, for organosomatic indices) • Organosomatic indices are ratios comparing the weight of an organ to body weight (eg, hepatosomatic index [liver:body weight, HSI], gonadosomatic index [gonads:body weight, GSI], viscerosomatic index [entire viscera:body weight, VSI], and splenosomatic index [spleen:body weight, SSI], Barton et al., 2002) • Values that are lower or higher than normal suggest that energy allocation to organs has been affected by stress (Kebus et al., 1992) • Necropsy-based methods (eg, health assessment index) requires autopsy of the sacrificed fish, whereby condition of internal organs is compared to published standards that outline criteria of an organ of normal condition (Adams et al., 1993)	detect chronic stress; Barton et al., 2002) • Some nonlethal options (eg, length–weight analysis, condition factor, relative weight)	• Some critiques regarding the use of condition indices given that they can lead to inappropriate conclusions based on inherent limitations of the various methods (Cone, 1989) • More involved measures (organosomatic and necropsy) are lethal • Condition indices can be influenced by seasons, stage of development, sexual maturation, and disease state • Often require large effect size to detect stress

Fluctuating asymmetry

Description			
• Differential development of a structure on one side of an organism (Jagoe and Haines, 1985) • Symmetrical structures should result from the same genetic material, so deviations from symmetry can represent stress in the form of genetic mutation, or environmental stress (Leary and Allendorf, 1989)	• Quantifying the presence of fluctuating asymmetry can indicate stress within a population • Can be used as an early warning indicator prior to population-level declines in abundance or adverse environmental conditions (Jagoe and Haines, 1985)	• Can be inexpensive and straightforward to measure as meristic/morphometric characters are measured or counted	• Many meristic characters (eg. fin rays) are variable, making them unreliable as stress indicators (Leary and Allendorf, 1989) • Not all species or characters lend themselves to studies of fluctuating asymmetry as a stress index • May require large sample sizes to identify trends/patterns (Jagoe and Haines, 1985) • May be challenging to define a causal mechanism or link between asymmetry and stress (Jagoe and Haines, 1985)

Reproduction and fitness

Description			
• Reproduction and reproductive output is a process critical to survival and persistence of an individual and a species • Reduced reproduction can result from stress	• Can assess extent of intersex as an indicator of environmental stress (Bortone and Davis, 1994) • Impacts of stress may be visible in a range of reproduction-related factors such as gamete quality and/or timing of reproduction (Schreck et al., 2001)	• Very ecologically relevant when relating stress to population-level parameters • Can measure a suite of gonadal indices (mass, size, egg stage, hormones, etc.) as proxies for reproduction • Can perform artificial crosses with known parents to define offspring survival and viability (Campbell et al., 1994)	• May need to collect data on reproduction only at certain times of the year • May be difficult to assign a cause-and-effect relationship between a stressor and reduced reproduction

(Continued)

Table 11.3 (Continued)

Indicator	Sampling and analytical considerations	Strengths of indicator	Weaknesses of indicator
Survival • The most extreme response to a stressor is death, whereby homeostasis cannot be maintained (Wood et al., 1983)	• Mortality can be measured by holding fish in nets, cages, pens, or tanks and simply counting the dead (Gutowsky et al. 2015) • Often instructive to mark/tag individual fish to determine which individuals died and their history • Increasing set of tools (eg, biotelemetry) available to study mortality in free-swimming fish in the wild (Pine et al., 2003)	• Often simple, inexpensive approach • Mortality is absolute and a clear, ecologically relevant fitness indicator • Most powerful when combined with other indicators that reveal the mechanistic basis for mortality (eg, Cooke et al., 2006)	• May be difficult to obtain animal care and use approvals to use death as an endpoint • Subject to bias from method used to assess mortality (eg, net or cage effects; Gutowsky et al., 2015) • Fish can die for many reasons (eg, senescence) such that it is necessary to have appropriate controls

necessarily equate to a less stressed or unstressed fish. Readers are encouraged to consider this caveat when investigating and selecting measures of fish stress.

3.1. Cellular and Molecular Indicators

Acting in concert with the responses that occur at higher levels of biological organization, the cellular and molecular suite of stress responses help to temporarily tolerate stressors (Kultz, 2005, Table 11.1). Cortisol is involved in molecular responses to stressors as it stimulates the expression of metallothionein, ubiquitin, and HSPs by interacting with heat shock factors (HSFs), which affect transcriptional regulation (Vamvakopoulos and Chrousos, 1994; Kassahn et al., 2009). Cortisol also binds with glucocorticoid receptors and interacts with the c-Jun component of the activation protein-1 (AP-1) transcription factor (Iwama et al., 2006). The transcriptional effects of cortisol binding to glucocorticoid receptors depend on tissue type and HSP 90 expression levels (Basu et al., 2001; Vijayan et al., 2003). A range of stressors induce a common set of responses, which can include repair of DNA and protein damage, cell cycle arrest or apoptosis, the removal of cellular and molecular debris, and changes in cellular metabolism that reflect the transition from anabolic to catabolic states (Iwama et al., 2004). Cumulatively, these cellular and molecular responses are triggered by the eukaryote minimal stress proteome, a set of evolutionarily conserved proteins (Kultz, 2005) (Faught et al., 2016; Chapter 4 in this volume).

Stressor exposure results in the production of reactive oxygen species (ROS), resulting in oxidative stress inside the cell. Oxidative stress leads to increased levels of protein damage. The amount of ubiquitin-labeled protein can indicate the level of protein damage in the cell (Iwama et al., 1998; intracellular enzymes also indicate cell damage or death, see Table 11.1). The damaged and ubiquitinated protein in the cell induces a heat shock response, aimed at repairing protein damage (Wu, 1995). The magnitude of the heat shock response depends on the magnitude and duration of the stressor and acclimation state (Iwama et al., 2004; Somero and Hofmann, 1996). HSP expression is regulated by heat shock factor 1 (HSF1), which dissociates HSPs from HSF1 following activation of the hypothalamic–pituitary–interrenal (HPI) axis. HSF1 then migrates to the nucleus and begins the transcription of HSPs (Kassahn et al., 2009). ROS can also accelerate the reduction in telomere length resulting from oxidative stress. Telomeres are the caps at the end of the chromosome that are essential for genome stability, and, when reduced at an accelerated rate can, in turn, hasten cell senescence (Richter and von Zglinicki, 2007). The concept that telomere length can be correlated with cellular senescence, and possibly organismal senescence, has become an emerging field of research (Ricklefs, 2008; Monaghan et al., 2009). A range of environmental stressors, including psychological stress (Epel et al., 2004) or elevated reproductive effort (Kotrschal et al., 2007), can affect telomere length in vertebrates. So there is evidence that environmental stressors can cause oxidative stress, which if sustained, may trigger cellular senescence and ultimately organismal senescence (Monaghan et al., 2009).

Changes in gene expression (ie, quantitative, qualitative, and changes in reaction coefficients) can be linked with a range of stressors (Krasnov et al., 2005). Genomics tools such as microarrays and gene expression profiling are now commonly used to understand the responses of fish to a range of stressors. Commonly examined stressors include temperature (eg, Jeffries et al., 2012), hypoxia (eg, Gracey et al., 2001), handling stress (eg, Donaldson et al., 2014), and toxicants (Williams et al., 2003). cDNA microarrays enable thousands of genes to be screened simultaneously to identify groups of differentially expressed genes related to biochemical pathways involved in a range of responses. Relatively few studies have assessed gene expression in relation to acute stressor exposure, compared to the literature assessing primary, secondary, and tertiary stress indicators (Caipang et al., 2008; Prunet et al., 2008). Instead, one of the main objectives of functional genomics studies is broadening our understanding of how gene expression is influenced by environmental conditions (Buckley, 2007; Miller et al., 2009). Microarrays using nonlethal tissue biopsies (ie, muscle and gill tissue) have identified potential genes involved in an unhealthy signature of migrating sockeye salmon, where individuals that are less likely to reach spawning grounds are characterized by indices of poor health, including downregulation of blood clotting factors and genes related to aerobic respiration, as well as expression of genes linked with immune function (Miller et al., 2011). Still, understanding the functional significance of many genes and gene families remains a challenge, and the implications of how genes respond following exposure to a stressor is not always clear.

In fish, a number of genes have been investigated as potential biomarkers for various stressors. Genes linked to cell apoptosis, such as cytochrome c and transcription factor JUNB, are upregulated in response to elevated temperatures in sockeye salmon (*Oncorhynchus nerka*; Jefferies et al., 2012). JUNB is likewise upregulated following low-water and air exposure stressors in rainbow trout (*O. mykiss*; Momoda et al., 2007). Transcription factor NUPR1, which is involved in the regulation of cell growth and apoptosis (Mallo et al., 1997), is also responsive to stressors in rainbow trout, and can remain upregulated several hours poststressor (Momoda et al., 2007). Changes in the gene expression of biological pathways related to inflammation, protein degradation, and the immune response have been observed in rainbow trout (*O. mykiss*) liver in relation to handling stress (Momoda et al., 2007; Wiseman et al., 2007). Also in rainbow trout, following repeated exposure to a netting stressor, Krasnov et al. (2005) observed changes in expression of genes related to immune responses, cell proliferation and growth, apoptosis and protein biosynthesis in the brain, and changes in expression of genes related to cellular biochemical processes in the kidney. The onset of recovery poststress may be evidenced by the expression of genes related to gluconeogenesis, glycogenolysis, and energy metabolism in the liver (Momoda et al., 2007; Wiseman et al., 2007). Donaldson et al. (2014) identified species- and sex-specific genomic responses in sockeye salmon related to the stress response and recovery following exposure to an exercise stressor. A complicating factor in understanding the behavior of these genes during the stress response is the fact that studies conducted on fish to date often focus on different time courses (Krasnov et al., 2005), species, tissues (Kassahn et al., 2009), techniques (Prunet et al., 2008), and genes of interest.

3.2. Primary and Secondary Physiological Indicators

Physiological indicators of stress include all the responses between the cellular and molecular level and the whole-animal level (Table 11.2). The stress response involves a wide range of physiological responses including, and beyond, HPI axis activation (Schreck and Tort, 2016; Chapter 1 in this volume, Figure 1.6). There are a suite of primary indicators (eg, catecholamines and stress hormones), as well as secondary responses (eg, changes in glucose, ion balance, acid–base balance, immunological functions, or other indicators of energetic metabolism) that can be used to assess stress in fish. Secondary changes happen over a slower timescale than the primary responses. The use of the different primary and secondary stress indicators has relative advantages or disadvantages depending on the stressor of interest, and the level of background information available about the species, population, or individual (Table 11.2).

Changes in certain primary and secondary stress indicators are notably useful when assessing responses to specific aquaculture or handling practices, or acute disturbances in the field. Catecholamines provide the fastest primary response to stressors, but are difficult to measure because they respond quickly to stressor exposure (Reid et al., 1998; Pottinger, 2008). Catecholamines are appropriate and powerful indicators in laboratory settings, but are often not logistically possible to measure in the field because they are highly responsive to capture and handling. Cortisol, the other measure of the primary stress response, is among the most commonly measured stress indicators (Mommsen et al., 1999; Barton, 2002; Pottinger,

2008). Because cortisol responds more slowly than catecholamines to specific stressors, it can be quantified in laboratory or field settings to obtain both baseline and poststressor levels, as long as animals can be sampled within a few minutes of capture (Romero and Reed, 2005). Secondary stress indicators commonly measured include (1) glucose elevation, a result of the increase in catabolism and release of glucose into the blood stream following stressor exposure (Barton, 2002); (2) lactate elevation, an indicator of anaerobic metabolism following hypoxia or exercise stressors (Wood et al., 1983); (3) osmolality or specific ions, which are altered as a result of the increase in catecholamines and subsequent increase in heart rate and gill permeability (McDonald and Milligan, 1997); and (4) leukocytes, which can not only reflect response to acute stressors, but may also predict future survival or performance (Davis et al., 2008).

Many of these physiological indicators are measured in plasma as they provide a snapshot of the circulating levels of the hormone or metabolite (Barton, 2002; Pottinger, 2008). The benefit of sampling plasma (or red blood cells for indicators such as hematocrit) is that, for most fishes, samples can be collected nonlethally. Also, because plasma is a conventional tissue for measuring physiological parameters, there are a number of studies that exist to aid in the interpretation of values, and assays or meters are commercially available to facilitate data generation (Wells and Pankhurst, 1999; Beecham et al., 2006). The main disadvantage of measuring indicators in plasma is that plasma is not always the most relevant tissue, and some indicators do not make sense for plasma measurements. For example, bioenergetics indicators measured following chronic stressor exposure such as glycogen liver stores could not obtained by taking only a plasma sample. Similarly, as phosphocreatine (PCr) and adenosine triphosphate (ATP) are indicators of acute exercise stress, these metrics are best measured in muscle tissue. Therefore, the specific tissue sampled will be dependent on the research question, and the primary/secondary physiological indicator of interest. Taking these caveats into account, primary and secondary stress indicators provide information about how individuals are perceiving and responding to environmental challenges, and can identify the extent to which animals are stressed by these challenges, as well as potentially predict future performance and survival (Breuner et al., 2008).

3.3. Whole-Organism Indicators

Whole-organism (or tertiary) responses to stressors include a number of aspects of fish performance such as changes in growth, condition, disease resistance, metabolism (Sadoul and Vijayan, 2016; Chapter 5 in this volume), cardiac activity, swimming performance, behavior, fitness, and even survival (Wedemeyer et al., 1990, Table 11.3). These responses can serve as indicators of stress, and are also generally considered to have ecological relevance. For example, growth rates can directly influence population models (Power, 2002) and bioenergetics models (Beyers et al., 2002), while fitness and survival influence demographic processes (McCallum, 2000). Behaviors related to food acquisition, predator avoidance, and habitat selection also have direct ecological relevance with a physiological underpinning (Godin, 1997). Many condition-related measures involve simple measures of mass (ie, usually the entire organism along with specific organs), length, or both (Goede and Barton, 1990). Traditional measures of fish condition (eg, relative weight, condition factor) are also simple to calculate, but are criticized for their lack of specificity and breadth (Bolger and Connolly, 1989). So-called organosomatic indices (eg, ratio of organ mass to body mass) are widely used as an index of stress and condition given their simplicity, but they require lethal sampling. Some researchers have generated indices that are a composite of various measures related to body condition and health, and perhaps the most commonly used is the Health Assessment Index (HAI), first proposed by Adams et al. (1993). The HAI can be modified to meet the needs of researchers and their study questions.

Swimming performance represents another tertiary endpoint that is functionally simple to both collect and interpret, and is also ecologically meaningful, but proper measurement requires custom-designed swim tunnels or flumes that can be technically challenging to build, or expensive to purchase. Critical swimming speed is an example of a metric that has been measured for decades (Beamish, 1978), and although it is now regarded as somewhat limited in value (Plaut, 2001), it does have use when making relative comparisons of fish swimming ability among stressor treatments. Portz (2007) reported that if the goal of a study is to simply make treatment-level comparisons in swimming ability, then simply defining the time to exhaustion for fish chased by tail pinching in a round tank generated values that correlate with more formal measures of critical swimming speed. Recognizing inherent variation in swimming ability among individual fish, Jain et al. (1997) refined swimming protocols to create a new method that compared the ability of fish to swim a second time shortly after doing an initial swim. Termed the "recovery ratio," the approach has served as a sensitive indicator of overall organismal status in the face of different stressors (see Jain et al., 1998). When we combine measures of cardiac activity (eg, heart rate or cardiac output) and metabolic rate (oxygen consumption) with active swimming challenges (see Webber et al., 1998), it is possible to obtain a multilevel understanding of stress and its influence on an ecologically relevant index of organism performance. In fact, laboratory-derived relationships between metabolism, cardiac function, swimming ability, and water temperature can be scaled to field environments to predict mortality (Farrell et al., 2008). Together, the measurement of fish swimming ability, using a range of possible techniques, has the ability to provide valuable information on how stress can influence an important aspect of fish ecology.

Tertiary stress indicators are more diverse than data on the primary or secondary stress response, which tend to fall almost exclusively within the realm of the physiologist. Collecting data related to the tertiary stress response, as well as interpreting those data, may require the expertise of ethologists, field ecologists, as well as physiologists. Nonetheless, there certainly are strong physiological underpinnings, some more direct than others, to tertiary stress indicators. Indicators related to cardiac activity and metabolism are the domain of physiologists, but also link to activities such as swimming (both swimming performance and general locomotor activity) and other elements of behavior. Metabolism and locomotor activity are large drivers of energy budgets (Boisclair and Leggett, 1989). Energy budgets define somatic growth rates (and condition), reproductive investment, and behavior. In that sense, many of the tertiary indicators are linked to bioenergetics (Beyers et al., 2002). Even the growing interest of using reflex impairment as indicators of fish vitality is linked to metabolism and neurological function (Davis, 2010).

As we move from measures focused on direct indicators of homeostasis (ie, primary or secondary stress indicators) to more whole-organism level indicators that represent integration of various mechanisms (eg, neuroendocrine processes), the ability to infer a stressed state becomes more challenging, and may be best studied in the context of comparisons among groups (eg, control vs stressor A and stressor B). For example, something as straightforward as mortality is a natural phenomenon, and a dead fish does not immediately imply that the fish was stressed; senescence is a natural process, and a fish can be predated upon independent of whether it was in a stressed state. Similarly, reduced feeding behavior of a fish does not indicate that the fish is experiencing stress because it may simply not be hungry, which could be influenced by previous feeding history, metabolic demands, food availability, seasonality, or even genetics. This is unlike primary and secondary stress indicators, such as cortisol or osmolality, for which there are clear reference ranges or where a specific threshold (eg, X ng/mL or X mmol/L) is meaningful. Also, circulating cortisol for an individual fish may be elevated relative to reference ranges, yet there are no organism-level endpoints evident. Does that mean that the fish is stressed? Context clearly matters, which means comparisons using appropriate controls must be used when inferring stress from tertiary endpoints.

Despite these challenges, the elegance of using tertiary indices to define stress is their simplicity, both in collection as well as interpretation, particularly in relation to other physiological measures of stress that can only be quantified with laboratory work. For example, two metrics that are simple, elegant, as well as highly relevant when identifying whole-organism stress, are reflex indicators (eg, is a specific reflex present or absent) and survival (eg, did an organism live or die). What is remarkable with these metrics is that something as simple as reflex impairment has been shown to correlate with survival (Davis, 2010; Raby et al., 2012), and is now being widely embraced by the research community to define the physiological state of animals in the field, and to predict mortality. Interestingly, however, the mechanistic basis for reflex impairment is not directly clear. Similarly, many tertiary indices of stress can be measured outside the traditional laboratory with wild, free-swimming fish. For example, some measures such as fine-scale locomotor activity (Cooke et al., 2004), cardiac activity (Cooke et al., 2004; Clark et al., 2010), and plasma glucose (Endo et al., 2009) can be measured remotely and either transmitted (ie, biotelemetry/biosensors) or stored aboard electronic tags for later downloading (ie, biologging). Reflex impairment measures can also be conducted in field settings, as they do not require any specialized equipment (Raby et al., 2012).

4. CONSIDERATIONS FOR MEASURING AND INTERPRETING STRESS

With regard to the stressor, study animal, and stress indicator, there are a number of factors to consider when quantifying and interpreting stress in fishes (Fig. 11.3). A selection of these factors is highlighted next.

4.1. Interspecific Differences

It is perhaps not surprising that different species of fish respond to the same stressors differently. Given evolutionary, ecological (ie, predation), environmental (ie, temperature), and life history differences that contribute to the divergence of species, HPI activity and other indicators of stress are apt to vary. Comparing stress across species is largely limited by variation among studies in stressor type, severity, and duration, which can dramatically affect the magnitude of the HPI stress response (see Section 4.4). When exposed to an identical stressor, plasma cortisol (Barton, 2000, 2002; Pottinger, 2010) and glucose (Jentoft et al., 2005) levels, gene expression (Jeffries et al., 2014b), immune function (Cnaani et al., 2004), habitat preference (Jacobsen et al., 2014), and avoidance behavior (Hansen et al., 1999) all can vary among species, including closely related species (eg, pink salmon, *Oncorhynchus gorbuscha* and sockeye salmon, *O. nerka*;

Donaldson et al., 2014). However, under certain conditions, different species can exhibit similar responses to the same stressor. Campbell et al. (1994) noted that repeated air exposure reduces progeny survival to a similar degree in rainbow and brown trout (*Salmo trutta*). Ryer et al. (2004) showed that juvenile sablefish, *Anoplopoma fimbria*, known to be robust against fisheries capture mortality, display similar behavioral impairment following capture as the more fragile walleye pollock, *Theragra chalcogramma*. Design of experiments whereby multiple species (and where possible, hybrids; eg, Noga et al., 1994) are exposed to identical stressors, and identical indicators are measured, can maximize collection of data that aid in understanding factors driving species-specific differences in stress.

4.2. Intraspecific Differences

Measures of stress can also vary within species. Often the existence of unique populations motivates examination of stress. Environmental and ecological factors linked to geographically distinct populations are predicted to drive divergence in stress responsiveness. Indeed, along a longitudinal gradient, populations of killifish, *Fundulus heteroclitus* (northern vs southern) vary in their stressor-induced plasma cortisol responses (DeKoning et al., 2004). Stressor-induced changes in ventilation rate of tropical poeciliids, *Brachyrhaphis episcopi* (confinement stressor, Brown et al., 2005) and three-spined stickleback, *Gasterosteus aculeatus* (predator stressor, Bell et al., 2010) vary depending on whether fish are collected from low- or high-predation habitats. In other instances, differences in physiological parameters of stress (eg, plasma cortisol, lactate, glucose) are not present among populations but fitness outcomes (ie, survival) can still vary poststressor exposure (Donaldson et al., 2012). Population-specific differences in the stress response are intriguing from an evolutionary standpoint, and should not be overlooked when making specieswide conclusions on stressor sensitivity.

Other intraspecific factors driving variation in stress indicators include sex, size, social status, and domestication. Sex differences can be especially pronounced in certain species. Adult female Pacific salmon (*Oncorhynchus* spp.) have higher stressor-induced plasma cortisol levels (Donaldson et al., 2014) as well as a greater likelihood of mortality (Martins et al., 2012) in response to temperature stressors. Timing of peak stressor-induced plasma cortisol is affected by body size in European sea bass (*Dicentrarchus labrax*; Fatira et al., 2014). Poststressor HPI activity is known to vary between dominant and subordinate rainbow trout (Jeffrey et al., 2014a). Variation in measures of stress are also present between hatchery and wild trout (Woodward and Strange, 1987; Lepage et al., 2000) and salmon (Johnsson et al., 2001), which in combination with all intraspecific considerations has implications with regard to maximizing performance of fishes reared for captivity or release into the wild for stock enhancement.

Finally, there is a rapidly developing appreciation for heritability of, and interindividual/interfamily differences in, stress-responsiveness (eg, Kittilsen et al., 2009; Pottinger, 2010; Hori et al., 2012), as well as coupling of physiological and behavioral stress indicators within an individual (ie, coping style, Castanheira et al., 2015). Aquaculture has utilized this variation for over a decade to breed genetically divergent lines of rainbow trout (*O. mykiss*) with high and low plasma cortisol responses to a 3 h confinement stressor (Pottinger and Carrick, 1999). This variability is now associated with consistent variability in myriad of behavioral indicators of stress (eg, activity following a conspecific intruder, Øverli et al., 2007). High and low responders have also been characterized in Atlantic salmon (*Salmo salar*, Fevolden et al., 1991), gilthead sea bream (*Sparus aurata*, Tort et al., 2001), Atlantic cod (*Gadus morhua*, Hori et al., 2012), and striped bass (*Morone saxatilis*, Wang et al., 2004). These individual-level differences may also influence individual tolerance to the sampling methodology to quantify stress (eg, recovery from cannulation, Bry and Zohar, 1980). Together, there are multiple aspects of intraspecific variation that can influence stress indicators; accounting for and describing this variation is warranted.

4.3. Context-Specific Differences

HPI axis activity, in addition to auxiliary molecular, physiological, and whole-animal responses can vary in magnitude depending on the environmental and ecological context of the stressor an individual is exposed to. Activation of the HPI axis should be consistent across contexts. Ultimately, any threat to an animal's fitness must be endured and overcome via HPI-mediated changes to physiology and behavior (Schreck and Tort, 2016; Chapter 1 in this volume). The changes to physiological and behavioral processes may nonetheless differ if the individual is faced with a predator, resource competition, restrictive feeding, hypoxia, aquatic pollution, or elevated water temperature.

Context-dependent variation in stress measures is observed at all organizational levels. At the neuroendocrine level (primary indicator), changes in brain monoamine (eg, norepinephrine, serotonin) concentrations differed in three-spined stickleback based on stressor type (unfamiliar conspecific vs predator; Bell et al., 2007). At later stages of the HPI axis, predator-induced whole-body cortisol response of winter flounder (*Pseudopleuronectes americanus*) differed depending on species of predator (Breves and Specker, 2005). More subtle differences in stressor type may not elicit different plasma responses (eg, intra- vs interspecific intruder, Ros et al.,

2014). Interestingly, two very different stressors with regard to ecological relevance, and potentially severity, can elicit similar responses. Visible implant elastomer tagging elicits a similar plasma cortisol response in three-spined stickleback as that following a simulated predator attack (Fürtbauer et al., 2015). Secondary, physiological processes are also influenced by stressor context. Degree of change in metabolism varied in *Galaxias maculatus* (Milano et al., 2010) depending on whether individuals were exposed to visual or olfactory predator cues. At the whole-animal level, change in ventilation rates varies among stressors (eg, confinement vs conspecific vs electroshock, Barreto and Volpato, 2006). Likely interlaced with stressor severity (see Section 4.4) the environmental and ecological context of a stressor matters when making decisions about which indicator to select, predicting the response of the indicator, and interpreting the results.

When assessing stress, exposure of animals to multiple stressors is arguably the most biologically relevant context. Experimental design incorporating different stressors (eg, elevated water temperature and fisheries capture, aquatic pollution, or immune challenge) may detect magnification of stress indicators (eg, Marcogliese et al., 2005; Jacobsen et al., 2014). Fish are also exposed to the same stressors more than once, and previous experience with a particular stressor and the individual's capacity to learn may influence stress indicators (Barcellos et al., 2010; also see discussion on habituation in Section 4.4). Housing conditions prior to stressor exposure (ie, isolation vs groups) can also influence stress indicators (Giacomini et al., 2015), potentially via communication of olfactory cues of conspecific stress (Barcellos et al., 2014a). Testing fish singly versus in groups can be motivated by the schooling and social tendencies of the species. Incorporating additional experimental variables to increase biological relevance can introduce other logistical challenges such as requirement of larger sample sizes. If the scientific priority is for biological relevance of stress indicators, such nuances of ecological and environmental context can be considered.

For some contexts, identifying suitable indicators is intuitive (eg, measuring ventilation frequency and avoidance behaviors following a simulated predator attack). Other scenarios allow for use of many suitable indicators. Exposure to toxicological stressors affects gene expression (Jeffries et al., 2015), gamete quality (Khan and Weis, 1987), and suites of physiological processes and behaviors across life stages (Scott and Sloman, 2004; Sloman and McNeil, 2012). Approaches using multiple indicators (eg, Woodley and Peterson, 2003) provide opportunities to discover or exclude indicators for a particular stressor. However, it is necessary to keep in mind that primary and secondary indicators of stress may not correlate with whole-animal or tertiary indicators (eg, HSP expression increasing in response to a thermal challenge but swimming performance remaining constant). Even within a class of indicators, responses can vary (eg, increases in lactate concentration indicating insufficient oxygen supply to muscles, but no change in HSP expression indicating absence of protein damage). Accordingly, the nature of the stressor or stressors, and relationship among indicators, is important to consider when determining the appropriate indicators of stress to measure.

4.4. Stressor Severity

As discussed by Schreck and Tort (2016; Chapter 1 in this volume), embedded within context-specific differences is whether the stressor will be applied in an acute/single, repeated, or chronic/prolonged manner. An acute or single exposure may be typified by brief durations (seconds to minutes), and associated with physiological responses that are adaptive. Variation in duration of an acute stressor can alter stress indicator levels in a gradated manner (Gesto et al., 2013, 2015). Acute stressors can also vary in intensity with regard to expected level of impairment (eg, 2-minute confinement vs 5-minute chase with net), again resulting in variation in stress indicator levels (Geslin and Auperin, 2004). For contaminant stressors, there are often clear patterns between dose concentration and extent of indicator response or mortality (ie, LC_{50}). However, as mentioned in Section 4.3, stressors that may be considered distinct based on severity can elicit similar responses, as well as identical stressors differing in duration (Fatira et al., 2014).

Definitions of what constitutes a chronic stressor exposure are not always consistent; chronic stress can be repeated, sequential exposure to an acute stressor (eg, daily handling over multiple weeks), or continuous prolonged exposure to a stressor (eg, continuous exposure to elevated water temperature over multiple weeks). How an indicator responds to these exposures may depend on its role in immediate (eg, avoidance behavior) versus long-term (eg, growth) effects on fitness (Schreck, 2000). Also, for some species, chronic stressor exposure is reality (as in Boonstra, 2013b). Although indicators may imply a chronically stressed state (ie, allostatic overload, McEwen and Wingfield, 2003), the responses may be adaptive (Boonstra, 2013b). For repeated exposures, the interval between stressor application can affect the response of the stress indicator as well (Schreck, 2000). Thus, habituation must be considered in order to validate that changes to stress indicators are indeed due to "disrupted negative feedback" (Romero et al., 2009) or allostatic overload (Schreck and Tort, 2016; Chapter 1 in this volume, Figures 1.3, 1.5, and 1.6).

Discerning habituation from desensitization and exhaustion using hormonal indices of stress is discussed by Schreck (2000) and Cyr and Romero (2009). Defining habituation using performance/behavioral indicators may not be as straightforward and measurement of an accompanying physiological parameter (ie, HPI axis activity) is needed. A subset of animals can be sampled midway through repeated exposure to an acute stressor (eg, immediately following daily handling stressor being applied over multiple weeks) to confirm that the HPI axis and stress indicator is still responsive to the stressor. A subset of animals can be exposed to an acute stressor midway through a sustained stressor exposure (eg, single, acute chase stressor during continuous exposure to elevated water temperature over multiple weeks) to confirm that the HPI axis and target stress indicator is still responsive. Diminution of the endocrine stress response is detected in a number of species chronically stressed (Barton et al., 1987; Jentoft et al., 2005, but see Barcellos et al., 2006). Wingfield et al. (2011) proposed that there are thresholds after which mounting a stress response is no longer adaptive. When these thresholds are surpassed, resistance potential to the stressor increases via attenuation of the stress response (Wingfield et al., 2011). Multiple samplings throughout the chronic exposure may identify tipping points whereby the stress indicator no longer responds (resilience) or recovers to baseline, prestressor levels (exhaustion).

Phenotypic plasticity should also be considered, particularly when quantifying stress in wild populations inhabiting environments with fluctuating conditions that are also chronic stressors (eg, climate change, aquatic pollution, Silvestre et al., 2012; Crozier and Hutchings, 2014). Laboratory-based indicators of stress may not be reliable indicators in wild animals (Dickens and Romero, 2013), and this may be due to shifting coping strategies. Alternatively, laboratory-based indicators of chronic stress may be detected in wild populations but without population-level consequences. Populations of invasive round goby from highly contaminated areas in Lake Ontario demonstrate indices of toxicological stress (eg, endocrine disruption, Marentette et al., 2010; impaired behaviors, Sopinka et al., 2010). Yet, populations of round goby in polluted areas are stable and populations in reference areas are declining (McCallum et al., 2014). This example illustrates how new organismal steady states may emerge and be interpreted as stress, but do not affect population-level processes (see Section 5).

4.5. Field Versus Laboratory

Laboratory and field measurements of stress each have their advantages and disadvantages. Certain indicators of stress must be measured in the laboratory due to the complexity of equipment used to obtain samples (eg, serial blood sampling via cannulations, Fig. 11.1). Other indicators can only be measured in the field due to the inability to replicate the behavior in the laboratory (eg, stressor-induced changes in migration rates, Donaldson

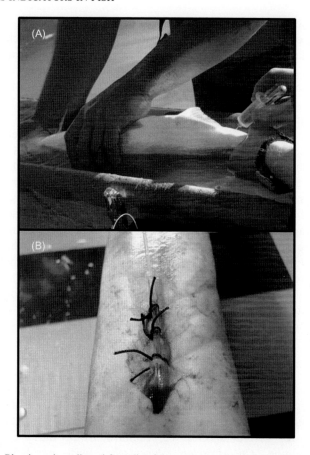

Figure 11.1. Blood can be collected from live fish to measure stress indicators by (A) drawing blood from the caudal vasculature using a syringe or Vacutainer; or (B) by implanting a cannula in the vasculature (often aortic), which also enables serial sampling. Photo credits: (A) Michael Donaldson, (B) Michael Lawrence.

et al., 2011). Laboratory studies allow for control of variables that can confound measurement of stress indicators, and otherwise are difficult to control in the wild (eg, water temperature). This degree of control is crucial and necessary for research targeting mechanistic connections between stressor and indicator. Still, mimicking stressors that fish will encounter in the wild in a laboratory setting does not truly encompass the entire stress response that an animal would elicit under ecologically relevant conditions. For example, latent effects of sublethal stress indicators (eg, postrelease predation following a fisheries capture stressor that impairs equilibrium,

Danylchuk et al., 2007) are not adequately accounted for in the laboratory. Depending on study hypothesis and goals, measurement of the chosen stress indicator may be more suitable under laboratory versus field conditions, or vice versa.

Development of remotely sensing devices (Cooke et al., 2004), as well as validation of point-of-care devices in fishes (Stoot et al., 2014), allows for field measurement of indicators once restrained to the laboratory. Specifically, with the advancement of biologging and biotelemetry technology, changes in activity levels and energetics (Burnett et al., 2014), foraging (Brownscombe et al., 2014), and heart rate (Clark et al., 2008) of free-swimming fish can be monitored before, during, and after a stressor (Donaldson et al., 2010; Raby et al., 2015). Limitations of this field-oriented mechanistic approach to measuring stress include cost, surgical requirements for tag implantation, detection efficiency, and tag retrieval, which can compromise study sample size. Measuring indicators in wild fishes naturally exposed to unpredictable, labile stressors (eg, cyclones, floods) can be incredibly revealing but is also highly opportunistic (Wingfield, 2013). A holistic, collaborative approach of field-based research coupled with complementary laboratory research focusing on mechanism(s) can provide the most complete assessment of the stress response.

4.6. Temporal Aspects

There are multiple levels of timing that can be considered when investigating animal stress. First, what life stage will the indicator of stress be measured? The hyporesponsive period early in development is well established for several species (eg, rainbow trout, Barry et al., 1995; Chinook salmon, *Oncorhynchus tshawytscha*, Feist and Schreck, 2001; yellow perch, *Perca flavescens*, Jentoft et al., 2002; lake sturgeon, *Acipenser fulvescens*, Zubair et al., 2012); endogenous stressor-induced cortisol production is not detected prior to hatch. Stressor-induced and resting plasma cortisol levels can shift (1) as sexual maturation progresses in salmonids (Pottinger and Carrick, 2000; Cook et al., 2011) and catfish (Barcellos et al., 2014b) and (2) during parental care in largemouth bass (Jeffrey et al., 2014b). Hyperresponsive periods are also present in smolting salmon (vs parr, Carey and McCormick, 1998). A longitudinal study by Koakoski et al. (2012) found that concentration and timing of peak stressor-induced plasma cortisol varied among fingerling, juvenile, and adult jundiá (*Rhamdia quelen*). To date, age effects largely focus on cortisol as the indicator of stress (Schreck and Tort, 2016; Chapter 1 in this volume, Figure 1.4). Life stage shifts in baseline levels of stress indicators may also confound quantification of stress. This caveat is relevant for other physiological (eg, ontogeny of antioxidant defenses, Otto and Moon, 1996), performance (eg, rapid growth of larval coral reef species), and behavioral (eg, ontogeny of predator avoidance, Brown, 1984) stress indicators that can vary across life stage.

Second, what time of the day will the indicator of stress be measured? Resting heart rates (Aissaoui et al., 2000), plasma cortisol (Cousineau et al., 2014), and various behaviors (eg, activity, Bayarri et al., 2004) fluctuate on a diel cycle. Inconsistent timing of collection can skew data and comparison between studies with different sampling times can compromise validity of conclusions. It is noted, however, that some species may exhibit plasticity in traits typically associated with circadian rhythms (Reebs, 2002).

Third, what time poststressor will the indicator of stress be measured? As mentioned previously, time-course sampling is the most comprehensive approach to ensure capture of rise, peak, and recovery of the indicator. Induction and recovery times will vary, however, for different indicators (Gesto et al., 2015). Catecholamines (and other sympathetic nervous system processes such as heart rate and ventilation) are elevated instantaneously (seconds), whereas cortisol takes longer to elevate (minutes to hours) above prestressor levels. A lag between stressor exposure and changes in mRNA abundance (ie, transcription) is expected. Induction and recovery times may also vary depending on stressor type and severity (see Section 4.4). Identifying and describing temporal influences on the stress response is itself a topic of interest in stress biology; still, when treatment-level impacts (eg, stressor type or severity) are of interest, temporal influences should also be accounted for.

5. FROM INDIVIDUAL INDICATORS TO ECOSYSTEM HEALTH

Thus far this chapter has reviewed indicators of stress at the individual level, but an interesting extension to this work is to ask how the stress of an individual scales to population- and ecosystem-level processes. It is important to note that the molecular, physiological, and whole-animal indicators of stress described in this chapter are responsive over shorter timescales (minutes to days) relative to the response of populations and whole ecosystems to environmental stressors (months to years, see Figure 1 of Adams and Greeley, 2000). Carryover effects and intergenerational components of stress are especially integral when linking individual stress indicators to downstream population effects. Harsh overwintering conditions, episodes of low resource availability, and other environmental stressors can have carryover effects on a fish's phenotype even if stress indicators suggest recovery (O'Connor et al., 2014; O'Connor and Cooke, 2015). Latent effects of stress on populations may be shaped by maternal match/mismatch (see Sheriff and Love, 2013; Love et al., 2013), and may not manifest for several generations. For example, using a 30-year dataset, Venturelli et al. (2010) found that maternally mediated effects on egg size in walleye (*Sander vitreus*) have the capacity to modulate population dynamics. Older, larger females produce larger eggs, which are apt to produce offspring with higher survival. Indeed, the authors detected higher population reproductive rates during years when older females were more abundant (Venturelli et al., 2010). If a stressor associated with a fisheries targeting older females compromises reproduction within this cohort, population stability could fluctuate via maternally mediated mechanisms (eg, changes in egg size, number, or energy content). Connecting individual stress indicators to larger scale processes can be achieved through modeling (Calow and Forbes, 1998; Fefferman and Romero, 2013) or correlation with population- and ecosystem-level metrics.

There are a number of commonly reported population- and ecosystem-level stress indicators including changes in population abundance, habitat use, age and size structure, sex ratios, and age at maturity (Shuter, 1990; Adams and Greeley, 2000; Bartell, 2006). At the ecosystem level, changes in indices of biotic integrity or species richness, food web structure, and productivity can all indicate environmental stress (Karr, 1981; Odum, 1985). Just as using multiple indicators is a robust way to define stress within an individual, an approach that incorporates many levels of biological organization to generate an ecosystem health assessment can be most informative (Attrill and Depledge, 1997; Adams and Greeley, 2000; Bartell, 2006; Yeom and Adams, 2007). For example, measuring indicators of thermal stress in captive adult salmon (eg, Jeffries et al., 2012) can be linked with fitness metrics of fishes migrating in the wild that are naturally experiencing higher water temperatures (Martins et al., 2012). These laboratory and field findings can then be scaled up to population and species survival trends using stock assessment data collected by government agencies. Changes in salmon population abundance can then be linked to health of other taxa (Bryan et al., 2013) and ecosystem-level processes (Gende et al., 2002). Connecting individual traits to an ecosystem is possible following pairing of laboratory recorded stress indicators of an individual, field-derived stress indicators, and population and ecosystem attributes amalgamated from an array of sources.

Drawing associations from individual to population or ecosystem, however, does require longitudinal datasets encompassing both individual-level stress indicators and field-based monitoring. Upon establishment that individual-level stress indicators correlate with population-level change, the more rapidly responding stress indicators can be utilized as early warning signals of forthcoming population and ecosystem effects (Adams and Greeley, 2000). Focusing on sentinel or ecologically important organisms as bioindicators significantly aids in implementing individual stress profiles with assessments of ecosystem stability (Adams and Greeley, 2000). Such endeavors highlight the value and necessity of collaboration among academics, government, and user groups in order to develop stress indicators into useful tools for conservation and management.

6. STRESS INDICATORS OF THE FUTURE

Work outlined in this chapter, as well as in this volume, has provided a comprehensive understanding of how fish detect and respond to stressors. Despite the breadth in tools available to measure responses of fish to stressors (Fig. 11.2), and the wealth of questions that can be answered, there are a number of new directions that research in this area can take moving forward. With human populations projected to grow, and impacts to the planet anticipated to continue or intensify, improving our understanding of the response of fish to stressors, particularly in response to multiple stressors, is critical. We feel that there are five main areas that researchers should consider as targets for future work in hopes of both developing novel stress indicators and refining existing stress indicators to maximize benefits and predictive values.

First, and most importantly, there is a need to better link indices of stress and disturbance with metrics of reproductive output and fitness. The primary stress response is in itself a characteristic that is under selection (Wingfield et al., 1998; Ricklefs and Wikelski, 2002). Studies have used magnitude of the acute stress response as suggestion that populations may be at risk of experiencing declines (Romero and Wikelski, 2001). Despite the certainty of the relationship between chronic stress and reduced reproduction, links between activation of the acute stress response and fitness outcomes currently exist, to some extent, as correlative relationships rather than causative mechanisms. Identifying activation of the stress response, either following acute or chronic stressor exposure, does not necessarily guarantee that an animal will experience reduced fitness relative to unexposed animals. The ability to confidently link acute or chronic stress responses to reductions in fitness, or negative changes to other population parameters, would represent a monumental leap forward in our understanding of the importance of the stress response. As well, our ability to predict the outcomes of exposure to stressful stimuli and use of the cortisol stress response as an early warning system for conservation would be greatly

Figure 11.2. There are a growing number of stress indicators for use in fish, including some that are reasonably novel such as: (A) use of high throughput omic techniques; (B) use of point-of-care handheld meters for measuring blood chemistry in the field; (C) whole-body extraction of cortisol from small fish; (D) evaluation of reflex status; (E) assessment of swimming performance and metabolic status; and (F) evaluation of the locomotory activity and energetics of free-swimming fish using accelerometer biologgers. Photo credits: (A) Katrina Cook, (B) Steven Cooke, (C) Julia Redfern, (D) Vivian Nguyen, (E) Zach Zucherman, (F) Jacob Brownscombe.

enhanced. A metric relating stress to reproduction could also be coupled with the concept of carryover effects allowing the impacts of a current stressor across multiple reproductive bouts, or across generations. Indices linked to reproduction could then be associated with landscape-level stressors such as habitat modifications or climate change to discern the impact of these broad challenges on populations.

Second, similar to links between stress and fitness, we feel there is a need to strengthen links between stress and fish performance. The performance of a fish is a broad concept that includes metrics such as swimming ability, aerobic scope, and scope for a stress response (reactive scope). Links between these different metrics and outcomes such as survival and food acquisition are well established in the literature (Plaut, 2001; Farrell et al., 2008). However, stress and fitness/reproduction may also be linked indirectly through declines in organism performance (ie, decreased swimming performance can lead to reduced feeding and/or an inability to escape predation). There are a number of performance metrics or reflex impairments (eg, body flex, gag response) that correlate positively with stress and can be used to predict individual mortality (Davis, 2010). Because these reflex impairments can easily and reliably be collected in the field, it would be advantageous to link these performance metrics to outcomes beyond survival, including concepts such as reduced investment in reproduction or lowered fitness. Therefore, understanding how stress impacts organismal performance, over both short and long terms, can aid in our ability to predict impacts of stress on individuals, and, ultimately, on populations.

Third, to facilitate connection between individual stress indicators and population-level processes, there is a need to further our understanding of the indicators mediating intergenerational effects. Egg size, fertilization success, and embryonic survival are established indicators of parental stress. Accompaniment of these metrics with the evaluation of physiological and behavioral traits of progeny is becoming more prevalent, and can reveal latent indicators of intergenerational stress. Still presently lacking is knowledge of gametic stress indicators driving changes to offspring phenotype. Elevated levels of egg cortisol are thought to be a reliable stress indicator that also serves as a mechanism of offspring change (Gingerich and Suski, 2011). However, whether maternal stressor exposure alters cortisol levels in eggs remains equivocal (eg, Stratholt et al., 1997 and Sopinka et al., 2014). Further experimentation is required to confirm if concentration of egg cortisol (or other hormones such as thyroid and sex steroids) is a reliable indicator of maternal stress. Also, with the advancement of molecular technologies (see later), quantification of epigenetic changes in the transcriptome of eggs, embryos, and sperm (Cabrita et al., 2014; Mommer and Bell, 2014) has the potential to serve as a valuable stress indicator. Expanding the repertoire of reproductive-based stress indicators will aid in predicting the cascading effects of stress from one generation to the next.

Fourth, research has recently demonstrated that stress hormones can be deposited and archived in structures such as fur or feathers (Bortolotti et al., 2009; Sheriff et al., 2011). The ability to extract stress hormones from structures such as fur or feathers provides a unique, long-term, integrated history of the activity of the stress axis, and can serve as a catalog of past events in the life of an animal. In addition, these structures can be collected nonlethally, and are often freely shed by animals. At present, we are aware of a single study that has measured stress hormones (cortisol) from scales (Aerts et al., 2015), demonstrating the potential to use elasmoid scales as a stress biomarker for fishes. We would encourage exploration into the area of cortisol deposition in scales as this could provide a valuable tool for both nonlethally defining the stress history of a free-swimming animal, and potentially relating stressful events in the past to reproductive output or fitness.

Finally, the last decade has seen the emergence of a number of new technologies for quantifying the molecular responses of animals to various disturbances, including tools such as transcriptomics, gene expression, and protein generation, which has provided a powerful new way to quantify how organisms interact with their environment (Evans and Hofmann, 2012). These techniques can provide reliable indices of stress, assay a number of different physiological systems simultaneously, and importantly, link changes in gene expression to ecologically relevant outcomes such as survival and fitness (Abzhanov et al., 2006). Fish are ideally suited for studies using these molecular tools because they possess a number of different nucleated tissues that can be collected nonlethally (eg, gills, Jeffries et al., 2014a; red blood cells, Dennis et al., 2015). These tools become particularly valuable when they are linked to whole-organism metrics of performance, intergenerational effects (ie, epigenetics), landscape-level challenges, or demographic patterns to define population-level trends, rather than simply cataloging the stress response of an animal. We therefore encourage the continued development and proliferation of these novel tools.

7. CONCLUSION

Exposure of fish to a stressor evokes activation of the HPI axis (primary response) and subsequent secondary and tertiary responses. Both HPI axis activity and whole-animal responses are quantified to indicate stress (see Section 3 and Tables 11.1–11.3). Quantification of stress under laboratory

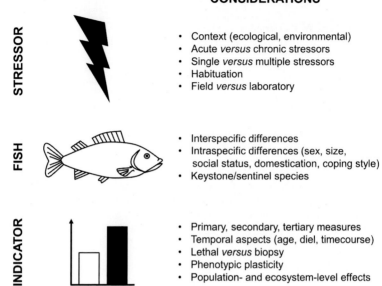

CONSIDERATIONS

STRESSOR
- Context (ecological, environmental)
- Acute *versus* chronic stressors
- Single *versus* multiple stressors
- Habituation
- Field *versus* laboratory

FISH
- Interspecific differences
- Intraspecific differences (sex, size, social status, domestication, coping style)
- Keystone/sentinel species

INDICATOR
- Primary, secondary, tertiary measures
- Temporal aspects (age, diel, timecourse)
- Lethal *versus* biopsy
- Phenotypic plasticity
- Population- and ecosystem-level effects

Figure 11.3. Considerations when quantifying and interpreting indicators of stress in fishes.

and field conditions is fundamental to understanding how a fish responds to changes in its internal and external environment. This knowledge is continuing to be implemented into the management and conservation of wild populations, as well as maintaining welfare of captive fishes. The number of different stress indicators reported in the literature is matched by the number of different factors that must be considered when quantifying and interpreting the indicators themselves (Fig. 11.3). Be it the severity or contemporaneous nature of the stressors, the variation in response within populations and among species, or the time of day the indicator is measured, our understanding of stress is challenged by many subtle and significant variables. The future of stress indicators will entail a combination of (1) optimizing experimental approaches to ensure indicators are reliable and ecologically-relevant; (2) examining stress across levels of organization within individuals (ie, molecular to whole-animal responses), as well as integrating individual responses with population- and ecosystem-level stress indices; (3) accounting for and determining indicators relevant for carryover and intergenerational effects; (4) enhancing new technologies such as biologging, telemetry, and noninvasive assessments of genomic and endocrine stress profiles; and (5) continued efforts to collaborate among disciplines.

REFERENCES

Abzhanov, A., Kuo, W. P., Hartmann, C., Grant, B. R., Grant, P. R. and Tabin, C. J. (2006). The calmodulin pathway and evolution of elongated beak morphology in Darwin's finches. *Nature* **442**, 563–567.

Adams, S. M. (1999). Ecological role of lipids in the health and success of fish populations. In Lipids in *Freshwater* Ecosystems (eds. M. T. Arts and B. C. Wainmann), pp. 132–160. New York, NY: Springer.

Adams, S. M. (2002). *Biological Indicators of Aquatic Ecosystem Stress*. Maryland: American Fisheries Society.

Adams, S. M. and Greeley, M. S. (2000). Ecotoxicological indicators of water quality: using multi-response indicators to assess the health of aquatic ecosystems. *Water Air Soil Poll.* **123**, 103–115.

Adams, S. M., Brown, A. M. and Goede, R. W. (1993). A quantitative health assessment index for rapid evaluation of fish condition in the field. Trans. Am. Fish. Soc. **122**, 63–73.

Aerts, J., Metz, J. R., Ampe, B., Decostere, A., Flik, G. and De Saeger, S. (2015). Scales tell a story on the stress history of fish. *PLoS One* **10**, e0123411.

Aissaoui, A., Tort, L. and Altimiras, J. (2000). Circadian heart rate changes and light-dependence in the Mediterranean seabream *Sparus aurata. Fish. Physiol. Biochem.* **22**, 89–94.

Al-Murrani, W. K., Al-Rawi, I. K. and Raof, N. M. (2002). Genetic resistance to *Salmonella typhimurium* in two lines of chickens selected as resistant and sensitive on the basis of heterophil/lymphocyte ratio. *Brit. Poult. Sci.* **43**, 501 507.

Alonso-Alvarez, C., Bertrand, S., Devevey, G., Prost, J., Faivre, B. and Sorci, G. (2004). Increased susceptibility to oxidative stress as a proximate cost of reproduction. *Ecol. Lett* **7**, 363–368.

Altimiras, J. and Larsen, E. (2000). Non–invasive recording of heart rate and ventilation rate in rainbow trout during rest and swimming. Fish go wireless!. *J. Fish Biol.* **57**, 197–209.

Anderson, W. G., Booth, R., Beddow, T. A., McKinley, R. S., Finstad, B., Økland, F., et al. (1998). Remote monitoring of heart rate as a measure of recovery in angled Atlantic salmon, *Salmo salar* (L.). In *Advances in Invertebrates and Fish Telemetry* (eds. J. P. Lagardere, M.-L. Begout Anras and G. Claireaux), pp. 233–240. Netherlands: Springer.

Armstrong, J. D. (1998). Relationships between heart rate and metabolic rate of pike: integration of existing data. *J. Fish Biol* **52**, 362–368.

Attrill, M. J. and Depledge, M. H. (1997). Community and population indicators of ecosystem health: targeting links between levels of biological organisation. *Aquat. Toxicol.* **38**, 183–197.

Aubin-Horth, N., Deschênes, M. and Cloutier, S. (2012). Natural variation in the molecular stress network correlates with a behavioural syndrome. *Horm. Behav.* **61**, 140–146.

Bacanskas, L. R., Whitaker, J. and Di Giulio, R. T. (2004). Oxidative stress in two populations of killifish (*Fundulus heteroclitus*) with differing contaminant histories. *Marine Environ. Res.* **58**, 597–601.

Baker, M. R. and Vynne, C. H. (2014). Cortisol profiles in sockeye salmon: sample bias and baseline values at migration, maturation, spawning, and senescence. *Fish Res.* **154**, 38–43.

Barcellos, L. J. G., Kreutz, L. C. and Quevedo, R. M. (2006). Previous chronic stress does not alter the cortisol response to an additional acute stressor in jundiá (*Rhamdia quelen*, Quoy and Gaimard) fingerlings. *Aquaculture* **253**, 317–321.

Barcellos, L. J. G., Ritter, F., Kreutz, L. C. and Cericato, L. (2010). Can zebrafish *Danio rerio* learn about predation risk? The effect of a previous experience on the cortisol response in subsequent encounters with a predator. *J. Fish Biol.* **76**, 1032–1038.

Barcellos, L. J., Koakoski, G., da Rosa, J. G., Ferreira, D., Barreto, R. E., Giaquinto, P. C., et al. (2014a). Chemical communication of predation risk in zebrafish does not depend on cortisol increase. *Sci. Rep.* **4**, 5076.

Barcellos, L. J., Woehl, V. M., Koakoski, G., Oliveira, T. A., Ferreira, D., da Rosa, J. G. S., et al. (2014b). Puberty influences stress reactivity in female catfish *Rhamdia quelen*. Physiol. Behav. **128**, 232–236.

Barreto, R. E. and Volpato, G. L. (2004). Caution for using ventilatory frequency as an indicator of stress in fish. *Behav. Process.* **66**, 43–51.

Barreto, R. E. and Volpato, G. L. (2006). Ventilatory frequency of Nile tilapia subjected to different stressors. *J. Exp. Anim. Sci.* **43**, 189–196.

Barry, T. P., Malison, J. A., Held, J. A. and Parrish, J. J. (1995). Ontogeny of the cortisol stress response in larval rainbow trout. *Gen. Comp. Endocrinol.* **97**, 57–65.

Bartell, S. M. (2006). Biomarkers, bioindicators, and ecological risk assessment- a brief review and evaluation. *Environ. Bioindic.* **1**, 60–73.

Barton, B. A. (2000). Salmonid fishes differ in their cortisol and glucose responses to handling and transport stress. *N. Am. J. Aquacul.* **62**, 12–18.

Barton, B. A. (2002). Stress in fishes: a diversity of responses with particular reference to changes in circulating corticosteroids. *Integr. Comp. Biol.* **42**, 517–525.

Barton, B. A. and Schreck, C. B. (1987). Metabolic cost of acute physical stress in juvenile steelhead. *Trans. Am. Fish. Soc.* **116**, 257–263.

Barton, B. A., Peter, R. E. and Paulencu, C. R. (1980). Plasma cortisol levels of fingerling rainbow trout (*Salmo gairdneri*) at rest, and subjected to handling, confinement, transport, and stocking. *Can. J. Fish. Aquat. Sci.* **37**, 805–811.

Barton, B. A., Schreck, C. B. and Barton, L. D. (1987). Effects of chronic cortisol administration and daily acute stress on growth, physiological conditions, and stress responses in juvenile rainbow trout. *Dis. Aquat. Organ.* **2**, 173–185.

Barton, B. A., Morgan, J. D., Vijayan, M. M. and Adams, S. M. (2002). Physiological and condition-related indicators of environmental stress in fish. In *Biological Indicators of Aquatic Ecosystem Stress* (ed. S. M. Adams), pp. 111–148. Maryland: American Fisheries Society.

Basu, N., Nakano, T., Grau, E. G. and Iwama, G. K. (2001). The effects of cortisol on heat shock protein 70 levels in two fish species. *Gen. Comp. Endocrinol.* **124**, 97–105.

Bayarri, M. J., Munoz-Cueto, J. A., López-Olmeda, J. F., Vera, L. M., De Lama, M. R., Madrid, J. A., et al. (2004). Daily locomotor activity and melatonin rhythms in Senegal sole (*Solea senegalensis*). *Physiol. Behav.* **81**, 577–583.

Beamish, F. W. H. (1978). Swimming capacity. In *Fish Physiology - Locomotion*, Vol. 7 (eds. W. S. Hoar and D. J. Randall), pp. 101–187. New York, NY: Academic Press.

Beamish, R. and McFarlane, G. A. (1983). The forgotten requirement for age validation in fisheries biology. *Trans. Am. Fish. Soc.* **112**, 735–743.

Beecham, R. V., Small, B. C. and Minchew, C. D. (2006). Using portable lactate and glucose meters for catfish research: acceptable alternatives to laboratory methods? N. Am. J. Aquacul. **68**, 291–295.

Bell, A. M., Backström, T., Huntingford, F. A., Pottinger, T. G. and Winberg, S. (2007). Variable neuroendocrine responses to ecologically-relevant challenges in sticklebacks. *Physiol. Behav.* **91**, 15–25.

Bell, A. M., Henderson, L. and Huntingford, F. A. (2010). Behavioral and respiratory responses to stressors in multiple populations of three-spined sticklebacks that differ in predation pressure. *J. Comp. Biochem. Physiol. B* **180**, 211–220.

Beyers, D. W., Rice, J. A. and Adams, S. M. (2002). Evaluating stress in fish using bioenergetics-based stressor-response models. In *Biological Indicators of Aquatic Ecosystem Stress* (ed. S. M. Adams), pp. 289–320. Maryland: American Fisheries Society.

Boisclair, D. and Leggett, W. C. (1989). The importance of activity in bioenergetics models applied to actively foraging fishes. *Can. J. Fish. Aquat. Sci.* **46**, 1859–1867.

Bolger, T. and Connolly, P. L. (1989). The selection of suitable indices for the measurement and analysis of fish condition. *J. Fish Biol.* **34**, 171–182.

Bonier, F., Martin, P. R., Moore, I. T. and Wingfield, J. C. (2009). Do baseline glucocorticoids predict fitness? *Trends Ecol. Evol.* **24**, 634–642.

Boonstra, R. (2013a). The ecology of stress: a marriage of disciplines. *Funct. Ecol.* **27**, 7–10.

Boonstra, R. (2013b). Reality as the leading cause of stress: rethinking the impact of chronic stress in nature. *Funct. Ecol.* **27**, 11–23.

Bortolotti, G. R., Marchant, T., Blas, J. and Cabezas, S. (2009). Tracking stress: localisation, deposition and stability of corticosterone in feathers. *J. Exp. Biol.* **212**, 1477–1482.

Bortone, S. A. and Davis, W. P. (1994). Fish intersexuality as indicator of environmental stress. *Bioscience* 165–172.

Breuner, C. W., Patterson, S. H. and Hahn, T. P. (2008). In search of relationships between the acute adrenocortical response and fitness. *Gen. Comp. Endocrinol.* **157**, 288–295.

Breves, J. P. and Specker, J. L. (2005). Cortisol stress response of juvenile winter flounder (*Pseudopleuronectes americanus*, Walbaum) to predators. *J. Exp. Mar. Biol. Ecol.* **325**, 1–7.

Brown, J. A. (1984). Parental care and the ontogeny of predator-avoidance in two species of centrarchid fish. *Anim. Behav.* **32**, 113–119.

Brown, C., Gardner, C. and Braithwaite, V. A. (2005). Differential stress responses in fish from areas of high-and low-predation pressure. *J. Comp. Biochem. Physiol. B* **175**, 305–312.

Brownscombe, J. W., Gutowsky, L. F., Danylchuk, A. J. and Cooke, S. J. (2014). Foraging behaviour and activity of a marine benthivorous fish estimated using tri-axial accelerometer biologgers. *Mar. Ecol. Prog. Ser.* **505**, 241–251.

Bry, C. and Zohar, Y. (1980). Dorsal aorta catheterization in rainbow trout (*Salmo gairdneri*). II. Glucocorticoid levels, hematological data and resumption of feeding for five days after surgery. *Reprod. Nutr. Dev.* **20**, 1825–1834.

Bryan, H. M., Darimont, C. T., Paquet, P. C., Wynne-Edwards, K. E. and Smits, J. E. (2013). Stress and reproductive hormones in grizzly bears reflect nutritional benefits and social consequences of a salmon foraging niche. *PloS One* **8**, e80537.

Brydges, N. M., Boulcott, P., Ellis, T. and Braithwaite, V. A. (2009). Quantifying stress responses induced by different handling methods in three species of fish. Appl. Anim. *Behav. Sci.* **116**, 295–301.

Buckley, B. A. (2007). Comparative environmental genomics in non-model species: using heterologous hybridization to DNA-based microarrays. *J. Exp. Biol.* **210**, 1602–1606.

Burnett, N. J., Hinch, S. G., Braun, D. C., Casselman, M. T., Middleton, C. T., Wilson, S. M., et al. (2014). Burst swimming in areas of high flow: delayed consequences of anaerobiosis in wild adult sockeye salmon. *Physiol. Biochem. Zool.* **87**, 587–598.

Burton, T., Killen, S. S., Armstrong, J. D. and Metcalfe, N. B. (2011). What causes intraspecific variation in resting metabolic rate and what are its ecological consequences? *Proc. Roy. Soc. B* **278**, 3465–3473.

Cabrita, E., Martínez-Páramo, S., Gavaia, P. J., Riesco, M. F., Valcarce, D. G., Sarasquete, C., et al. (2014). Factors enhancing fish sperm quality and emerging tools for sperm analysis. *Aquaculture* **432**, 389–401.

Caipang, C. M. A., Brinchmann, M. F., Berg, I., Iversen, M., Eliassen, R. and Kiron, V. (2008). Changes in selected stress and immune–related genes in Atlantic cod, *Gadus morhua*, following overcrowding. *Aquac. Res.* **39**, 1533–1540.

Calow, P. and Forbes, V. E. (1998). How do physiological responses to stress translate into ecological and evolutionary processes? *Comp. Biochem. Physiol. A* **120**, 11–16.

Campbell, P. M., Pottinger, T. G. and Sumpter, J. P. (1994). Preliminary evidence that chronic confinement stress reduces the quality of gametes produced by brown and rainbow trout. *Aquaculture* 120, 151–169.

Carey, J. B. and McCormick, S. D. (1998). Atlantic salmon smolts are more responsive to an acute handling and confinement stress than parr. *Aquaculture* 168, 237–253.

Castanheira, M. F., Conceição, L. E., Millot, S., Rey, S., Bégout, M. L., Damsgard, B., et al. (2015). Coping styles in farmed fish: consequences for aquaculture. *Rev. Aquac* 7, 1–19.

Cawthon, R. M. (2002). Telomere measurement by quantitative PCR. *Nucl. Acids Res.* 30, e47–e47.

Claireaux, G., Webber, D., Kerr, S. and Boutilier, R. (1995). Physiology and behaviour of free-swimming Atlantic cod (*Gadus morhua*) facing fluctuating salinity and oxygenation conditions. *J. Exp. Biol.* 198, 61–69.

Clark, T. D., Taylor, B. D., Seymour, R. S., Ellis, D., Buchanan, J., Fitzgibbon, Q. P., et al. (2008). Moving with the beat: heart rate and visceral temperature of free-swimming and feeding bluefin tuna. *Proc. R. Soc. B* 275, 2841–2850.

Clark, T. D., Sandblom, E., Hinch, S. G., Patterson, D. A., Frappell, P. B. and Farrell, A. P. (2010). Simultaneous biologging of heart rate and acceleration, and their relationships with energy expenditure in free-swimming sockeye salmon (*Oncorhynchus nerka*). *J. Comp. Biochem. Physiol. B* 180, 673–684.

Cnaani, A., Tinman, S., Avidar, Y., Ron, M. and Hulata, G. (2004). Comparative study of biochemical parameters in response to stress in *Oreochromis aureus*, *O. mossambicus* and two strains of *O. niloticus*. *Aquac. Res.* 35, 1434–1440.

Cone, R. S. (1989). The need to reconsider the use of condition indices in fishery science. *Trans. Am. Fish. Soc.* 118, 510–514.

Congleton, J. L. and Wagner, T. (2006). Blood-chemistry indicators of nutritional status in juvenile salmonids. *J. Fish Biol* 69, 473–490.

Cook, K. V., McConnachie, S. H., Gilmour, K. M., Hinch, S. G. and Cooke, S. J. (2011). Fitness and behavioral correlates of pre-stress and stress-induced plasma cortisol titers in pink salmon (*Oncorhynchus gorbuscha*) upon arrival at spawning grounds. *Horm. Behav.* 60, 489–497.

Cook, K. V., Crossin, G. T., Patterson, D. A., Hinch, S. G., Gilmour, K. M. and Cooke, S. J. (2014). The stress response predicts migration failure but not migration rate in a semelparous fish. *Gen. Comp. Endocrinol.* 202, 44–49.

Cooke, S. J., Hinch, S. G., Wikelski, M., Andrews, R. D., Kuchel, L. J., Wolcott, T. G., et al. (2004). Biotelemetry: a mechanistic approach to ecology. *Trends Ecol. Evol.* 19, 334–343.

Cooke, S. J., Hinch, S. G., Crossin, G. T., Patterson, D. A., English, K. K., Healey, M. C., et al. (2006). Mechanistic basis of individual mortality in Pacific salmon during spawning migrations. *Ecology* 87, 1575–1586.

Cooke, S. J., Sack, L., Franklin, C. E., Farrell, A. P., Beardall, J., Wikelski, M., et al. (2013). What is conservation physiology? Perspectives on an increasingly integrated and essential science. *Conserv. Physiol.* 1, cot001.

Cooke, S. J., Blumstein, D. T., Buchholz, R., Caro, T., Fernández-Juricic, E., Franklin, C. E., et al. (2014). Physiology, behavior, and conservation. *Physiol. Biochem. Zool.* 87, 1–14.

Costantini, D. (2008). Oxidative stress in ecology and evolution: lessons from avian studies. *Ecol. Lett.* 11, 1238–1251.

Cousineau, A., Midwood, J. D., Stamplecoskie, K., King, G., Suski, C. D. and Cooke, S. J. (2014). Diel patterns of baseline glucocorticoids and stress responsiveness in a teleost fish (bluegill, *Lepomis macrochirus*). *Can. J. Zool.* 92, 417–421.

Crossin, G. T. and Hinch, S. G. (2005). A nonlethal, rapid method for assessing the somatic energy content of migrating adult Pacific salmon. *Trans. Am. Fish. Soc.* 134, 184–191.

Crozier, L. G. and Hutchings, J. A. (2014). Plastic and evolutionary responses to climate change in fish. *Evol. App.* **7**, 68–87.

Cyr, N. E. and Romero, L. M. (2009). Identifying hormonal habituation in field studies of stress. *Gen. Comp. Endocrinol.* **161**, 295–303.

Dale, V. H. and Beyeler, S. C. (2001). Challenges in the development and use of ecological indicators. *Ecol. Indic.* **1**, 3–10.

Danylchuk, S. E., Danylchuk, A. J., Cooke, S. J., Goldberg, T. L., Koppelman, J. and Philipp, D. P. (2007). Effects of recreational angling on the post-release behavior and predation of bonefish (*Albula vulpes*): the role of equilibrium status at the time of release. *J. Exp. Mar. Biol. Ecol.* **346**, 127–133.

Davis, M. W. (2010). Fish stress and mortality can be predicted using reflex impairment. *Fish Fish.* **11**, 1–11.

Davis, M. W., Olla, B. L. and Schreck, C. B. (2001). Stress induced by hooking, net towing, elevated sea water temperature and air in sablefish: lack of concordance between mortality and physiological measures of stress. *J. Fish Biol.* **58**, 1–15.

Davis, A. K., Maney, D. L. and Maerz, J. C. (2008). The use of leukocyte profiles to measure stress in vertebrates: a review for ecologists. *Funct. Ecol.* **22**, 760–772.

DeKoning, A. L., Picard, D. J., Bond, S. R. and Schulte, P. M. (2004). Stress and interpopulation variation in glycolytic enzyme activity and expression in a teleost fish *Fundulus heteroclitus*. *Physiol. Biochem. Zool.* **77**, 18–26.

Dennis, C. E., Kates, D. F., Noatch, M. R. and Suski, C. D. (2015). Molecular responses of fishes to elevated carbon dioxide. *Comp. Biochem. Physiol. A* **187**, 224–231.

Dhabhar, F. S., Miller, A. H., McEwen, B. and Spender, R. L. (1996). Stress-induced changes in blood leukocyte distribution – role of adrenal steroid hormones. *J. Immunol.* **157**, 1638–1644.

Dickens, M. J. and Romero, L. M. (2013). A consensus endocrine profile for chronically stressed wild animals does not exist. *Gen. Comp. Endocrinol.* **191**, 177–189.

Donaldson, M. R., Cooke, S. J., Patterson, D. A. and Macdonald, J. S. (2008). Cold shock and fish. *J. Fish Biol.* **73**, 1491–1530.

Donaldson, M. R., Clark, T. D., Hinch, S. G., Cooke, S. J., Patterson, D. A., Gale, M. K., et al. (2010). Physiological responses of free-swimming adult coho salmon to simulated predator and fisheries encounters. *Physiol. Biochem. Zool.* **83**, 973–983.

Donaldson, M. R., Hinch, S. G., Patterson, D. A., Hills, J., Thomas, J. O., Cooke, S. J., et al. (2011). The consequences of angling, beach seining, and confinement on the physiology, post-release behaviour and survival of adult sockeye salmon during upriver migration. *Fish Res.* **108**, 133–141.

Donaldson, M. R., Hinch, S. G., Raby, G. D., Patterson, D. A., Farrell, A. P. and Cooke, S. J. (2012). Population-specific consequences of fisheries-related stressors on adult sockeye salmon. *Physiol. Biochem. Zool.* **85**, 729–739.

Donaldson, M. R., Hinch, S. G., Jeffries, K. M., Patterson, D. A., Cooke, S. J., Farrell, A. P., et al. (2014). Species-and sex-specific responses and recovery of wild, mature pacific salmon to an exhaustive exercise and air exposure stressor. *Comp. Biochem. Physiol. A* **173**, 7–16.

Eaton, L., Edmonds, E. J., Henry, T. B., Snellgrove, D. L. and Sloman, K. A. (2015). Mild maternal stress disrupts associative learning and increases aggression in offspring. *Horm. Behav.* **71**, 10–15.

Ellerby, D. J. and Herskin, J. (2013). Swimming flumes as a tool for studying swimming behavior and physiology: current applications and future developments. In Swimming Physiology of *Fish* (eds. A. P. Palstra and J. V. Planas), pp. 345–375. Springer Berlin Heidelberg.

Ellis, T., James, J. D., Stewart, C. and Scott, A. P. (2004). A non–invasive stress assay based upon measurement of free cortisol released into the water by rainbow trout. *J. Fish Biol.* **65**, 1233–1252.

Endo, H., Yonemori, Y., Hibi, K., Ren, H., Hayashi, T., Tsugawa, W., et al. (2009). Wireless enzyme sensor system for real-time monitoring of blood glucose levels in fish. *Biosens. Bioelectron.* **24**, 1417–1423.

Epel, E. S., Blackburn, E. H., Lin, J., Dhabhar, F. S., Adler, N. E., Morrow, J. D., et al. (2004). Accelerated telomere shortening in response to life stress. *Proc. Natl. Acad. Sci. U. S. A.* **101**, 17312–17315.

Evans, T. G. and Hofmann, G. E. (2012). Defining the limits of physiological plasticity: how gene expression can assess and predict the consequences of ocean change. *Philos. Trans. R. Soc. B* **367**, 1733–1745.

Fangue, N. A., Hofmeister, M. and Schulte, P. M. (2006). Intraspecific variation in thermal tolerance and heat shock protein gene expression in common killifish, *Fundulus heteroclitus. J. Exp. Biol.* **209**, 2859–2872.

Farrell, A. P. (1991). From hagfish to tuna: a perspective on cardiac function in fish. *Physiol. Zool.* **64**, 1137–1164.

Farrell, A. P., Lee, C. G., Tierney, K., Hodaly, A., Clutterham, S., Healey, M., et al. (2003). Field–based measurements of oxygen uptake and swimming performance with adult Pacific salmon using a mobile respirometer swim tunnel. *J. Fish Biol.* **62**, 64–84.

Farrell, A. P., Hinch, S. G., Cooke, S. J., Patterson, D. A., Crossin, G. T., Lapointe, M., et al. (2008). Pacific salmon in hot water: applying aerobic scope models and biotelemetry to predict the success of spawning migrations. *Physiol. Biochem. Zool.* **81**, 697–709.

Fatira, E., Papandroulakis, N. and Pavlidis, M. (2014). Diel changes in plasma cortisol and effects of size and stress duration on the cortisol response in European sea bass (*Dicentrarchus labrax*). *Fish. Physiol. Biochem.* **40**, 911–919.

Faught, E., Aluru, N. and Vijayan, M. M. (2016). The Molecular Stress Response. In *Fish Physiology - Biology of Stress in Fish*, Vol. 35 (eds. C. B. Schreck, L. Tort, A. P. Farrell and C. J. Brauner), San Diego, CA: Academic Press.

Fefferman, N. H. and Romero, L. M. (2013). Can physiological stress alter population persistence? A model with conservation implications. *Conserv. Physiol.* **1**, cot012.

Feist, G. and Schreck, C. B. (2001). Ontogeny of the stress response in chinook salmon, *Oncorhynchus tshawytscha. Fish. Physiol. Biochem.* **25**, 31–40.

Fevolden, S. E., Refstie, T. and Røed, K. H. (1991). Selection for high and low cortisol stress response in Atlantic salmon (*Salmo salar*) and rainbow trout (*Oncorhynchus mykiss*). *Aquaculture* **95**, 53–65.

Fürtbauer, I., King, A. J. and Heistermann, M. (2015). Visible implant elastomer (VIE) tagging and simulated predation risk elicit similar physiological stress responses in three–spined stickleback *Gasterosteus aculeatus. J. Fish Biol.* **86**, 1644–1649.

Gallo, V. P. and Civinini, A. (2003). Survey of the adrenal homolog in teleosts. *Inter. Rev. Cytol.* **230**, 89–187.

Gende, S. M., Edwards, R. T., Willson, M. F. and Wipfli, M. S. (2002). Pacific salmon in aquatic and terrestrial ecosystems. *Bioscience* **52**, 917–928.

Geslin, M. and Auperin, B. (2004). Relationship between changes in mRNAs of the genes encoding steroidogenic acute regulatory protein and P450 cholesterol side chain cleavage in head kidney and plasma levels of cortisol in response to different kinds of acute stress in the rainbow trout (*Oncorhynchus mykiss*). *Gen. Comp. Endocrinol.* **135**, 70–80.

Gesto, M., López-Patiño, M. A., Hernández, J., Soengas, J. L. and Míguez, J. M. (2013). The response of brain serotonergic and dopaminergic systems to an acute stressor in rainbow trout: a time course study. *J. Exp. Biol.* **216**, 4435–4442.

Gesto, M., López–Patiño, M. A., Hernández, J., Soengas, J. L. and Míguez, J. M. (2015). Gradation of the stress response in rainbow trout exposed to stressors of different severity: the role of brain serotonergic and dopaminergic systems. *J. Neuroendocrinol.* **27**, 131–141.

Giacomini, A. C. V. V., de Abreu, M. S., Koakoski, G., Idalêncio, R., Kalichak, F., Oliveira, T. A., et al. (2015). My stress, our stress: blunted cortisol response to stress in isolated housed zebrafish. *Phys. Behav.* **139**, 182–187.

Gingerich, A. J. and Suski, C. D. (2011). The role of progeny quality and male size in the nesting success of smallmouth bass: integrating field and laboratory studies. *Aquat. Ecol.* **45**, 505–515.

Gingerich, A. J., Philipp, D. P. and Suski, C. D. (2010). Effects of nutritional status on metabolic rate, exercise and recovery in a freshwater fish. *J. Comp. Biochem. Physiol. B* **180**, 371–384.

Godin, J. G. J. (1997). *Behavioural Ecology of Teleost Fishes*. Oxford: Oxford University Press.

Goede, R. W. and Barton, B. A. (1990). Organismic indices and an autopsy-based assessment as indicators of health and condition of fish. *Am. Fish. Soc. Symp.* **8**, 93–108.

Gorissen, M. and Flik, G. (2016). Endocrinology of the Stress Response in Fish. In *Fish Physiology - Biology of Stress in Fish*, Vol. 35 (eds. C. B. Schreck, L. Tort, A. P. Farrell and C. J. Brauner), San Diego, CA: Academic Press.

Gracey, A. Y., Troll, J. V. and Somero, G. N. (2001). Hypoxia-induced gene expression profiling in the euryoxic fish *Gillichthys mirabilis*. *Proc. Natl. Acad. Sci. U.S.A.* **98**, 1993–1998.

Gutowsky, L. F., Aslam, W., Banisaeed, R., Bell, L. R., Bove, K. L., Brownscombe, J. W., et al. (2015). Considerations for the design and interpretation of fishing release mortality estimates. *Fish. Res.* **167**, 64–70.

Hammer, C. (1995). Fatigue and exercise tests with fish. *Comp. Biochem. Physiol. A* **112**, 1–20.

Hansen, J. A., Marr, J. C., Lipton, J., Cacela, D. and Bergman, H. L. (1999). Differences in neurobehavioral responses of chinook salmon (*Oncorhynchus tshawytscha*) and rainbow trout (*Oncorhynchus mykiss*) exposed to copper and cobalt: behavioral avoidance. *Environ. Toxicol. Chem.* **18**, 1972–1978.

Henderson, R. J. and Tocher, D. R. (1987). The lipid composition and biochemistry of freshwater fish. *Prog. Lipid Res.* **26**, 281–347.

Henry, J. B. (1996). *Clinical Diagnosis and Management by Laboratory Methods*. Philadelphia: Saunders.

Hori, T. S., Gamperl, A. K., Hastings, C. E., Vander Voort, G. E., Robinson, J. A. B., Johnson, S. C., et al. (2012). Inter-individual and-family differences in the cortisol responsiveness of Atlantic cod (*Gadus morhua*). *Aquaculture* **324**, 165–173.

Hughes, P. and Dragunow, M. (1995). Induction of immediate-early genes and the control of neurotransmitter-regulated gene expression within the nervous system. *Pharmacol. Rev.* **47**, 133–178.

Inuzuka, H., Nanbu-Wakao, R., Masuho, Y., Mmrmatsu, M., Tojo, H. and Wakao, H. (1999). Differential regulation of immediate early gene expression in preadipocyte cells through multiple signaling pathways. *Biochem. Biophys. Res. Commun.* **265**, 664–668.

Iwama, G. K. (2007). The welfare of fish. *Dis. Aquat. Organ.* **75**, 155–158.

Iwama, G. K., McGeer, J. C. and Pawluk, M. P. (1989). The effects of five fish anaesthetics on acid-base balance, hematocrit, blood gases, cortisol, and adrenaline in rainbow trout. *Can. J. Zool.* **67**, 2065–2073.

Fish stress and health. In *Aquaculture* Society for Experimental Biology Seminar Series, vol. 62 (G. K. Iwama, A. D. Pickering, J. P. Sumpter and C. B. Schreck), Cambridge: Cambridge University Press.

Iwama, G. K., Thomas, P. T., Forsyth, R. B. and Vijayan, M. M. (1998). Heat shock protein expression in fish. *Rev. Fish Biol. Fish.* **8**, 35–56.

Iwama, G. K., Afonso, L. O. B., Todgham, A., Ackerman, P. and Nakano, K. (2004). Are hsps suitable for indicating stressed states in fish? *J. Exp. Biol.* **207**, 15–19.

Iwama, G. K., Afonso, L. O. B. and Vijayan, M. M. (2006). Stress in fishes. In *The Physiology of Fishes* (eds. D. H. Evans and J. B. Claiborne), pp. 319–342. Boca Raton: Taylor & Francis.

Jacobsen, L., Baktoft, H., Jepsen, N., Aarestrup, K., Berg, S. and Skov, C. (2014). Effect of boat noise and angling on lake fish behaviour. *J. Fish Biol.* **84**, 1768–1780.

Jagoe, C. H. and Haines, T. A. (1985). Fluctuating asymmetry in fishes inhabiting acidified and unacidified lakes. *Can. J. Zool.* **63**, 130–138.

Jain, K. E., Hamilton, J. C. and Farrell, A. P. (1997). Use of a ramp velocity test to measure critical swimming speed in rainbow trout (*Onchorhynchus mykiss*). *Comp. Biochem. Physiol. A* **117**, 441–444.

Jain, K. E., Birtwell, I. K. and Farrell, A. P. (1998). Repeat swimming performance of mature sockeye salmon following a brief recovery period: a proposed measure of fish health and water quality. *Can. J. Zool.* **76**, 1488–1496.

Jeffrey, J. D., Gollock, M. J. and Gilmour, K. M. (2014a). Social stress modulates the cortisol response to an acute stressor in rainbow trout (*Oncorhynchus mykiss*). *Gen. Comp. Endocrinol.* **196**, 8–16.

Jeffrey, J. D., Cooke, S. J. and Gilmour, K. M. (2014b). Regulation of hypothalamic–pituitary–interrenal axis function in male smallmouth bass (*Micropterus dolomieu*) during parental care. *Gen. Comp. Endocrinol.* **204**, 195–202.

Jeffries, K. M., Hinch, S. G., Sierocinski, T., Clark, T. D., Eliason, E. J., Donaldson, M. R., et al. (2012). Consequences of high temperatures and premature mortality on the transcriptome and blood physiology of wild adult sockeye salmon (*Oncorhynchus nerka*). *Ecol. Evol.* **2**, 1747–1764.

Jeffries, K. M., Hinch, S. G., Gale, M. K., Clark, T. D., Lotto, A. G., Casselman, M. T., et al. (2014a). Immune response genes and pathogen presence predict migration survival in wild salmon smolts. *Mol. Ecol.* **23**, 5803–5815.

Jeffries, K. M., Hinch, S. G., Sierocinski, T., Pavlidis, P. and Miller, K. M. (2014b). Transcriptomic responses to high water temperature in two species of Pacific salmon. *Evol. Appl.* **7**, 286–300.

Jeffries, K. M., Brander, S. M., Britton, M. T., Fangue, N. A. and Connon, R. E. (2015). Chronic exposures to low and high concentrations of ibuprofen elicit different gene response patterns in a euryhaline fish. *Environ. Sci. Pollut. Res.* **22** (22), 17397–17413. In press.

Jentoft, S., Held, J. A., Malison, J. A. and Barry, T. P. (2002). Ontogeny of the cortisol stress response in yellow perch (*Perca flavescens*). *Fish. Physiol. Biochem.* **26**, 371–378.

Jentoft, S., Aastveit, A. H., Torjesen, P. A. and Andersen, Ø. (2005). Effects of stress on growth, cortisol and glucose levels in non-domesticated Eurasian perch (*Perca fluviatilis*) and domesticated rainbow trout (*Oncorhynchus mykiss*). *Comp. Biochem. Physiol. A* **141**, 353–358.

Johnsson, J. I., Höjesjö, J. and Fleming, I. A. (2001). Behavioural and heart rate responses to predation risk in wild and domesticated Atlantic salmon. *Can. J. Fish. Aquat. Sci.* **58**, 788–794.

Karr, J. R. (1981). Assessment of biotic integrity using fish communities. *Fisheries* **6**, 21–27.

Kassahn, K. S., Crozier, R. H., Pörtner, H. O. and Caley, M. J. (2009). Animal performance and stress: responses and tolerance limits at different levels of biological organisation. *Biol. Rev.* **84**, 277–292.

Kebus, M. J., Collins, M. T., Brownfield, M. S., Amundson, C. H., Kayes, T. B. and Malison, J. A. (1992). Effects of rearing density on the stress response and growth of rainbow trout. *J. Aquat. Anim. Health* **4**, 1–6.

Khan, A. T. and Weis, J. S. (1987). Toxic effects of mercuric chloride on sperm and egg viability of two populations of mummichog, *Fundulus heteroclitus. Environ. Pollut.* **48**, 263–273.

Kilgas, P., Tilgar, V. and Mand, R. (2006). Hematological health state indices predict local survival in a small passerine bird, the great tit (*Parus major*). *Physiol. Biochem. Zool.* **79**, 565–572.

Kittilsen, S., Ellis, T., Schjolden, J., Braastad, B. O. and Øverli, Ø. (2009). Determining stress-responsiveness in family groups of Atlantic salmon (*Salmo salar*) using non-invasive measures. *Aquaculture* **298**, 146–152.

Koakoski, G., Oliveira, T. A., da Rosa, J. G. S., Fagundes, M., Kreutz, L. C. and Barcellos, L. J. G. (2012). Divergent time course of cortisol response to stress in fish of different ages. *Physiol. Behav.* **106**, 129–132.

Kotrschal, A., Ilmonen, P. and Penn, D. J. (2007). Stress impacts telomere dynamics. *Biol. Lett.* **3**, 128–130.

Krasnov, A., Koskinen, H., Pehkonen, P., Rexroad, C. E., Afanasyev, S. and Mölsä, H. (2005). Gene expression in the brain and kidney of rainbow trout in response to handling stress. *BMC Genomics* **6**, 3.

Kultz, D. (2005). Molecular and evolutionary basis of the cellular stress response. *Annu. Rev. Physiol.* **67**, 225–257.

Kushner, R. F. (1992). Bioelectrical impedance analysis: a review of principles and applications. *J. Am. Coll. Nutr.* **11**, 199–209.

Landsman, S. J., Wachelka, H. J., Suski, C. D. and Cooke, S. J. (2011). Evaluation of the physiology, behaviour, and survival of adult muskellunge (*Esox masquinongy*) captured and released by specialized anglers. *Fish Res.* **110**, 377–386.

Leary, R. F. and Allendorf, F. W. (1989). Fluctuating asymmetry as an indicator of stress: implications for conservation biology. *Trends Ecol. Evol.* **4**, 214–217.

Lefrançois, C., Claireaux, G. and Lagardere, J. P. (1998). Heart rate telemetry to study environmental influences on fish metabolic expenditure. In Advances in Invertebrates *and Fish Telemetry* (eds. J. P. Lagardere, M.-L. Begout Anras and G. Claireaux), pp. 215–224. Netherlands: Springer.

Lepage, O., Overli, O., Petersson, E., Jarvi, T. and Winberg, S. (2000). Differential stress coping in wild and domesticated sea trout. *Brain Behav. Evol.* **56**, 259–268.

Lesser, M. P. (2006). Oxidative stress in marine environments: biochemistry and physiological ecology. *Annu. Rev. Physiol.* **68**, 253–278.

Lobato, E., Moreno, J., Merino, S., Sanz, J. J. and Arriero, E. (2005). Haeomatological variables are good predictors of recruitment in nestling pied flycatchers (*Ficedula hypoleuca*). *Ecoscience* **12**, 27–34.

Love, O. P., McGowan, P. O. and Sheriff, M. J. (2013). Maternal adversity and ecological stressors in natural populations: the role of stress axis programming in individuals, with implications for populations and communities. *Funct. Ecol.* **27**, 81–92.

MacDougall-Shackleton, S. A., Dindia, L., Newman, A. E. M., Potvin, D. A., Stewart, K. A. and MacDougall-Shackleton, E. A. (2009). Stress, song and survival in sparrows. *Biol. Lett.* **5**, 746–748.

Madliger, C. L. and Love, O. P. (2014). The need for a predictive, context–dependent approach to the application of stress hormones in conservation. *Conserv. Biol.* **28**, 283–287.

Mallo, G. V., Fiedler, F., Calvo, E. L., Ortiz, E. M., Vasseur, S., Keim, V., et al. (1997). Cloning and expression of the rat p8 cDNA, a new gene activated in pancreas during the acute phase of pancreatitis, pancreatic development, and regeneration, and which promotes cellular growth. *J. Biol. Chem.* **272**, 32360–32369.

Marçalo, A., Mateus, L., Correia, J. H. D., Serra, P., Fryer, R. and Stratoudakis, Y. (2006). Sardine (*Sardina pilchardus*) stress reactions to purse seine fishing. *Mar. Biol.* **149**, 1509–1518.

Marcogliese, D. J., Brambilla, L. G., Gagné, F. and Gendron, A. D. (2005). Joint effects of parasitism and pollution on oxidative stress biomarkers in yellow perch *Perca flavescens*. *Dis. Aquat. Organ.* **63**, 77–84.

Marentette, J. R., Gooderham, K. L., McMaster, M. E., Ng, T., Parrott, J. L., Wilson, J. Y., et al. (2010). Signatures of contamination in invasive round gobies (*Neogobius melanostomus*): a double strike for ecosystem health?. *Ecotoxicol. Environ. Saf.* **73**, 1755–1764.

Martins, E. G., Hinch, S. G., Patterson, D. A., Hague, M. J., Cooke, S. J., Miller, K. M., et al. (2012). High river temperature reduces survival of sockeye salmon (*Oncorhynchus nerka*) approaching spawning grounds and exacerbates female mortality. *Can. J. Fish. Aquat. Sci.* **69**, 330–342.

Mazeaud, M. M. and Mazeaud, F. (1981). Adrenergic responses to stress in fish. In Stress and Fish (ed. A. D. Pickering), pp. 49–75. London: Academic Press.

Mazeaud, M. M., Mazeaud, F. and Donaldson, E. M. (1977). Primary and secondary effects of stress in fish: some new data with a general review. *Trans. Am. Fish. Soc.* **106**, 201–212.

McCallum, H. (2000). *Population Parameters: Estimation for Ecological Models*. Oxford: Blackwell Publishing Ltd.

McCallum, E. S., Charney, R. E., Marenette, J. R., Young, J. A., Koops, M. A., Earn, D. J., et al. (2014). Persistence of an invasive fish (*Neogobius melanostomus*) in a contaminated ecosystem. *Biol. Invas.* **16**, 2449–2461.

McDonald, G. and Milligan, L. (1997). Ionic, osmotic and acid-base regulation in stress. In *Fish Stress and Health in Aquaculture* (eds. G. K. Iwama, A. D. Pickering, J. P. Sumpter and C. B. Schreck), pp. 119–145. Cambridge: Cambridge University Press.

McEwen, B. S. and Wingfield, J. C. (2003). The concept of allostasis in biology and biomedicine. *Horm. Behav.* **43**, 2–15.

Milano, D., Lozada, M. and Zagarese, H. E. (2010). Predator-induced reaction patterns of landlocked *Galaxias maculatus* to visual and chemical cues. *Aquat. Ecol.* **44**, 741–748.

Miller, K. M., Schulze, A. D., Ginther, N., Li, S., Patterson, D. A., Farrell, A. P., et al. (2009). Salmon spawning migration: metabolic shifts and environmental triggers. *Comp. Biochem. Physiol. D* **4**, 75–89.

Miller, K. M., Li, S., Kaukinen, K. H., Ginther, N., Hammill, E., Curtis, J. M., et al. (2011). Genomic signatures predict migration and spawning failure in wild Canadian salmon. *Science* **331**, 214–217.

Mommer, B. C. and Bell, A. M. (2014). Maternal experience with predation risk influences genome-wide embryonic gene expression in threespined sticklebacks (*Gasterosteus aculeatus*). *PloS One* **9**, e98564.

Mommsen, T. P., Vijayan, M. M. and Moon, T. W. (1999). Cortisol in teleosts: dynamics, mechanisms of action, and metabolic regulation. *Rev. Fish Biol. Fish.* **9**, 211–268.

Momoda, T. S., Schwindt, A. R., Feist, G. W., Gerwick, L., Bayne, C. J. and Schreck, C. B. (2007). Gene expression in the liver of rainbow trout, *Oncorhynchus mykiss*, during the stress response. *Comp. Biochem. Physiol. D* **2**, 303–315.

Monaghan, P., Metcalfe, N. B. and Torres, R. (2009). Oxidative stress as a mediator of life history trade-offs: mechanisms, measurements and interpretation. *Ecol. Lett.* **12**, 75–92.

Monaghan, P. and Haussmann, M. F. (2006). Do telomere dynamics link lifestyle and lifespan? *Trends Ecol. Evol* **21**, 47–53.

Moreno, J., Merino, S., Martinez, J., Sanz, J. J. and Arriero, E. (2002). Heterophil/lymphocyte ratios and heat-shock protein levels are related to growth in nestling birds. *Ecoscience* **9**, 434–439.

Morrissey, M. B., Suski, C. D., Esseltine, K. R. and Tufts, B. L. (2005). Incidence and physiological consequences of decompression in smallmouth bass after live-release angling tournaments. *Trans. Am. Fish. Soc.* **134**, 1038–1047.

Moseley, P. L. (1997). Heat shock proteins and heat adaptation of the whole organism. *J. Appl. Physiol.* **83**, 1413–1417.

Nelson, J. A. and Chabot, D. (2011). General energy metabolism. In *Encyclopedia of Fish Physiology: From Genome to Environment*, vol. 3 (ed. A. P. Farrell), pp. 1566–1572. San Diego: Academic Press.

Nelson, J. A., Gotwalt, P. S., Reidy, S. P. and Webber, D. M. (2002). Beyond U_{crit}: matching swimming performance tests to the physiological ecology of the animal, including a new fish 'drag strip'. *Comp. Biochem. Physiol. A* **133**, 289–302.

Noga, E. J., Kerby, J. H., King, W., Aucoin, D. P. and Giesbrecht, F. (1994). Quantitative comparison of the stress response of striped bass (*Morone saxatilis*) and hybrid striped bass (*Morone saxatilis* x *Morone chrysops* and *Morone saxatilis* x *Morone americana*). *Am. J. Vet. Res.* **55**, 405–409.

O'Connor, C. M. and Cooke, S. J. (2015). Ecological carryover effects complicate conservation. *Ambio* **44** (6), 582–591.

O'Connor, C. M., Gilmour, K. M., Arlinghaus, R., Hasler, C. T., Philipp, D. P. and Cooke, S. J. (2010). Seasonal carryover effects following the administration of cortisol to a wild teleost fish. *Physiol. Biochem. Zool.* **83**, 950–957.

O'Connor, C. M., Gilmour, K. M., Arlinghaus, R., Matsumura, S., Suski, C. D., Philipp, D. P., et al. (2011). The consequences of short-term cortisol elevation on individual physiology and growth rate in wild largemouth bass. *Can. J. Fish. Aquat. Sci.* **68**, 693–705.

O'Connor, C. M., Norris, D. R., Crossin, G. T. and Cooke, S. J. (2014). Biological carryover effects: linking common concepts and mechanisms in ecology and evolution. *Ecosphere* **5**, art28.

Odum, E. P. (1985). Trends expected in stressed ecosystems. *Bioscience* **35**, 419–422.

Oswald, R. L. (1978). The use of telemetry to study light synchronization with feeding and gill ventilation rates in *Salmo trutta*. *J. Fish Biol.* **13**, 729–739.

Otto, D. M. and Moon, T. W. (1996). Endogenous antioxidant systems of two teleost fish, the rainbow trout and the black bullhead, and the effect of age. *Fish. Physiol. Biochem.* **15**, 349–358.

Øverli, Ø., Sørensen, C., Pulman, K. G., Pottinger, T. G., Korzan, W., Summers, C. H., et al. (2007). Evolutionary background for stress-coping styles: relationships between physiological, behavioral, and cognitive traits in non-mammalian vertebrates. *Neurosci. Biobehav. Rev.* **31**, 396–412.

Palmisano, A. N., Winton, J. R. and Dickhoff, W. W. (2000). Tissue-specific induction of Hsp90 mRNA and plasma cortisol response in chinook salmon following heat shock, seawater challenge, and handling challenge. *Mar. Biotechnol.* **2**, 329–338.

Pankhurst, N. W. and Van Der Kraak, G. (1997). Effects of stress on reproduction and growth of fish. In *Fish Stress and Health in Aquaculture (Society for Experimental Biology Seminar Series*, vol. 62 (eds. G. K. Iwama, A. D. Pickering, J. P. Sumpter and C. B. Schreck), pp. 73–93. Cambridge: Cambridge University Press.

Pickering, A. D., Pottinger, T. G. and Christie, P. (1982). Recovery of the brown trout, *Salmo trutta* L., from acute handling stress: a time–course study. *J. Fish. Biol.* **20**, 229–244.

Pine, W. E., Pollock, K. H., Hightower, J. E., Kwak, T. J. and Rice, J. A. (2003). A review of tagging methods for estimating fish population size and components of mortality. *Fisheries* **28**, 10–23.

Plaut, I. (2001). Critical swimming speed: its ecological relevance. *Comp. Biochem. Physiol. A* **131**, 41–50.

Portz, D. E. (2007). *Fish-Holding-Associated Stress in Sacramento River Chinook Salmon (*Oncorhynchus tshawytscha*) at South Delta Fish Salvage Operations: Effects on Plasma Constituents, Swimming Performance, and Predator Avoidance, PhD Dissertation*. Davis: University of California.

Portz, D. E., Woodley, C. M. and Cech, J. J., Jr (2006). Stress-associated impacts of short-term holding on fishes. *Rev. Fish Biol. Fish.* **16**, 125–170.

Pottinger, T. G. (2008). The stress response fish: mechanisms, effects and measurements. In *Fish Welfare* (ed. E. J. Branson), pp. 32–48. Oxford: Blackwell Publishing Ltd.

Pottinger, T. G. (2010). A multivariate comparison of the stress response in three salmonid and three cyprinid species: evidence for inter–family differences. *J. Fish Biol.* **76**, 601–621.

Pottinger, T. G. and Carrick, T. R. (1999). Modification of the plasma cortisol response to stress in rainbow trout by selective breeding. *Gen. Comp. Endo* **116**, 122–132.

Pottinger, T. G. and Carrick, T. R. (2000). Contrasting seasonal modulation of the stress response in male and female rainbow trout. *J. Fish Biol.* **56**, 667–675.

Pottinger, T. G. and Carrick, T. R. (2001). Stress responsiveness affects dominant–subordinate relationships in rainbow trout. *Horm. Behav.* **40**, 419–427.

Power, M. (2002). Assessing fish population responses to stress. In *Biological Indicators of Aquatic Ecosystem Stress* (ed. S. M. Adams), pp. 379–429. Maryland: American Fisheries Society.

Priede, I. G. (1983). Heart rate telemetry from fish in the natural environment. *Comp. Biochem. Physiol. A* **76**, 515–524.

Prunet, P., Cairns, M. T., Winberg, S. and Pottinger, T. G. (2008). Functional genomics of stress responses in fish. *Rev. Fish. Sci.* **16**, 157–166.

Rabben, H. and Furevik, D. M. (1993). Application of heart rate transmitters in behaviour studies on Atlantic halibut (*Hippoglossus hippoglossus*). *Aquac. Eng.* **12**, 129–140.

Raby, G. D., Donaldson, M. R., Hinch, S. G., Patterson, D. A., Lotto, A. G., Robichaud, D., et al. (2012). Validation of reflex indicators for measuring vitality and predicting the delayed mortality of wild coho salmon bycatch released from fishing gears. *J. App. Ecol.* **49**, 90–98.

Raby, G. D., Clark, T. D., Farrell, A. P., Patterson, D. A., Bett, N. N., Wilson, S. M., et al. (2015). Facing the river gauntlet: understanding the effects of fisheries capture and water temperature on the physiology of coho salmon. *PLoS One* e0124023.

Randall, D. J. and Perry, S. F. (1992). In *Catecholamines. In Fish Physiology, vol. 12 Part B* (eds. W. S. Hoar, D. J. Randall and A. P. Farrell), pp. 255–300. New York, NY: Academic Press.

Rankin, M. A. and Burchsted, J. C. A. (1992). The cost of migration in insects. *Annu. Rev. Entomol* **37**, 533–559.

Reebs, S. G. (2002). Plasticity of diel and circadian activity rhythms in fishes. *Rev. Fish Biol. Fish.* **12**, 349–371.

Reid, S. G., Bernier, N. J. and Perry, S. F. (1998). The adrenergic stress response in fish: control of catecholamine storage and release. *Comp. Biochem. Physiol. C* **120**, 1–27.

Richter, T. and von Zglinicki, T. (2007). A continuous correlation between oxidative stress and telomere shortening in fibroblasts. *Exp. Gerontol.* **42**, 1039–1042.

Ricklefs, R. E. (2008). The evolution of senescence from a comparative perspective. *Funct. Ecol.* **22**, 379–392.

Ricklefs, R. E. and Wikelski, M. (2002). The physiology/life-history nexus. *Trends Ecol. Evol.* **17**, 462–468.

Romero, L. M. and Wikelski, M. (2001). Corticosterone levels predict survival probabilities of Galapagos marine iguanas during El Nino events. *Proc. Natl. Acad. Sci. U. S. A.* **98**, 7366–7370.

Romero, L. M. and Reed, J. M. (2005). Collecting baseline corticosterone samples in the field: is under 3 min good enough?. *Comp. Biochem. Physiol. A* **140**, 73–79.

Romero, L. M., Dickens, M. J. and Cyr, N. E. (2009). The reactive scope model—a new model integrating homeostasis, allostasis, and stress. *Horm. Behav.* **55**, 375–389.

Ros, A. F., Vullioud, P., Bruintjes, R., Vallat, A. and Bshary, R. (2014). Intra-and interspecific challenges modulate cortisol but not androgen levels in a year-round territorial damselfish. *J. Exp. Biol.* **217**, 1768–1774.

Ryer, C. H., Ottmar, M. L. and Sturm, E. A. (2004). Behavioral impairment after escape from trawl codends may not be limited to fragile fish species. *Fish Res.* **66**, 261–269.

Sadoul, B. and Vijayan, M. M. (2016). Stress and Growth. In *Fish Physiology - Biology of Stress in Fish*, Vol. 35 (eds. C. B. Schreck, L. Tort, A. P. Farrell and C. J. Brauner), San Diego, CA: Academic Press.

Sapolsky, R. M., Romero, L. M. and Munck, A. U. (2000). How do glucocorticoids influence stress responses? Integrating permissive, suppressive, stimulatory, and preparative actions 1. *Endocr. Rev* **21**, 55–89.

Schreck, C. B. (2000). Accumulation and long-term effects of stress in fish. In *The Biology of Animal Stress* (eds. G. Moberg and J. A. Mench), pp. 147–158. Oxon: CABI Publishing.

Schreck, C. B. (2010). Stress and fish reproduction: the roles of allostasis and hormesis. *Gen. Comp. Endocrinol.* **165**, 549–556.

Schreck, C. B., Olla, B. L. and Davis, M. W. (1997). Behavioral responses to stress. In *Fish Stress and Health in Aquaculture (Society for Experimental Biology Seminar Series)*, vol. 62 (eds. G. K. Iwama, A. D. Pickering, J. P. Sumpter and C. B. Schreck), pp. 145–170. Cambridge: Cambridge University Press.

Schreck, C. B., Contreras-Sanchez, W. and Fitzpatrick, M. S. (2001). Effects of stress on fish reproduction, gamete quality, and progeny. *Aquaculture* **197**, 3–24.

Schreck, C. B. and Tort, L. (2016). The Concept of Stress in Fish. In *Fish Physiology - Biology of Stress in Fish*, Vol. 35 (eds. C. B. Schreck, L. Tort, A. P. Farrell and C. J. Brauner), San Diego, CA: Academic Press.

Schulte, P. M. (2014). What is environmental stress? Insights from fish living in a variable environment. *J. Exp. Biol.* **217**, 23–34.

Scott, G. R. and Sloman, K. A. (2004). The effects of environmental pollutants on complex fish behaviour: integrating behavioural and physiological indicators of toxicity. *Aquat. Toxicol.* **68**, 369–392.

Semenza, G. L. (1998). Hypoxia-inducible factor 1: master regulator of O_2 homeostasis. *Curr. Opin. Gen. Dev.* **8**, 588–594.

Sheriff, M. J. and Love, O. P. (2013). Determining the adaptive potential of maternal stress. *Ecol. Lett.* **16**, 271–280.

Sheriff, M. J., Dantzer, B., Delehanty, B., Palme, R. and Boonstra, R. (2011). Measuring stress in wildlife: techniques for quantifying glucocorticoids. *Oecologia* **166**, 869–887.

Shuter, B. J. (1990). Population-level indicators of stress. *Am. Fish. Soc. Symp.* **8**, 145–166.

Silvestre, F., Gillardin, V. and Dorts, J. (2012). Proteomics to assess the role of phenotypic plasticity in aquatic organisms exposed to pollution and global warming. *Integr. Comp. Biol.* **52**, 681–694.

Sloman, K. A. and McNeil, P. L. (2012). Using physiology and behaviour to understand the responses of fish early life stages to toxicants. *J. Fish Biol.* **81**, 2175–2198.

Sloman, K. A., Motherwell, G., O'Connor, K. I. and Taylor, A. C. (2000). The effect of social stress on the standard metabolic rate (SMR) of brown trout, *Salmo trutta*. *Fish. Physiol. Biochem.* **23**, 49–53.

Somero, G. N. and Hofmann, G. E. (1996). Temperature thresholds for protein adaptation: when does temperature change start to 'hurt'? In *Global Warming: Implications for Freshwater and Marine Fish (Society for Experimental Biology Seminar Series)*, vol. 61 (eds. C. M. Wood and D. G. McDonald), pp. 1–24. Cambridge: Cambridge University Press.

Sopinka, N. M., Marentette, J. R. and Balshine, S. (2010). Impact of contaminant exposure on resource contests in an invasive fish. *Behav. Ecol. Sociobiol.* **64**, 1947–1958.

Sopinka, N. M., Hinch, S. G., Middleton, C. T., Hills, J. A. and Patterson, D. A. (2014). Mother knows best, even when stressed? Effects of maternal exposure to a stressor on offspring performance at different life stages in a wild semelparous fish. *Oecologia* **175**, 493–500.

Stoot, L. J., Cairns, N. A., Cull, F., Taylor, J. J., Jeffrey, J. D., Morin, F., et al. (2014). Use of portable blood physiology point-of-care devices for basic and applied research on vertebrates: a review. *Conserv. Physiol.* **2**, cou011.

Stratholt, M. L., Donaldson, E. M. and Liley, N. R. (1997). Stress induced elevation of plasma cortisol in adult female coho salmon (*Oncorhynchus kisutch*), is reflected in egg cortisol content, but does not appear to affect early development. *Aquaculture* **158**, 141–153.

Struthers, D. P., Danylchuk, A. J., Wilson, A. D. M. and Cooke, S. J. (2015). Action cameras: bringing aquatic and fisheries research into view. *Fisheries* **40**, 502–512.

Suski, C. D. and Philipp, D. P. (2004). Factors affecting the vulnerability to angling of nesting male largemouth and smallmouth bass. *Trans. Am. Fish. Soc.* **133**, 1100–1106.

Tomanek, L. and Somero, G. N. (1999). Evolutionary and acclimation-induced variation in the heat-shock responses of congeneric marine snails (genus *Tegula*) from different thermal habitats: implications for limits of thermotolerance and biogeography. *J. Exp. Biol.* **202**, 2925–2936.

Tort, L., Montero, D., Robaina, L., Fernández–Palacios, H. and Izquierdo, M. S. (2001). Consistency of stress response to repeated handling in the gilthead sea bream *Sparus aurata* Linnaeus, 1758. *Aquac. Res.* **32**, 593–598.

Torres, R. and Velando, A. (2007). Male reproductive senescence: the price of immune-induced oxidative damage on sexual attractiveness in the blue-footed booby. *J. Anim. Ecol* **76**, 1161–1168.

Trushenski, J. T., Bowker, J. D., Gause, B. R. and Mulligan, B. L. (2012). Chemical and electrical approaches to sedation of hybrid striped bass: induction, recovery, and physiological responses to sedation. *Trans. Am. Fish. Soc.* **141**, 455–467.

Valavandis, A., Vlahogianni, T., Dassenakis, M. and Scoullos, M. (2006). Molecular biomarkers of oxidative stress in aquatic organisms in relation to toxic environmental pollutants. *Ecotoxicol. Environ. Saf.* **64**, 178–189.

Vamvakopoulos, N. C. and Chrousos, G. P. (1994). Hormonal regulation of human corticotropin-releasing hormone gene expression: implications for the stress response and immune/inflammatory reaction. *Endocr. Rev.* **15**, 409–420.

Venturelli, P. A., Murphy, C. A., Shuter, B. J., Johnston, T. A., van Coeverden de Groot, P. J., Boag, P. T., et al. (2010). Maternal influences on population dynamics: evidence from an exploited freshwater fish. *Ecology* **91**, 2003–2012.

Vijayan, M. M., Raptis, S. and Sathiyaa, R. (2003). Cortisol treatment affects glucocorticoid receptor and glucocorticoid-responsive genes in the liver of rainbow trout. *Gen. Comp. Endocrinol.* **132**, 256–263.

Wagner, T. and Congleton, J. L. (2004). Blood chemistry correlates of nutritional condition, tissue damage, and stress in migrating juvenile chinook salmon (*Oncorhynchus tshawytscha*). *Can. J. Fish. Aquat. Sci.* **61**, 1066–1074.

Wang, C., King, W. and Woods, L. C. (2004). Physiological indicators of divergent stress responsiveness in male striped bass broodstock. *Aquaculture* **232**, 665–678.

Webber, D. M., Boutilier, R. G. and Kerr, S. R. (1998). Cardiac output as a predictor of metabolic rate in cod *Gadus morhua*. *J. Exp. Biol.* **201**, 2779–2789.

Wedemeyer, G. A., Barton, B. A. and Mcleay, D. J. (1990). Stress and acclimation. In Methods for *Fish* Biology (eds. C. B. Schreck and P. B. Moyle), pp. 451–489. Maryland: American Fisheries Society.

Wells, R. M. G. and Pankhurst, N. W. (1999). Evaluation of simple instruments for the measurement of blood glucose and lactate, and plasma protein as stress indicators in fish. *J. World Aquacul. Soc.* **30**, 276–284.

Wendelaar-Bonga, S. (1997). The stress response in fish. *Physiol. Rev.* **77**, 591–625.

White, A. J., Schreer, J. F. and Cooke, S. J. (2008). Behavioral and physiological responses of the congeneric largemouth (*Micropterus salmoides*) and smallmouth bass (*M. dolomieu*) to various exercise and air exposure durations. *Fish Res.* **89**, 9–16.

Wikelski, M. and Cooke, S. J. (2006). Conservation physiology. *Trends Ecol. Evol.* **21**, 38–46.

Williams, T. D., Gensberg, K., Minchin, S. D. and Chipman, J. K. (2003). A DNA expression array to detect toxic stress response in European flounder (*Platichthys flesus*). *Aquat. Toxicol.* **65**, 141–157.

Wingfield, J. C. (2013). Ecological processes and the ecology of stress: the impacts of abiotic environmental factors. *Funct. Ecol.* **27**, 37–44.

Wingfield, J. C., Maney, D. L., Breuner, C. W., Jacobs, J. D., Lynn, S., Ramenofsky, M., et al. (1998). Ecological bases of hormone-behaviour internactions: the 'emergency life history stage'. *Am. Zool.* **38**, 191–206.

Wingfield, J. C., Kelley, J. P. and Angelier, F. (2011). What are extreme environmental conditions and how do organisms cope with them. *Curr. Zool.* **57**, 363–374.

Wiseman, S., Osachoff, H., Bassett, E., Malhotra, J., Bruno, J., VanAggelen, G., et al. (2007). Gene expression pattern in the liver during recovery from an acute stressor in rainbow trout. *Comp. Biochem. Physiol. D* **2**, 234–244.

Wood, C. M., Turner, J. D. and Graham, M. S. (1983). Why do fish die after severe exercise? *J. Fish Biol.* **22**, 189–201.

Woodley, C. M. and Peterson, M. S. (2003). Measuring responses to simulated predation threat using behavioral and physiological metrics: the role of aquatic vegetation. *Oecologia* **136**, 155–160.

Woodward, J. J. (1982). Plasma catecholamines in resting rainbow trout, *Salmo gairdneri* Richardson, by high pressure liquid chromatography. *J. Fish Biol.* **21**, 429–432.

Woodward, C. C. and Strange, R. J. (1987). Physiological stress responses in wild and hatchery-reared rainbow trout. *Trans. Am. Fish. Soc.* **116**, 574–579.

Wu, C. (1995). Heat shock transcription factors: structure and regulation. *Annu. Rev. Cell Dev. Biol.* **11**, 441–469.

Yeom, D. H. and Adams, S. M. (2007). Assessing effects of stress across levels of biological organization using an aquatic ecosystem health index. *Ecotoxicol. Environ. Saf.* **67**, 286–295.

Zubair, S. N., Peake, S. J., Hare, J. F. and Anderson, W. G. (2012). The effect of temperature and substrate on the development of the cortisol stress response in the lake sturgeon, *Acipenser fulvescens*, Rafinesque (1817). *Environ. Biol. Fish.* **93**, 577–587.

12

STRESS MANAGEMENT AND WELFARE

LYNNE U. SNEDDON

DAVID C.C. WOLFENDEN

JACK S. THOMSON

Stress poses a significant challenge to the health and welfare of fish in a variety of contexts. Preserving fish well-being has obvious benefits for aquaculture, fisheries management practices, large-scale fisheries, recreational fishing, research, and the ornamental fish industry. Healthy fish provide better economic return, contribute to population size, provide experimentally sound data, are attractive and pleasing to watch, and pose no risk to public health. However, many practices in each of these areas where fish are used cause stress and as such may impair fish welfare. The impact of routine procedures that fish are subject to is discussed to better understand how stress can be managed in captivity. The means of reducing stress and its

Biology of Stress in Fish: Volume 35
FISH PHYSIOLOGY

deleterious effects on fish behavior, development, growth, reproduction, and immune function are considered as practical management tools that could be employed by those using fish if they wish to minimize stress and improve health. The opportunity to have a sense of control over stress or being able to anticipate and prepare for stress improves the ability of fish to cope with any stressors in their environment. Inescapable, unpredictable, or chronic stress leads to loss of control and allostatic overload. This can result in behavioral abnormalities leading to displaced aggression and stereotypical behavior. Thus, allowing fish to have other behavior options such as hiding or redirecting behavior can be provided by environmental enrichment. Conditioning fish to associate cues with the onset of a stress allows them to anticipate and prepare for stress, which can be beneficial. Operant conditioning, where fish can operate self-feeders, allows fish to control their own foraging behavior and also has positive effects on fish welfare. Providing the right kind of environmental and cognitive stimulation along with optimal environmental conditions, appropriate feeding regimes and social contact appear to be key to reducing stress in captive contexts if this is logistically possible. Practices in a variety of fish industries are considered where stress may be elevated with countermeasures suggested. Future studies should investigate implementing these factors to understand their impact in different circumstances and in different species if reducing stress is important. The development of robust stress indicators and automated alert systems based on behavior or environmental parameters to detect fish health will be vital for the assessment and alleviation of stress.

1. INTRODUCTION

From the preceding chapters, it is clear that stress has a substantial impact upon fish biology and influences overall health. Short-term acute stress can be considered an adaptive response and actually allows the individual fish to cope with a stressful event and further to regain homeostasis (Schreck and Tort, 2016; Chapter 1 in this volume). Thus acutely stressful events, which should be avoided where possible, may not necessarily compromise the welfare of the fish. However, long-term chronic stress has detrimental effects upon fish growth (Sadoul and Vijayan, 2016; Chapter 5 in this volume), the immune system (Yada and Tort, 2016; Chapter 10 in this volume), and reproductive output (Pankhurst, 2016; Chapter 8 in this volume), which also has negative consequences for aquaculture, the fisheries, and ornamental fish industries and research using fish as experimental models (Spagnoli et al., 2016; Chapter 13 in this volume) (Davis, 2006). It is in the interest of those managing and using fish to keep them in good health and free from stress as far as possible. This ensures that fish perform well biologically, grow to a marketable size, reproduce effectively, and benefit aquaculture and fisheries, but they also will perform well if liberated into the wild, look attractive, and exhibit normal behaviors in the ornamental fish industry. Additionally unstressed fish provide robust, sound experimental data increasing scientific validity of experiments. This chapter considers the impact of stress upon fish in terms of their health and welfare, how we can alleviate stress in managed populations, as well as adopt strategies for improving welfare, and finally discusses gaps in our knowledge to identify where future research is needed.

1.1. Defining Welfare

Welfare can be defined as a state in which an animal feels well according to Volpato et al. (2007), and these authors question the validity of phrases such as poor welfare, due to the apparent oxymoronic nature of such terms. Semantics aside, various definitions of welfare have been proposed, although the subject is a highly complex one, with no clear consensus on how welfare should actually be defined or measured. Fish, as animals, present interesting challenges for the measurement of welfare due to the differences in their behavior and physiology compared with terrestrial mammals. Nevertheless, fish welfare is an area that can be investigated scientifically, in spite of the difficulties in identifying and measuring welfare (Table 12.1; Martins et al., 2012; Sneddon, 2011). Several approaches have been proposed to measure the welfare of animals, although three main methodologies have been extensively developed (Stamp Dawkins, 2012). Specifically, these definitions are feelings-based, function-based, or use comparison with the natural lives of wild animals. These have varying emphases along a continuum from philosophical arguments to those concerned purely with empirically measurable biological factors, with some authors adopting a synthesis of more than one approach. Broadly, welfare can be defined as, "is the animal well?" and "does it have what it needs?" (Stamp Dawkins, 2012).

Definitions employing the subjective experiences of the animal are generally termed feelings-based. The premise underlying feelings-based definitions is that the animal should feel well for good welfare to apply. However, this also means the individual can experience adverse welfare states such as pain, fear, and stress (Fraser, 2008). Assessing the internal subjective

Table 12.1

Examples of behavioral changes in response to procedures or stressors that could be used as species-specific welfare indicators in farmed fish showing both positive and negative welfare consequences

Welfare state	Measure	Procedures/stressors	Species	Welfare indicator	Affect wild fish
Positive	Exploratory activity	Stocking density	Rainbow trout, *Oncorhynchus mykiss*	Low stocking densities: ↑ use of self-feeders	Unlikely to affect natural populations. May affect native species if large populations of farmed fish are added for restocking due to competition, hybridization, and predation by introduced fish
	Food-anticipatory activity	Feeding method	Gilthead Sea Bream, *Sparus aurata*	Scheduled feeding (vs random): ↑ food anticipatory activity	Unlikely unless native food supply is affected by anthropogenic factors; for example, mortality by pollution or overfishing
Negative	Aggression	Feeding method	African catfish, *Clarias gariepinus*	Hand feeding (vs self-feeding): ↑ % of fish bitten	Unlikely
			Atlantic salmon, *Salmo salar*	Fixed ration (vs demand feeding): ↑ aggressive acts	
			Atlantic salmon, *S. salar*	Underfeeding (vs satiation): ↑ aggressive acts	
		Photoperiod and light intensity	African catfish, *C. gariepinus*	Long periods of light and high light intensity: ↑ skin lesions	May be affected by anthropogenic lighting in urban areas

	Species	Finding	Implications for wild populations
Stocking density	African catfish, C. gariepinus	Low stocking densities: ↑ skin lesions	Unlikely to affect natural populations. May affect native species if large populations of farmed fish are added for restocking due to competition, hybridization, and predation by introduced fish
	Rainbow trout O. mykiss	High stocking densities: ↑ aggressive behavior	
Grading	White sea bream, Diplodus sargus	High stocking densities: ↑ social interactions	Not applicable
	African catfish, C. gariepinus	Homogenous groups of small fish: ↑ % of fish bitten	
Foraging behavior Feeding method	Atlantic halibut, Hippoglossus hippoglossus	Floating pellets (vs sinking): ↓ number of pellets eaten	Unlikely unless native food supply is affected by anthropogenic factors; for example, mortality by pollution or overfishing of prey
Water exchange rate in RAS	Nile tilapia, Oreochromis niloticus	High water exchange rates: ↓ feeding latency in small size classes	Not applicable
Grading	African catfish, C. gariepinus	Homogenous groups of small fish: ↓ total feeding time and feeding rate	Not applicable
Fasting periods	Sea bass, Dicentrarchus labrax	↑ Feeding rate and daily feeding times after feeding is resumed	Not applicable
Cleaning protocols	Sea bass, D. labrax	↓ Self-feeding	Not applicable

(Continued)

Table 12.1 (Continued)

Welfare state	Measure	Procedures/stressors	Species	Welfare indicator	Affect wild fish
	Swimming activity	Hypoxia	Atlantic cod, *Gadus morhua*	↓ Swimming speed	Occurs due to heat wave increasing temperature thereby reducing dissolved oxygen, algal blooms, and pollution
			White sturgeon, *Acipenser transmontanus*	↓ Swimming speed	
			Brook charr, *Salvelinus fontinalis*	↑ Swimming speed	
			Sockeye salmon, *Oncorhynchus nerka*	↓ Recovery test	
			Nile tilapia, *O. niloticus*	↑ and ↓ Swimming activity of schools	
		Hyperoxia/supersaturation	Atlantic salmon, *S. salar*	↓ Swimming speed	Can naturally occur and native populations have adapted strategies to cope with this. In sudden events this poses a risk to wild fish
		Pollutants	Sockeye salmon, *O. nerka*	Dehydroabietic acid: ↓ recovery test	Risk increased due to anthropogenic inputs into water bodies, leaching and run-off from industrial/agricultural land
			Brown trout, *Salmo trutta*	Cu: ↓ recovery test	
			Juvenile lake chubsuckers, *Erimyzon sucetta*	Ash: ↓ U_{crit}	
		Infections	Delta smelt, *Hypomesus transpacificus*	↓ U_{crit}	Naturally occurring infection worsened by environmental and biological disturbance that causes stress

Parasites	Atlantic salmon, *S. salar*	$\downarrow U_{crit}$	Naturally occurring infection worsened by environmental and biological disturbance that causes stress
Temperature	Largemouth bass, *Micropterus salmoides*	Low temperature: $\downarrow U_{crit}$	Occurs due to heat wave increasing temperature thereby reducing dissolved oxygen. Reduced temperature due to severely cold weather events below freezing may mean water bodies become hypoxic under ice
Transportation	Rainbow trout, *O. mykiss*	$\uparrow$ EMG	Only poses a risk to wild caught fish if used in fisheries management or scientific studies; for example, transport to laboratory
Underfeeding	Atlantic cod, *G. morhua* Rainbow trout, *O. mykiss*	$\uparrow$ Swimming speed $\uparrow$ EMG	Unlikely unless native food supply is affected by anthropogenic factors; for example, mortality by pollution or overfishing of prey
	Gilthead sea bream, *S. aurata*	$\uparrow$ Swimming speed; $\uparrow$ school maneuver complexity	
	Turbot, *Scophthalmus maximus*	$\uparrow$ Swimming speed	
		$\uparrow$ Swimming speed	Not applicable

(Continued)

Table 12.1 (Continued)

Welfare state	Measure	Procedures/ stressors	Species	Welfare indicator	Affect wild fish
		Short- and long-term cage submergence	Atlantic salmon, *S. salar*		
		Photoperiod and light intensity	African catfish, *C. gariepinus*	Long periods of light and high light intensity: ↑ swimming activity	May be affected by anthropogenic lighting in urban areas
		Photoperiod	African catfish, *C. gariepinus*	Long periods of light: ↑ swimming activity	May be affected by anthropogenic lighting in urban areas
		Stocking density	African catfish, *C. gariepinus* European sea bass, *D. labrax*	High stocking density: ↑ swimming activity High stocking density: ↓ swimming speed	Unlikely to affect natural populations. May affect native species if large populations of farmed fish are added for restocking due to competition, hybridization, and predation by introduced fish
			European sea bass, *D. labrax*	High stocking density: ↑ muscular activity measured via electromyogram (EMG)	
			Atlantic halibut, *H. hippoglossus* Atlantic salmon, *S. salar*	High stocking density: ↑ swimming activity Changes in space use and patchiness	
	Environmental gradients Feeding method				Wild fish naturally adapted to gradients Not applicable

	African catfish, *C. gariepinus*	Hand feeding (vs self-feeding): ↑ swimming activity	Unlikely to affect natural populations. May affect native species if large populations of farmed fish are added for restocking due to competition, hybridization, and predation by introduced fish
	Atlantic salmon, *S. salar*	Scheduled (vs on demand): ↑ swimming speed and turning angles	
Ventilatory activity			
Stocking density	African catfish, *C. gariepinus*	High stocking density: ↑ air breathing	
Hypoxia	Nile tilapia, *O. niloticus*	↑ Ventilator frequency	Can naturally occur and native populations have adapted strategies to cope with this. In sudden events this poses a risk to wild fish
Confinement	Nile tilapia, *O. niloticus*	↑ Ventilator frequency	Occurs during catch-and-release angling in keep nets or during fisheries management practices/scientific research. May also occur when anthropogenic structures are built; for example, dams

(*Continued*)

Table 12.1 (Continued)

Welfare state	Measure	Procedures/stressors	Species	Welfare indicator	Affect wild fish
		Handling	Rainbow trout, *O. mykiss*	↑ Ventilator frequency	Occurs during catch-and-release angling or during fisheries management practices/scientific research
	Stereotypic behavior and abnormal behaviors	Stocking density	African catfish, *C. gariepinus*	High stocking density: ↑ escape attempts	Unlikely to affect natural populations. May affect native species if large populations of farmed fish are added for restocking due to competition, hybridization, and predation by introduced fish
			Atlantic halibut, *H. hippoglossus*	High stocking density: ↑ loops of vertical swimming; ↑ surface swimming	
		Feed characteristics	Atlantic halibut, *H. hippoglossus*	Floating pellets (vs sinking): ↑ loops of vertical swimming	Unlikely unless native food supply is affected by anthropogenic factors; for example, mortality by pollution or overfishing of prey

Source: Adapted from Martins, C. I. M., Galhardo, L., Noble, C., Damsgård, B., Spedicato, M. T., Zupa, W., et al. (2012). Behavioural indicators of welfare in farmed fish. *Fish Physiol. Biochem.* **38**, 17–41, with kind permission from Springer; ↑ indicates an increase and ↓ indicates a decrease in the parameter measured; critical swimming speed, U_{crit}). These procedures/stressors may also affect wild fish as indicated by Sneddon, Thomson, and Wolfenden in the last column. The measures are yet to be tested in wild fish.

state of any animal is obviously very difficult but generally these are linked to choice experiments where the animal self-selects a resource it either values or would improve its state (Flecknell et al., 1999) or approaches where the animal must pay a cost to accessing a valued resource (Mason et al., 2001). Behavior is measured to give an indication of the animal's emotional state; for example, fear may be expressed by animals, and this has evolved to protect the animal in times of danger (Fendt and Fenselow, 1999). Studies in fish have demonstrated a consistent response to threatening stimuli such as avoidance of novel objects, freezing to reduce conspicuousness; escape or fleeing behaviors; thigmotaxis where the fish swims next to tank walls avoiding open, central areas; sinking to depth; fast-start swimming and diving responses; and antipredator behaviors (review in Maximino et al., 2010). Brain architecture innervating fear responses in zebrafish are similar to that in mammals (eg, the habenula, Agetsuma et al., 2010). Additionally, antianxiety drugs reduce fear responses in zebrafish (Grossman et al., 2011). Although such approaches may give some insight into how an animal feels, there are difficulties with a feelings-based approach. It has been criticized for being highly subjective and open to observer bias or anthropomorphic interpretations, and it is often difficult to interpret behaviors in terms of feelings, although the results may suggest lapses in welfare (Dawkins, 1980). There exists a range of opinions on the value of feelings-based definitions of welfare. Some scientists argue (eg, McGlone, 1993), for instance, that it is fundamentally impossible to know how an animal is feeling, and that feelings have no place in assessment of animal welfare. This viewpoint relies upon the measurement of purely physiological criteria such as growth and reproduction to assess an animal's welfare status. Others, however (eg, Duncan, 1993, 1996), maintain that feelings-based judgments are at the very heart of any discussion on animal welfare concepts.

Function-based definitions of animal welfare measurement concern an animal's ability to cope with its environment. Various aspects can be measured; for example, growth, incidence of disease, fecundity, and behavioral indicators, as well as physiological factors such as stress hormones. These parameters provide robust, experimental measurements of welfare and as such provide convincing evidence on the state of an animal. The premise here is that if the animal is in good health, it should grow well, have a healthy immune system, and be free of disease, be actively reproducing, and have low stress indicators. Thus, biological functioning definitions of welfare provide empirical measurements by which welfare assessment can be informed. For example, a study by Sneddon et al. (2003) investigated the response of rainbow trout when acetic acid and bee venom were administered to the lips. The fish subjected to this potentially painful stimulus demonstrated a range of atypical behaviors including rocking on the pectoral fins, rubbing of the affected area on the aquarium substrate, cessation of feeding, and increased respiratory rate (Sneddon et al., 2003) as well as increased cortisol (Ashley et al., 2009). These abnormal behaviors were reduced by administering morphine, a painkiller (Sneddon, 2003a). However, whereas function-based measurement of welfare may provide objective criteria, it has been argued, for example, by Dawkins (1980) that this may not provide the complete picture. For example, a shoaling fish maintained in isolation may appear to be perfectly healthy and fulfill all function-based criteria for acceptable welfare. However, such individuals' welfare may in fact be compromised if access to conspecifics is denied and the individual cannot shoal. Altered stress indicators are generally informative of homeostatic challenges but often acute stress responses help the animal to cope with a novel threat. Therefore, in this case high cortisol production would reflect a natural coping mechanism and does not necessarily mean poor welfare. Alternatively, chronic stress does result in impaired immune function, reduction in reproductive output, and reduced growth potentially leading to mortality. Thus, biological functioning measures do need to be interpreted cautiously in the context that they are being measured (Keeling and Jensen, 2009).

The natural lives approach to quantifying welfare involves a direct comparison between the observed behaviors of captive animals and their wild counterparts that may experience variable conditions (eg, food availability, predators, environmental disturbance) compared to captivity, making a comparison between wild and captive environments difficult. The premise in natural lives definitions of animal welfare are that if they are in a good welfare state then the animal's behavior should be similar to that seen in a natural environment. We should also provide all the resources the animals need to perform their natural behavioral repertoire (or artificial analogs of those resources). Therefore, animals should lead a natural life and should be able to exhibit natural behaviors shown by their wild counterparts or ancestors (Rollin, 1993). However, in captivity the motivation to perform normal feeding behavior such as foraging may be reduced since food is provided to animals, so caution may be necessary when inferring from the comparison of captive animal behavior with wild animal behavior. Conditions in production systems may also remove the motivation to perform some behaviors. For example, adult Atlantic salmon (*Salmo salar*) may migrate many thousands of miles in the oceans searching for food, yet farmed Atlantic salmon are held in relatively small sea cages compared with the open ocean but are provided with plentiful food. Does this remove the motivation (or simply the opportunity) for these fish to migrate, or does this explain why they constantly swim in a circular fashion in the cage?

Irrespective of what definition is employed it is important to maintain good welfare from a moral, ethical, or legal perspective, but it also benefits the fish-based industries and science as well as maintaining biodiversity and conserving wild fish populations. Many authors have suggested simpler definitions such as whether the animal appears in good health or well-being and whether it has what it needs (Curtis, 1985; Stamp Dawkins, 2012). The needs of an animal can be considered holistically in terms of physiology, safety, and behavioral needs (Curtis, 1985). Physiological needs consist of providing appropriate nutrition, environmental conditions, and keeping the animal in good health, whereas safety needs are comprised of those that prevent the animal from succumbing to adverse weather conditions, predation, and accidents involving equipment and facilities. Behavioral needs include keeping the animal free from maltreatment (abuse, neglect, or deprivation) and providing the animal with all the external stimuli and resources to perform its natural behavioral repertoire. For example, the male three-spined stickleback (*Gasterosteus aculeatus*) becomes highly territorial in the breeding season and is motivated to build a nest (Mayer and Pall, 2007). Thus, in captivity we must consider providing males in breeding condition (signaled by their red throat color) with enough space to establish a territory and materials to build a nest to accommodate innate behaviors, thus reducing stress. However, such accommodations are species specific. For example, salmonids in aquaculture and under laboratory conditions are aggressive and form dominance hierarchies (eg, Sneddon et al., 2011). Fighting, a normal part of the salmonid behavioral repertoire, can result in injuries and long-term stress for subordinates whereas dominants flourish. We can then question whether we should provide these fish the opportunity to act aggressively and meet this behavioral need or whether conditions should be sought to reduce aggression since subordinates are experiencing poorer well-being. This would be difficult to answer but as a resolution in aquaculture fish are regularly size graded to remove larger individuals. Therefore, for the benefit of the fish and for the humans using the fish it is in our interest to minimize and alleviate stress to ensure good welfare; thus it would be advantageous to provide appropriate housing and conditions if the goal is to hold the animals in ideal conditions.

2. MANAGING STRESS IN FISH

For fish introduced from the wild into a captive environment or those captively reared and destined for release into natural environments, the disparity in the conditions they face are likely to cause problems with survival and stress management. In general, captive environments may provide numerous benefits to fish: food is often supplied in abundance (Garner, et al. 2010; although in some instances, such as in ornamental aquaria where owners forget to feed their fish, food provision can be low, Gronquist and Berges, 2013), there are usually no predators, fish have shelter, and mates are often provided and so reproductive success secured. In contrast, compared to wild environments, fish are also kept in inappropriately high (eg, intensive farming) or low densities (such as individuals from schooling species kept alone), often with restricted space (captive environments are often structurally simple with little to no enrichment), the food is of uniform quality and artificial, the method of food delivery often leads to aggressive interactions, and fish are regularly disturbed and often handled (Benhaïm et al., 2013; Garner et al., 2010; Huntingford, 2004; Kulczykowska and Vázquez, 2010; Newman et al., 2015). These challenges will often cause physiological stress, particularly through increased hypothalamic–pituitary–interrenal (HPI) axis activity.

Stress is a fundamental and adaptive aspect of animal biology that initiates a suite of physiological and behavioral responses that have been selected to ameliorate the effects of poor quality environments. Initial responses to stress, such as increased vigilance and arousal followed by freezing or the fight-or-flight response (Galhardo and Oliveira, 2009), are followed by longer-term HPI activity, which returns the organism to homeostasis (Pankhurst, 2011) (Schreck and Tort, 2016; Chapter 1 in this volume). The magnitude of stress can be measured through change in HPI axis activity as measured through, for example, plasma cortisol. However, baseline and poststress plasma cortisol concentrations are known only for a few commercially or experimentally important species and can vary considerably both inter- and intraspecifically (Pankhurst, 2011); therefore the relative physiological impact of a stressor on an individual is difficult to determine and, also, difficult to place into context considering variation in HPI reactivity between species. Furthermore, some poststress cortisol responses to the same stressor may vary by as much as two orders of magnitude between species (Barton, 2002; Clearwater and Pankhurst, 1997). Stress responses can also vary with method of sampling particularly since fish are sensitive to the various disturbances arising from the various sampling procedures (Clearwater and Pankhurst, 1997; Ellis et al., 2012; Pankhurst, 2011). Furthermore, within captivity stressors are often chronic or repeated (Huntingford et al., 2006). Chronic stress can have a negative impact on HPI reactivity, resulting in responses of lower magnitude and overall desensitization of the axis to further stressors (eg, Huntingford et al., 2006; Pankhurst, 2011; Zuberi et al., 2014). Behavioral markers of chronic stress may include alteration to swimming patterns, inappropriate responses to predators, disruption to feeding patterns, and social inhibition (Schreck et al., 1997).

Many fish brought into captivity are unable to cope with their new conditions and die (see Section 3.4, this chapter), and thus the captive environment provides selective pressure favoring animals that are able to survive in these conditions (Huntingford, 2004; Pankhurst, 1998; Zuberi et al., 2014). For instance, among rainbowfish (*Melanoteania duboulayi*) the physiological response to both confinement stress and to chasing with a simulated predator was attenuated in captive compared to wild fish (Zuberi et al., 2011, 2014). In both cases the cause of the physiological change was attributed to an environment where fish are exposed to repeated stressors, and where a physiological response to all such challenges would be maladaptive. The differences were considered derived from ontogenetic desensitization to the environmental challenge combined with long-term selection pressure (Zuberi et al., 2011). Captivity therefore promotes different coping mechanisms to those that might have been selected for in the wild (Huntingford, 2004; LePage et al., 2000; Pampoulie et al., 2006). Often behavioral and physiological changes in fish moved from the wild into captivity can be observed in as little as just one generation (Huntingford, 2004; Zuberi et al., 2011). Such habituation is useful for fish being retained within captivity and allows for appropriate responses to a new set of challenges.

Stress physiology, measured as HPI activity, appears to be a heritable trait in its own right (although heritability varies widely between species; Øverli et al., 2005). As such, adaptations toward captive living include increased stress tolerance in some populations (Pankhurst, 1998; Douxfils et al., 2011), particularly to those stressors common in captivity such as handling and confinement (Cleary et al., 2000; Douxfils et al., 2011). Furthermore, though chronic stress has been implicated in immunosuppression (Ashley, 2007; Huntingford et al., 2006), selection in some captive fish populations appears to have resulted in improved immune function (in both innate and acquired immune parameters) relative to previous generations (eg, Douxfils et al., 2011). Though stress-related immunosuppression is linked to cortisol, it is likely also modulated by various other hormones including catecholamines, endogenous opioids, pituitary hormones, and serotonin, among others (Douxfils et al., 2011 and references therein) (Gorissen and Flik, 2016; Chapter 3 in this volume) (Yada and Tort, 2016; Chapter 10 in this volume).

Variation in reproductive strategy also influences the genetic health of captive and wild populations. In the wild, reproductive success is variable and not guaranteed, and pairings tend to be nonrandom (Garner et al., 2010). In contrast, captive populations are often generated from a small number of parents, generating a significant founder effect that limits the available gene pool, and are also exposed to greater levels of inbreeding, resulting in reduced genetic diversity and allelic richness, genetic drift, and reduced heterozygosity compared to wild conspecifics (Douxfils et al., 2011; Doyle et al., 2001; Garner et al., 2010). Such a loss in genetic variability per se is theorized to have the potential to harm the ability of the population to adapt to a natural environment, particularly where such environments are not stable (Pampoulie et al., 2006; Douxfils et al., 2011).

Stress during reproduction can also have profound impacts on health of progeny, although the negative impacts are seemingly reduced in hatchery-reared fish compared to wild-caught and trapped (Pankhurst, 2016; Chapter 8 in this volume). Stress to egg-bearing mothers can affect reproductive output through reductions in plasma gonadal steroids (eg, 17β-estradiol and testosterone; Cleary et al., 2000; Clearwater and Pankhurst, 1997) with wide-ranging effects including increased atresia, reductions in egg size, delayed ovulation, and reduced survivability and growth in progeny (Campbell et al., 1992). High cortisol in embryos transferred from the ovarian fluid of stressed mothers can also result in unhealthy larvae with reduced growth and reduced survival, and this influences behavior in adult progeny (Barton, 2002; Sloman, 2010). However, some of these negative effects appear to differ between wild and captive fish, likely due to genetically and experientially derived tolerance to stress (Cleary et al., 2000; Pankhurst, 2011).

Furthermore, the rearing environment may directly influence variations in somatic growth, such as in brain size and structure. Conditions at early life stages may initiate a developmental trajectory resulting in physical, physiological, and behavioral differences in adulthood (Kihslinger and Nevitt, 2006). This may arise from habitat utilization in larval or juvenile stages, with more complex habitats encouraging complex behaviors that promote development of particular brain regions such as the telencephalon and optic tectum associated with these behaviors (Marchetti and Nevitt, 2003). Other studies have focused on neuronal proliferation, particularly within the telencephalon, and noted differences in the brains of fish reared in complex or simple environments (reviewed in Ebbesson and Braithwaite, 2012). Furthermore, neural plasticity, along with the ability to navigate mazes, was improved in three-spined sticklebacks (*Gasterosteus aculeatus*) reared in an enriched compared to an impoverished environment (Salvanes et al., 2013). Thus habitat complexity may have profound impacts on brain development with, consequently, effects on relevant behaviors.

2.1. Considerations for Care of Wild Fish in Captivity

Bringing wild fish into captivity or restocking wild populations with captive-reared stock engenders a need to ensure that the welfare of fish transferred between environments is the best it can be. Changing environment—whether from wild to captive or vice versa—is inherently stressful: for instance, the survival rate of hatchery-reared fish released into natural habitats compared to wild fish is generally poor (McNeil, 1991), and related to lack of experience in obtaining food and avoiding predators (Brown et al., 2003). Likewise, wild fish introduced into captivity also exhibit high mortality. Provision must therefore be made for fish to adjust to their new conditions. This includes care in methods of general husbandry but also an awareness of the potential stressors that differ between the two environments and how fish can (1) maintain appropriate physiological and behavioral responses to natural challenges if they are destined for restocking, or (2) adjust to the highly regulated and stressful conditions associated with captivity.

Feeding in wild fish may consist of continuous grazing or intermittent bouts separated by periods of enforced fasting; however, the periodicity and timing of feeding often varies between species based largely on sensory apparatus (Kulczykowska and Vázquez, 2010). Feeding may be exclusively nocturnal or diurnal, while some are capable of shifting between the two. For instance, sea bass (*Dicentrarchus labrax*) are able to rapidly switch feeding times to coincide with an adjustment to prevailing light:dark cycles (Sánchez-Vázquez et al., 1995). Likewise, salmon switch feeding periodicity based on environmental temperature, with diurnal feeding occurring where temperature was $> 10°C$ (Fraser et al., 1993). However, in captive environments feeding is often regulated and delivered at predictable times during the day. The ability of fish to adjust to this regularity is, again, species specific. Some fish are incapable of adjusting natural rhythmicity (eg, nocturnal feeding in tench (*Tinca tinca*) Herrero et al., 2005; diurnal feeding in seabream (*Diplodus puntazzo*) Vera et al., 2006) and disruption of these rhythms is inherently stressful. Random feeding events, as may be experienced in natural environments where food sources are unpredictable, tend to also induce stress. Higher cortisol and glucose was recorded in sea bream (*Sparus aurata*) fed at random intervals, and was associated with greater activity and being continuously alert. Ultimately this resulted in reduced growth rate. A consequence of the predictable delivery of feed is that fish become more alert and aggressive as feeding time approaches. In some species predictability of feeding time reduces stress compared to random feeding (Kulczykowska and Vázquez, 2010); however, the competition for feed can be extremely stressful and cause damage to fish through fighting, and the distribution of feed will favor the larger, dominant fish.

Predictability can also limit the likelihood of fish retaining the ability to search, forage, and capture their own food if they are released into a wild environment, potentially resulting in reduced growth rate and condition factor (Brown et al., 2003; Garner et al., 2010; although since aquaculture strains are usually selected for improved growth rate the opposite may be true, eg, Zuberi et al., 2011). These responses can be relearned with social learning improving the speed with which individuals obtain these skills and thus supplementary feeding with live feed may invoke natural foraging (Brown et al., 2003; Sneddon, 2003b). Some hatchery-reared fish, either through boldness (ie, reduced inhibition when confronted with new objects) or through lack of experience, approach novel food items more quickly in the wild (Huntingford, 2004), though the ability to differentiate between suitable and unsuitable prey once released appears more difficult (Brown et al., 2003). For these fish, enriched rearing conditions may be the most beneficial environment for development of appropriate foraging and ability to discriminate between prey (Brown et al., 2003). A further consideration of potentially increased boldness among hatchery-reared fish is a propensity to ignore dangers such as predators when foraging (see later).

Clearly an understanding of the individual needs of species under care is important in limiting stress. One method of resolving such issues associated with predictability and different feeding regimes would be to switch to automated self-demand feeders that deliver feed when fish interact with them; such methods can limit the aggressive interactions and provide equal opportunities for weaker fish to obtain food or to do so at less competitive times (Huntingford and Kadri, 2014).

Fish also vary their diet dependent on individual needs, which may vary temporally and on a species-specific basis (in terms of nutrient content or even just shape and the amount of movement exhibited by the food; Huntingford, 2004), and will depend on environmental variables and often the life-stage of the fish (Kulczykowska and Vázquez, 2010). Dietary requirements for particular species, how they make use of particular nutrients within their diets, and how these nutrients interact may not, however, be known (Oliva-Teles, 2012). Within captive environments fish are often fed a formulated and uniform monotype diet that given the preceding, therefore, poses the risk of failing to meet individual dietary requirements (Oliva-Teles, 2012) and is potentially a distasteful or unattractive diet for the fish (Huntingford and Kadri, 2014). These are challenges that must be accounted for on a species-by-species basis and clearly require further study to meet the demands for the wide variety of species in captivity.

Captive fish are considered inexperienced, particularly in terms of antipredator strategies, and this is reflected in poor or no response to danger. Released fish have been observed to take greater risks while foraging and more frequently suffer from predation than fish that developed in the wild (eg, Benhaïm et al., 2012; Garner et al., 2010; Zuberi et al., 2011). This may in part be due to an inability in some fish to recognize natural predators (Zuberi et al., 2011), though some domesticated strains retain the ability to recognize olfactory cues from historic predators (Huntingford, 2004). The nature of food provision also encourages surface-feeding, which may increase the likelihood of avian predation (Maynard et al., 2001).

In hatchery-reared fish, the issues associated with predation can be somewhat ameliorated by providing experience through conditioning with a combination of olfactory stimuli (eg, predator scent in combination with conspecific alarm signal), providing close-contact experiences, or rearing with fish capable of exhibiting appropriate antipredator responses (Huntingford, 2004). Furthermore some domesticated salmonids retain appropriate antipredator responses and these could be a particular focus for restocking or accompanying fish destined for restocking (Benhaïm et al., 2012).

Captive environments, particularly stock tanks and in aquaculture facilities, often make use of high density culture, which is inherently stressful and can negatively impact welfare (Ashley, 2007). However, the social needs of fish also depend on their species: some are solitary living whereas others prefer living in groups, though the predictability of the hatchery or captive environment may inhibit social cues (Garner, et al. 2010). Even among group-living species the aquarium environment provides stressful challenges. Isolation as part of husbandry practices can cause stress in facultative schoolers, such as rainbowfish, *Melanoteania duboulayi*, with the response exaggerated in wild fish that have not become accustomed to the practice (Zuberi et al., 2014). Aggression may also be common, particularly in high-density/low-space enclosures. Some species develop dominant–subordinate relationships or hierarchies (eg, Atlantic salmon, *Salmo salar*; rainbow trout, *Oncorhynchus mykiss*) in which subordinates exhibit chronic increases in HPI and serotonergic activity but reduced hypothalamic–pituitary–gonadal (HPG) activity (Cubitt et al., 2008) with associated impacts on welfare. In combination with spatially and temporally predictable food provision, these relationships can lead to contest behavior, injury, and differential feed intake between fish, enhancing growth/size disparity between individuals (eg, Cubitt et al., 2008; Montero et al., 2009).

Aggression itself varies between captive and wild populations, with increased or reduced aggressiveness selected for in captive fish dependent upon the rearing environment and method of feeding (Huntingford, 2004). Increased aggression may arise through increased competitiveness over food, increased boldness (linked strongly with aggression), or even an inability for an individual to communicate its decision to withdraw from a contest (Huntingford, 2004). Context may influence the outcome of fights, particularly where wild fish are able to utilize experience of, for example, territory defense to outcompete hatchery-reared fish without these skills (Huntingford, 2004).

Solutions to these problems may be difficult, particularly in industries focusing on high output. Minimizing density and maximizing space may improve conditions, although suggested densities to minimize aggression may vary between studies (Gronquist and Berges, 2013), indicating other factors may also be important. Providing shelter for animals can also minimize the impact of aggressive interactions, as can ensuring that space within enclosures is fully utilized (Huntingford and Kadri, 2014).

Habitat complexity can be an issue in many captive environments, particularly in laboratories where reduction in extraneous stimuli is often necessary. Home and public aquaria usually provide enrichment in terms of decoration, but aquaculture and hatchery environments are usually devoid of these forms of enrichment. Lack of habitat complexity can have profound effects on fish living within these environments, restricting the capacity to exhibit the natural range of physiological and behavioral responses to stimuli (Brydges and Braithwaite, 2009).

Enrichment provides social and physical stimulation that is otherwise lacking in some captive environments and, at the very least, minimizes monotony, which can prevent stress. The positive effects of enrichment appear numerous with effects on temperament behavior and cognition, resulting in animals exhibiting behavior more commonly associated with wild animals (Brown et al., 2003; Huntingford, 2004; Brydges and Braithwaite, 2009). As discussed previously, structural complexity can also enhance the development of young fish through behavioral interactions (Kihslinger and Nevitt, 2006; Marchetti and Nevitt, 2003). Even simple provisions such as lighting to highlight space can reduce crowding and, therefore, density-related stress (Huntingford and Kadri, 2014). The benefits of providing enrichment are therefore clear, but consideration must again be made of species-specific needs in constructing structures and preferred substrate.

Though not explicitly related to phenotypic differences between wild and captive fish, transportation of fish can also induce stress responses. The process of transportation involves a variety of invasive practices including handling for capture and unloading in addition to the transport itself, and such responses may last for a protracted period of time (Ashley, 2007; Olla et al., 1998). If fish are unable to recover sufficiently quickly then they may be unable to respond appropriately to their new habitats (eg, through increased risk of predation in the wild; Olla et al., 1998). If, however, transport-related stress is itself of sufficient magnitude, then fish may be unable to regain homeostasis resulting directly in mortality (Olla et al., 1998). One interesting result of transportation stress was observed in two lines of rainbow trout bred for divergent stress responses: although divergent poststress plasma cortisol profiles were maintained following transportation, behavioral differences were observed and attributed to differential loss in body mass ostensibly caused by stress (Ruiz-Gomez et al., 2008). This may itself indicate intraspecific and individual differences in the ability of fish to cope with this particular stress.

Despite being stressful in its own right, and therefore requiring measures to minimize the impact on fish, transport is also the first opportunity to mitigate the negative effects of introduction into a new environment (Ashley, 2007). High quality transport conditions can result in more rapid recovery posttransport (Tang et al., 2009); indeed, the loading procedures themselves may be responsible for the majority of stress involved in transport and should be a focus for future research on welfare during transport (Iversen et al., 2005; Tang et al., 2009). Fish should be carefully allowed to acclimate to their new conditions through gradual change in water conditions (Walster et al., 2014), including temperature and chemical/osmotic composition, but also taking other environmental factors such as light and sound levels into account. Provision of a recovery period following transport to allow homeostasis to be restored will assist fish in dealing with subsequent stressors (eg, predators, obtaining food; Ashley, 2007). Furthermore, immediate after-care following transport, principally to minimize the risk of new tank syndrome where spikes in waste products occur (Walster et al., 2014), will further minimize stress and potential mortality. Finally, recommendations for transport include sedation with light anesthesia, with an appropriate anesthetic dependent on species (Ashley, 2007). Additionally, dietary supplements and even preparation through the administration of acute, mild stressors may all inhibit transport-related stress (Ashley, 2007). Such procedures must necessarily be performed in advance of transport or handling. Temporary fasting of fish prior to transport evacuates the guts, thereby reducing waste production and excretion (Handy and Poxton, 1993) and metabolic rate during transport. Physiological stress is thereby reduced and welfare improved. However, recommended duration of prehandling starvation periods varies between species; small species, for example, may only require a day (eg, Froese, 1988) whereas Farmed Animal Welfare Council (FAWC) guidelines for salmon indicate a maximal limit of 72 h (FAWC, 1996).

The natural environment is replete with complex colors and lights of varying wavelength, which vary on a continuous basis. In contrast, the timing and quality of lighting in captive environments is often based on human needs, particularly in laboratories where strong lighting and single-color high-contrast backgrounds may be common. Consideration of the particular parts of the light spectrum must be made bearing the fish in mind, since different wavelengths of light are utilized to different extents dependent on the ecology of the animal. Many fish, for example, utilize ultraviolet (UV) wavelengths (Losey et al., 1999), but studies into excess UV as a cause of stress in captivity provide inconsistent results (Gronquist and Berges, 2013, and references therein). Since red wavelengths are attenuated rapidly in water some fish are unable to perceive this color, exhibiting behavior and physiology similar to fish kept in the dark (eg, Owen et al., 2010), but bright white light or background, typical of aquaculture, is a known stressor (Owen et al., 2010). Light quality is influential even during egg-stage and larval development; for instance, eggs reared under conditions most closely simulating natural environments (eg, with day–night cycles of light of blue wavelengths) developed more rapidly compared to those under red lights or without light cycles (Blanco-Vives et al., 2011). Furthermore, fish derived from and preferring turbid environments may be stressed by clear water (Owen et al., 2010) and the welfare of these fish must be considered against intended outcomes of holding these species in captivity.

The use of sound for communication is important among aquatic organisms, and high volume noise in the environment is considered stressful across many species (Kight and Swaddle, 2011). Captive environments, including aquaculture, labs, and ornamental aquaria, are noisy environments, producing loud sounds within the hearing range of many fish species and are typically louder than natural environments devoid of anthropogenic sounds (Bart et al., 2001; Kight and Swaddle, 2011). Noise is generated by, for example, aeration and pumps, producing high frequency sound (above 500 Hz), and water recirculation systems, motors, and any machinery used to automate processes such as feeding producing low frequency sound (below 400 Hz; Bart et al., 2001). Sound pressure levels (SPLs) are generally higher in the lower frequency range with SPL generally in the range 90–130 dB compared to 110–115 dB at higher frequencies (Bart et al., 2001); however, SPLs of 153 dB (Bart et al., 2001) or ranging from 70 to 160 dB (Clark et al., 1996; Bart et al., 2001) have been detected in aquaculture and lab environments. There is, therefore, a distinct likelihood for SPL within captive environments to be greater than natural habitats (eg, above 110 dB re 1 µPa in streams and similar high-energy environments, 100 dB re 1 µPa in lakes and similar low-energy environments, and concentrated in the frequency range below 500 Hz; Wysocki et al., 2007). Hearing capabilities of fish occur across a broad range and vary across species and taxonomic groups, and is often classified according to the presence and utilization of anatomical structures: hearing specialists, with higher sensitivity (particularly in an optimal frequency range of approximately 100–1000 Hz), utilize gas-filled structures (eg, a swim bladder) connected to the inner ear whereas hearing generalists/nonspecialists lack these structures (Popper et al., 2003).

In between these extremes are fish with gas-filled structures that are not linked to the inner ear. The impact of noise on fish is highly variable, dependent on complex interactions of these factors, and a complete account cannot be provided here. However, the range of effects can be broadly summarized, and include not just stress but temporary or permanent physical damage with associated loss of sensitivity.

For hearing specialists such as goldfish (*Carassius auratus*), even relatively low SPL sounds associated with aquaria (above 110 dB re 1 μPa, Gutscher et al., 2011; above 130 dB re 1 μPa, Smith et al., 2004) prevent the fish from fully utilizing their sense of hearing and mask sounds important in communication (Gutscher et al., 2011). Hearing specialists may also show startle responses to the onset of noise (Smith et al., 2004) where hearing generalists do not (although cf. Davidson et al., 2009) and, although behavioral change diminished within minutes, this may be important in systems where fish are exposed to acute bursts of noise. For hearing generalists, such as rainbow trout (*Oncorhynchus mykiss*) noise levels of 130 dB re 1 μPa are known to be capable of causing stress (Kight and Swaddle, 2011). Noise produces a variety of stress impacts on fish, resulting in increases in plasma glucose and cortisol, and can cause physical damage to HPI-axis-related structures with numerous downstream effects (Kight and Swaddle, 2011). Fatalities in developing eggs/embryos and long-term stress impacts in the surviving fry can also be caused by high noise levels (see Kight and Swaddle, 2011 for review). Critically, if occurring at an early age or in developing juveniles, many of the effects of loud noise may not be reversible (Kight and Swaddle, 2011). Furthermore, exceptionally loud noises can cause temporary hearing loss (a temporary threshold shift, resulting from noise above 140 dB re 1 μPA) or physical damage to the inner ear or swim bladder (Kight and Swaddle, 2011).

The presence of sounds of particular frequency per se, rather than the power of the sound, may be stressful or induce maladaptive behaviors in exposed fish. Low frequency sounds in particular are considered to be similar to those produced by predators and cause fright and escape responses in fish (Knudsen et al., 1992; Bui et al., 2013). Since the stimulus of infrasound is often not coupled with predatory activity within captive environments it is possible that the behavioral responses (such as escape swimming) may diminish through habituation but continuous exposure remains a welfare concern (Bui et al., 2013). In ornamental aquaria the tapping of glass by visitors can also be stressful, resulting in reduced foraging and elevated plasma cortisol (Bart et al., 2001; Gronquist and Berges, 2013). Since fish are sensitive to infrasound (< 20 Hz), even down to below 1 Hz infrasound devices have been used to repel fish in the environment to ensure natural populations stay away from damaging anthropogenic structures such as dams (Sand et al., 2001). Infrasound is generated by the motors of boats and ships, sonar, and construction (Slabbekoorn et al., 2010). Larval Atlantic cod (*Gadus morhua*) exposed to playback of the sound of underwater ship motors during development had a lower body width:length ratio and were easier to catch in a predator avoidance experiment (Nedelec et al., 2015).

The effects of sound on fish rely on extremely complex interactions of power and frequency of sound, whether it is acute or chronic, and the hearing capabilities and anatomy of the fish, among other factors. Though these cannot be sufficiently summarized in this section it is clear that anthropogenic noises are capable of disturbing or damaging fish and thus care should be taken to ensure that such noise is minimized in captive environments. Ensuring noises are kept at a minimum is extremely difficult in the environments inhabited by fish in captivity but may be an important issue in reducing welfare and stress.

Not only are both anthropogenic light and sound a problem for fish, other senses, such as magneto- and electroreception, may be affected by metal structures in natural environments and in aquarium facilities where tanks may be placed on metal racks. Hooks with additional magnets used in long line fisheries can actually attract blue sharks, *Prionace glauca*, increasing the bycatch of this species, which has a near-threatened conservation status (Porsmoguer et al., 2015). These hooks remain magnetized even after magnet removal. Indeed, permanent magnets and electropositive metal alloys are being investigated as shark repellents to reduce bycatch with mixed results (O'Connell et al., 2014). However, the impact of metal structures such as wind turbines and racking in aquaria as a potential source of stress has yet to be investigated.

2.2. The Impact of Psychological Stress

Although stress and mental state have been discussed at great length by several authors, only recently has the concept of psychological stress been considered in fish (Galhardo and Oliviera, 2009). The Cognitive Activation Theory of Stress formally defines some concepts linked to the experience of stress with positive (eg, coping) and negative valence (eg, hopelessness, helplessness; Ursin and Eriksen, 2004). The stress response comprises both neurophysiological activation and arousal and if acute it can be considered as normal. In contrast long-term arousal and stress is undesirable as previously discussed. The primary alarm response is vital for the behavioral response to acute stress (Ursin and Eriksen, 2004). The stress arousal response modulates sensory input, perceptual processing, and cognitive mechanisms to draw the animal's full attention to the stressor

(Steckler, 2005). When confronted by a challenge the animal stops current activity and focuses upon the stressor. Often behavioral inhibition is observed: in fish, the most idiosyncratic response is freezing behavior where an individual remains entirely motionless (Vilhunen and Hirvonen, 2003). Freezing is observed during fear responses (Yue et al., 2004), or as an antipredator strategy (Vilhunen and Hirvonen, 2003). The fight-or-flight response is also enlisted with the function of removing the threat by attacking or by retreating (Schreck et al., 1997; Steckler, 2005). A variety of different behavioral patterns have been related to fight-or-flight (Ashley and Sneddon, 2008; Sneddon, 2013) such as fleeing, hiding, and seeking shelter or increased shoal cohesion. Aggressive behavior includes chases and attacks (eg, Oliveira and Almada, 1998). Whole body rocking and rubbing against surfaces are only seen during nociception or pain (Sneddon et al., 2003; Sneddon, 2015). Swimming or activity patterns are also sensitive to many stressors (Huntingford et al., 2006; Martins et al., 2012).

If the stressor is chronic or inescapable, behavioral responses can be altered; these reflect an inability to regain homeostasis and the animal being unable to remove itself from danger. There are a variety of behaviors that fish perform during chronic stress. Substantial alterations in the swimming patterns, disruption to antipredator behavior, anorexia, increased refuge use, modified social behavior, and impaired learning ability are seen in response to chronic stress (Schreck et al., 1997). We could infer from these elevated responses that the fish exposed to chronic stress is also in a negative psychological state. If maintaining good fish health and welfare is the goal, we should consider ways of reducing any stressful situations, but further, how can we implement strategies to assist the fish in preparing for stressors that are necessary in the management of fish?

The extent to which fish have internal subjective experiences is subject to debate (eg, experience of pain, Rose, 2002; Sneddon et al., 2014; Sneddon, 2015). However, there is growing scientific evidence to suggest that fish do have some form of negative affective state associated with poor welfare. Studies tend to focus upon poor welfare states such as pain, fear, and stress (Table 12.1; Martins et al., 2012); however, there are a few intriguing studies that suggest fish experience positive states where there are no signs of stress and fish are engaging in a rewarding behavior. Cleaner wrasse (*Labroides dimidiatus*) engage in interspecific interactions in which cleaner fish remove parasites from visiting client reef fish. However, there is no conclusive evidence yet that cleaning organisms significantly improve the health of their clients. The stress response of wild-caught individuals of two client species, the chocolate dip chromis (*Chromis dimidiata*) and the lyretale anthias (*Pseudanthias squamipinnis*), was lower after a cleaning interaction where the cleaner wrasse massaged the client with its pelvic fins (Bshary et al., 2007).

Indeed when a client, the surgeonfish (*Ctenochaetus striatus*), was exposed to a stationary or moving model of a cleaner wrasse the surgeonfish voluntarily engaged with the moving model and had lower plasma cortisol concentrations (Soares et al., 2011). This suggests the cleaning or tactile action of the model and the cleaner wrasse have a calming effect that fish will seek out. Some authors have suggested this is an example of pleasure in fish. Fish have intelligence on a par with mammals (Brown, 2015) and are capable of highly complex behaviors that are challenging dogma on brain evolution and size (Abbott, 2015). Complicated behaviors have been observed such as cooperation within species and between species. For example, moray eels go hunting with groupers and signal to one another. Studies have shown fish have the ability to recognize related individuals and others, can adopt Machiavellian manipulation of clients in cleaner wrasse relationships, have bold and shy personalities where bold, aggressive extraverted fish contrast with shy, timid, cautious fish. Fish can display fear responses, use tools, learn complicated navigational tasks, and make future behavioral decisions based upon their experiences and learn skills from others, have long-term memories, and can recognize themselves through smell. These studies suggest fish are sentient and display awareness (review in Sneddon, 2011). Therefore, it would be sensible to presume that the state of being stressed results in a negative psychological state in fish. Caution should be applied here not to anthropomorphize and apply the feelings associated with human stress to fish; it is possible their experience may be more primitive in nature and but less important from a health and well-being perspective.

The amount of stress an animal perceives will naturally depend on the intensity and duration of the stressor and will be influenced by the individual's psychological and physiological state at any given moment (Curtis, 1985). Accounting for individual variation in sex, age, breeding condition, stress coping style (Fig. 12.1) (Winberg et al., 2016; Chapter 2 in this volume), as well as the diurnal cycle, the resources that promote positive well-being in the environment, whether the environment is predictable and controllable as opposed to more stressful unpredictable situations should all be considered (Curtis, 1985; Sneddon, 2011). Interpretation of results from studies on stress in fish must have biological relevance and extrapolation between species should be done with caution. Territorial species such as rainbow trout grow well in isolation from others whereas the gregarious common carp (*Cyprinus carpio*) can become ill when individually housed. Schooling or shoaling species, such as the zebrafish (*Danio rerio*) show little aggression in large groups but when held as groups of four or less, highly aggressive interactions causes stress in subordinates. Therefore, social context is particularly important with contrasting situations needed for gregarious and territorial species (Volpato et al., 2007).

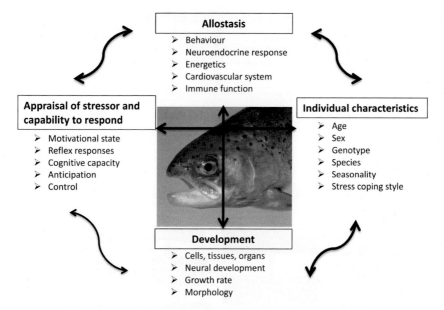

Figure 12.1. The various factors that affect allostasis in fish and the response to environmental stress. Both appraisal of the positive or negative valence of the external event and the capacity to respond is dependent upon the degree of allostatic load available to an individual that will all be shaped by that individual's state at a given time as well as its previous experience. Curved and straight arrows show that the four main contributory factors—allostasis, appraisal of the stressor and capacity to respond, development, and individual characteristics—all interplay in shaping the stress response.

2.3. Controlling and Preparing for Stress

The traditional concept of homeostasis has been challenged by various authors. Sapolsky (2004) termed this phenomenon allostasis, which is an animal achieving "stability through change." Rather than homeostasis being fixed at a certain state, allostasis is based upon the idea that internal balance is more flexible (McEwen, 1998). In this case internal balance is molded according to the specific requirements of the animal. This concept assumes animals are able to anticipate stressors that are predictable through physiological mechanisms (eg, seasonality) and respond to unpredictable changes through physiological mechanisms (eg, fight-or-flight). These two mechanisms are employed in the appropriate situation in accordance to the type of stressors the animal encounters and these have differing costs attached to them (allostatic load) (McEwen and Wingfield, 2003). If an animal is able to overcome a challenging situation its allostatic load is manageable (Fig. 12.2). However, when an individual cannot cope

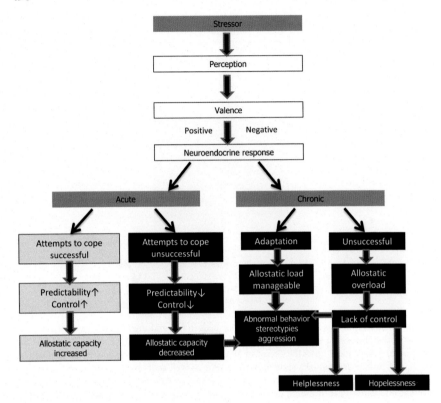

Figure 12.2. A diagrammatic representation of different scenarios when a fish is subject to a stressor and how the valence, type of stressor (acute or chronic) can result in a number of different scenarios including increased allostatic capacity, decreased allostatic capacity, and allostatic overload. Overload can lead to hopelessness or helplessness and possibly mortality.

or is subject to chronic stress this results in allostatic overload, which results in pathological changes (Fig. 12.2). In this case stress is maladaptive and is detrimental to welfare (Moberg, 1985; McEwen, 1998; Broom, 2008). Two key issues can be addressed when considering allostasis: (1) whether the stress is predictable or (2) whether the animal can exert control over its situation. This should then allow animals to regain internal balance without allostatic overload. When considering fish this assumes that they can anticipate and learn when an event will occur or can perform behaviors that allow some form of control or escape. If welfare is based upon allostasis then stress has an inverted U-shaped relationship; relatively little (hypostimulation) or chronic stress (allostatic load) results in negative welfare (Korte et al., 2007). This concept allows discrimination between normal acute stress responses and chronic stress where fish experience sustained stress over days, weeks, months, and more. If chronic stress results in impaired immune and reproduction function, reduced or negative growth rate, and possibly mortality, then this is defined as allostatic overload. In turn positive welfare is regarded as an animal that possesses a dynamic and flexible predictive physiological, cognitive, and behavioral capacity to anticipate and respond to abiotic and biotic challenges concurrent with meeting environmental demands (allostatic load is discussed by Schreck and Tort, 2016; Chapter 1 in this volume).

Control can be defined as the capability to avoid or decrease exposure to a challenge possibly influencing the outcome and as such can moderate stress responsiveness (Galhardo and Oliveira, 2009). Is it possible to give a fish some degree or sense of control over stress? An example would be to provide a refuge for the fish to actively seek cover or hide during disturbances such as tank cleaning and maintenance. Another approach is to give the animal a prior experience of the stress so it can anticipate how to respond in the future (Ursin and Eriksen, 2004). An individual's previous response, if effective in reducing stress, can allow a sense of control where arousal can be lowered and thus it can cope with subsequent challenges (Eriksen et al., 2005). This may explain why dominant fish tend to have lower cortisol levels than fish of lower status (Earley et al., 2006) and may explain the winner-and-loser effect (Eriksen et al., 2005), where winners' probability of winning fights in the future increases (Rutte et al., 2006). For example, bold rainbow trout became shyer after negative prior experiences (losing pair-wise contests; Frost et al., 2007). In contrast to acquiring control, an animal may learn that previous strategies to cope with stress were futile (total lack of control), thus learned helplessness may develop (Ursin and Eriksen, 2004; Lovallo, 2005). Helplessness is distinct from hopelessness, where some form of negative control exists and attempts to cope may increase stress (Fig. 12.2). Both helplessness and hopelessness elicit cortisol activation (Ursin and Eriksen, 2004; Lovallo, 2005), and low-ranking subordinate fish usually end up in this state due to the chronic stress of subordination (Earley et al., 2006).

Is there evidence for predictability reducing stress in fish? Cichlid fish (*Oreochromis mossambicus*) can appraise a predictable versus unpredictable situation and this modified their stress response. The impact of predictability depended upon the positive or negative valence of the experimental context where confinement or feeding were used. When stressed by confinement, the predictable group were more attentive to the visual anticipatory signal, exhibited reduced freezing behavior and a lower cortisol response relative to baseline. Predictability of food presentation triggered higher performance of anticipatory behavior and activity. Reduced cortisol levels were recorded in the predictable group but the

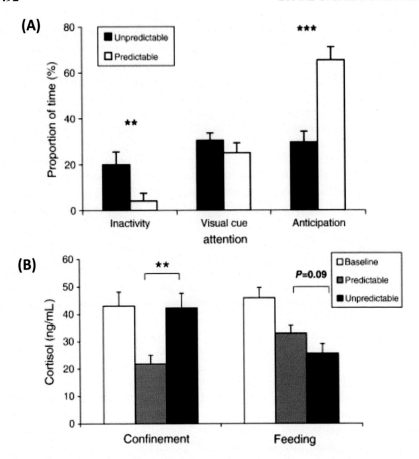

Figure 12.3. (A) Proportion of time (%) the cichlid, *Oreochromis mossambicus*, spent in inactivity, paying attention to a visual cue or engaged in anticipatory behavior when feeding was unpredictable or predictable. Fish spent more time engaged in anticipatory behavior when the feeding regime was predictable. (B) Plasma cortisol concentrations in the baseline, predictable and unpredictable treatments after (A) confinement and (B) feeding events. **p < 0.01; ***p < 0.001. Reproduced from Galhardo, L., Vital, J, and Oliveira, R. F. (2011). The role of predictability in the stress response of a cichlid fish. *Physiol. Behav.* **102**, 367–372 by kind permission from Elsevier.

unpredictable group during food presentation also exhibited a reduction in cortisol (Fig. 12.3; Galhardo et al., 2011). Studies investigating predictability in feed provision have actually produced conflicting results and it would seem predictability can be negative for species such as sea bream (*Sparus aurata*; Sánchez et al., 2009); this will have to be investigated on a species-by-species basis. Other types of stress can be reduced by classical conditioning, allowing fish to associate a neutral cue with an event or reward. Schreck et al. (1995) trained Chinook salmon (*Oncorhynchus tshawytscha*) to associate food with handling and transport. These fish had a reduced physiological response to stress, improved well-being during transport, and a better coping response when exposed to further stressors. Thus, positive reinforcement training can modulate the deleterious effects of stressors with benefits for fish welfare (Laule and Desmond, 1998).

Another successful strategy to provide control to fish is operant conditioning, where animals perform a novel behavior such as pressing a key to receive a reward (eg, rodents). This approach has been used in self-feeders that are operated by the fish who can then control the provision of the food they receive. This promotes growth and improved feed conversion ratios in rainbow trout (Alanärä, 1996). The ability for choosing between behavioral options also bestows control. Having a choice may result in a positive perception of an individual's coping ability (Eriksen et al., 2005). For example, carp had lower stress parameters and performed avoidance behavior if a refuge was provided when exposed to a threatening olfactory cue (Hoglund et al., 2005). Enrichments can allow animals to redirect their behavior when stressed, providing an outlet for frustration. Frustration-like behavior has been observed in rainbow trout when an expected food reward was denied where trout increased the performance of aggressive behavior; this was correlated with changes in brain neurochemistry comparable with that seen in mammals (Vindas et al., 2014). Provision of substrate can allow subordinate individuals of *Oreochromis mossambicus* to perform digging behaviors, avoiding aggressive interactions with dominant males, and provides a behavioral outlet for lack of breeding opportunities and low social status resulting in better welfare (Galhardo et al., 2008). Social context provides stability in captivity, thus fish should be held in an appropriate manner according to their life history (Carlstead, 1996; Galhardo and Oliveira, 2009). Isolation in gregarious species has a negative impact whereas small groups can be detrimental to territorial species. Social contact, however, does appear to reduce fear in some individuals, thus social enrichment can be used to improve welfare (Galhardo and Oliveira, 2009). If fish are to learn to cope and be able to anticipate stress they must then be exposed to stressors. This will allow them use their prior experience of coping or to learn to anticipate a stressful situation and improve their allostatic capability (Fig. 12.2).

An environment deficient of stimulation may impair cognitive and physiological coping mechanisms. Thus it is important for animals to be exposed to an appropriate level of environmental stimuli, but excessive stimulation should be avoided since it could lead to exhaustion and allostatic overload, which is contrary to positive welfare (Meehan and

Mench, 2007; McEwen and Wingfield, 2003; Galhardo and Oliveira, 2009). How much stress is excessive needs to be determined on a case-by-case basis. Pathological changes in the nervous system are seen in other animals where frustration and boredom can result in abnormal (including self-directed behaviors and displaced aggression) or stereotypical behaviors (Sapolsky, 2004). In mammals, victims of violence often direct aggressive behavior toward another individual or object that was not originally involved in the violence. This psychological condition is termed displaced aggression and appears to be an evolutionary conserved mechanism that can also be demonstrated in fish. Rainbow trout that received aggression from a larger, aggressive fish responded subsequently by directing aggression toward smaller individuals (Øverli et al., 2004). In the forebrain of these fish exhibiting displaced aggression patterns of serotonin turnover and plasma cortisol suggest that the aggression directed against socially subordinate fish inhibited neuroendocrine stress responses. These findings suggest subordinate individuals may serve as an outlet to reduce stress, and displaced aggression toward lower ranking individuals is possible as a behavioral stress coping strategy. Similarly when confronted with a competitor, a video display of a dominant male, territorial male *Astatotilapia burtoni* performed directed aggression toward the video images whereas nonterritorial males directed aggression toward other conspecifics in the tank possibly as an outlet for the stress or frustration at the situation (Clement et al., 2005). Stereotypes are repetitive behaviors that have no obvious function and often develop when animals are held in impoverished (barren) conditions or are subject to chronic stress (Mason and Latham, 2004). Performance of these behaviors is believed to be indicative of a detrimental mental state although some would argue these repetitive behaviors are actually a coping mechanism for dealing with chronic stress (Mason and Latham, 2004). Stereotypical behavior in fish has not received much attention but fish are known to perform repetitive swimming movements (eg, Atlantic salmon and Atlantic halibut, *Hippoglossus hippoglossus*, Ashley, 2007) that may resemble pacing in mammals.

3. THE IMPACT OF STRESS ON FISH WELFARE

When considering the empirical evidence (the adverse impact stress has on a variety of biological and behavioral measures in fish, for the capacity of fish to respond to and learn to avoid and anticipate negative events, and for their ability to choose between resources so that clear preferences can be identified, for the evidence of pain and fear in fish), these complex cognitive abilities suggests that fish can be considered capable of experiencing poor welfare (Sneddon, 2011). Even if we disagree with the concept of fish experiencing adverse states, all scientists concur with the idea that we must keep fish healthy for improved economic return, sustainability of fisheries, conservation, scientific validity, and to prevent transmission of zoonotic diseases from unhealthy, stressed fish to the general public. Fish are subject to practices that cause stress with large numbers harvested in aquaculture and fisheries as well as fish caught as a recreational pastime or for sport. Fish are also subject to stressful procedures in laboratory research although there are humane guidelines with respect to the treatment of experimental fish in most countries. Finally, ornamental fish are a popular companion animal easily purchased by the public who may have no experience in fish husbandry. These matters will be discussed later, with reference to stress and welfare, with suggestions for improvements to current practice if the goal of those who use fish is to maintain them in good health.

3.1. Stress in Fisheries

Fisheries exploiting wild populations of fish employ techniques that may significantly compromise fish welfare. These are discussed by Sneddon and Wolfenden (2012) and include many factors involved in capture methods, bycatch discards where unwanted species are returned to the water usually damaged or dead, and postcapture procedures including treatment and handling of fish as well as slaughter (Sneddon and Wolfenden, 2012). It has been estimated that 2.7 trillion fish were caught annually between 1999 and 2007 (Fishcount.org, 2015) compared with 3 billion mammals (cattle, sheep, pigs, goats) and 57 billion birds (chickens, turkeys, ducks) used in farming. Wild-caught fish are used as a foodstuff for humans but a considerable proportion is utilized as fishmeal and oil used to feed farmed animals and for fish oil nutritional supplements. Catching fish in large-scale commercial fisheries involves a variety of invasive and destructive techniques that have received media attention. The campaign in the EU, Fish Fight (Fish Fight, 2015), has 0.87 million supporters that wanted a ban on bycatch discard and encouraged the public to broaden the type of fish that they eat rather than focusing on a few overfished species. This campaign led the European Union to eventually ban bycatch discards (Fish Fight, 2015). Fish are caught in a variety of ways including trawling over a long time where fish may swim to exhaustion, can suffer broken spines and other injuries, are crushed during boarding by the weight of others, and may experience barometric trauma from being brought up rapidly from depth. Damage may also occur in other fishing methods (eg, gill and seine nets). Long-line fishing often involves hooked fish trapped for hours or days and nontarget species such as sharks, turtles, dolphins, and sea birds are also caught and suffer mortality. Live bait where smaller species are hooked on fishing gear to catch larger species is also practiced. The range of techniques utilized to capture fish can lead to significant stress (Sneddon and Wolfenden, 2012). These include the use of hook and line as in long-lining, trolling, and pole-and-line, as well as netting techniques such as trawls, gill nets, and seines (review in Sneddon and Wolfenden, 2012). Solutions to ameliorate the stress of capture could be investigated by improving fishing gear, capture methods, and refining the procedure of capture and slaughter to enhance welfare where logistically practical (Metcalfe, 2009).

The discarding of nontarget species (bycatch) is a controversial practice, necessitated in many countries due to quota enforcement, regulatory discards, and economic discards (lower value fish discarded so they do not take up space in the hold). Bycatch fish may suffocate on board the vessel and/or be subject to the injuries described earlier as a result of capture (Metcalfe, 2009). Higher water temperature increases postrelease mortality of discards with significant impact upon stress physiology (Cooke and Suski, 2005; Gale et al., 2013; Raby et al., 2015). Exhaustive exercise, hypoxia and air emersion, injury during capture and barotrauma from being brought up from depth too fast, and a neuroendocrine stress response are seen in caught fish (Davis and Olla, 2001, 2002; Davis et al., 2001; Farrell et al., 2001; Raby et al., 2012). The use of bycatch discards can be thought of as wasteful if perfectly marketable fish are thrown overboard; on welfare grounds alone, banning the discard of fish bycatch would influence an improvement in welfare; however, there are also proven positive impacts for the overall health of fish populations as a result of prohibiting discards. Trials of discard bans (eg, the Norwegian ban of discards in Northeast Arctic waters) resulted in recovery in fish stocks, and despite the initial dip in economic return, the market soon adapts to the more diverse range of species captured (Diamond and Beukers-Stewart, 2011). Other strategies to reduce stress in fisheries are being proposed such as slowing speed of trawls, allowing nontarget species to escape or reducing incidence of injury and exhaustion; bringing fish up from depth slowly to avoid barotrauma and rapid temperature change; and fishing outside of warmer temperatures to reduce stress (review in Metcalfe, 2009). Many scientists are now working with the fisheries industry to reduce wildlife bycatch (Consortium for Wildlife Bycatch Reduction, 2016). For example, employing LED lights in ocean shrimp trawling reduced the bycatch of fish species with between 56 to 91% less fish caught (Hannah et al., 2015). Adding a divider to provide a lower and upper cod end in trawls allowed Norwegian lobster to be separated from fish such that the fish were in better condition with fewer injuries (Karlsen et al., 2015).

Another issue of concern is that capture is stressful in some cases and this may result in delayed mortality so even if the fish swim away after release they may be so physiologically compromised that mortality occurs some time afterward. Plasma measurements from Coho salmon that had been caught in beach seine demonstrated that the physiological stress profile was similar to exhaustive exercise and during exposure to hypoxia, with significantly elevated cortisol and lactate concentrations for fish entangled for a longer duration in the fishing gear (Raby et al., 2012). Fish towed in a net in the laboratory to simulate trawling were then held for up to 60 days to determine delayed mortality, which occurred up to 20 days after towing (Davis, 2007). Method of capture can influence delayed mortality with post-caught cod faring better (9%) than long-line caught animals (39%; Humborstad et al., 2016). Studies in sablefish, *Anoplopoma fimbria*, have suggested that the elevated stress response during simulated trawling and hooking in laboratory conditions impairs immune function that may contribute to delayed mortality (Lupes et al., 2006). Factors enhancing delayed mortality are thought to be capture during warmer temperatures and lengthy exposure to air (eg, sable fish, Davis and Parker, 2004). Predicting the propensity for delayed mortality may be useful when deciding whether to discard fish or euthanize when fish are unlikely to survive. Reflex action mortality predictors (RAMP) is a technique that tests for the presence or absence of a reflex response (eg, movement to a light tail pinch) to generate a condition RAMP score after capture to predict fate. The method has been employed effectively in laboratory and field-based holding studies to indicate capture stress and to predict delayed mortality (Davis and Ottmar, 2006; Lupes et al., 2006; Davis, 2007; Raby et al., 2012; Humborstad et al., 2009, 2016). If the aim of releasing fish is to ensure they survive, the use of RAMP may allow an informed decision to be made as to the likelihood of any given fish to survive postrelease.

There are considerable variations in the techniques employed in fisheries to dispatch fish after they are caught. These range from simply allowing the fish to suffocate on the boat's deck, which anecdotally has been reported to take up to 250 min (fishcount.org.uk, 2013), put on ice, bleeding by cutting of the gills (exsanguination) to live gutting. Approximately 75% of cod caught by Danish seine that were left in air took 10 min to die with some taking up to 20 min (Humborstad et al., 2009). Robb et al. (2000) found bleeding takes over 7 min for brain death to occur in contrast to concussion followed by brain destruction, which was almost immediate. However, concussing the large numbers of fish caught in a trawl is a logistical challenge, thus, different sized fisheries may use different methods. Each of the techniques presents a challenge to the welfare of the fish concerned, although research is currently lacking in the welfare status of subjects from wild fisheries during slaughter compared to their farmed counterparts. Death by suffocation can take from 55 to 250 min and those fish that were gutted alive were still responsive for up to 65 min after (Fishcount.org, 2015). Turbot (*Scophthalmus maximus*) immersed in ice water were still responsive after 75 min of immersion (Lambooij et al., 2015). The Humane Slaughter Association (HSA, 2014) has produced recommendations for farmed fish whereby fish should be rendered insensible or unconscious before a more invasive method is applied to ensure brain destruction prior to processing. Therefore, it has been proposed that fisheries adopt percussive stunners. However, this has not been widely adopted due to the sheer numbers of fish caught that it may be impractical to stun every fish. Future research should investigate more humane methods of slaughter on board fishing vessels to reduce stress and improve welfare if logistically possible.

Clearly, there are improvements that could be explored to reduce stress during current fishing methods, specifically, reduce the time spent fishing so that fish are landed more quickly, reduce the injuries sustained to the fish by improving equipment; use quick, efficient humane killing techniques on board; and reduce bycatch (Metcalfe, 2009). These may be conveniently considered under the following categories and some countries may legally require these:

- Reducing the initial numbers of nontarget species (bycatch) captured
- Increasing the survival chances of discarded bycatch
- Reducing the duration of the capture experience
- Mitigating the stressful experience of slaughter for target species
- Adjusting fishing practices to exclude the use of live bait fish

3.2. Stress in Aquaculture

There are a range of welfare issues in aquaculture where routine husbandry and management procedures can cause stress (Table 12.2; Ashley, 2007; Conte, 2004). Exposure to unnatural stocking densities or social contexts can present a challenge since fish are reared intensively. Many studies indicate that stocking density can have a negative impact on the welfare of fish, and as discussed by Turnbull et al. (2008), water quality and the effect of social interaction are closely linked to the welfare of farmed fish. However, the situation is a complex one, and further studies are needed to determine the true impact of stocking density on fish welfare. For example, Ellis et al. (2005) suggest that both high and low stocking densities may compromise welfare. It is suggested that high stocking density and the resulting suboptimal water quality may cause decreased growth and lack of

Table 12.2

The major welfare issues fish experience in aquaculture that can lead to acute and chronic stress with suggested improvements

Area of welfare concern	Welfare issues involved	Improvements
Winter diseases		
Several diseases associated with low temperatures	Although many are clearly associated with specific bacterial pathogens, immunosuppression during winter may play a large role	Immunization Adapted diet providing a supplementary dosage of vitamins and trace minerals to assist the immune system and altered feeding regime controlling level of nutrients available to the pathogen
Fin rot		
Abrasion with the environment and/or aggressive interactions cause fin damage and secondary infection may follow	Injectable vaccines have superseded antibiotics although vaccines and adjuvants are associated with inflammation and granuloma, as well as the stress of handling anesthesia and injection	Vaccines with improved efficacy and reduced side effects as well as oral application
Sea lice		
Parasitic copepods may cause severe tissue damage	Lice have developed resistance to traditional chemical treatments	Potential alternative controls include vaccination and selective breeding toward louse resistance Biological control with cleaner wrasse but should consider wrasse welfare
Viral diseases		
Examples: infectious pancreatic necrosis, infectious hematopoietic necrosis, viral hemorrhagic septicemia, infectious salmon anemia, sleeping disease	Traditional vaccines developed over the past 20 years have shown only moderate success and there are relatively few commercial vaccines and specific therapeutics with adequate efficacy	The development of alternative antiviral treatments such as DNA vaccines and selection for disease resistance
Noninfectious production-related deformities		
Deformities of the heart, swim bladder, and spine	Fish with heart deformities show a high mortality rate during stress due to impaired cardiovascular function, cardiac failure, or heart rupture. Both genetic and environmental factors may contribute to spinal deformities	High temperatures during incubation of salmon should be avoided. Spinal deformities may be reduced by increasing smolt weight at seawater introduction, vaccinating, and reducing salinity and temperature variations. Fish from families showing a high incidence of deformities should not be used for breeding
Grading, handling, crowding		
Inherently stressful	Many procedures, such as grading, are aimed at improving welfare. There is a	Appropriate supplementation of dietary vitamins C and E, and glucan may protect

(*Continued*)

Table 12.2 (Continued)

Area of welfare concern	Welfare issues involved	Improvements
	large variation between species in stress response to procedures and handling stressors can affect subsequent stress response	against the adverse effects of chronic stress The appropriate use of good crowd management, suitable nets, careful handling, recovery periods, and movement using fish pumps and transfer pipes is preferable Appropriate feeding technique and stocking densities may avoid frequent grading
Transportation Inherently stressful as may involve capture, loading, transport, unloading, and stocking	Transport stressors can affect fish over a prolonged period	Adverse effects may be reduced by suitable acclimation and recovery periods as well as species-appropriate use of anesthesia and dilute salt solutions
Food withdrawal Starvation prior to slaughter, transportation, and other management practices	May benefit welfare by reducing metabolism, oxygen demand, and waste production. Atlantic salmon and rainbow trout show long anorexic periods in the wild, so the welfare effect of food deprivation in aquaculture is not known. Deprivation for short periods under appropriate conditions may not diminish welfare	Starvation for up to 72 h for Atlantic salmon and 48 h for rainbow trout should only occur where beneficial to welfare and empirical studies on the effects of starvation on stress physiology or behavior are required
Slaughter Slaughter should be as humane as possible—fish should be stunned prior to slaughter, causing an immediate loss of consciousness that lasts until death	Dewatering followed by asphyxiation in ice slurry of rainbow trout and gilt head sea bream; immersion in CO_2 saturated water followed by gill cut or gill cutting alone for Atlantic salmon and rainbow trout; and desliming followed by evisceration of eels do not meet the criteria for humane slaughter	Percussive or electrical stunning methods appear to achieve humane slaughter in Atlantic salmon, gilt-head sea bream, turbot, and rainbow trout The use of electric stunning tongues or electrically stunning batches of eels in freshwater combined with nitrogen flushing can cause immediate unconsciousness
Stocking density Pivotal factor affecting welfare in a number of different ways (eg, through aggression, water quality, and activity/feeding patterns)	The effect of stocking density comprises of numerous interacting and case-specific factors. Sea bass show high stress levels at high densities. Arctic charr show low growth and food intake at low and very high densities. Halibut tolerance for high stocking	Feeding pattern and floor space may be altered to improve the effect of density on welfare in halibut; also see aggression, to follow Salmon swimming depth and shoal density can be manipulated by artificial light levels, and feeding patterns can

Table 12.2 (Continued)

Area of welfare concern	Welfare issues involved	Improvements
	density appears to be stage-dependent. Rainbow trout show a decrease in welfare at high densities, water quality being a key factor. High stocking densities, above a given threshold, are associated with reduced welfare in Atlantic salmon in sea cages. Site-specific factors also have an effect on welfare	alter aggressive interactions in several species including Atlantic salmon
Aggression		
Formation of social hierarchies may lead to injuries, chronic social stress, and size heterogeneity	Sociobiology, stocking density, and feeding technique have strong influences on the levels of social interactions	Feeding technique should be species-appropriate to avoid excess competition and aggression Consider use of self-feeders to reduce aggression The presence of a small number of larger fish may reduce aggression within groups of smaller fish Increased dietary levels of L-tryptophan has been shown to suppress aggressive activity Substrate or background color may be used to influence aggressive behavior in some species
Abnormal behavior and the freedom to express normal behavior		
Abnormal behavior includes repetitive behavior and abnormal swimming activity/patterns	Understanding the functional origin of apparently abnormal behavior is important Empirical studies are required to establish whether abnormal behaviors represent diminished welfare or adaptive responses with no effect on welfare	Enriched rearing environments may improve welfare following release to augment wild populations Without empirical studies the importance of a given behavioral pattern to a given species is unclear. Studies of the mechanism of control and/or the behavioral and physiological consequences of denial of expression of key behaviors are required Choice studies may allow assessment of the value associated with a given behavior or resource. Ability to make behavioral choices (eg, shelter during stressful events or conditioning to neutral cues) so fish can anticipate the onset of a stressor

Source: Adapted from Ashley, P. J. (2007). Fish welfare: current issues in aquaculture. *Appl. Anim. Behav. Sci.* **104**, 199–235, with kind permission from Elsevier.

body condition as well as physical symptoms such as fin erosion. Fin erosion, often termed fin rot in an aquacultural context, refers to loss of, or damage to, the tissue of the fins of fish, and is a common problem affecting farmed salmonids (Ellis et al., 2008). Exposure to suboptimal water quality is one of the primary stressors in fish. The range of water quality parameters that can have deleterious effects on fish in aquaculture have been extensively evaluated, due to the high economic value of such fish (MacIntyre et al., 2008). This is particularly important in recirculating aquaculture systems (RAS), which may have as little as 10% of water replenished each day (Martins et al., 2011). If water quality is not monitored or action to remedy the situation is not taken quickly enough this can pose a significant threat to fish. Factors including pH, ammonia, nitrite, nitrate, salinity, oxygenation, carbon dioxide, and temperature should ideally be kept within an optimal range for the species concerned. Automatic monitoring systems encompass an alert system that can inform managers of any water quality issues. Some species undergo rapid changes in phenotype during smoltification; for example, Atlantic salmon go from freshwater to a marine phenotype that has to be managed carefully in aquaculture. A comprehensive analysis of production parameters demonstrated that higher stocking densities and exposing juvenile Atlantic salmon seawater warmer than freshwater resulted in better growth and survivorship (Kristensen et al., 2012). The management of these species is complex and at all life stages more research is needed to reduce the impact of stressors during key transformation periods.

Fish in aquaculture may be exposed to handling stress, and this may be repeated a number of times for each individual, for grading, vaccination, slaughter, and other procedures. Handling has been shown to elicit a stress response in numerous studies; for example, in the red sea bream *Pagrus major* (Biswas et al., 2006), Atlantic salmon (McCormick et al., 1998), and common carp (Saeji et al., 2003). Minimal handling and the development of handling techniques that limit stress will assist in the improvement of fish welfare. For example, the type of light fish are held under when handled can reduce stress responses. Blue light was most beneficial for sea bass, *Dicentrarchus labrax*, compared with white light during a stressor (Karakatsouli et al., 2012). In contrast yellow perch, *Perca flavescens*, grew slower under blue light compared with red and full spectrum light (Head and Malison, 2000). Sudden changes in lighting should also be avoided as this leads to stress in salmonids, thus light intensity should be gradually altered to simulate dawn and dusk (Mork and Gulbrandsen, 1994). Caged Atlantic salmon will avoid high light intensity and prefer swimming at depth during the day (Johansson et al., 2006), thus handling of these fish would be desirable at lower light intensities to prevent light-induced stress.

The range of diseases affecting teleosts is vast, and they may be precipitated by, and their effects exacerbated by, stressors such as poor water quality, excessive handling, and crowding. Management of factors such as water quality assist in the management of disease, thus promoting welfare, although chemical control of many pathogens is widely practiced in the event of outbreaks. Additionally, vaccination is now commonplace for many diseases and is a vital aspect of modern salmonid farming (Sommerset et al., 2005). Biological control has also been successfully trialed against certain parasitic diseases, notably through the use of cleaning wrasse, primarily the goldsinny (*Ctenolabrus rupestris*) to control sea lice in salmonids (Costello, 1993). Vaccination presents its own challenges; the adjuvants used to act as vehicle for the vaccine cause abdominal peritonitis and stress in Atlantic salmon (Bjørge et al., 2011), thus other vehicles should be explored.

In many countries legislation that regulates the transport of animals is in place, and this applies to the transport of fish as well as mammals. The stress response of fish to transport appears to vary according to the technique used. Whereas road transport may elicit measurable stress due to confinement and deterioration in water quality, Erikson et al. (1997) suggest that the use of modern well boats for the transport of Atlantic salmon—an established practice in Norwegian salmonid fisheries—results in various physiological parameters typically used in measurement of stress (including muscle tissue pH and adenylate energy charge) not changing significantly from those of unstressed fish. This is explained by the well boat's ability to maintain optimal water quality, as well as the rapid netting facilitated at the place of slaughter. Thus maintaining temperature, good water quality, and minimizing transport is key to reducing stress in fish that need to be moved. Other improvements include use of very low anesthesia to keep fish sedated during the journey and use of 5–10 ppm saline to prevent bacterial and fungal disease outbreaks (Ashley, 2007). Other strategies include using commercial products such as Slime Coat or Stress Coat to reduce stress and replace any lost mucus layer from the fish's body; however, this has been anecdotally proposed and has not been rigorously tested. Monitoring of temperature and oxygen is a must, which can lead to stress via a variety of factors such as large temperature fluctuations, hypoxia or supersaturation, and the build-up of carbon dioxide (Tang et al., 2009). Fish also experience stress during crowding, netting, and pumping during loading and unloading (Nomura et al., 2009).

Aggression is a long-standing problem in aquaculture and as discussed previously can lead to allostatic overload for subordinates due to chronic stress of low status. Juvenile salmonids are naturally aggressive defending feeding territories, which can present problems in the confined aquaculture environment especially during feed restriction (Brännäs and Alanärä, 1994).

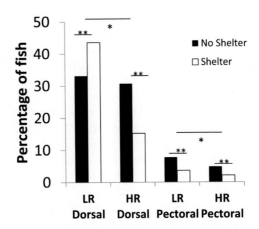

Figure 12.4. Percentage of Atlantic salmon, *Salmo salar*, possessing damage to the pectoral and dorsal fins when held under a low ration (LR) or high ration (HR) of feeding and whether shelters were present or absent. Fin damage was higher in fish on lower food rations and without shelter, suggesting an increase in aggression $*p < 0.05$; $**p < 0.005$. Adapted from Persson, L. and Alanara, A. (2014). The effect of shelter on welfare of juvenile Atlantic salmon *Salmo salar* reared under a feed restriction regimen. *J. Fish. Biol.* **85**, 645–656, with kind permission from John Wiley and Sons.

Increased aggression can result in more severe fin damage (eg, Cañon Jones et al., 2010), with the dorsal and caudal fin most affected (Turnbull et al., 2008; Persson and Alanara, 2014; Fig. 12.4). Dorsal fin injuries and stress after stocking can increase mortality rates (Petersson et al., 2013). To reduce this problem scientists have explored increasing the heterogeneity of the environment and providing enrichment in the otherwise barren, monotonous tank or raceway. For example, cobblestones used as substrate in raceways decreases fin erosion in rainbow trout and cutthroat trout, *Oncorhynchus clarkii* (Bosakowski and Wagner, 1995). Fin quality of rainbow trout and Atlantic salmon markedly improved in a more heterogeneous hatchery rearing environment since this provided refuge and reduced aggressive interactions (eg, Näslund et al., 2013). Several attempts have been made to increase the structural complexity of the rearing environment (eg, Brockmark et al., 2007), but whether structural complexity reduces aggression, stress, and as a result, improves long-term well-being is currently unknown for many species.

Foraging behavior defined as seeking and finding food is another potential stressor in aquaculture. When food is limited, animals will fight over access to food, therefore, it is incredibly important to feed at the right time, provide an adequate amount of appropriate nutritional value, and ensure all fish have the opportunity to satiate their hunger

(Martins et al., 2012). Foraging strategies are as diverse as fish are and are influenced by the aquaculture system, age, gender, genetics, seasonality, and other factors. As discussed earlier, fish adopt various feeding strategies (see Section 1.1; bottom feeders, surface feeders, and feeding from the water column; eg, nocturnal vs diurnal vs crepuscular; regular vs random; predators, scavengers, or more passive feeders; piscivorous, vegetarian, or omnivorous). Thus it is vital fish are provided with the right kind of food, in the right quantity, and in the right manner (Section 2.1). Without this, aggression and stress can ensue, and stress can elicit anorexia resulting in reduced growth (Wendelaar Bonga, 1997). Bottom-feeding flatfish fed sinking pellets compared with floating pellets had improved welfare (Kristiansen and Fernö, 2007). Sea bream subject to alterations in the feeding time (regular timed vs random) exhibited negative changes in behavior and stress physiology when the expected food was not provided (Sánchez et al., 2009). This may suggest that regularly timed feed provided on schedule allows fish a sense of control as they can then prepare themselves for further foraging bouts. Sea bream increased swimming activity before feeding (ie, food-anticipation), when a regularly timed feeding regime was employed in contrast to randomly assigned feeding (Sánchez et al., 2009). However, caution should be applied since regular feeding times increased fin damage when compared with demand feeding (Noble et al., 2007, 2008), and Atlantic salmon displayed signs of aggression and frustration when the expected feed was absent with greater variability in individual growth rates compared with random feeding (Vindas et al., 2012). Thus aquaculture managers and carers should carefully consider what feeding regime to employ in each case since this will be species specific and also depend upon the type of facility and management strategy. Employing self-feeders that fish can operate can improve feed conversion ratio and growth in rainbow trout (Alanärä and Brännäs, 1996), suggesting lower stress (Suzuki et al., 2008). This strategy allows the fish complete control over their feeding and as such may be beneficial for welfare.

Many routine husbandry procedures cause stress and disrupt foraging behavior indicating reduced welfare. Altered water quality, cleaning protocols providing disturbance to the tank, vaccination, aggression, and size grading procedures impact feeding behavior. All of these stressors could be coupled with a neutral cue to allow fish to anticipate them and prepare for future exposure. Another strategy would be to provide enrichment or refugia so fish have alternate behavioral options when faced with disturbance. Many of these management practices are necessary and beneficial for the health of the fish so consideration as to how to reduce the timing, intensity, and duration of these stressors should be investigated in future studies.

There are also welfare concerns over the fate of fish during harvesting and slaughter. Regulations in some countries state fish should be killed rapidly and humanely to reduce stress (eg, HSA, 2014). To this end humane methods of killing fish have been developed as mentioned previously, such as percussive stunners and electrical stunners so the fish are unconscious during secondary methods used to ensure brain destruction (Lines and Spence, 2014). Harvesting of fish often employs crowding and physical methods of removing fish after a period of feed restriction to reduce waste products and prevent food in the gut affecting flesh quality (Lines and Spence, 2014). Both the lack of feeding opportunities and the harvesting procedure are stressful so this must be done rapidly. Recent research in rainbow trout showed that these fish cleared the gut within 24 h but typically they would be fasted for up to 72 h where indicators of stress were elevated (Lopez-Luna et al., 2013). Thus shorter fasting periods should be explored. Harvesting involves crowding followed by pumping the fish out of the water or brailing (Lines and Spence, 2014). Reduction in oxygen and subjecting fish to reduced space as well as the fish swimming against turbulence are all stressors (Ortuño et al., 2001; Brown et al., 2010). In brailing a net is used to remove fish from the water—in wet brailing the fish are held in water preventing air emersion whereas in dry brailing there are risk to the fish through physical damage and abrasion from contact with other fish, the net, and being placed on hard surfaces (Lines and Spence, 2014), thus it is advisable to keep brailing times to a minimum to reduce stress. There is no doubt these transport methods cause acute stress so identifying better ways of performing transport such as well boats where fish are kept in optimal water quality should be a priority.

Methods of stunning and killing fish vary from asphyxiation in air or on ice, live chilling, bleeding out by severing the gills, beheading, live gutting, and the use of carbon dioxide saturated water, all of which significantly compromise welfare and are not considered humane approaches (reviewed in Lines and Spence, 2014). More welfare-friendly developments include the use of automated percussive stunners and electrical stunning. In percussive stunning, mainly used by the salmonid industry, a rapid mechanical blow to the top of the head results in rapid and prolonged or permanent insensibility. This can be done very effectively if the blow is strong enough and if it is applied at the correct location. Manual percussive stunning can be affected by the ability, training, and tiredness of the operator. This approach is mainly used in round species such as the salmonids and is ineffective in flatfish and those species with hard bony plates in the skull like carp, catfish, and tilapia (Lambooij et al., 2007). As an alternative, electrical stunning has been proposed to be effective in stunning fish prior to physical methods of killing (bleeding). Advantages of wet electrical stunning where electrical stimuli is applied to a whole water body is that large numbers of fish can be stunned simultaneously; the fish do not have to be exposed to air, crowded, or handled until after they are insensible. This obviously is an improvement on welfare. However, fish are not permanently stunned and so the operators must employ a secondary method quickly to ensure the fish do not come round after stunning (Lines and Spence, 2014). A more recent study has shown that applying electrical stimuli produces more stress in *Oreochromis mossambicus* than handling alone (Roques et al., 2012) so the method of slaughter requires careful consideration. Another method being proposed as more humane is the use of anesthesia in certain countries such as Australia, Chile, New Zealand, Korea, Costa Rica, and Honduras. In this case fish are anesthetized in their tank in large numbers using isoeugenol, a clove oil derivative, but this substance is not yet approved for use in many other countries. Anesthesia would be a more favorable way of killing fish (Kiessling et al., 2009) assuming a stress response is avoided (Zahl et al., 2010), the drug is not aversive to the fish (Readman et al., 2013), and it is fit for human consumption. The dose used for humane killing should be higher than the dose used for anesthesia to ensure death occurs (Sneddon, 2011).

3.3. Stress in Recreational Fishing

Fish are caught by recreational anglers and by sport fishers. Many of the issues described under fisheries (see Section 3.1) are also relevant here except that in many cases this is one individual human catching fish as opposed to large-scale fishing. Therefore, we do not provide an exhaustive review but outline some of the specific issues related to angling and possible improvements to this practice that may minimize stress. It is an ethical question for the individual engaged in angling/sport fishing who may judge their enjoyment of catching fish outweighs any cost (stress physiology, behavioral, injury, etc.) to the fish as to what practices they adopt in line with local guidelines and licensing. Often the fish are killed, and we could argue they provide a benefit to the human as food that outweighs any costs to the fish if the capture method and killing are humane. Fish are also released in a process called "catch-and-release," which may be down to the undesirable size or species of the fish as inappropriate for food or to meet local harvest regulations or other local requirements (Cooke and Sneddon, 2007). If catch-and-release angling is beneficial for the conservation and sustainability of fish stocks then it is assumed most of the fish survive after release. However, catching fish involves hooking injuries, exposing the fish to air emersion, and during sport if held in keep nets fish experience hypoxia (Cooke and Sneddon, 2007). There are numerous studies that provide analyses of the consequences of catch-and-release where fish experience

Table 12.3
Proposed improvement to catch-and-release angling proposed by Cooke and
Sneddon (2007) with additional comments on the benefit to the fish (all approaches
would reduce stress and improve fish welfare)

Proposed improvement	Benefit
Barbless hooks, circle hooks	Reduced injury
Cut fishing line if deeply hooked	Reduced air emersion since it is not possible to retrieve hook
	Consider euthanasia
Avoid contact with gills or eyes	Reduced damage
Play fish minimally	Reduced stress and exhaustion; energy for subsequent behavior
Keep fish in water	Reduce stress of air emersion
Euthanasia of injured fish where legal	Improved welfare in fish that would not survive release
Use knotless nets	Reduced net abrasion
Raise fish slowly from depth	Reduced barometric trauma
Risk of fishing at warm temperatures	Reduced stress and mortality

Source: Adapted from Cooke, S. J., and Sneddon, L. U. (2007). Animal welfare
perspectives on recreational angling. *Appl. Anim. Behav. Sci.* **104**, 176–198, with kind
permission from Elsevier.

physical injury; sublethal alterations in behavior, physiology, or fitness; and
mortality. Research suggests that all recreational fishing results in some
form of injury and stress to an individual fish (review in Cooke et al., 2013).
However, the severity of injury, magnitude of stress, and potential for
mortality varies extensively in response to a variety of environmental factors
as well as the behavior of the angler. However, this information can be used
to positively identify strategies that could reduce stress through changes in
either gear (eg, type of hook, bait, or net) or angling practices (eg, duration
of fight or play, limiting air exposure, no fishing during extreme
environmental conditions, avoiding the reproductive period; Table 12.3).
Naturally the catch-and-release angling experience involves some stress and
injury (eg, the need to physically hook the fish), but perhaps adopting less
invasive, stressful practices is beneficial to fish health, will improve
survivorship after release and overall welfare (Table 12.3). This will have
an obvious benefit on the fish population and fishery to enhance
conservation and management.

3.4. Stress in Ornamental Fish

Ornamental fish are kept or reared for their attractiveness. The breeding
of ornamental fish began over 1000 years ago with the domestication of
goldfish in China, and nowadays many fish species are highly prized for their

distinct markings (eg, koi carp). It is estimated 1 in 10 households possess pet fish in the United Kingdom (Balon, 2004; The Telegraph, 2012) and in the United States (Davenport, 1996), with an estimated 20–25 million kept in aquaria and 20 million in ponds in the United Kingdom alone (PFMA, 2014). Approximately 4000 freshwater and marine species are held as pets or in public exhibits. The global trade in ornamental fish is substantial and estimated to be worth around GBP£3–4 billion (OATA, 2014). The commercial (retail) value of individual pet fish ranges from around one pound (GBP) (eg, goldfish) to over 10,000 GBP for a top-quality koi.

Pet fish are diverse and from a variety of natural habitats, thus all the welfare issues already mentioned in keeping wild fish in captivity (Section 2) and in aquaculture (Section 3.2) also apply to ornamental fish. Many species are wild caught and transported for up to 3 days from areas such as the Amazon and coral reefs (Walster, 2008). Mortality rates can be from 1–30% in shipments of wild-caught fish (Ploeg, 2005 since the fish experience poor conditions and both transportation and acclimation can result in stress as the animals pass through various stages in the live-fish chain. Mortality rates can also be high after delivery (Ploeg, 2005). Generally, 90% of freshwater ornamentals are farmed, whereas only 10% are collected in the wild. Farmed ornamental fish are often produced in large-scale facilities in, for example, Israel and the Far East, and these present perhaps the greatest welfare concerns in terms of disease and mortality. For example, often potentially devastating viral outbreaks have been reported from several of these intensive facilities (Chua, 1996; Paperna et al., 2001). For marine species around 95% are wild caught and just 5% are reared (Oliver, 2001, 2003; Sea Shepherd, 2014). So the tropical marine fish trade may affect natural populations and has negative consequences for conservation and biodiversity. Various extreme morphologies have been produced by ornamental fish breeders such as fancy bubble eye goldfish (UFAW, 2013) and balloon mollies (mollies with stunted body shape and distended abdomens (UFAW, 2013), all of which compromise welfare due to stress and ill health, plus a number of hybridized species have been produced (eg, the parrot cichlid) for their ornamental appeal, and the welfare consequences of these are unknown so more evidence is needed. This is a similar situation to pedigree dog and cat breeding where certain breeds selected for extreme traits has led to veterinary health problems (eg, snout shortening in some breeds of dog leads to brachycephalic obstructive airway syndrome; Packer et al., 2015).

Fish are relatively inexpensive pets and, generally speaking, any adult can purchase an aquarium or pond fish knowing nothing about keeping fish except for water and food requirements (in contrast to keeping mammalian pets since humans are mammals so have a better understating of mammal husbandry and care). Reputable pet stores often enquire about prospective owner's knowledge and can provide advice on tank size, water conditions, and any other factors relevant to the specific species. However, not all vendors exhibit such good practice and can result in stress and poor welfare in the hands of inexperienced owners.

Legislation banning trade in mutilated pet fish (eg, dye-tattooed, dye-injected; "cosmetic" surgery) is advisable as well as a ban or restrictions on certain man-made strains that are of welfare concern (eg, celestial eyed goldfish; scoliotic parrot cichlid hybrids; balloon mollies that experience swimming problems due to unnatural body shape) if fish welfare is considered important. Restrictions on the import of unsuitable large species that grow too large for the home aquaria such as pacu sp. and several large catfish species (such as *Pangassius,* red-tailed catfish and shovel-nosed catfish) would prevent them being held in stressful circumstances where often these fish die prematurely or end up being rehomed at public aquaria.

In wild-caught fish, mortality rates can be high (Townsend, 2011), up to 30% (Ploeg, 2005), due to the stressful conditions under which the fish are shipped and due to acclimation stress in the animals. Transportation times of more than 24 h are stressful for common clownfish even though guidelines allow up to 3 days for transport (Wolfenden et al., unpub. data). Rather than taking fish from the wild, sustainable breeding programs in the importer's country may reduce such problems and lower transport times thereby improving fish welfare.

3.5. Stress in Research Within a Laboratory Context

In terms of their role in biological and medical research, the use of fish as experimental models is increasing. Zebrafish are a significant model species (Clark and Ekker, 2015) (Spagnoli et al., 2016; Chapter 13 in this volume). Fish are second only to mice in terms of the numbers used in regulated procedures in the United Kingdom (Home Office, 2014) and their use is increasing in Europe (eg, 28% increase from 2008 to 2011 to approximately 1.3 million, Europa, 2015). The increasing use of fish species is likely to reflect global experimental animal use (eg, USDA, 2013). The precise nature of stress and its impact upon fish welfare is covered by Spagnoli and colleagues (2016; Chapter 13 in this volume), but again all the aforementioned husbandry, environmental, nutritional, and social stressors apply to experimental fish as they do to those held in other captive contexts (Sections 2.1 and 3.2) (Reinhardt, 2004). For example, laboratory aquaria can lack any type of enrichment and this is especially true of zebrafish holding facilities, yet zebrafish given a choice of enriched areas over barren areas choose substrate and plant enrichment even though they had been reared in barren conditions for 7 months (Schroeder et al., 2014). The health and welfare of fish is tightly legislated in Europe with clear guidelines to improve the well-being and health of captive laboratory fish subject to procedures that may cause stress, pain, or lasting harm (European Directive, 2010/63/EU). Many other countries also have clear guidelines to promote good welfare since this is beneficial for scientific data collection. For example, the American Fisheries Society has published guidelines for the use of fish in research (AFS, 2014) that is a useful resource for scientists and carers (technical staff) in North America. Although laboratory fish welfare is far behind that of mammals, particularly rodents, much can be learned from reviewing the literature on rodent housing and husbandry (Singhal et al., 2014). Appropriate social contact, sensory stimulation, tank design, nutrition, physical enrichment, and novel objects that provide mental stimulation are key factors in improving well-being. However, more research is needed to truly understand the needs of laboratory fish species combined with an understanding of their life history, environmental tolerances, and behavior from a species-specific perspective. This will ensure the integrity of the data collected and the relevance of the scientific outputs (Spence et al., 2008).

3.6. Stress and Welfare in Wild Fish

3.6.1. FISH SURVEYS

Biological research and fisheries management surveys often require capture and/or collection from wild populations, and any such interventions are likely to cause stress in the animals. Minimizing stress or its effects should not only be a prerequisite for ethical reasons but is also likely to maximize success in studies utilizing mark-recapture techniques, and ensures maximal survival rates in fish that are rereleased or transferred elsewhere (eg, laboratories, aquaria). Furthermore, behavioral and physiological measurements of fish exposed to improper or poor techniques may be compromised by the procedure yielding poor quality data, and thus care should be provisioned to ensure fish surveyed in the wild are exposed to minimal stress.

The process of trapping involves a variety of discrete stressors throughout the process, which may involve an initial stage of interruption of natural fish behavior (eg, prevention of migration along its route), the process of actual trapping, and a final stage of processing (Clements et al., 2002). The method of capture itself has direct impact on fish health and survival, though the harmful effects of different capture techniques may not be comprehensively understood (Clements et al., 2002). This section does not attempt to provide an exhaustive list of trap types but summarizes some of the primary sources of stress and welfare issues among the most common methods of capture. Trapping using nets or wire cages, either passively (ie, planting the device and allowing fish to swim in) or actively (dragging nets in a targeted manner), is an extremely common method particularly among commercial fishing. Stress can be induced even before fish are actually trapped; prevention of fish from achieving a desired goal (such as movement along its migration route by a barrier) can result in increased burst activity and activation of the stress response (Clements et al., 2002). Subsequent entrainment within the trap is similarly associated with stress, likely due to confinement with any escape route not obvious (Clements et al., 2002)

Electrofishing is an alternative and popular method of collecting numerous fish rapidly for sampling by utilizing strong electric fields in water to disorientate and stun the fish, which are then easily captured and processed. Reorientation and recovery of fish is rapid once the electric field is removed, often with few external indicators of damage, suggesting limited impact on fish biology. However, spinal injuries including misalignment and fracture are apparent among 44–67% of fish X-rayed following capture through electrofishing (Sharber and Carothers, 1988). Lethality and magnitude of nonlethal damage of electrofishing procedures varies according to a variety of poorly understood factors including the nature of the electric field and the biology of the fish, including size and species (Snyder, 2003). Spinal injuries and muscular hemorrhaging are thought to occur during convulsions as fish are exposed to sudden voltage differentials across the body, which may occur as the apparatus is switched on or off, as the electric field is pulsed, or as fish exit and reenter the water (eg, by leaping as an escape response or removed and replaced by net), though the nature and quality of the evidence limits any direct methods of ameliorating potential injury and stress. Recommendations include use of currents that are continuous (eg, direct current) and of low power, and improved postcapture care, including a reduction in handling time and ensuring holding-tank conditions are of high quality (Snyder, 2003).

3.6.2. HOLDING CONDITIONS

Holding conditions subsequent to capture should prioritize fish welfare, and therefore an understanding of how to improve the conditions of holding should be paramount for practitioners. This is particularly important where, for instance, ongoing survival and health are important (eg, supply of stock, research, capture-release, etc.); here, the primary concern is to ensure stress during both capture and holding are minimized. In contrast, in commercial fisheries where capture is quickly followed by slaughter, holding welfare may be considered less important and, furthermore, facilities to ensure good standards of welfare may be unsuitable on board vessels. Even within the fishery industry, however, practitioners should be mindful of the impact of premortality stress on flesh quality (eg, Poli et al., 2005) and, therefore, from both an ethical and commercial perspective, high welfare standards should be maintained where possible. This section, however, will focus on instances where ongoing welfare is important though, clearly, these methods are applicable to all practices where improvements in welfare are to be made.

Holding, which may occur within a trap itself or subsequently in a holding tank or chamber, can cause stress, evidenced by increases in plasma cortisol, which is likely due to confinement (Clements et al., 2002; Portz et al., 2006; Sharpe et al., 1998). Indeed, confinement is considered a standardized stressor for measuring poststress cortisol responses in fish (eg, Pottinger et al., 1992; Sharpe et al., 1998) and the basis is thought to be due to overcrowding and, additionally, the resulting impact on water quality (Portz et al., 2006). The magnitude of any negative impacts on fish caused by confinement are linked closely with the duration and intensity of the confinement; extremely high fish density (as in, for example, holding tanks while fish are waiting to be processed) can cause highly pronounced stress responses with effects equivalent to those caused by chronic stress (Portz et al., 2006). In holding tanks, therefore, fish should be maintained at as low a density as possible and reasonable for the particular species (eg, see Sections 2.1 and 3.2) and for as short a period of time as possible.

Water oxygen content, temperature, light intensity, stocking density (see Section 2.1 for further details on stocking density), and duration of capture have all been implicated as among the most important features of any holding environment where stress is a concern (Oldenburg et al., 2011), with some dependent upon the species being held. Dissolved oxygen at levels near saturation (but not above since this can cause gas-bubble disease) is critical, particularly since oxygen consumption rates in stressed fish are increased (eg, Barton and Schreck, 1987). Fish should also be maintained at species-specific optimal temperatures. While minimizing thermal stress, those fish in particular that are exposed to surgical procedures (eg, for tagging, see Section 3.6.3) will benefit from improved healing rate and reduced chance of immunosuppression when held at appropriate temperatures (Oldenburg et al., 2011). Furthermore, rapid fluctuations in temperature are detrimental to fish health; thus, the temperature in holding tanks or pens should be constant and similar to that from which the fish was captured (Oldenburg et al., 2011).

Finally, while duration of capture should be minimized since maintenance in artificial conditions may inherently be stressful (Portz et al., 2006), provision of a recovery period following any invasive procedure should be provided (Oldenburg et al., 2011). The period of recovery should be sufficient for fish to return to homeostasis and is therefore dependent on the magnitude and duration of the stress response, itself dependent on the procedure. However, a clear suggestion is to ensure that fish are returned to the wild as soon as possible, and are thus processed quickly, allowed to recover in a suitable environment, and released upon recovery if we wish to reduce stress (Clements et al., 2002; Oldenburg et al., 2011)

3.6.3. TAGGING AND INVASIVE PROCEDURES

Tagging has become an important aspect in monitoring fish populations and/or individuals. However, in addition to any stress induced by capture and holding, the process of implanting tags can elicit stress. There are a number of types of tag that can be used and that will have various impacts on the fish both in terms of the difficulty of the procedure, but also in short-term effects through immediate recovery, and longer-term effects over the span of the life of the fish or until the tag is lost. Tags can be external, such as pop-up satellite archival tags and anchor/T-bar tags (which, though considered external, often require a gun to implant within tissue) or simple procedures such as fin clipping; can be implanted subcutaneously such as visible implant elastomer; or could be internal via gastric insertion or through surgery and implantation into the coelomic cavity, such as with passive integrated transponder tags (Fürtbauer et al., 2015; Hoolihan et al., 2011; Newby et al., 2007). Researchers have long known that tags can cause both short-term (through physiological responses to the capture and handling procedure) and long-term (through injury, acclimation to the tag or poorly attached tags) modifications to behavior, although the extent of stress and behavioral change is not always known or completely understood (Hoolihan et al., 2011).

External tags are often quick and simple to apply, and in relation to surgery certainly less intrusive. However, they can still cause significant stress and/or pain in the fish. In Nile tilapia, *Oreochromis niloticus*, and carp, *Cyprinus carpio*, caudal fin clipping did not produce a cortisol response any greater than that of the associated handling procedure; however, the fin clip did appear to cause changes in gill physiology and pain that required at least 6 h of recovery (Roques et al., 2010). Experimental fish exhibited altered behavior compared to controls, including increased swimming activity and aversion to areas of the tank frequented by control individuals, with these differences abolished after 6 h. Furthermore the presence of C and A-δ nerve fibers also indicated the capability for nociception with the behavioral recovery time consistent with that observed for dull pain in mammals (Roques et al., 2010). T-bar tags may be applied with or without anesthesia, depending upon the circumstances. Tagged fish may exhibit greater mortality (Jolivet et al., 2009). In this study, any effect of the tags on growth were masked by the high mortality rate among tagged fish. However, other studies indicate no effect of external tagging on feeding or swimming behavior (Berumen and Almany, 2009). Thus the effects of external tags on fish could be wide-ranging and the determinants of survivability posttagging are not yet clear.

VIE tags consist of a fluid elastomer of various colors, which are implanted subcutaneously. To do so fish are often anesthetized though, again, not all VIE tagging protocols include anesthesia. Application of the tag appears to be stressful with stickleback, *Gasterosteus aculeatus*, responding with a physiological change similar to that after exposure to a predator (Fürtbauer et al., 2015). However, this does not necessarily imply that the tagging process is considered an attempted predation event and the rate of recovery postimplantation may differ from a survived predation event.

Internal tags are necessarily invasive and the effects on fish stress and health appear variable. Gastric tags do not require surgery; instead the tag is introduced through the mouth/esophagus and ingested, either forced or voluntarily swallowed (although this still requires coaxing and the introduction of the tag into the mouth). The requirements for handling, whether forced or voluntarily swallowed, generate stress with forced feeding potentially more stressful, resulting in more rapid regurgitation (c. 30 days rather than 60), reduced appetite, and lower survival rates compared to fish that are allowed to swallow voluntarily (Winger and Walsh, 2001). Extremely careful handling of fish with minimal contact and good teamwork between handlers can ensure that intragastric tags can be implanted with minimal stress or behavioral change relative to untagged conspecifics (Smith et al., 2009). In contrast, intraperitoneal/intracoelomic tags must be implanted surgically, often necessitating the use of anesthesia (although, for food fish particularly, the use of chemical anesthetics may be governed by various regulations often requiring lengthy withdrawal periods post-surgery; Trushenski et al., 2013). The process generates a stress response with higher plasma cortisol in fish having undergone tag-implantation compared to control fish that are handled or fish both handled and held under anesthesia (Lower et al., 2005). In the long term there have been notable issues in health of the tissue associated with the site of surgery (Caputo et al., 2009) although reports on physiological health vary, with some studies indicating poor physiological health in the long term (Montoya et al., 2012) and others suggesting no long-term physiological impacts on the fish (Caputo et al., 2009; Loher and Rensmeyer, 2011).

The use of tags therefore requires careful consideration regarding the most appropriate type of tag to be utilized, taking into account not just experimental factors such as the type of data that needs to be collected and the recovery rates of different types of tag, but also the immediate and long-term impact on fish and fish health if we consider this an important issue.

Vaccination is also an important procedure, particularly in aquaculture, which provides protection from disease and minimizes the impact that may occur on local ecosystems by the provision of antibiotics (see Section 3.2). Again, however, vaccinations require the temporary handling and containment of fish as well as injection with the associated impact on welfare and stress. Furthermore, intraperitoneal vaccination can cause some damage at the injection site with resultant stress and associated welfare impacts (Poli, 2009). Fish may behave aberrantly and have reduced appetite and, therefore, impacted growth. Obviously, the benefit of vaccination (improved immunocompetence and reduced reliance on antibiotics) must be weighed against the costs of stress on a case-by-case basis. Importantly the levels of care throughout the entire procedure, from trapping through to subsequent release, should be carefully considered to minimize impact on fish.

Processing fish—a series of actions involving measurements, fin clips, and checking for tags/clips—often occurs as part of any survey and itself can cause significant stress thought to be separate from that of confinement as might be found in traps/holding tanks (Clements et al., 2002). In addition to handling (see Section 3.2), emersion itself is considered an acute stressor, generating very high plasma cortisol concentrations in fish (eg, Pickering and Pottinger, 1989). Furthermore, attempts to capture fish (eg, with a net from a holding tank) may simulate chasing by a predator and incur a physiological response (Brown et al., 2007). Additionally, any handling can cause disruption of the mucous epithelial layer of the fish, leading to injury and infection. Thus handling at this point must be rapid, with minimal contact, and must be performed a minimum number of times to minimize any negative physiological effects. Effective organization and coordination between handlers can help minimize any stress caused through handling (Smith et al., 2009). The use of particular nets (which may be species-specific) or handling with wet towels/hands will also minimize physical damage to the tissue or mucous layer of the fish (Ashley, 2007).

3.6.4. DAMS AND OTHER BARRIERS

Fish may often come into contact with barriers that inhibit movement along their intended path. These may be anthropogenic, such as dams and hydropower turbines, but may also include natural barriers such as waterfalls and high gradient slopes or debris (including, for example, beaver dams). Particularly for migrating species these can act as significant barriers to progress, inhibiting movement both upstream and downstream. In most cases, however, barriers can induce stress and injury in fish with the effect varying according to the availability of alternative routes and, across dams, to the types of turbine used (Brown et al., 2012). Fish communities are likely to have developed around natural barriers such that particular species/populations are geographically segregated whereas others are capable of traversing these barriers (eg, Torrente-Vilara et al., 2011). However, the impact of natural barriers on fish migration appears to have been largely ignored in the literature favor of the study of the impact of anthropogenic barriers (Thorstad et al., 2008), which, in contrast, can cause significant fragmentation of stream habitats, though our knowledge of the impact of such fragmentation is currently limited (Bourne et al., 2011). Furthermore, the effect of dams on fish will inherently depend both on the species of fish and its life stage (for instance migratory vs nonmigratory, and direction of travel at different ontogenetic stages; Coutant and Whitney, 2000); a similar variability is likely for both natural and artificial barriers. Nevertheless this section will focus primarily on the amelioration of the effects of artificial barriers.

Dams are often built with passage provided (eg, spills and trash racks), which allow for almost complete survival (<2% mortality rate; Muir et al., 2001, Whitney et al., 1997) compared to when fish attempt to make passage through the turbine itself (where mortality rates are approximately 5–15%; Muir et al., 2001; Whitney et al., 1997). Mortality and damage generally arises from abrasion with hard structures (the walls of the dam) or with the turbine itself as fish are unable to maintain orientation and position within the water column (Coutant and Whitney, 2000; Brown et al., 2012) and can include hemorrhaging and rupture of the swim bladder (Brown et al., 2012). The ability of fish to travel through the turbine itself is largely dependent on their biology: the vertical position of the fish in the water column, influenced by their buoyancy, will influence their route of passage, and thus method of buoyancy can directly impact on safe passage. Fish with swim bladders will be unable to adjust buoyancy quickly enough to account for the downward pull through the turbine (Alexander, 1993), resulting in disorientation. These fish generally attempt to regain neutral buoyancy by swimming upward, reorientating themselves relative to the turbine (Coutant and Whitney, 2000). In contrast, fish utilizing dynamic lift through the turbine will sink since they will be unable to maintain forward propulsion through the turbulent water (Coutant and Whitney, 2000). Additionally, any fish implanted with intracoelomic telemetry tags may face increased risk of damage or mortality, which increases with size of the tag relative to the fish (Brown et al., 2012). This danger can be largely eradicated through the use of neutrally buoyant, externally attached acoustic transmitters (Brown et al., 2012). Though these tags may cause tissue damage and are associated with reduced swimming capabilities, reduced growth and reduced survival rates, use of carefully selected tagging methods appropriate to the species can inhibit negative impacts (Jepsen et al., 2015). Furthermore, tags developed specifically for the purpose of passage through dams provide promising initial results, although with further development necessary (Brown et al., 2013).

The safest passage is likely, therefore, through routes that bypass the turbine, but the efficacy of these routes likewise depends on the biology and, particularly, swimming patterns of fish. For instance, juvenile salmon migrating downstream often make use of the upper portions of the water column and thus have a greater chance of making use of spills than adults (Giorgi and Stevenson, 1995; Coutant and Whitney, 2000). These fish will generally descend to deeper waters in search of passage if such a route is not available (Coutant and Whitney, 2000). The efficacy of spills also depends on their structure. Increasing spill volume (the amount of water passing over the spill rather than through the turbine) will increase the number of fish making use of the spill in a curvilinear manner (Coutant and Whitney, 2000). Greater rates of flow of water can also improve successful passage, and this may vary seasonally. Many migrating fish will align themselves with the strongest flow of the water, which is usually in the center of the river and toward the surface where friction with the channel is lowest (Coutant and Whitney, 2000). Finally, spills aligned with the flow of the river also maximize the number of fish utilizing them (Willis, 1982; Coutant and Whitney, 2000). Thus positioning spills adequately with respect to how fish use them will minimize the number of fish resorting to passage through the turbine and maximize survival across the dam. Additionally, guidance mechanisms such as intake screens, electric fields, and underwater sounds or lights can also encourage fish along a particular route.

Adult salmonids returning to the natal river to spawn will face a different range of issues. Passage past turbines and dams is provided through fishways or ladders (Pon et al., 2009). Traversing dams in such a manner still represents a significant difficulty. Differences in stress may be difficult to measure in fish that have undergone excursions across dams since the plasma cortisol response is transient or simply low relative to the overall stress of migration (Pon et al., 2009; Roscoe et al., 2011); however, significant mortality can be observed in spawning salmon migrating upriver across a dam compared to those that do not (Pon et al., 2009; Roscoe et al., 2011). Despite this there appear to be no clear physiological or behavioral correlates indicating success or failure in breaching dams, at least that are yet known (Pon et al., 2009) although sex may be a factor, with fewer female sockeye salmon, *Oncorhynchus nerka*, surviving than males in one study (Roscoe et al., 2011). Stress may also arise through an inability of fish to navigate the dam. Fish unable to progress in their journey if a river is blocked with a net may show increased burst swimming behavior and increases in plasma cortisol (Clements et al., 2002). Thus exhaustive anaerobic exercise associated with this swimming may inhibit further progress. However, for adult fish navigating a dam but unable to find passage, no evidence was found of any burst swimming or compromised metabolic state except for somewhat reduced plasma Na^+ concentrations that were still within accepted standards for migrating salmon (Pon et al., 2009). Dams, weirs, and power stations can also cause delays of days to weeks; Atlantic salmon, *Salmo salar*, attempting to maneuver past often ignore bypasses either since they have imprinted to travelling through the dam/station itself or because the greater water discharge is a more attractive route than the gentler discharge of the bypass (Thorstad et al., 2008). Furthermore, fish that are not migrating but are found in the presence of dams circumstantially are also at risk, particularly young fish with poor or weak swimming ability that can be drawn into the turbines accidentally (Coutant and Whitney, 2000). It would be extremely difficult to provide mechanisms to minimize risk to these fish.

The data suggest that dams provide a significant barrier to fish passing up- and downstream and pose a threat to fish inhabiting nearby areas. The development of safe passage through the dams appears to ameliorate much of the damage caused to fish and does reduce the mortality rates observed. However, improvements are still clearly necessary, particularly for fish travelling upstream.

3.6.5. AQUACULTURE AND THE IMPACT ON WILD FISH

Often interactions occur between farmed fish and the local natural population near which they are held (Skaala et al., 1990), with important ramifications for the health of the natural stocks. Such interactions can occur due to the presence of wild fish near farms, drawn by the availability of uneaten feed that has passed through the cage (Dempster et al., 2009), or due to escapes from farms through failure of the nets or human error (Grigorakis and Rigos, 2011; Johanesen et al., 2011). The nature of these interactions can be varied, including reproduction and, thereby, genetic alterations in both groups. Loss of genetic variation among captively reared fish through inbreeding, and the inheritance of maladaptive characteristics (such as reduced disease resistance, reduced or excessive growth rate, reduced or increased fecundity, and poor quality gametes/offspring) can be detrimental to wild stocks if the genetic material is introduced. This may occur through restocking programs or through escapes although the factors that can drive or limit genetic mixing between captive and natural populations remain unclear (Skaala et al., 1990). However, other interactions, such as competition and predation and the introduction of pathogens (Skaala et al., 1990), can have much more direct impacts on wild populations. The passage of pathogens, for instance, is fairly unrestrained due to the passage of water through open net pens (Johanesen et al., 2011), with pathogens then potentially easily spread as wild fish are able to disperse.

The presence of wastes, a combination of uneaten feed, incompletely digested feed, and excreta, can cause deteriorations in water quality through organic loading (Grigorakis and Rigos, 2011), including soluble nitrogen and phosphorous compounds in fish farms in natural environments. The impact is most strongly felt on local ecology, with increases in microbial load and fouling organisms and alterations to benthic communities. Also of considerable importance is the use of persistent antibiotics, which have been implicated in increased drug-resistance among pathogens prevalent in fishery-reared fish; however, toxicity of drugs to local communities appears to be limited (Grigorakis and Rigos, 2011). However, these effects can feed directly back into stocked animals within the pens and, particularly where water quality is concerned, cause welfare issues.

Within aquaculture and fish farms clear management strategies must be in place to ensure that clean water is provided and allowed to flow through the system, removing wastes and replenishing oxygen. Disease and pathogens must also be dealt with appropriately, but in all cases care must be made to ensure that local environmental health is not damaged.

3.7. Surgery and Anesthesia

Anesthesia (review in Sneddon, 2011) and surgery elicit a stress response in fish. In wild fish anesthesia and surgery is typically conducted to implant invasive tags (see Section 3.6.3) from which a wealth of important biological knowledge is obtained for fisheries management and fish biological research (Loher and Rensmeyer, 2011). In flat fishes internal tags can result in abdominal adhesions forming that could be potentially painful, but these fish grew at the same rate as controls so this is unlikely to be stressful (Loher and Rensmeyer, 2011). Intracoelomic tagging procedures should include good surgical practice including pre- and postoperative care, sterile instruments, anesthesia, wound closure using absorptive sutures, and the use of antibiotics (although debated as to their utility; Brown et al., 2011). In the laboratory fish may be subject to a range of invasive surgeries such as tag implantation, fin clipping, cannulation, spinal and optic nerve crush, brain lesions, intestinal resection, and heart surgery (eg, Itou et al., 2014; Saera-Vila et al., 2015; Schall et al., 2015). Again it is vital to use good surgical practice to promote healing and to reduce secondary infections. Analgesia is not routinely adopted in fish surgical procedures but recent studies have shown the utility of analgesia in ameliorating behavioral and physiological changes in fish during a potentially painful event (Sneddon, 2011, 2015).

Anesthesia should be carefully considered due to the impact of many agents upon stress physiology (Sneddon, 2011). A variety of anesthetic drugs are applied to fish via immersion. Correct dosing can result in effective anesthesia for acute procedures as well as loss of consciousness for surgical procedures. However, it is important to note that dose and anesthetic drug vary between species of fish and their utility can be affected by a variety of physiological factors (eg, body weight, stress status) as well as environmental conditions (eg, temperature). A range of drugs are currently in use in laboratory, veterinary, and aquaculture contexts. The most common are buffered MS-222 (Tricaine), benzocaine, isoeugenol, metomidate, 2-phenoxyethanol, and quinaldine (Ackerman et al., 2005; Ross and Ross, 2008; Neiffer and Stamper, 2009). Induction, recovery rate, and pharmacokinetics as well as undesirable or adverse side effects have been explored (Sneddon, 2011). However, these are limited to a relatively small number of species and using any of these agents on a nontested species should be done carefully. Recent research has demonstrated MS-222 and benzocaine are aversive to zebrafish and that metomidate and 2,2,2 tribromoethanol (TBE) do not elicit an avoidance response (Readman et al., 2013). However, these nonaversive drugs have no analgesic properties whereas MS222 and benzocaine are local anesthetics with pain-relieving properties, so the question remains whether these would be favorable to use in invasive procedures. Other methods of anesthesia such as carbon dioxide and live chilling are controversial. The use of carbon dioxide should be avoided as it causes a peripheral tissue acidosis that excites nociceptors in rainbow trout and is banned in the European Union (Mettam et al., 2012). Live chilling may only cause paralysis but this has been shown to be less stressful then MS222 anesthesia in small tropical species such as zebrafish (Wilson et al., 2009). This method must keep water temperature at 4°C and ice should not be allowed to touch the fish in case of ice crystal formation within the body. To ensure good welfare in fish the choice of anesthesia method should be intelligently made by ensuring adequate knowledge of the species to be used and its responses to the particular method or drug.

4. CONCLUSIONS AND FUTURE DIRECTIONS

Clearly the environments in which captive and wild-reared fish are exposed engender multiple physiological and behavioral differences. Attenuated stress responses and behavioral repertoire in captive fish, derived from developmental, experiential, and genetic factors, may reduce the likelihood of survival or success if released into natural environments. Likewise, fish brought in from the wild into laboratory, ornamental, or aquaculture facilities face a very different social and physical environment that may favor different phenotypes than those that would be successful in the wild. Managing stress in fish held under different contexts provides a plethora of benefits to the researchers or carers of those fish. Advances in our understanding of the allostatic response linked to the psychological impact of stress on well-being and health means we can now begin to explore different strategies to improve the welfare of fish in captivity. Clearly reducing allostatic load is of crucial importance especially before a stressful event is about to commence to avoid the additive effects of stress; thus, the timing of procedures should be considered. The effects of additive stress is seen in Atlantic salmon subject to stress before and after vaccination (Figure 12.5) where both elevated HPI axis activity and mortality was observed leading to allostatic overload (Iversen and Eliassen, 2014). If minimizing stress is the goal, ensuring fish have a sufficient amount of time to recover from stressors is also imperative. Further allowing fish to have control or be able to anticipate stressful events should be the goal of scientists using fish. Environmental enrichment in captivity can provide an outlet for frustration and displacement behaviors or a means of allowing an individual to escape or hide from the stress. Simple classical conditioning approaches (eg, light cue) can be used to signal the onset of a stressor, which allows fish to anticipate and prepare for the advent of stress in captivity. Additionally, operant conditioning employed as a means of bestowing control allowing fish to operate self-feeders can have positive welfare benefits. These cognitive strategies do need to be investigated in different contexts and in different species. This is also true for the impact of stocking density and its influence on foraging behavior due to the diversity of fish behavior, ecology, and life history, which is modulated by the captive environment. Identifying what is an acceptable allostatic load would be a major step forward since too little stress may result in negative welfare as well as too much stress.

Many other factors impinge upon an individual's ability to cope with stress including genotype, stress coping style, previous experience, and also transfer of stress history from parents to offspring, which alters development and behavior into adulthood (Andersson et al., 2011). Studies are needed to understand the role these factors play in shaping phenotype. The generation of strains of fish tolerant to stress (eg, the breeding of high and low stress-responding lines of rainbow trout; Pottinger and Carrick, 1999) may be an alternative. Otherwise, care in husbandry, particularly during transport and the point of change between wild and captive habitats and vice versa, is required to maximize survival and optimize welfare if stress and possible mortality are to be avoided. This is particularly important when ensuring captive laboratory fish are able to exhibit wild-type responses when scientific data are intended to be generalized from laboratory trials into wild populations in case domestication of laboratory-held fish affects any of their responses, making it difficult to extrapolate to the wild situation (eg, Christie et al., 2012).

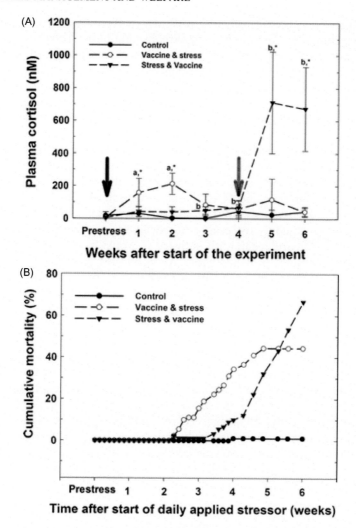

Figure 12.5. (A) The effects of 4 weeks of daily applied stressor on baseline levels of plasma cortisol either before ("vaccine and stress") or after ("stress and vaccine") secondary vaccination in Atlantic salmon, *Salmo salar*. Note that control group was not subjected to either secondary vaccination or daily stressor. Values are expressed in mean ± SD ($n = 6$). Significant changes ($p < 0.05$) from prestress are indicated with letters. Significant changes ($p < 0.05$) between groups are indicated with asterisk. Black arrow indicates time of second vaccination for "vaccine and stress" group, while gray arrow indicates time of second vaccination for "stress and vaccine" group. (B) The effect of 4 weeks of daily applied stressor on overall cumulative mortality (%) either before ("vaccine and stress") or after ("stress and vaccine") secondary vaccination. Note that control group was not subjected to either secondary vaccination or daily stressor. Reproduced from Iversen, M. H. and Eliassen, R. A. (2014). The effect of allostatic load on hypothalamic-pituitary-interrenal (HPI) axis before and after secondary vaccination in Atlantic salmon postsmolts (*Salmo salar* L.). *Fish Physiol. Biochem.* **40**, 527–538, with kind permission from Springer.

Finally, a means of identifying stress early on is required so that remedial action can be taken. For example, oxygen consumption and behavior was used to monitor the health of Atlantic salmon during transport (Farrell, 2006). Automatic monitoring systems that can provide an alert when behavior strays from the norm would be a valuable tool. This would involve the use of cameras at the side, above, or even inside holding tanks to record behavior, but knowledge would be required of the normal behavior of each species. Additionally physiological measurements could also be made. This could assist in understanding when fish have recovered from a stressful event so that the next procedure can be effectively timed to reduce impact on the fish. Behavioral monitoring systems have been used to monitor welfare of rainbow trout in flow-through and RAS systems (Colson et al., 2015) although numbers were low at 50 per tank (stocking density: 30 kg/m^3). Development of such systems would be especially beneficial for aquaculture, laboratory aquaria, and the ornamental fish industry so would have a wide impact on reducing stress and improving welfare. Alongside this, providing fish with the optimal environment, water quality, social contact, correct feeding opportunities, and employing disturbing practices in husbandry in a manner that allows anticipation, preparation, and control should all reduce stress and enhance the welfare of fish. Further, when subjecting fish to stressors in aquaculture, fisheries, catch and release, experimentation, and in ornamental fish, continuing to employ the most humane methods or alleviating poststress manipulation will ensure the health and well-being of the fish we use. For any captive population, the development of robust stress indicators to safeguard stock health is important to retain the desired behavioral and physiological responses to stressful challenges. These indicators can be used to alert investigators or those involved in husbandry as to when deviations from expected standards occur, allowing intervention. Thus, future studies should focus on developing meaningful, practical indicators of welfare in fish if the goal is to minimize stress to ensure good health.

REFERENCES

Abbott, A. (2015). Clever fish. *Nature* **521**, 412–414.
Ackerman P.A., Morgan J.D., Iwama G.K. (2005) Anesthetics. CCAC guidelines on: The care and use of fish in research, teaching and testing, Canadian Council on Animal Care, Ottawa Canada. Available at: ⟨http://www.ccac.ca/Documents/Standards/Guidelines/Fish.pdf⟩.
AFS. (2014). ⟨http://fisheries.org/guide-for-the-use-of-fishes-in-research⟩.
Agetsuma, M., Aizawa, H., Aoki, T., Nakayama, R., Takahoko, M., Goto, M., et al. (2010). The habenula is crucial for experience-dependent modification of fear responses in zebrafish. *Nat. Neurosci.* **13**, 1354–1356.
Alanärä, A. (1996). The use of self-feeders in rainbow trout (*Oncorhynchus mykiss*) production. *Aquaculture* **145**, 1–20.

Alanärä, A. and Brännäs, E. (1996). Dominance in demand-feeding behaviour in Arctic charr and rainbow trout: the effect of stocking density. *J. Fish. Biol.* **48**, 242–254.

Alexander, R. M. (1993). Buoyancy. In *The Physiology of Fishes* (ed. D. H. Evans), pp. 75–97. Boca Raton, FL: CRC Press.

Andersson, M. A., Silva, P. I. M., Steffensen, J. F. and Hoglund, E. (2011). Effects of maternal stress coping style on offspring characteristics in rainbow trout (*Oncorhynchus mykiss*). *Horm. Behav.* **60**, 699–705.

Ashley, P. J. (2007). Fish welfare: current issues in aquaculture. *Appl. Anim. Behav. Sci.* **104**, 199–235.

Ashley, P. J. and Sneddon, L. U. (2008). Pain and fear in fish. In *Fish Welfare* (ed. E. J. Branson), pp. 49–77. Oxford: Blackwell Publishing.

Ashley, P. J., Ringrose, S., Edwards, K. L., McCrohan, C. R. and Sneddon, L. U. (2009). Effect of noxious stimulation upon antipredator responses and dominance status in rainbow trout. *Anim. Behav.* **77**, 403–410.

Balon, E. K. (2004). About the oldest domesticates among fishes. *J. Fish. Biol.* **65**, 1–27.

Bart, A. N., Clark, J., Young, J. and Zohar, Y. (2001). Underwater ambient noise measurements in aquaculture systems: a survey. *Aquacul. Eng.* **25**, 99–110.

Barton, B. A. (2002). Stress in fishes: a diversity of responses with particular reference to changes in circulating corticosteroids. *Integr. Comp. Biol.* **42**, 517–525.

Barton, B. A. and Schreck, C. N. (1987). Metabolic cost of acute physical stress in juvenile steelhead. *Trans. Am. Fish. Soc.* **116**, 257–263.

Benhaïm, D., Péan, S., Lucas, G., Blanc, N., Chatain, B. and Bégout, M.-L. (2012). Early life behavioural differences in wild caught and domesticated sea bass (*Dicentrarchus labrax*). *Appl. Anim. Behav. Sci.* **141**, 79–90.

Benhaïm, D., Guyomard, R., Chatain, B., Quillet, E. and Bégout, M.-L. (2013). Genetic differences for behaviour in juveniles from two strains of brown trout suggest an effect of domestication history. *Appl. Anim. Behav. Sci.* **147**, 235–242.

Berumen, M. L. and Almany, G. R. (2009). External tagging does not affect the feeding behavior of a coral reef fish, *Chaetodon vagabundus* (Pisces: Chaetodontidae). *Environ. Biol. Fish.* **86**, 447–450.

Biswas, A. K., Seoka, M., Takii, K., Maita, M. and Kumai, H. (2006). Stress response of red sea bream *Pagrus major* to acute handling and chronic photoperiod manipulation. *Aquaculture* **252** (2–4), 566–572.

Bjørge, M. H., Nordgreen, J., Janczak, A. M., Poppe, T., Ranheim, B. and Horsberg, T. E. (2011). Behavioural changes following intraperitoneal vaccination in Atlantic salmon (*Salmo salar*). *Appl. Anim. Behav. Sci.* **133**, 127–135.

Blanco-Vives, B., Aliaga-Guerrero, M., Cañavate, J. P., Muñoz-Cueto, J. A. and Sánchez-Vázquez, F. J. (2011). Does lighting manipulation during incubation affect hatching rhythms and early development of sole? *Chronobiol. Int.* **28** (4), 300–306.

Bosakowski, T. and Wagner, E. J. (1995). Experimental use of cobble substrates in concrete raceways for improving fin condition of cutthroat (*Oncorhynchus clarkii*) and rainbow trout (*O. mykiss*). *Aquaculture* **130**, 159–165.

Bourne, C. M., Kehler, D. G., Wiersma, Y. F. and Cote, D. (2011). Barriers to fish passage and barriers to fish passage assessments: the impact of assessment methods and assumptions on barrier identification and quantification of watershed connectivity. *Aquat. Ecol.* **45**, 389–403.

Brännäs, E. and Alanärä, A. (1994). Effect of reward level on individual variability in demand feeding activity and growth rate in Arctic charr and rainbow trout. *J. Fish Biol* **45**, 423–434.

Brockmark, S., Neregard, L., Bohlin, T., Bjornsson, B. T. and Johnsson, J. I. (2007). Effects of rearing density and structural complexity on the pre- and postrelease performance of Atlantic salmon. *Trans. Am. Fish. Soc.* **136**, 1453–1462.

Broom, D. M. (2008). Welfare assessment and relevant ethical decisions: key concepts. *Annu. Rev. Biomed. Sci.* **10**, T79–T90.

Brown, C. (2015). Fish intelligence, sentience and ethics. *Anim. Cogn.* **18**, 1–17.

Brown, C., Davidson, T. and Laland, K. (2003). Environmental enrichment and prior experience improve foraging behaviour in hatchery-reared Atlantic salmon. *J. Fish. Biol.* **63** (S1), 187–196.

Brown, C., Burgess, F. and Braithwaite, V. A. (2007). Heritable and experiential effects on boldness in a tropical poeciliid. *Behav. Ecol. Sociobiol.* **62**, 237–243.

Brown, J. A., Watson, J., Bourhill, A. and Wall, A. (2010). Physiological welfare of commercially reared cod and effects of crowding for harvesting. *Aquaculture* **298**, 315–324.

Brown, R. S., Eppard, M. B., Murchie, K. J., Nielsen, J. L. and Cooke, S. J. (2011). An introduction to the practical and ethical perspectives on the need to advance and standardize the intracoelomic surgical implantation of electronic tags in fish. *Rev. Fish Biol. Fisher* **21**, 1–9.

Brown, R. S., Pflugrath, B. D., Carlson, T. J. and Deng, Z. D. (2012). The effect of an externally neutrally buoyant transmitter on mortal injury during simulated hydroturbine passage. *J. Renew. Sustain. Energy* **4** (1), 013107.

Brown, R. S., Deng, Z. D., Cook, K. V., Pflugrath, B. D., Li, X., Fu, T., et al. (2013). A field evaluation of an external and neutrally buoyant acoustic transmitter for juvenile salmon: implications for estimating hydroturbine passage survival. *PLoS One* **8** (10), e77744.

Brydges, N. M. and Braithwaite, V. A. (2009). Does environmental enrichment affect the behaviour of fish commonly used in laboratory work? *Appl. Anim. Behav. Sci.* **118**, 137–143.

Bshary, R., Oliveira, R. F., Oliveira, T. S. F. and Canário, A. V. M. (2007). Do cleaning organisms reduce the stress response of client reef fish? *Front. Zool.* **4**, 21.

Bui, S., Oppedal, F., Korsøen, Ø. J., Sonny, D. and Dempster, T. (2013). Group behavioural responses of Atlantic salmon (*Salmo salar* L.) to light, infrasound and sound stimuli. *PLoS One* **8** (5), e63696.

Campbell, P. M., Pottinger, T. G. and Sumpter, J. P. (1992). Stress reduces the quality of gametes produced by rainbow trout. *Biol. Reprod.* **47**, 1140–1150.

Cañon Jones, H. A., Hansen, L., Noble, C., Damsgård, B., Broom, D. M. and Pearce, G. P. (2010). Social network analysis of behavioural interactions influencing fin damage development in Atlantic salmon (*Salmo salar*) during feed-restriction. *Appl. Anim. Behav. Sci.* **127**, 139–151.

Caputo, M., O'Connor, C. M., Hasler, C. T., Hanson, K. C. and Cooke, S. J. (2009). Long-term effects of surgically implanted telemetry tags on the nutritional physiology and condition of wild freshwater fish. *Dis. Aquat. Organ.* **84** (1), 35–41.

Carlstead, K. (1996). Effects of captivity on the behavior of wild mammals. In *Wild Mammals in Captivity: Principles and Techniques* (eds. D. G. Kleiman, M. E. Allen, K. V. Thompson and S. Lumpkin). Chicago: The University of Chicago Press.

Christie, M. R., Marine, M. L., French, R. A. and Blouin, M. S. (2012). Genetic adaptation to captivity can occur in a single generation. *PNAS* **109**, 238–242.

Chua, F. H. C. (1996). Aquaculture health management in Singapore: current status and future directions. In *Health Management in Asian Aquaculture. Proceedings of the Regional Expert Consultation on Aquaculture Health Management in Asia and the Pacific* (eds. R. P. Subasinghe, J. R. Arthur and M. Shariff), pp. 115–126. Fisheries Technical paper No. 360, Rome: FAO.

Clark, K. J. and Ekker, S. C. (2015). How zebrafish genetics informs human biology. *Nat. Educ.* **8** (4), 3.

Clark, J., Young, J., Bart, A. N., Zohar, Y. (1996). Underwater ambient noise measurements. In: 30th Proceedings of the Acoustical Society of America, St. Louis, MO.

Clearwater, S. J. and Pankhurst, N. W. (1997). The response to capture and confinement stress of plasma cortisol, plasma sex steroids and vitellogenic oocytes in the marine teleost, red gurnard. *J. Fish. Biol.* **50**, 429–441.

Cleary, J. J., Pankhurst, N. W. and Battaglene, S. C. (2000). The effect of capture and handling stress on plasma steroid levels and gonadal condition in wild and farmed snapper *Pagrus auratus*. *J. World Aquacul. Soc.* **31**, 558–569.

Clement, T. S., Parikh, V., Schrumpf, M. and Fernald, R. D. (2005). Behavioral coping strategies in a cichlid fish: the role of social status and acute stress response in direct and displaced aggression. *Horm. Behav.* **47**, 336–342.

Clements, S. P., Hicks, B. J., Carragher, J. F. and Dedual, M. (2002). The effect of a trapping procedure on the stress response of wild rainbow trout. *N. Am. J. Fish. Manage.* **22** (3), 907–916.

Colson, V., Sadoul, B., Valotaire, C., Prunet, P., Gaume, M. and Labbe, L. (2015). Welfare assessment of rainbow trout reared in a recirculating aquaculture system: comparison with a flow-through system. *Aquaculture* **436**, 151–159.

Consortium for Wildlife Bycatch Reduction. (2016). <http://www.bycatch.org/publications/>.

Conte, F. S. (2004). Stress and the welfare of cultured fish. *Appl. Anim. Behav. Sci.* **86**, 205–223.

Cooke, S. J. and Sneddon, L. U. (2007). Animal welfare perspectives on recreational angling. *Appl. Anim. Behav. Sci.* **104**, 176–198.

Cooke, S. J. and Suski, C. D. (2005). Do we need species-specific guidelines for catch-and release recreational angling to effectively conserve diverse fishery resources? *Biodiv. Cons* **14**, 1195–1209.

Cooke, S. J., Raby, G. D., Donaldson, M. R., Hinch, S. G., O'Connor, C. M., Arlinghaus, R., et al. (2013). The physiological consequences of catch-and-release angling: perspectives on experimental design, interpretation, extrapolation and relevance to stakeholders. *Fish. Manage. Ecol.* **20**, 268–287.

Costello, M. J. (1993). Review of methods to control sea lice (Caligidae: Crustacea) infestations on salmon (*Salmo salar*) farms. In *Pathogens of Wild and Farmed Fish: Sea Lice* (eds. G. A. Boxshall and D. Defaye), pp. 219–254. Chichester: Ellis Horwood.

Coutant, C. C. and Whitney, R. R. (2000). Fish behaviour in relation to passage through hydropower turbines: a review.. *Trans. Am. Fish. Soc.* **192**, 351–380.

Cubitt, K. F., Winberg, S., Huntingford, F. A., Kadri, S., Crampton, V. O. and Øverli, Ø. (2008). Social hierarchies, growth and brain serotonin metabolism in Atlantic salmon (*Salmo salar*) kept under commercial rearing conditions. *Physiol. Behav.* **94**, 529–535.

Curtis, S. E. (1985). What constitutes animal well-being? In *Animal Stress* (ed. G. P. Moberg), pp. 1–14. Bethesda: American Physiological Society.

Davenport, K. E. (1996). Characteristics of the current international trade in ornamental fish, with special reference to the European Union. *Rev. Sci. Tech. de l'Office International des Epizooties* **15**, 435–443.

Davidson, J., Bebak, J. and Mazik, P. (2009). The effects of aquaculture production noise on the growth, condition factor, feed conversion and survival of rainbow trout, *Oncorhynchus mykiss*. *Aquaculture* **288**, 337–343.

Davis, K. B. (2006). Management of physiological stress in finfish aquaculture. *N. Am. J. Aquacul.* **68**, 116–121.

Davis, M. W. (2007). Simulated fishing experiments for predicting delayed mortality rates using reflex impairment in restrained fish. *ICES J. Mar. Sci.* **64**, 1535–1542.

Davis, M. W. and Olla, B. L. (2001). Stress and delayed mortality induced in Pacific halibut by exposure to hooking, net towing, elevated seawater temperature and air: Implications for management of bycatch. *North Am. J. Fisheries Manag* **21**, 725–732.

Davis, M. W. and Olla, B. L. (2002). Mortality of lingcod towed in a net as related to fish length, seawater temperature, and air exposure: A laboratory bycatch study. *North Am. J. Fisheries Manag* **22**, 1095–1104.

Davis, M. W., Olla, B. L. and Schreck, C. B. (2001). Stress induced by hooking, net towing, elevated sea water temperature and air in sablefish: lack of concordance between mortality and physiological measures of stress. *J. Fish Biol* **58**, 1–15.

Davis, M. W. and Parker, S. J. (2004). Fish size and exposure to air: potential effects on behavioral impairment and mortality rates in discarded sablefish. *N. Am. J. Fish. Manage.* **24**, 518–524.

Davis, M. W. and Ottmar, M. L. (2006). Wounding and reflex impairment may be predictors for mortality in discarded or escaped fish. *Fish. Res.* **82**, 1–6.

Dawkins, M. S. (1980). *Animal Suffering: The Science of Animal Welfare*. Netherlands: Springer.

Dempster, T., Uglem, I., Sanchez-Jerez, P., Fernandez-Jover, D., Bayle-Sempere, J., Nilsen, R., et al. (2009). Coastal salmon farms attract large and persistent aggregations of wild fish: an ecosystem approach. *Mar. Ecol. Prog. Ser.* **385**, 1–14.

Diamond, B. and Beukers-Stewart, B. D. (2011). Fisheries discards in the north sea: waste of resources or necessary evil? *Rev. Fish. Sci.* **19**, 231–245.

Douxfils, J., Mathieu, C., Mandiki, S. N. M., Milla, S., Henrotte, E., Wang, N., et al. (2011). Physiological and proteomic evidences that domestication process differentially modulates the immune status of juvenile Eurasian perch (*Perca fluviatilis*) under chronic confinement stress. *Fish. Shellfish Immunol.* **31**, 1113–1121.

Doyle, R. W., Perez-Enriquez, R., Takagi, M. and Taniguchi, M. (2001). Selective recovery of founder genetic diversity in aquacultural broodstocks and captive, endangered fish population. *Genetica* **111**, 291–304.

Duncan, J. H. (1993). Welfare is to do with what animals feel. *J. Agric. Environ. Ethics* **6** (Suppl. 2), 8–14.

Duncan, J. H. (1996). Animal welfare is defined in terms of feelings. Acta Agricul. Scand. *A Anim. Sci.* **27** (Suppl.), 29–35.

Earley, R. L., Edwards, J. T., Aseem, O., Felton, K., Blumer, L. S., Karom, M., et al. (2006). Social interactions tune aggression and stress responsiveness in a territorial cichlid fish (*Archocentrus nigrofasciatus*). *Physiol. Behav.* **88**, 353–363.

Ebbesson, L. O. E. and Braithwaite, V. A. (2012). Environmental effects on fish neural plasticity and cognition. *J. Fish. Biol.* **81** (7), 2151–2174.

Ellis, T., James, J. D. and Scott, A. P. (2005). Branchial release of free cortisol and melatonin by rainbow trout. *J. Fish. Biol.* **67**, 535–540.

Ellis, T., Oidtmann, B., St-Hilaire, S., Turnbull, J. F., North, B. P., MacIntyre, C. M., et al. (2008). Fin erosion in farmed fish. In *Fish Welfare* (ed. E. J. Branson), pp. 121–149. Oxford: Blackwell.

Ellis, T., Yildiz, H. Y., López-Olmeda, J., Spedicato, M. T., Tort, L., Øverli, Ø., et al. (2012). Cortisol and finfish welfare. *Fish Physiol. Biochem.* **38**, 163–188.

Eriksen, H. R., Murison, R., Pensgaard, A. M. and Ursin, H. (2005). Cognitive activation theory of stress (CATS): from fish brains to the Olympics. *Psychoneuroendocrinology* **30**, 933–938.

Erikson, U., Sigholt, T. and Seland, A. (1997). Handling stress and water quality during live transportation and slaughter of Atlantic salmon (*Salmo salar*). *Aquaculture* **149** (3), 243–252.

Europa. (2015). <http://eur-lex.europa.eu/legal-content/EN/TXT/?uri=CELEX:52013SC0497/>.

European Directive. (2010/63/EU). <http://eur-lex.europa.eu/legal-content/EN/TXT/?uri=CE-LEX:32010L0063/>.

Farrell, A. P. (2006). Bulk oxygen uptake measured with over 60,000 kg of adult salmon during live-haul transportation at sea. *Aquaculture* **254**, 646–652.

Farrell, A. P., Gallaugher, P. E. and Routledge, R. (2001). Rapid recovery of exhausted adult coho salmon after commercial capture by troll fishing. *Can. J. Fish. Aquat. Sci* **58**, 2319–2324.

FAWC. (1996). FAWC (Farmed Animal Welfare Council), 1996. Report on the Welfare of Farmed Fish. Surbiton, Surrey.

Fendt, M. and Fenselow, M. S. (1999). The neuroanatomical and neurochemical basis of conditioned fear. *Neurosci. Biobehav. Rev.* **23**, 743–760.

Fishcount.org. (2015). <http://fishcount.org.uk/>.

Fish Fight. (2015). <http://www.fishfight.net/story.html/>.

Flecknell, P. A., Roughan, J. V. and Stewart, R. (1999). Use of oral buprenorphine ('buprenorphine jello') for postoperative analgesia in rats: a clinical trial. *Lab. Anim.* **33**, 169–174.

Fraser, D. (2008). *Understanding Animal Welfare, The Science in its Cultural Context.* Chichester: Wiley Blackwell.

Fraser, N. H. C., Metcalfe, N. B. and Thorpe, J. E. (1993). Temperature-dependent switch between diurnal and nocturnal foraging in salmon. *Proc. R. Soc. Lond. B Biol.* **252** (1334), 135–139.

Froese, R. (1988). Relationship between body weight and loading densities in fish transport using the plastic bag method. *Aquacul. Fish. Manage.* **19**, 275–281.

Frost, A. J., Winrow-Giffen, A., Ashley, P. J. and Sneddon, L. U. (2007). Plasticity in animal personality traits: does prior experience alter the degree of boldness? *Proc. R. Soc. Lond. B Biol. Sci.* **274**, 333–339.

Fürtbauer, I., King, A. J. and Heistermann, M. (2015). Visible implant elastomer (VIE) tagging and simulated predation risk elicit similar physiological stress responses in three-spined stickleback *Gasterosteus aculeatus. J. Fish. Biol.* **86**, 1644–1649.

Gale, M. K., Hinch, S. G. and Donaldson, M. R. (2013). The role of temperature in the capture and release of fish. *Fish Fish* **14**, 1–33.

Galhardo, L. and Oliveira, R. F. (2009). Psychological stress and welfare in fish. *ARBS Ann. Rev. Biomed. Sci.* **11**, 1–20.

Galhardo, L., Correia, J. and Oliveira, R. F. (2008). The effect of substrate availability on behavioural and physiological indicators of welfare in the African cichlid (*Oreochromis mossambicus*). *Anim. Welfare* **17**, 239–254.

Galhardo, L., Vital, J. and Oliveira, R. F. (2011). The role of predictability in the stress response of a cichlid fish. *Physiol. Behav.* **102**, 367–372.

Garner, S. R., Madison, B. N., Bernier, N. J. and Neff, B. D. (2010). Behavioural interactions and hormones in naturally and hatchery-spawned Chinook salmon. *Ethology* **117**, 37–48.

Giorgi, A. E. and Stevenson, J. R. (1995). *A Review of Biological Investigations Describing Smolt Passage Behavior at Portland District Corps of Engineers Projects: Implications to Surface Collection Systems.* Boise, Idaho: Don Chapman Consultants.

Gorissen, M. and Flik, G. (2016). Endocrinology of the Stress Response in Fish. In *Fish Physiology - Biology of Stress in Fish*, Vol. 35 (eds. C. B. Schreck, L. Tort, A. P. Farrell and C. J. Brauner), San Diego, CA: Academic Press.

Grigorakis, K. and Rigos, G. (2011). Aquaculture effects on environmental and public welfare – the case of Mediterranean mariculture. *Chemosphere* **855**, 899–919.

Gronquist, D. and Berges, J. A. (2013). Effects of aquarium-related stressors on the zebrafish: a comparison of behavioural, physiological, and biochemical indicators. *J. Aquatic Health* **25**, 53–65.

Grossman, L., Stewart, A., Gaikwad, S., Utterback, E., Wu, N., DiLeo, J., et al. (2011). Effects of piracetam on behavior and memory in adult zebrafish. *Brain Res. Bull.* **85**, 58–63.

Gutscher, M., Wysocki, L. E. and Ladich, F. (2011). Effects of aquarium and pond noise on hearing sensitivity in an otophysine fish. *Bioacoustics* **20**, 117–136.

Handy, R. D. and Poxton, M. G. (1993). Nitrogen pollution in mariculture: toxicity and excretion of nitrogenous compounds by marine fish. *Rev. Fish. Biol. Fisher* **3**, 205–241.

Hannah, R. W., Lomeli, M. J. M. and Jones, S. A. (2015). Tests of artificial light for bycatch reduction in an ocean shrimp (*Pandalus jordani*) trawl: strong but opposite effects at the footrope and near the bycatch reduction device. *Fish. Res.* **170**, 60–67.

Head, A. B. and Malison, J. A. (2000). Effects of lighting spectrum and disturbance level on the growth and stress responses of yellow perch Perca flavescens. *J. World Aquacult. Soc* **31**, 73–80.

Herrero, M. J., Madrid, J. A. and Sánchez-Vázquez, F. J. (2005). Demand-feeding rhythms and feeding-entrainment of locomotor rhythms in tench (*Tinca tinca*). *Physiol. Behav.* **84**, 595–605.

Hoglund, E., Weltzien, F. A., Schjolden, J., Winberg, S., Ursin, H. and Doving, K. B. (2005). Avoidance behavior and brain monoamines in fish. *Brain Res.* **1032**, 104–110.

Home Office. (2014). <https://www.gov.uk/government/statistics/statistics-of-scientific-procedures-on-living-animals-great-britain-2013/>.

Hoolihan, J. P., Luo, J., Abascal, F. J., Campana, S. E., De Metrio, G., Dewar, H., et al. (2011). Evaluating post-release behavior modification in large pelagic fish deployed with pop-up satellite archival tags. *ICES J. Mar. Sci.* **68** (5), 880–889.

HSA. (2014). Humane Harvesting of Fish <http://www.hsa.org.uk/downloads/related-items/harvesting-of-fish.pdf>.

Humborstad, O. B., Davis, M. W. and Lokkeborg, S. (2009). Reflex impairment as a measure of vitality and survival potential of Atlantic cod (*Gadus morhua*). *Fish. Bull.* **107**, 395–402.

Humborstad, O. B., Breen, M., Davis, M. W., Lokkeborg, S., Mangor-Jensen, A., Midling, K. O., et al. (2016). Survival and recovery of longline- and pot-caught cod (*Gadus morhua*) for use in capture-based aquaculture (CBA). *Fish. Res.* **174**, 103–108.

Huntingford, F. A. (2004). Implications of domestication and rearing conditions for the behaviour of cultivated fishes. *J. Fish. Biol.* **65** (Suppl. A), 122–142.

Huntingford, F. A. and Kadri, S. (2014). Defining, assessing and promoting the welfare of farmed fish. *Rev. Sci. Tech. de L'Office International des Epizooities* **33**, 233–244.

Huntingford, F. A., Adams, C., Braithwaite, V. A., Kadri, S., Pottinger, T. G., Sandoe, P., et al. (2006). Current issues in fish welfare. *J. Fish. Biol.* **68**, 332–372.

Itou, J., Akiyama, R., Pehoski, S., Yu, X., Kawakami, H. and Kawakami, Y. (2014). Regenerative responses after mild heart injuries for cardiomyocyte proliferation in zebrafish. *Developmental Dynamics* **243**, 1477–1486.

Iversen, M. H. and Eliassen, R. A. (2014).). The effect of allostatic load on hypothalamic-pituitary-interrenal (HPI) axis before and after secondary vaccination in Atlantic salmon postsmolts (*Salmo salar* L.). *Fish Physiol. Biochem.* **40**, 527–538.

Iversen, M., Finstad, B. and Nilssen, K. J. (2005). Recovery from loading and transport stress in Atlantic salmon (*Salmo salar* L.) smolts. *Aquaculture* **168**, 387–394.

Jepsen, N., Thorstad, E. B., Havn, T. and Lucas, M. C. (2015). The use of external electronic tags on fish: an evaluation of tag retention and tagging effects. *Anim. Biotelem.* **3**, 49.

Johansson, D., Ruohonen, K., Kiessling, A., Oppedal, F., Stiansen, J. E., Kelly, M., et al. (2006). Effect of environmental factors on swimming depth preferences of Atlantic salmon (Salmo salar L.) and temporal and spatial variations in oxygen levels in sea cages at a fjord site. *Aquacult* **254**, 594–605.

Johanesen, L.-H., Jensen, I., Mikkelsen, H., Bjørn, P. A., Jansen, P. A. and Bergh, Ø. (2011). Disease interaction and pathogens exchange between wild and farmed fish populations with special reference to Norway. *Aquaculture* **315**, 167–186.

Jolivet, A., de Pontual, H., Garren, F. and Bégout, M. L. (2009). Effects of T-bar and DST tagging on survival and growth of European hake. In *Tagging and Tracking of Marine Animals with Electronic Devices* (eds. J. L. Nielsen, H. Arrizabalaga, N. Fragosos, A. Hobday, M. Lutcavage and L. Sibert), pp. 181–193. Dordrecht, Netherlands: Springer.

Karakatsouli, N., Katsakoulis, P., Leondaritis, G., Kalogiannis, D., Papoutsoglou, S. E., Chadio, S., et al. (2012). Acute stress response of European sea bass Dicentrarchus labrax under blue and white light. *Aquacult* **364**, 48–52.

Karlsen, J. D., Krag, L. A., Albertsen, C. M. and Frandsen, R. P. (2015). From fishing to fish processing: separation of fish from crustaceans in the Norway lobster-directed multispecies trawl fishery improves seafood quality. *PLoS One* **10**, e0140864.

Keeling, L. and Jensen, P. (2009). Behavioural disturbances, stress and welfare. In *The Ethology of Domestic Animals: an Introductory Text* (ed. P. Jensen), second ed., pp. 85–101. Wallingford, Oxfordshire: CABI Publishing.

Kiessling, A., Johansson, D., Zahl, L. H. and Samuelsen, O. B. (2009). Pharmacokinetics, plasma cortisol and effectiveness of benzocaine, MS-222 and isoeugenol measured in individual dorsal aorta-cannulated Atlantic salmon (*Salmo salar*) following bath administration. *Aquaculture* **286**, 301–308.

Kight, C. R. and Swaddle, J. P. (2011). How and why environmental noise impacts animals: an integrative, mechanistic review. *Ecol. Lett.* **14**, 1052–1061.

Kihslinger, R. L. and Nevitt, G. A. (2006). Early rearing environment impacts cerebellar growth in juvenile salmon. *J. Exp. Biol.* **209**, 504–509.

Knudsen, F. R., Enger, P. S. and Sand, O. (1992). Awareness reactions and avoidance responses to sound in juvenile Atlantic salmon, *Salmo salar* L. *J. Fish. Biol.* **40**, 523–534.

Korte, S. M., Olivier, B. and Koolhaas, J. M. (2007). A new animal welfare concept based on allostasis. *Physiol. Behav.* **92** (3), 422–428.

Kristensen, T., Haugen, T. O., Rosten, T., Fjellheim, A., Atland, A. and Rosseland, B. O. (2012). Effects of production intensity and production strategies in commercial Atlantic salmon smolt (*Salmo salar* L.) production on subsequent performance in the early sea stage. *Fish Physiol. Biochem.* **38**, 273–282.

Kristiansen, T. S. and Fernö, A. (2007). Individual behaviour and growth of halibut (*Hippoglossus hippoglossus* L.) fed sinking and floating feed: evidence of different coping styles. *Appl. Anim. Behav. Sci.* **104**, 236–250.

Kulczykowska, E. and Vázquez, F. J. S. (2010). Neurohormonal regulation of feed intake and response to nutrients in fish: aspects of feeding rhythm and stress. *Aquacul. Res.* **41**, 654–667.

Lambooij, E., Eilarczyk, M., Bialowas, H., van den Boogaart, J. G. M. and van de Vis, J. W. (2007). Electrical and percussive stunning of the common carp (*Cyprinus carpio* L.): neurological and behavioural assessment. *Aquacul. Eng.* **37**, 171–179.

Lambooij, B., Bracke, M., Reimert, H., Foss, A., Imsland, A. and van de Vis, H. (2015). Electrophysiological and behavioural responses of turbot (*Scophthalmus maximus*) cooled in ice water. *Physiol. Behav.* **149**, 23–28.

Laule, G. and Desmond, T. (1998). Positive reinforcement training as an enrichment strategy. In *Second Nature: Environmental Enrichment for Captive Animals* (eds. D. J. Shepherdson, J. D. Mellen and M. Hutchins), Washington: Smithsonian Institution Press.

LePage, O., Øverli, Ø., Petersson, E., Järvi, T. and Winberg, S. (2000). Differential stress coping in wild and domesticated sea trout. *Brain Behav. Evol.* **56**, 259–268.

Lines, J. A. and Spence, J. (2014). Humane harvesting and slaughter of farmed fish. *Rev. Sci. Tech. Off. Int. des Epizooties* **33**, 255–264.

Loher, T. and Rensmeyer, R. (2011). Physiological responses of Pacific halibut, *Hippoglossus stenolepis*, to intracoelomic implantation of electronic archival tags, with a review of tag implantation techniques employed in flatfish. *Rev. Fish Biol. Fisher* **21** (1), 97–115.

Lopez-Luna, J., Yásquez, L., Torrent, E. and Villarroel, M. (2013). Short-term fasting and welfare prior to slaughter in rainbow trout, *Oncorhynchus mykiss. Aquaculture* **400–401**, 142–147.

Losey, G. S., Cronin, T. W., Goldsmith, T. H., Hyde, D., Marshall, N. J. and McFarland, W. N. (1999). The UV visual world of fishes: a review. *J. Fish. Biol.* **54**, 921–943.

Lovallo, W. R. (2005). *Stress & Health: Biological and Psychological Interactions.* Thousand Oaks: Sage Publications.

Lower, N., Moore, A., Scott, A. O., Ellis, T., James, J. D. and Russell, I. C. (2005). A non-invasive method to assess the impact of electronic tag insertion on stress levels in fishes. *J. Fish. Biol.* **67** (5), 1202–1212.

Lupes, S. C., Davis, M. W., Olla, B. L. and Schreck, C. B. (2006). Capture-related stressors impair immune system function in sablefish. *Trans. Am. Fish. Soc.* **135**, 129–138.

MacIntyre, C. M., Ellis, T., North, B. P. and Turnbull, J. F. (2008). The influences of water quality on the welfare of farmed rainbow trout: a review. In *Fish Welfare* (ed. E. J. Branson), pp. 150–184. Oxford: Blackwell.

Marchetti, M. P. and Nevitt, G. A. (2003). Effects of hatchery rearing practices on brain structures of rainbow trout (*Oncorhynchus mykiss*). *Environ. Biol. Fishes* **66**, 9–14.

Martins, C. I. M., Eding, E. H. and Verreth, J. A. J. (2011). Stressing fish in recirculating aquaculture systems (RAS): does stress induced in one group of fish affect the feeding motivation of other fish sharing the same RAS? *Aquacul. Res.* **42**, 1378–1384.

Martins, C. I. M., Galhardo, L., Noble, C., Damsgård, B., Spedicato, M. T., Zupa, W., et al. (2012). Behavioural indicators of welfare in farmed fish. *Fish Physiol. Biochem.* **38**, 17–41.

Mason, G. J. and Latham, N. R. (2004). Can't stop, won't stop: is stereotypy a reliable animal welfare indicator? *Anim. Welfare* **13**, S57–S69.

Mason, G. J., Cooper, J. J. and Clareborough, C. (2001). Frustrations of fur-farmed mink. *Nature* **410**, 35–36.

Maximino, C., Marques de Brito, T., Waneza da Silva Batista, A., Herculano, A. M., Morato, S. and Gouveia, A., Jr. (2010). Measuring anxiety in zebrafish: a critical review. *Behav. Brain Res.* **215**, 157–171.

Mayer, I. and Pall, M. (2007). Hormonal control of reproductive behaviour in the stickleback. In *Biology of the Three-Spined Stickleback. [Marine Biology Series.]* (eds. S. Ostlund-Nilsson, I. Mayer and F. A. Huntingford), pp. 249–269. Boca Raton, FL: CRC Press.

Maynard, D. J., Berejikian, B. A., Flagg, T. A., and Mahnken, C. V. W. (2001). Development of a natural rearing system to improve supplemental fish quality, 1996–1998 Progress Report, Project No. 199105500, 174 electronic pages (BPA Report DOE/BP-00004768-1). <http://www.nwfsc.noaa.gov/assets/26/5432_04302014_123559_Maynard.et.al.2001-NATURES-1996-1998.pdf>.

McCormick, S. D., Shrimpton, J. M., Carey, J. B., O'Dea, M. F., Sloan, K. E., Moriyama, S., et al. (1998). Repeated acute stress reduces growth rate of Atlantic salmon parr and alters plasma levels of growth hormone, insulin-like growth factor I and cortisol. *Aquaculture* **168**, 221–235.

McEwen, B. S. (1998). Stress, adaptation, and disease–allostasis and allostatic load. *Neuroimmunomodulation* **840**, 33–44.

McEwen, B. S. and Wingfield, J. C. (2003). The concept of allostasis in biology and biomedicine. *Horm. Behav.* **43**, 2–15.

McGlone, J. J. (1993). What is animal welfare? *J. Agric. Environ. Ethics* **6** (Suppl. 2), 22–36.

McNeil, W. J. (1991). Expansion of cultured Pacific salmon into marine ecosystems. *Aquaculture* **98**, 173–183.

Meehan, C. L. and Mench, J. A. (2007). The challenge of challenge: can problem solving opportunities enhance animal welfare? *Appl. Anim. Behav. Sci.* **102**, 246–261.

Metcalfe, J. D. (2009). Welfare in wild-capture marine fisheries. *J. Fish. Biol.* **75**, 2855–2861.

Mettam, J. J., McCrohan, C. R. and Sneddon, L. U. (2012). Characterisation of chemosensory trigeminal receptors in the rainbow trout (Oncorhynchus mykiss): responses to irritants and carbon dioxide. *J. Exp. Biol.* **215**, 685–693.

Moberg, G. P. (1985). Biological response to stress: key to assessment of animal well-being? In *Animals Stress* (ed. G. P. Moberg), pp. 27–49. Bethesda: American Physiological Society.

Montero, D., Lalumera, G., Izquierdo, M. S., Caballero, M. J., Saroglia, M. and Tort, L. (2009). Establishment of dominance relationships in gilthead sea bream *Sparus aurata* juveniles during feeding: effects on feeding behavior, feed utilization and fish health. *J. Fish. Biol.* **74**, 790–805.

Montoya, A., López-Olmeda, J. F., Lopez-Capel, A., Sánchez-Vázquez, F. J. and Pérez-Ruzafa, A. (2012). Impact of a telemetry-transmitter implant on daily behavioural rhythms and physiological stress indicators in gilthead seabream (*Sparus aurata*). *Mar. Environ. Res.* **79**, 48–54.

Mork, O. I. and Gulbrandsen, J. (1994). Vertical activity of 4 salmonid species in response to changes between darkness and 2 intensities of light. *Aquacult* **127**, 317–328.

Muir, W. D., Smith, S. G., Williams, J. G. and Sandford, B. P. (2001). Survival of juvenile salmonids passing through bypass systems, turbines, and spillways with and without flow detectors at Snake River dams. *N. Am. J. Fish. Manage.* **21** (1), 135–146.

Näslund, J., Rosengren, M., Del Villar, D., Gansel, L., Norrgård, J. R., Persson, L., et al. (2013). Hatchery tank enrichment affects cortisol levels and shelter-seeking in Atlantic salmon (*Salmo salar*). *Can. J. Fish. Aquat. Sci.* **70**, 585–590.

Nedelec, S. L., Simpson, S. D., Morley, E. L., Nedelec, B. and Radford, A. N. (2015). Impacts of regular and random noise on the behavior, growth and development of larval Atlantic cod (*Gadus morhua*). *Proc. R. Soc. B Biol.* **282** (1817), 20151943.

Neiffer, D. L. and Stamper, M. A. (2009). Fish sedation, anesthesia, analgesia, and euthanasia: considerations, methods, and types of drugs. *ILAR J* **50**, 343–360.

Newby, N., Binder, T. R. and Stevens, D. (2007). Passive integrated transponder (PIT) tagging did not negatively affect the short-term feeding behavior or swimming performance of juvenile rainbow trout. *Trans. Am. Fish. Soc.* **136** (2), 341–345.

Newman, A. E. M., Edmunds, N. B., Ferraro, S., Heffell, Q., Merritt, G. M., Pakkala, J. J., et al. (2015). Using ecology to inform physiology studies: implications of high population density in the laboratory. *Am. J. Physiol.* **308**, R449–R454.

Noble, C., Kadri, S., Mitchell, D. F. and Huntingford, F. A. (2007). Influence of feeding regime on intraspecific competition, fin damage and growth in 1+Atlantic salmon parr (*Salmo salar* L.) held in freshwater production cages. *Aquacul. Res.* **38**, 1137–1143.

Noble, C., Kadri, S., Mitchell, D. F. and Huntingford, F. A. (2008). Growth, production and fin damage in cage-held 0+Atlantic salmon pre-smolts (*Salmo salar* L.) fed either (a) on-demand, or (b) to a fixed satiation-restriction regime: data from a commercial farm. *Aquaculture* **275**, 163–168.

Nomura, M., Sloman, K.A., von Keyserlingk, M.A.G. and Farrell, A.P. (2009). Physiology and behaviour of Atlantic salmon (Salmo salar) smolts during commercial land and sea transport. Physiol. Behav. 96, 233–243.

OATA. (2014). Available from: < http://www.ornamentalfish.org/useful-links/> (retrieved 09.02.14.).

O'Connell, C. P., Gruber, S. H., O'Connell, T. J., Johnson, G., Grudecki, K. and He, P. (2014). The Use of Permanent Magnets to Reduce Elasmobranch Encounter with a Simulated Beach Net. 1. The Bull Shark (Carcharhinus leucas). *Ocean & Coastal Management* **97**, 12–19.

Oldenburg, E. W., Colotelo, A. H., Brown, R. S. and Eppard, M. B. (2011). Holding of juvenile salmonids for surgical implantation of electronic tags: a review and recommendations. *Rev. Fish Biol. Fisher* **21**, 35–42.

Oliva-Teles, A. (2012). Nutrition and health of aquaculture fish. *J. Fish. Dis.* **35**, 83–108.

Oliveira, R. F. and Almada, V. C. (1998). Dynamics of social interactions during group formation in males of the cichlid fish *Oreochromis mossambicus. Acta Ethol.* **1**, 57–70.

Oliver, K. (2001). *The ornamental fish marketFAO/GLOBEFISH Research Programme*, vol. 67. Rome: FAO.

Oliver, K. (2003). World trade in ornamental species. In *Collection, Culture and Conservation* (eds. J. C. Cato and C. L. Brown), pp. 49–64. Iowa: Blackwell Publishing.

Olla, B. L., Davis, M. W. and Ryer, C. H. (1998). Understanding how the hatchery environment represses or promotes the development of behavioral survival skills. *Bull. Mar. Sci.* **62**, 531–550.

Ortuño, J., Esteban, M. and Meseguer, J. (2001). Effects of short-term crowding stress on the gilthead seabream (*Sparus aurata* L.) innate immune response. *Fish Shellfish Immun.* **11** (2), 187–197.

Øverli, Ø., Korzan, W. J., Larson, E. T., Winberg, S., Lepage, O., Pottinger, T. G., et al. (2004). Behavioral and neuroendocrine correlates of displaced aggression in trout. *Horm. Behav.* **45** (5), 324–329.

Øverli, Ø., Winberg, S. and Pottinger, T. G. (2005). Behavioral and neuroendocrine correlates of selection for stress responsiveness in rainbow trout – a review. *Integrat. Comp. Biol.* **45**, 463–474.

Owen, M. A. G., Davies, S. J. and Sloman, K. A. (2010). Light colour influences the behaviour and stress physiology of captive tench (*Tinca tinca*). *Rev. Fish Biol. Fisheries* **20**, 375–380.

Packer, R. M. A., Hendricks, A. and Burn, C. C. (2015). Impact of facial conformation on canine health: brachycephalic obstructive airway syndrome. *PLoS One* **10**, e0137496.

Pampoulie, C., Jörunsdóttir, T. D., Steinarsson, A., Pétursdóttir, G., Stefánsson, M. O. and Daníelsdóttir, A. K. (2006). Genetic comparison of experimental farmed strains and wild Icelandic populations of Atlantic cod (*Gadus morhua* L.). *Aquaculture* **261**, 556–564.

Pankhurst, N. W. (1998). Reproduction. In *Biology of Farmed Fish* (eds. K. Black and A. D. Pickering), pp. 1–26. Sheffield: Sheffield Academic Press.

Pankhurst, N. W. (2011). The endocrinology of stress in fish: an environmental perspective. *Gen. Comp. Endocrinol.* **170**, 265–275.

Pankhurst, N. W. (2016). Reproduction and Development. In *Fish Physiology - Biology of Stress in Fish*, Vol. 35 (eds. C. B. Schreck, L. Tort, A. P. Farrell and C. J. Brauner), San Diego, CA: Academic Press.

Paperna, I., Vilenkin, M. and Alves de Matos, P. (2001). Iridovirus infections in farm-reared tropical ornamental fish. *Dis. Aquatic Orgs.* **48**, 17–25.

Persson, L. and Alanara, A. (2014). The effect of shelter on welfare of juvenile Atlantic salmon *Salmo salar* reared under a feed restriction regimen. *J. Fish. Biol.* **85**, 645–656.

Petersson, E., Karlsson, L., Ragnarsson, B., Bryntesson, M., Berglund, A., Stridsman, S., et al. (2013). Fin erosion and injuries in relation to adult recapture rates in cultured smolts of Atlantic salmon and brown trout. *Can. J. Fish. Aquat. Sci.* **70**, 915–921.

PFMA. (2014). < http://www.pfma.org.uk/pet-population-2014/>.

Pickering, A. D. and Pottinger, T. G. (1989). Stress responses and disease resistance in salmonid fish: effects of chronic elevation of plasma cortisol. *Fish Physiol. Biochem.* **7**, 253–258.

Ploeg, A. (2005). Facts on mortality with shipments of ornamental fish. In: Ornamental Fish International. <http://www.ornamental-fish-int.org/files/files/mortality.pdf/>.

Poli, B. M. (2009). Farmed fish welfare-suffering assessment and impact on product quality. *Ital. J. Anim. Sci* **8** (Suppl. 1), 139–160.

Poli, B. M., Parisi, G., Scappini, F. and Zampacavallo, G. (2005). Fish welfare and quality as affected by pre-slaughter and slaughter management. *Aquacul. Int.* **13**, 29–49.

Pon, L. B., Hinch, S. G., Cooke, S. J., Patterson, D. A. and Farrel, A. P. (2009). Physiological, energetic and behavioural correlates of successful fishway passage of adult sockeye salmon *Oncorhynchus nerka* in the Seton River, British Columbia. *J. Fish. Biol.* **74**, 1323–1336.

Popper, A. N., Fay, R. R., Platt, C. and Sand, O. (2003). Sound detection mechanisms and capabilities of teleost fishes. In *Sensory Processing in Aquatic Environments* (eds. S. P. Collin and N. J. Marshall), pp. 3–38. New York, NY: Springer-Verlag.

Porsmoguer, S. B., Banaru, D., Boudouresque, C. F., Dekeyser, I. and Almarcha, C. (2015). Hooks equipped with magnets can increase catches of blue shark (Prionace glauca) by longline fishery. *Fish. Res* **172**, 345–351.

Portz, D. E., Woodley, C. M. and Cech, J. J., Jr. (2006). Stress-associated impacts of short-term holding on fishes. *Rev. Fish. Biol. Fisher.* **16**, 125–170.

Pottinger, T. G. and Carrick, T. R. (1999). Modification of the plasma cortisol response to stress in rainbow trout by selective breeding. *Gen. Comp. Endocrinol.* **116**, 122–132.

Pottinger, T. G., Pickering, A. D. and Hurley, M. A. (1992). Consistency in the stress response of individuals of two strains of rainbow trout, *Oncorhynchus mykiss*. *Aquaculture* **103**, 275–289.

Raby, G. D., Donaldson, M. R., Hinch, S. G., Patterson, D. A., Lotto, A. G., Robichaud, D., et al. (2012). Validation of reflex indicators for measuring vitality and predicting the delayed mortality of wild coho salmon bycatch released from fishing gears. *J. Appl. Ecol.* **49**, 90–98.

Raby, G. D., Clark, T. D., Farrell, A. P., Patterson, D. A., Bett, N. N., Wilson, S. M., et al. (2015). Facing the river gauntlet: understanding the effects of fisheries capture and water temperature on the physiology of coho salmon. *PLoS ONE* **10**, e0124023.

Readman, G. D., Owen, S. E., Murrell, J. G. and Knowles, T. G. (2013). Do fish perceive anaesthetics as aversive? *PLoS One* **8** (9), e73773.

Reinhardt, V. (2004). Common husbandry-related variables in biomedical research with animals. *Lab. Anim.* **38**, 213–235.

Robb, D. H. F., Wotton, S. B., McKinstry, J. L., Sorensen, N. K. and Kestin, S. C. (2000). Commercial slaughter methods used on Atlantis salmon: determination of the onset of brain failure by electroencephalography. *Vet. Rec.* **147**, 298–303.

Rollin, B. E. (1993). Animal welfare, science and value. *J. Agricult. Environ. Ethics* **6** (Suppl. 2), 44–50.

Roques, J. A. C., Abbink, W., Geurds, F., van de Vis, H. and Flik, G. (2010). Tailfin clipping, a painful procedure: studies on Nile tilapia and common carp. *Physiol. Behav.* **101**, 533–540.

Roques, J. A. C., Abbink, W., Chereau, G., Fourneyron, A., Spanings, T., Burggraaf, D., et al. (2012). Physiological and behavioral responses to an electrical stimulus in Mozambique tilapia (*Oreochromis mossambicus*). *Fish Physiol. Biochem.* **38**, 1019–1028.

Roscoe, D. W., Hinch, S. G., Cooke, S. J. and Patterson, D. A. (2011). Fishway passage and post-passage mortality of up-river migrating sockeye salmon in the Seton River, British Columbia. *River Res. Appl.* **27**, 693–705.

Rose, J. D. (2002). The neurobehavioral nature of fishes and the question of awareness and pain. *Rev. Fish. Sci.* **10**, 1–38.

Ross, L. G. and Ross, B. (2008). *Anaesthetic and Sedative Techniques for Aquatic Animals* (3rd Edition.). Oxford: Blackwells.

Ruiz-Gomez, M. deL., Kittilsen, S., Höglund, E., Huntingford, F. A., Sørensen, C., Pottinger, T. G., et al. (2008). Behavioral plasticity in rainbow trout (*Oncorhynchus mykiss*) with divergent coping styles: when doves become hawks. *Horm. Behav.* **54**, 534–538.

Rutte, C., Taborsky, M. and Brinkhof, M. W. (2006). What sets the odds of winning and losing? *Trends. Ecol. Evol.* **21**, 16–21.

Saeji, J. P. J., Vervurg-van-Kemenade, L. B. M., Van Muiswinkel, W. B. and Wiegertjes, G. F. (2003). Daily handling stress reduces resistance of carp to *Trypanoplasma borreli*: in vitro modulatory effects of cortisol on leukocyte function and apoptosis. *Dev. Comp. Immunol.* **27** (3), 233–245.

Saera-Vila, A., Kasprick, D. S., Junttila, T. L., Grzegorski, S. J., Louie, K. W., Chiari, E. F., et al. (2015). Myocyte dedifferentiation drives extraocular muscle regeneration in adult zebrafish. *Investigative Ophthalm. Visual Sci* **56**, 4977–4993.

Sadoul, B. and Vijayan, M. M. (2016). Stress and Growth. In Fish Physiology - Biology of Stress in Fish, Vol. 35 (eds. C. B. Schreck, L. Tort, A. P. Farrell and C. J. Brauner), San Diego, CA: Academic Press.

Salvanes, A. G. V., Moberg, O., Ebbesson, L. O. E., Nilsen, T. O., Jensen, K. H. and Braithwaite, V. A. (2013). Environmental enrichment promotes neural plasticity and cognitive ability in fish. *Proc. R. Soc. B. Biol.* **280**, 20131331.

Sánchez, J. A., López-Olmeda, J. F., Blanco-Vives, B. and Sánchez-Vázquez, F. J. (2009). Effects of feeding schedule on locomotor activity rhythms and stress response in sea bream. *Physiol. Behav.* **98**, 125–129.

Sánchez-Vázquez, F. J., Madrid, J. A. and Zamora, S. (1995). Circadian rhythms of feeding activity in sea bass, *Dicentrarchus labrax* L.: dual phasing capacity of diel demand-feeding pattern. *J. Biol. Rhythm.* **10**, 256–266.

Sand, O., Enger, P. S., Karlsen, H. E. and Knudsen, F. R. (2001). Detection of infrasound in fish and behavioral responses to intense infrasound in juvenile salmonids and European silver eels: a minireview. *Am. Fish. Soc. Symp.* **26**, 183–193.

Sapolsky, R. M. (2004). *Why Zebras Don't Get Ulcers.* New York, NY: Henry Holt and Company.

Schall, K. A., Holoyda, K. A., Grant, C. N., Levin, D. E., Torres, E. R., Maxwell, A., et al. (2015). Adult zebrafish intestine resection: a novel model of short bowel syndrome, adaptation, and intestinal stem cell regeneration. *Am. J. Physiol.-Gastrointestinal Liver Physiol* **309**, G135–G145.

Schreck, C. B., Jonsson, L., Feist, G. and Reno, P. (1995). Conditioning improves performance of juvenile Chinook salmon, *Oncorhynchus tshawytscha*, to transportation stress. *Aquaculture* **135**, 99–110.

Schreck, C. B., Olla, B. L. and Davis, M. W. (1997). Behavioral responses to stress. In *Fish Stress and Health in Aquaculture* (eds. G. K. Iwama, A. D. Pickering, J. P. Sumpter and C. B. Schreck), pp. 145–170. Cambridge: Cambridge University Press.

Schreck, C. B. and Tort, L. (2016). The Concept of Stress in Fish. In *Fish Physiology - Biology of Stress in Fish*, Vol. 35 (eds. C. B. Schreck, L. Tort, A. P. Farrell and C. J. Brauner), San Diego, CA: Academic Press.

Schroeder, P., Jones, S., Young, I. Y. and Sneddon, L. U. (2014). What do zebrafish want? Impact of social grouping, dominance and gender on preference for enrichment. *Lab. Anim.* **48**, 328–337.

Sea Shepherd. (2014). < http://www.seashepherd.org/reef-defense/aquarium-trade.html/ > .

Sharber, N. G. and Carothers, S. W. (1988). Influence of electrofishing pulse shape on spinal injuries in adult rainbow trout. *N. Am. J. Fish. Manage.* **8**, 117–122.

Sharpe, C. S., Thompson, D. A., Blankenship, H. L. and Schreck, C. B. (1998). Effects of routine handling and tagging procedures on physiological stress responses in juvenile Chinook salmon. *Prog. Fish Cult.* **60** (2), 81–87.

Singhal, G., Jaehne, E. J., Corrigan, F. and Baune, B. T. (2014). Cellular and molecular mechanisms of immunomodulation in the brain through environmental enrichment. *Front. Cell. Neurosci.* Available from: <http://dx.doi.org/10.3389/fncel.2014.00097>.

Skaala, Ø., Dahle, G., Jørstad, K. E. and Nævdal, G. (1990). Interactions between natural and farmed fish populations: information from genetic markers. *J. Fish. Biol.* **36**, 449–460.

Slabbekoorn, H., Bouton, N., van Opzeeland, I., Coers, A., ten Cate, C. and Popper, A. N. (2010). A noisy spring: the impact of globally rising underwater sound levels on fish. *Trends. Ecol. Evol.* **25** (7), 419–427.

Sloman, K. A. (2010). Exposure of ova to cortisol pre-fertilisation affects subsequent behaviour and physiology of brown trout. *Horm. Behav.* **58**, 433–439.

Smith, M. E., Kane, A. S. and Popper, A. N. (2004). Acoustical stress and hearing sensitivity in fishes: does the linear threshold shift hypothesis hold water? *J. Exp. Biol.* **207**, 3591–3602.

Smith, J. M., Mather, M. E., Frank, H. J., Muth, R. M., Finn, J. T. and McCormick, S. D. (2009). Evaluation of a gastric radio tag insertion technique for anadromous river herring. *N. Am. J. Fish. Manage.* **29** (2), 367–377.

Sneddon, L. U. (2003a). The evidence for pain in fish: the use of morphine as an analgesic. *Appl. Anim. Behav. Sci.* **83** (2), 153–162.

Sneddon, L. U. (2003b). The bold and the shy: individual differences in rainbow trout. *J. Fish. Biol.* **62**, 971–975.

Sneddon, L. U. (2011). Cognition and welfare. In *Fish Cognition and Behavior* (eds. C. Brown, K. Laland and J. Krause), second ed., pp. 405–434. Oxford: Wiley-Blackwell.

Sneddon, L. U. (2013). Do painful sensations and fear exist in fish? In *Animal Suffering: From Science to Law, International Symposium* (eds. T. A. van der Kemp and M. Lachance), pp. 93–112. Toronto: Carswell.

Sneddon, L. U. (2015). Pain in aquatic animals. *J. Exp. Biol.* **218**, 967–976.

Sneddon, L. U. and Wolfenden, D. (2012). How are fish affected by large scale fisheries: pain perception in fish? In *See the Truth* (ed. K. Soeters, Amsterdam: Nicolaas G. Pierson Foundation.

Sneddon, L. U., Braithwaite, V. A. and Gentle, M. J. (2003). Do fish have nociceptors? Evidence for the evolution of a vertebrate sensory system. *Proc. R. Soc. Lond. B* **270**, 1115–1121.

Sneddon, L. U., Schmidt, R., Fang, Y. and Cossins, A. R. (2011). Molecular correlates of social dominance: a novel role for ependymin in aggression. *PLoS One* **6** (4), e18181.

Sneddon, L. U., Elwood, R. W., Adamo, S. A. and Leach, M. C. (2014). Defining and assessing pain in animals. *Anim. Behav.* **97**, 201–212.

Snyder, D. (2003). Inited overview: conclusions from a review of electrofishing and its harmful effects on fish. *Rev. Fish Biol. Fish.* **13**, 445–453.

Soares, M. C., Oliveira, R. F., Ros, A. F. H., Grutter, A. S. and Bshary, R. (2011). Tactile stimulation lowers stress in fish. *Nat. Commun.* **2**, 534.

Sommerset, I., Krossøy, B., Biering, E. and Frost, P. (2005). Vaccines for fish in aquaculture. *Expert. Rev. Vaccines* **4** (1), 89–101.

Spagnoli, S., Lawrence, C. and Kent, M. L. (2016). Stress in Fish as Model Organisms. In *Fish Physiology - Biology of Stress in Fish*, Vol. 35 (eds. C. B. Schreck, L. Tort, A. P. Farrell and C. J. Brauner), San Diego, CA: Academic Press.

Spence, R., Gerlach, G., Lawrence, C. and Smith, C. (2008). The behaviour and ecology of the zebrafish. *Danio Rerio. Biol. Rev. Cam. Philos. Soc.* **83**, 13–34.

Stamp Dawkins, M. (2012). *Why animals matter: animal consciousness, animal welfare, and human well-being.* Oxford: Oxford University Press.

Steckler, T. (2005). The neuropsychology of stress. In *Handbook of Stress and the Brain. Part 1: The Neurobiology of Stress* (eds. T. Steckler, N. H. Kalin and J. M. H. N. Reul), Amsterdam: Elsevier.

Suzuki, K., Mizusawa, K., Noble, C. and Tabata, M. (2008). The growth, feed conversion ratio and fin damage of rainbow trout *Oncorhynchus mykiss* under self-feeding and hand-feeding regimes. *Fish. Sci.* **74**, 941–943.

Tang, S., Brauner, C. J. and Farrell, A. P. (2009). Using bulk oxygen uptake to assess the welfare of adult Atlantic salmon, *Salmo salar*, during commercial live-haul transport. *Aquaculture* **286**, 318–323.

The Telegraph. (2012). Available from: <http://www.telegraph.co.uk/lifestyle/pets/9217643/One-in-ten-Britons-now-have-pet-fish.html/> (retrieved 09.02.14.).

Thorstad, E. B., Økland, F., Aarestrup, K. and Heggberget, T. G. (2008). Factors affecting the within-river spawning migration of Atlantic salmon, with emphasis on human impacts. *Rev. Fish Biol. Fisher* **18**, 345–371.

Torrente-Vilara, G., Zuanon, J., Leprieur, F., Oberdorff, T. and Tedesco, P. A. (2011). Effects of natural rapids and waterfalls on fish assemblage structure in the Madeira River (Amazon Basin). *Ecol. Freshw. Fish* **20**, 588–597.

Townsend, D. (2011). Sustainability, equity and welfare: a review of the tropical marine ornamental fish trade. *SPC Live Reef Fish Inform. Bull.* **20**, 2–12.

Trushenski, J. T., Bowker, J. D., Cooke, S. J., Erdahl, D., Bell, T., MacMillan, J. R., et al. (2013). Issues regarding the use of sedatives in fisheries and the need for immediate-release options. *Trans. Am. Fish. Soc.* **142** (1), 156–170.

Turnbull, J. F., North, B. P., Ellis, T., Adams, C. E., Bron, J., MacIntyre, C. M., et al. (2008). Stocking density and the welfare of farmed salmonids. In *Fish Welfare* (ed. E. J. Branson), pp. 111–120. Oxford: Blackwell.

UFAW. (2013). Available from: <http://www.ufaw.org.uk/fish/fish> (retrieved 31.08.14.).

Ursin, H. and Eriksen, H. R. (2004). The cognitive activation theory of stress. *Psychoneuroendocrinology* **29**, 567–592.

USDA. (2013). <https://speakingofresearch.files.wordpress.com/2008/03/usda-animal-research-use-2011-13.pdf/>.

Vera, L. M., Madrid, J. A. and Sánchez-Vázquez, F. J. (2006). Locomotor, feeding and melatonin daily rhythms in sharpsnout seabream (*Diplodus puntazzo*). *Physiol. Behav.* **88** (1–2), 167–172.

Vilhunen, S. and Hirvonen, H. (2003). Innate antipredator responses of Arctic charr (*Salvelinus alpinus*) depend on predator species and their diet. *Behav. Ecol. Sociobiol.* **55**, 1–10.

Vindas, M. A., Folkedal, O., Kristiansen, T. S., Stien, L. H., Braastad, B. O., Mayer, I., et al. (2012). Omission of expected reward agitates Atlantic salmon (*Salmo salar*). *Anim. Cogn.* **15**, 903–911.

Vindas, M. A., Johansen, I. B., Vela-Avitua, S., Nørstrud, K. S., Aalgaard, M., Braastad, B. O., et al. (2014). Frustrative reward omission increases aggressive behaviour of inferior fighters. *Proc. R. Soc. B* **281**, 20140300.

Volpato, G. L., Gonçalves-de-Freitas, E. and Fernandes-de-Castilho, M. (2007). Insights into the concept of fish welfare. *Dis. Aquat. Orgs.* **75**, 165–171.

Walster, C. (2008). The welfare of ornamental fish. In *Fish Welfare* (ed. E. J. Branson), pp. 271–290. Oxford: Blackwell.

Walster, C., Rasidi, E., Saint-Erne, N. and Loh, R. (2014). The welfare of ornamental fish in the home aquarium. *Companion Anim.* **20**, 302–306.

Wendelaar Bonga, S. E. (1997). The stress response in fish. *Physiol. Rev.* **77**, 591–625.

Wilson, J.M., Bunte, R.M. and Carty, A.J. Evaluation of rapid cooling and tricaine methanesulfonate (MS222) as methods of euthanasia in zebrafish (Danio rerio). JAALAS 48, 785–789.

Whitney, R. R., Calvin, L. D., Erho Jr., M. W., and Coutant, C. C. (1997). Downstream passage for salmon at hydroelectric projects in the Columbia River basin: development, installation, evaluation. North-West Power Planning Council, NPCC Resport 97-15, Portland, Oregon.

Willis, C. F. (1982). Indexing of juvenile salmonids migrating past The Dalles Dam, 1982. Report of Oregon Department of Fish and Wildlife (Contract DACW57-78-C-0056) to U.S. Army Coprs of Engineers, Portland, Oregon.

Winger, P. D. and Walsh, S. J. (2001). Tagging of Atlantic cod (*Gadus morhua*) with intragastric transmitters: effects of forced insertion and voluntary ingestion on retention, food consumption and survival. *J. Appl. Ichthyol.* **17** (5), 234–239.

Winberg, S., Höglund, E. and Øverli, Ø. (2016). Variation in the Neuroendocrine Stress Response. In *Fish Physiology - Biology of Stress in Fish*, Vol. 35 (eds. C. B. Schreck, L. Tort, A. P. Farrell and C. J. Brauner), San Diego, CA: Academic Press.

Wysocki, L. E., Amoser, S. and Ladich, F. (2007). Diversity in ambient noise in European freshwater habitats: noise levels, spectral profiles, and impact on fishes. *J. Acoust. Soc. Am.* **121** (5), 2559–2566.

Yada, T. and Tort, L. (2016). Stress and Disease Resistance: Immune System and Immunoendocrine Interactions. In *Fish Physiology - Biology of Stress in Fish*, Vol. 35 (eds. C. B. Schreck, L. Tort, A. P. Farrell and C. J. Brauner), San Diego, CA: Academic Press.

Yue, S., Moccia, R. D. and Duncan, I. J. H. (2004). Investigating fear in domestic rainbow trout, Oncorhynchus mykiss, using an avoidance learning task. *Appl. Anim. Behav. Sci* **87**, 343–354.

Zahl, H. L., Kiessling, A., Samuelsen, O. B. and Olsen, R. E. (2010). Anaesthesia induces stress in Atlantic salmon (*Salmo salar*), Atlantic cod (*Gadus morhua*) and Atlantic halibut (*Hippoglossus hippoglossus*). *Fish Physiol. Biochem.* **36**, 719–730.

Zuberi, A., Ali, S. and Brown, C. (2011). A non-invasive assay for monitoring stress responses: a comparison between wild and captive-reared rainbowfish (*Melanoteania duboulayi*). *Aquaculture* **321**, 267–272.

Zuberi, A., Brown, C. and Ali, S. (2014). Effect of confinement on water-borne and whole body cortisol in wild and captive-reared rainbowfish (*Melanoteania duboulayi*). *Int. J. Agricul. Biol.* **16**, 183–188.

13

STRESS IN FISH AS MODEL ORGANISMS

SEAN SPAGNOLI

CHRISTIAN LAWRENCE

MICHAEL L. KENT

With the advent of the zebrafish, the three-spined stickleback, and the medaka as laboratory animals, the emergence of fish as model organisms has provided a wide variety of potential experimental subjects, simultaneously introducing new challenges to both researchers and aquaculturists. With regard to stress in these fishes, we must shift the emphasis beyond the traditional definitions of production in terms of fecundity and growth and toward the goals of experimental consistency, animal welfare, and model robustness. In order to improve fish as model organisms, aquaculturists and researchers must use each organism's natural history as a template for developing appropriate husbandry practices. In this chapter, we review published data regarding stress in laboratory fishes in order to provide a foundation for building better husbandry protocols for the purposes of improving fish as model organisms.

Biology of Stress in Fish: Volume 35
FISH PHYSIOLOGY

1. INTRODUCTION

The use of aquatic models as part of an integrative approach to biomedical research has expanded rapidly in the last few decades. The popularity of the zebrafish as a laboratory organism is due to its plethora of advantages regarding high-throughput research: a known genome; rapid life cycle; decreased cost of rearing compared to mice; transparent body in larvae and, occasionally, also in adults. Furthermore, as a vertebrate, they provide utility and efficiency of so-called lower organisms while maintaining physiological parallels to man and so-called higher organisms (Lieschke and Currie, 2007; Allen and Neely, 2010; Harland and Grainger, 2011; Phillips and Westerfield, 2014). This is perhaps most strikingly exemplified by the zebrafish (*Danio rerio*), as the Zebrafish Model Organism Database website (http://zfin.org) now lists approximately 1000 laboratories that employ zebrafish, and a 2014 search of the US National Institutes of Health RePORTER website using the term "zebrafish" revealed a list of 735 grants using this model. In fact, zebrafish recently surpassed *Drosophila* in PubMed listings (2,279 vs 2,265 PubMed listings in 2011).

It is important to keep both the researchers' and aquaculturists' primary goals in mind when shifting between food production, conservation, and laboratory settings. For much of this book, stress has been discussed mostly as it relates to fish performance, which is a single word with numerous meanings depending on context. In a production setting, performance can be measured in terms of fecundity, growth, and survival. In a conservation setting, performance can be measured in terms of population stability, recovery, and homing and straying rates. In the laboratory, the definition of performance becomes entirely fluid, changing based on the objectives of the research at hand. Fortunately, there are some fundamental goals that apply to all experimental organisms: similarity of subjects apart from experimental manipulations, animal welfare, and robustness of the model under study—both in general and as it pertains to specific research end points.

For many years, researchers have used adult trout (Bailey et al., 1996; Williams et al., 2003), medaka (Takeda and Shimada, 2010), and swordtail X platyfish hybrids (Patton et al., 2010) as models for basic biomedical science. Zebrafish experimentation was initially led by developmental geneticists whose primary interests concerned embryos or larval fish. As researchers have explored the utility of the zebrafish as a model for aging, chronic disease, and complex neurobehavioral syndromes, increasing focus has been placed on juvenile and adult-stage zebrafish (Dooley and Zon, 2000). Challenges in experiments utilizing embryos or newly hatched larvae have many similarities with research using postlarval through adult fish because both require appropriate rearing and population maintenance. However, the latter type of research requires additional diligence in order to avoid nonprotocol-induced variation in in vivo experiments. The emergent challenges involved in rearing and maintaining populations of juvenile and adult zebrafish compared to embryos and larvae has led to a heightened need for greater involvement by veterinary specialists trained in comparative pathology and laboratory animal medicine.

Specific areas of study utilizing juvenile and adult zebrafish are varied and include infectious disease susceptibility and immune system function (Novoa and Figueras, 2012), aging (Gerhard, 2003; Kishi et al., 2009), toxicology (Truong et al., 2011), oncology (Liu and Leach, 2011; Ceol et al., 2011), and behavior (Sisson et al., 2006; Wong et al., 2010; Stewart et al., 2012). Regarding infectious diseases, some 30 different bacterial pathogens have been studied using the zebrafish model (Meijer and Spaink, 2011). A number of programs rely on experimental infection of zebrafish with *Mycobacterium* spp. (eg, Prouty et al., 2003; Meijer et al., 2005; Swaim et al., 2006), which could be confounded by underlying, preexisting infections. Diverse work on fish in the field of cancer research includes the establishment of xenotransplantation models, where human tumors are transplanted into immune-compromised zebrafish (Patton et al., 2011; White et al., 2008; Patton and Zon, 2005; Taylor and Zon, 2009).

Other aquatic species commonly used as model organisms include three-spined sticklebacks (*Gasterosteus aculeatus*), Japanese medaka (*Oryzias latipes*), fathead minnow (*Pimephales promelas*), and several *Xipophorus* species. Sticklebacks are considered a robust model for studies concerning evolution and ecology, specifically phenotypic and genomic variation, speciation, and natural selection (Hendry et al., 2013). Medaka have been indispensable to the fields of genomics and toxicology, and were the first fish in which transgenesis was achieved (Shima and Mitani, 2004). Fathead minnows have been used in numerous toxicology studies (Ankley and Villeneuve, 2006) and are used as bioindicator species. *Xipophorus* species are excellent models for melanomas and sexual development disorders (Schartl, 2014). Other common research species include common carp (*Cyprinus carpio*) in environmental toxicology studies, channel catfish (*Ictalurus punctatus*) in studies of evolutionary genomics (Liu, 2003), and Mozambique tilapia (*Oreochormis mossambicus*) in osmoregulatory research (Gardell et al., 2013).

Despite the wide utility of these organisms, investigators are only just now becoming aware of the potential impact that underlying stress and chronic disease may have on research utilizing postlarval fish in medium- to long-term experiments. With the dramatic rise of fish as model organisms for neurobehavioral experiments (Rihel and Schier, 2012; Spence, 2011; Bailey et al., 2013), it is likely that chronic stress may exert unwanted effects on studies with behavioral or psychological end points. Kent et al. (2012) reviewed the impacts of chronic diseases as causes of nonprotocol-induced variation in zebrafish research. This chapter will expand on this theme to highlight potential and demonstrated impacts of stress in model laboratory fish and to provide a general overview of stress in these animals.

2. INDICATORS OF STRESS IN LABORATORY FISH

The relationship between researchers and aquaculturists is a paradox: Researchers house animals adapted to natural environments in totally artificial habitats to eliminate potentially confounding variables, including environmental olfactory cues, local environmental microbiota, and wild-type nutritional profiles, which causes stress (Wedemeyer, 1997). Aquaculturists consider that what may be best for the fish in terms of health, welfare, and environmental adaptations may introduce confounding variables in the forms of unspecified bacterial flora, altered behavioral syndromes, and environmental variability associated with more natural settings. When working with fish in the laboratory, we must integrate the knowledge of the animals' natural histories and potential adaptation to the laboratory environment over generations of artificial rearing with the demands of the research at hand in order to produce robust scientific models without sacrificing animal welfare. Indeed, close attention to animal welfare can reduce or eliminate stress as a confounding variable, and the first step in reducing stress for fish in the laboratory is the ability to measure and monitor stress responses (Sneddon et al., 2016; Chapter 12 in this volume).

The most commonly used indicator of stress in fish is cortisol, a readily measured component of the primary neuroendocrine stress response (Ellis et al., 2012) (Sopinka et al., 2016; Chapter 11 in this volume). Elevation of plasma cortisol is a well-documented indicator of stress in fishes, although there are many other hormonal and physiological changes associated with stress (Martinez-Porchas et al., 2009; Ellis et al., 2012) (Schreck and Tort, 2016; Chapter 1 in this volume). Whole-body cortisol concentrations have been used in small fish when blood volumes are inadequate to provide measurements for plasma cortisol (Barry et al., 1995; Feist and Schreck, 2001). This approach has been extended to zebrafish (Canavello et al., 2011; Ramsay et al., 2006, 2009a,b,c; Egan et al., 2009). Pavlidis et al. (2011, 2013) used a similar approach with zebrafish, but confined the analysis to the trunk portion of the body. More recently, Gesto et al. (2015) proposed the use of gill biopsies as an in vivo alternative to whole or partial body homogenates, the advantages being that gill biopsy can be obtained without euthanasia of the fish and that sampling induces far less stress than blood draws. Laboratory fishes, such as zebrafish, are usually held in relatively small tanks, and assays have been developed to test cortisol levels in the water (Ellis et al., 2004; Gronquist and Berges, 2013), which correlate well with plasma cortisol values (Félix et al., 2013). This nonlethal approach also has the advantage of providing an overall assessment of whole-tank stress using only a single test. These studies evaluating cortisol/stress associations have shown that crowding, brief handling (netting), or underlying chronic infections by two zebrafish pathogens commonly found in research facilities (*Pseudoloma neurophilia* or *Mycobacterium marinum*) are all associated with elevated cortisol levels.

Recently, Aerts, and coauthors demonstrated that the cortisol content in both ontogenetic and regenerated scales of common carp (*Cyprinus carpio*) could be used as a reliable and quantitative biomarker of chronic stress in this species (Aerts et al., 2015). As conventional methods for quantification of cortisol in fishes from plasma, feces, and water are only able to provide information relative to the short-term response of a given animal or set of animals, this new approach will allow for assessment of stress levels in fish over much longer periods of time.

Cortisol, while easily the most quantifiable indicator of stress reactions, does not necessarily tell the whole story. Stress, from perception to processing to response, involves the coordination of dozens of neural circuits and endocrine loops (Maximino et al., 2010). While each strand in the web is still being explored, behavior, one of the end products of the stress response, can be used as both a quantitative and qualitative indicator. Behavior is a summation and synthesis of all facets of the stress response, and because zebrafish have relatively robust and repeatable responses to certain stimuli, behavior can be employed as a soft indicator of stress. Whereas not as specific as cortisol, behavioral responses may be a better way of determining fish preferences in vivo, leading to improved welfare and experimental consistency.

Most stress-associated behaviors tend to be conserved among verte-brates, with species-specific natural histories guiding individual reactions. In the case of zebrafish, having evolved as a prey species in its range in the major river drainages in India, Bangladesh, and Nepal, it is no great stretch to surmise that the animal prefers darkness and depth—the better to hide from predators—under stressful conditions (Engeszer et al., 2007). There-fore, Egan et al. (2009) found that behaviors such as reduced tank exploration, scototaxis (light avoidance), thigmotaxis (wall hugging), erratic swimming, diving, and freezing (sudden, brief immobility) are correlated with increased cortisol levels and are indicative of stress. The authors also showed that different strains of zebrafish exhibited different inherent anxiety levels, which is particularly important in genetically modified animals: Knowing the background strain of a fish may help to account for differential levels of stress. To make matters even more complicated, there can be a great deal of variability in behavioral syndromes (sometimes referred to as personalities) between individual fish, leading to a potentially high level of variability in stress responses (Dugatkin et al., 2004; Conrad et al., 2011).

Although impractical as diagnostic tests for stress in the laboratory, the use of pharmacological substances that are specifically associated with stress and anxiety allow researchers to identify stress-induced behavioral preferences and to correlate these behaviors with stress-associated cortisol levels. For example, in a preference experiment where a zebrafish must choose between a white-walled and a black-walled compartment, fish treated with anxiolytic (anxiety-reducing) compounds such as selective serotonin reuptake inhibitors spend more time in the white compartment while fish treated with anxiogenic (anxiety-enhancing) compounds such as caffeine and substituted amphetamines spend more time in the black compartment than controls (Araujo et al., 2012). Extending the correlation between behavior and cortisol levels to larval zebrafish, Tudorache et al. (2015) showed 8-day-old larvae that hatched early exhibited higher cortisol peaks, but faster recovery times, than those hatching later. The close evaluation of stress-related behaviors in laboratory animals is particularly important in the use of zebrafish as model organisms for drug discovery, particularly in the arena of pharmaceutical screening. If certain behaviors associated with stress are confirmed, then behavioral responses following pharmaceutical exposure can be used to characterize the neurobehavioral effects of the drug (Stewart et al., 2015). Behavioral indicators of stress also have the advantage of being noninvasive to both the individual and to the tank as a whole when measurements are taken.

3. FACTORS IMPACTING STRESS IN LABORATORY FISH HANDLING

Handling of fish by humans, which usually entails removing them from water, is probably the greatest stressor most laboratory fish will face during their lifetimes (Ghisleni et al., 2012). Hence, handling events should be kept to a minimum. Because experimentation requires handling at various times during a laboratory fish's life, understanding responses to handling can help reduce stress and increase posthandling survival. Fortunately, because the perception and anticipation of stress make up a large component of the stress response, manipulation of perception and anticipation can greatly reduce overall stress in some species. For example, anesthetization of yearling chinook salmon prior to netting can reduce cortisol levels and increase postnetting survival. Regarding nonsalmonids commonly used in research, serum cortisol levels in goldfish are not affected by rapid capture with a net, ice immobilization, tricaine methanesulfonate (MS-222) anesthesia, or electric immobilization (Singley and Chavin, 1975).

In contrast, zebrafish exhibit rapid cortisol elevation and subsequent recovery from acute netting stress (Ramsay et al., 2009a,b,c; Tran et al., 2014). Brydges and Braithwait (2009) found that scooping fish in water, rather than netting, resulted in an attenuated cortisol response. There are several reports of handling associated with reduced fecundity, which are discussed later.

Fish handling in a laboratory setting frequently involves more elaborate and stressful events than netting: laboratory fishes are often exposed to pathogens or chemical agents by immersion, gavage, or various routes of injection. Therefore, understanding the stress involved in these procedures is important, but data are extremely limited and can only be inferred from overall recovery and/or postprocedure survival rates. Kahl et al. (2001), in one of the few such studies available, found that anesthesia with MS-222 and intraperitoneal injections with corn oil and 10% ethanol was not correlated with behavioral or secondary sex changes, reduced survival, or fecundity.

4. HOUSING

Fish used in research are most typically housed in aquaria that are on either flow-through or recirculating aquaculture systems (RAS) (Lawrence and Mason, 2012). While literature documenting differential effects of different types of aquatic systems on stress is sparse, it has been suggested that animals held in RAS might experience elevated stress as a result of metabolite, cortisol, and alarm substance accumulation in intensively stocked systems (Martins et al., 2011). This hypothesis was tested in Nile tilapia, but the results of the study were inconclusive (Martins et al., 2009).

Because of the limited data available, the subject deserves further examination, especially because laboratory species like zebrafish are almost exclusively maintained in RAS.

Parker et al. (2012) examined the effect of varying prehousing conditions on stress reactivity measured by the novel tank exploration test. They found that group size, water changes, and especially the ability to make visual contact with conspecifics influenced novel tank exploration in zebrafish. They also showed that basal cortisol levels were lower in individually housed than in group-housed fish. These data support the idea that housing conditions mediate the stress response in this species, and therefore need to be standardized in order to improve study design.

Other housing characteristics have been evaluated relative to their impact on stress and anxiety in zebrafish. Blaser and Rosemberg (2012) found that tank depth and wall color elicited different responses in depth preference and scototaxis assays. An aspect of housing vital to fish husbandry that generally goes unnoticed due to its constant presence is the sound of air and water pumps. Some work has been done on the subject of noise in salmonids: specific frequencies produce avoidance responses in juvenile Atlantic salmon (*Salmo salar* L.) (Knudsen et al., 1992). While specific research regarding the effects of husbandry-associated noise on zebrafish behavior is in its infancy, Smith et al. (2003) demonstrated that goldfish (*Carassius auratus*), a cyprinid like *D. rerio*, may be susceptible to stress and hearing loss in the face of chronic noise exposure. More recently, Neo et al. (2015) found that moderate sound levels alter group cohesion, swimming velocity, and depth preference in laboratory zebrafish. Interestingly, the same study found that there was no avoidance response to chronic, loud sound levels. These data provide further evidence that housing configuration can be an important contributor to zebrafish stress and behavioral responses.

The effects of external acoustics on fish behavior have been explored in a number of model and nonmodel organisms aside from zebrafish, with overall results indicating a subtle and potentially insidious influence of sound on survival and ecology. Purser and Radford (2011) found that while captive three-spined sticklebacks exposed to brief and prolonged noise had no reduction in total feed consumption, the fish had increased incidences of food-handling errors and decreased discrimination between food- and nonfood items, indicating a distraction effect. Similarly, European eels (*Anguilla anguilla*) showed reduced antipredator behaviors following exposure to anthropogenic noise (Simpson et al., 2014). Furthermore, there is ample evidence to suggest that anthropogenic noise interferes with acoustic communication between fish, another important survival strategy associated with social behavior (Radford et al., 2014).

4.1. Density

Most studies utilizing fish in the laboratory are not concerned with production in terms of fecundity or feed efficiency, except in nutrition or husbandry studies. However, the number of animals present per unit volume along with the water exchange rate is closely related to fish stress levels, and may contribute to unexpected nonprotocol-induced variation in studies measuring other criteria, including immune, metabolic, and behavioral responses.

Relationships between fish number per unit volume, water flow, and stress may be neither direct nor intuitive. Generally, reduction of animal numbers among zebrafish housed at initially high densities correlates with reduced cortisol levels (Ramsay et al., 2006; Pavlidis et al., 2013), but only to a certain point. There exists a lower limit of stocking density past which cortisol levels begin to increase as the number of fish decreases (Fig. 13.1). This may be due to the establishment of dominance hierarchies among small shoals lacking the capacity to spread chasing and aggression behavior throughout a large population of individuals (Filby et al., 2010; Pavlidis et al., 2013). Strikingly, despite the fact that zebrafish are a shoaling species, Parker et al. (2012) and Giacomini et al. (2015) showed that individuals held in isolation actually had reduced cortisol, presumably due to reduced interfish aggression, The latter authors also suggested that this reduced stress response may be related to the absence of behavioral and chemical cues from other fish held in the same housing unit. Correspondingly, zebrafish exposed to chemical cues released from dead fish showed elevated cortisol (Oliveira et al., 2014). This should be of particular note in zebrafish husbandry, since the vast majority of research laboratories use recirculating

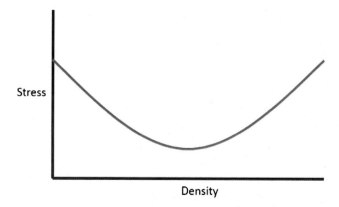

Figure 13.1. Cortisol levels in zebrafish related to density in adult zebrafish. Adapted from Harper and Lawrence (2010). *The Laboratory Zebrafish. CRC Press, Boca Raton, FL.*

systems and pheromones such as alarm substance could persist in and spread through these systems, causing elevated levels of intertank stress.

A survey of the literature shows that the relationship between stocking density and stress is extremely variable and complex. For example, in contrast to all the observations just mentioned, Pangnussat et al. (2013) found that zebrafish held in isolation had higher cortisol levels than those held in small groups of three. Likewise, another study found no correlation between stocking density and tank water cortisol levels (Gronquist and Berges, 2013). Although densities were the same (4 fish/L), Ramsay et al. (2006) found that fish crowded within 76 L aquaria resulted in higher cortisol levels than those held in 4 L tanks. These studies show that differences in environmental conditions, husbandry methods, and study design may result in very different stress responses, especially when cortisol is used as the primary indicator.

As discussed earlier, cortisol levels may not tell the whole story of stress and social interactions. A behavioral shoaling preference is well described in zebrafish, which, when presented with a choice between an empty tank and a tank full of conspecifics, tend to spend more time near the shoal (Stewart et al., 2012). Furthermore, minimal numbers of conspecifics are considered environmentally enriching by many animal care and use protocols. Even though individual fish may have lower cortisol levels in isolation, there is a behavioral preference for the shoal, which may indicate that some degree of social stress may be a positive, rather than a negative source of stress from the fish's point of view.

Interestingly, there is evidence that stocking density may have an effect on zebrafish behavior as early as the egg stage. Steenbergen et al. (2011) found that embryos raised in isolation had different responses to a dark pulse assay than embryos raised in batches of 25. This effect was rather elegantly proven to be due to contact between embryonic eggs as the behavioral effects of isolation could be reversed by artificial tactile stimulation during the first 2 days of embryogenesis. Therefore, consistency in embryo stocking densities may be important to the development of larval, and possibly even adult, behavioral phenotypes.

4.2. Enrichment

A considerable proportion of this chapter has been devoted to discussing practices concerning the optimization of production in laboratory fish. It is necessary to remember that a "happy" fish is often a productive fish and that animal welfare is the cornerstone of improved production, especially in a laboratory setting. Optimizing husbandry goes beyond simply meeting an animal's physiological needs such as stocking density, nutrition, and water quality. In this regard, animal care committees and regulatory oversight bodies in research institutions require that laboratory mammals be provided with various forms of environmental enrichment (Baumans and Van Loo, 2013; Hutchinson et al., 2005; Institute of Laboratory Animal Resources Committee, 2011). The advantages of providing structural enrichment—for all animals—is very much species- and context-dependent and a number of studies have shown that enriched environments affect certain behavioral syndromes in fish (Brydges and Braithwait, 2009). For example, Kistler et al. (2011) found that zebrafish were attracted to structures (artificial plants), but that the attraction response was stronger in checkered barbs (*Puntius oligolepis*). In zebrafish, spatial complexity tends to reduce aggression, and decreased aggression is associated with increased fecundity (Carfagnini et al., 2009). However, a direct correlation between environmental enrichment and stress reduction has not been demonstrated. For example, Wilkes et al. (2012) found that glass rods simulating plant stems did not reduce cortisol levels in zebrafish. In another study, Keck and coauthors (2015) found that while the provision of vegetation in the form of plastic plants did not reduce cortisol levels in pair-housed zebrafish, it did reduce mortality rates associated with aggression in these settings. In contrast, Pavlidis et al. (2013) showed that darker tanks (dark backgrounds) resulted in reduced cortisol compared to zebrafish held in tanks with light backgrounds. The findings of these, and similar studies, serve to highlight the complexity of fish behavior as it relates to husbandry. More research is needed to explore the specific environmental enrichment needs of laboratory fish in order to optimize their well-being and their utility as model organisms.

4.3. Light/Dark Cycle

Fish, like all animals, are profoundly influenced by photoperiod (Pavlidis et al., 1999; Valenzuela et al., 2012; Boeuf and Le Bail, 1999), so it is critically important to understand and regulate light cycles in the housing environment of captive fishes. Given that photoperiod exerts such profound influence on the fish physiology, it is not surprising that variations in "day" length beyond the animal's evolved preferences can be a source of distress. For example, African catfish (*Clarias gariepinus*) exhibit elevated cortisol levels when held under extended daylight conditions (Almazán-Rueda et al., 2005). Furthermore, Mustapha et al. (2014) found that Nile tilapia and African catfish exhibited increased mortality when maintained under constant light without a dark period (Mustapha et al., 2014). Female *Astyanax bimaculatus* exposed to longer light cycles showed elevated cortisol, higher aggression rates, and decreased survival than fish held under constant darkness.

Whereas it is well established that zebrafish are photoperiodic breeders (Lawrence, 2007; Westerfield, 2007), there is little data available concerning the effects of day length on stress. Existing photoperiod studies do not address stress specifically, but their results tend to be relatively consistent: larval fish show increased rates of swimming activity in darkness as opposed to light (De Esch et al., 2012) and both larval and adult fish show elevated activity levels when lights are abruptly turned off during light/dark challenge tests (Vignet et al., 2013). It can be inferred from these studies that light and darkness have neurobehavioral consequences for zebrafish, but more research is needed to specifically address potential correlations between photoperiod and stress.

5. FEEDING AND STRESS

Many animal care committees consider daily feeding of fish to be a requirement for humane treatment. With small tanks, especially static aquaria, this may cause issues with water quality and increase the adverse effects of high stocking density. Salmonids may exhibit increased cortisol when feeding is abruptly ceased, probably due to increased aggression (Brännäss et al., 2003). In contrast, Gronquist and Berges (2013) showed a clear reduction in cortisol levels in zebrafish that were subjected to food deprivation for upto 9 days, possibly due to increased water quality outstripping metabolic needs in relatively sedentary fish. Another possibility is that captive fish without natural activity patterns are chronically overfed, which could also interfere with experimentation. Conversely, stress inhibits food intake in many species (Steenbergen et al., 2011). Interestingly, the timing of feeding for groups of cichlid fish accustomed to being fed at predictable intervals induced a rise in cortisol levels (Galhardo et al., 2011), suggesting that changes in feeding routines might be a source of stress in fish held under laboratory conditions.

6. SEX AND HIERARCHIES

Zebrafish in aquaria, as well as a number of other species including salmon, establish dominance hierarchies, which are more pronounced in males (Spence et al., 2008; Paull et al., 2010; Ejike and Schreck, 1980). Filby et al. (2010) showed that subordinate males exhibited a greater rise in cortisol than subordinate females, potentially due to the more prevalent dominance hierarchies observed in males.

In laboratory zebrafish, physiological factors associated with dominance include animal size and aggressive behavior (Paull et al., 2010). Subordinate animals overall appear to experience more stress than dominants across species. For example, nonterritorial male African cichlids (*Haplochromis burtoni*) tend to have higher cortisol levels than territorial males (Fox et al., 1997). This is also true of juvenile rainbow trout, in which the submissive behavioral phenotype is considered to be a chronic stressor. Interestingly, elevated cortisol and submissive behavior likely have an interactive relationship of positive reinforcement wherein submission causes stress and the resulting elevated cortisol results in increased submissive interactions due to poor health, small size, and so on (Gilmour et al., 2005). The question of what makes a fish more likely to be dominant or submissive may be related to genetic predispositions to elevated cortisol responses to external stimuli, although this hypothesis is still being explored (Ramsay et al., 2009a,b,c).

In order to improve behavioral consistency and reduce hierarchy-related stress, adequate feeding has been suggested as a potential mitigating factor in zebrafish as it promotes consistent size distributions within a tank while reducing the need for aggression to compete for food. Crowding, of course, goes hand in hand with resource management and hierarchy formation and so appropriate stocking density may help to reduce hierarchy-induced stress (Ramsay et al., 2009a,b,c).

7. SEX DETERMINATION AND REVERSAL

Fish are often used in studies wherein gonadal phenotype is an end point, as in the evaluation of endocrine-disrupting compounds. The Japanese medaka (*Oryzias latipes*) is an important species in toxicology research. Hayashi et al. (2010) showed that cortisol caused female-to-male sex reversal and that metyrapone (an inhibitor of cortisol synthesis) inhibited HT-induced masculinization of female medaka. As cortisol has also been implicated in influencing gonadal fate in other species, including pejerrey (*Odontesthes bonariensis*) (Hattori et al., 2009) and Japanese flounder (*Paralichthys olivaceus*), the possibility that the hormone may have similar effects in laboratory species such as zebrafish should be considered.

8. STRESS, CORTISOL, AND REPRODUCTION

The tremendous increase in the use of fish in biomedical research has been primarily focused on early developmental stages, particularly the embryo. This is especially true of the zebrafish, which is well suited to high-throughput experimental protocols due to its high fecundity, external fertilization, rapid development, and larval whole-body transparency compared to traditional mammalian models (Murphey and Zon, 2006). Many laboratories, therefore, use adult fish solely for embryo production and the effects of stress on fecundity and embryogenesis are of particular concern in this setting.

Because fecundity is one of the major production end points in aquaculture, most studies regarding the relationship between stress and fecundity have been done with rainbow trout and other salmonids. These studies were extensively reviewed by Schreck et al. (2001). Similar studies on laboratory fish species are far less common; however, Cloud (1981) found that hatching rates in medaka were accelerated when eggs were incubated with deoxycorticosterone (DOC).

A discussion of stress and fecundity must focus on the concept of maternal stress effects on offspring, most of which are epigenetic. Schreck and Tort (Schreck and Tort, 2016; Chapter 1 in this volume) have already discussed the data concerning the maternal effects of stress on salmonid offspring, but data on its effects on laboratory fish are sparse. Nesan and Vijayan (2012) found that prehatching exposure to cortisol in zebrafish embryos induced cardiac dysfunction and other developmental defects. In this study, injection of cortisol into single-celled eggs was meant to mimic maternal cortisol delivery to eggs because independent cortisol production does not occur in zebrafish until after hatching. In three-spined sticklebacks, maternal exposure to predation risk was associated with increased embryonic transcription of genes associated with metabolism, epigenetic inheritance, and neuronal development. The embryos of predator risk-exposed mothers also tended to be larger than control embryos (Mommer and Bell, 2014).

The consequences of larval and embryonic stress can extend into adulthood: Rainbow trout exposed to stress at early embryonic stages, both at stages involving maternal and endogenous cortisol, actually resulted in lowered stress responses as adults (Steenbergen et al., 2011). These studies highlight that consistency in maternal sourcing and husbandry, as well as consistent care for embryos and larvae are necessary for reducing the impact of stress as a source of nonprotocol-induced variation.

9. ANESTHETICS

Laboratory fishes are often anesthetized to allow for various procedures, including injections, tagging, and implants. Palić et al. (2006) evaluated the effects of MS-222, metomidate and eugenol on cortisol levels in fathead minnows. Fish anesthetized with metomidate or eugenol (but not MS-222) showed reduced cortisol compared to controls following handling and crowding. Recently, two separate studies found that zebrafish perceive the most commonly used anesthetic agent in fish research, MS-222, as aversive. Aversive in this case means that fish attempt to avoid the substance under study—given the choice, they prefer water with lower concentrations of the substance or places where it is absent entirely. Using behavioral analysis software that quantified spatial preference based on swimming behaviors, Readman et al. (2013) showed that adult zebrafish displayed aversive behavior when exposed to seven of nine commonly used anesthetics, including MS-222 and benzocaine (Readman et al., 2013). They suggested the use of etomidate as a possible alternative for this species. In a separate study, Wong et al. (2014) used a conditioned place avoidance paradigm test to demonstrate zebrafish aversion to MS-222 (Wong et al., 2014). Whereas these data apply only to zebrafish, they clearly demonstrate that the choice of anesthetic should be carefully considered depending on the species and experimental application. Also worth consideration is the balance of a brief aversive stimulus in the form of anesthesia compared to the pain and stress of handling or surgery without anesthesia. We do not suggest that anesthesia should not be used prior to experimental manipulation—in fact it is necessary to survival in many cases—merely that broader options for less aversive or stressful anesthesia methods should be explored.

10. UNDERLYING DISEASES

Kent et al. (2012) reviewed the documented and potential impacts that underlying lesions and infections have or may have on research using zebrafish. In line with the topic of this book, stress and underlying diseases have a symbiotic relationship, with disease potentially elevating incipient stress levels and stress exacerbating pathogenicity in turn. For example, *Pseudoloma neurophilia* is extremely common in zebrafish facilities (Sanders et al., 2012), and we recently showed that this microsporidium, which targets the central nervous system, can infect many laboratory fishes, such as medaka and fathead minnows. Most infections are subclinical (Spagnoli et al., 2015a,b), but fish subjected to crowding stress develop more severe infections (Ramsay et al., 2009a,b,c). *Pseudoloma neurophilia* infections have been associated with altered responses to assessments of stress and anxiety, indicating that infections may result in high-responding behavioral phenotypes (Spagnoli et al., 2015a,b). Zebrafish with underlying *Mycobac-terium marinum* infections subjected to crowding stress exhibited higher infection-associated mortality than unstressed controls. This was not due to increasing horizontal transmission, as this pathogen requires ingestion of infected tissue or invertebrates for transmission among laboratory fishes (Peterson et al., 2013). Rather, the increased mortality was attributed to the exacerbation of preexisting infections by stress in individual fish. There are numerous examples of stress, crowding, and elevated cortisol enhancing the proliferation and virulence of pathogens within the fish host, since cortisol is notoriously immunosuppressive (Cortés et al., 2013). Conversely, the presence of underlying parasite infections can be stressors themselves, resulting in increased cortisol levels (Hoole and Williams, 2004). The importance of cortisol as an immune suppressor is particularly important in studies concerning inflammatory and immune responses with regard to underlying fish stress (Yada and Tort, 2016; Chapter 10 in this volume).

11. CONSISTENCY

Whereas all the specific aspects of husbandry discussed previously are important individually, there is one factor that should unite them all and direct the decision-making process of anyone working with these animals: consistency in care and husbandry. Changes are obviously necessary as the fish grows from a larva to an adult, but steps should be taken to minimize drastic changes in factors like feeding, temperature, ambient noise, and water quality once the specific husbandry practices of that life stage are established. A fascinating study by Piato et al. (2011) found that zebrafish have the potential to be an excellent model for unpredictable chronic stress in humans. When exposed to unpredictable multiple stressors given at irregular intervals and unpredictable time points, zebrafish showed increased anxiety and decreased cognitive function as well as increased cortisol after 7 days of exposure. Whereas this is excellent news for the zebrafish as a model for human neurobehavioral disease, it could mean that successive, random changes in husbandry practices could both produce undue stress and alter responses to neurobehavioral research. Therefore, husbandry changes or frequent random stress events such as equipment repair or replacement, as well as alarms or lighting changes, should be avoided whenever possible.

12. CONCLUSION AND KEY UNKNOWNS

Future research should explore the exact effects of specific husbandry manipulations on stress and its capacity to influence the fish as a model organism. It is likely that, as more data are gathered on the subject, the scientific community will begin to unearth some heretofore unknown complicating factors. For example, domestication selection in salmonids is a well-established phenomenon (Araki et al., 2008) with marked genetic and behavioral changes occurring as quickly as a single generation. It is likely, therefore, that in a totally artificial laboratory environment, domestication selection could strongly influence behavioral syndromes of model fishes. To support this concept, behavioral differences have been observed between AB and TU/TL zebrafish lines (Varga, 2011). This artificial selection may actually be beneficial in that it could reduce genetic variability among fish within a laboratory, however, inadvertent selection for distinct behavioral phenotypes could produce variable stress responses between different laboratories and breeders. For example, the behavioral shifts associated with *Pseudoloma neurophilia* infection observed by Spagnoli et al. (2015a,b) may have been due to the selective pressure of the parasite killing off fish with a particular behavioral phenotype. If this is the case, then it is possible that laboratories contaminated by this very common parasite could be inadvertently selecting for zebrafish with high-responding, nonmortality-associated behavioral phenotypes.

This chapter has provided an extensive discussion of the sources, measurability, and effects of stress in laboratory fishes. It is important here to reiterate the three basic goals of animal husbandry for research: consistency of subjects, robustness to the model under study, and animal welfare. Stress is a fundamental neurophysiological state that links the brain and body in all vertebrates. Therefore, the myriad of experimental end points utilizing laboratory fish, whether they are neurobehavioral, pathologic, molecular, or metabolic, can all be influenced by stress. A thorough understanding of the causes and consequences of stress in laboratory fish will ultimately lead to the development of husbandry protocols that marry animal welfare with experimental consistency.

REFERENCES

Aerts, J., Metz, J. R., Ampe, B., Decostere, A., Flik, G. and De Saeger, S. (2015). Scales tell a story on the stress history of fish. *PLoS One* **10**, e0123411.

Allen, J. P. and Neely, M. N. (2010). Trolling for the ideal model host: zebrafish take the bait. *Future Microbiol.* **5**, 563–569.

Almazán-Rueda, P., Van Helmond, A. T. M., Verreth, J. A. J. and Schrama, J. W. (2005). Photoperiod affects growth, behaviour and stress variables in *Clarias gariepinus*. *J. Fish. Biol.* **67**, 1029–1039.

Ankley, G. T. and Villeneuve, D. L. (2006). The fathead minnow in aquatic toxicology: past, present and future. *Aquat. Toxicol.* **78**, 91–102.

Araki, H., Berejikian, B. A., Ford, M. J. and Blouin, M. S. (2008). Fitness of hatchery-reared salmonids in the wild. *Evol. Appl.* **1**, 342–355.

Araujo, J., Maximino, C., de Brito, T. M., da Salva, A. W. B., Oliveira, K. R. M., Batista, E. J. O., et al. (2012). Behavioral and pharmacological aspects of anxiety in the light/dark preference test. In *Zebrafish Protocols for Neurobehavioral Research* (eds. A. V. Kalueff and A. M. Stewart), pp. 191–203. New York, NY: Springer.

Bailey, G. S., Williams, D. E. and Hendricks, J. D. (1996). Fish models for environmental carcinogenesis: the rainbow trout. *Environ. Health Perspect.* **104**, 5–21.

Bailey, J., Oliveri, A. and Levin, E. D. (2013). Zebrafish model systems for developmental neurobehavioral toxicology. *Birth Defects Res. C Embryo Today* **99**, 14–23.

Barry, T. P., Malison, J. A., Held, J. A. and Parrish, J. J. (1995). Ontogeny of the cortisol stress response in larval rainbow trout. *Gen. Comp. Endocrinol* **97**, 57–65.

Baumans, V. and Van Loo, P. L. P. (2013). How to improve housing conditions of laboratory animals: the possibilities of environmental refinement. *Vet. J.* **195**, 24–32.

Blaser, R. E. and Rosemberg, D. B. (2012). Measures of anxiety in zebrafish (*Danio rerio*): dissociation of black/white preference and novel tank test. *PLoS One* **7**, e36931.

Boeuf, G. and Le Bail, P. Y. (1999). Does light have an influence on fish growth? *Aquaculture* **177**, 129–152.

Brännäss, E., Jonsson, S. and Lundqvist, H. (2003). Influence of food abundance in individual behaviour strategy and growth rate in juvenile Brown Trout (*Salmo trutta*). *Can. J. Zool.* **81**, 684–692.

Brydges, N. M. and Braithwait, V. A. (2009). Does environmental enrichment affect the behaviour of fish commonly used in laboratory work? *Appl. Anim. Behav. Sci.* **118**, 137–143.

Canavello, P. R., Cachat, J. M., Beeson, E. C., Laffoon, A. L., Grimes, C., Haymore, W. A. M., et al. (2011). Measuring endocrine (cortisol) responses of zebrafish to stress. In *Zebrafish Neurobehavioral Protocols* (eds. A. V. Kalueff and A. M. Stewart), pp. 135–142. New York, NY: Springer.

Carfagnini, A. G., Rodd, F. H., Jeffers, K. B. and Bruce, A. E. E. (2009). The effects of habitat complexity on aggression and fecundity in zebrafish (*Danio rerio*). *Environ. Biol. Fish* **86**, 403–409.

Ceol, C. J., Houvras, Y., Jane-Valbuena, J., Bilodeau, S., Orlando, D. A., Battisti, V., et al. (2011). The histone methyltransferase SETDB1 is recurrently amplified in melanoma and accelerates its onset. *Nature* **471**, 513–517.

Cloud, J. G. (1981). Deoxycorticosterone-induced precocious hatching of teleost embryos. *J. Exp. Zool.* **216**, 197–199.

Conrad, J. L., Weinersmith, K. L., Brodin, T., Saltz, J. B. and Sih, A. (2011). Behavioural syndromes in fishes: a review with implications for ecology and fisheries management. *J. Fish. Biol.* **78**, 395–435.

Cortés, R., Teles, M., Tridico, R., Acerete, L. and Tort, L. (2013). Effects of cortisol administered through slow release implants on innate immune responses in rainbow trout (*Onchorhynchus mykiss*). *Int. J. Genomics* **2013**, 619714. Online edition.

De Esch, C., van der Linde, H., Slieker, R., Willemsen, R., Wolterbeek, A., Woutersen, R., et al. (2012). Locomotor activity assay in zebrafish larvae: influence of age, strain and ethanol. *Neurotoxicol. Teratol.* **34**, 425–433.

Dooley, K. and Zon, L. I. (2000). Zebrafish: a model for the study of human disease. *Curr. Opin. Genet. Dev.* **10**, 252–256.

Dugatkin, L., McCall, M. A., Gregg, R. G., Cavanaugh, A., Christensen, C. and Unseld, M. (2004). Zebrafish (*Danio rerio*) exhibit individual differences in risk-taking behavior during predator inspection. *Ethol. Ecol. Evol.* **17**, 77–81.

Egan, R. J., Bergner, C. L., Hart, P. C., Cachat, J. M., Canavello, P. R., Elegante, M. F., et al. (2009). Understanding behavioral and physiological phenotypes of stress and anxiety in zebrafish. *Behav. Brain Res.* **205**, 38–44.

Ejike, C. and Schreck, C. B. (1980). Stress and social hierarchy rank in coho salmon. *Trans. Am. Fish. Soc.* **109**, 423–426.

Ellis, T., James, J. D., Stewart, C. and Scott, A. P. (2004). A non-invasive stress assay based upon measurement of free cortisol released into the water by rainbow trout. *J. Fish. Biol.* **65** (5), 1233–1252.

Ellis, T., Yildiz, H. Y., Lopez-Olmeda, J., Spedicato, M. T., Tort, L., Overli, O., et al. (2012). Cortisol and finfish welfare. *Fish. Physiol. Biochem.* **38**, 163–188.

Engeszer, R. E., Patterson, L. B., Rao, A. A. and Parichy, D. M. (2007). Zebrafish in the wild: a review of natural history and new notes from the field. *Zebrafish* **4**, 21–40.

Feist, G. and Schreck, C. B. (2001). Ontogeny of the stress response in chinook salmon, Oncorhynchus tshawytscha. *Fish. Physiol. Biochem* **25**, 31–40.

Félix, A. S. I., Faustino, A. I., Cabral, E. M. and Oliveira, R. F. (2013). Noninvasive measurement of steroid hormones in zebrafish holding-water. *Zebrafish* **10**, 110–115.

Filby, A. L., Paull, G. C., Bartlett, E. J., Van Look, K. J. W. and Tyler, C. R. (2010). Physiological and health consequences of social status in zebrafish (*Danio rerio*). *Physiol. Behav.* **101**, 576–587.

Fox, H. E., White, S. A., Kao, M. H. F. and Fernald, R. D. (1997). Stress and dominance in a social fish. *J. Neurosci.* **17**, 663–6469.

Galhardo, L., Vital, J. and Oliveira, R. F. (2011). The role of predictability in the stress response of a cichlid fish. *Physiol. Behav.* **102**, 367–372.

Gardell, A. M., Yang, J., Sacchi, R., Fangue, N. A., Hammock, B. D. and Kultz, D. (2013). Tilapia (*Oreochromis mossambicus*) brain cells respond to hyperosmotic challenge by inducing myo-inositol biosynthesis. *J. Exp. Biol.* **216**, 4615–4625.

Gerhard, G. S. (2003). Comparative aspects of zebrafish (*Danio rerio*) as a model for aging research. *Exp. Gerontol.* **38**, 1333–1341.

Gesto, M., Hernandez, J., Lopez-Patino, M. A., Soengas, J. L. and Miguez, J. M. (2015). Is gill cortisol concentration a good acute stress indicator in fish? A study in rainbow trout and zebrafish. *Comp. Biochem. Physiol. A Mol. Integr. Physiol.* **188**, 65–69.

Ghisleni, G., Capiotti, K. M., Da Silva, R. S., Oses, J. P., Piato, A. L., Soares, V., et al. (2012). The role of CRH in behavioral responses to acute restraint stress in zebrafish. *Prog. Neuro-Psychopharmacol. Biol. Psychiatry* **36**, 176–182.

Giacomini, A. C., de Abreu, M. S., Koakoski, G., Idalencio, R., Kalichak, F., Oliveira, T. A., et al. (2015). My stress, our stress: blunted cortisol response to stress in isolated zebrafish. *Physiol. Behav.* **139**, 182–187.

Gilmour, K. M., DiBattista, J. D. and Thomas, J. B. (2005). Physiological causes and consequences of social status in salmonid fish. *Integ. Comp. Biol.* **45**, 263–273.

Gronquist, D. and Berges, J. A. (2013). Effects of aquarium-related stressors on the zebrafish: a comparison of behavioral, physiological, and biochemical indicators. *J. Aquat. Anim. Health* **25**, 53–65.

Harland, R. M. and Grainger, R. N. (2011). *Xenopus* research: metamorphosed by genetics and genomics. *Trends Gen.* **27**, 507–515.

Hattori, R. S., Fernandino, J. I., Kishii, A., Kimura, H., Kinno, T., Oura, M., et al. (2009). Cortisol-induced masculinization: does thermal stress affect gonadal fate in pejerrey, a teleost fish with temperature-dependent sex determination? *PLoS One* **4** (8), e6548.

Hayashi, Y., Kobira, J., Yamaguchi, T., Shiraishi, E., Yazawa, T., Hirai, T., et al. (2010). High temperature causes masculinization of genetically female medaka by elevation of cortisol. *Mol. Rep. Dev.* **77**, 679–686.

Hendry, A. P., Peichel, C. L., Matthews, B., Boughman, J. W. and Nosil, P. (2013). Stickleback research: the now and the next. *Evol. Ecol. Res.* **15**, 111–141.

Hoole, D. and Williams, G. T. (2004). The role of apoptosis in non-mammalian host-parasite relationships. In *Host-Parasite Interactions* (eds. G. Fli and G. Wiegertjes), pp. 13–45. BIOS Scientific Publishers.

Hutchinson, E., Avery, A. and VandeWoude, S. (2005). Environmental enrichment for laboratory rodents. *ILAR J.* **46**, 148–161.

Institute of Laboratory Animal Resources Committee. (2011). Guide for the Care and Use of Laboratory Animals.

Kahl, M. D., Jensen, K. M., Korte, J. J. and Ankley, G. T. (2001). Effects of handling on endocrinology and reproductive performance of the fathead minnow. *J. Fish. Biol.* **59**, 515–523.

Keck, V. A., Edgerton, D. S., Hajizadeh, S., Swift, L. L., Dupont, W. D., Lawrence, C. and Boyd, K. L. (2015). Effects of habitat complexity on pair-housed zebrafish. *J. Am. Assoc. Lab. Anim. Sci.* **54**, 378–383.

Kent, M. L., Harper, C. and Wolf, J. C. (2012). Documented and potential research impacts of subclinical diseases in zebrafish. *ILAR J.* **53**, 126–134.

Kishi, S., Slack, B. E., Uchiyama, J. and Zhdanova, I. V. (2009). Zebrafish as a genetic model in biological and behavioral gerontology: where development meets aging in verterates – a mini-review. *Gerontology* **55**, 430–441.

Kistler, C., Hegglin, D., Würbel, H. and König, B. (2011). Preference for structured environment in zebrafish (*Danio rerio*) and checker barbs (*Puntius oligolepis*). *Appl. Anim. Behav. Sci.* **135**, 318–327.

Knudsen, F. R., Enger, P. S. and Sand, O. (1992). Awareness reactions and avoidance responses to sound in juvenile Atlantic salmon, Salmo salar L. *J. Fish. Biol* **40**, 523–534.

Lawrence, C. (2007). The husbandry of zebrafish (*Danio rerio*): a review. *Aquaculture* **269**, 1–4.

Lawrence, C. and Mason, T. (2012). Zebrafish housing systems: a review of basic operating principles and considerations for design and functionality. *ILAR J.* **53**, 179–191.

Lieschke, G. J. and Currie, P. D. (2007). Animal models of human disease: zebrafish swim into view. *Nat. Rev. Genet.* **8**, 353–367.

Liu, Z. (2003). A review of catfish genomics: progress and perspectives. *Comp. Funct. Genom.* **4**, 259–265.

Liu, S. and Leach, S. D. (2011). Zebrafish models for cancer. *Ann. Rev. Pathol. Mech. Dis.* **6**, 71–93.

Martinez-Porchas, M., Martinez-Cordova, L. T. and Ramos-Enriquez, R. (2009). Cortisol and glucose: reliable indicators of fish stress? *J. Aquat. Sci.* **4**, 158–178.

Martins, C. I. M., Ochola, D., Ende, S. S. W., Eding, E. H. and Verreth, J. A. J. (2009). Is growth retardation present in Nile tilapia Oreochromis niloticus cultured in low water exchange recirculating aquaculture systems? *Aquaculture* **298**, 1–2.

Martins, C. I. M., Eding, E. H. and Verreth, J. A. J. (2011). Stressing fish in recirculating aquaculture systems (RAS): does stress induced in one group of fish affect the feeding motivation of other fish sharing the same RAS? *Aquac. Res.* **42**, 1378–1384.

Maximino, C., deBrito, T. M., da Silva Batista, A. W., Herculano, A. M., Morato, S. and Gouveia, A., Jr. (2010). Measuring anxiety in zebrafish: a critical review. *Behav. Brain Res.* **214**, 157–171.

Meijer, A. H. and Spaink, H. P. (2011). Host-pathogen interactions made transparent with the zebrafish model. *Curr. Drug. Targets* **12**, 1000–1017.

Meijer, A. H., Verbeek, F. J., Salas-Vidal, E., Corredor-Adamez, M., Bussman, J., van der Sar, A. M., et al. (2005). Transcriptome profiling of adult zebrafish at the late stage of chronic tuberculosis due to *Mycobacterium marinum* infection. *Mol. Immunol.* **42**, 1185–1203.

Mommer, B. C. and Bell, A. M. (2014). Maternal experience with predation risk influences genome-wide embryonic gene expression in threespined sticklebacks (*Gasterosteus aculeatus*). *PLoS One* **9**.

Murphey, R. D. and Zon, L. I. (2006). Small molecule screening in the zebrafish. *Methods* **39**, 255–261.

Mustapha, M. K., Oladokun, O. T., Salman, M. M., Adeniyi, I. A. and Ojo, D. (2014). Does light duration (photoperiod) have an effect on the mortality and welfare of cultured *Oreochromis niloticus* and *Clarias gariepinus*? *Turk. J. Zool.* **38**, 466–470.

Neo, Y. Y., Parie, L., Bakker, F., Snelderwaard, P., Tudorache, C., Schaaf, M., et al. (2015). Behavioral changes in response to sound exposure and not special avoidance of noisy conditions in captive zebrafish. *Front. Behav. Neurosci.* **9**, 28.

Nesan, D. and Vijayan, M. M. (2012). Embryo exposure to elevated cortisol level leads to cardiac performance dysfunction in zebrafish. *Mol. Cell. Endocrinol.* **363**, 85–91.

Novoa, B. and Figueras, A. (2012). Zebrafish: model for the study of inflammation and the innate immune response to infectious diseases. *Adv. Exp. Med. Biol.* **946**, 253–275.

Oliveira, T. A., Koakoski, G., da Motta, A. C., Piato, A. L., Barreto, R. E., Volpato, G. L., et al. (2014). Death-associated odors induce stress in zebrafish. *Horm. Behav.* **65**, 340–344.

Palić, D., Herolt, D. M., Andreasen, C. B., Menzel, B. W. and Roth, J. A. (2006). Anesthetic efficacy of tricaine methanesulfonate, metomidate and eugenol: effects on plasma cortisol concentration and neutrophil function in fathead minnows (*Pimephales promelas* Rafinesque, 1820). *Aquaculture* **254**, 675–685.

Pangnussat, N., Piato, A. L., Schaefer, I. C., Blank, M., Tamborski, A. R., Guerim, L. D., et al. (2013). One for all and all for one: the importance of shoaling on behavioral and stress responses in zebrafish. *Zebrafish* **10**, 338–342.

Parker, M. O. I., Millington, M. E., Combe, F. J. and Brennan, C. H. (2012). Housing conditions differentially affect physiological and behavioural stress responses of zebrafish, as well as the response to anxiolytics. *PLoS One* **7** (4), e34992.

Patton, E. E. and Zon, L. I. (2005). Taking human cancer genes to the fish: a transgenic model of melanoma in zebrafish. *Zebrafish* **1**, 363–368.

Patton, E. E., Mitchell, D. L. and Nairn, R. S. (2010). Genetic and environmental melanoma models in fish. *Pigment. Cell. Melanoma. Res.* **23**, 314–337.

Patton, E. E., Mathers, M. E. and Schartl, M. (2011). Generating and analyzing fish models of melanoma. *Methods Cell. Biol.* **105**, 339–366.

Paull, G. C., Filby, A. L., Giddins, H. G., Coe, T. S., Hamilton, P. B. and Tyler, C. R. (2010). Dominance hierarchies in zebrafish (*Danio rerio*) and their relationship with reproductive success. *Zebrafish* **7**, 109–117.

Pavlidis, M., et al. (1999). The effect of photoperiod on diel rhythms in serum melatonin, cortisol, glucose, and electrolytes in the common dentex, *Dentex dentex*. *Gen. Comp. Endocrinol.* **113**, 240–250.

Pavlidis, M., Digka, N., Theodoridi, A., Campo, A., Brasakis, K., Skouradakis, G., et al. (2013). Husbandry of zebrafish, *Danio rerio*, and the cortisol stress response. *Zebrafish* **10**, 524–531.

Peterson, T. S., Ferguson, J. A., Watral, V. G., Mutoji, K. N., Ennis, D. G. and Kent, M. L. (2013). *Paramecium caudatum* enhances transmission and infectivity of *Mycobacterium marinum* and *Mycobacterium chelonae* in zebrafish (*Danio rerio*). *Dis. Aquat. Org.* **106**, 229–239.

Phillips, J. B. and Westerfield, M. (2014). Zebrafish models in translational research: tipping the scales towards advancements in human health. *Dis. Model Mech.* **7**, 739–743.

Piato, A. L., Capiotti, K. M., Tamborski, A. R., Oses, J. P., Barcellos, L. J. G., Bogo, M. R., et al. (2011). Unpredictable chronic stress model in zebrafish (*Danio rerio*): behavioral and physiological responses. *Prog. Neuro-Psychopharmacol. Biol. Psychiatry* **35**, 561–567.

Prouty, M. G., Correa, N. E., Barker, L. P., Jagadeeswaran, P. and Klose, K. E. (2003). Zebrafish-*Mycobacterium marinum* model for mycobacterial pathogenesis. *FEMS Microbiol. Lett.* **225**, 177–182.

Purser, J. and Radford, A. N. (2011). Acoustic noise induces attention shifts and reduces foraging performance in three-spined sticklebacks (*Gasterosteus aculeatus*). *PLoS One* ⟨http://dx.doi.org/10.1371/journal.pone.0017478⟩.

Radford, A. N., Kerridge, E. and Simpson, S. D. (2014). Acoustic communication in a noisy world: can fish compete with anthropogenic noise? *Behav. Ecol.* **25** (5), 1022–1030.

Ramsay, J. M., Feist, G. W., Varga, Z. M., Westerfield, M., Kent, M. L. and Schreck, C. B. (2006). Whole-body cortisol is an indicator of crowding stress in adult zebrafish, *Danio rerio. Aquaculture* **258**, 565–574.

Ramsay, J. M., Feist, G. W., Varga, Z. M., Westerfield, M., Kent, M. L. and Schreck, C. B. (2009a). Rapid cortisol response of zebrafish to acute net handling stress. *Aquaculture* **297**, 157–162.

Ramsay, J. M., Watral, V., Schreck, C. B. and Kent, M. L. (2009b). Husbandry stress exacerbates mycobacterial infections in adult zebrafish, *Danio rerio* (Hamilton). *J. Fish. Dis.* **32**, 931–941.

Ramsay, J. M., Watral, V., Schreck, C. B. and Kent, M. L. (2009c). *Pseudoloma neurophilia* (Microsporidia) infections in zebrafish (*Danio rerio*): effects of stress on survival, growth and reproduction. *Dis. Aquat. Org.* **88**, 69–84.

Readman, G. D., Owen, S. F., Murrell, J. C. and Knowles, T. G. (2013). Do fish perceive anaesthetics as aversive? *PLoS One* **8** (9), e73773.

Rihel, J. and Schier, A. F. (2012). Behavioral screening for neuroactive drugs in zebrafish. *Dev. Neurobiol.* **72**, 373–385.

Sanders, J. L., Watral, V. and Kent, M. L. (2012). Microsporidiosis in zebrafish research facilities. *ILAR J.* **53**, 106–113.

Schartl, M. (2014). Beyond the zebrafish: diverse fish species for modeling human disease. *Dis. Model Mech.* **7**, 181–192.

Schreck, C. B. and Tort, L. (2016). The Concept of Stress in Fish. In *Fish Physiology - Biology of Stress in Fish*, Vol. 35 (eds. C. B. Schreck, L. Tort, A. P. Farrell and C. J. Brauner), San Diego, CA: Academic Press.

Schreck, C. B., Contreas-Sanchez, W. and Fitzbatrick, M. S. (2001). Effects of stress on fish reproduction, gamete quality, and progeny. *Aquaculture* **197**, 3–24.

Shima, A. and Mitani, H. (2004). Medaka as a research organism: past, present and future. *Mech. Dev.* **121**, 599–604.

Simpson, S. D., Purser, J. and Radford, A. N. (2014). Anthropogenic noise compromises antipredator behavior in European eels. *Glob. Change Biol.* **21** (2), 586–593.

Singley, J. S. and Chavin, W. (1975). Serum cortisol in normal goldfish (*Carassius auratus L.*). *Comp. Biochem. Physiol. A Physiol.* **50**, 77–82.

Sisson, M., Cawker, J., Buske, C. and Gerlai, R. (2006). Fishing for genes influencing vertebrate behavior: zebrafish making headway. *Lab Anim (NY)* **35** (5), 33–39.

Smith, M. E., Kane, A. S. and Popper, A. N. (2003). Noise-induced stress response and hearing loss in goldfish (*Carassius auratus*). *J. Exp. Biol.* **207**, 427–435.

Sneddon, L. U., Wolfenden, D. C. C. and Thomson, J. S. (2016). Stress Management and Welfare.. In *Fish Physiology - Biology of Stress in Fish*, Vol. 35 (eds. C. B. Schreck, L. Tort, A. P. Farrell and C. J. Brauner), San Diego, CA: Academic Press.

Spagnoli, S., Xue, L. and Kent, M. L. (2015a). The common neural parasite *Pseudoloma neurophilia* is associated with altered startle response habituation in adult zebrafish (*Danio rerio*): implications for the zebrafish as a model organism. *Behav. Brain Res.* **291:**, 351–360.

Spagnoli, S., Xue, L., Murray, K. M., Chow, F. and Kent, M. L. (2015b). *Pseudoloma neurophilia*: a retrospective and descriptive study with new implications for pathogenesis and behavioral phenotypes. *Zebrafish* **12**, 189–201.

Spence, R. (2011). Zebrafish models in neurobehavioral research. *Neuromethods* **52**, 211–222.

Spence, R., Gerlach, G., Lawrence, C. and Smith, C. (2008). The behaviour and ecology of the zebrafish, *Danio rerio*. *Biol. Rev. Camb. Philos. Soc.* **83**, 13–34.

Steenbergen, P. J., Richardson, M. K. and Champagne, D. L. (2011). The use of the zebrafish model in stress research. *Prog. Neuro-Psychopharmacol. Biol. Psychiatry* **35**, 1432–1451.

Stewart, A., Gaikwad, S., Kyzar, E., Green, J., Roth, A. and Kalueff, A. V. (2012). Modeling anxiety using adult zebrafish: a conceptual review. *Neuropharmacology* **62**, 135–143.

Stewart, A. M., Gerlai, R. and Kalueff, A. V. (2015). Developing higher throughput zebrafish screens for in-vivo CNS drug discovery. *Front Behav. Neurosci.* **9**, 1–8.

Swaim, L. E., Connoly, L. E., Volkman, H. E., Humbert, O., Born, D. E. and Ramakrishnan, L. (2006). *Mycobacterium marinum* infection of adult zebrafish causes caseating granulomatous tuberculosis and is moderated by adaptive immunity. *Infect. Immun.* **74**, 6108–6117.

Takeda, H. and Shimada, A. (2010). The art of medaka genetics and genomics: what makes them so unique? *Annu. Rev. Genet.* **44**, 217–241.

Taylor, A. M. and Zon, L. I. (2009). Zebrafish tumor assays: the state of transplantation. *Zebrafish* **6**, 339–346.

Tran, S., Chatterjee, D. and Gerali, R. (2014). Acute net stressor increases whole-body cortisol levels without altering whole-brain monoamines in zebrafish. *Behav. Neurosci.* **128**, 621–624.

Truong, L., Harper, S. L. and Tanguay, R. L. (2011). Evaluation of embryo toxicity using the zebrafish model. *Methods Mol. Biol.* **691**, 271–279.

Tudorache, C., Ter Braake, A., Tromp, M., Slabbekoom, H. and Schaaf, M. J. (2015). Behavioral and physiological indicators of stress coping styles in larval zebrafish. *Stress* **18**, 121–128.

Valenzuela, A., Campos, V., Yanez, F., Alveal, K., Gutierrez, P., Rivas, M., et al. (2012). Application of artificial photoperiod in fish: a factor that increases susceptibility to infectious diseases? *Fish Physiol. Biochem.* **38**, 943–950.

Varga, Z. M. (2011). Aquaculture and husbandry at the zebrafish international resource center. *Methods Cell. Biol.* **104**, 453–478.

Vignet, C., Begout, M. L., Pean, S., Lyphout, L., Leguay, D. and Cousin, X. (2013). Systematic screening of behavioral responses in two zebrafish strains. *Zebrafish* **10**, 365–375.

Yada, T. and Tort, L. (2016). Stress and Disease Resistance: Immune System and Immunoendocrine Interactions. In Fish Physiology - Biology of Stress in Fish, Vol. 35 (eds. C. B. Schreck, L. Tort, A. P. Farrell and C. J. Brauner), San Diego, CA: Academic Press.

Wedemeyer, G. A. (1997). Rearing conditions: effects on fish in intensive culture. In *Fish Stress and Health in Aquaculture* (eds. G. K. Iwama, A. D. Pickering, J. P. Sumpter and C. B. Schreck), pp. 35–73. New York, NY: Cambridge University Press.

Westerfield, M. (2007). The Zebrafish Book. A Guide for the Laboratory Use of Zebrafish (Danio rerio) (fifth ed.). Eugene: University of Oregon Press.

White, R. M., Sessa, A., Burke, C., Bowman, T., LeBlanc, J., Ceol, C., et al. (2008). Transparent adult zebrafish as a tool for in vivo transplantation analysis. *Cell. Stem. Cell.* **2**, 183–189.

Wilke, L., Owen, S. F., Readman, G. D., Sloman, K. A. and Wislon, R. W. (2012). Does structural enrichment for toxicology studies improve zebrafish welfare? *Appl. Anim. Behav. Sci.* **129**, 143–150.

Williams, D. E., Bailey, G. S., Reddy, A., Hendricks, J. D., Oganesian, A., Orner, G. A., et al. (2003). The rainbow trout (*Oncorhynchus mykiss*) tumor model: recent applications in low-dose exposures to tumor initiators and promoters. *Toxicol. Pathol.* **31**, 58–61.

Wong, K., Eegante, M., Bartels, B., Elkhayat, S., Tien, D., Roy, S., et al. (2010). Analyzing habituation responses to novelty in zebrafish (*Danio rerio*). *Brain. Res.* **208**, 450–457.

Wong, D., von Keyserlingk, M. A., Richards, J. G. and Weary, D. M. (2014). Conditioned place avoidance of zebrafish (*Danio rerio*) to three chemicals used for euthanasia and anaesthesia. *PLoS One* **9** (2), e88030.

INDEX

Note: Page numbers followed by "*f*" and "*t*" refer to figures and tables, respectively.

OTHER VOLUMES IN THE
FISH PHYSIOLOGY SERIES

Printed in the United States
By Bookmasters